Spatial Tessellations
Concepts and Applications
of Voronoi Diagrams

Probability and Mathematical Statistics (Continued)

ROBERTSON, WRIGHT, and DYKSTRA • Order Restricted Statistical Inference
ROGERS and WILLIAMS • Diffusions, Markov Processes, and Martingales, Volume II: Îto Calculus
ROSS • Stochastic Processes
RUBINSTEIN • Simulation and the Monte Carlo Method
RUZSA and SZEKELY • Algebraic Probability Theory
SCHEFFE • The Analysis of Variance
SEBER • Linear Regression Analysis
SEBER • Multivariate Observations
SEBER and WILD • Nonlinear Regression
SERFLING • Approximation Theorems of Mathematical Statistics
SHORACK and WELLNER • Empirical Processes with Applications to Statistics
STAUDTE and SHEATHER • Robust Estimation and Testing
STOYANOV • Counterexamples in Probability
STYAN • The Collected Papers of T. W. Anderson: 1943–1985
WHITTAKER • Graphical Models in Applied Multivariate Statistics
YANG • The Construction Theory of Denumerable Markov Processes

Applied Probability and Statistics

ABRAHAM and LEDOLTER • Statistical Methods for Forecasting
AGRESTI • Analysis of Ordinal Categorical Data
AGRESTI • Categorical Data Analysis
ANDERSON and LOYNES • The Teaching of Practical Statistics
ANDERSON, AUQUIER, HAUCK, OAKES, VANDAELE, and WEISBERG • Statistical Methods for Comparative Studies
ASMUSSEN • Applied Probability and Queues
* BAILEY • The Elements of Stochastic Processes with Applications to the Natural Sciences
BARNETT • Interpreting Multivariate Data
BARNETT and LEWIS • Outliers in Statistical Data, *Second Edition*
BARTHOLOMEW, FORBES, and McCLEAN • Statistical Techniques for Manpower Planning, *Second Edition*
BATES and WATTS • Nonlinear Regression Analysis and Its Applications
BELSLEY • Conditioning Diagnostics: Collinearity and Weak Data in Regression
BELSLEY, KUH, and WELSCH • Regression Diagnostics: Identifying Influential Data and Sources of Collinearity
BHAT • Elements of Applied Stochastic Processes, *Second Edition*
BHATTACHARYA and WAYMIRE • Stochastic Processes with Applications
BIEMER, GROVES, LYBERG, MATHIOWETZ, and SUDMAN • Measurement Errors in Surveys
BLOOMFIELD • Fourier Analysis of Time Series: An Introduction
BOLLEN • Structural Equations with Latent Variables
BOX • R. A. Fisher, the Life of a Scientist
BOX and DRAPER • Empirical Model-Building and Response Surfaces
BOX and DRAPER • Evolutionary Operation: A Statistical Method for Process Improvement
BOX, HUNTER, and HUNTER • Statistics for Experimenters: An Introduction to Design, Data Analysis, and Model Building
BROWN and HOLLANDER • Statistics: A Biomedical Introduction
BUCKLEW • Large Deviation Techniques in Decision, Simulation, and Estimation
BUNKE and BUNKE • Nonlinear Regression, Functional Relations and Robust Methods: Statistical Methods of Model Building
CHATTERJEE and HADI • Sensitivity Analysis in Linear Regression
CHATTERJEE and PRICE • Regression Analysis by Example, *Second Edition*

*Now available in a lower priced paperback edition in the Wiley Classics Library.

CLARKE and DISNEY • Probability and Random Processes: A First Course with Applications, *Second Edition*
COCHRAN • Sampling Techniques, *Third Edition*
COCHRAN and COX • Experimental Designs, *Second Edition*
CONOVER • Practical Nonparametric Statistics, *Second Edition*
CONOVER and IMAN • Introduction to Modern Business Statistics
CORNELL • Experiments with Mixtures, Designs, Models, and the Analysis of Mixture Data, *Second Edition*
COX • A Handbook of Introductory Statistical Methods
COX • Planning of Experiments
CRESSIE • Statistics for Spatial Data
DANIEL • Applications of Statistics to Industrial Experimentation
DANIEL • Biostatistics: A Foundation for Analysis in the Health Sciences, *Fifth Edition*
DAVID • Order Statistics, *Second Edition*
DEGROOT, FIENBERG, and KADANE • Statistics and the Law
* DEMING • Sample Design in Business Research
DILLON and GOLDSTEIN • Multivariate Analysis: Methods and Applications
DODGE and ROMIG • Sampling Inspection Tables, *Second Edition*
DOWDY and WEARDEN • Statistics for Research, *Second Edition*
DRAPER and SMITH • Applied Regression Analysis, *Second Edition*
DUNN • Basic Statistics: A Primer for the Biomedical Sciences, *Second Edition*
DUNN and CLARK • Applied Statistics: Analysis of Variance and Regression, *Second Edition*
ELANDT-JOHNSON and JOHNSON • Survival Models and Data Analysis
FLEISS • The Design and Analysis of Clinical Experiments
FLEISS • Statistical Methods for Rates and Proportions, *Second Edition*
FLEMING and HARRINGTON • Counting Processes and Survival Analysis
FLURY • Common Principal Components and Related Multivariate Models
GALLANT • Nonlinear Statistical Models
GROSS and HARRIS • Fundamentals of Queueing Theory, *Second Edition*
GROVES • Survey Errors and Survey Costs
GROVES, BIEMER, LYBERG, MASSEY, NICHOLLS, and WAKSBERG • Telephone Survey Methodology
HAHN and MEEKER • Statistical Intervals: A Guide for Practitioners
HAND • Discrimination and Classification
HEIBERGER • Computation for the Analysis of Designed Experiments
HELLER • MACSYMA for Statisticians
HOAGLIN, MOSTELLER, and TUKEY • Exploratory Approach to Analysis of Variance
HOAGLIN, MOSTELLER, and TUKEY • Exploring Data Tables, Trends and Shapes
HOAGLIN, MOSTELLER, and TUKEY • Fundamentals of Exploratory Analysis of Variance
HOAGLIN, MOSTELLER, and TUKEY • Understanding Robust and Exploratory Data Analysis
HOCHBERG and TAMHANE • Multiple Comparison Procedures
HOEL • Elementary Statistics, *Fifth Edition*
HOGG and KLUGMAN • Loss Distributions
HOLLANDER and WOLFE • Nonparametric Statistical Methods
HOSMER and LEMESHOW • Applied Logistic Regression
IMAN and CONOVER • Modern Business Statistics
JACKSON • A User's Guide to Principal Components
JOHN • Statistical Methods in Engineering and Quality Assurance
JOHNSON • Multivariate Statistical Simulation

*Now available in a lower priced paperback edition in the Wiley Classics Library.

Continued on back

Spatial Tessellations Concepts and Applications of Voronoi Diagrams

ATSUYUKI OKABE
University of Tokyo, Japan

BARRY BOOTS
Wilfrid Laurier University, Ontario, Canada

KOKICHI SUGIHARA
University of Tokyo, Japan

With a Foreword by
D.G. KENDALL

JOHN WILEY & SONS
Chichester · New York · Brisbane · Toronto · Singapore

Other Wiley Editorial Offices

John Wiley & Sons, Inc., 605 Third Avenue,
New York, NY 10158-0012, USA

Jacaranda Wiley Ltd, G.P.O. Box 859, Brisbane,
Queensland 4001, Australia

John Wiley & Sons (Canada) Ltd, 22 Worcester Road,
Rexdale, Ontario M9W 1L1, Canada

John Wiley & Sons (SEA) Pte Ltd, 37 Jalan Pemimpin #05-04,
Block B, Union Industrial Building, Singapore 2057

Library of Congress Cataloging-in-Publication Data

Boots, B.N.
Spatial tessellations : concepts and applications of voronoi diagrams / Barry Boots, Atsuyuki Okabe, Kokichi Sugihara ; with a foreword by D.G. Kendall.
p. cm.—(Wiley series in probability and mathematical statistics)
Includes bibliographical references and index.
ISBN 0 471 93430 5
1. Voronoi polygons. 2. Spatial analysis (Statistics)
3. Geometry—Data processing. I. Okabe, Atsuyuki, 1945–
II. Sugihara, Kōkichi, 1948– . III. Title. IV. Series.
QA278.2.B664 1992
519.5′36—dc20 91-47687
CIP

British Library Cataloguing in Publication Data

A catalogue record for this book is available from the British Library

ISBN 0 471 93430 5

Typeset in 10/12pt Times by Photo·graphics Ltd, Honiton
Printed and bound in Great Britain by Biddles Ltd, Guildford, Surrey

Contents

Foreword **ix**

Acknowledgements **xi**

Chapter 1 Introduction **1**

1.1 Outline 3
1.2 History of the concept of the Voronoi diagram 6
1.3 Mathematical preliminaries 10
1.3.1 Sets 10
1.3.2 Matrices 13
1.3.3 Geometry 16
1.3.4 Derivatives 28
1.3.5 Integration 32
1.3.6 Graphs 35
1.3.7 Probabilities and statistical tests 43
1.3.8 Spatial stochastic point processes 56

Chapter 2 Definitions and Basic Properties of Voronoi Diagrams **65**

2.1 Definitions of the ordinary Voronoi diagram 65
2.2 Definitions of the Delaunay tessellation (triangulation) 72
2.3 Basic properties of the Voronoi diagram 76
2.4 Basic properties of the Delaunay triangulation 89
2.5 Graphs related to the Delaunay triangulation 115

Chapter 3 Generalizations of the Voronoi Diagram **123**

3.1 Weighted Voronoi diagrams 128
3.1.1 Multiplicatively weighted Voronoi diagram 129
3.1.2 Additively weighted Voronoi diagram 133
3.1.3 Compoundly weighted Voronoi diagram 136
3.1.4 Power diagram 138
3.1.5 Sectional Voronoi diagram 140
3.2 Higher-order Voronoi diagrams 142
3.2.1 Order-k Voronoi diagram 142
3.2.2 Ordered order-k Voronoi diagram 151

3.3 Farthest-point Voronoi diagram and kth nearest-point Voronoi diagram 158
3.3.1 Farthest-point Voronoi diagram 159
3.3.2 kth nearest-point Voronoi diagram 162
3.4 Voronoi diagrams with obstacles 164
3.4.1 Shortest-path Voronoi diagram 164
3.4.2 Voronoi diagram in a simple polygon 168
3.4.3 Visibility-shortest-path Voronoi diagram 168
3.5 Voronoi diagrams for lines 170
3.5.1 Voronoi diagrams for a set of points and straight line segments 172
3.5.2 Voronoi diagrams for a set of points, straight line segments and circular arcs 178
3.5.3 Voronoi diagrams for a set of circles 179
3.5.4 Medial axis 182
3.6 Voronoi diagrams for areas 183
3.7 Voronoi diagrams with V-distances 184
3.7.1 Voronoi diagrams with the Minkowski metric L_p 185
3.7.2 Voronoi diagram with the Karlsruhe metric 188
3.7.3 Voronoi diagrams with the Hausdorff distance 190
3.7.4 Voronoi diagrams on a sphere 192
3.7.5 Voronoi diagrams on a cylinder 194
3.7.6 Voronoi diagrams on a cone 196
3.7.7 Voronoi diagrams on a polyhedral surface 197
3.7.8 Miscellany 198
3.8 Network Voronoi diagrams 200
3.8.1 Network Voronoi node-diagram 200
3.8.2 Network Voronoi link-diagram 202
3.8.3 Network Voronoi area-diagram 204
3.9 Notes on applications 205

Chapter 4 Algorithms for Computing Voronoi Diagrams 209
4.1 Computational preliminaries 209
4.2 Data structure for representing a Voronoi diagram 215
4.3 Incremental method 223
4.4 Divide-and-conquer method 232
4.5 Plane sweep method 239
4.6 Practical techniques for implementing the algorithms 246
4.6.1 Inconsistency caused by numerical errors 246
4.6.2 Construction of an error-free world 247
4.6.3 Topology-oriented approach 251
4.7 Algorithms for higher-dimensional Voronoi diagrams 257
4.8 Algorithms for generalized Voronoi diagrams 261
4.9 Approximation algorithms 267

Chapter 5 Poisson Voronoi Diagrams 273
5.1 Properties of Poisson Voronoi diagrams 276
5.2 Uses of Poisson Voronoi diagrams 278

5.3 Simulating Poisson Voronoi cells and Poisson Delaunay cells 281
5.4 Properties of Poisson Voronoi cells 286
5.5 Sectional Poisson Voronoi diagrams 307
5.6 Additively weighted Poisson Voronoi diagrams: the Johnson–Mehl model 315
5.7 Higher order Poisson Voronoi diagrams 318
5.8 Poisson Voronoi diagrams on the surface of a sphere 322
5.9 Properties of Poisson Delaunay cells 323
5.10 Other random Voronoi diagrams 329

Chapter 6 Spatial Interpolation **335**

6.1 Polygonal methods 337
6.1.1 Nearest neighbour interpolation 340
6.1.2 Natural neighbour interpolation 343
6.2 Triangular methods 351
6.3 Modifying Delaunay triangulations 356
6.4 Approximating surfaces 361
6.5 Ordering multivariate data 362

Chapter 7 Models of Spatial Processes **365**

7.1 Assignment models 366
7.2 Growth models 381
7.3 Spatial–temporal processes 385
7.3.1 Spatial competition models: the Hotelling model 385
7.3.2 Adjustment models 392
7.4 Two-species models 394
7.5 Recognizing Voronoi diagrams 397
7.5.1 Reconstructing Voronoi diagrams 397
7.5.2 Approximating Voronoi diagrams 399

Chapter 8 Point Pattern Analysis **405**

8.1 Polygon-based methods 409
8.1.1 Direct approach 409
8.1.2 Information theory approaches 412
8.1.3 Estimating spatial intensity 415
8.2 Triangle-based methods 417
8.3 Nearest neighbour distance methods 422
8.3.1 Nearest neighbour distance method for point-like objects 425
8.3.2 Nearest neighbour distance method for line-like objects 428
8.3.3 Nearest neighbour distance method for area-like objects 431
8.3.4 Multi nearest neighbour distance method 432

Chapter 9 Locational Optimization through Voronoi Diagrams **433**
9.1 Preliminaries 434
9.1.1 The non-linear non-convex programming problem 434
9.1.2 The descent method 437
9.1.3 The penalty function method 441
9.2 Locational optimization of points 443
9.2.1 Locational optimization of point-like facilities used by independent users 444
9.2.2 Locational optimization of points in a three-dimensional space 451
9.2.3 Locational optimization of point-like facilities used by groups 452
9.2.4 Locational optimization of a hierarchical facility 454
9.2.5 Locational optimization of observation points for estimating the total quantity of a spatial variable continuously distributed over a plane 458
9.2.6 Locational optimization of service points of a mobile facility 462
9.2.7. Locational optimization of terminal points through which users go to a central point 463
9.3 Locational optimization of lines 467
9.3.1 Locational optimization of a service route 467
9.3.2 Locational optimization of a network 470
9.3.3 Euclidean Steiner tree 473
9.4 Locational optimization over time 478
9.4.1 Multi-stage locational optimization 478
9.4.2 Periodic locational optimization 481
9.5 Voronoi fitting and its application to locational optimization problems 485
9.5.1 Method of fitting a Voronoi diagram to a polygonal tessellation 485
9.5.2 Locational optimization for minimizing restricted areas 488
References **489**
Index **523**

Foreword

I was delighted to be asked to write a preface to this beautiful and outstandingly original book. It is the unique treatise on its subject, it fills a serious gap in the literature and it covers the theory and the huge range of applications in a masterly way.

The authors are right to distinguish Voronoi *diagrams* and Delone *tessellations*. The Delone construction decomposes a Euclidean space of m dimensions, containing a given set of points, into non-overlapping space-filling *simplexes* (not, of course, all of the same shape and size), so that it tessellates the space using tiles that are identical with one another up to linear transformations. The Voronoi construction also splits up the space into polyhedral cells, but now they are much less uniform in character—the number of faces will vary from one cell to another, so that it would be wrong to call the result a tessellation.

These mutually dual procedures give fascinating but different insights into the structure of a set of points in m dimensions, and they have found numerous applications. At the time of writing there is a new application on the largest of all possible scales which throws light on the structure of the Universe as we see it. This will be seen as a particularly interesting development when one recalls that most of the earlier applications (for example, to the study of the structure of metallic composites, and other such aggregates) were on the microscopic scale. The reader of this book is strongly urged to look at a review paper just published by Icke and van de Weygaert (*Quart. Journal, Royal Astronomical Society*, **32**, 85–112). There it is shown that the Voronoi construction not only gives insight into the distribution of galaxies, but also permits a new approach to the dynamics that mould the shape of the universe we live in.

My own contributions have been in the Delone tradition, and are concerned (for example) with the way in which high-dimensional Delone simplexes pack together around a common vertex. Thus in 15 dimensions the number of such locally associated simplexes turns out to be of the order of 44 million million. This implies a related statement about the Voronoi polyhedra, and there tells us something about the number of faces of an individual cell. It seems likely that the huge number of Delone simplexes in such a local 'fan' can be roughly partitioned into a moderate number of 'chunky' simplexes (substantial faces in the Voronoi case), and a vast number of 'needle-like' ones (tiny faces), but we have no precise information on this matter at the moment.

It is a great pleasure to welcome this book to the Wiley series.

David Kendall

Acknowledgments

So many people helped in so many ways during the preparation of this book that it is only possible to acknowledge a few of them individually. First, we are deeply grateful to D.G. Kendall, who read through the draft and encouraged its publication; and to Y. Asami, C. M. Hoffmann, M. Iri, K. Murota and A. Suzuki, who suggested or commented on parts of the draft. Our special thanks also go to D. A. Aboav, F. Aurenhammer, H. Edelsbrunner, S. Egginton, J. D. Embury, M. F. Goodchild, M. Hori, H.-C. Imhof, G. Le Caer, U. Lorz, J. Mecke, R. E. Miles, J. Moller, L. Muche, Y. Ohsawa, N. Rivier, Y. M. Seoung, D. Stoyan, T. Suzuki, M. Tanemura, G. Toussaint, D. S. Wilkinson, H. Yomono and L. Zanietti, who provided material. We must also express our debt to A. Dawkins, S. Henry, H. Honkers, J. Horton, T. Kaneko, O. Kurita, R. Metcalfe, P. Schaus, M. Stone and T. Yoshikawa, among others, who assisted in production. For the help that they have given us, we are indebted to the staff of the publisher, in particular, C. Farmer, S. Gale, J. Narain and H. Ramsey. We should also acknowledge the award of a Book Preparation Grant from Wilfrid Laurier University which helped meet costs incurred during the preparation of the manuscript. Finally we are grateful for academic e-mail networks which made us feel as if we had been working in the same office.

CHAPTER 1

Introduction

On initial consideration the following problems, which concern a variety of phenomena at disparate scales, would appear to have little in common:

- an astronomer studying the structure of the Universe;
- an archaeologist attempting to identify the parts of a region under the influence of different neolithic clans;
- a meteorologist estimating precipitation at a gauge which has failed to operate;
- an urban planner locating public schools in a city;
- a physicist studying the behaviour of liquid argon;
- a physiologist examining capillary supply to muscle tissue.

However, these problems (all of which, together with many others, are dealt with below) can be resolved by approaches developed from a single concept which forms the subject of this book.

The concept is a simple but intuitively appealing one. Given a finite set of distinct, isolated points in a continuous space, we associate all locations in that space with the closest member of the point set. The result is a partitioning of the space into a set of regions. The solid lines in Figure 1.0.1 show a simple instance for two-dimensional Euclidean space. Given its widespread occurrence it is not surprising that this concept has been 'discovered' many times in many different places and as a result both the diagram and its constituent regions are known by a plethora of aliases. One immediate problem we faced in writing this book was the choice of an appropriate name which would be recognized by those interested in the concept. We selected the term *Voronoi diagram*, which seems to be the one most extensively used, although equivalent terms are identified in Section 1.2.

A second diagram can be constructed from the Voronoi diagram in m-dimensional space by joining those points whose regions share an $(m-1)$-dimensional face (see the broken lines in Figure 1.0.1). We refer to this dual diagram as the *Delaunay tessellation*, although it too has a number of aliases (see Section 1.2). The Delaunay tessellation may also be constructed

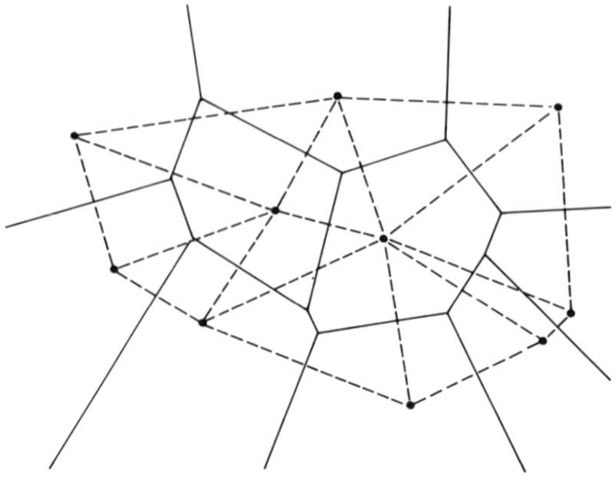

Figure 1.0.1 A planar Voronoi diagram (solid lines) and its dual Delaunay tessellation (dashed lines).

directly from the point set by taking each $(m + 1)$-ad of points and examining its circumsphere. If the interior of this does not contain a point of the set, we construct the simplex determined by the $(m + 1)$ points, but if it is not empty we do nothing. After all possible $(m + 1)$-ads have been considered in this way, the result is the Delaunay tessellation.

Voronoi diagrams and Delaunay tessellations are two of a few truly interdisciplinary concepts with relevant material to be found in, but not limited to, anthropology, archaeology, astronomy, biology, cartography, computationl geometry, crystallography, ecology, forestry, geography, geology, marketing, metallography, meteorology, operations research, physics, physiology, statistics and urban and regional planning. An unfortunate consequence of this is that the material is extremely diffused, of varying mathematical sophistication and quite often idiosyncratic. The amount of material relating to Voronoi diagrams has also been subject to continuing escalation in recent years as witnessed by the dominance of post 1980 references in the References. Much of this proliferation results from the basic concept being extended in a variety of ways (see Chapter 3). Consequently, our major motive for writing the book was to synthesize and to unify as much of the material as possible and to present it in a structured form. This seemed particularly important because, to our knowledge, no other book exists that is devoted exclusively to the topic. The one which comes closest is Aurenhammer (1988) but this is a short (87 pages), highly selective survey which emphasizes computational algorithms and only offers brief descriptions of applications. A number of recent books also provide limited treatments of Voronoi diagrams as part of more extensive coverage, for example Ahuja and Schachter (1983), Preparata and Shamos (1985), Iri and Koshizuka (1986), and Stoyan *et al.* (1987) but in most cases Voronoi diagrams account for less than ten percent of the books' contents. In addition, the second-last work is in Japanese which restricts its exposure to an English language audience.

In the works mentioned above, material relating to Voronoi diagrams has been presented by adopting a particular perspective, say that of the natural or social scientist, the mathematician or the computer scientist, with the last perspective being particularly favoured. With the exception of Stoyan *et al.* (1987), little consideration has been given to applications. Here we deliberately avoid emphasizing a particular perspective and instead attempt to provide a comprehensive and balanced treatment of all aspects of Voronoi diagrams including their definition, history, computation, statistical properties and applications. It is our hope and intention that the book should appeal equally to those whose concern with Voronoi diagrams is theoretical, practical or both. However, in view of the on-going and rapid development of Voronoi diagrams, we feel that this book is best regarded as a report on the current state of affairs from a most dynamic area rather than an exhaustive study.

1.1 OUTLINE

We have already noted that the concept of the Voronoi diagram is used extensively in a variety of disciplines and has independent roots in many of them. In Section 1.2 we begin our treatment of the Voronoi diagram by tracing its origins and that of the Delaunay tessellation from the nineteenth century up to the present. In the course of this exploration we identify the other names by which both structures are known. Before proceeding further, in order to make the book as self-contained as possible, in Section 1.3 we briefly review some basic features of a number of topics involved in the subsequent discussion of Voronoi diagrams. These include sets, matrices, geometry, calculus, graphs, probability theory and stochastic point processes. The material in this section is not intended to be exhaustive but rather to provide a sufficient introduction for those who may be unfamiliar with such topics.

Chapter 2 begins with a formal definition of both the Voronoi diagram (Section 2.1) and the Delaunay tessellation (Section 2.2). Sections 2.3 and 2.4 present properties of both structures, most of which are exploited in applications described elsewhere in the book. This chapter concludes with a discussion of a number of other spatial structures which are related to the Delaunay tessellation and which have been used in a variety of applications including diffusion modelling, numerical taxonomy and the exploratory display and analysis of data.

A major reason for the continuing success of the Voronoi diagram is that it can be generalized in a variety of ways. Such generalizations are the subject of Chapter 3 and include weighting the points in various ways (Section 3.1), considering regions associated with subsets of points rather than individual points (Sections 3.2 and 3.3), including obstacles in the space (Section 3.4), considering regions associated with sets of geometric features other than points (Sections 3.5 and 3.6), examining Voronoi diagrams involving non-Euclidean distances (Section 3.7) and on networks (Section 3.8).

In Chapter 4, after an initial consideration of data structures for representing the Voronoi diagram (Section 4.2), we present major algorithms for constructing it (Sections 4.3–4.5). For application purposes it is necessary to be able to implement these algorithms and in Section 4.6 we consider various practical techniques for doing this. Up to this point in the chapter the discussion concentrates on the planar Voronoi diagram, but in Section 4.7 we consider Voronoi diagrams in higher dimensions and in Sections 4.8 and 4.9 we examine algorithms for generating the various generalized diagrams introduced in Chapter 3.

The Poisson Voronoi diagram (PVD) refers to the situation in which the points are located in space 'at random' according to the homogeneous Poisson point process. This diagram has been used extensively both as a descriptive and a normative model in the investigation of a wide range of empirical situations in both the natural and social sciences (Section 5.2) as diverse as the nesting territories of Royal terns and mouthbreeder fish, bus service districts in England and Wales and carbon particles in steels. Chapter 5 presents the major properties of the PVD (Section 5.1) and its constituent regions (Section 5.4) as well as those of the dual Poisson Delaunay tessellation (Section 5.9), sections of the three-dimensional PVD (Section 5.5) and generalizations of the PVD produced by weighting the points (Section 5.6), considering subsets of points rather than individual points (Section 5.7) and the PVD on the surface of a sphere (Section 5.8). We also consider briefly other types of 'random' Voronoi diagrams (Section 5.10). In view of the importance of the PVD and associated Voronoi diagrams to applications, the emphasis in this chapter is on the presentation of results relating to characteristics of Voronoi regions in such structures. Although some of these are exact results derived analytically, the majority are estimates obtained from Monte Carlo simulation procedures, the different types of which are described in Section 5.3.

In the remaining chapters the emphasis shifts towards applications of Voronoi diagrams. We identify four major areas, spatial interpolation, models of spatial processes, point pattern analysis and locational optimization, each of which is the subject of a separate chapter. In Chapter 6, under the general heading of spatial interpolation, we consider how the Voronoi diagram has been used to facilitate the presentation and analysis of spatial data (values of one or more variables observed at a set of data sites in space). Such uses range from employing Voronoi diagrams for defining various types of neighbour relationships (Section 6.1), through ordering multivariate data (Section 6.5) to the construction of continuous surfaces from a set of data values observed at punctiform data sites (Sections 6.1–6.3), including the approximation of such surfaces (Section 6.4). In this chapter the material is organized in terms of whether the Voronoi diagram (Section 6.1) or the Delaunay tessellation (Sections 6.2–6.5) is used.

Chapter 7 considers how, by equating the procedures involved in defining the Voronoi diagram with assumptions involved in spatial processes, we may produce simple spatial models. At least four types of spatial processes, assignment (Section 7.1), growth (Section 7.2), some spatial temporal processes (Section 7.3) and two-species competition (Section 7.4), can be represented in this way. The application of such models is extensive and examples from areas as diverse as crystallography, solid state physics, biology, astronomy, physiology, archaeology and urban economics are presented. To determine if such models are appropriate for a given phenomenon we need to know to what degree the spatial structure of the phenomenon resembles that of a Voronoi diagram and so we conclude this chapter (Section 7.5) by considering techniques for determining if a given structure is a Voronoi diagram and, if not, the extent to which it approximates a Voronoi diagram.

Use of the Voronoi diagram and the Delaunay tessellation in point pattern analysis is described in Chapter 8. Section 8.1 considers various methods using the Voronoi diagram while Section 8.2 examines those that make use of the Delaunay tessellation. In both of these sections our concern is with how the points in a given set are positioned with respect to each other and the space in which they are located. In the final section of this chapter (Section 8.3) our attention shifts to a consideration of how the members of the point set are located with respect to other objects, not members of the set, which are also located in the same space. These objects may be other points, lines or areas or a combination of all three.

Finally, Chapter 9 deals with the various ways in which the Voronoi diagram can be applied in the solutions to locational optimization problems. Although problems relating to locational optimization on a network have been the subject of extensive prior investigation, less work has been undertaken on problems in the plane and it is such problems that are emphasized in this chapter. Planar optimization problems are important in a number of natural and social sciences. Examples range from locating observation sites for estimating a continuous spatial random variable to the location of various types of public facilities such as health clinics, fire stations and public transportation stops. All of these problems involve objects which can be represented as points and they, and other situations, form the subject matter of Section 9.2. Section 9.3 considers problems relating to the location of linear features, while Section 9.4 extends the discussion to situations where the objects under study are not located synchronously in the space.

The final section (Section 9.5) illustrates how locational optimization procedures relate to fitting a Voronoi diagram to a given tessellation, a topic that is treated more generally in Section 7.5.

1.2 HISTORY OF THE CONCEPT OF THE VORONOI DIAGRAM

It is quite possible that the concept of the Voronoi diagram is of considerable antiquity. As we will show elsewhere in the book (but especially in Chapter 7), many different kinds of natural structures closely resemble Voronoi diagrams and it seems unlikely that such structures would have gone unnoticed by early scientists and observant laymen alike. In his treatment of cosmic fragmentation in both *Le Monde de Mr Descartes, ou Le Traité de la Lumière* published in 1644 (but written between 1629 and 1633) and in Part III of *Principia Philosophiae*, also published in 1644, Descartes uses Voronoi-like diagrams to show the disposition of matter in the solar system and its environs (see Figure 1.2.1). Since these diagrams are not accompanied by any special commentary relating to their construction, it is possible that such figures were not uncommon at that time. In fact diagrams such as that in Figure 1.2.1 are actually more akin to a generalized version of a Voronoi diagram, known as a weighted Voronoi diagram (discussed in Section 3.1).

The first undisputed presentations of the concept that we are aware of appear in the work of Dirichlet (1850) and Voronoi (1908) who, in their studies on positive definite quadric forms, considered a special form of the Voronoi diagram. Dirichlet treated two- and three-dimensional cases, whereas Voronoi examined the general m-dimensional case. Their conern was with the distribution of points with integer coordinates that give minima of the values of a given quadric form. In that context they considered the set of points regularly placed in the m-dimensional space generated by linear combinations of m linearly independent vectors with integer coefficients. This set contains infinitely many points, and the Voronoi diagram generated by this set of points gives the partition of the space into mutually congruent polyhedra. Using this concept Voronoi, for example, gave a simple proof of the upper bound $2(2^m-1)$ of the number of $(m-1)$-dimensional facets of an m-dimensional polyhedron, the translational repetition of which can fill the whole space without overlap. As a result of this early work two of the many labels used to describe the diagram, the *Voronoi diagram* and the *Dirichlet tessellation*, came to be adopted.

Given that the first developments of the Voronoi diagram concept involved sets of points that were regularly placed in space, it is not surprising that its first extension was to the area of crystallography. Work in this area began in the late nineteenth and early twentieth centuries (see Nowacki, 1965, for a review) and was dominated by German researchers. It was in this context that Niggli (1927) introduced the label *Wirkungsbereich* (domain of action, field of activity, area of influence) to refer to a Voronoi region.

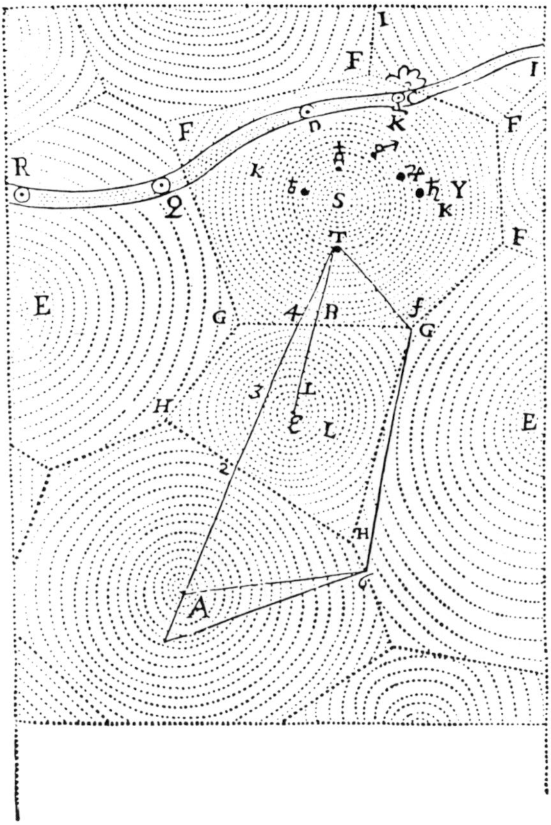

Figure 1.2.1 The disposition of matter in the Solar System and its environs by Descartes (S is the sun; ϵ is a star; RQD represents the path of a comet; polygonal areas represent heavens). (*Source*: Copy of the original in Descartes (1644) reproduced in Mahoney (1979).)

Much of this early work was involved with the broader concern of identifying congruent space-filling domains, a line of research that has continued to the present (see Section 7.1). In fact, although they were aware of the previous development of Voronoi regions, Grünbaum and Shephard (1980) coined the term *plesiohedra* to reflect that they were part of a larger set of regions known as stereohedra, congruent copies of which are capable of filling three-dimensional space without overlap except possibly for their boundaries.

At the same time the Voronoi diagram concept was being applied in crystallography, it appears to have been developed independently in at least two other areas, both involving spatial interpolation. The first of these occurred in meteorology when Thiessen (1911) used Voronoi regions as an aid to computing more accurate estimates of regional rainfall averages (see Section 6.1.1). While acknowledging Thiessen's paper, Horton (1917) claims to have developed the procedure independently in exactly the same context. Notwithstanding this, Whitney (1929) referred to the procedure as 'Thiessen's method' and since then the term *Thiessen polygon* has been the preferred one in two-dimensional applications in both meteorology and geography.

The second area involved the estimation of ore reserves in a deposit using information obtained from bore holes (see Section 6.1.1). The earliest work on this topic by Boldyrev appeared in Russia in 1909 (Popoff, 1966) and was independently developed in the United States by Davis and Harding (Harding, 1920–21, 1923). Here the Voronoi regions are termed simply *area of influence polygons*, a name that remains current in this area of application (Hayes and Koch, 1984).

Despite this expansion of the use of Voronoi diagrams, further independent rediscoveries continued. In their study of the chemical properties in metallic sodium the physicists Wigner and Seitz (1933) described Voronoi regions for points arranged on a lattice in three-dimensional space. Although they did not label these polyhedra, the term *Wigner–Seitz regions* has come into quite common usage amongst physicists even when they are aware of the precedence of Voronoi diagrams, with the former term being used when the regions are regular polytypes and the latter when they are not. Somewhat ironically, over twenty years later, Voronoi diagrams were rediscovered in a similar context by Frank and Kasper (1958) modelling complex alloy structures as packings of spheres. Their concern focused on the neighbour relationships defined by the Voronoi polyhedra of atoms contiguous to that of the Voronoi polyhedron of a given atom, which they labelled the *domain of an atom*.

Later still, and in the span of only two years, Voronoi diagrams were discovered not once but twice more in the field of ecology. First, in his work in estimating the intensity of trees in a forest, Brown (1965) defined a Voronoi region for an individual tree, calling it the *area potentially available (APA)* to a tree. One year later Mead's paper appeared (Mead, 1966) using the same concept for plants and labelling the resulting Voronoi regions as *plant polygons*. This name has remained reasonably common in this area of application despite the early recognition by Jack (1967) that such polygons are equivalent to Voronoi regions.

The final rediscovery that we are aware of is by Hoofd *et al.* (1985) who defined Voronoi regions with respect to the centres of capillaries in sections of tissue, labelling the resulting polygons *capillary domains*. It is also possible that there have been more obscure discoveries made at other times and places and that there might even be future rediscoveries. Indeed, as late as 1987 Icke (Icke and Van de Weygaert, 1987) admits to developing them yet again, this time in an astronomical context, only to be saved from claiming such work as novel by his co-author's discovery of the vast field of literature on Voronoi diagrams.

Although the development and initial applications of Voronoi diagrams occurred in the natural sciences, the concept also spread into the social sciences. The earliest application that we are aware of was by Bogue (1949) who used the Voronoi polygons defined about US metropolitan centres (represented as points on a map) as surrogates for their market areas. However, prior to this, work by a number of researchers, some of it dating back to the mid-nineteenth century, had considered the nature of the boundary between the market areas of two competing centres under various conditions of market prices and transportation costs. This material, which is now usually referred to as the economic law of market areas, involves the same concepts used in the definition of weighted Voronoi diagrams (see Section 3.1). Shieh (1985) reviewed the development of this material. Aided by other initial studies by Snyder (1962) and Dacey (1965), the definition of market areas has remained the main area of application with work extending down to the level of individual retail stores, although geographers, anthropologists and archaeologists have also used the concept to model other types of human territorial systems. Following the suggestion of Evans (1967), geographers were also the first to use Voronoi (and Delaunay) concepts in the analysis of two-dimensional point patterns (Boots, 1974).

So by the 1960s knowledge of the Voronoi concept was current in both the natural and social sciences. However, empirical applications remained somewhat limited because of the lack of simple and efficient means of constructing them, with practitioners relying on methods involving compass and ruler and fraught with ambiguities (Kopec, 1963). This situation motivated solutions from the rapidly developing area of computer science and by the early 1970s a number of algorithms had been developed to construct Voronoi diagrams in both two and three dimensions. In turn, this work stimulated other developments in computer science which contributed to the now flourishing field of computational geometry. This endeavour is typified by the seminal paper of Shamos and Hoey (1975). Not only did they present an algorithm for constructing the Voronoi diagram, they also illustrated how it could be used to solve a series of what were then seen as essentially independent problems relating to a finite set of distinct points in Euclidean space, such as finding the minimum spanning tree, identifying the nearest neighbour of each point and finding the largest circle, containing no points of the set, the centre of which is inside the convex hull of the point set. In addition, they also suggested ways of generalizing the Voronoi diagram by considering Voronoi regions associated with subsets of k points of the entire point set rather than individual points. As a consequence of this and other contemporary initiatives, in the past 15 years the basic concept has been extended in a great variety of ways, which are detailed in Chapter 3.

The dual concept of the Delaunay tessellation also has a history marked by rediscoveries, although these are not as frequent as in the case of the Voronoi diagram. The concept has its origins with Voronoi (1908) who defined it by way of the neighbour relationships in the Voronoi diagram. However, it was the Russian number theorist, Boris Nikolaevitch Delone (1890–1980), who first defined the concept using the empty sphere method (Delaunay, 1934), thus explaining our choice of the name for the concept. As we show in Chapter 2, one of the properties of the Delaunay tessellation in two dimensions when the tessellation consists exclusively of triangles is that the triangles are as equilateral as possible, a property used by Whitney (1929) to establish a triangulation (and hence a Delaunay tessellation) for use in spatial interpolation.

Smith (1964) seems to have rediscovered the concept in his work on three-dimensional aggregates, labelling the structure the *simplical graph*. Nevertheless, a year later (Smith, 1965) he recognized the precedence of Delaunay's work and re-labelled the constituent regions as *Delaunay simplices*, a term also used by Rogers (1964) in his comprehensive work on tessellations. A recent rediscovery of the concept is that of Christ *et al.* (1982) in the context of random lattice field theory. Although they recognized that the structure is the dual of the Voronoi diagram, they did not give a name to it.

1.3 MATHEMATICAL PRELIMINARIES

In this section we present some mathematical notions that will be commonly used in this text. The major notions are those related to sets, matrices, geometry, derivatives, integration, graphs, probabilities, statistical tests and spatial stochastic point processes. Since space is limited, some developments may not be in full detail. The reader who wishes to understand these notions with full formal derivations should consult the relevant mathematical textbooks (such as Royden, 1963 (real analysis); Busacker and Saaty, 1965 (graph theory); Feller, 1957 (probability) and Wilks, 1962 (statistics)).

1.3.1 Sets

We naïvely understand a *set of elements* as a collection of any description. We allow a set with no elements, called an *empty set*, denoted by $\varnothing$. A set with at least one element is called a *non-empty set*. A non-empty set consisting of n elements, $a_1, \ldots, a_n$, is written as $\{a_1, \ldots, a_n\}$, which is sometimes denoted by $\{a_i\}$ in brief when n is understood. The notation $\mathbb{R}$ is used for the set of all real numbers; I_∞ for the set of all natural numbers, $I_\infty = \{1, 2, \ldots, n, \ldots\}$; and I_m for the set of natural numbers from 1 to m ($m < \infty$), where the suffix m indicates the maximum integer in the set, e.g. $I_m = \{1, \ldots, m\}$, $I_n = \{1, \ldots, n\}$. If the elements of a set can be indexed with I_n, such as $\{a_1, \ldots, a_n\}$, the set is called a *finitely countable set*. A set whose elements can be indexed with I_∞ is called an *infinitely countable set*. Finitely and infinitely countable sets are called *countable sets*. Sets that are not countable sets are called *uncountable sets*. An example is the set of all real numbers in $\mathbb{R}$.

A set of elements satisfying some conditions is written as $\{x|$ conditions$\}$. For example, the set of all real numbers that are greater than 1 is written as $A = \{x| \ x > 1, x \in \mathbb{R}\}$. If element x belongs to set A, we denote it by $x \in A$; if it does not belong to A, then $x \notin A$. For example, $4 \in A$ and $0 \notin A$ for the set A given in the above example. If all elements in set A belong to set B, i.e. $a \in B$ for all $a \in A$, we call set A a *subset* of B, or we say that set A is *contained* in B (set B *contains* set A), and denote it by $A \subset B$. For example, if $B = \{x| \ x > 0, x \in \mathbb{R}\}$, then $A \subset B$, where the set A is given in the above.

For two sets A and B, we define three set operations: intersection, union and complement. The *intersection* of A and B, denoted by $A \cap B$, is the set of all elements that belong to both A and B, i.e.

$$A \cap B = \{x| x \in A \quad \text{and} \quad x \in B\} . \tag{1.3.1}$$

For example, if set A is given by points in the square in Figure 1.3.1(a) and set B is given by points in the triangle, then the intersection $A \cap B$ is given by points in the shaded area. If $A \cap B = \varnothing$, sets A and B are said to be *disjoint*. The *union* of A and B, denoted by $A \cup B$, is the set of all elements that belong either to A or B (Figure 1.3.1(b)), i.e.

$$A \cup B = \{x| x \in A \quad \text{or} \quad x \in B\} . \tag{1.3.2}$$

The *complement* of B with respect to A, denoted by $A \backslash B$, is the set of all elements of A that do not belong to B (Figure 1.3.1(c)), i.e.

$$A \backslash B = \{x| x \in A \quad \text{and} \quad x \notin B\} . \tag{1.3.3}$$

The above set operations satisfy quite a number of laws. Here we mention only three major ones: *commutative laws, associative laws* and *distributive laws*, which are defined by

$$A \cap B = B \cap A, \qquad A \cup B = B \cup A, \tag{1.3.4}$$

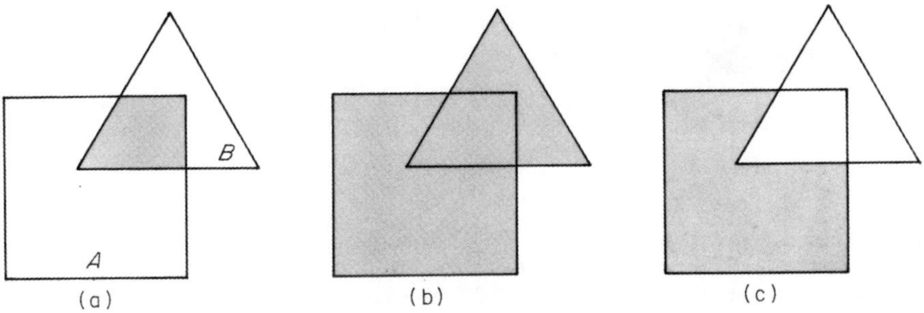

Figure 1.3.1 Set operations: (a) intersection; (b) union; (c) complement.

$$A \cap (B \cap C) = (A \cap B) \cap C, \qquad A \cup (B \cup C) = (A \cup B) \cup C \tag{1.3.5}$$

$$A \cap (B \cup C) = (A \cap B) \cup (A \cap C), \qquad A \cup (B \cap C) = (A \cup B) \cap (A \cup C) \tag{1.3.6}$$

respectively.

Using equation (1.3.5), we define $A_1 \cap A_2 \cap A_3 \cap \cdots \cap A_n$ by $(((A_1 \cap A_2) \cap A_3) \cap \cdots) \cap A_n$, which is written as $\cap_{i=1}^{n} A_i$ in brief, i.e.

$$\bigcap_{i=1}^{n} A_i = \{x | x \in A_1, \text{ and } \ldots, \text{ and } x \in A_n\} . \tag{1.3.7}$$

Similarly, we write $(((A_1 \cup A_2) \cup A_3) \cup \cdots \cup A_n)$ as

$$\bigcup_{i=1}^{n} A_i = \{x | x \in A_1, \text{ or } \ldots, \text{ or } x \in A_n\} . \tag{1.3.8}$$

If A and B are non-empty sets, then the *Cartesian product* of A and B, denoted by $A \times B$, is the set of all ordered pairs (a, b) with $a \in A$ and $b \in B$ (the shaded area in Figure 1.3.2(a)). More generally, if $A_1, \ldots, A_m$ are non-empty sets, then the *m-dimensional Cartesian product*, $A_1 \times \cdots \times A_m$, is the set of ordered m-tuples $(a_1, \ldots, a_m)$ with $a_i \in A_i$, $i \in I_m$. We sometimes refer to $(a_1, \ldots, a_m)$ as *Cartesian coordinates* and call a_i the *i*th *coordinate* or *component*. When $A_i = \mathbb{R}$ for all $i \in I_m$, the Cartesian product is called the m-dimensional *real Cartesian space*. Specifically, we sometimes refer to $\mathbb{R}$ as the *real line* and $\mathbb{R}^2$ as the *real plane*.

For two non-empty sets A and B, we define a set, f, of ordered pairs (a, b) in $A \times B$ which satisfy the condition that if $(a, b) \in f$ and $(a, b') \in f$, then $b = b'$. We call the set f a *function* from $\{a| (a, b) \in f\}$ to B (Figure 1.3.2(a)). We refer to element b as the *value* of f at point a, or we say that element b is the *image* under f of point a. Customarily, we write $b = f(a)$

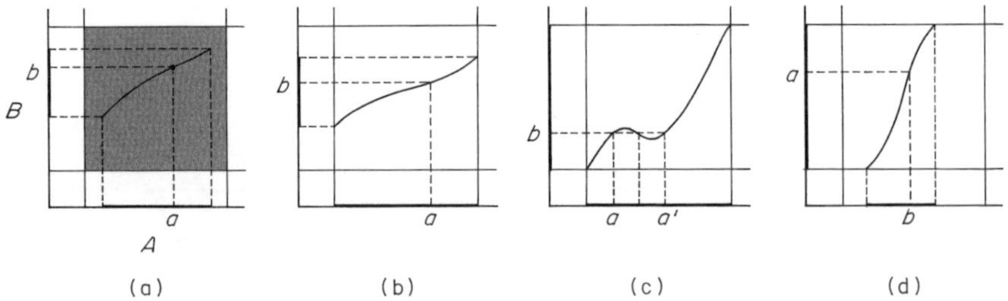

Figure 1.3.2 Cartesian product $A \times B$ and functions. (a) Cartesian product $A \times B$ (the shaded area), coordinates (a, b), a function $b = f(a)$, the domain of f (the heavy solid line in A), and the range of f (the heavy solid line in B). (b) Mapping of A into B. (c) Mapping of A onto B. (d) The function inverse to the function shown in (b).

for $(a, b) \in f$. The set $\{a|\ (a, b) \in f\}$ is called the *domain* of f (the heavy line in A in Figure 1.3.2(a)) and the set $\{b|\ (a, b) \in f\}$ is called the *range* of f (the heavy line in B in Figure 1.3.2(a)). If the domain of f is A (Figure 1.3.2(a)), we sometimes say that f *maps A into B* or f is a *mapping* of A *into B*. If, in addition, the range of f is B (Figure 1.3.2(c)), we say that f *maps A onto B* or f is a mapping of A *onto B*. It should be noted that the condition for a function mentioned above does not imply the condition that if $(a, b) \in f$ and $(a', b) \in f$, then $a = a'$ (observe Figure 1.3.2(c)). If this condition does hold, we say that f is *one-to-one*. The functions in Figures 1.3.2(a) and 1.3.2(b) are one-to-one, but the function in Figure 1.3.2(c) is not. Suppose that f is a one-to-one function with the domain in A and the range in B, and let $g = \{(b, a) \in B \times A|\ (a, b) \in f\}$. Then g is a one-to-one function with the domain given by the range of f and with the range given by the domain of f. We call the function g the *function inverse to f* or the *inverse function* of f, and we ordinarily denote g by f^{-1}. Figure 1.3.2(d) shows the function inverse to the function shown in Figure 1.3.2(b).

1.3.2 Matrices

A *matrix* is an array of elements which has a particular shape. It is always a rectangular (including square) arrangement of elements in rows and columns. For example, a matrix with m rows and n columns, called an $m \times n$ matrix, is written as

$$A = \begin{bmatrix} a_{11} & a_{12} & \cdots & a_{1n} \\ a_{21} & a_{22} & \cdots & a_{2n} \\ \vdots & \vdots & \ddots & \vdots \\ a_{m1} & a_{m2} & \cdots & a_{mn} \end{bmatrix}. \tag{1.3.9}$$

Specifically, we call an $m \times 1$ matrix a *column vector*, and a $1 \times n$ matrix a *row vector*. In this text, we denote a column vector by a bold face letter, such as

$$\mathbf{x} = \begin{bmatrix} x_1 \\ x_2 \\ \vdots \\ x_m \end{bmatrix}. \tag{1.3.10}$$

To save space, we may write the matrix of equation (1.3.9) as $[a_{ij}]$ when $m \times n$ is understood.

A matrix in which the number of rows and columns are the same (an $n \times n$ matrix) is called a *square matrix*. A square matrix $[a_{ij}]$ whose elements are given by $a_{ij} = 1$ for $j = i \in I_n$ and $a_{ij} = 0$ for $j \neq i$, $i, j \in I_n$, i.e.

$$I = \begin{bmatrix} 1 & 0 & \cdots & 0 \\ 0 & 1 & \cdots & 0 \\ \vdots & \vdots & \ddots & \vdots \\ 0 & 0 & \cdots & 1 \end{bmatrix}, \tag{1.3.11}$$

is called an *identity matrix*, and denoted by I (distinguish I from I_n defined in Section 1.3.1).

A square matrix $[a_{ij}]$ whose elements are given by $a_{ij} = 0$ for all $j \neq i$, i, $j \in I_n$, i.e.

$$\begin{bmatrix} a_{11} & 0 & \cdots & 0 \\ 0 & a_{22} & \cdots & 0 \\ \vdots & \vdots & \ddots & \vdots \\ 0 & 0 & \cdots & a_{nn} \end{bmatrix}, \tag{1.3.12}$$

is called a *diagonal matrix*. Obviously, the identity matrix is a diagonal matrix.

The *transpose* of an $m \times n$ matrix A, denoted by A^T, is an $n \times m$ matrix whose rows are, in order, the columns of A. For example, the transpose of the matrix A of equation (1.3.9) is written as

$$A^T = \begin{bmatrix} a_{11} & a_{21} & \cdots & a_{m1} \\ a_{12} & a_{22} & \cdots & a_{m2} \\ \vdots & \vdots & \ddots & \vdots \\ a_{1n} & a_{2n} & \cdots & a_{mn} \end{bmatrix}. \tag{1.3.13}$$

If a matrix A is a square matrix and if $A = A^T$, the matrix A is said to be a *symmetric matrix*.

It follows from the definition of A^T that the transpose of a column vector is a row vector, and the transpose of a row vector is a column vector. For example, the transpose of the column vector of equation (1.3.10) is written as

$$\mathbf{x}^T = (x_1, x_2, \ldots, x_m). \tag{1.3.14}$$

The *matrix addition* of two $m \times n$ matrices $A = [a_{ij}]$ and $B = [b_{ij}]$ is performed by

$$A + B = \begin{bmatrix} a_{11} + b_{11} & a_{12} + b_{12} & \cdots & a_{1n} + b_{1n} \\ a_{21} + b_{21} & a_{22} + b_{22} & \cdots & a_{2n} + b_{2n} \\ \vdots & \vdots & \ddots & \vdots \\ a_{m1} + b_{m1} & a_{m2} + b_{m2} & \cdots & a_{mn} + b_{mn} \end{bmatrix}, \quad (1.3.15)$$

Similarly, the *matrix subtraction* is performed by $A - B = [a_{ij} - b_{ij}]$.

The *scalar multiplication* or *scalar product* of a matrix A by α is performed by

$$\alpha A = \begin{bmatrix} \alpha a_{11} & \alpha a_{12} & \cdots & \alpha a_{1n} \\ \alpha a_{21} & \alpha a_{22} & \cdots & \alpha a_{2n} \\ \vdots & \vdots & \ddots & \vdots \\ \alpha a_{m1} & \alpha a_{m2} & \cdots & \alpha a_{mn} \end{bmatrix}. \quad (1.3.16)$$

The *matrix multiplication* of an $m \times n$ matrix A and an $n \times k$ matrix B is performed by

$$AB = \begin{bmatrix} \Sigma_{l=1}^{n} a_{1l}b_{l1} & \Sigma_{l=1}^{n} a_{1l}b_{l2} & \cdots & \Sigma_{l=1}^{n} a_{1l}b_{lk} \\ \Sigma_{l=1}^{n} a_{2l}b_{l1} & \Sigma_{l=1}^{n} a_{2l}b_{l2} & \cdots & \Sigma_{l=1}^{n} a_{2l}b_{lk} \\ \vdots & \vdots & \ddots & \vdots \\ \Sigma_{l=1}^{n} a_{ml}b_{l1} & \Sigma_{l=1}^{n} a_{ml}b_{l2} & \cdots & \Sigma_{l=1}^{n} a_{ml}b_{lk} \end{bmatrix} \quad (1.3.17)$$

(an $m \times k$ matrix). Note that $AB \neq BA$ in general even if both A and B are square matrices. Thus, we say that multiplying B from the right on A for AB, and multiplying B from the left on A for BA.

In a specific case in which A and B are vectors, i.e. $\boldsymbol{x}_1^{\mathrm{T}} = (x_{11}, \ldots, x_{1n})$ and $\boldsymbol{x}_2^{\mathrm{T}} = (x_{21}, \ldots, x_{2n})$, respectively, we call $A^{\mathrm{T}} B$, i.e.

$$\boldsymbol{x}_1^{\mathrm{T}} \boldsymbol{x}_2 = x_{11} x_{21} + \cdots + x_{1n} x_{2n} = \sum_{k=1}^{n} x_{1k} x_{2k}, \quad (1.3.18)$$

the *inner product* or sometimes the *scalar product* of $\boldsymbol{x}_1$ and $\boldsymbol{x}_2$.

For a given square matrix A, if a matrix B satisfies $AB = BA = I$, we call the matrix B the *inverse matrix* of the matrix A. Usually the matrix B is denoted by A^{-1}. An inverse matrix is used in various contexts. In our text it appears in the form $A^{-1} \boldsymbol{y}$. In theory, we can directly calculate this vector with an explicit form of A^{-1}. In practice, however, we usually carry it out through solving $A\boldsymbol{x} = \boldsymbol{y}$ with respect to $\boldsymbol{x}$, because the computational time is shorter. (Multiplying A^{-1} from the left on both sides of $A\boldsymbol{x} = \boldsymbol{y}$, we obtain $A^{-1} A\boldsymbol{x} = A^{-1} \boldsymbol{y}$, and so $\boldsymbol{x} = A^{-1} \boldsymbol{y}$.) This problem is known as a system of *simultaneous linear equations*. Many computational methods are proposed for this problem, such as the sweep-out method and the LU-decomposition method. The computer programs of these methods are now available in the ordinary software library.

1.3.3 Geometry

Consider an m-dimensional real Cartesian space $\mathbb{R}^m$ defined in Section 1.3.1. In this space an m-tuple or the Cartesian coordinates $(x_1, \ldots, x_m)$ indicate a *point*, denoted by p (Figure 1.3.3). If there are n points in $\mathbb{R}^m$, these points are denoted by $p_1, \ldots, p_n$ with the Cartesian coordinates $(x_{i1}, \ldots, x_{im})$, $i \in I_n$. Points p and p_i may be also represented by vectors $\mathbf{x}^T = (x_1, \ldots, x_m)$ and $\mathbf{x}_i^T = (x_{i1}, \ldots, x_{im})$, respectively. Sometimes it is more convenient to represent point p as the arrow from the origin o of $\mathbb{R}^m$ to point p, denoted by $\overrightarrow{op}$. The length of the arrow in $\mathbb{R}^2$ is, using the Pythagoras theorem, given by $\sqrt{(x_1^2 + x_2^2)} = \sqrt{(\mathbf{x}^T \mathbf{x})}$. Similarly, in $\mathbb{R}^m$, the length of the arrow, or the *norm* of vector $\mathbf{x}$, denoted by $\|\mathbf{x}\|$, is given by

$$\|\mathbf{x}\| = \sqrt{\mathbf{x}^T \mathbf{x}} = \left\{ \sum_{i=1}^{m} x_i^2 \right\}^{1/2}. \tag{1.3.19}$$

If a vector is designated as an arrow, it may be shifted provided that its length and direction are maintained, that is, an arrow may be shifted in a parallel way. In Figure 1.3.3, for example, $\overrightarrow{p_2p_3}$ and $\overrightarrow{op}$ have the same length and the same direction, and so these two arrows indicate the same vector. It should be noted, however, that if a vector is designated as a point, then the point cannot be shifted. For example, point p is indicated by $\overrightarrow{op}$ but not by $\overrightarrow{p_2p_3}$. We sometimes call a vector indicating a point a *location vector* or *a point vector*.

The scalar multiplication of $\mathbf{x}$ by a positive scalar λ, i.e. $\lambda\mathbf{x}$, is represented by the arrow obtained from extending (shrinking) $\overrightarrow{op}$ λ times in the same direction (Figure 1.3.4(a)). The scalar multiplication of $\mathbf{x}$ by a negative scalar $-\lambda$, i.e. $-\lambda\mathbf{x}$, is represented by the arrow obtained from extending

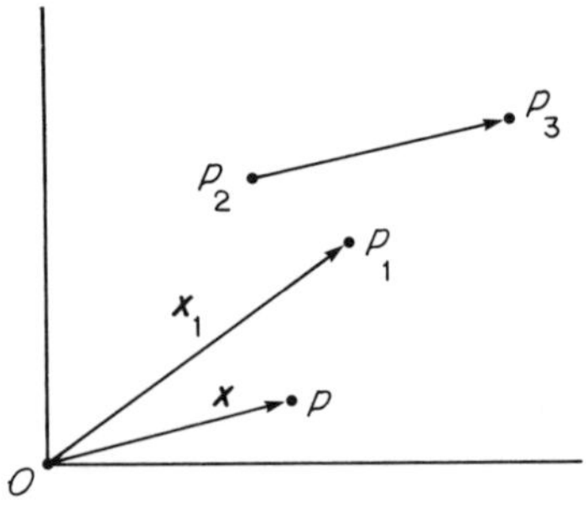

Figure 1.3.3 Vectors.

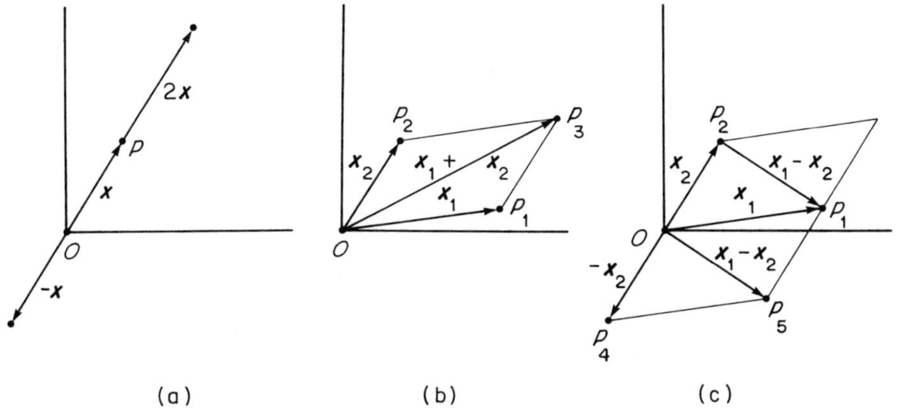

Figure 1.3.4 Vector operations: (a) scalar multiplication; (b) addition; (c) subtraction.

(shrinking) $\overrightarrow{op}$ λ times in the opposite direction (Figure 1.3.4(a)). The vector addition, $\mathbf{x}_1 + \mathbf{x}_2$, is represented by the diagonal $\overrightarrow{op_3}$ of the parallelogram constructed from $\overrightarrow{op_1}$ and $\overrightarrow{op_2}$ (Figure 1.3.4(b)). The vector subtraction, $\mathbf{x}_1 - \mathbf{x}_2$, is obtained from the addition $\mathbf{x}_1 + (-\mathbf{x}_2)$, where the vector $-\mathbf{x}_2$ is obtained from the vector $\mathbf{x}_2$ multiplied by a negative scalar -1, that is, $\mathbf{x}_1 - \mathbf{x}_2$ is indicated by the diagonal $\overrightarrow{op_5}$ of the parallelogram constructed from $\overrightarrow{op_1}$ and $\overrightarrow{op_4}$ (Figure 1.3.4(c)). Alternatively, the vector subtraction, $\mathbf{x}_1 - \mathbf{x}_2$, is represented by the diagonal $\overrightarrow{p_2p_1}$ of the parallelogram constructed from $\overrightarrow{op_1}$ and $\overrightarrow{op_2}$ (observe that $\overrightarrow{op_5}$ and $\overrightarrow{p_2p_1}$ represent the same vector $\mathbf{x}_1 - \mathbf{x}_2$).

Suppose that two vectors $\mathbf{x}_1$ and $\mathbf{x}_2$ are given by those in Figure 1.3.5(a). Then, we can find a scalar λ that satisfies $\mathbf{x}_2 = \lambda \mathbf{x}_1$. On the other hand, we cannot find such a λ for the vectors shown in Figure 1.3.5(b). We say that the vectors in panel (a) are *linearly dependent* and those in panel (b) *linearly*

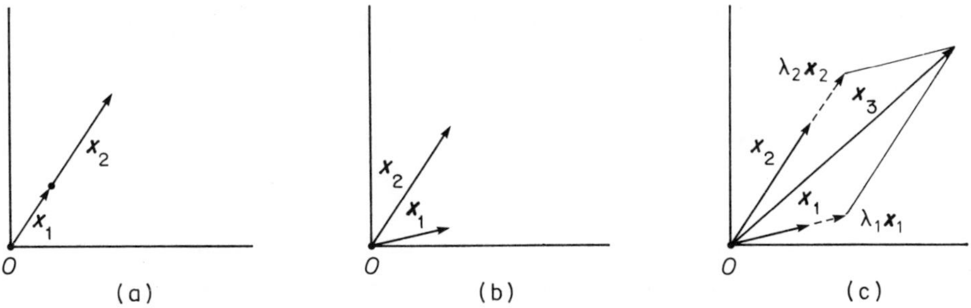

Figure 1.3.5 (a) Two vectors being linearly dependent; (b) linearly independent vectors; (c) three vectors being linearly dependent.

independent. This notion can be extended to n vectors in $\mathbb{R}^m$. We say that vectors $\mathbf{x}_1, \ldots, \mathbf{x}_n$ are *linearly dependent* in $\mathbb{R}^m$ if there exists a set of n scalars $\Lambda = \{\lambda_1, \ldots, \lambda_n\}$ which are not all zero such that

$$\lambda_1 \mathbf{x}_1 + \cdots + \lambda_n \mathbf{x}_n = 0 . \tag{1.3.20}$$

On the other hand, if the only Λ for which equation (1.3.20) holds is $\lambda_1 = \cdots = \lambda_n = 0$, then the vectors are said to be *linearly independent*. The number of linearly independent vectors in $\mathbb{R}^m$ is at most m. For example, in $\mathbb{R}^2$, as is shown in Figure 1.3.5(b), two vectors $\mathbf{x}_1$ and $\mathbf{x}_2$ are linearly independent, but three vectors $\mathbf{x}_1$, $\mathbf{x}_2$ and $\mathbf{x}_3$ are linearly dependent, because, as is indicated by the broken arrows in Figure 1.3.5(c), $\lambda_1 \mathbf{x}_1 + \lambda_2 \mathbf{x}_2 - \mathbf{x}_3 = 0$.

When two vectors $\mathbf{x}_1$ and $\mathbf{x}_2$ form the right angle as in Figure 1.3.6(a), we say that the vectors $\mathbf{x}_1$ and $\mathbf{x}_2$ are *orthogonal*, and denote it by $\mathbf{x}_1 \perp \mathbf{x}_2$. From Figure 1.3.6(a) it is evident that vectors $\mathbf{x}_1$ and $\mathbf{x}_2$ are orthogonal if and only if $\|\mathbf{x}_1 + \mathbf{x}_2\|$ equals $\|\mathbf{x}_1 - \mathbf{x}_2\|$, i.e. $(\mathbf{x}_1 + \mathbf{x}_2)^{\mathrm{T}} (\mathbf{x}_1 + \mathbf{x}_2) = (\mathbf{x}_1 - \mathbf{x}_2)^{\mathrm{T}} (\mathbf{x}_1 - \mathbf{x}_2)$. After a few steps of calculation, we obtain the following relation:

$$\mathbf{x}_1 \perp \mathbf{x}_2 \quad \text{if and only if} \quad \mathbf{x}_1^{\mathrm{T}} \mathbf{x}_2 = 0 . \tag{1.3.21}$$

In Figure 1.3.6(b), we perpendicularly project point p_2 on the line containing $\overrightarrow{op_1}$. Let p_2' be the projected point and $\mathbf{x}_2'$ be its location vector. Then we say that p_2' is the *perpendicular projection* or *orthogonal projection* of p_2 on the line containing $\overrightarrow{op_1}$. Since $\mathbf{x}_2' = \lambda \mathbf{x}_1$, and $(\lambda \mathbf{x}_1 - \mathbf{x}_2)^{\mathrm{T}} (\lambda \mathbf{x}_1) = 0$ (from equation (1.3.21)), the vector $\mathbf{x}_2'$ is written as

$$\mathbf{x}_2' = \frac{\mathbf{x}_2^{\mathrm{T}} \mathbf{x}_1}{\|\mathbf{x}_1\|^2} \mathbf{x}_1 . \tag{1.3.22}$$

Equation (1.3.22) becomes simply when point p_i with coordinates (x_{i1}, x_{i2}) in $\mathbb{R}^2$ is projected orthogonally on the x_1 axis. In this case the orthogonal projection of p_i on the x_1 axis is given by $(x_{i1}, 0)$. Similarly, the orthogonal projection of p_i with coordinates (x_{i1}, x_{i2}, x_{i3}) in $\mathbb{R}^3$ on the x_1–x_2 plane is given by $(x_{i1}, x_{i2}, 0)$. Furthermore, we can project a set S in $\mathbb{R}^3$ orthogonally on the x_1–x_2 plane by $S' = \{(x_1, x_2, 0) \mid (x_1, x_2, x_3) \in S\}$. In this case we

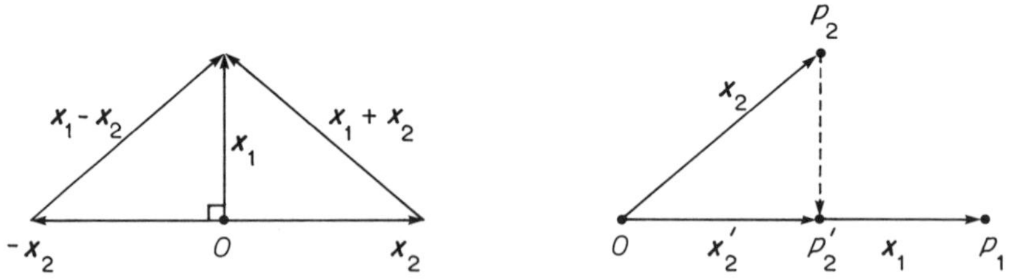

Figure 1.3.6 (a) Orthogonal vectors; (b) an orthogonal projection.

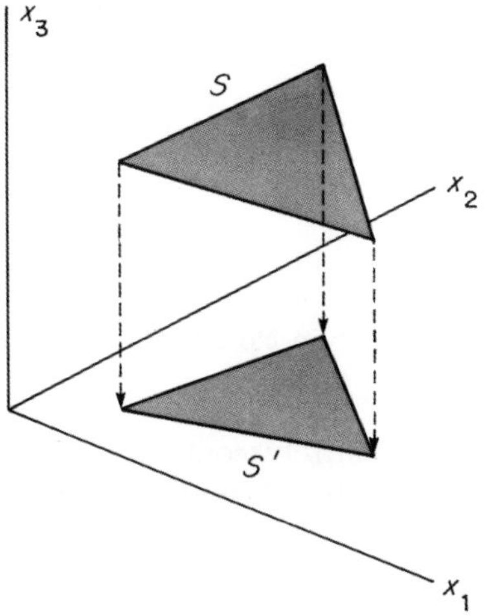

Figure 1.3.7 Orthographic projection.

say that S' is the *orthographic projection* of S on the x_1–x_2 plane. Figure 1.3.7 depicts the orthographic projection of a triangle placed in $\mathbb{R}^3$ on the x_1–x_2 plane.

As is seen in Figure 1.3.4(c), $\overrightarrow{p_2p_1}$ and $\overrightarrow{op_5}$ both represent the vector $\mathbf{x}_2 - \mathbf{x}_1$. Since the length of $\overrightarrow{op_5}$ is given by equation (1.3.19), where $\mathbf{x} = \mathbf{x}_2 - \mathbf{x}_1$, we understand that the *Euclidean distance* from point $\mathbf{x}_i^{\mathrm{T}} = (x_{i1}, \ldots, x_{im})$ to point $\mathbf{x}_j^{\mathrm{T}} = (x_{j1}, \ldots, x_{jm})$ is given by

$$\|\mathbf{x}_i - \mathbf{x}_j\| = \sqrt{(\mathbf{x}_i - \mathbf{x}_j)^{\mathrm{T}}(\mathbf{x}_i - \mathbf{x}_j)}$$
$$= \left[\sum_{k=1}^{m} (x_{ik} - x_{jk})^2\right]^{1/2} . \qquad (1.3.23)$$

We call the m-dimensional real Cartesian space with the Euclidean distance the m-dimensional *Euclidean space*.

In the Euclidean space, we now show a few topological concepts. For a point c in $\mathbb{R}^m$ and $\epsilon > 0$, we define the set

$$N_\epsilon(\mathbf{c}) = \{\mathbf{x} \mid \|\mathbf{x} - \mathbf{c}\| < \epsilon\} , \qquad (1.3.24)$$

which is called an *open ball* (in particular, an *open disk* for $m = 2$; Figure 1.3.8(a)) with radius ϵ centred at $\mathbf{c}$. Using this term, we say that the set A in $\mathbb{R}^m$ is *open* if every point in A is the centre of some open ball entirely contained in A (Figure 1.3.8(c)); the set A in $\mathbb{R}^m$ is *closed* if $\mathbb{R}^m \backslash A$ is open in $\mathbb{R}^m$; a point $\mathbf{x}$ in $\mathbb{R}^m$ is a *boundary point* of A if every open ball centred at $\mathbf{x}$ contains a point of A and a point of $\mathbb{R}^m \backslash A$ (Figure 1.3.7(d)). The set of all boundary points of A is called the *boundary* of A, and it is denoted by ∂A. If the set A includes the boundary of A, every point in $\mathbb{R}^m \backslash A$ is the

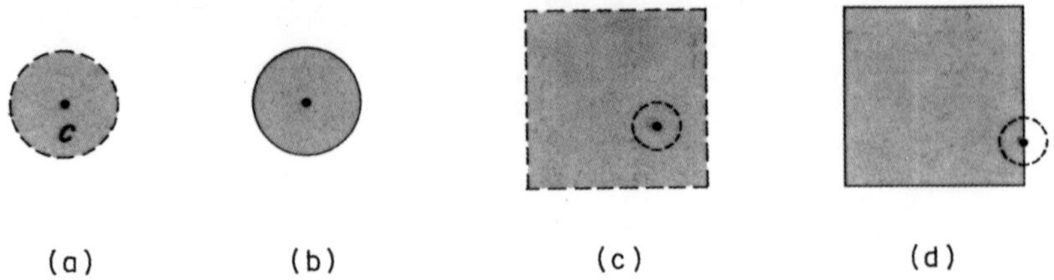

Figure 1.3.8 (a) An open disk; (b) a closed disk; (c) an open set; (d) a closed set.

centre of some open ball entirely contained in $\mathbb{R}^m \backslash A$. Thus the set A is closed if it contains its boundary. A point $\boldsymbol{x}$ in $\mathbb{R}^m$ is called an *interior point* of the set A if there exists an open ball centred at $\boldsymbol{x}$ that contains only points in the set A (point $\boldsymbol{x}$ in Figure 1.3.8(c)). If $\mathbf{c}$ is a point in $\mathbb{R}^m$, then any set that contains an open set containing $\mathbf{c}$ is called a *neighbourhood* of $\mathbf{c}$ in $\mathbb{R}^m$. The open ball defined by equation (1.3.24) is a neighbourhood of $\mathbf{c}$, and is sometimes referred to as an ϵ-neighbourhood.

In conjunction with an open ball, we define the sets

$$\bar{N}_\epsilon(\mathbf{c}) = \{\boldsymbol{x} | \, \|\boldsymbol{x} - \mathbf{c}\| \leq \epsilon\} \quad \text{for } \mathbf{c} \in \mathbb{R}^m, \epsilon > 0 \tag{1.3.25}$$

(notice that $<$ in equation (1.3.24) is replaced by $\leq$), and

$$\partial \bar{N}_\epsilon(\mathbf{c}) = \{\boldsymbol{x} | \, \|\boldsymbol{x} - \mathbf{c}\| = \epsilon\} \quad \text{for } \mathbf{c} \in \mathbb{R}^m, \epsilon > 0 \, , \tag{1.3.26}$$

which is a hypersphere in $\mathbb{R}^m$ (specifically, a circle in $\mathbb{R}^2$, a sphere in $\mathbb{R}^3$). The set $\partial \bar{N}_\epsilon(\mathbf{c})$ is the boundary of $\bar{N}_\epsilon(\mathbf{c})$. Since the set $\bar{N}_\epsilon(\mathbf{c})$ includes its boundary, the set $\bar{N}_\epsilon(\mathbf{c})$ is a closed set. Thus we call $\bar{N}_\epsilon(\mathbf{c})$ a *closed ball* (specifically, a *closed disk* for $m = 2$; Figure 1.3.8(b)) centred at $\mathbf{c}$ with radius ϵ.

For a set C, if there exist two open sets A and B such that $(A \cap C) \cap (B \cap C) = \varnothing$ and $(A \cap C) \cup (B \cap C) = C$, the set C is said to be *disconnected* (Figure 1.3.9(a)). If a set is not disconnected, the set is said to be *connected* (Figure 1.3.9(b)).

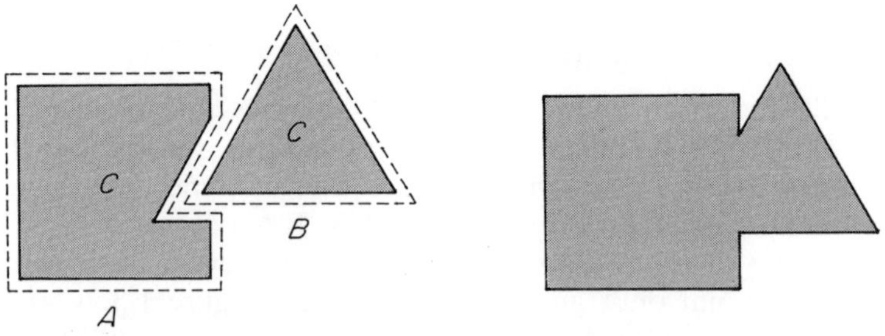

Figure 1.3.9 (a) A disconnected set C; (b) a connected set C.

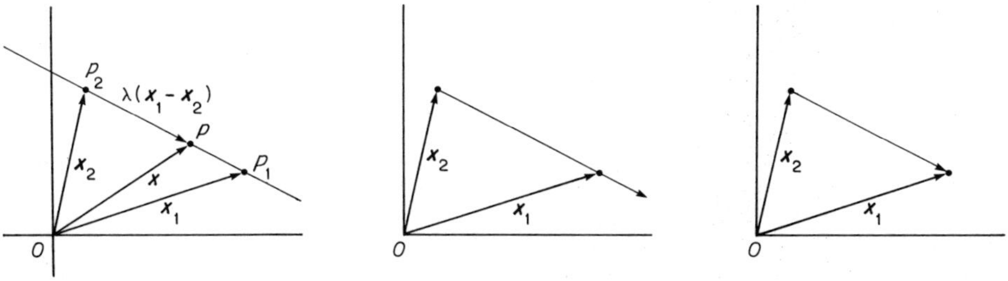

Figure 1.3.10 Lines (bold lines) defined by vectors: (a) a line; (b) a half line; (c) a line segment.

As we showed in Figure 1.3.5(c), $\overrightarrow{p_2p_1}$ represents the vector $\boldsymbol{x}_1 - \boldsymbol{x}_2$. We extend or shrink this arrow by multiplying a scalar λ keeping its initial point at p_2 (Figure 1.3.10(a)). This extended or shrunken arrow represents the vector $\lambda(\boldsymbol{x}_1 - \boldsymbol{x}_2)$. Let $\boldsymbol{x}$ be the location vector indicating the head point p, which is represented by $\overrightarrow{op}$. Recalling the vector addition depicted in Figure 1.3.4(b), we notice that $\overrightarrow{op}$ is obtained by adding $\overrightarrow{op_2}$ and $\overrightarrow{p_2p}$ (Figure 1.3.10(a)). Since these arrows represent the vectors $\boldsymbol{x}$, $\boldsymbol{x}_2$ and $\lambda(\boldsymbol{x}_1 - \boldsymbol{x}_2)$, respectively, we obtain the equation $\boldsymbol{x} = \lambda(\boldsymbol{x}_1 - \boldsymbol{x}_2) + \boldsymbol{x}_2 = \lambda\boldsymbol{x}_1 + (1 - \lambda)\boldsymbol{x}_2$. From this equation the (*straight*) *line* passing through p_1 and p_2 ($p_1 \neq p_2$) is written as

$$L_1 = \{\boldsymbol{x} | \boldsymbol{x} = \lambda\boldsymbol{x}_1 + (1 - \lambda)\boldsymbol{x}_2, \lambda \in \mathbb{R}\} . \qquad (1.3.27)$$

Observing that $\boldsymbol{x} = \boldsymbol{x}_2$ if $\lambda = 0$ and $\boldsymbol{x} = \boldsymbol{x}_1$ if $\lambda = 1$, we notice that the *half line* radiating from p_2 passing through p_1 (Figure 1.3.10(b)) is given by the set

$$L_2 = \{\boldsymbol{x} | \boldsymbol{x} = \lambda\boldsymbol{x}_1 + (1 - \lambda)\boldsymbol{x}_2, \lambda \geq 0\} , \qquad (1.3.28)$$

and the (*straight*) *line segment* connecting p_1 and p_2 (Figure 1.3.10(c)) is given by the set

$$L_3 = \{\boldsymbol{x} | \boldsymbol{x} = \lambda\boldsymbol{x}_1 + (1 - \lambda)\boldsymbol{x}_2, 0 \leq \lambda \leq 1\} . \qquad (1.3.29)$$

Points p_1 and p_2 are called the *end points* of L_3. Line segment L_3 contains its end points p_1 and p_2.

Let us next consider a plane in $\mathbb{R}^3$ passing through points p_1, p_2 and p_3 which are not all on the same line, or equivalently, $\boldsymbol{x}_2 - \boldsymbol{x}_1$ and $\boldsymbol{x}_3 - \boldsymbol{x}_1$ are linearly independent. A point $\boldsymbol{v}$ on the line passing through p_1 and p_2 is, from equation (1.3.27), given by $\boldsymbol{v} = \alpha\boldsymbol{x}_1 + (1 - \alpha)\boldsymbol{x}_2$; a point $\boldsymbol{w}$ on the line passing through p_1 and p_3 is given by $\boldsymbol{w} = \beta\boldsymbol{x}_1 + (1 - \beta)\boldsymbol{x}_3$ (Figure 1.3.11). Thus a point $\boldsymbol{x}$ on the plane is given by $\boldsymbol{x} = \gamma\boldsymbol{v} + (1 - \gamma)\boldsymbol{w} = (\alpha\gamma + (1 - \alpha)\beta)\boldsymbol{x}_1 + \gamma(1 - \alpha)\boldsymbol{x}_2 + (1 - \beta)(1 - \gamma)\boldsymbol{x}_3$ (Figure 1.3.11). It follows from this equation that if we define the set by

$$A_1 = \{\boldsymbol{x} | \boldsymbol{x} = \lambda_1\boldsymbol{x}_1 + \lambda_2\boldsymbol{x}_2 + (1 - \lambda_1 - \lambda_2)\boldsymbol{x}_3, \lambda_1, \lambda_2 \in \mathbb{R}\} , \qquad (1.3.30)$$

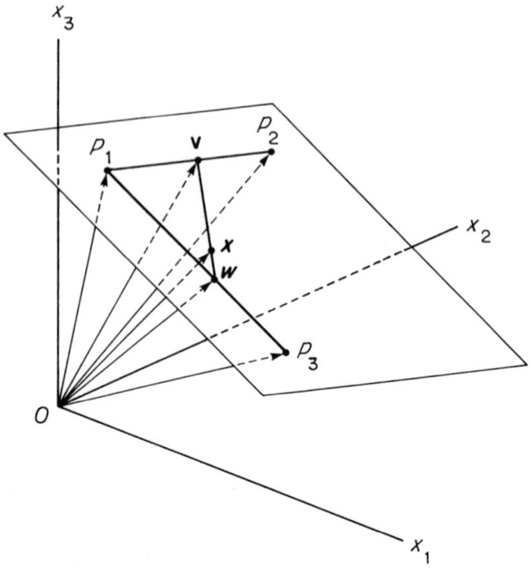

Figure 1.3.11 A plane in $\mathbb{R}^3$.

then the set A_1 represents a *plane*. More generally, we define the set in $\mathbb{R}^m$ by

$$A_2 = \left\{ \mathbf{x} \middle| \ \boldsymbol{x} = \sum_{i=1}^{m-1} \lambda_i \mathbf{x}_i + \left(1 - \sum_{i=1}^{m-1} \lambda_i\right) \mathbf{x}_m, \ \lambda_i \in \mathbb{R}, \ i \in I_{m-1} \right\}, \tag{1.3.31}$$

where $\mathbf{x}_2 - \mathbf{x}_1, \ldots, \mathbf{x}_m - \mathbf{x}_1$ are linearly independent. We call the set A_2 a *hyperplane* in $\mathbb{R}^m$.

In the Cartesian plane $\mathbb{R}^2$, a line is given by equation (1.3.27). The equation in the right-hand side of equation (1.3.27) is written as $x_1 = \lambda x_{11} + (1 - \lambda)x_{21}$ and $x_2 = \lambda x_{21} + (1 - \lambda)x_{22}$. Cancelling λ from these equations, we obtain $(x_{12} - x_{22})x_1 + (x_{21} - x_{11})x_2 = (x_{12} - x_{22})x_{21} + (x_{21} - x_{11})x_{22}$. Thus, a line is alternatively written as $L_1 = \{\mathbf{x} | a_1x_1 + a_2x_2 = b\}$, where a_1 and a_2 are constants, at least one of which is non zero, and b is a constant. Similarly, in $\mathbb{R}^3$, a plane is alternatively written as $A_1 = \{\mathbf{x} |\ a_1x_1 + a_2x_2 + a_3x_3 = b\}$, where a_1, a_2 and a_3 are constants, at least one of which is non-zero. Extending this expression to $\mathbb{R}^m$, we can alternatively write the hyperplane A_2 as

$$A_2 = \{\mathbf{x} |\ \mathbf{a}^{\mathrm{T}}\,\mathbf{x} = b, \mathbf{a}, \mathbf{x} \in \mathbb{R}^m, \|\mathbf{a}\| \neq 0\}\ . \tag{1.3.32}$$

A line splits a plane into two disjoint regions (Figure 1.3.10(a)); a plane splits a space into two disjoint regions (Figure 1.3.11); generally, a hyperplane splits a space into two disjoint regions. We call one of the regions with the hyperplane, i.e. the region defined by

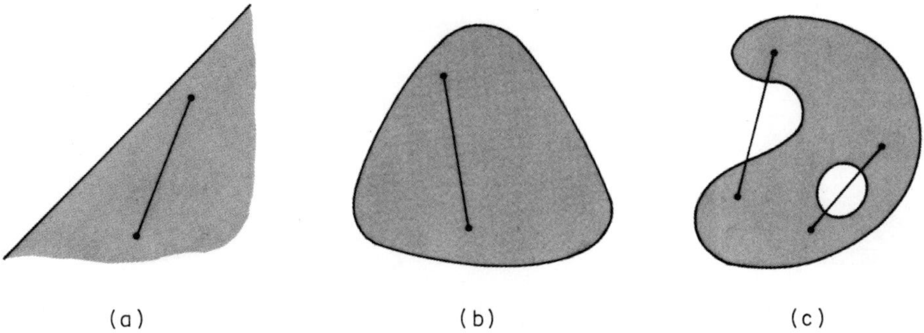

Figure 1.3.12 Convex and non-convex sets: (a) a half plane; (b) a convex set; (c) a non-convex set.

$$H = \{\boldsymbol{x} | \boldsymbol{a}^{\mathrm{T}} \boldsymbol{x} \leq b, \boldsymbol{a}, \boldsymbol{x} \in \mathbb{R}^m, \|\boldsymbol{a}\| \neq 0\} \ , \tag{1.3.33}$$

a *(closed) half space* (in particular, a *(closed) half plane* for $m = 2$; Figure 1.3.12(a)).

Observing Figure 1.3.12(a), we readily understand that the line segment between any two points in a half plane is included in the half plane. This property holds not only for a half plane but also other figures, such as Figure 1.3.12(b). Unlike Figure 1.3.12(c), Figure 1.3.12(b) does not have holes or its boundary does not bend into the figure. In such a case we say that a geometrical figure is *convex*. Formally, if for any two points $\boldsymbol{x}_1$ and $\boldsymbol{x}_2$ in A, the line segment joining the points is contained in A, i.e.

$$\lambda \boldsymbol{x}_1 + (1 - \lambda)\boldsymbol{x}_2 \in A \quad \text{for all } 0 \leq \lambda \leq 1 \ , \tag{1.3.34}$$

then we call the set A a *convex* set; otherwise, a *non-convex* set. In particular, if for any two points $\boldsymbol{x}_1$ and $\boldsymbol{x}_2$ in A, the relation

$$\lambda \boldsymbol{x}_1 + (1 - \lambda)\boldsymbol{x}_2 \in A \backslash \partial A \quad \text{for all } 0 < \lambda < 1 \ , \tag{1.3.35}$$

holds, we call the set A a *strictly convex* set.

A half space (equation (1.3.33)) is convex, because $\boldsymbol{a}^{\mathrm{T}}(\lambda \boldsymbol{x}_1 + (1 - \lambda)\boldsymbol{x}_2) = \lambda \boldsymbol{a}^{\mathrm{T}} \boldsymbol{x}_1 + (1 - \lambda)\boldsymbol{a}^{\mathrm{T}} \boldsymbol{x}_2 \leq \lambda b + (1 - \lambda)b = b$ for $\boldsymbol{x}_1, \boldsymbol{x}_2 \in H_2$ and $0 < \lambda < 1$. The intersection of two convex sets is also convex. To prove it, let $\boldsymbol{x}_1$ and $\boldsymbol{x}_2$ be arbitrary points in the intersection of two convex sets A_1 and A_2. Then, $\boldsymbol{x}_1, \boldsymbol{x}_2 \in A_1$ and $\boldsymbol{x}_1, \boldsymbol{x}_2 \in A_2$. Since A_1 and A_2 are convex sets, from relation (1.3.34), the point $\lambda \boldsymbol{x}_1 + (1 - \lambda)\boldsymbol{x}_2$, $0 \leq \lambda \leq 1$, is in A_1 and in A_2. Therefore, point $\lambda \boldsymbol{x}_1 + (1 - \lambda)\boldsymbol{x}_2$, $0 \leq \lambda \leq 1$, is in $A_1 \cap A_2$. Applying this proof, we can see that the intersection of convex sets is also convex. It is an exercise to show that a convex polygon A with vertices $\boldsymbol{x}_1, \ldots, \boldsymbol{x}_n$ is given by

$$A = \left\{ \boldsymbol{x} \,\middle|\, \sum_{i=1}^{n} \lambda_i \boldsymbol{x}_i \text{ where } \sum_{i=1}^{n} \lambda_i = 1, \lambda_i \geq 0, i \in I_n \right\} \tag{1.3.36}$$

(recall the derivation leading to equation (1.3.30)).

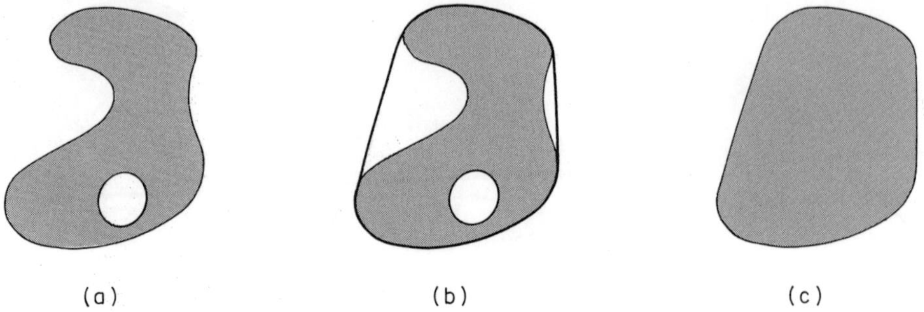

Figure 1.3.13 A convex hull: (a) a non-convex figure A; (b) the boundary of the convex hull of A (the heavy line); (c) the convex hull of A.

To define another type of a convex set, let us consider the geometric figure in Figure 1.3.13(a). Suppose that this figure is encircled by a rubber band. Then the rubber band becomes like the heavy line in Figure 1.3.13(b). We fill the inside of this rubber band with points. Then we obtain Figure 1.3.13(c), which we call a *convex hull*. To give a formal definition of the convex hull, consider all possible convex sets that contain the set A. The number of such sets is infinite. We take the intersection of all such convex sets. Then the resulting set is called the *convex hull* of the set A. Since the intersection of convex sets is a convex set, the convex hull of the set A is a convex set. Moreover, it is the 'smallest' convex set that contains the set A. Obviously, if the set A is a convex set, the convex hull of A is A itself.

As a special type of a convex hull, we define a simplex. A *simplex* in $\mathbb{R}^m$ is the convex hull of any set of $m + 1$ points which do not all lie on one hyperplane in $\mathbb{R}^m$. If $m = 0$, the simplex is a point itself, called the zeroth-order simplex; if $m = 1$, the simplex is the straight line segment connecting the two points, called the first-order simplex (Figure 1.3.14(a)); if $m = 2$, the simplex is a triangle, called the second-order simplex (Figure 1.3.14(b)); if $m = 3$, the simplex is a tetrahedron, called the third-order simplex (Figure 1.3.14(c)), and so forth. The first-order simplex is written as equation (1.3.29). Similarly, the second-order simplex is written as equation (1.3.36),

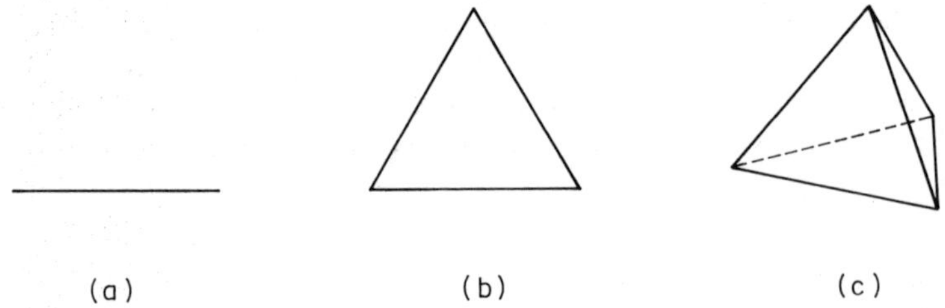

Figure 1.3.14 Simplices.

where $n = 3$ and $\boldsymbol{x}_2 - \boldsymbol{x}_1$ and $\boldsymbol{x}_3 - \boldsymbol{x}_1$ are linearly independent. In general, the mth-order simplex is written as equation (1.3.36), where $\boldsymbol{x}_2 - \boldsymbol{x}_m, \ldots, \boldsymbol{x}_{m+1} - \boldsymbol{x}_1$ are linearly independent. If a set is contained in a simplex, we say that the set is *bounded*; otherwise *unbounded*. We sometimes say that the region given by a bounded set is a *finite region* and the region given by an unbounded set is an *infinite region*.

Let A be a subset of $\mathbb{R}^m$. We call an element $\boldsymbol{a}_u$ in $\mathbb{R}^m$ an *upper bound* of A if $\boldsymbol{a}_u \geq \boldsymbol{x}$ for all $\boldsymbol{x}$ in A. Similarly, we call an element $\boldsymbol{a}_\ell$ in $\mathbb{R}^m$ a *lower bound* of A if $\boldsymbol{a}_\ell \leq \boldsymbol{x}$ for all $\boldsymbol{x}$ in A. When the set A has an upper bound, we say that A is *bounded above*. Similarly, if the set A has a lower bound, we say that A is *bounded below*. If the set A is bounded above in $\mathbb{R}^m$, we say that an upper bound of A is a *supremum* of A if it is less than any other upper bound of A. Similarly, we say that a lower bound of A is an *infimum* of A if it is greater than any other lower bound of A.

The concept of convexity can be introduced into a function. Let $y = f(\boldsymbol{x})$ be a function from $\mathbb{R}^n$ to $\mathbb{R}$, and $\{(\boldsymbol{x}, y) | \, y \geq f(\boldsymbol{x}), \, \boldsymbol{x} \in S\}$, where S is the domain of the function f. This set is called the *epigraph* of the function f. Two examples of the epigraph are shown in Figure 1.3.15. If the epigraph of the function f is convex, we call the function f a *convex function* (Figure 1.3.15(a)); otherwise, a *non-convex function* (Figure 1.3.15(b)).

Suppose, as is shown in Figure 1.3.16, that five half planes, $H_1, \ldots, H_5$, are given. Using a finite number of intersection and union operations with respect to these half planes, we may construct a connected region. For example, $(H_1 \cap H_2 \cap H_3) \cup (H_1 \cap H_4 \cap H_5)$ in Figure 1.3.16(a). We call this region a *polygon*. Stated a little more generally, if a region constructed from a finite number of operations of intersection and union with respect to a finite number of half planes is a non-empty connected set, then we call the region a *polygon*. The boundary of a polygon consists of straight line

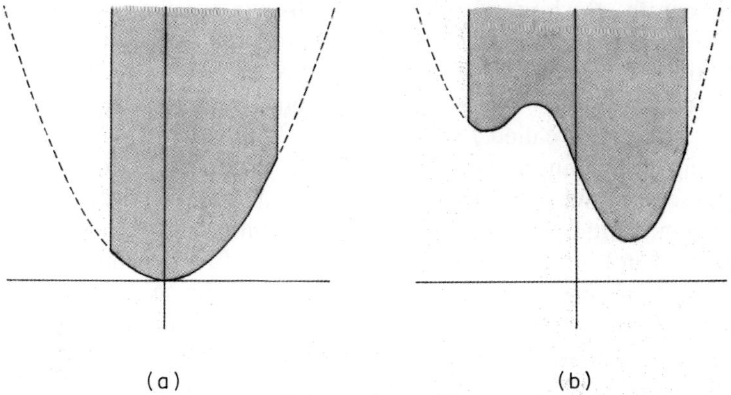

Figure 1.3.15 (a) The epigraph of a convex function, and (b) that of a non-convex function.

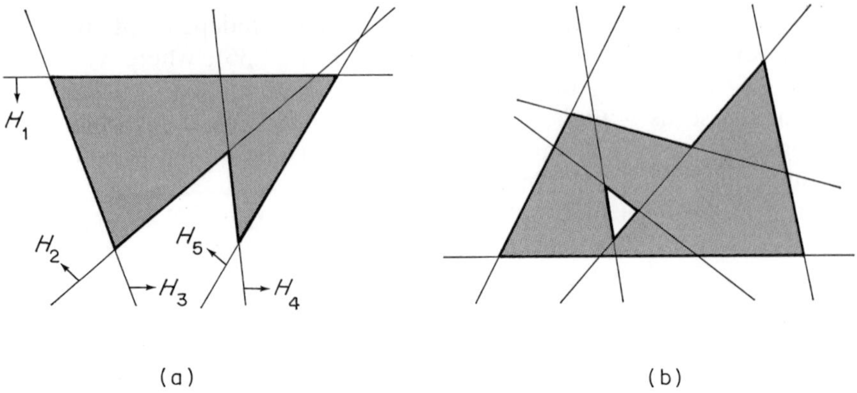

Figure 1.3.16 A polygon constructed from half planes: (a) a simple polygon; (b) a polygon that is not simple.

segments. These line segments are called *edges* and their end points are called *vertices*. If a polygon does not have a hole in it, and each vertex of a polygon has exactly two edges and every edge intersects only at vertices, we call the polygon a *simple polygon*. The polygon in Figure 1.3.16(a) is a simple polygon, but that in Figure 1.3.5(b) is not.

The notion of a polygon defined in $\mathbb{R}^2$ can be extended to $\mathbb{R}^m$. If a set constructed from a finite number of operations of intersection and union on a finite number of half spaces is a non-empty connected bounded set, we call the set a *polyhedron*. A polyhedron (polygon) may be convex or non-convex. If a polyhedron is convex and bounded, it is sometimes called a *polytope*. A simplex (for example, the tetrahedron shown in Figure 1.3.14(c)) is a polytope.

The boundary of a polyhedron in $\mathbb{R}^3$ consists of polygons, which are called *faces*. Generally, the boundary of a polyhedron in $\mathbb{R}^m$ consists of polyhedrons in $\mathbb{R}^{m-1}$, which are called $(m-1)$-*faces*; the boundary of a $(m-1)$-face consists of polyhedrons in $\mathbb{R}^{m-2}$, which are called $(m-2)$-*faces*; and so forth. Note that 0-faces are *vertices*, and 1-faces are *edges*, and that $(m-1)$-faces are sometimes called *facets*.

Recalling the definition of a convex set (equation (1.3.34)), we notice that the sets shown in Figure 1.3.17 are non-convex, because the line segment connecting points p_1 and p_2 is not contained in the set. However, if we fix one end point at p_0, as in Figure 1.3.17(a), the line connecting the point p_0 and any point in the set is contained in the set. We call a set with this geometric property *star-shaped*. Stated precisely, the set A is *star-shaped* if there exists a point $\mathbf{x}_0$ in A for which $\lambda\mathbf{x}_0 + (1-\lambda)\mathbf{x} \in A$ holds for all $0 \le \lambda \le 1$ and for all $\mathbf{x} \in A$. The set of such points, $\{\mathbf{x}_0 | \lambda\mathbf{x}_0 + (1-\lambda)\mathbf{x} \in A, 0 \le \lambda \le 1, \mathbf{x} \in A\}$, is called the *kernel* of the set A. The kernel of the star-shaped polygon of Figure 1.3.17(a) is shown by the densely shaded region

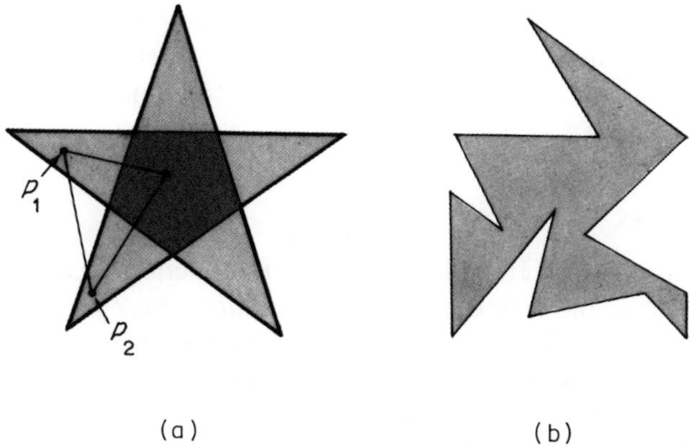

Figure 1.3.17 (a) A star-shaped polygon (the shaded region) and its kernel (the densely shaded region); (b) a polygon that is not star-shaped.

in the figure. The figure in Figure 1.3.17(b) is not star-shaped. Obviously, a convex polygon is star-shaped, and its kernel is the convex polygon itself.

Let S be a closed subset of $\mathbb{R}^m$ and $\mathcal{S} = \{S_1, \ldots, S_n\}$, where S_i is a closed subset of S. If elements in the set $\mathcal{S}$ satisfy

$$[S_i \backslash \partial S_i] \cap [S_j \backslash \partial S_j] = \varnothing, \quad i \neq j, i, j \in I_n , \tag{1.3.37}$$

and

$$\bigcup_{i=1}^{n} S_i = S , \tag{1.3.38}$$

then we call the set $\mathcal{S}$ a *tessellation* of S. Specifically, we call a tessellation in $\mathbb{R}^2$ a *planar tessellation* of S. Figure 1.3.18 shows two planar tessellations. The tessellation in panel (a) consists of polygons with three or more vertices,

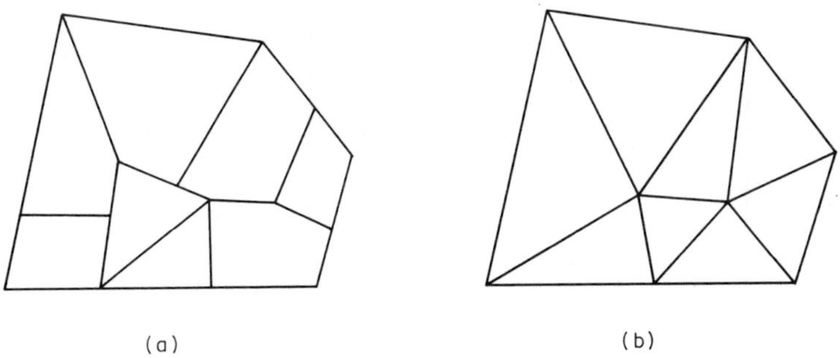

Figure 1.3.18 Tessellations: (a) a tessellation that is not a triangulation; (b) a triangulation.

whereas that in panel (b) consists of only triangles. We call a planar tessellation like panel (b), i.e. a tessellation in which S_i in $\mathcal{S}$ is a triangle for all $i \in I_n$, a *triangulation* of S. In this text, a triangulation will be denoted by $\mathcal{T} = \{T_1, \ldots, T_n\}$.

1.3.4 Derivatives

Let f be a function from $\mathbb{R}^n$ to $\mathbb{R}$. If $n = 1$, the function is called a *univariate function*; if $n \geq 2$, the function is called a *multivariate function*. In panels (a) and (b) of Figure 1.3.19, two contrasting univariate functions are depicted (the filled circle indicates a closed end point and the unfilled circle indicates an open end point). The difference between these functions is visually distinct; the function in Figure 1.3.19(b) has a gap at point c, whereas the function in Figure 1.3.19(a) does not. Informally, we may say that a function without a gap is a continuous function. Formally, however, it is not so simple to define a gap.

To define a gap formally, we shall employ the so-called ϵ–δ method. Suppose that a positive number ϵ is given by the ϵ indicated in Figure 1.3.19(a). Then we can find a positive number δ (shown in Figure 1.3.19(a)) that satisfies

$$f(c) - \epsilon < f(x) < f(c) + \epsilon \quad \text{for all } x \text{ in } c - \delta < x < c + \delta \,. \qquad (1.3.39)$$

Of course, if we choose δ' in Figure 1.3.19(a), this relation does not hold. Hence, a positive number δ depends upon ϵ, and so we write δ as $\delta(\epsilon)$. As can be seen in Figure 1.3.19(a), for a given ϵ we can always find a $\delta(\epsilon)$ that satisfies the relation (1.3.39). Moreover, we can find such a $\delta(\epsilon)$ not only for a given ϵ but also for any ϵ. In this case the function f is said to be *continuous* at c. On the other hand, in Figure 1.3.19(b) there exists an ϵ for which we cannot find a $\delta(\epsilon)$ that satisfies the relation (1.3.39). This implies that the function in Figure 1.3.19(b) has a gap and is not continuous.

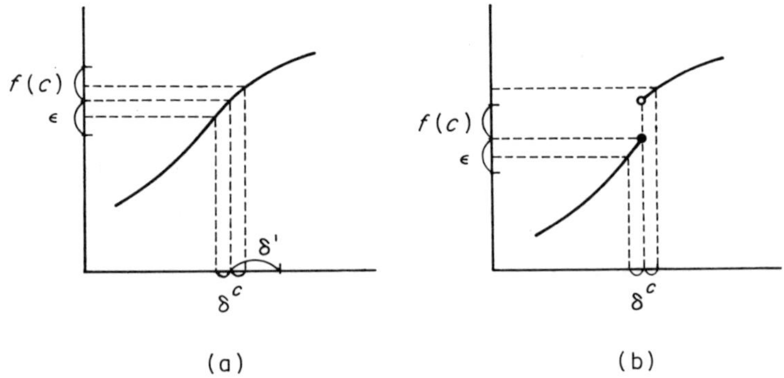

Figure 1.3.19 (a) A function continuous at c; (b) a function discontinuous at c.

To sum up the function $f(x)$ is said to be *continuous* at c if for any positive number ϵ there exists a positive number $\delta(\epsilon)$ such that if x is in the domain of the function f and $|x - c| < \delta(\epsilon)$, then $|f(x) - f(c)| < \epsilon$; otherwise, the function $f(x)$ is *discontinuous* at c. The same definition can be applied to a multivariate function. Generally, the function $y = f(\mathbf{x})$ from $\mathbb{R}^m$ to $\mathbb{R}$ is said to be *continuous* at $\mathbf{c}$ if, for any positive number ϵ, there exists a positive number $\delta(\epsilon)$ such that if $\mathbf{x}$ is in the domain of the function f and $\|\mathbf{x} - \mathbf{c}\| < \delta(\epsilon)$, then $|f(\mathbf{x}) - f(\mathbf{c})| < \epsilon$.

In Figure 1.3.20 a continuous univariate function $y = f(x)$ is shown with the line passing through points $(c, f(c))$ and $(x, f(x))$ (the fine line). The gradient of this line is given by $(f(x) - f(c))/(x - c)$, which is regarded as a function of x. The value of $(f(x) - f(c))/(x - c)$ changes as x approaches c (the fine lines in Figure 1.3.20). In the limit, as is indicated by the heavy line in Figure 1.3.20, the value of $(f(x) - f(c))/(x - c)$ appears to approach a certain value D. If it actually approaches D, we write

$$\lim_{x \to c} \frac{f(x) - f(c)}{x - c} = D\,. \qquad (1.3.40)$$

Stated more formally, if c is in the domain of function f, we say that a real number D is the *(first) derivative* of f at c if, for every positive number ϵ, there exists a positive number $\delta(\epsilon)$ such that if x belongs to the domain of f and if $0 < |x - c| < \delta(\epsilon)$, then

$$\left| \frac{f(x) - f(c)}{x - c} - D \right| < \epsilon\,. \qquad (1.3.41)$$

In this case, D is written as $\mathrm{d}f(c)/\mathrm{d}x$ or $f'(c)$ (or simply $\mathrm{d}f/\mathrm{d}x$ or f' when c is understood). This definition is sometimes written as

$$\frac{\mathrm{d}f(c)}{\mathrm{d}x} = \lim_{x \to c} \frac{f(x) - f(c)}{x - c}\,. \qquad (1.3.42)$$

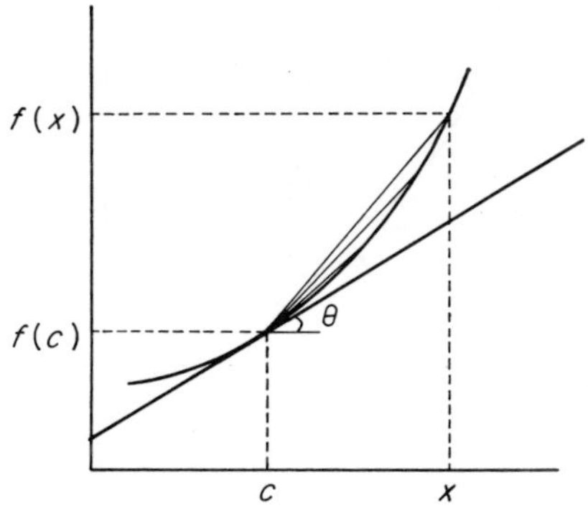

Figure 1.3.20 Derivative of the function f at c.

A continuous function and its derivative defined above are closely related. It is shown that if the function f has a derivative at c, then f is continuous at c. Note, however, that the converse is not true. For example, $f(x) = |x - 1|$ is continuous at $x = 1$ but it does not have a derivative there.

Suppose that the function f has a derivative at any x in the neighbourhood of c. Then, $y = f'(x)$ can be regarded as a function in that neighbourhood. Applying the above definition to this function, we define

$$\frac{d^2 f(c)}{dx^2} = \lim_{x \to c} \frac{f'(x) - f'(c)}{x - c}, \tag{1.3.43}$$

which is called the *second derivative* of f at c. Sometimes, $d^2 f(c)/dx^2$ is written as $f''(c)$. In this manner, we can define the kth *derivative*, which is denoted by $d^k f(c)/dx^k$ or $f^{(k)}(c)$.

To extend the above definition to a multivariate function, let us first consider a simple example, $y = f(x_1, x_2) = x_1^2 + x_2^2$. The set $\{(x_1, x_2, y) \mid y = x_1^2 + x_2^2\}$ forms a surface called a paraboloid in $\mathbb{R}^3$, which is depicted in Figure 1.3.21. We cut this surface by the vertical plane parallel to the x_2 axis passing through a fixed point $x_1 = c_1$. The resulting intersection is a parabola (Figure 1.3.21), which is mathematically written as $y = f(c_1, x_2) = c_1^2 + x_2^2$. Since $y = c_1^2 + x_2^2$ can be regarded as a univariate function of x_2 with constant c_1, we can apply the above definition to the function $y = f(c_1, x_2) = c_1^2 + x_2^2$, i.e.

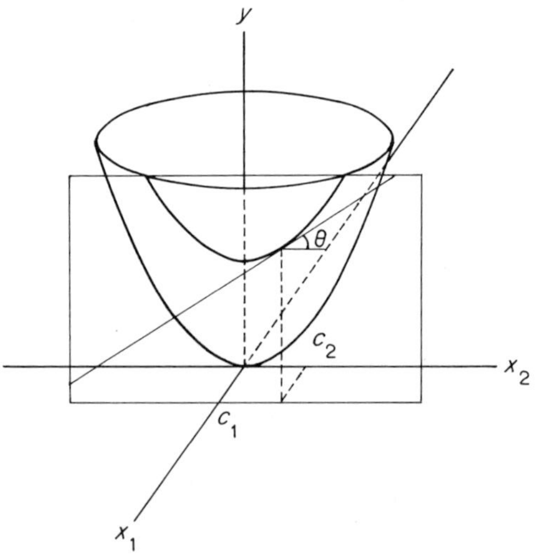

Figure 1.3.21 Partial derivative of the function $f(x_1, x_2) = x_1^2 + x_2^2$ with respect to x_2.

$$\frac{\partial f(c_1, c_2)}{\partial x_2} = \lim_{x_2 \to c_2} \frac{f(c_1, x_2) - f(c_1, c_2)}{x_2 - c_2}, \tag{1.3.44}$$

which is called a *partial derivative* of the function f at point (c_1, c_2) with respect to x_2. Similarly, we can define $\partial f(c_1, c_2)/\partial x_1$. In Figure 1.3.21, $\partial f(c_1, c_2)/\partial x_2$ is given by the tangent of the θ indicated in Figure 1.3.21.

The definition of a partial derivative in the above simple example can be readily extended to that of a general multivariate function $y = f(x_1, \ldots, x_n)$. The function $y = f(c_1, \ldots, c_{i-1}, x_i, c_{i+1}, \ldots, c_n)$ can be regarded as a univariate function of x_i with constants $c_1, \ldots, c_{i-1}, c_{i+1}, \ldots, c_n$. Applying the definition of the derivative of a univariate function, we can define

$$\frac{\partial f(c_1, \ldots, c_n)}{\partial x_i} = \lim_{x_i \to c_i} \frac{f(c_1, \ldots, x_i, \ldots, c_n) - f(c_1, \ldots, c_i, \ldots, c_n)}{x_i - c_i}, \tag{1.3.45.}$$

which is called a *partial derivative* of f with respect to x_i at $(c_1, \ldots, c_n)$.

Suppose that there exists a partial derivative of the function f with respect to x_i at every point in a neighbourhood of $(c_1, \ldots, c_j, \ldots, c_n)$ in $\{(c_1, \ldots, x_j, \ldots, c_n) | -\infty < x_j < \infty\}$. Then $\partial f(c_1, \ldots, x_j, \ldots, c_n)/\partial x_i$ can be regarded as a univariate function of x_j in that neighbourhood. Applying the definition of the derivative of a univariate function, we can define

$$\frac{\partial^2 f(c_1, \ldots, c_n)}{\partial x_j \partial x_j} = \lim_{x_j \to c_j} \frac{\dfrac{\partial f(c_1, \ldots, x_j, \ldots, c_n)}{\partial x_i} - \dfrac{\partial f(c_1, \ldots, c_j, \ldots, c_n)}{\partial x_i}}{x_j - c_j}, \tag{1.3.46}$$

which is called the *second partial derivative* of f at $(c_1, \ldots, c_n)$. In this manner, we can define the kth partial derivatives.

In Figure 1.3.20 the heavy line shows the line tangent to the curve $y = f(x)$ at c. The gradient of this line is indicated by $\mathrm{d}f(c)/\mathrm{d}x$. If $\mathrm{d}f(c)/\mathrm{d}x$ exists, the line tangent to the curve $y = f(x)$ at c always exists. In Figure 1.3.22 the shaded plane shows the plane tangent at the surface $y = x_1^2 + x_2^2$ at (c_1, c_2). Like $\mathrm{d}f(c)/\mathrm{d}x$ of a univariate function, $(\partial f(c)/\partial x_1, \partial f(c)/\partial_2)$ indicates the gradient of the plane, which is sometimes written as $\mathrm{grad}f(c)$ or $\nabla f(c)$. Unlike a univariate function, however, even if both $\partial f(c_1, c_2)/\partial x_1$ and $\partial f(c_1, c_2)/\partial x_1$ exist, the plane tangent to the surface $y = f(x_1, x_2)$ at (c_1, c_2) is not necessarily unique. To guarantee a unique plane, we need one more condition.

Generally, if the partial derivatives of $f(\mathbf{x})$ exist in the neighbourhood of $\mathbf{c}$ and, in addition, they are continuous at $\mathbf{c}$, then we say that the function f is *differentiable* at $\mathbf{c}$.

If the function $y = f(x_1, x_2)$ is differentiable at (c_1, c_2), there exists a unique plane tangent to the surface given by $\{(x_1, x_2, y) \mid y = f(x_1, x_2)\}$ at (c_1, c_2). The plane is written as

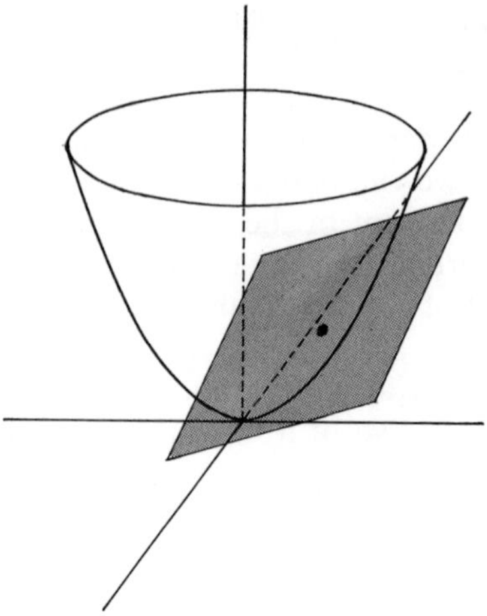

Figure 1.3.22 A plane tangent to the surface in $\mathbb{R}^3$.

$$y = \left(\frac{\partial f(c_1, c_2)}{\partial x_1}, \frac{\partial f(c_1, c_2)}{\partial x_2}\right)\begin{pmatrix} x_1 - c_1 \\ x_2 - c_2 \end{pmatrix} + f(c_1, c_2)\ . \tag{1.3.47}$$

Generalizing equation (1.3.47), if the function $y = f(\boldsymbol{x})$ is differentiable at $\mathbf{c}$, we can define the hyperplane tangent to the surface $\{(\boldsymbol{x}, y) \mid y = f(\boldsymbol{x}), \boldsymbol{x} \in \mathbb{R}^n\}$ at $\mathbf{c}$, which is written as

$$y = \nabla f(\mathbf{c})(\boldsymbol{x} - \mathbf{c}) + f(\mathbf{c})\ , \tag{1.3.48}$$

where

$$\operatorname{grad} f(\mathbf{c}) = \nabla f(\mathbf{c}) = \left(\frac{\partial f(\mathbf{c})}{\partial x_1}, \ldots, \frac{\partial f(\mathbf{c})}{\partial x_n}\right) \tag{1.3.49}$$

(observe equation (1.3.47) written as equation (1.3.48)).

1.3.5 Integration

When we measure an area or a volume, we evaluate an integral. In this section we give the definition of the Riemann integral in $\mathbb{R}^m$ (first, $m = 1$, next $m \geq 2$).

Let $y = f(x)$ be a function defined on the closed interval $C = [a, b] = \{x \mid a \leq x \leq b\}$ of the real line $\mathbb{R}$. We assume that the function f is bounded, i.e. $-\infty < K_1 < f(x) < K_2 < \infty$. We partition C into a finite number of subintervals $[x_{i-1}, x_i]$, $i = 1, \ldots, n$, which are collectively exhaustive and mutually exclusive except for boundary points in C (note that $x_0 = a$, $x_n = b$). The set $T = \{[x_{i-1}, x_i], i \in I_n\}$ is called a *partition* of

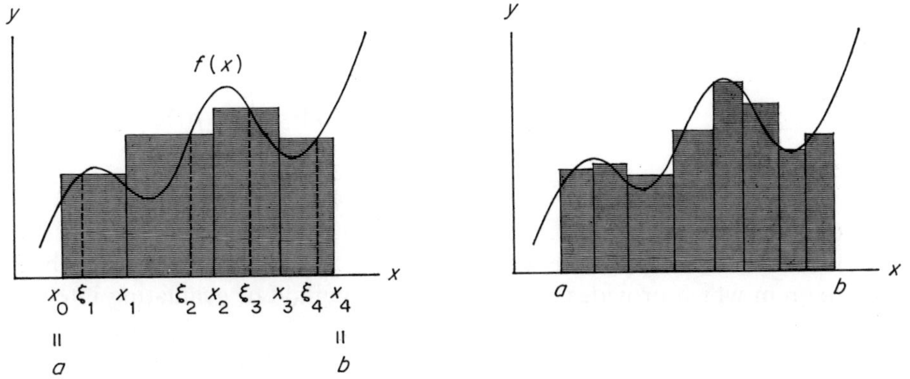

Figure 1.3.23 (a) Riemann sum with partition T; (b) Riemann sum with the partition of a refinement of T.

C (an example is shown in Figure 1.3.23). For a given partition T of C, we define

$$S(f, T) = \sum_{i=1}^{n} f(\xi_i)(x_i - x_{i-1}) , \tag{1.3.50}$$

where $x_{i-1} \leq \xi_i \leq x_i$, $i \in I_n$. We call the real number $S(f, T)$ the *Riemann sum* of f corresponding to T. In Figure 1.3.23 the Riemann sum is indicated by the shaded area. Obviously, a Riemann sum depends upon the partition T and the choice of ξ_i.

If T_1 and T_2 are partitions of C, we say that T_2 is a *refinement* of T_1 if every subinterval of T_2 is contained in some subinterval of T_1 (Figure 1.3.23(b)). Let T_i be a refinement of T_{i-1}, $i = 1, 2 \ldots$ ($T_0 = C$). If we take an 'appropriate' series of $T_1, T_2, T_3, \ldots$, for example $T_i = \{[0, (b-a)/i^2], [(b-a)/i^2, 2(b-a)/i^2], \ldots, [(i^2-1)(b-a)/i^2, b]\}$, then $S(f, T_i)$ approaches a certain value R as i increases (the partition becomes finer). Formally, we say that the function f is *(Riemann) integrable* on C if there exists a real number R such that for every positive number ϵ, there exists a partition T_ϵ of $[a, b]$ such that if T is any refinement of T_ϵ and $S(f, T)$ is any Riemann sum corresponding to T, then $|S(f, T) - R| < \epsilon$. The number R is uniquely determined and is denoted by

$$R = \int_a^b f(x)\mathrm{d}x = \int_C f(x)\mathrm{d}x . \tag{1.3.51}$$

We call the real number R the *(Riemann) integral* of f over C.

Suppose that the function f is continuous in C, and let

$$F(x) = \int_a^x f(t)\mathrm{d}t . \tag{1.3.52}$$

Then $F(x)$ can be regarded as a continuous function. If $\delta > 0$ is such that $x + \delta \in C$, then

$$F(x+\delta) - F(x) = \int_a^{x+\delta} f(t)\mathrm{d}t - \int_a^x f(t)\mathrm{d}t = \int_x^{x+\delta} f(t)\mathrm{d}t$$
$$= f(\xi)(x + \delta - x) \qquad (1.3.53)$$

for some ξ in $x \le \xi \le x + \delta$. Since f is continuous at x, $\lim_{\delta\to 0}(F(x+\delta) - F(x))/(x + \delta - x) = f(x)$. Hence $F'(x) = f(x)$. From this result, we obtain the theorem which provides the basis for the method of evaluating integrals in calculus.

Fundamental theorem of integral calculus Let $f(x)$ be continuous on $[a, b] = \{x | \, a \le x \le b\}$. A function F on $[a, b]$ satisfies

$$F(x) - F(a) = \int_a^x f(x)\mathrm{d}x \quad \text{for } x \in [a, b] \, , \qquad (1.3.54)$$

if and only if $F' = f$ on $[a, b]$.

The Riemann integral in $\mathbb{R}^2$ can be defined almost in the same manner. Let C be a bounded closed area in $\mathbb{R}^2$ (the shaded region in Figure 1.3.24), and the function $y = f(x_1, x_2)$ is assumed to be bounded. For convenience, $f(x_1, x_2) = 0$ for $(x_1, x_2) \notin C$. Let $a_i = \min\{x_i | \, \mathbf{x} \in C\}$, $b_i = \max\{x_i | \, \mathbf{x} \in C\}$, $i = 1, 2$, and $U = [a_1, b_1] \times [a_2, b_2]$ (Figure 1.3.24). Let $T_{u_i} = \{[x_{ij-1}, x_{ij}], j \in I_{n_i}\}$ be a partition of $U_i = [a_i, b_i]$, ($x_{i0} = a_i$, $x_{in_i} = b_i$, $i = 1, 2$). Then the partitions T_{u_1} and T_{u_2} generate a partition $T^{(2)}$ of U

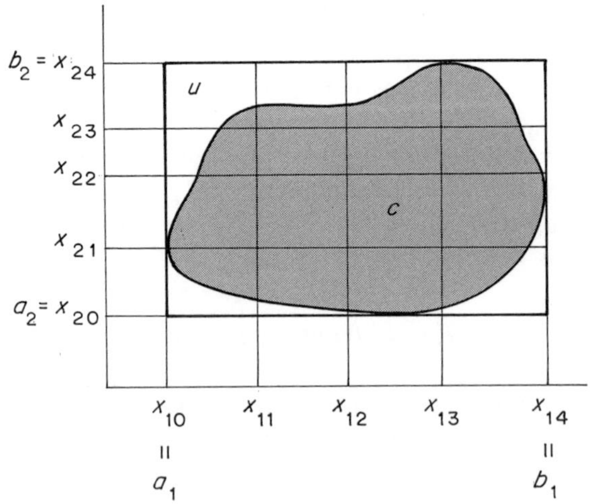

Figure 1.3.24 Partition in $\mathbb{R}^2$ for a Riemann sum.

consisting of a finite number of non-overlapping (except for boundaries) closed subintervals whose union is U, i.e. $T^{(2)} = \{[x_{1k-1}, x_{1k}] \times [x_{2l-1}, x_{2l}], k \in I_{n_1}, l \in I_{n_2}\}$ (Figure 1.3.24). For the partition $T^{(2)}$, we define

$$S(f, T^{(2)}) = \sum_{k=1}^{n_1} \sum_{l=1}^{n_2} f(\xi_k, \eta_l)(x_{1k} - x_{1k-1})(x_{2l} - x_{2l-1}) , \qquad (1.3.55)$$

where $x_{1k-1} \leq \xi_k \leq x_{1k}$, $x_{2l-1} \leq \eta_l \leq x_{2l}$. We call the real number $S(f, T^{(2)})$ the *Riemann sum* of f corresponding to the partition $T^{(2)}$.

For two partitions $T_1^{(2)}$ and $T_2^{(2)}$ of U, if each subinterval of $T_1^{(2)}$ is contained in some subinterval of $T_2^{(2)}$, the partition $T_1^{(2)}$ is called a *refinement* of partition $T_2^{(2)}$. With this notion, we define a Riemann integrable function. The function f is *Riemann integrable* on C if, for every positive real number ϵ, there exist a real number R and a partition $T_\epsilon^{(2)}$ of U such that if $T^{(2)}$ is a refinement of $T_\epsilon^{(2)}$ and $S(f, T^{(2)})$ is any Riemann sum corresponding to $T^{(2)}$, then $|S(f, T^{(2)}) - R| < \epsilon$. The number R is uniquely determined and is denoted by

$$R = \int\int_C f(x_1, x_2)\mathrm{d}x_1\mathrm{d}x_2 = \int_C f(x_1, x_2)\mathrm{d}x_1\mathrm{d}x_2 = \int_C f(\mathbf{x})\mathrm{d}\mathbf{x} . \qquad (1.3.56)$$

We call the real number R the *Riemann integral* of f over C.

As is understood from the definitions of the Riemann integral in $\mathbb{R}$ and $\mathbb{R}^2$, it is routine to define the Riemann integral in $\mathbb{R}^m$. For this integral, we may employ one of the notations

$$\int \cdots \int_C f(x_1, \ldots, x_m)\mathrm{d}x_1 \ldots \mathrm{d}x_m = \int_C f(x_1, \ldots, x_m)\mathrm{d}x_1 \ldots \mathrm{d}x_m$$

$$= \int_C f(\mathbf{x})\mathrm{d}\mathbf{x} . \qquad (1.3.57)$$

1.3.6 Graphs

Suppose, as is shown in Figure 1.3.25(a), that there are four towns $p_{g1}, \ldots, p_{g4}$ (indicated by the filled circles) in a region and those towns are connected with roads $L_{g1}, \ldots L_{g4}$ (indicated by lines). We are living in town p_{g1} and wish to visit towns p_{g3} and p_{g4} in this order without going through the other towns, and return to town p_{g1}. To examine this type of trip feasibility, the distance between towns or the shape of roads are not essential; the essential information is whether or not there exists a direct road between towns. This information is summarized in Table 1.3.1. Mathematically, we can treat the incidence of roads with towns listed in Table 1.3.1 in graph theory, which is to be introduced here.

To define a graph formally, let $P = \{p_1, \ldots, p_{n_p}\}$ be a non-empty set, and $\{p_i, p_j\}$ be an unordered pair, where $p_i, p_j \in P$. The set of all distinct

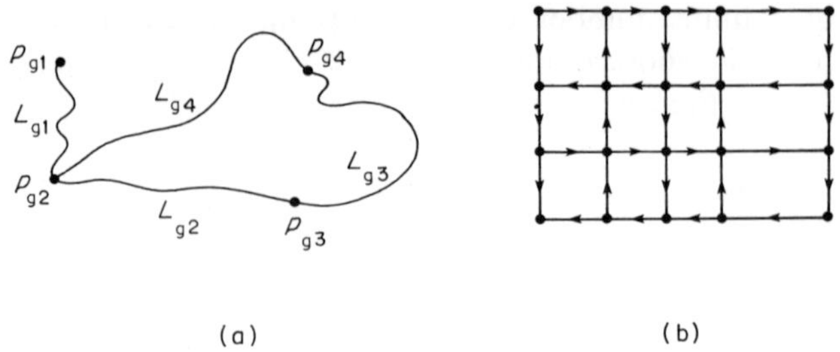

Figure 1.3.25 (a) Towns and roads between towns, and (b) one-way streets, which can be regarded as geometric graphs.

Table 1.3.1 Roads and their incident towns

Road L_{gi}	Towns incident with road L_{gi}
L_{g1}	p_{g1}, p_{g2}
L_{g2}	p_{g2}, p_{g3}
L_{g3}	p_{g3}, p_{g4}
L_{g4}	p_{g4}, p_{g2}

unordered pairs $\{\{p_i, p_j\} | p_i, p_j \in P\}$ is called the *unordered product* of P, and denoted by $P * P$ (note that $\{p_i, p_i\}$ is allowed). Let $L = \{L_1, \ldots, L_{n_l}\}$ be a set which is possibly empty, and f be a mapping of L into $P * P$, i.e. $f(L_i) = \{p_j, p_k\}$. Just for notational simplicity, we write $L_i \sim \{p_j, p_k\}$ for $f(L_i) = \{p_j, p_k\}$. We call the paired sets P and L with mapping f an *abstract graph*, or simply a *graph*, and denote it by $G(P, L, f)$, $G(P, L)$ or G. To distinguish a geometric graph to be defined below, we sometimes use G_a for an abstract graph. An example of an abstract graph is given by

$$G_a(P, L, f) = G_a(\{p_1, p_2, p_3, p_4\}, \{L_1, L_2, L_3, L_4\}, \{L_1 \sim \{p_1, p_2\},$$
$$L_2 \sim \{p_2, p_3\}, L_3 \sim \{p_3, p_4\}, L_4 \sim \{p_2, p_4\}\}) \, . \quad (1.3.58)$$

If the sets P and L in G are finite, we call the graph a *finite graph*. In this text, we shall deal with only finite graphs.

Given a graph $G(P, L, f)$, we can construct a graph, called a subgraph, consisting of selected links and nodes of G with the same incidences as those of G satisfying that the selected nodes include all of the end points of the selected links. Mathematically, we call $G(P_s, L_s, f_s)$ a *subgraph* of G if and only if $G(P_s, L_s, f_s)$ satisfies the following three conditions: (i) $P_s \subset P$ and $L_s \subset L$; (ii) $f(L_i) = f_s(L_i)$ for every $L_i \in L_s$; (iii) if $L_i \in L_s$ and $f(L_i) = \{p_j, p_k\}$, then $p_j \in P_s$ and $p_k \in P_s$.

Inspecting the similarity between the function f of equation (1.3.58) and Figure 1.3.25(a) or Table 1.3.1 (which actually shows a function), we notice that the above abstract graph may be represented by a set of geometric points, and a set of geometric line segments connecting those points. For this representation, let us introduce the notion of a geometric graph.

Let $P_g = \{p_{g1}, \ldots, p_{gn_p}\}$ be a set of n distinct points in the m-dimensional Euclidean space and $L_g = \{L_{g1}, \ldots, L_{gn_\ell}\}$ be a set of non-self-intersecting lines with end points in P_g. The line segments in L_g may be straight or may be curved, but they should satisfy the following three properties.

(i) Every line segment that forms a loop (a non-self-intersecting line segments whose end points coincide) contains exactly one point of P_g (for example, a loop with point p_{g1} in Figure 1.3.26).

(ii) Every line segment that does not form a loop contains exactly two points of P_g, and these points agree with its end points (for example, line segment L_{g2} with points p_{g2} and p_{g3} in Figure 1.3.26).

(iii) Every line segment in L_g has no common points except for points of P_g (for example, line segments L_{g2} and L_{g3} share the point p_{g3} in Figure 1.3.26).

We call the paired sets (P_g, L_g) satisfying the above properties a *geometric graph* and denote it by $G(P_g, L_g)$ or G. When we distinguish a geometric graph from an abstract graph G_a, we sometimes use G_g. As is noticed from this definition, the end points of every line segment in L_g are included in P_g, but a point in P_g is not necessarily an end point of a line segment in L_g (such as p_{g8} in Figure 1.3.26).

From the definition of a geometric graph, we notice that the geometric points and the geometric line segments shown in Figure 1.3.25(a) actually form a geometric graph, i.e.

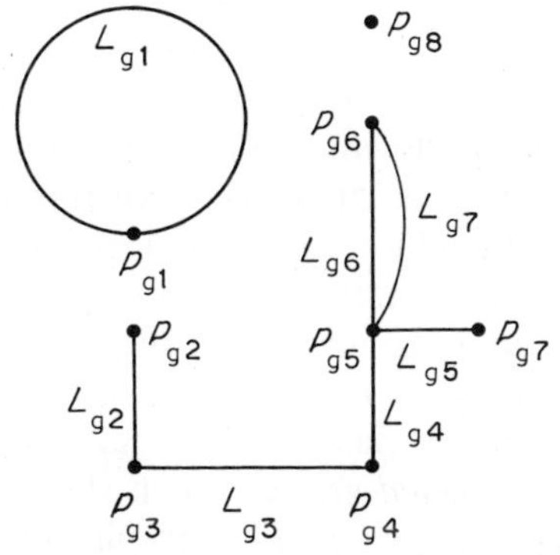

Figure 1.3.26 A geometric graph $G_g = G_g(\{p_{g1}, \ldots, p_{g8}\}, \{L_{g1}, \ldots, L_{g7}\})$.

$$G_g(P_g, L_g, f_g) = G_g(\{p_{g1}, p_{g2}, p_{g3}, p_{g4}\}, (L_{g1}, L_{g2}, L_{g3}, L_{g4}\}, \{L_{g1} \sim \{p_{g1}, p_{g2}\}, L_{g2} \sim \{p_{g2}, p_{g3}\}, L_{g3} \sim \{p_{g3}, p_{g4}\}, L_{g4} \sim \{p_{g2}, p_{g4}\}\}) \,. \quad (1.3.59)$$

Moreover, we notice that there is a one-to-one correspondence between the vertices in the abstract graph G_a of equation (1.3.58) and those in the geometric graph G_g of equation (1.3.59) such that the adjacency relations are preserved.

To state this correspondence more precisely, let us consider two graphs $G_\alpha(P_\alpha, L_\alpha, f_\alpha)$ and $G_\beta(P_\beta, L_\beta, f_\beta)$. First, when $G_\alpha(P_\alpha, L_\alpha, f_\alpha)$ is an abstract graph and $G_\beta(P_\beta, L_\beta, f_\beta)$ is an abstract graph or a geometric graph, if there exist a one-to-one mapping Φ_p of P_α onto P_β, and a one-to-one mapping Φ_ℓ of L_α onto L_β such that $f_\alpha(L_i) = \{p_j, p_k\}$, $L_i \in L_\alpha$, p_j, $p_k \in P_\alpha$, if and only if $f_\beta(\Phi_\ell(L_i)) = \{\Phi_p(p_j), \Phi_p(p_k)\}$, then we say that graphs $G_\alpha(P_\alpha, L_\alpha, f_\alpha)$ and $G_\beta(P_\beta, L_\beta, f_\beta)$ are *isomorphic*. If an abstract graph G_a is isomorphic to a geometric graph G_g, then we say that graph G_g is a *geometric realization* of graph G_a. If a graph has a geometrical realization in $\mathbb{R}^2$, we call the graph a *planar graph*; otherwise, a *non-planar graph*. Although graphs are not always planar graphs, every finite graph has a geomtric realization in $\mathbb{R}^3$. Comaring equations (1.3.58) and (1.3.59), we readily notice that the abstract graph G_a and the geometric graph G_g are isomorphic, and hence the geometric graph G_g is a geometric realization of the abstract graph G_a. Moreover, since the geometric graph G_g can be drawn on a plane (Figure 1.3.25(a)), it is a planar graph.

Second, when two graphs $G(P_\alpha, L_\alpha, f_\alpha)$ and $G(P_\beta, L_\beta, f_\beta)$ are both geometric graphs, the notion of isomorphic becomes a little stronger than the above. If $G(P_\beta, L_\beta, f_\beta)$ is obtained by deforming $G(P_\alpha, L_\alpha, f_\alpha)$ by moving the vertices continuously in $\mathbb{R}^2$ in such a way that the edges are also moved as their end points are moved (the edges may be extended or shrunken) and that no edges hit a vertex other than its own end points, then we say that graphs $G(P_\alpha, L_\alpha, f_\alpha)$ and $G(P_\beta, L_\beta, f_\beta)$ are *isomorphic*.

In the example of Figure 1.3.25(a) we implicitly assume that every road is two-ways; we can drive, say from town p_{g1} to town p_{g2} as well as from town p_{g2} to town p_{g1}. In the downtown of a city, however, streets are often one-way such as in Figure 1.3.25(b). To take this directional feature into account in a graph, we introduce a direction on links $L_i \in L$.

Let $P = \{p_1, \ldots, p_{n_p}\}$ be a non-empty set, $L = \{L_1, \ldots, L_{n_\ell}\}$ be a set which is possibly empty, and f be a mapping of L into $P \times P$, i.e. $f(L_i) = (p_j, p_k)$, which is simply written as $L_i \sim (p_j, p_k)$. Since $P \times P$ is the Cartesian product, (p_j, p_k) is an ordered pair (cf. $\{p_j, p_k\}$). We call the paired sets P and L with mapping f a *directed graph*, which should be distinguished from a *undirected graph* defined above. The graphs in Figures 1.3.27(a) and 1.3.27(b) are an undirected graph and a directed graph, respectively.

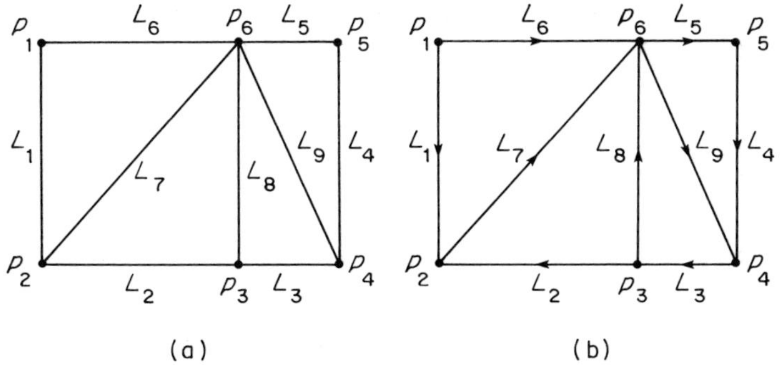

Figure 1.3.27 (a) An undirected graph and (b) a directed graph.

The elements of P and L in graph G are called *nodes* and *links*, respectively, and a mapping f is called an *incidence mapping*. In particular, in a directed graph the nodes p_j and p_k in $L_i \sim (p_j, p_k)$ are called an *initial* node and a *terminal* node, respectively. For example, in the directed graph in Figure 1.3.27(b), node p_1 is the initial node of link L_1, and node p_2 is the terminal node of link L_1. If $L_i \sim \{p_j, p_k\}$ (for an undirected graph) or $L_i \sim (p_j, p_k)$ (for a directed graph), then link L_i is said to be *incident* with each of nodes p_j and p_k. In particular, in a directed graph, link L_i is said to be *positively incident* with node p_j and *negatively incident* with node p_k. The nodes incident with a link are called *end points*. For $L_i \sim \{p_j, p_k\}$ or $L_i \sim (p_j, p_k)$, if $p_j = p_k$, then link L_i is called a *loop* (L_{g1} in Figure 1.3.26 is a loop). If $L_i \sim \{p_j, p_k\}$ or $L_i \sim (p_j, p_k)$ and $L_\ell \sim \{p_j, p_k\}$ or $L_\ell \sim (p_j, p_k)$, then links L_i and L_ℓ are called *parallel links* (L_{g6} and L_{g7} are parallel links in Figure 1.3.26).

If $L_i \sim \{p_j, p_k\}$ or $L_i \sim (p_j, p_k)$, then nodes p_j and p_k are called *adjacent nodes* (for example, nodes p_2 and p_3 in Figure 1.3.27(a),(b) are adjacent nodes). If links L_i and L_j share at least one common end point, then links L_i and L_j are called *adjacent links* (for example, links L_2 and L_3 in Figure 1.3.27(a),(b) are adjacent links). The number of links incident with a node p_i is called the *degree* of node p_i, and denoted by $\delta(p_i)$ (for example, $\delta(p_2) = 3$ in Figure 1.3.27(a),(b)). In particular, in a directed graph, the number of positively incident links with node p_i, denoted by $\delta^+(p_i)$, is called the *positive degree* of node p_i, and the number of negatively incident links with node p_i, denoted by $\delta^-(p_i)$, is called the *negative degree* of node p_i (for example, $\delta^+(p_2) = 1$ and $\delta^-(p_2) = 2$ in Figure 1.3.27(b). Obviously, the equation $\delta(p_i) = \delta^+(p_i) + \delta^-(p_i)$ holds. A node in P whose degree is zero is called an *isolated node* (for example, p_{g8} in Figure 1.3.26).

The adjacency of nodes may be indicated in terms of a matrix, called an adjacency matrix. To be explicit, consider a graph $G(\{p_1, \ldots, p_n\}, L)$ and let $C = [c_{ij}]$ be an $n \times n$ matrix. When graph G is a undirected graph, let

c_{ij} be the number of links incident with both nodes p_i and p_j. When graph G is a directed graph, let c_{ij} be the number of links directed from node p_i to node p_j. Then, the adjacency of nodes in graph G can be indicated by the matrix C, which we call an *adjacency matrix* of nodes. The adjacency matrix C_1 of the undirected graph shown in Figure 1.3.27(a) and that C_2 of the directed graph shown in Figure 1.3.27(b) are given by

$$C_1 = \begin{bmatrix} 0 & 1 & 0 & 0 & 0 & 1 \\ 1 & 0 & 1 & 0 & 0 & 1 \\ 0 & 1 & 0 & 1 & 0 & 1 \\ 0 & 0 & 1 & 0 & 1 & 1 \\ 0 & 0 & 0 & 1 & 0 & 1 \\ 1 & 1 & 1 & 1 & 1 & 0 \end{bmatrix}, \quad C_2 = \begin{bmatrix} 0 & 1 & 0 & 0 & 0 & 1 \\ 0 & 0 & 0 & 0 & 0 & 1 \\ 0 & 1 & 0 & 0 & 0 & 1 \\ 0 & 0 & 1 & 0 & 0 & 0 \\ 0 & 0 & 0 & 1 & 0 & 0 \\ 0 & 0 & 0 & 1 & 1 & 0 \end{bmatrix},$$

respectively.

From the definition of c_{ij} and that of $\delta(p_i)$, it follows that the equation $\delta(p_i) = \sum_{j=1}^{n} c_{ij}$ holds for an undirected graph, and the equations $\delta^+(p_i) = \sum_{j=1}^{n} c_{ij}$ and $\delta^-(p_i) = \sum_{j=1}^{n} c_{ji}$ hold for a directed graph.

For a given graph, if we can traverse from a node to a node on successively adjacent links without visiting the same links more than once, we call such a set of links a *chain* in an undirected graph, and a *path* in a directed graph (note that in a directed graph, we are supposed to follow directions). For example, L_7, L_9, L_4, L_5, L_6 in Figure 1.3.27(a) is a chain, and L_7, L_9, L_3, L_8, L_5 in Figure 1.3.27(b) is a path. In traversing a chain or a path, if we visit every node only once, the chain or the path is called a *simple chain* or a *simple path*. The chain or the path in the above example is not a simple chain nor a simple path, because we visit p_6 twice. An example of a simple chain in Figure 1.3.27(a) is L_7, L_9, L_3, and that of a simple path in Figure 1.3.27(b) is L_7, L_5, L_4. Note that we allow a chain or a path that returns to the same node, for example L_1, L_7, L_9, L_4, L_5, L_6 in Figure 1.3.27(a), and L_7, L_9, L_3, L_2 in Figure 1.3.27(b). Such a chain and a path are called a *circuit* and a *cycle*, respectively. If we can traverse a circuit and a cycle visiting every node only once, we call them a *simple circuit* and *simple cycle*, respectively. The circuit in the above example is not a simple circuit because we visit p_6 twice, but the cycle in the above example is a simple cycle.

In the graph of Figure 1.3.27(a) we can traverse from any node in P to any node in P through links. In this sense, the graph is connected. Stated a little more precisely, an undirected graph $G(P, L)$ is *connected* if there exists at least one chain between every pair of nodes in P; graph G is *disconnected* if there exists no chain between at least one pair of nodes in P. For example, the graph in Figure 1.3.27(a) is connected; the graph in Figure 1.3.26 is disconnected. In a directed graph $G(P, L)$, if there exists at least one path from any node to any node in P, we say that the graph is *strongly connected*. The directed graph in Figure 1.3.27(b) is not strongly connected because there is no path from node p_2 to node p_1. A connected subgraph which is not contained in any larger connected subgraph is called a *component*. The graph in Figure 1.3.26(a) consists of three components.

The graph shown in Figure 1.3.28 is a connected graph, which looks like a tree (we shall often show a tree upside down). Obviously, an actual tree does not have looped twigs. In graph theory, if a graph is connected and does not have circuits, the graph is called a *tree* (*graph*). When a graph is a tree, the following equation holds:

$$n_p - n_\ell = 1 , \tag{1.3.60}$$

where n_p and n_ℓ are the number of nodes and that of links in a tree. To prove this equation, consider a tree graph, G_0, with n_p nodes and n_ℓ links (an example is shown in Figure 1.3.28(a), where $n_p = 8$, $n_\ell = 7$). From tree G_0, we delete one link and one node keeping the remaining tree, G_1, connected (Figure 1.3.28(b)). As a result, tree G_1 has $n_p - 1$ nodes and $n_\ell - 1$ links (in Figure 1.3.28(b), $n_p = 7$, $n_\ell = 6$). The difference between those numbers is $(n_p - 1) - (n_\ell - 1)$, which is the same as that of tree G_0, i.e. $n_p - n_\ell$. From tree G_1, we delete one link and one node keeping the remaining tree, G_2, connected (Figure 1.3.28(c)). As a result, tree G_2 has $n_p - 2$ nodes and $n_\ell - 2$ links (in Figure 1.3.28(c), $n_p = 6$, $n_\ell = 5$). Again, the difference is $(n_p - 2) - (n_\ell - 2) = n_p - n_\ell$. In this manner, we delete the links and nodes until the remaining tree becomes a tree with only one link, $G_{n_\ell - 1}$. Tree $G_{n_\ell} - 1$ has two nodes and one link (Figure 1.3.28(d)), and hence the difference is $2 - 1$. Therefore $n_p - n_\ell = (n_p - 1) - (n_\ell - 1) = \cdots = 2 - 1 = 1$, proving equation (1.3.60).

When a graph is not a tree, equation (1.3.60) does not necessarily hold, but we have a more general equation, called Euler's formula. When a graph is a planar graph, links of the graph may partition the plane into disjoint regions one of which is unbounded (Figure 1.3.29(a)). Let n_r be the number of those regions. Note that when a graph is a tree, the links of the tree do

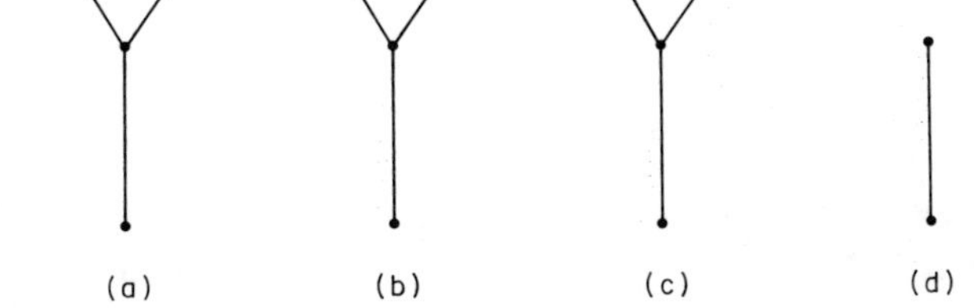

Figure 1.3.28 Tree graphs.

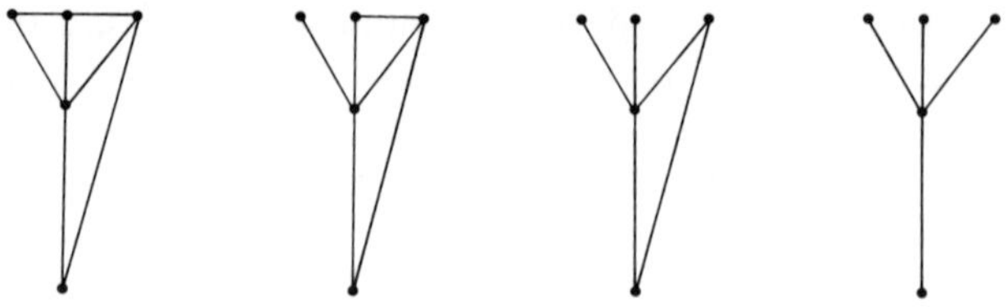

Figure 1.3.29 Deleting links from circuits in a non-tree graph.

not partition the plane and hence we have only one (unbounded) region ($n_r = 1$). In terms of n_p, n_ℓ and n_r, Euler's formula is stated as follows.

Euler's formula For a connected planar graph with n_p nodes, n_ℓ links and n_r regions, the equation

$$n_p - n_\ell + n_r = 2 \qquad (1.3.61)$$

holds.

We can prove this equation through the following procedure. If a graph is a tree, as mentioned above, we have only one (unbounded) region, $n_r = 1$, and hence equation (1.3.61) becomes $n_p - n_\ell + 1 = 2$. This equation is equivalent to equation (1.3.60) already proved above. When a graph is not a tree, we delete one link from a circuit in the original graph, G_0, keeping the resulting graph, G_1, connected (Figure 1.3.29(b)). Graph G_1 may be a tree or may not. If graph G_1 is not a tree, we delete one link from a circuit in graph G_1 keeping the resulting graph, G_2, connected (Figure 1.3.29(c)). We continue this procedure until the resulting graph becomes a tree (Figure 1.3.29(d)). Since graph G_0 has n_r regions, a tree is obtained by deleting $n_r - 1$ links from graph G_0. Since the number of nodes does not change by this deletion, the resulting tree has n_p nodes and $n_\ell - (n_r - 1)$ links. For this tree, as shown above, equation (1.3.60) holds. Upon substituting $n_\ell - n_r$ into this equation, we obtain $n_p - (n_\ell - (n_r - 1)) = 1$, from which equation (1.3.61) is derived.

Euler's formula can be generalized in $\mathbb{R}^m$. Let $\mathcal{S}$ be a tessellation of a bounded set S in $\mathbb{R}^m$, and n_i be the number of i-faces in $\mathcal{S}$. Then the equation

$$\sum_{i=0}^{m} (-1)^i n_i = 1 + (-1)^m \qquad (1.3.62)$$

holds. We call this equation the *Euler–Schlaefli formula*.

A planar graph is always associated with a 'dual graph'. To define it explicitly, let $G_g(P_g, L_g)$ be a planar graph with k regions R_i, $i \in I_k$. We take a point q_i in each region R_i (open circles in Figure 1.3.30(a)). For each

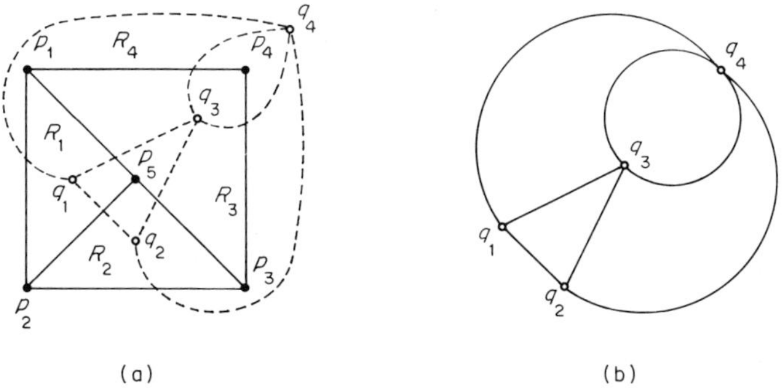

Figure 1.3.30 The primal graph and the dual graph.

link which is shared by regions R_i and R_j, we generate a line segment combining q_i and q_j which crosses the common link only once and has no point in common with any other links of L (the broken line in Figure 1.3.30(a)). Let L_g^* be the resulting links. Then we obtain a new geometric graph G_g^* with nodes $P_g^* = \{q_i, i \in I_k\}$ and L_g^* (Figure 1.3.30(b)). We call graph $G_g^*(P_g^*, L_g^*)$ the *dual graph* of graph $G_g(P_g, L_g)$. For the dual graph G_g^*, graph G_g is called the *primal graph*. Note that the dual and primal are a relative notion. We can say that the dual graph of G_g^* is graph G_g. In this case, the primal graph is graph G_g^*.

1.3.7 Probabilities and statistical tests

In statistics, any process of observation is referred to as an *experiment*, and a result obtained from an experiment is called an *event*. For example, an experiment may consist of throwing two dice. The sum of the numbers on the two faces being four is an event. This event can be further decomposed into three events: face 1 and face 3; face 3 and face 1, face 2 and face 2. Each of these events, however, cannot be decomposed any more. If an event cannot be decomposed into more elementary events in the context of a given experiment, we call the event a *simple event*, and an event consisting of two or more simple events a *compound event*.

Suppose that some experiment can result in exactly one of n events A_1, ..., A_n which are mutually exclusive in the sense that the occurrence of one event precludes that of any other. To every event we assign some number $P(A_i)$ between zero and one. If an event is impossible, we assign zero for that event. If an event certainly happens, we assign unity for that event. If we assign the number $P(A_i)$ to event A_i and the number $P(A_j)$ to event A_j, then we assign the number $P(A_i) + P(A_j)$ to the event in which event A_i or event A_j is to be realized. Mathematically, $0 \leq P(A_i) \leq 1$; $\Pr(A_i$ and $A_j) = 0$, $i \neq j$; $\Pr(A_i$ or $\cdots$ or $A_n) = 1$; $\Pr(A_i$ or $A_j) = P(A_i) + P(A_j)$, where Pr(conditions) indicates the probability that the event satisfying the conditions occurs. We may call the number $P(A_i)$ characterized by these properties the *probability* of event A_i.

To define the probability more precisely, we introduce one more notion. Let S be a set of all possible events of an experiment. If S is a countable set, we call the set S a *discrete sample space*. In a discrete sample space S, a *probability* is defined as a real number $P(A)$ associated with an event A that satisfies the following three postulates.

Postulate 1 $P(A) \geq 0,\ A \subset S$.

Postulate 2 $P(S) = 1$.

Postulate 3 If $A_1, A_2, A_3, \ldots$ are a countable number of disjoint subsets of S, then $P(A_1 \cup A_2 \cup A_3 \cdots) = P(A_1) + P(A_2) + P(A_3) + \cdots$.

The real-valued function $P(A)$ is called a *probability measure*. From these postulates, it follows that

$$P(A) \leq 1 \quad \text{for } A \subset S \,, \tag{1.3.63}$$

$$P(\varnothing) = 0 \,, \tag{1.3.64}$$

$$P(A) + P(S \backslash A) = 1, \quad A \subset S \,. \tag{1.3.65}$$

The formula $P(A_1 \cup A_2) = P(A_1) + P(A_2)$, $A_1 \cap A_2 = \varnothing$ of Postulate 3 is called the *special rule of addition*, because the condition $A_1 \cap A_2 = \varnothing$ is imposed. If we relax this condition, we have the *general rule of addition*: for any $A_1 \subset S$, $A_2 \subset S$,

$$P(A_1 \cup A_2) = P(A_1) + P(A_2) - P(A_1 \cap A_2) \,. \tag{1.3.66}$$

Informally, if the realization of an event A is not affected by an event B, we say that event A is independent of event B. For instance, when we throw a die twice, the event that face 1 appears in the second throw is independent of the event that face 1 appears in the first throw. Formally, we say that an event A and an event B are *independent* if and only if

$$P(A \cap B) = P(A)P(B) \,, \tag{1.3.67}$$

that is, the probability of realizing both event A and event B is given by the product of the probability of realizing event A and that of realizing event B.

Consider again the experiment in which we throw two dice. We are now concerned with the sum of the numbers appearing on the two faces. The sample space for this experiment is then given by $S = \{(i, j) |\ i \in I_6, j \in I_6\}$, where (i, j) indicates (the number on the first face, that of the second face).

For each element (i, j) of S, we assign $i + j$. This assignment can be regarded as a function $x((i, j)) = i + j$, and we call x a *random variable*. Stated generally, if S is a discrete sample space with a probability measure and if x is a real-valued function given by $x(A)$, $A \subset S$, then x is called a *discrete random variable*.

In the above experiment, since $P((i, j)) = 1/36$ for all $i, j \in I_6$, the probability of x being equal to a certain number is given by Table 1.3.2. This table may be regarded as a function, which is called a *probability function*. Generally, a *probability function* is defined as a function $f(x)$ which assigns a probability to a discrete random variable x.

From the probability function shown in Table 1.3.2, we can obtain the probability that the total number is not greater than a certain number. For instance, the probability that the total number is not greater than four is 1/6. Generally, if we define

$$F(x) = \sum_{t \leq x} f(t) \tag{1.3.68}$$

with a probability function $f(x)$, then $F(x)$ shows the probability that a random variable t is not greater than x. We call this function a *(cumulative) distribution function*. For example, if the probability function is given by the function indicated in Table 1.3.2, then its distribution function is given by Figure 1.3.31.

In statistics, various kinds of probability functions are known (a good encyclopaedia is provided by Johnson and Kotz (1969).) Here we show only three of them, called the binomial distribution, the multinomial distribution and the Poisson distribution.

First, to derive a binomial distribution, consider the experiment in which we throw a die five times, and we wish to know the probability that face 1 appears exactly twice in this experiment. To obtain this probability, we first consider the number of ways in which face 1 appears exactly twice in five trials. We denote the event that face 1 appears only in the first and second trials by (11 * * *), the event that face 1 appears only in the first and third trials by (1 * 1 * *), and so forth. Then possible combinations are written

Table 1.3.2 The sum of the numbers appearing on the faces of two dice

Total number	Probability	Total number	Probability
2	1/36	8	5/36
3	2/26	9	4/36
4	3/36	10	3/36
5	4/36	11	2/36
6	5/36	12	1/36
7	6/36		

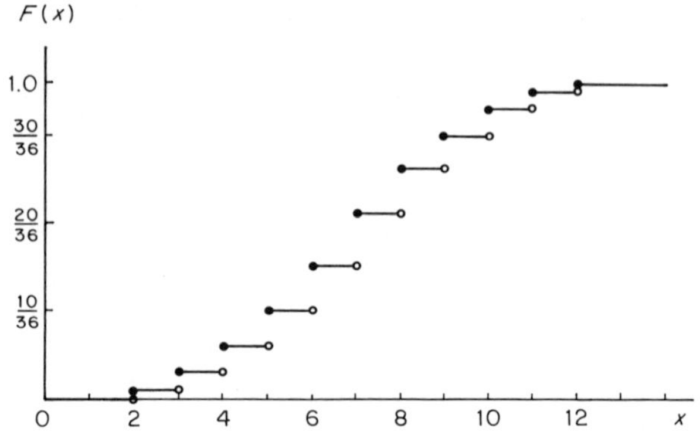

Figure 1.3.31 The distribution function of the probability function given by Table 1.3.2.

as (11 * * *), (1 * 1 * *), (1 * * 1 *), (1 * * * 1), (* 11 * *), (* 1 * 1 *), (* 1 * * 1), (* * 11 *), (* * 1 * 1), (* * * 11).There are ten different ways in which face 1 appears exactly twice in five trials.

Generally the number of combinations of x elements out of n elements is given by

$$_nC_x = \binom{n}{x} = \frac{n!}{x!(n-x)!} , \tag{1.3.69}$$

where $n! = n(n-1) \cdots 1$. In the above case, $_5C_2 = 5!/(2!3!) = 10$.

The probability that face 1 appears in the first trial is 1/6 and the probability that face 1 appears in the second trial is 1/6; the probabilities that face 1 does not appear in the third, fourth, and fifth trials are 5/6, 5/6 and 5/6, respectively. Hence the probability that face 1 appears only in the first and second trials is given by $(1/6)^2\ (1-1/6)^3$. From this probability and the number of ways obtained in the above, the probability that face 1 appears exactly twice in five trials is given by $10(1/6)^2\ (1-1/6)^3$. This probability is a specific value of a binomial distribution. In general, the *binomial distribution* is defined by

$$b(x; n, \theta) = \binom{n}{x} \theta^x (1-\theta)^{n-x} , \quad x = 0, 1, \ldots, n , \tag{1.3.70}$$

where $1 \leq n < \infty$ and $0 \leq \theta \leq 1$. In the above example, $n = 5$, $x = 2$ and $\theta = 1/6$.

The binomial distribution is extended to the *multinomial distribution*, which is defined by

$$g(x_0, x_1, \ldots, x_m) = \frac{n!}{x_0!x_1! \cdots x_m!} \theta_0^{x_0}\ \theta_1^{x_1} \cdots \theta_m^{x_m} ,$$

$$x_i = 0, 1, \ldots, n , \tag{1.3.71}$$

where $1 \leq \Sigma_{i=0}^{m} x_i = n < \infty$ and $\Sigma_{i=0}^{m} \theta_i = 1$, $\theta_i > 0$. If $\theta_0 = \theta$, $\theta_1 = 1 - \theta$, equation (1.3.71) becomes equation (1.3.70). To give an example, consider the experiment in which we throw a die five times. Then, the probability that face 1 appears once, face 2 appears twice and the other faces appear twice is given by $g(1, 2, 2) = (5!/(1!2!2!))(1/6)^1(1/6)^2(4/6)^2$.

The binomial distribution is also closely related to the *Poisson distribution*, which is defined by

$$p(x; \lambda) = \frac{\lambda^x e^{-\lambda}}{x!}, \quad x = 0, 1, 2, \ldots . \tag{1.3.72}$$

It is shown that the binomial distribution approaches the Poisson distribution if we increase n to a sufficiently large number and decrease θ to a sufficiently small number keeping $n\theta = \lambda$ constant.

As is observed in Figure 1.3.31, a distribution function of a discrete random variable is a step function. A distribution function, however, is not necessarily a discontinuous function. We may consider a continuous distribution function like Figure 1.3.32. When we consider continuous distribution functions, we shall limit ourselves to distribution functions $F(x)$ that are continuous and have derivative $\mathrm{d}F(x)/\mathrm{d}x$ except at a finite number of points. As can be seen from Figure 1.3.32, the variable x of a continuous distribution function takes a value on a continuum in $\mathbb{R}$, and hence we call a variable x a *continuous random variable* in contrast to a discrete random variable in Figure 1.3.31. Noticing that $F(-\infty) = 0$ and $F(\infty) = 1$ for any distribution function $F(x)$, and using the fundamental theorem of integral calculus (equation (1.3.54)), we can write

$$\int_{-\infty}^{x} f(t)\mathrm{d}t = F(x) - F(-\infty)$$

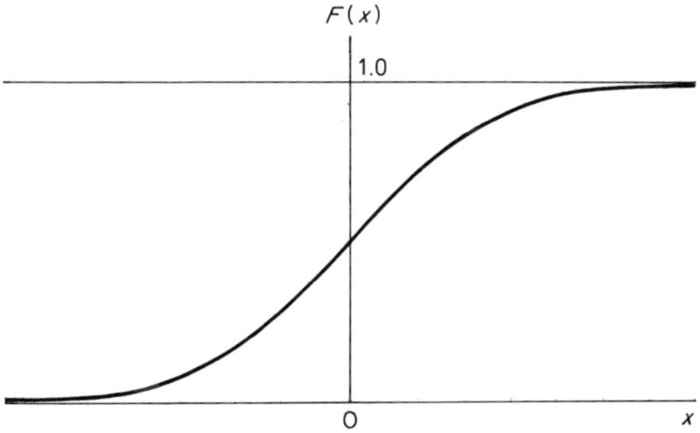

Figure 1.3.32 The distribution function of a continuous random variable.

or

$$F(x) = \int_{-\infty}^{x} f(t)\mathrm{d}t \ , \qquad (1.3.73)$$

where $f(x) = \mathrm{d}F(x)/\mathrm{d}x$. We call the function f a *probability density function*, which is sometimes abbreviated to pdf. A probability density function of a continuous variable is the counterpart of a probability function of a discrete random variable.

The function $F(x)$ shows the probability that a continuous random variable takes on a value not greater than x, which corresponds to $F(x)$ of equation (1.3.68) of a discrete random variable. As a consequence of equation (1.3.73), $F(b) - F(a)$ shows the probability that a random variable x takes on a value on the interval from a to b ($a < b$).

To show an example of the derivation from a distribution function to a density function, consider the distribution function given by

$$F(x) = \begin{cases} 0 & \text{for } x \leq a \ , \\ x & \text{for } a < (x-a)/(b-a) < b \ , \\ 1 & \text{for } x \geq b \ , \end{cases} \qquad (1.3.74)$$

which is depicted in Figure 1.3.33(a). Taking the derivative of $F(x)$ at x, we obtain its corresponding probability density function as

$$f(x) = \begin{cases} 0 & \text{for } x < a \ , \\ 1/(b-a) & \text{for } a < x < b \ , \\ 0 & \text{for } x > b \ , \end{cases} \qquad (1.3.75)$$

which is depicted in Figure 1.3.33(b). This probability distribution function is called the *uniform distribution*.

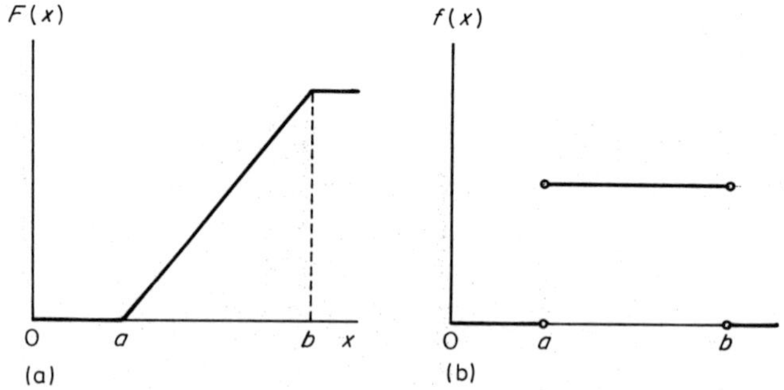

Figure 1.3.33 The uniform distribution: (a) the distribution function; (b) the probability density function.

One of the most frequently used probability density functions in statistics is that of the *normal distribution*, which is defined by

$$f(x) = \frac{1}{\sigma\sqrt{2\pi}} \exp\left(-\frac{1}{2}\left(\frac{x-\mu}{\sigma}\right)^2 \right) \quad \text{for } -\infty < x < \infty \qquad (1.3.76)$$

(Figure 1.3.34(a)). This density function is sometimes referred to as $N(x; \mu, \sigma^2)$. Specifically, the normal distribution with $\mu = 0$ and $\sigma = 1$, i.e. $N(x; 0, 1^2)$, is called the *standard normal distribution.*

We can construct several distributions from the normal distribution. One example is the chi-square distribution, which is often used in this text. If random variables $x_1, \ldots, x_k$ independently follow the standard normal

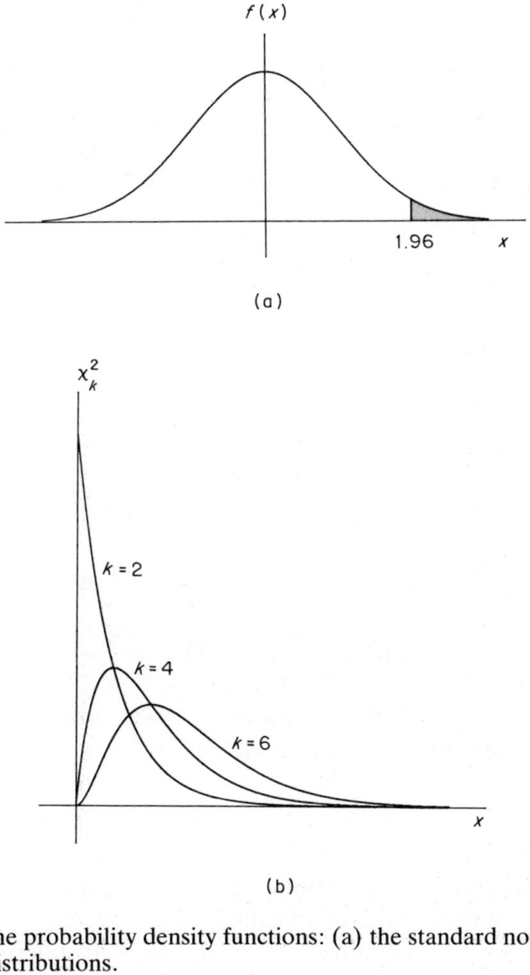

Figure 1.3.34 The probability density functions: (a) the standard normal distribution; (b) chi-square distributions.

distribution, then $x_1^2 + \cdots + x_k^2$ follows the *chi-square distribution* with k degrees of freedom, which is written as

$$\chi_k^2(x) = \begin{cases} \dfrac{1}{2^{k/2}\,\Gamma(k/2)}\, \mathrm{e}^{-x/2}\, x^{(k/2)-1} & \text{if } x > 0\ , \\ 0 & \text{if } x \leq 0\ . \end{cases} \tag{1.3.77}$$

The characteristics of a probability function or a probability density function are often indicated by a few parameter values. To define these parameter values, let us consider a dice game using the experiment shown in Table 1.3.2. Suppose that we join a dice game with \$8 in which a dice is thrown twice, and we can gain the dollars of the sum of numbers on two faces. As listed in Table 1.3.2, we have a chance of receiving \$2 with probability 1/36; \$3 with probability 2/36; ...; \$12 with probability 1/36. Then, what is our expectation of this game? In statistics, a mathematical expectation is the sum of the products of potential gains and the probability that this gain will actually be realized. In the above game, our mathematical expectation is, from Table 1.3.2, obtained as \$7, which is smaller than \$8. If we are 'rational' enough, we should not join this game.

In general, the mathematical *expectation*, *expected value* or *mean* of a discrete random variable x, denoted by $\mathrm{E}(x)$, is defined by

$$\mathrm{E}(x) = \sum_{x \in S} x\, f(x)\ , \tag{1.3.78}$$

where $f(x)$ is the value of a probability function at x. For example, when a discrete random variable follows the binomial distribution given by equation (1.3.70), its expected value is obtained as

$$\mathrm{E}(x) = \sum_{x=0}^{n} x\, b(x;\, n,\, \theta) = n\theta\ . \tag{1.3.79}$$

In the case of the Poisson distribution given by equation (1.3.72), we obtain:

$$\mathrm{E}(x) = \sum_{x=0}^{\infty} x\, p(x;\, \lambda) = \lambda\ . \tag{1.3.80}$$

Correspondingly, the mathematical *expectation*, *expected value* or *mean* of a continuous random variable x is defined by

$$\mathrm{E}(x) = \int_{-\infty}^{\infty} x\, f(x)\mathrm{d}x\ , \tag{1.3.81}$$

where $f(x)$ is the value of a probability density function at x. The expected values of x that follow the uniform distribution of equation (1.3.75), the normal distribution of equation (1.3.76) and the chi-square distribution of equation (1.3.77) are given by

$$\mathrm{E}(x) = \int_a^b \frac{x}{b-a}\mathrm{d}x = \frac{a+b}{2} , \tag{1.3.82}$$

$$\mathrm{E}(x) = \int_{-\infty}^{\infty} x\, N(x; \mu, \sigma^2)\mathrm{d}x = \mu , \tag{1.3.83}$$

$$\mathrm{E}(x) = \int_0^{\infty} x\, \chi_k^2(x)\mathrm{d}x = k , \tag{1.3.84}$$

respectively.

From equations (1.3.78) and (1.3.81), we obtain the following formulae for discrete as well as a continuous random variables x and x_i, $i \in I_n$:

$$\mathrm{E}(ax + b) = a\mathrm{E}(x) + b , \tag{1.3.85}$$

$$\mathrm{E}\left(\sum_{i=1}^{n} x_i\right) = \sum_{i=1}^{n} \mathrm{E}(x_i) , \tag{1.3.86}$$

where a and b are constants.

Like the expected value of x, we can calculate the expected value of x^r by

$$\mathrm{E}(x^r) = \sum_{x \in S} x^r f(x) , \tag{1.3.87}$$

$$\mathrm{E}(x^r) = \int_{-\infty}^{\infty} x^r f(x)\mathrm{d}x . \tag{1.3.88}$$

We call $\mathrm{E}(x^r)$ the *rth moment about the origin* of the distribution of a random variable x. The rth moment can be calculated not only about the origin but also about the mean, i.e.

$$\mathrm{E}((x - \mathrm{E}(x))^r) = \sum_{x \in S} (x - \mathrm{E}(x))^r f(x) , \tag{1.3.89}$$

$$\mathrm{E}((x - \mathrm{E}(x))^r) = \int_{-\infty}^{\infty} (x - \mathrm{E}(x))^r f(x)\mathrm{d}x . \tag{1.3.90}$$

We call $\mathrm{E}((x - \mathrm{E}(x))^r)$ the *rth moment about the mean*. In particular, we call the second moment about the mean the *variance* and denote it by $\mathrm{Var}(x)$, and we call $\sqrt{\mathrm{Var}(x)}$ the *standard deviation*. It follows from equations (1.3.85)–(1.3.90) that

$$\mathrm{Var}(x) = \mathrm{E}(x^2) - \mathrm{E}(x)^2 , \tag{1.3.91}$$

$$\mathrm{Var}(ax + b) = a^2\mathrm{Var}(x) , \tag{1.3.92}$$

where a and b are constants. One may expect that the equation corresponding to equation (1.3.86) holds with respect to a variance. In general, this expectation is false, but when variables x_i, $i \in I_n$, are independent, it is true; that is, for independent random variables x_i, $i \in I_n$, the equation

$$\mathrm{Var}\left(\sum_{i=1}^{n} x_i\right) = \sum_{i=1}^{n} \mathrm{Var}(x_i) \tag{1.3.93}$$

holds.

It is an exercise to obtain the variances of the binomial, Poisson, uniform, normal and chi-square distributions. They are given by

$$\mathrm{Var}(x) = \sum_{x=0}^{n} (x - n\theta)^2\, b(x; n, \theta) = n\theta(1-\theta)\ , \tag{1.3.94}$$

$$\mathrm{Var}(x) = \sum_{x=0}^{\infty} (x - \lambda)^2\, p(x; \lambda) = \lambda\ , \tag{1.3.95}$$

$$\mathrm{Var}(x) = \int_a^b \left(x - \frac{a+b}{2}\right)^2 \frac{1}{b-a}\,\mathrm{d}x = \frac{1}{12}(b-a)^2\ , \tag{1.3.96}$$

$$\mathrm{Var}(x) = \int_{-\infty}^{\infty} (x - \mu)^2\, N(x; \mu, \sigma^2)\mathrm{d}x = \sigma^2\ , \tag{1.3.97}$$

$$\mathrm{Var}(x) = \int_0^{\infty} (x - k)^2\, \chi_k^2(x)\mathrm{d}x = 2k\ , \tag{1.3.98}$$

respectively.

From equations (1.3.86) and (1.3.93), if n random variables x_i, $i \in I_n$, independently follow the normal distribution of equation (1.3.76), then the average of these random variables, $\bar{x} = \Sigma_{i=1}^{n} x_i/n$, has $\mathrm{E}(\bar{x}) = \mu$ and $\mathrm{Var}(\bar{x}) = \sigma^2/n$. Moreover, it can be shown that the random variable $\bar{x}$ follows the normal distribution, $N(\bar{x}; \mu, \sigma^2/n)$. One might hopefully expect that the average value of independent random variables follows a normal distribution regardless of a distribution function. This expectation is false, but in some sense it is true, which we shall show with an example.

Consider the uniform distribution of equation (1.3.75) which is depicted in Figure 1.3.33(b), and suppose that n random variables independently follow this distribution. If $n = 1$, then $\bar{x} = x_1$, and so the probability density function of $\bar{x}$ is given by Figure 1.3.33(b) or 1.3.35(a). Comparing Figure 1.3.35(a) and Figure 1.3.34(a), we readily notice that the shape in Figure 1.3.35(a) is distinctively different from that of the normal distribution. When $n = 2$, the probability density function of $\bar{x} = (x_1 + x_2)/2$ is given by Figure 1.3.35(b). Again, the shape in Figure 1.3.35(b) is quite different from that of the normal distribution, but these two functions are both unimodal. When $n = 3$, the probability density function of $\bar{x} = (x_1 + x_2 + x_3)/3$ is given by Figure 1.3.35(c). It is quite surprising to see that the shape of this function is quite similar to that of the normal distribution. In this fashion we draw

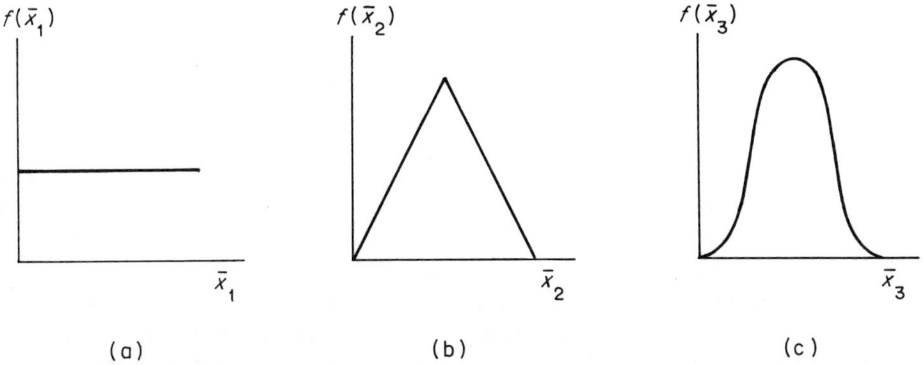

Figure 1.3.35 Probability density functions of (a) $\bar{x}_1$, (b) $\bar{x}_2$ and (c) $\bar{x}_3$, where $\bar{x}_n = \Sigma_{j=1}^{n} x_j/n$, and x_i independently follows the uniform distribution.

the distribution of $\bar{x} = (x_1 + \cdots + x_n)/n$ for $n = 1, 2, 3, 4, \ldots$, and we find that the shape of the probability density function of $\bar{x}$ becomes more similar to that of the normal distribution as n increases. From this observation we may expect that the probability density function of $\bar{x}$ approaches the normal distribution as n increases. This expectation is proved to be true by the following theorem. More surprisingly, the probability distribution function of $\bar{x}$ approaches the normal distribution as n increases not only for the uniform distribution but also for a very broad class of distribution functions.

The central limit theorem Let $x_1, \ldots, x_n$ be independent random variables from the same probability density function with the expected value μ and the finite variance σ^2. Let $\bar{x} = \Sigma_{i=1}^{n} x_i/n$ and

$$z = \frac{\bar{x} - \mathrm{E}(\bar{x})}{\sqrt{\mathrm{Var}(\bar{x})}} = \frac{\bar{x} - \mu}{\sigma/\sqrt{n}}. \tag{1.3.99}$$

Then, the probability density function of z approaches the standard normal distribution $N(z; 0, 1^2)$ as n approaches infinity.

The proof is shown in Gnedenko and Kolmogorov (1954).

Suppose that we generate 100 real numbers using Program U, which is designed to generate independent random variables following the uniform distribution with $a = 0$, $b = 1$ in equation (1.3.75), and that we obtain the average of those numbers as 0.56. Since 0.56 is a little higher than the expected value 0.50 (equation (1.3.82)), we might wonder if Program U is appropriately designed. If Program U independently generates random variables x_i, $i \in I_{100}$, according to the uniform distribution, then, from the central limit theorem with equations (1.3.82) and (1.3.96), the random variable $z = 20\sqrt{3}(\bar{x} - 0.5)$ (where $\bar{x} = \Sigma_{i=1}^{100} x_i/100$ approximately follows the standard normal distribution. From the numerical table of the standard normal distribution, the probability of $z = 20\sqrt{[3(\bar{x} - 0.5)]} \geq 1.96$ is 0.05 (Figure 1.3.34(a)). This means that $20\sqrt{[3(0.56 - 0.50)]} = 2.08 > 1.96$ or the probability of $\bar{x} \geq 0.56$ is less than 0.05. Thus if Program U truly generates the uniform distribution with $a = 0$, $b = 1$, we rarely observe (with probability less than 0.05) that the average value is 0.56. It is hence more natural to consider another possibility, say to consider that Program U has parameter values different from $a = 0$, $b = 1$ rather than to consider that Program U has parameter values $a = 0$, $b = 1$ and it happens to give a rare event. This type of statistical examination is called a *statistical test.*

Let us formalize the above statistical test generally. Suppose we know that random variables $x_1, \ldots, x_n$ are independently distributed according to the same distribution $f(x_i) = f(x)$ with known variance σ^2. We do not know, however, the expected value of x_i. We are provided with observed values $\mathrm{x}_1, \ldots, \mathrm{x}_n$, where x_i is a realization of the random variable x_i, $i \in I_n$. Given $\bar{\mathrm{x}} = \Sigma_{i=1}^{n} \mathrm{x}_i/n$, we are concerned with whether the true expected value μ is μ_0 (in the above example, $\bar{\mathrm{x}} = 0.56$, $\mu_0 = 0.50$). To examine it, we place the hypothesis, called the *null hypothesis*, that $\mu = \mu_0$ against an alternative hypothesis that $\mu > \mu_0$. If the null hypothesis is true and if n is large enough, then, from the central limit theorem, $z = (\bar{x} - \mu_0)/(\sigma/\sqrt{n})$ (where $\bar{x} = \Sigma_{i=1}^{n} x_i/n$) approximately follows the standard normal distribution (if $f(x)$ is a normal distribution, z exactly follows the standard normal distribution). For the observed value $\bar{\mathrm{x}}$, we obtain $\mathrm{z} = (\bar{\mathrm{x}} - \mu_0)/(\sigma/\sqrt{n})$ (in the above example, $\mathrm{z} = 20\sqrt{[3(0.56 - 0.50)]}$). Let z_α be $F(z_\alpha) = -\alpha$, where $F(x)$ is the probability distribution function of the standard normal distribution (in the above example, $z_{0.05} = 1.96$). The value of α is called a *significance level*, and usually $\alpha = 0.05$ or 0.01 is adopted. If the observed value z is greater than z_α, then the probability that the random variable z takes a value greater than or equal to z is smaller than α. This implies that if the null hypothesis is true, we rarely observe the event $z > z_\alpha$. It is hence more natural to consider that the hypothesis is false rather than true. To sum up, if $\mathrm{z} > z_\alpha$, we reject the null hypothesis with significance level α; otherwise, we cannot reject the null hypothesis. We call this test the *upper one-sided statistical test on a mean with a known variance.* This test can be generalized in a few respects. First, almost in the same manner, we can test with $\mathrm{z} < -z_\alpha$ or with $\mathrm{z} > z_{\alpha/2}$ or $\mathrm{z} < -z_{\alpha/2}$; these tests are called the *lower one-sided statistical test* and *two-sided statistical test*, respectively. Furthermore, we can test a mean with an unknown variance. Regarding these tests, the reader should consult a textbook of statistics (such as Wilks, 1962).

Last we show another statistical test which is often used in this text, called the *goodness-of-fit test.* This test is used to determine whether a set of data may be regarded as values of a random variable having a given distribution. To illustrate the procedure of this test, suppose that we carry out $n = 100$ experiments each of which consists of five throws of a die. The unfilled

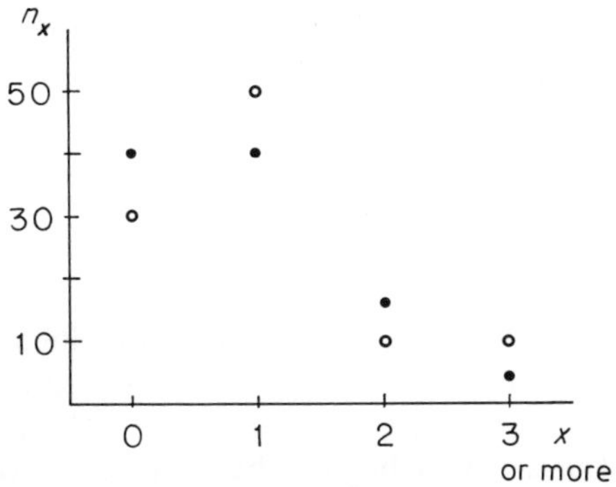

Figure 1.3.36 The observed frequencies and the expected frequencies.

circles in Figure 1.3.36 shows the observed frequencies n_x of face 1 appearing exactly x (= 0, 1, 2 and more than 2) times in five throws over $n = 100$ experiments. Recalling the experiment of five throws of a dice discussed in the binomial distribution, we notice that the probability θ_x of face 1 appearing x times in five throws is given by the binomial distribution with the parameter values 5 and $\theta = 1/6$. Therefore the expected frequency is given by 100 $b(x; 5, 1/6)$, which is shown by the filled circles in Figure 1.3.36. If we place the hypothesis that the die is correctly made, we can expect that the unfilled circles are close to the filled circles. In Figure 1.3.36, however, we observe some discrepancies, and wonder if the hypothesis is true. To examine this hypothesis, we calculate the value of

$$X^2 = \sum_{x=0}^{m} \frac{(n_x - n\theta_x)^2}{n\theta_x}. \tag{1.3.100}$$

Note that we usually aggregate some x's into one x to avoid a small probability, for example $\Sigma_{x=m}^{\infty} \theta_x$, and reset θ_m as $\Sigma_{x=m}^{\infty} \theta_x$. In the above example, since θ_3, θ_4, θ_5 are small (each is less than 0.04), we aggregate these probabilities into θ_3 (= $\theta_3 + \theta_4 + \theta_5$). With this modification, we calculate the value of equation (1.3.100) and obtain $X^2 = 13.25$. We compare this value with the critical value of the chi-square distribution of $m = 3$ degrees of freedom with a small significance level, say 0.05. Consulting the numerical table of the chi-square distribution, we find that this critical value is $\chi^2_{3,0.05} = 7.81$. Since $X^2 > \chi^2_{3,0.05}$, we reject the hypothesis that the die is correctly made; it is more natural to consider that the die is deformed.

The theory of the goodness-of-fit test underlying the above procedure is as follows. Suppose that n discrete random numbers are independently generated according to a probability function $f(x)$, $x = 0, 1, \ldots, m$. Let n_x be the frequency of a random number being x over n random numbers. Then $(n_0, \ldots, n_m)$ forms a multivariate random variable. Recalling equation

(1.3.71), we notice that this random variable follows the multinomial distribution with the parameter values $\theta_x = f(x)$. With this multivariate random variable, we construct a variable

$$\chi^2 = \sum_{x=0}^{m} \frac{(n_x - n\theta_x)^2}{n\theta_x} . \tag{1.3.101}$$

This is also a random variable. We can prove, after a fairly long derivation, that the random variable of equation (1.3.101) in which $(n_0, \ldots, n_m)$ follows the multinomial distribution is approximately distributed according to the chi-square distribution with m degrees of freedom (see Wilks, 1962, pp. 338–339). Let $(\mathrm{n}_0, \ldots, \mathrm{n}_m)$ be an observed value of $(n_0, \ldots, n_m)$. The statistical examination is to test whether or not this observed value is a random realization under the probability function $f(x)$. The logic of the test is just the same as that of the test on a mean shown above. If the observed value X^2 of χ^2 (which is obtained from substituting n_x into n_x in equation (1.3.101)) is greater than the critical value $\chi^2_{m,\alpha}$ with the degree of freedom $m-1$ and the significance level α, we reject the null hypothesis that random numbers are generated according to the distribution $f(x)$; otherwise, we cannot reject it.

In the above goodness-of-fit test we treat a discrete distribution. We can also treat a continuous distribution by transforming it into a 'discrete' distribution. Let $F(x)$ be a continuous distribution function, and $\theta_i = F(x_{i+1}) - F(x_i)$, where $-\infty = x_0 < x_1 < x_2 < \cdots < x_m < x_{m+1} = \infty$. Let n_i be the observed number of events whose values are in $x_i \leq x < x_{i+1}$. Then we can proceed with the goodness-of-fit test for n_i, $i = 0, \ldots, m$, in the same manner as mentioned above.

1.3.8 Spatial stochastic point processes

In the previous section we defined the uniform distribution in $\mathbb{R}$ by equation (1.3.75). Similarly, we can define the uniform distribution in $\mathbb{R}^m$ by

$$f(\mathbf{x}) = \begin{cases} \dfrac{1}{|S|}, & \mathbf{x} \in S , \\ 0 , & \mathbf{x} \notin S , \end{cases} \tag{1.3.102}$$

where $S \subset \mathbb{R}^m$ and $|S|$ is the area or volume of S ($|S| < \infty$). We call the process in which points are generated over S according to the uniform distribution a *binomial point process*. Figure 1.3.37 shows the binomial point process with $n = 50$ points in a unit square. From the term 'process', the reader might call to mind a process over time. In the study of stochastic point processes, however, a time process is usually implicit. When we explicitly deal with a stochastic point process over time, we call it a *spatial–temporal stochastic point process*.

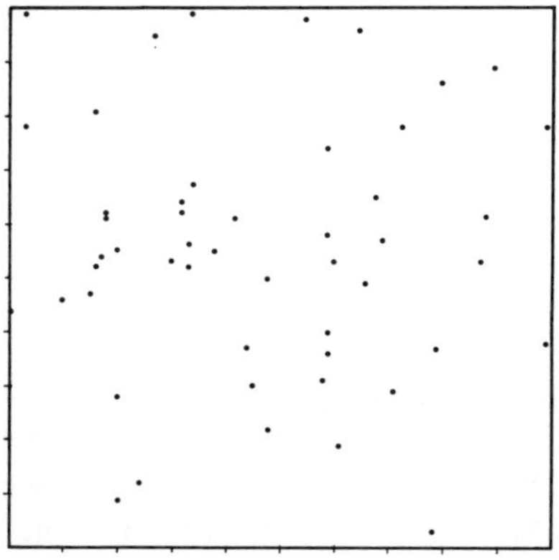

Figure 1.3.37 A binomial point process ($n = 50$).

The reader may be curious to know why the above point process is called the binomial point process. To see this reason, let A be a subset of S, and $N(A)$ be the number of points placed in A when n points are generated according to the uniform distribution. Since the probability of a point being placed in A is given by $|A|/|S|$, the probability, $\Pr(N(A) = x)$, that x points are placed in A is given by

$$\Pr(N(A) = x) = {}_nC_x \left(\frac{|A|}{|S|}\right)^x \left(1 - \frac{|A|}{|S|}\right)^{n-x} \tag{1.3.103}$$

(recall the derivation leading to equation (1.3.70)). This distribution is the binomial distribution with the parameter values n and $\theta = |A|/|S|$ (denoted by $b(x; n, |A|/|S|)$ in Section 1.3.7), after which the above process is named.

In the above we defined the binomial point process with the uniform distribution. Alternatively, we can directly define the binomial point process by the binomial distribution. The *binomial point process* is defined by the process in which, for any subset A in S, the probability distribution of $N(A)$ is given by equation (1.3.103).

In a similar way, we can define a general stochastic point process. Let S be a non-empty subset of $\mathbb{R}^m$, and $N(A)$ be the number of points in a subset, A, of S. Then a *stochastic point process* is defined by the process in which points are generated according to the probability distribution $\Pr(N(A) = x)$, $x = 0, 1, \ldots$, for any A in S. Specifically, if $m = 2$, the process is called a *planar stochastic point process*. If the probability of two points being coincident is zero under a stochastic point process, we say that the process is *simple*. If any finite region A in $\mathbb{R}^m$ contains a finite number of points under a stochastic point process, the process is said to be *locally finite*. In this text, we deal with stochastic point processes that satisfy these two conditions.

Among many simple and locally finite stochastic point processes, the most fundamental process is the *Poisson point process*, which is defined by the process in which for any subset A in $S = \mathbb{R}^m$, $\Pr(N(A) = x)$ is given by

$$\Pr(N(A) = x) = \frac{\lambda|A|\, e^{-\lambda|A|x}}{x!}, \quad x = 0, 1, 2, \dots . \tag{1.3.104}$$

We can also define the Poisson point process as a limit of the binomial point process. The limit means the expansion of the finite region S to an infinite region keeping $n/|S|$ constant. Since $\lambda = n/|S|$ is the density of points, this expansion implies the expansion of the region S keeping the density of points constant. In the limit, as we remarked in Section 1.3.7, the binomial distribution of equation (1.3.103) approaches the Poisson distribution of equation (1.3.104). We understand from this derivation that the parameter λ in the Poisson point process shows the density of points in a unit volume. The density λ is sometimes referred to as *intensity*.

In the above, we derived the Poisson point process directly from the Poisson distribution. More fundamentally, we can derive the Poisson point process from the following four postulates, through which we can see the characteristics of the Poisson point process explicitly.

Postulate 1 $0 < \Pr(N(A) = 0) < 1$ if $0 < |A| < \infty$.

Postulate 2 The probability distribution of $N(A)$ depends on $|A|$, and $\Pr(N(A) \geq 1)$ approaches zero as $|A|$ approaches 0.

Postulate 3 If $A_1, \dots, A_n$ are disjoint, then $N(A_1), \dots, N(A_n)$ are mutually independent and $N(A_1 \cup \cdots \cup A_n) = N(A_1) + \cdots + N(A_n)$.

Postulate 4

$$\lim_{|A| \to 0} \frac{\Pr(N(A) \geq 1)}{\Pr(N(A) = 1)} = 1 . \tag{1.3.105}$$

Postulate 1 asserts that if region A has a positive volume, then the probability of a point being placed in region A is not zero. Postulate 2 asserts that $\Pr(N(A) = x)$ does not depend upon the location of A but only upon its volume (or its area for $m = 2$). There is no tendency for points to occupy particular regions in $\mathbb{R}^m$. Postulate 3 means no interactions between points. We call this property *independent scattering*. It should be noted that the binomial point process is not independent scattering, because if x points are placed in A, then $n - x$ are inevitably placed in $S \backslash A$. Postulate 4 implies that the process is simple. If random variable $N(A)$ satisfies the above four postulates for $A \subset \mathbb{R}^m$, then $N(A)$ is proved to follow the distribution of equation (1.3.104). Since this proof requires much space, we omit it here; the reader should consult a textbook of stochastic processes, for example Karlin (1969, pp. 337–43).

As is seen in Postulate 2, the Poisson point process has the invariance property in the sense that $P(N(A_1) = x) = P(N(A_2) = x)$ for $|A_1| = |A_2|$. Generally, we can examine the characteristics of stochastic point processes in terms of invariance properties. First, let us consider the transformation of A in S defined by

$$\varphi_c(A) = \{\mathbf{x} + \mathbf{c} | \mathbf{x} \in A\}, A \subset S, \varphi_c(A) \subset S , \tag{1.3.106}$$

where $\mathbf{c} \in \mathbb{R}^m$ is a constant vector. If we regard A as a figure in $\mathbb{R}^m$, we may say that φ_c moves the figure A without rotation in $\mathbb{R}^m$ (see Figure 1.3.38(a), $m = 2$). We call this transformation *translation*. In observing equation (1.3.104), we readily notice that $\Pr(N(A) = x) = \Pr(N(\varphi_c(A)) = x)$ holds, that is, the Poisson point process is invariant under translation. Generally if a stochastic point process satisfies

$$\Pr(N(A) = x) = \Pr(N(\varphi_c(A)) = x) \; A \subset S, \varphi_c(A) \subset S , \tag{1.3.107}$$

we say that the stochastic point process is *homogeneous* or *stationary*. The binomial point process and the Poisson point process are homogeneous. This may be intuitively understood from the fact that the processes are derived from the uniform distribution.

We next consider the transformation of A in S defined by

$$\varphi_\theta(A) = \{B\mathbf{x} | \mathbf{x} \in A, |B| = 1, B^T B = I\}, \quad A \subset S, \varphi_\theta(A) \subset S , \tag{1.3.108}$$

where B is an $m \times m$ matrix and $|B|$ is its determinant. The transformation φ_θ rotates a figure A around the origin in $\mathbb{R}^m$ as is shown in Figure 1.3.38(b) ($m = 2$). We call this transformation *rotation*. Since $|A| = |\varphi_\theta(A)|$ holds,

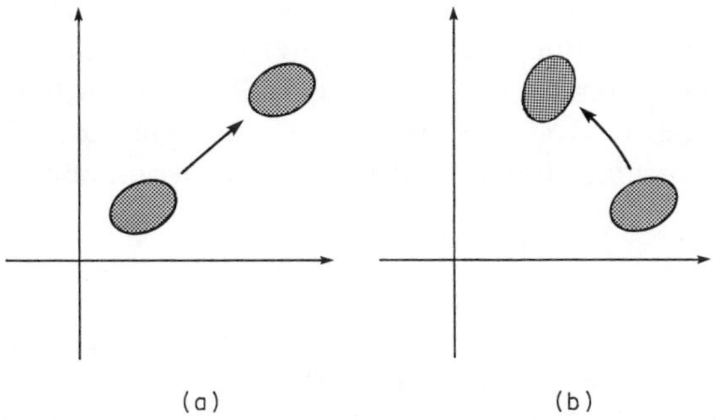

Figure 1.3.38 (a) Translation; (b) rotation.

$\Pr(N(A) = x) = \Pr(N(\varphi_\theta(A)) = x)$ holds for equation (1.3.104). Generally, if a stochastic point process satisfies

$$\Pr(N(A) = x) = \Pr(N(\varphi_\theta(A)) = x), A \subset S, \varphi_\theta(A) \subset S , \qquad (1.3.109)$$

we say that the stochastic point process is *isotropic*. If a stochastic point process is homogeneous and isotropic, we call the stochastic point process *motion-invariant*. The binomial point process and the Poisson point process are motion-invariant.

The binomial point process and the Poisson point process have a stronger property than the motion-invariant property. If a transformation, φ_v, of A in S satisfies

$$|\varphi_v(A)| = |A|, \quad A \subset S, \varphi_v(A) \subset S , \qquad (1.3.110)$$

we call φ_v the *volume-preserving transformation* (*area-preserving transformation* for $m = 2$). Obviously, translation and rotation are volume-preserving transformations. Generally, if a stochastic point process satisfies

$$\Pr(N(A) = x) = \Pr(N(\varphi_v(A)) = x), \quad A \subset S, \varphi_v(A) \subset S , \qquad (1.3.111)$$

we say that the stochastic point process is *volume-preserving*. From equation (1.3.103), we notice that the binomial point process is volume-preserving. The Poisson point process is also volume-preserving because Postulate 2 is satisfied in that process.

The volume-preserving transformation can be extended to a measure-preserving transformation. To be explicit, consider a set S and a set, $\mathcal{A}$, of subsets of S which satisfies the following three conditions: (i) $S \in \mathcal{A}$; (ii) if $A_i \in \mathcal{A}$, then $S \backslash A_i \in \mathcal{A}$; (iii) if $A_i \in \mathcal{A}$, $i = 1, 2, \ldots$, then $\cup_{i=1}^{\infty} A_i \in \mathcal{A}$. We call the pair $(S, \mathcal{A})$ satisfying these conditions a *measurable space*. If a function, m, from $\mathcal{A}$ to a non-negative real satisfies $m(\emptyset) = 0$ and $m(\cup_{i=1}^{\infty} A_i) = \Sigma_{i=1}^{\infty} m(A_i)$, where $A_1, A_2, \ldots$ are mutually disjoint, we say that m is a *measure* on $(S, \mathcal{A})$. A measurable space $(S, \mathcal{A})$ with a measure m is called a *measure space* and is denoted by a triple $(S, \mathcal{A}, m)$. In particular, if $m(S) = 1$, we call $(S, \mathcal{A}, m)$ a *probability space* (recall the probability space defined by Postulates 1–3 in Section 1.3.7). If a transformation satisfies $\varphi^{-1}(A_i) \in \mathcal{A}$ for $A_i \in \mathcal{A}$, we say that φ is *measurable*. If φ is one-to-one, if $\varphi(S) = S$ and if $A_i \in \mathcal{A}$ implies $\varphi^{-1}(A_i) \in \mathcal{A}$, then we say that φ is *invertible*. For a measurable transformation φ in $(S, \mathcal{A}, m)$, if

$$m(\varphi^{-1}(A)) = m(A), \quad A \in \mathcal{A} , \qquad (1.3.112)$$

we say that φ is a *measure-preserving* transformation; if φ is invertible, equation (1.3.112) is equivalent to $m(\varphi(A)) = m(A)$. In the following exposition, we assume, for simplicity, that a measure-preserving transformation is invertible. Obviously the volume-preserving transformation is a measure-preserving transformation. Thus the binomial point process and the Poisson point process are measure-preserving point processes.

To sum up, the Poisson point process is simple, independent scattering and invariant under the measure-preserving transformation; consequently it is motion-invariant, i.e. it is homogeneous and isotropic. Because of these properties, we sometimes say that a pattern of points resulting from the Poisson point process is a *complete spatial random pattern* or *complete spatial randomness*, and abbreviated to CSR (Diggle, 1983).

In conjunction with a measure-preserving transformation, we define two more important properties, called ergodicity and mixing, which will be referred to in Chapter 5. For a measure-preserving transformation φ in a probability space $(S, \mathscr{A}, m)$ we call set $A \in \mathscr{A}$ an *invariant set* if

$$\varphi(A) = A \tag{1.3.113}$$

holds. For an invariant set A, the equation $\varphi(S\backslash A) = S\backslash A$ holds as well as $\varphi(A) = A$. Hence if $0 < m(A) < 1$, then $0 < m(S\backslash A) < 1$. This implies that the whole space S is decomposed into A and $S\backslash A$ with φ. On the other hand, if $m(A) = 0$ or $m(S\backslash A) = 0$ holds for any invariant set A, the whole space is indecomposable with φ (note that we neglect a set of zero measure in measure theory). We call indecomposable measure-preserving transformations ergodic. To be precise, for a probability space $(S, \mathscr{A}, m)$, if any invariant set $A \in \mathscr{A}$ satisfies $m(A) = 0$ or $m(A) = 1$, then the measure-preserving function φ is called *ergodic*.

Ergodicity has many important implications. To see them, first consider the sequence $\varphi(A)$, $\varphi^2(A) = \varphi(\varphi(A))$, $\varphi^3(A) = \varphi(\varphi^2(A))$, ..., which we call the orbit of A with φ. Regarding this orbit, the following theorem is proved. A measure-preserving transformation φ is ergodic if and only if for every $A \in \mathscr{A}$ with $m(A) > 0$, the equation $m(\cup_{n=1}^{\infty} \varphi^n(A)) = 1$ holds (Walters, 1982. Theorem 1.5(iii)). This implies that the orbit $\{\varphi(A), \varphi^2(A), ..., \varphi^n(A), ...\}$ of any set with non-zero measure A sweeps out the whole space S. Furthermore, it is shown that the orbit $\{x, \varphi(x), \varphi^2(x), ..., \varphi^n(x), ...\}$ of $x \in S$ enters A with relative frequency $m(A)$. If these properties are taken for granted (in fact, they can be proved to be true), the following theorem may be understood intuitively.

Ergodic theorem Suppose that φ is an ergodic measure-preserving transformation in a probability space $(S, \mathscr{A}, m)$, $\int_S |f(x)| \mathrm{d}m < \infty$, and there exists $\lim_{n\to\infty} \sum_{i=0}^{n-1} f(\varphi^i(x))/n$. Then the following equation holds almost everywhere (i.e. except for measure-zero sets):

$$\lim_{n\to\infty} \frac{1}{n} \sum_{i=0}^{n-1} f(\varphi^i(x)) = \int_S f(x) \mathrm{d}m \; . \tag{1.3.114}$$

The proof is shown in Walters (1982, Section 1.6). This intuitively means that the arithmetic mean of f over the orbit x, $\varphi(x)$, $\varphi^2(x)$, $\varphi^3(x)$, ... (the left-hand side of equation (1.3.114)) equals to the spatial mean of f over the whole space S (the right-hand-side).

To see another implication of ergodicity, the following theorem is suggestive. A measure-preserving function φ is ergodic if and only if for every A and B with $m(A) > 0$ and $m(B) > 0$ in a probability space $(S, \mathcal{A}, m)$, there exists n with $m(\varphi^n(A) \cap B) > 0$. This intuitively means that the orbit of A obtained by the ergodic measure-preserving transformation φ is 'mixed' with B for large n. Then in the limit as $n \to \infty$, to what extent is the transformed A mixed with B? For this question, we have the following theorem (Walters, 1982, Corollary 1.14.2). Suppose that φ is a measure-preserving transformation of a probability space $(S, \mathcal{A}, m)$. The transformation φ is ergodic if and only if

$$\lim_{n \to \infty} \frac{1}{n} \sum_{i=0}^{n-1} m(\varphi^i(A) \cap B) = m(A)m(B) \quad \text{for any } A, B \in \mathcal{A} \,. \qquad (1.3.115)$$

This intuitively means that the probability of the orbit of A in a given B is, on average, asymptotically the same as the probability of A in the whole probability space. Stated differently, ergodicity implies that for any set A, $\varphi^n(A)$ becomes, asymptotically, independent of B on average.

The condition given by equation (1.3.115) becomes stronger if we write

$$\lim_{n \to \infty} m(\varphi^n(A) \cap B) = m(A)m(B) \quad \text{for any } A, B \in \mathcal{A} \,. \qquad (1.3.116)$$

We say that a measure-preserving transformation φ in a probability space $(S, \mathcal{A}, m)$ is *(strongly) mixing* if and only if equation (1.3.116) holds. This implies that $\varphi^n(A)$ becomes, asymptotically, independent of set B (not 'on average'). Obviously, mixing implies ergodicity, but not conversely. Since ergodic and mixing transformations are both measure-preserving transformations, the Poisson point process is invariant for ergodic and mixing transformations.

The Poisson point process may be extended in several directions. First, we may relax homogeneity, and define the point process by

$$\Pr(N(A) = x) = \frac{\lambda(A)|A| \, \mathrm{e}^{-\lambda(A)|A|x}}{x!} \,, \quad x = 0, 1, 2, \ldots \,, \qquad (1.3.117)$$

where

$$\lambda(A) = \int_A \lambda(\boldsymbol{x}) \mathrm{d}\boldsymbol{x} \,. \qquad (1.3.118)$$

We call this point process the *general Poisson point process* with the intensity function $\lambda(\boldsymbol{x})$. An example is shown in Figure 1.3.39, where $\lambda(\boldsymbol{x}) = c_1 \exp(-c_2 \sqrt{(x_1^2 + x_2^2)})$, $c_1, c_2 > 0$).

The general Poisson point process reduces to the Poisson point process of equation (1.3.104) if $\lambda(\boldsymbol{x}) = \lambda$ (a constant). Since λ in equation (1.3.104)

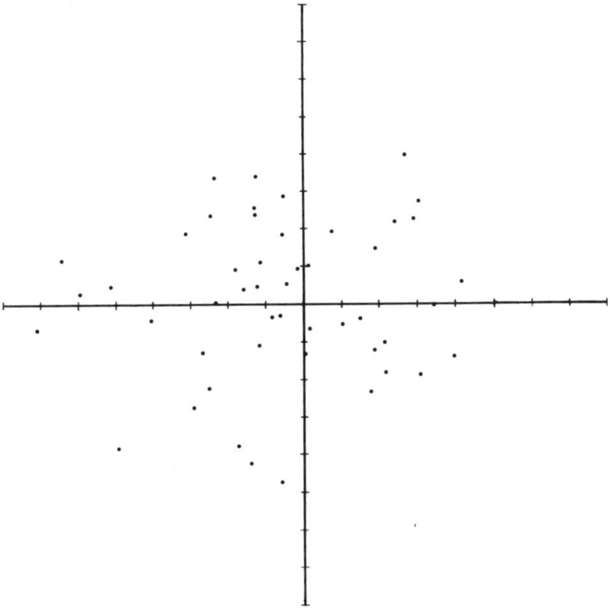

Figure 1.3.39 A general Poisson point process with $\lambda(\boldsymbol{x}) = c_1 \exp(-c_2\sqrt{x_1^2 + x_2^2})$.

indicates the density of points, $\lambda(\boldsymbol{x})$ implies that the density depends upon where A is placed. Thus the density of points may vary from location to location in a general Poisson point process. This property distinguishes from that of the Poisson point process defined by equation (1.3.104) which we call a *homogeneous Poisson point process*.

In the homogeneous Poisson point process the parameter λ is deterministic. We may relax this assumption, and regard it as probabilistic. We call the Poisson point process with a probabilistic parameter λ a *doubly stochastic Poisson point process* or a *compound Poisson point process*.

We may also relax the property of independent scattering. One of the well-known processes is the Gibbs point processes. The reader who is interested in general point processes, including the Gibbs point processes, should consult Bartlett (1975), Hagett *et al.* (1977, Chapter 13), Pielou (1977), Getis and Boots (1978), Cliff and Ord (1981), Diggle (1983), Upton and Fingleton (1985), Stoyan *et al.* (1987), and Ripley (1988), among others.

CHAPTER 2

Definitions and Basic Properties of Voronoi Diagrams

In Chapter 1 we discussed the concept of a Voronoi diagram from a historical view point. In this chapter we wish to fully develop this concept from a geometric view point.

The chapter consists of five sections. In Section 2.1 we define a Voronoi diagram and introduce notations to be commonly used in this text. In Section 2.2 we define a Delaunay tessellation, in particular a Delaunay triangulation which is a 'dual tessellation' of a Voronoi diagram. Voronoi diagrams and Delaunay tessellations are, as we overviewed in Section 1.2, quite useful in various fields because of their nice geometric properties. In Sections 2.3 and 2.4 we show the basic properties of Voronoi diagrams and Delaunay triangulations, respectively (specific properties only used in each chapter will be shown in that chapter). In Section 2.5 we refer to geometric graphs closely related to a Delaunay triangulation, such as a Gabriel graph, a relative neighbourhood graph and a minimum spanning tree.

2.1 DEFINITIONS OF THE ORDINARY VORONOI DIAGRAM

For ease of exposition, we first give a fairly intuitive definition of a Voronoi diagram in a plane. We next restate this definition more precisely in mathematical terms. Lastly, extending the definition in $\mathbb{R}^2$ to $\mathbb{R}^m$, we define an m-dimensional Voronoi diagram.

Suppose that a set of points is given in the Euclidean plane (for example, the filled circles in Figure 2.1.1). The number of points is assumed to be two or more but finite and they are all distinct in the sense that no points coincide in the plane. Given this point set, we assign every location in the plane to the closest member in the point set (in Figure 2.1.1, for example, the point p is assigned to the filled circle incident to the heavy broken line). If a location happens to be equally close to two or more members of the point set, we assign the location to those members (point p' in Figure 2.1.1

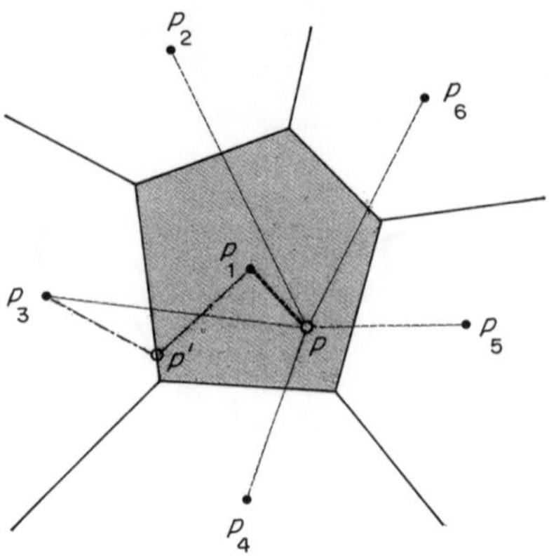

Figure 2.1.1 An ordinary Voronoi diagram.

is assigned to the two filled circles incident to the dash-dot lines in Figure 2.1.1). As a result, the set of locations assigned to each member in the point set forms its own region (the shaded region in Figure 2.1.1). The resulting regions are collectively exhaustive in the plane because every location is assigned to at least one member in the point set. The set of locations assigned to two or more members in the point set forms the boundaries of the regions (the continuous lines in Figure 2.1.1). Adjacent regions are hence overlapped only on their boundaries. Thus the set of the regions is collectively exhaustive and mutually exclusive except for boundaries, namely, the set of the regions forms a tessellation. We call this tessellation a *planar ordinary Voronoi diagram*, and the regions constituting the Voronoi diagram *ordinary Voronoi polygons*.

To sum up, we have the following definition of a Voronoi diagram.

Definition V1 (a planar ordinary Voronoi diagram) Given a set of two or more but a finite number of distinct points in the Euclidean plane, we associate all locations in that space with the closest member(s) of the point set with respect to the Euclidean distance. The result is a tessellation of the plane into a set of the regions associated with members of the point set. We call this tessellation the *planar ordinary Voronoi diagram* generated by the point set, and the regions constituting the Voronoi diagram *ordinary Voronoi polygons*.

Note that when the Euclidean plane (space) is understood, we shall ofen refer to a planar ordinary Voronoi diagram simply as a *Voronoi diagram* and an ordinary Voronoi polygon as a *Voronoi polygon*. Also note that we exclude from our discussion the point set consisting of only one member, because such a Voronoi diagram is too trivial (it consists of only one Voronoi polygon which is the whole plane.)

Let us next restate Definition V1 in mathematical terms. We consider a finite number, n, of points in the Euclidean plane, and assume that $2 \leq n < \infty$. The n points are labelled by $p_1, \ldots, p_n$ with the Cartesian coordinates $(x_{11}, x_{12}), \ldots, (x_{n1}, x_{n2})$ or location vectors $\mathbf{x}_1, \ldots, \mathbf{x}_n$. The n points are *distinct* in the sense that $\mathbf{x}_i \neq \mathbf{x}_j$ for $i \neq j$, $i, j \in I_n = \{1, \ldots, n\}$. Let p be an arbitrary point in the Euclidean plane with coordinates (x_1, x_2) or location vector $\mathbf{x}$. Then the Euclidean distance from p to p_i is given by $d(p, p_i) = \|\mathbf{x} - \mathbf{x}_i\| = \sqrt{[(x_1 - x_{i1})^2 + (x_2 - x_{i2})^2]}$. If p_i is the nearest point from p or p_i is one of the nearest points from p, we have the relation $\|\mathbf{x} - \mathbf{x}_i\| \leq \|\mathbf{x} - \mathbf{x}_j\|$ for $j \neq i$, $j \in I_n$ (recall the broken and dash-dot lines in Figure 2.1.1). In this case, p is assigned to p_i. Therefore, Definition V1 is written mathematically as follows.

Definition V2 (a planar ordinary Voronoi diagram) Let $P = \{p_1, \ldots, p_n\}$, where $2 \leq n < \infty$ and $\mathbf{x}_i \neq \mathbf{x}_j$ for $i \neq j$, $i, j \in I_n$. We call the region given by

$$V(p_i) = \{\mathbf{x} \mid \|\mathbf{x} - \mathbf{x}_i\| \leq \|\mathbf{x} - \mathbf{x}_j\| \text{ for } j \neq i, j \in I_n\} \tag{2.1.1}$$

the *(ordinary) Voronoi polygon* associated with p_i (or the Voronoi polygon of p_i), and the set given by

$$\mathcal{V} = \{V(p_1), \ldots, V(p_n)\} \tag{2.1.2}$$

the *(planar ordinary) Voronoi diagram* generated by P (or Voronoi diagram of P).

We call the p_i of $V(p_i)$ the *generator point* or *generator* of the ith Voronoi polygon, and the set $P = \{p_1, \ldots, p_n\}$ the *generator set* of the Voronoi diagram $\mathcal{V}$. For brevity we may write V_i for $V(p_i)$. Also we may use $V(x_{i1}, x_{i2})$ or $V(\mathbf{x}_i)$ when we want to emphasize the coordinates or location vector of the generator p_i. In addition, we may use $\mathcal{V}(P)$ when we want to explicitly indicate the generator set P of $\mathcal{V}$.

In Definition V2 the reader should notice that the relation in equation (2.1.1) is defined in terms of $\leq$, but not $<$. A Voronoi polygon is hence a closed set. Alternatively, we may define a Voronoi polygon as

$$V^{\circ}(p_i) = \{\mathbf{x} \mid \|\mathbf{x} - \mathbf{x}_i\| < \|\mathbf{x} - \mathbf{x}_j\| \text{ for } j \neq i, j \in I_n\}\ , \tag{2.1.3}$$

which is an open set. Both definitions are acceptable, but in this text, we define a Voronoi polygon as a closed set. When we want to emphasize this property, we call $V(p_i)$ a *closed Voronoi polygon*, and $V^{\circ}(p_i)$ an *open Voronoi polygon*.

Since a Voronoi polygon is a closed set, it contains its boundary denoted by $\partial V(p_i)$. The boundary of a Voronoi polygon may consist of line segments, half lines or infinite lines, which we call *Voronoi edges*. We denote a Voronoi edge by e_i. Noticing that $=$ is included in the relation of equation

(2.1.1), we may alternatively define a Voronoi edge as a line segment, a half line or an infinite line shared by two Voronoi polygons with its end points. Mathematically, if $V(p_i) \cap V(p_j) \neq \emptyset$, the set $V(p_i) \cap V(p_j)$ gives a Voronoi edge (which may be degenerate into a point). Corresponding to notation e_i, we use $e(p_i, p_j)$ for $V(p_i) \cap V(p_j)$, which is read as the Voronoi edge generated by p_i and p_j. Note that $e(p_i, p_j)$ may be empty. If $e(p_i, p_j)$ is neither empty nor a point, we say that the Voronoi polygons $V(p_i)$ and $V(p_j)$ are *adjacent*.

An end point of a Voronoi edge is called a *Voronoi vertex*. Alternatively, a Voronoi vertex may be defined as a point shared by three or more Voronoi polygons. We denote a Voronoi vertex by q_i. When there exists at least one Voronoi vertex at which four or more Voronoi edges meet in the Voronoi diagram $\mathcal{V}$ (the unfilled circle in Figure 2.1.2(a)), we say that $\mathcal{V}$ is *degenerate*; otherwise, we say that $\mathcal{V}$ is *non-degenerate*. The Voronoi diagram in Figure 2.1.2(a) is degenerate and that in Figure 2.1.1 is non-degenerate. A degenerate Voronoi diagram often appears when the generator points are regularly spaced, such as in Figure 2.1.2(b). In some derivations, a degenerate Voronoi diagram requires special lengthy treatments which are not always essential. To avoid this difficulty, we shall often make the following assumption.

Assumption V1 (the non-degeneracy assumption) Every Voronoi vertex in a Voronoi diagram has exactly three Voronoi edges.

In Definition V1 or V2, we defined a Voronoi diagram in an unbounded plane. In practical applications, however, we often deal with a bounded region S, where generators are placed (the heavy lines Figure 2.1.3). In this case we consider the set given by

$$\mathcal{V} \cap S = \{V(p_1) \cap S, \dots, V(p_n) \cap S\} . \quad (2.1.4)$$

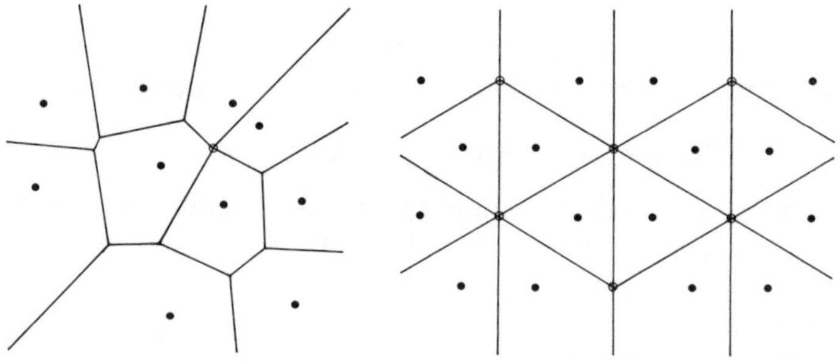

(a) (b)

Figure 2.1.2 Degenerate Voronoi diagrams.

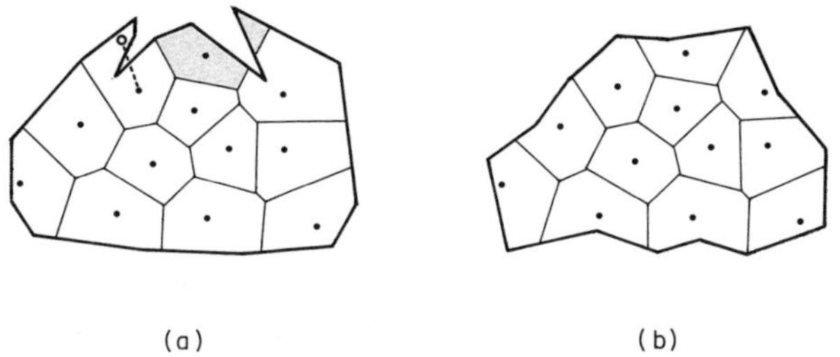

Figure 2.1.3 Bounded Voronoi diagrams: (a) a disconnected boundary Voronoi region (the shaded region) and a non-star-shaped boundary Voronoi polygon with respect to its generator (the broken line segment); (b) star-shaped boundary Voronoi polygons with respect to their generators.

We call this set a *bounded Voronoi diagram* or the *Voronoi diagram bounded by S*. If the Voronoi polygon $V(p_i)$ shares the boundary of S, we call the region $V(p_i) \cap S$ a *boundary Voronoi polygon* or *region* (the term 'region' is used when ∂S is curved or when $V(p_i) \cap S$ is not connected).

We should note that a boundary Voronoi region may be disconnected (the shaded region in Figure 2.1.3(a)), and that the line segment joining a point in $V(p_i)$ and p_i may not be contained in $V(p_i) \cap S$ (the broken line segment in Figure 2.1.3(a)). In practical applications, such boundary Voronoi polygons are often problematic, and so we have to define a more appropriate Voronoi diagram. We shall show such an alternative Voronoi diagram in Section 3.4. A bounded Voronoi diagram may be meaningful in practice if every boundary Voronoi region is star-shaped with respect to its generator point (Figure 2.1.3(b)). In practice, we usually treat such a well-formed bounded Voronoi diagram.

As we observed in Figure 2.1.1, the ordinary Voronoi diagram consists of polygons. Recalling that a polygon is defined in terms of half planes (Section 1.3), we may alternatively define a Voronoi diagram in terms of half planes. To show this alternative definition, we consider the line perpendicularly bisecting the line segment $\overline{p_ip_j}$ joining two generators p_i and p_j (Figure 2.1.4). We call this line the *bisector* between p_i and p_j and denote it by $b(p_i, p_j)$. Since a point on the bisector $b(p_i, p_j)$ is equally distant from the generators p_i and p_j, $b(p_i, p_j)$ is written as

$$b(p_i, p_j) = \{\mathbf{x} \mid \|\mathbf{x} - \mathbf{x}_i\| = \|\mathbf{x} - \mathbf{x}_j\|\}\ , \quad j \neq i\ . \tag{2.1.5}$$

The bisector divides the plane into two half planes and gives

$$H(p_i, p_j) = \{\mathbf{x} \mid \|\mathbf{x} - \mathbf{x}_i\| \leq \|\mathbf{x} - \mathbf{x}_j\|\}\ , \quad j \neq i\ . \tag{2.1.6}$$

We call region $H(p_i, p_j)$ the *dominance region of p_i over p_j*. In Figure 2.1.4 we indicate the dominance regions of p_1 over p_2, p_3 and p_4 by the horizontally, diagonally and vertically hatched regions, respectively. Obviously, in

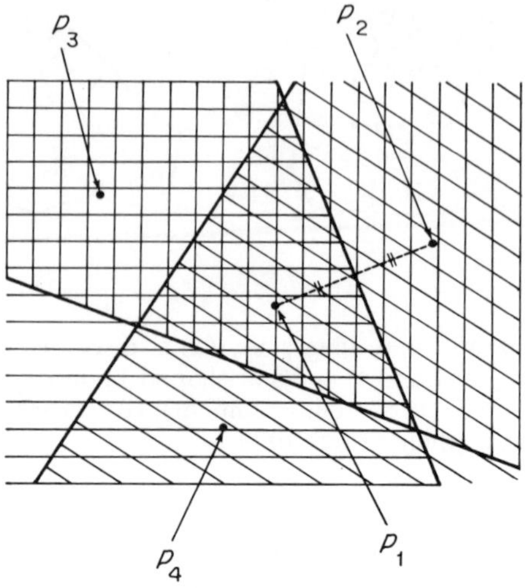

Figure 2.1.4 A Voronoi polygon obtained from half planes.

the dominance region $H(p_i, p_j)$ the distance to generator p_i is shorter than or equal to the generator p_j. In Figure 2.1.4, therefore, the distance from a point p in the intersection of the vertically, horizontally and diagonally hatched regions to the generator p_1 is shorter than or equal to the distance from p to the generator p_j, $j = 2, 3, 4$. This relation is equivalent to equation (2.1.1), and hence the intersection $H(p_1, p_2) \cap H(p_1, p_3) \cap H(p_1, p_4)$ gives the Voronoi polygon associated with p_1. From this example, we understand that the following definition is an alternative to Definition V2.

Definition V3 (a planar ordinary Voronoi diagram defined with half planes) Let $P = \{p_1, \ldots, p_n\} \subset \mathbb{R}^2$, where $2 \leq n < \infty$ and $\mathbf{x}_i \neq \mathbf{x}_j$ for $i \neq j$, $i, j \in I_n$. We call the region

$$V(p_i) = \bigcap_{j \in I_n \setminus \{i\}} H(p_i, p_j) \tag{2.1.7}$$

the *(ordinary) Voronoi polygon* associated with p_i and set $\mathcal{V}(P) = \{V(p_1), \ldots, V(p_n)\}$ the *(planar ordinary) Voronoi diagram* generated by P.

The equivalence of Definition V3 to Definition V2 is apparent, because $\|\mathbf{x} - \mathbf{x}_i\| \leq \|\mathbf{x} - \mathbf{x}_j\|$ if and only if $\mathbf{x} \in H(p_i, p_j)$ for $j \neq i$.

We can readily extend the above definition to the m-dimensional Euclidean space.

Definition V4 (an ordinary Voronoi diagram in $\mathbb{R}^m$) Let $P = \{p_1, \ldots, p_n\} \subset \mathbb{R}^m$, where $2 \leq n < \infty$ and $\mathbf{x}_i \neq \mathbf{x}_j$ for $i \neq j$, $i, j \in I_n$. We call the region given by

$$V(p_i) = \{\mathbf{x} | \|\mathbf{x} - \mathbf{x}_i\| \leq \|\mathbf{x} - \mathbf{x}_j\| \text{ for } j \neq i, j \in I_n\} \tag{2.1.8}$$

$$= \bigcap_{j \in I_n \setminus \{i\}} H(p_i, p_j) \tag{2.1.9}$$

the *m-dimensional ordinary Voronoi polyhedron* associated with p_i, and set $\mathcal{V}(P) = \{V(p_1), \ldots, V(p_n)\}$ the *m-dimensional ordinary Voronoi diagram* generated by P, where $H(p_i, p_j)$ is given by equation (2.1.6) for $\mathbb{R}^m$.

Note that when $\mathbb{R}^m$ is understood, we may simply call $\mathcal{V}(P)$ an *ordinary Voronoi diagram* or just a *Voronoi diagram.*

For the three-dimensional Voronoi diagram, the boundary of a Voronoi polyhedron consists of facets, which we call *Voronoi facets*. The boundary of a Voronoi facet consists of line segments, half lines or infinite lines, which we call *Voronoi edges*. The boundary of a Voronoi edge consists of points, which we call *Voronoi vertices.* Figure 2.1.5 shows a stereographic view of a three-dimensional Voronoi diagram obtained by Gen (1983) (see also Bowyer, 1981).

A visual presentation of an m-dimensional Voronoi diagram becomes difficult when $m \geq 4$. Formally, we call the boundary of an m-dimensional Voronoi polyhedron $(m-1)$-*dimensional Voronoi faces*; the boundary of an $(m-1)$-dimensional Voronoi face $(m-2)$-*dimensional Voronoi faces*; ...; the boundary of a three-dimensional Voronoi face *two-dimensional Voronoi faces* or simply *Voronoi faces*; the boundary of a two-dimensional Voronoi face *one-dimensional Voronoi faces* or *Voronoi edges*; the boundary of a one-dimensional Voronoi face *zero-dimensional Voronoi faces* or *Voronoi vertices*. An $(m-1)$-dimensional Voronoi face is also called a *Voronoi facet.*

Obviously we allow $m = 1$, a one-dimensional Voronoi diagram, or a Voronoi diagram on a line (Figure 2.1.6). In this case, a 'one-dimensional Voronoi polyhedron' is a half line or a line segment called a *Voronoi line*, and Voronoi vertices are end points of Voronoi lines. We easily notice that

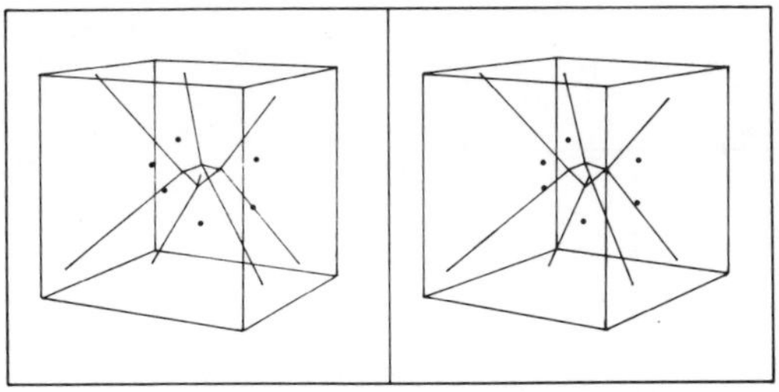

Figure 2.1.5 A stereographic view of a three-dimensional Voronoi diagram (view the left panel with the left eye and the right panel with the right eye at the same time, or use a stereographic viewer). (*Source*: Gen (1983, p. 178).)

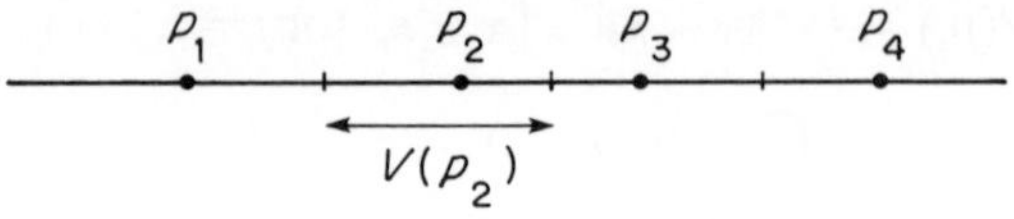

Figure 2.1.6 A Voronoi diagram on a line.

the boundary point between two adjacent Voronoi lines is the mid-point of the generators of those Voronoi lines; the number of unbounded Voronoi lines is always two; the number of Voronoi lines adjacent to a Voronoi line is one or two. Because of these simple geometric properties, a line Voronoi diagram is often adopted in a simplified theoretical context to avoid complicated geometric arguments (for example, the original Hotelling model in Section 7.3). In this text we shall mainly treat a Voronoi diagram whose dimension is two or more.

2.2 DEFINITIONS OF THE DELAUNAY TESSELLATION (TRIANGULATION)

In the same way that a planar graph has its dual graph, a Voronoi diagram has its 'dual tessellation', called the Delaunay tessellation. In this section we define the Delaunay tessellation, in particular the Delaunay triangulation.

We consider a Voronoi diagram in the Euclidean plane, and assume that generators of the Voronoi diagram are not on the same line like Figure 2.2.1(a). Since this assumption will be adopted often in this text, we shall refer to it as:

Assumption D1 (the non-collinearity assumption) For a given set of points $P = \{p_1, \ldots, p_n\} \subset \mathbb{R}^m$ $(n \geq 2)$, $p_1, \ldots, p_n$ are not on the same line.

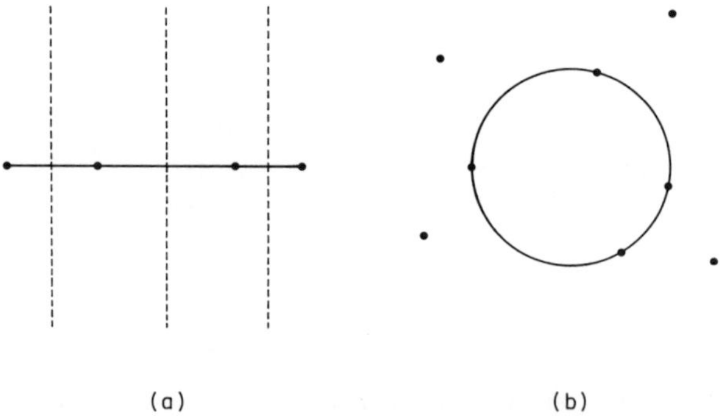

Figure 2.2.1 (a) Collinear generators; (b) cocircular generators.

In addition to this assumption we assume that the number of points in P is three or more but finite. Note that the non-collinearity assumption implicitly implies $n \neq 2$ because two points always can be placed on a line. Also note that these two assumptions were not made in the preceding section, but they are made here because, as will be shown later, without these assumptions, we cannot obtain a triangulation (we can obtain a Voronoi diagram even when $n = 2$ or generators are collinear (the broken lines in Figure 2.2.1(a)).

Under the non-collinearity assumption and $3 \leq n < \infty$, we now show how to obtain the Delaunay triangulation from a Voronoi diagram (the broken lines in Figure 2.2.2). First we choose a Voronoi edge in a Voronoi diagram (the heavy broken line in Figure 2.2.2(a)). This Voronoi edge is shared by two Voronoi polygons. We join the generators (the filled circles) of these Voronoi polygons by a line segment (the heavy solid line in Figure 2.2.2(a)). We carry out this line generation with respect to all Voronoi edges in the Voronoi diagram. As a result, we obtain a second tessellation of the convex hull of the generators (the solid lines in Figure 2.2.2(a)). In Figure 2.2.2(b) we have another tessellation of the convex hull of the generators (the solid

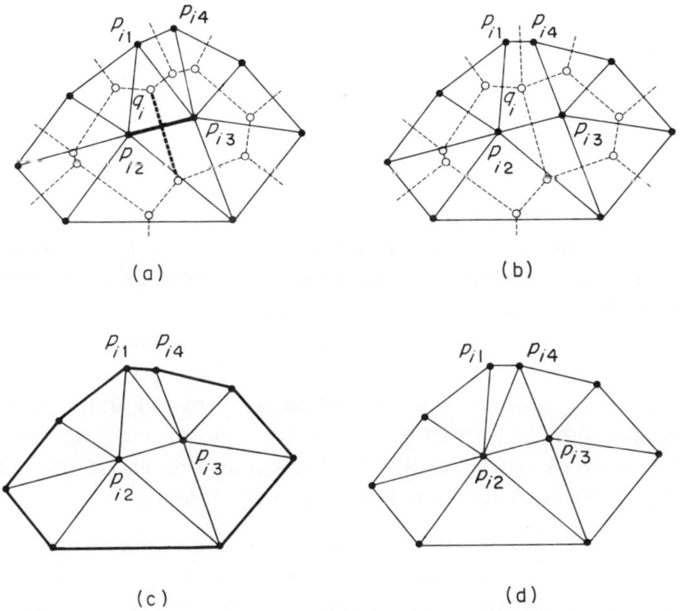

Figure 2.2.2 Voronoi diagrams (the broken lines) and triangulations (the solid lines): (a) a Delaunay triangulation; (b) a Delaunay pretriangulation; (c) a Delaunay triangulation obtained from (b); (d) another Delaunay triangulation obtained from (b).

lines) obtained from another Voronoi diagram (the broken lines). These two tessellations are different in that the tessellation in panel (a) consists of only triangles, whereas the tessellation in panel (b) contains a quadrangle. We call the former tessellation a *Delaunay triangulation*. The latter is not a triangulation, but since it is to be triangulated, we call it a *Delaunay pretriangulation*. Inspecting the Voronoi diagram indicated by the broken lines in Figure 2.2.2(b), we notice that a Delaunay pretriangulation occurs when the Voronoi diagram is degenerate (this condition will be proved in Property V7 in Section 2.3).

A Delaunay pretriangulation has polygons having four or more vertices. We partition these non-triangular polygons into triangles by non-intersecting line segments joining the vertices. For example, we partition the quadrangle with vertices p_{i1}, p_{i2}, p_{i3}, p_{i4} in Figure 2.2.2(b) by the line segment $\overline{p_{i1}p_{i3}}$. As a result, the Delaunay pretriangulation becomes a tessellation consisting of only triangles (Figure 2.2.2(c)). We also call this tessellation a *Delaunay triangulation* (and (sometimes) a *degenerate Delaunay triangulation* to distinguish it from the above Delaunay triangulation). We should note that, as can be seen in Figures 2.2.2(c) and 2.2.2(d), a Delaunay triangulation obtained from a Delaunay pretriangulation is not unique; the quadrangle $p_{i1}p_{i2}p_{i3}p_{i4}$ can be triangulated either by line segment $\overline{p_{i1}p_{i3}}$ or line segment $\overline{p_{i2}p_{i4}}$. Both triangulations are acceptable.

A Delaunay triangulation should consist of at least one triangle, and this condition is guaranteed by the non-collinearity assumption and $3 \leq n < \infty$. If the non-collinearity assumption is not satisfied, as in Figure 2.2.1(a), the above procedure does not produce triangles but line segments, or triangles which degenerate into line segments.

To sum up, we have the following definition.

Definition D1 (a Delaunay triangulation) Given a Voronoi diagram where generators are not collinear and its number is three or more but finite, we join all pairs of generators whose Voronoi polygons share a common Voronoi edge. As a result, we obtain a second tessellation. If this tessellation consists of only triangles, we call it a *Delaunay triangulation*. If not, we call it a *Delaunay pretriangulation*. For the Delaunay pretriangulation, we partition non-triangular polygons into triangles by non-intersecting line segments joining the vertices. As a result, the Delaunay pretriangulation becomes a triangulation. We also call it a *Delaunay triangulation*.

In Definition D1 we constructed a tessellation by generating line segments with respect to every Voronoi edge. Alternatively, we may construct a tessellation by generating line segments with respect to every Voronoi vertex. Although both methods are almost the same, we show the latter method here because it will be easily extended to the definition in $\mathbb{R}^m$ (Definition D3).

We consider a Voronoi diagram $\mathcal{V}(P)$ generated by a set of distinct points $P = \{p_1, \ldots, p_n\}$ $(3 \leq n < \infty)$ which satisfies the non-collinearity assumption. Consider a Voronoi vertex q_i and let $V(p_{i1}), \ldots, V(p_{ik_i})$ be the Voronoi polygons sharing the Voronoi vertex q_i. Note that the generators $p_{i1}, \ldots, p_{ik_i}$ are indexed counterclockwise (or clockwise) around q_i (Figure 2.2.2(a)). For these generators, we construct the polygon with line segments $\overline{p_{ij}p_{ij+1}}$, $j = 1, \ldots, k_i$ (p_{ik_i+1} is read as p_{i1}). If the resulting polygon is a triangle (such as $\triangle p_{i1}p_{i2}p_{i3}$ in Figure 2.2.2(a)), we leave it as it is. If it is not a triangle (such as the quadrangle $p_{i1}p_{i2}p_{i3}p_{i4}$ in Figure 2.2.2(b)), we partition the polygon into triangles by non-intersecting line segments joining the vertices (Figures 2.2.2(c) and 2.2.2(d)). In this manner we construct triangles with respect to all Voronoi vertices in $\mathcal{V}(P)$. As a result, we obtain the set of triangles that forms a tessellation of the convex hull of P, CH(P). We call this tessellation the *Delaunay triangulation* of CH(P) spanning P. Stated a little more mathematically, we have the following definition.

Definition D2 (a Delaunay Triangulation) Let $\mathcal{V}(P)$ be a Voronoi diagram generated by a set of n distinct points $P = \{p_1, \ldots, p_n\} \subset \mathbb{R}^2$ $(3 \leq n < \infty)$ that satisfies the non-collinearity assumption; $Q = \{q_1, \ldots, q_{n_v}\}$ be the set of Voronoi vertices in $\mathcal{V}(P)$; $\mathbf{x}_{i1}, \ldots, \mathbf{x}_{ik_i}$ be the location vectors of the generators whose Voronoi polygons share vertex q_i. We define the set by

$$T_i = \left\{ \mathbf{x} \middle| \, \mathbf{x} = \sum_{j=1}^{k_i} \lambda_j \mathbf{x}_{ij}, \text{ where } \sum_{j=1}^{k_i} \lambda_j = 1, \lambda_j \geq 0, j \in I_{k_i} \right\}, \tag{2.2.1}$$

and let

$$\mathcal{D} = \{T_1, \ldots, T_{n_v}\} . \tag{2.2.2}$$

If $k_i = 3$ for all $i \in I_{n_v}$, we call the set $\mathcal{D}$ the *Delaunay triangulation* of CH(P) spanning P. If there exists at least one $k_i \geq 4$, we call the set $\mathcal{D}$ the *Delaunay pretriangulation* of CH(P) spanning P. We partition T_i having $k_i \geq 4$ into $k_i - 2$ triangles by non-intersecting line segments joining the vertices, and denote the resulting triangles by $T_{i1}, \ldots, T_{ik_i-2}$ (note that $T_{ik_i-2} = T_{i1} = T_i$ for $k_i = 3$). Let

$$\mathcal{D} = \{T_{11}, \ldots, T_{1k_1-2}, \ldots\ldots, T_{n_v1}, \ldots, T_{n_v}, \ldots, T_{n_vk_{n_v}-2}\} . \tag{2.2.3}$$

We call the set $\mathcal{D}$ the *Delaunay triangulation* of CH(P) spanning P, and triangles in $\mathcal{D}$ *Delaunay triangles*.

As is seen in equation (2.2.1), a Delaunay triangle is defined as a closed set, and it contains the boundary consisting of line segments. We call these line segments *Delaunay edges*. Specifically, if a Delaunay edge is shared by two Delaunay triangles, we call it an *internal Delaunay edge* (the light lines in Figure 2.2.2(c)); otherwise, we call it an *external Delaunay edge* (the heavy lines in Figure 2.2.2(c)). When $\mathcal{V}$ is non-degenerate, a Voronoi edge and a Delaunay edge are in one-to-one correspondence. Hence the number of Voronoi edges in $\mathcal{D}$ is the same as that of Voronoi edges in $\mathcal{V}$. Unlike a Voronoi edge, a Delaunay edge is always finite. We call the end points of a Delaunay edge *Delaunay vertices*. Obviously, every Delaunay vertex is a generator of $\mathcal{V}(P)$, and hence the set of Delaunay vertices in $\mathcal{D}$ is given by P.

In Definitions D1 and D2 we defined a Delaunay triangulation using a Voronoi diagram. This does not imply that a Delaunay triangulation should always be defined with a Voronoi diagram. In fact, we shall show in Section 2.3 that a Delaunay triangulation can be defined without a Voronoi diagram. We can define a Delaunay triangulation once a set of points P is given. To indicate this property explicitly, we sometimes write $\mathcal{D}$ as $\mathcal{D}(P)$ like we write $\mathcal{V}$ as $\mathcal{V}(P)$.

Since triangulation is a two-dimensional geometric notion, a Delaunay triangulation is defined only in $\mathbb{R}^2$. We can, however, extend this notion (Definition D2) to the m-dimensional space as follows.

Definition D3 (an *m*-dimensional Delaunay tessellation) Let $\mathcal{V}(P)$ be a Voronoi diagram generated by $P = \{p_1, \ldots, p_n\} \subset \mathbb{R}^m$ $(3 \leq n < \infty)$, where the generators satisfy the non-collinearity assumption; $Q = \{q_1, \ldots, q_{n_v}\}$ be the set of Voronoi vertices (0-faces) in $\mathcal{V}(P)$; $V(p_{i1}), \ldots, V(p_{ik_i})$ be Voronoi $(m - 1)$-faces sharing q_i; T_i be the m-dimensional convex hull spanning generators $p_{i1}, \ldots, p_{k_i}$. If $k_i = m + 1$ for all $i \in I_{n_v}$, the set $\mathcal{D}(P) = \{T_1, \ldots, T_{n_v}\}$ consists of m-dimensional simplices. We call the set $\mathcal{D}(P)$ the *m-dimensional Delaunay tessellation* of CH(P) spanning P. If there exists at least one $k_i \geq m + 2$, we call the set $\mathcal{D}(P)$ the *m-dimensional Delaunay pretessellation* of CH(P) spanning P. We partition T_i having $k_i \geq m + 2$ into $k_i - m$ simplices by non-intersecting hyperplanes passing through the vertices of T_i. Let $T_{i1}, \ldots, T_{ik_i-m}$ be the resulting simplices ($T_{ik_i-m} = T_{i1} = T_i$ for $k_i = m + 1$), and $\mathcal{D}(P) = \{T_{11}, \ldots, T_{1k_1-m}, \ldots, T_{n_v1}, \ldots, T_{n_vk_{n_v}-m}\}$. We call the set $\mathcal{D}(P)$ the *m-dimensional Delaunay tessellation* of CH(P) spanning P, and a simplex in $\mathcal{D}(P)$ an *m-dimensional Delaunay simplex*.

As defined in Definition D1 or D2, a two-dimensional Delaunay tessellation is called a Delaunay triangulation. A three-dimensional Delaunay tessellation is called a *Delaunay tetrahedrization*. Figure 2.2.3 shows a stereographic view of a Delaunay tetrahedrization.

2.3 BASIC PROPERTIES OF THE VORONOI DIAGRAM

Having defined a Voronoi diagram in Section 2.1, we now wish to observe their geometric properties. We mainly deal with the properties of a planar

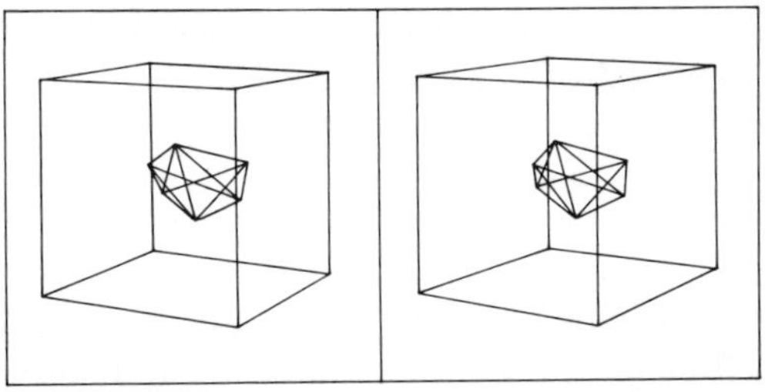

Figure 2.2.3 A stereographic view of a Delaunay tetrahedrization (view the left panel with the left eye and the right panel with the right eye, or use a stereographic viewer). (*Source*: Gen (1983, p. 78).)

Voronoi diagram, but some of them may be readily extended to an m-dimensional Voronoi diagram. (Note that in what follows, fairly technical or lengthy derivations are shown in **Proof** ... □. The reader who wishes to know the properties first may skip those derivations without losing the main stream.)

Although we have already stated in Definition V2 that the region $V(p_i)$ given by equation (2.1.1) is a polygon, this property is readily seen from Definition V3 which defines a Voronoi polygon in terms of half planes (equation (2.1.7)). Since a half plane is a convex set and the intersection of convex sets is a convex set, a Voronoi polygon is a convex set. Since generators are distinct, a Voronoi polygon is non-empty. Since every point in $\mathbb{R}^2$ is assigned to at least one of the generators (Definition V1), the Voronoi polygons in $\mathcal{V}(P)$ are collectively exhaustive in $\mathbb{R}^2$. Moreover, noticing that Definition V3 implies $V(p_i) \subset H(p_i, p_j)$, $V(p_j) \subset H(p_j, p_i)$ and $[H(p_i, p_j)\backslash\partial H(p_i, p_j)] \cap [H(p_j, p_i)\backslash\partial H(p_j, p_i)] = \emptyset$, we understand that Voronoi polygons are mutually exclusive except for boundaries. These properties are quite obvious for an ordinary Voronoi diagram, but some of them may not hold for generalized Voronoi diagrams (to be shown in Chapter 3). To draw attention to this contrast, we sum up these properties in Property V1.

Property V1 Let $P = \{p_1, \ldots, p_n\} \subset \mathbb{R}^2$ $(2 \leq n < \infty)$ be a set of distinct points. The set $V(p_i)$ defined by

$$V(p_i) = \{\mathbf{x} | \, \|\mathbf{x} - \mathbf{x}_i\| \leq \|\mathbf{x} - \mathbf{x}_j\| \text{ for } j \neq i, j \in I_n\} \tag{2.3.1}$$

is a non-empty convex polygon, and $\mathcal{V}(P) = \{V(p_1), \ldots, V(p_n)\}$ satisfies

$$\bigcup_{i=1}^{n} V(p_i) = \mathbb{R}^2 , \tag{2.3.2}$$

$$[V(p_i)\backslash\partial V(p_i)] \cap [V(p_j)\backslash\partial V(p_j)] = \emptyset, \quad i \neq j, i, j \in I_n . \tag{2.3.3}$$

The Voronoi diagram $\mathcal{V}(P)$ is thus a unique tessellation for P.

It is straightforward to see that Property V1 also holds in $\mathbb{R}^m$ if a polygon is replaced by a polyhedron and $\mathbb{R}^2$ is replaced by $\mathbb{R}^m$. Since a Voronoi polyhedron is a convex set, a finite Voronoi polyhedron is sometimes called a *Voronoi polytope*.

Since Voronoi polygons cover the Euclidean plane, it is obvious that at least one Voronoi polygon is infinite. From the assumption that $n \geq 2$, we have two or more Voronoi polygons, some of which may be finite. When generators are collinear, we readily notice from Figure 2.2.1(a) that all Voronoi polygons are infinite. When generators are non-collinear, we have Property V2.

Property V2 For a Voronoi diagram generated by a set of distinct points $P = \{p_1, \ldots, p_n\}$ $(2 \leq n < \infty)$ in $\mathbb{R}^2$, a Voronoi polygon $V(p_i)$ is infinite if and only if $p_i \in \partial \mathrm{CH}(P)$.

Proof Under the non-collinearity assumption, assume that some generators are in the interior of CH(P). We shall show that a Voronoi polygon whose generator is an interior point of CH(P) is finite. First, consider an interior generator point, p_i (p_i in Figure 2.3.1(a)). For p_i we can construct a triangle such that its vertices are generators on the boundary of CH(P), say p_{i1}, p_{i2}, $p_{i3} \in P$, and it contains p_i in its interior (see Figure 2.3.1(a)). We next construct the triangle by the intersection of the half planes $H(p_i, p_{i1})$, $H(p_i, p_{i2})$ and $H(p_i, p_{i3})$ (the dash-dot lines in Figure 2.3.1(a)). Since p_i is an interior point of $\triangle p_{i1}p_{i2}p_{i3}$, this triangle is finite and $H(p_i, p_{i1}) \cap H(p_i, p_{i2}) \cap H(p_i, p_{i3}) \supset \cap_{j=1, j\neq i}^{n} H(p_i, p_j) = V(p_i)$. The Voronoi polygon $V(p_i)$ is thus finite.

Conversely, assume that the Voronoi polygon $V(p_i)$ is bounded by Voronoi edges $e(p_i, p_j)$, $j \in J_i$, where J_i is the set of indices of Voronoi polygons adjacent to $V(p_i)$ (Figure 2.3.1(b)). It is then obvious that p_i is in the

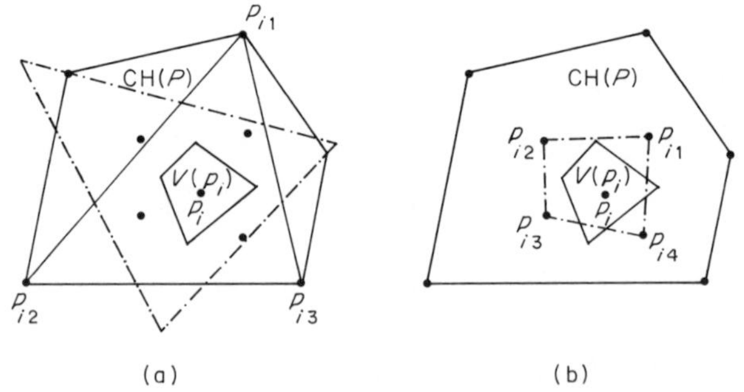

Figure 2.3.1 Illustration of the Proof of Property V2.

interior of $\mathrm{CH}(\{p_j, j \in J_i\})$ (the dash-dot lines in Figure 2.3.1(b)). Since $\mathrm{CH}(\{p_j, j \in J_i\}) \subset \mathrm{CH}(P)$, p_i is not on the boundary of $\mathrm{CH}(P)$.

When generators are collinear, all generators are on $\mathrm{CH}(P)$ which degenerate into a line segment, and all Voronoi polygons are infinite. □

Note that almost in the same manner, we can prove that Property V2 holds for an m-dimensional Voronoi polyhedron.

When a Voronoi diagram $\mathcal{V}(P)$ is given, we can obtain the convex hull of P using Property V2; line segments joining generators whose Voronoi polygons are infinite and share the boundary give $\mathrm{CH}(P)$.

It follows from Property V2 that:

Property V3 For the Voronoi diagram generated by a set of distinct points $P = \{p_1, \ldots, p_n\}$ $(2 \leq n < \infty)$:

(i) Voronoi edges are infinite straight lines if and only if P is collinear.

(ii) Voronoi edge $e(p_i, p_j)(\neq \emptyset)$ is a half line if and only if P is non-collinear and p_i and p_j are consecutive generators of the boundary of $\mathrm{CH}(P)$.

(iii) Suppose that p_i and p_j give a Voronoi edge $e(p_i, p_j)$. Then this edge is a finite line segment if and only if P is non-collinear and at least one of p_i, p_j is in the interior of $\mathrm{CH}(P)$.

From Definition V3, we notice that Voronoi edges are part of bisectors, but bisectors do not always generate Voronoi edges. The following property shows a sufficient condition for the generation of a Voronoi edge.

Property V4 The nearest generator point of p_i generates a Voronoi edge of $V(p_i)$.

To prove this property, let p_j be the nearest generator of p_i, and suppose that the bisector $b(p_i, p_j)$ does not generate a Voronoi edge. Let r_m be the mid point of $\overline{p_ip_j}$ $(r_m \in b(p_i, p_j)$, Figure 2.3.2). Since $b(p_i, p_j)$ does not generate a Voronoi edge, r_m is outside of $V(p_i)$. Hence $\overline{r_mp_i}$ intersects a Voronoi edge, say $e(p_i, p_k)$, of $V(p_i)$, and let r_v be this intersection point. Since $\overline{p_ip_k}/2 \leq \overline{p_ir_v} < \overline{p_ir_m} = \overline{p_ip_j}/2$, we have $\overline{p_ip_k} < \overline{p_ip_j}$, but this contradicts that p_j is the nearest point of p_i. Thus, $b(p_i, p_j)$ generates the Voronoi edge $e(p_i, p_j)$.

From Property V4, we readily notice Property V5.

Property V5 The nearest generator point from p_i exists in the generators whose Voronoi polygons share the Voronoi edges of $V(p_i)$.

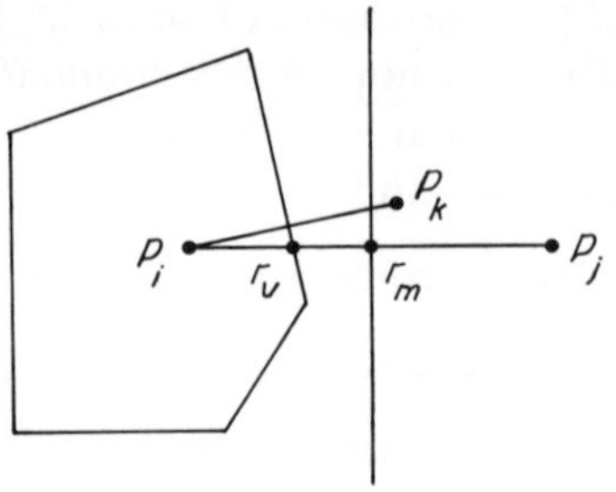

Figure 2.3.2 Illustration of the proof of Property V4.

This property is useful in solving the following well-known problems (Shamos and Hoey, 1975; Bentley *et al.*, 1980).

Problem P1 (the closest pair problem) For a set of distinct points P, if the distance between $p_i \in P$ and $p_j \in P$ is the minimum among the distances between all possible pairs of points in P, we say that the pair $\{p_i, p_j\}$ is the *closest pair*. Given P, find the closest pair.

Problem P2 (the all nearest neighbour problem) Given a set of distinct points P, find the nearest neighbour point of p_i, for every $p_i \in P$.

Property V5 suggests that when we want to solve the above problems, we need not compare all distances from p_i to the rest of the generators; we just compare the distances from p_i to the generators whose Voronoi polygons share the Voronoi edges of $V(p_i)$. Once Problem P2 is solved with this comparison, it is easy to find reciprocal pairs. A pair $\{p_i, p_j\}$ is a *reciprocal (nearest) pair* when p_i is the nearest neighbour point of p_j and p_j is the nearest neighbour point of p_i. The closest pair is found in the reciprocal pairs.

Properties V4 and V5 are concerned with the nearest generator point from a generator point. The next property is concerned with the nearest generator point from an arbitrary point.

Property V6 The generator p_i is the nearest generator point from point p if and only if $V(p_i)$ contains p.

Although this property is quite obvious (almost the definition of a Voronoi polygon itself), we state it in the above because it is very useful in solving the following problem (Knuth, 1973).

Problem P3 (the nearest-search problem) Given a set of distinct points P, find the nearest neighbour point among P from a given point p (p is not necessarily a point in P).

A naive method for solving Problem P3 is to find the minimum distance among n distances, $\{\|x - x_i\|, i \in I_n\}$. When we have only one probe point, this calculation is not so arduous. However, if the number of probe points becomes fairly large, the calculation with this naive method becomes time consuming. To unburden this computational load, Property V6 is useful. In general, when a tessellation is given, the problem of finding a region in which a given probe point is placed is called the *point-location problem.* Using efficient methods developed in computational geometry, such as the bucketing technique in Chapter 4, we can efficiently search to find in which Voronoi polygon a probe point exists. Once this polygon is known, say $V(p_i)$, then we immediately know from Property V6 that p_i is the nearest generator point.

A Voronoi vertex also has interesting properties. Let $Q = \{q_1, \ldots, q_{n_v}\}$ be the set of Voronoi vertices of a Voronoi diagram generated by P. Since a Voronoi vertex q_i is shared by three or more Voronoi polygons, $V(p_{i1})$, $\ldots$, $V(p_{ik})$, $k \geq 3$, $p_{ij} \in P$, $j \in I_k$, where j is indexed counterclockwise (Figure 2.3.3), three or more Voronoi edges, $e(p_{i1}, p_{i2}) \ldots, e(p_{ik}, p_{i1})$, meet at q_i. Recalling that $e(p_{ij}, p_{ij+1})$ is contained in the bisector $b(p_{ij}, p_{ij+1})$, we see that the Voronoi vertex q_i is equally distant from p_{ij} and p_{ij+1}, i.e. $\overline{q_ip_{ij}} = \overline{q_ip_{ij+1}}$. Since this holds for $j \in I_k$ ($k + 1$ is read as 1), we have $\overline{q_ip_{i1}} = \cdots = \overline{q_ip_{ik}}$ (the broken lines in Figure 2.3.3). Points $p_{i1}, \ldots, p_{ik}$ are hence on the same circle, which is denoted by C_i.

We next show that the circle C_i is an empty circle. Suppose that point $p_i \in P \backslash \{p_{i1}, \ldots, p_{ik}\}$ is inside C_i. Then $\overline{q_ip_i} < \overline{q_ip_{ij}}$ for $j = 1, \ldots, k$ (the dash-dot line in Figure 2.3.3), implying that q_i should be assigned to $V(p_i)$. This contradicts that q_i is shared only by $V(p_{i1}), \ldots, V(p_{ik})$. Hence C_i is an empty circle. From this derivation, we have Property V7.

Property V7 For every Voronoi vertex, $q_i \in Q$, in a Voronoi diagram, there exists a unique empty circle C_i centred at q_i which passes through

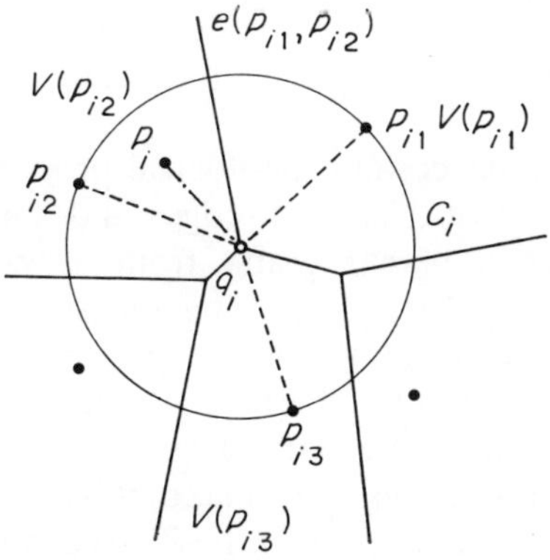

Figure 2.3.3 Illustration of the proof of Property V7.

three or more generators. Under the non-degeneracy assumption, C_i passes through exactly three generators.

From this property, the non-degeneracy assumption is equivalent to the following assumption.

Assumption V2 (the non-cocircularity assumption) Given a set of points $P = \{p_1, \ldots, p_n\} \subset \mathbb{R}^2$ $(4 \leq n < \infty)$, there does not exist a circle, C, such that $p_{i1}, \ldots, p_{ik} \in P$, $k \geq 4$, are on C, and all points in $P \backslash \{p_{i1}, \ldots, p_{ik}\}$ are outside C_0.

This assumption can be extended to the assumption in $\mathbb{R}^m$ if we replace a circle with a (hyper)sphere and 4 with $m + 2$. In this case, we may call the assumption the *non-cosphericity assumption*.

For a given set P of points, if a circle does not contain any points in its interior, the circle is called an *empty circle*. Immediately from Property V7, we obtain Property V8.

Property V8 Circle C_i in Property V7 is the largest empty circle among empty circles centred at the Voronoi vertex q_i.

Using this property with a slight modification, we can solve the following problem.

Problem P4 (the largest empty circle problem) Given a set of distinct points P, find the largest empty circle whose centre is in the convex hull of P.

Let Q' be the vertices of the bounded Voronoi diagram $\mathcal{V} \cap \mathrm{CH}(P)$ (here Q' also includes the vertices formed at the points of intersection between Voronoi edges and the boundary of the CH(P)), and C_i be the largest empty circle centred at $q_i \in Q'$. Then the largest circle among $\{C_i \mid q_i \in Q'\}$ gives the answer to Problem P4. An example is shown in Figure 2.3.4. This circle is useful to find the largest 'empty area' (an example will be given in Section 7.3).

A Voronoi diagram has certain topological properties with respect to the number of Voronoi vertices, Voronoi edges and Voronoi polygons. To see them, we first induce a planar graph from a two-dimensional Voronoi diagram, $\mathcal{V}(P)$. Let $Q = \{q_1, \ldots, q_{n_v}\}$ be the set of Voronoi vertices and $E = \{e_1, \ldots, e_{n_e}\}$ be the set of Voronoi edges in $\mathcal{V}(P)$ in which the first n_c edges are infinite. Since infinite edges are not allowed in a geometric graph, a Voronoi diagram cannot be regarded as a geometric graph; we have to modify infinite Voronoi edges. To make this modification, we place a dummy point q_0 sufficiently far from CH(P). We cut every infinite Voronoi edge at a certain point, and join its end point and q_0 with a line segment

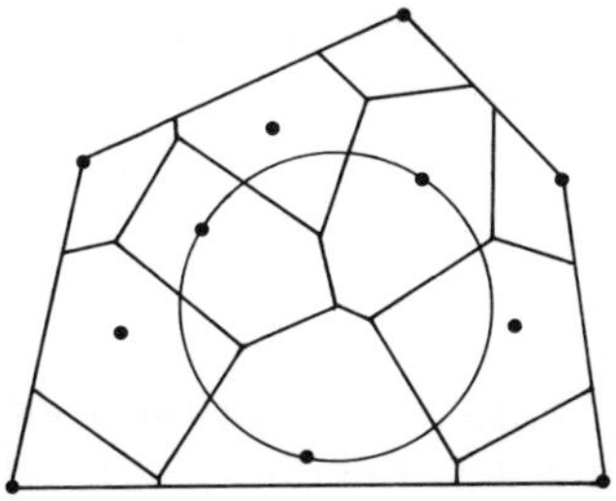

Figure 2.3.4 The largest empty circle whose centre is inside the convex hull of a set of points.

(Figure 2.3.5). Let $e_{b1}, \ldots, e_{bn_c}$ be the edges obtained from $e_1, \ldots, e_{n_c}$ with this modification, and $E_b = [E \backslash \{e_1, \ldots, e_{n_c}\}] \cup \{e_{b1} \ldots, e_{bn_c}\}$. Then, the pair of sets $Q_{+1} = Q \cup \{q_0\}$ and E_b actually form a planar graph, $G(Q_{+1}, E_b)$. We call $G(Q_{+1}, E_b)$ the *Voronoi graph* induced from $\mathcal{V}$(P). For a Voronoi graph, the Euler formula (equation (1.3.61)) holds, and hence we have the equation $(n_v + 1) - n_e + n = 2$ (n_r in equation (1.3.61) is n in $\mathcal{V}$), which is written as Property V9.

Property V9 Let n, n_e and n_v be the numbers of generators, Voronoi edges and Voronoi vertices of a Voronoi diagram in $\mathbb{R}^2$, respectively ($2 \le n < \infty$). Then

$$n_v - n_e + n = 1 \,. \tag{2.3.4}$$

As an extension of this property, we obtain Property V10 from the Euler–Schlaflie formula (equation (1.3.62)).

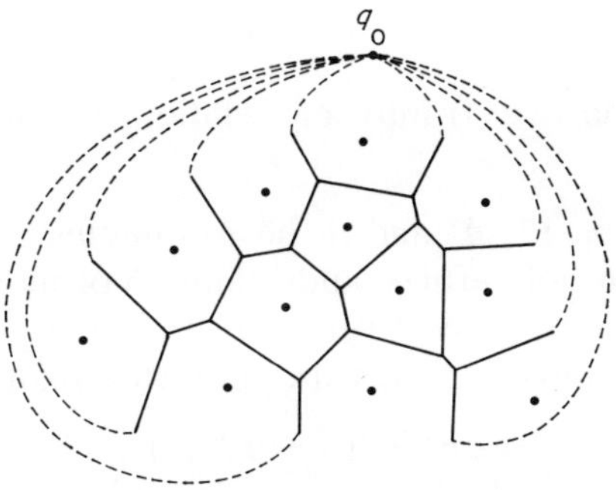

Figure 2.3.5 A Voronoi graph induced from a Voronoi diagram.

Property V10 Let n_k be the number of k-dimensional Voronoi faces in an m-dimensional Voronoi diagram. Then,

$$\sum_{k=0}^{m} (-1)^k n_k = (-1)^m . \tag{2.3.5}$$

Under the non-collinearity assumption and $3 \leq n < \infty$, every vertex in the Voronoi graph $G(Q_{+1}, E_b)$ has at least three edges. Hence the number of edges in $G(Q_{+1}, E_b)$ is not less than $3(n_v + 1)/2$ (a 2 in the denominator implies that every edge is counted twice), i.e.

$$n_e \geq \tfrac{3}{2}(n_v + 1) . \tag{2.3.6}$$

Substituting this relation into equation (2.3.4), we obtain Property V11.

Property V11 Let n, n_e and n_v be the numbers of generators, Voronoi edges and Voronoi vertices, of a Voronoi diagram in $\mathbb{R}^2$, respectively, and assume that $3 \leq n < \infty$. Then

$$n_e \leq 3n - 6 , \tag{2.3.7}$$

$$n_v \leq 2n - 5 . \tag{2.3.8}$$

Although we derived this property under the non-collinearity assumption, Property V11 does not require it, because if generators are collinear, then $n_v = 0$ and $n_e = n$, for which relations (2.3.7) and (2.3.8) hold for $n \geq 3$.

When a set P of generators satisfies the non-collinearity assumption with $n \geq 3$, every finite Voronoi polygon has at least three Voronoi edges, and every infinite Voronoi polygon has at least two Voronoi edges. From these facts and recalling that the number of finite Voronoi polygons is $n - n_c$ and that of infinite Voronoi polygons is n_c, we notice that the number of edges is not less than $[3(n - n_c) + 2n_c]/2$ (a 2 in the denominator implies that every edge is counted twice), i.e.

$$n_e \geq \frac{3(n - n_c) + 2n_c}{2} . \tag{2.3.9}$$

Substituting equation (2.3.4) into this relation, we obtain Property V12.

Property V12 Let n, n_e, n_v and n_c be the numbers of Voronoi polygons, Voronoi edges, Voronoi vertices and infinite Voronoi polygons of a planar Voronoi diagram $\mathcal{V}(P)$, respectively, where $3 \leq n < \infty$ and p satisfies the non-collinearity assumption. Then the following relations hold:

$$n_v \geq \tfrac{1}{2}(n - n_c) + 1 , \tag{2.3.10}$$

$$n_e \geq 3n_v + n_c - 3 \tag{2.3.11}$$

A Voronoi polygon is given by the intersection of $n - 1$ half planes (equation (2.1.9)). A Voronoi polygon thus has $n - 1$ Voronoi edges at the maximum. At the minimum when generators are collinear (Figure 2.2.1(a)), the leftmost and the rightmost Voronoi polygons have only one Voronoi edge. So, what is the average number of Voronoi edges of a Voronoi polygon? From relation (2.3.7) and the fact that every Voronoi edge is shared by exactly two Voronoi polygons, we notice that the average number of Voronoi edges per Voronoi polygon is less than or equal to $2(3n - 6)/n$. We hence obtain Property V13.

Property V13 The average number of Voronoi edges per Voronoi polygon does not exceed six.

Relation (2.3.8) shows that the possible maximum number of Voronoi vertices in a two-dimensional Voronoi diagram is $2n - 5$. For an m-dimensional Voronoi diagram, we have the following property.

Property V14 The maximum number, $n_{\max}$, of Voronoi vertices in an m-dimensional Voronoi diagram satisfies

$$\frac{n-1}{n-1-r}\binom{n-1-r}{r} \leq n_{\max} < 2\binom{n-1-r}{r}$$

$$\text{for even } m = 2r\ , n > m + 1\ , \tag{2.3.12}$$

$$\frac{(n-r-1)(nr-r-1)}{(r+1)(nr-r^2+1)}\binom{[nr/(r+1)]-r}{r-1} < n_{\max} < \frac{n}{n-r}\binom{n-r}{r}$$

$$\text{for odd } m = 2r - 1, n > m + 1\ . \tag{2.3.13}$$

The proof of this property is lengthy and is omitted here. The reader should consult Klee (1980).

When a generator set is given, we can of course construct Voronoi edges. Conversely, when Voronoi edges are given without generators, can we determine the locations of generators? This problem is referred to as the generator recognition problem.

Problem P5 (the generator recognition problem) Provided that the Voronoi edges of a non-degenerate Voronoi diagram $\mathcal{V}(P)$ are given, we recover the locations of generators P.

Note that this problem can be extended to generalized Voronoi diagrams (Chapter 3; see Ash and Bolker, 1985).

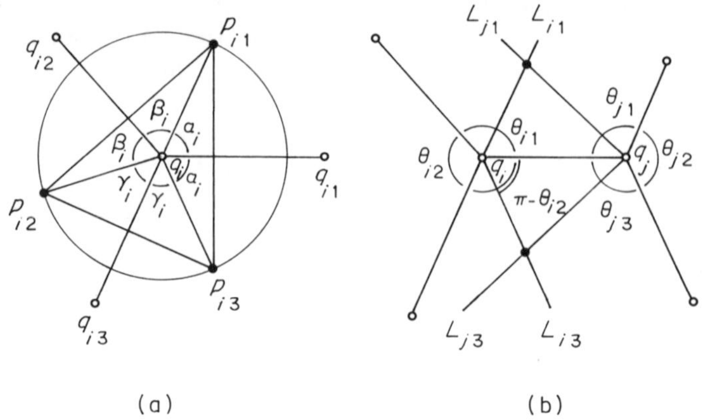

Figure 2.3.6 Recognition of generators from Voronoi edges.

To consider this problem, see Figure 2.3.6. Let q_i be a Voronoi vertex, p_{i1}, p_{i2}, p_{i3} be generators whose Voronoi polygons share q_i, and q_{i1}, q_{i2}, q_{i3} be Voronoi vertices of the Voronoi edges incident to q_i. From Property V7, q_i is the centre of the circle that passes through p_{i1}, p_{i2}, p_{i3}. Since Voronoi edges $e(p_{i1}, p_{i2})$, $e(p_{i2}, p_{i3})$ and $e(p_{i3}, p_{i1})$ perpendicularly bisect Voronoi edges $\overline{q_{i2}q_i}$, $\overline{q_{i3}q_i}$, $\overline{q_{i1}q_i}$, respectively, we have the equations $\angle p_{i1}q_iq_{i1} = \angle p_{i3}q_iq_{i1} = \alpha_i$, $\angle p_{i1}q_iq_{i2} = \angle p_{i2}q_iq_{i2} = \beta_i$, $\angle p_{i2}q_iq_{i3} = \angle p_{i3}q_iq_{i3} = \gamma_i$. Hence $2\alpha_i + 2\beta_i + 2\gamma_i = 2\pi$, i.e. $\alpha_i = \pi - \beta_i - \gamma_i = \pi - \angle q_{i2}q_iq_{i3}$. From this equation, we obtain Property V15.

Property V15 Let $\overline{q_iq_j}$ be a Voronoi edge in a non-degenerate Voronoi diagram, θ_{ik} and θ_{jk}, $k = 1, 2, 3$, be acute angles at q_i and q_j, respectively, where k is indexed counterclockwise from $\overline{q_iq_j}$ at q_i and clockwise at q_j (Figure 2.3.6(b)). Let L_{ik} (L_{jk}) be the half line radiating from q_i (q_j) with angle $\pi - \theta_{i2}$ ($\pi - \theta_{j2}$) with $\overline{q_iq_j}$ in the sector of θ_{ik} (θ_{jk}), $k = 1, 3$. Then the intersection point made by L_{i1} and L_{j1}, and that by L_{i3} and L_{j3} give the generators of the Voronoi diagram sharing $\overline{q_iq_j}$.

This property can be developed into a more general property with which we can examine whether or not a given planar tessellation, $\mathcal{S} = \{S_1, \ldots, S_n\}$, is a Voronoi diagram. Suppose that the tessellation $\mathcal{S}$ consists of convex polygons and every vertex has exactly three edges. Let $q_{i1}, \ldots, q_{ik_i}$ be the vertices of a polygon S_i indexed counterclockwise. Let p_{ij} be the intersection point in S_i obtained through the same procedure stated in Property V15, where $\overline{q_iq_j}$ is replaced by $\overline{q_{ij}q_{ij+1}}$, $j = 1, \ldots, k_i$ ($k_i + 1$ should be read as 1). Then we have the following property.

Property V16 A planar tessellation consisting of convex polygons whose vertices are all degree three is a Voronoi diagram if and only if $p_{i1} = p_{i2} = \cdots = p_{ik_i}$ holds for $i \in I_{n_V}$, where p_{ij} is defined in the above.

The proof of this property is provided by Ash and Bolker (1985).

A planar Voronoi diagram is of course defined in $\mathbb{R}^2$, but interestingly enough we can construct this Voronoi diagram through a polyhedron in $\mathbb{R}^3$. This construction uses a transformation from a point in $\mathbb{R}^2$ to a point $\mathbb{R}^3$, which is illustrated in Figure 2.3.7. Let p_i be a point on the x–y plane with coordinates (x_i, y_i). We lift this point vertically by height $z_i = x_i^2 + y_i^2$, and denote the lift-up point by p_i^*. The coordinates of p_i^* are given by $(x_i, y_i, x_i^2 + y_i^2)$. Recalling that the set

$$A = \{(x, y, z) | z = x^2 + y^2\} \tag{2.3.14}$$

represents the paraboloid of revolution along the z axis (Figure 1.3.21), we notice that the point p_i^* is on the paraboloid A. Thus the above transformation is to lift a point p_i on the x–y plane vertically to the point p_i^* on the paraboloid A. We call this transformation the lift-up transformation, that is, we call the transformation from (x_i, y_i) to (x_i, y_i, z_i) by

$$z_i = x_i^2 + y_i^2 \tag{2.3.15}$$

the *lift-up transformation* from point p_i to p_i^*, and call p_i^* the *lift-up* of point p_i. For a set of points $P = \{p_1, \ldots, p_n\}$, we also say that $P^* = \{p_1^*, \ldots, p_n^*\}$ is the *lift-up* of P.

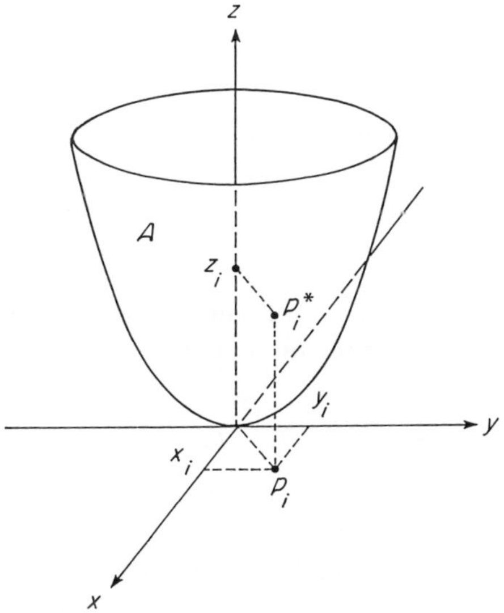

Figure 2.3.7 The lift-up transformation (p_i^* is the lift-up of p_i).

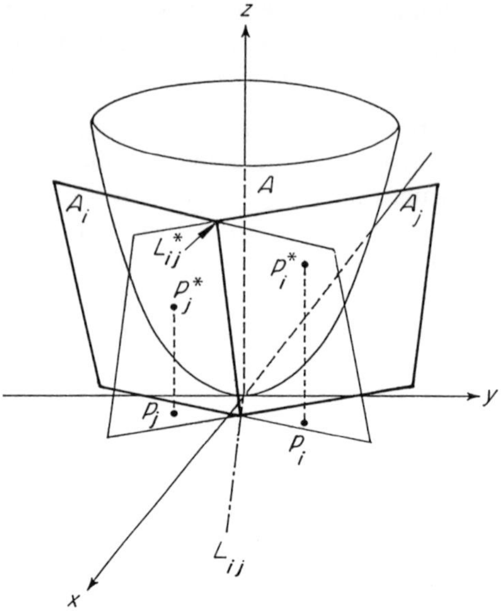

Figure 2.3.8 The construction of a planar Voronoi diagram through a polyhedron in $\mathbb{R}^3$.

For a given generator set P on the x–y plane, let P^* be the lift-up of P, and A_i be the plane tangential to the paraboloid A at p^*_i (Figure 2.3.8). Then, from equation (1.3.47), the plane A_i is given by

$$z = 2x_i x + 2y_i y - (x_i^2 + y_i^2) \, . \tag{2.3.16}$$

Notice that p^*_i is on the plane A_i and the points of P^*, other than p^*_i, are all above the plane A_i.

We next consider the intersection of two planes A_i and A_j, $i \neq j$ (Figure 2.3.8). Obviously this intersection, denoted by L^*_{ij}, is a line in $\mathbb{R}^3$. We project the line L^*_{ij} on the x–y plane orthographically, and let L_{ij} be the projected line (the dash-dot line in Figure 2.3.8). The line L_{ij} is algebraically obtained by eliminating z from the equations of (2.3.16) for i and j. As a result, we get

$$2(x_i - x_j)x + 2(y_i - y_j)y - (x_i^2 + y_i^2 - x_j^2 - y_j^2) = 0 \, . \tag{2.3.17}$$

This equation represents the line perpendicularly bisecting $\overline{p_i p_j}$, namely the bisector $b(p_i, p_j)$ between p_i and p_j.

The plane A_i divides $\mathbb{R}^3$ into two half spaces. Let H_i be the half space given by the region above the plane A_i with boundary A_i, i.e.

$$H_i = \{(x, y, z) | \, z \geq 2x_i x + 2y_i y - (x_i^2 + y_i^2)\} \, . \tag{2.3.18}$$

For two half spaces H_i and H_j $(i \neq j)$, we construct the intersection $H_i \cap H_j$ (the heavy solid lines in Figure 2.3.8), which forms a polyhedron in $\mathbb{R}^3$. Since the orthographic projection of the edge of the polyhedron $H_i \cap H_j$ on the x–y plane is the bisector between p_i and p_j, the orthographic projection of the facet of $H_i \cap H_j$ containing A_i onto the x–y plane gives the half plane $H(p_i, p_j)$. From this property we understand that the orthographic projection of the surface of the polyhedron given by $\cap_{i=1}^{n} H_i$ onto the x–y plane gives the Voronoi diagram $\mathcal{V}(P)$. To sum up, we obtain the following property.

Property V17 The Voronoi diagram $\mathcal{V}(P)$ is obtained as the orthographic projection of the facets of the polyhedron made by the intersection of all upper half spaces H_i, $i \in I_n$, onto the x–y plane, where the upper half space H_i is the region above the plane tangential to the paraboloid at the lift-up point of the generator p_i.

In conjunction with this property, we note that a planar Voronoi diagram can also be obtained through the inversion transformation in $\mathbb{R}^3$, which will be referred to in Section 2.4 (Property D7).

2.4 BASIC PROPERTIES OF THE DELAUNAY TRIANGULATION

Corresponding to Property V1 of a Voronoi diagram, we first state the following basic property which is almost obvious from Definition D2.

Property D1 The set T_i defined by equation (2.2.1) is a unique non-empty polygon, and the set $\mathcal{D}(P) = \{T_1, \ldots, T_{n_v}\}$ given by equation (2.2.2) satisfies

$$\bigcup_{i=1}^{n_v} T_i = \mathrm{CH}(P) , \tag{2.4.1}$$

$$[T_i \backslash \partial T_i] \cap [T_j \backslash \partial T_j] = \emptyset, \quad i \neq j, i, j \in I_{n_v} . \tag{2.4.2}$$

Hence the set $\mathcal{D}(P)$ forms a unique tessellation spanning P. If P satisfies the non-cocircularity assumption, $\mathcal{D}(P)$ is a Delaunay triangulation; otherwise, $\mathcal{D}(P)$ is a Delaunay pretriangulation. A Delaunay triangulation obtained from the Delaunay pretriangulation through the procedure mentioned in Definition D2 is not unique.

Although implicitly stated in equation (2.4.1) (and Figure 2.2.2(c)), it follows from Definition D1 and Property V2 that:

Property D2 The external Delaunay edges in $\mathcal{D}(P)$ constitute the boundary of the convex hull of P.

Thus a Delaunay triangulation is a triangulation of CH(P) spanning P. Since CH(P) is bounded, all Delaunay triangles and Delaunay edges are finite. This property is in contrast to Properties V2 and V3 of a Voronoi diagram.

In Properties V4–V6 we showed the properties concerning the nearest neighbour point in a Voronoi diagram. To find a similar property for a Delaunay triangulation, let us inspect the Delaunay triangulations in Figure 2.4.1, where the unfilled circles, the broken lines and the solid lines represent the Voronoi vertices, the Voronoi edges and the Delaunay triangulation spanning six generators $p_1, p_2, \ldots, p_6$, respectively. In observing Delaunay triangles T_1, T_3, T_4 in panel (a), we notice that the nearest generator from any point in a triangle is one of the three vertices of the triangle, because the whole triangle is covered with the three Voronoi polygons whose generators are the vertices of the triangle. On the other hand, the nearest generator from some points in T_2 (points in the shaded area) is a generator other than the three vertices of T_2, because T_2 is not exhaustively covered with the three Voronoi polygons whose generators are the vertices of T_2. If a triangulation consists of only the former type triangles, such as in Figure 2.4.1(b), we call the triangulation a Pitteway triangulation. Generally, for a triangulation $\mathcal{T}(P)$ of CH(P) spanning P, if for any point p in CH(P) the closest generator point from p is one of the three vertices of the triangle containing p, we call the triangulation $\mathcal{T}(P)$ a *Pitteway triangulation* spanning P (cf. McLain, 1976). The triangulation in Figure 2.4.1(b) is a Pitteway triangulation.

Between a Delaunay triangulation and a Pitteway triangulation, we have the following relationship.

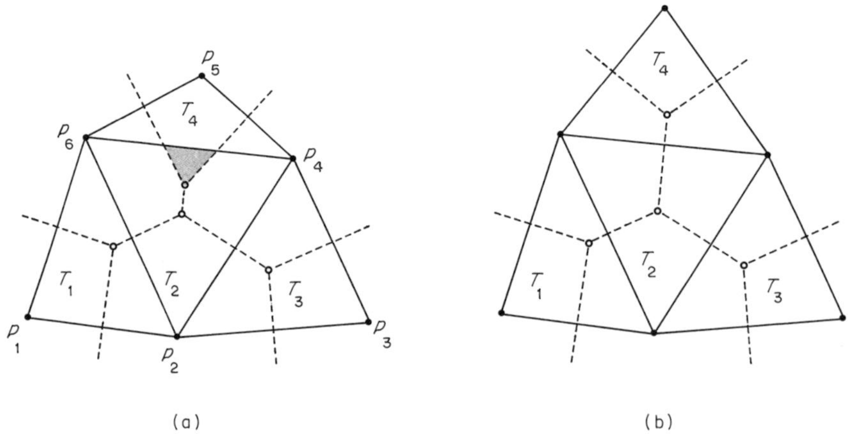

Figure 2.4.1 Delaunay triangulations: (a) is not a Pitteway triangulation, while (b) is a Pitteway triangulation.

Property D3 (Pitteway triangulation theorem) The Delaunay triangulation spanning P is a Pitteway triangulation spanning P if and only if every internal Delaunay edge crosses the associated Voronoi edge of the Voronoi diagram generated by P.

Proof First, suppose that the Delaunay triangulation $\mathcal{D}(P)$ is a Pitteway triangulation spanning P. Then, any Delaunay triangle is covered with the three Voronoi polygons whose generators are the vertices of the triangle. Let $\overline{p_ip_j}$ be an internal Delaunay edge shared by Delaunay triangles $\triangle p_ip_jp_k$ and $\triangle p_ip_jp_l$. Assume, contrary to the property, that $\overline{p_ip_j}$ does not cross the associated Voronoi edge. This means that if the four generators p_i, p_j, p_k and p_l are not cocircular, the Voronoi vertices generated by p_i, p_j, p_k and by p_i, p_j, p_l are both on the same side of $\overline{p_ip_j}$. If the four generators are cocircular, the Voronoi vertex shared by $V(p_i,)$ $V(p_j)$, $V(p_k)$ and $V(p_l)$ (in other words, the length-zero Voronoi edge associated with the Delaunay edge $\overline{p_ip_j}$) is not on $\overline{p_ip_j}$. Hence, the midpoint of $\overline{p_ip_j}$ belongs to $V(p_k)$ or $V(p_l)$ but not to the both, which contradicts that $\mathcal{T}(P)$ is a Pitteway triangulation. Thus, any internal Delaunay edge crosses the associated Voronoi edge.

Assume next that any internal Delaunay edge crosses the associated Voronoi edge. Let p_i, p_j and p_k be the three vertices forming a Delaunay triangle. The proof varies according to which edges are internal edges among $\overline{p_ip_j}$, $\overline{p_jp_k}$ and $\overline{p_kp_i}$.

Case 1. Suppose that $\overline{p_ip_j}$, $\overline{p_jp_k}$ and $\overline{p_kp_i}$ are all internal edges (such as T_2 in Figure 2.4.1(b)). Then, because of the above assumption, the triangle contains the Voronoi vertex generated by p_i, p_j and p_k, but does not contain any other Voronoi vertex. Since the line segment connecting the generator p and a point on the boundary of $V(p)$ belongs to $V(p)$, the triangle is covered with $V(p_i)$, $V(p_j)$ and $V(p_k)$.

Case 2. Suppose that $\overline{p_ip_j}$, $\overline{p_jp_k}$ and $\overline{p_kp_i}$ are all external Delaunay edges. This implies that there are exactly three generators in P, and hence the triangle (actually the whole plane) is covered with $V(p_i)$, $V(p_j)$ and $V(p_k)$.

Case 3. Suppose that one edge, say $\overline{p_ip_j}$, is an external Delaunay edge while the other two edges are internal. Then, the Voronoi vertex generated by p_i, p_j and p_k is either in the triangle or on the side opposite to the triangle with respect to $\overline{p_ip_j}$. In the former case, the external Delaunay edge $\overline{p_ip_j}$ also crosses the associated Voronoi edge, and hence the triangle is covered with $V(p_i)$, $V(p_j)$ and $V(p_k)$, as we saw in Case 1. In the latter case, the triangle contains no Voronoi vertex, and the Voronoi edge generated by p_j and p_k croses both $\overline{p_jp_k}$ and $\overline{p_ip_j}$; the Voronoi edge generated by p_i and p_k crosses both $\overline{p_ip_k}$ and $\overline{p_ip_j}$. No other Voronoi edge traverses the triangle. This means that the triangle is covered with the three Voronoi polygons generated by its vertices.

Case 4. Suppose that two edges, say $\overline{p_ip_j}$ and $\overline{p_jp_k}$, are external edges and the other is an internal edge (such as T_1 in Figure 2.4.1(b)). Then, the Voronoi vertex generated by p_i, p_j and p_k is either in the triangle or on the side opposite to the triangle with respect to one of the external edges. In both cases the configuration is the same as in Case 3, and hence the triangle is covered with the three Voronoi polygons generated by its vertices. □

Property D3 gives a necessary and sufficient condition for a Delaunay triangulation to be a Pitteway triangulation. The next question is: If the point set P admits a Pitteway triangulation, does it necessarily coincide with the Delaunay triangulation? The answer is 'no'. A counterexample is shown in Figure 2.4.2, where P consists of four points forming a rhombus; (a) is the Delaunay triangulation spanning P whereas (b) is not. However, both (a) and (b) are Pitteway triangulations. Thus, a Pitteway triangulation spanning P, if it exists, is not necessarily unique, and does not necessarily coincide with the Delaunay triangulation.

As we observed in Definitions D1 and D2, we have nice relationships between the Voronoi diagram $\mathcal{V}(P)$ and the Delaunay triangulation $\mathcal{D}(P)$ that share the same point set P. First, from Definition D1, the Delaunay vertices of $\mathcal{D}(P)$ are the generators of $\mathcal{V}(P)$; conversely, the generators of $\mathcal{V}(P)$ are the Delaunay vertices of $\mathcal{D}(P)$. Second, from Property V7, the Voronoi vertices of $\mathcal{V}(P)$ are the circumcentres of Delaunay triangles in $\mathcal{D}(P)$; conversely, the circumcentres of Delaunay triangles in $\mathcal{D}(P)$ are the Voronoi vertices in $\mathcal{V}(P)$. Lastly the number of Delaunay edges of $\mathcal{D}(P)$ is greater than or equal to that of Voronoi edges; under the non-degeneracy assumption, both numbers are the same. To sum up, we have the following property.

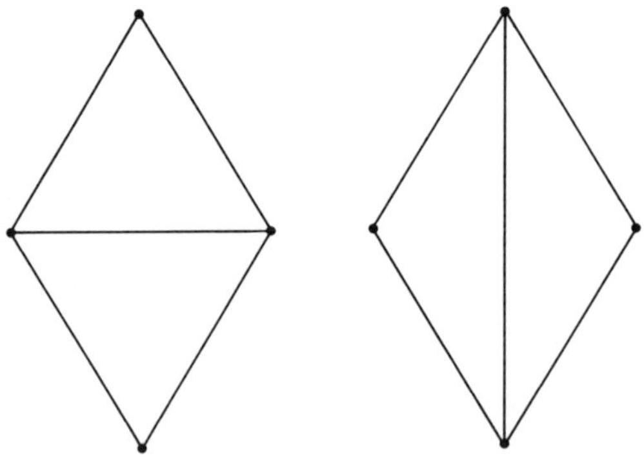

Figure 2.4.2 Two Pitteway triangulations spanning four points forming a rhombus; (a) is the Delaunay triangulation while (b) is not.

Property D4 For the Voronoi diagram $\mathcal{V}(P)$ generated by P, and the Delaunay triangulation spanning P, let Q and Q_d be the set of the Voronoi vertices and that of Delaunay vertices, respectively; E and E_d be the set of Voronoi edges and that of Delaunay edges, respectively; C_d be the set of circumcentres of Delaunay triangles. Then,

(i) $Q_d = P$.

(ii) $C_d = Q$.

(iii) $|E_d| \geq |E|$; $|E_d| = |E|$ if and only if $\mathcal{V}(P)$ is non-degenerate, where $|E|$ indicates the number of elements in the set E.

From Property D4 and Property V7, a circumcircle of a Delaunay triangle has the next property which corresponds to Property V8.

Property D5 All circumcircles of Delaunay triangles are empty circles.

An example is shown in Figure 2.4.3. Using this property, we can solve the largest empty circle problem (Problem P4) (note that a slight modification is necessary, as we remarked in Property V8).

Having observed Figure 2.4.3, the reader may question whether or not the converse relation in Property D5 holds, that is, if the circumcircles of all triangles in a triangulation are empty circles, then the triangulation is a Delaunay triangulation. To answer this question, we first introduce the following criterion.

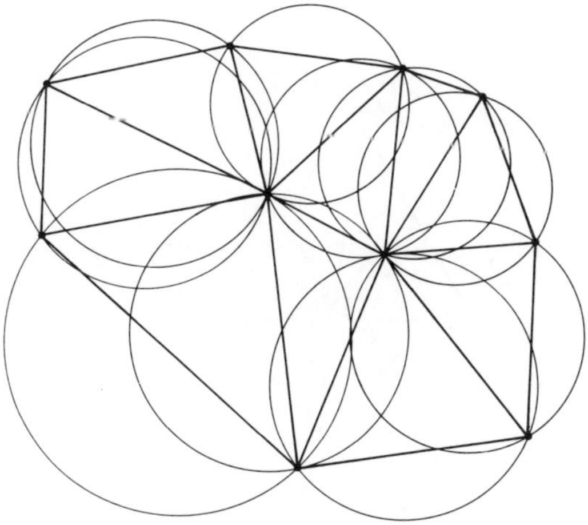

Figure 2.4.3 Empty circumcircles of Delaunay triangles in a Delaunay triangulation.

The empty circumcircle criterion Given a triangulation of CH(P) spanning P, if the circumcircle of a triangle in the triangulation is an empty circle, we say that the triangle satisfies the *empty circumcircle criterion*.

With this criterion, we can now show Property D6.

Property D6 (empty circumcircle theorem) Every triangle in a triangulation spanning P satisfies the empty circumcircle criterion if and only if the triangulation is the Delaunay triangulation spanning P.

Proof Since we have already obtained Property D5, what is left is to prove that a triangulation in which every triangle satisfies the empty circumcircle criterion is a Delaunay triangulation.

Consider triangles sharing the point $p_i \in P$ in a triangulation $\mathcal{T}$ spanning P (Figure 2.4.4). Let $p_{i1}, \ldots, p_{ik}$ be vertices of these triangles where the points are indexed counterclockwise, and $q_{i1}, \ldots, q_{ik}$ be the centres of the circumcircles, $C_{i1}, \ldots, C_{ik}$, of triangles $\triangle p_i p_{i1} p_{i2}, \ldots, \triangle p_i p_{ik} p_{i1}$ (the heavy solid lines in Figure 2.4.4). Since $\overline{p_i p_{ij}}$ is orthogonal to $\overline{q_{ij} q_{ij-1}}$, $j = 1, \ldots, k$ (0 is read as k), the polygon with vertices $q_{i1}, \ldots, q_{ik}$ is given by $\cap_{j=1}^{k} H(p_i\, p_{ij})$ (the broken lines in Figure 2.4.4). From the assumption, all points of P other than $p_i, p_{ij}, j = 1, \ldots, k$, are outside the circumcircles $C_{i1}, \ldots, C_{ik}$. We take a point, say p_h, from $P \backslash \{p_i, p_{ij}, \ldots, p_{ik}$, and draw the line segment $\overline{p_i p_h}$. Then $\overline{p_i p_h}$ intersects the arc of the circumcircle on the boundary of CH $(\cup_{j=1}^{k} C_{ij})$, say the arc $p_{ij} p_{ij+1}$ of C_{ij} ($j+1=5$ is read as 1). Let r_h be this intersection point. Then the boundary of $H(p_i, r_h)$ (the dash-dot line in Figure 2.4.4) passes through q_{ij} because r_h is on the circumcircle,

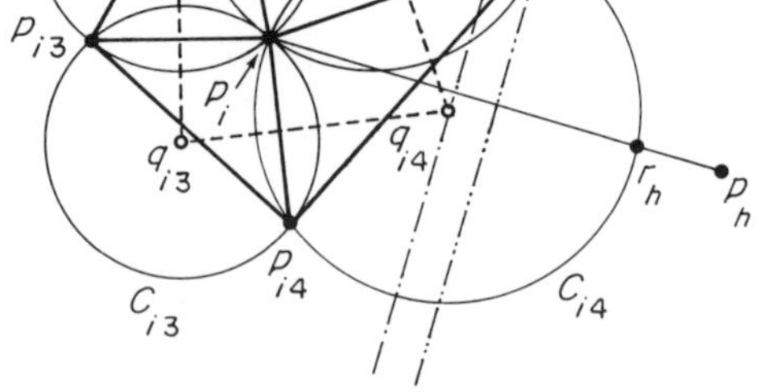

Figure 2.4.4 Illustration of the proof that a triangulation in which every triangle satisfies the empty circumcircle criterion is a Delaunay triangulation.

and $\cap_{j=1}^{k} \mathrm{H}(p_i, p_{ij}) \subset H(p_i, r_h) \subset H(p_i, p_h)$. This relation holds for all p_h in $P \backslash \{p_i, p_{ij}, \ldots, p_{ik}\}$. Therefore

$$\bigcap_{j=1}^{k} H(p_i, p_{ik}) = \bigcap_{j \in I_n \backslash \{i\}} H(p_i, p_j) = V(p_i) . \tag{2.4.3}$$

Thus the polygon with $q_{i1}, \ldots, q_{ik}$ is the Voronoi polygon of p_i. If we carry out the above procedure for every p_i in P, we obtain the Voronoi diagram $\mathcal{V}(P)$. Since triangulation $\mathcal{T}$ is the same as the Delaunay triangulation obtained from the Voronoi diagram $\mathcal{V}(P)$ (Definition D2), the given triangulation is the Delaunay triangulation spanning P. □

A Delaunay triangulation is of course a tessellation in $\mathbb{R}^2$, but like Property V17 we can also construct a Delaunay triangulation through a polyhedron in $\mathbb{R}^3$. Here we show two methods: one using the inversion transformation (to be defined below), and one using the lift-up transformation adopted in Property V17.

The *inversion transformation* with the inversion centre at the origin o and the inversion radius 1 is defined by

$$F(\mathbf{x}) = \frac{\mathbf{x}}{\|\mathbf{x}\|^2}, \quad \mathbf{x} \in \mathbb{R}^m \tag{2.4.4}$$

(Dodge, 1972, Chapter 5). This function has the property that $\|F(\mathbf{x})\| = 1/\|\mathbf{x}\|$, that is, the direction of the vector $F(\mathbf{x})$ is the same as that of the vector $\mathbf{x}$; the length of the vector $F(\mathbf{x})$ is the inverse of the length of the vector $\mathbf{x}$. We now consider the three-dimensional Euclidean space $\mathbb{R}^3$ with the Cartesian coordinate system (x_1, x_2, x_3), and let A be the plane parallel to the x_1–x_2 plane passing through the point $(0, 0, -1)$ whose location vector is given by $\mathbf{e}$ (see Figure 2.4.5). Suppose that p is on the plane A and $\mathbf{x}$ is the location vector of p. Then the vectors $\mathbf{e}$ and $\mathbf{x} - \mathbf{e}$ are orthogonal

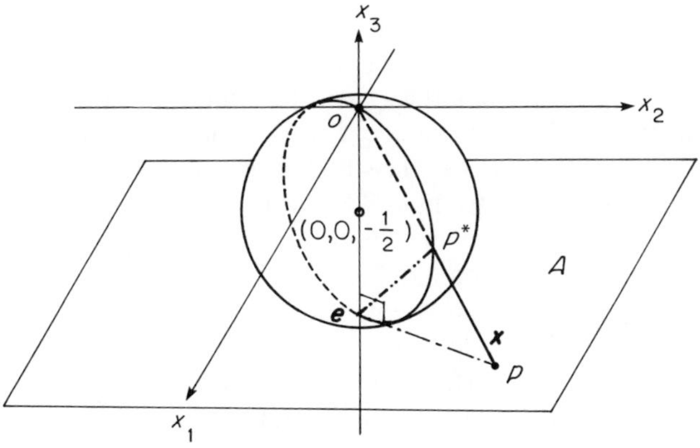

Figure 2.4.5 The inversion transformation.

(observe that the x_3-axis and the dash-dot line in Figure 2.4.5 are orthogonal), which is mathematically written as $\mathbf{e}^T(\mathbf{x} - \mathbf{e}) = 0$ or $\mathbf{e}^T \mathbf{x} = 1$. Let $\mathbf{x}^* = F(\mathbf{x}) = \mathbf{x}/\|\mathbf{x}\|^2$. Then $\mathbf{x}^{*T}(\mathbf{x}^* - \mathbf{e}) = (\mathbf{x}^T/\|\mathbf{x}\|^2)(\mathbf{x}/\|\mathbf{x}\|^2 - \mathbf{e}) = (1/\|\mathbf{x}\|^2)(1 - \mathbf{x}^T\mathbf{e}) = 0$, showing that the vectors $\mathbf{x}^*$ and $\mathbf{x}^* - \mathbf{e}$ are orthogonal (observe that the heavy broken line and the dash-dot-dot line in Figure 2.4.5 are orthogonal). This means that the point p^* indicated by the location vector $\mathbf{x}^*$ is on the circle centred at $(0, 0, -\frac{1}{2})$ with radius $\frac{1}{2}$ (see the great circle in Figure 2.4.5). Since this property holds for any point on the plane A, the set

$$S = \{\mathbf{x}^* | \mathbf{x}^* = F(\mathbf{x}), \mathbf{x} \in A\} \tag{2.4.5}$$

forms a sphere without the point o. Thus, the inversion transformation of equation (2.4.4) maps the plane A onto $S \backslash \{o\}$. Stated more specifically, the image of a point p on the plane A under the inversion transformation is the point p^* given by the intersection point of the sphere S and the line joining the point p and the origin (Figure 2.4.5). This mapping is indicated by $p^* = F(p)$.

The inversion transformation is a one-to-one function, and hence the inverse function F^{-1} exists. To obtain F^{-1} explicitly, we solve $\mathbf{x}^* = \mathbf{x}/\|\mathbf{x}\|^2$ with respect to $\mathbf{x}$ using the property $\|\mathbf{x}^*\| = 1/\|\mathbf{x}\|$. As a result, we get

$$F^{-1}(\mathbf{x}^*) = \frac{\mathbf{x}^*}{\|\mathbf{x}^*\|^2}. \tag{2.4.6}$$

We notice from equations (2.4.4) and (2.4.6) that $F^{-1} = F$. The inverse inversion transformation (i.e. the inversion transformation) maps $S \backslash \{o\}$ onto the plane A.

Using the inversion transformation, we now construct a triangulation according to the following procedure (Brown, 1979).

In $\mathbb{R}^3$ with the Cartesian coordinates (x_1, x_2, x_3) we consider the plane A which is parallel to the x_1–x_2 plane passing through $(0, 0, -1)$, and place a set of n distinct points $P = \{p_1, \ldots, p_n\}(3 \leq n < \infty)$ on A, where P satisfies the non-collinearity assumption (this assumption can be easily relaxed). Let $P^* = \{p_1^*, \ldots, p_n^*\} = \{F(p_1), \ldots, F(p_n)\}$, where F is given by equation (2.4.4) ($m = 3$) (the filled circles on the sphere in Figure 2.4.6). Then, CH(P^*) (the polyhedron in Figure 2.4.6) consists of triangular faces T_i^*, $i \in I_r$. These triangles can be classified into 'near-side' triangles and 'far-side' triangles. A *near-side* triangle T_i^* is a triangle that satisfies the condition that CH(P^*) and the origin are in the same half space produced by the plane containing T_i^* (the shaded face in Figure 2.4.6). We reindex T_i^* so that the first n_v triangles are near-side triangles. Let p_{ij}^*, $j = 1, 2, 3$, be vertices of a near-side triangle T_i^*, $i \in I_{n_v}$, where $j = 1, 2, 3$ are indexed counterclockwise on the surface of CH(P^*), and let $p_{ij} = F^{-1}(p_{ij}^*)$, $j = 1, 2, 3$. On plane A, we construct the triangle by $\overline{p_{ij}p_{ij+1}}$, $j = 1, 2, 3$ (where 4 is read as 1), and denote it by T_i. Then the set $\mathcal{T}(P) = \{T_1, \ldots, T_{n_v}\}$ gives

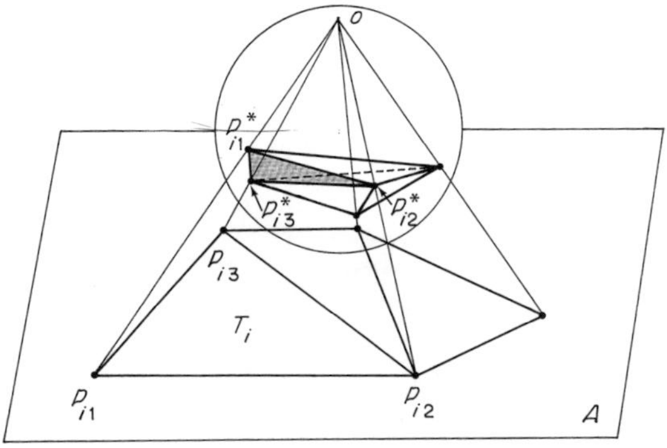

Figure 2.4.6 A Delaunay triangulation obtained through the inversion transformation.

a triangulation of CH(P) spanning P. We now demonstrate that this triangulation satisfies the next property.

Property D7 The triangulation obtained through the above inversion transformation procedure is a Delaunay triangulation.

Proof Let $\mathbf{x}^*_{ij} = F(\mathbf{x}_{ij}) = \mathbf{x}_{ij}/\|\mathbf{x}_{ij}\|^2$, $j = 1, 2, 3$, be the location vectors of the vertices $p^*_{i1}, p^*_{i2}, p^*_{i3}$ of a near-side triangular face T^*_i (the shaded face in Figure 2.4.7) of CH(P^*), where $P^* = \{p^*_1, \ldots, p^*_n\}, p^*_i = F(p_i), p^*_{ij} = F(p_{ij})$, $p_{ij} \in P$ and $\mathbf{x}_{ij}$ is the location vector of p_{ij}. Let A^*_i be the plane containing the face T^*_i, i.e.

$$A^*_i = \{\mathbf{z}^* |\, \mathbf{z}^* = \lambda_1 \mathbf{x}^*_{i1} + \lambda_2 \mathbf{x}^*_{i2} + \lambda_3 \mathbf{x}^*_{i3}, \text{ where } \lambda_1 + \lambda_2 + \lambda_3 = 1\}$$

$$= \left\{ \mathbf{z}^* |\, \mathbf{z}^* = \lambda_1 \frac{\mathbf{x}_{i1}}{\|\mathbf{x}_{i1}\|^2} + \lambda_2 \frac{\mathbf{x}_{i2}}{\|\mathbf{x}_{i2}\|^2} + \lambda_3 \frac{\mathbf{x}_{i3}}{\|\mathbf{x}_{i3}\|^2}, \right.$$

$$\left. \text{where } \lambda_1 + \lambda_2 + \lambda_3 = 1 \right\}. \tag{2.4.7}$$

Let S_i be the sphere passing through points $p_{i1}, p_{i2}, p_{i3}, o$; $\mathbf{r}$ be the location vector of the centre c_i of the sphere S_i; c'_i be the point indicated by $2\mathbf{r}$. If a point p (indicated by the location vector $\mathbf{x}$) is on S_i, points o, c'_i, p are on the great circle of S_i (the great circle in Figure 2.4.7) whose diameter is given by the line segment joining o and c'_i, and so $\angle\, opc'_i = \pi/2$, i.e. the vectors $\mathbf{x}$ and $\mathbf{x} - 2\mathbf{r}$ are orthogonal. Thus S_i is represented by the set of points

$$S_i = \{\mathbf{x} |\, \mathbf{x}^{\mathrm{T}}(\mathbf{x} - 2\mathbf{r}) = 0, \mathbf{x} \in \mathbb{R}^3\}. \tag{2.4.8}$$

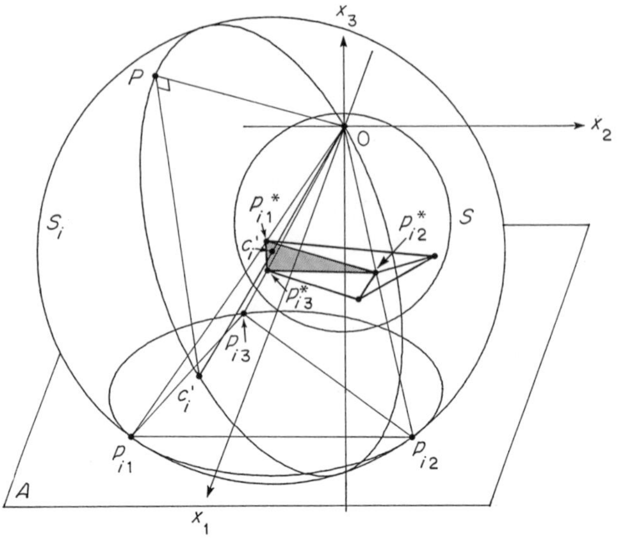

Figure 2.4.7 Illustration of the proof of Property D7.

Since points p_{i1}, p_{i2}, p_{i3} are on S_i, the equation $\mathbf{x}_{ij}^{\mathrm{T}}(\mathbf{x}_{ij} - 2\mathbf{r}) = 0$, i.e. $\mathbf{x}_{ij}^{\mathrm{T}}\mathbf{x}_{ij} = 2\mathbf{x}_{ij}^{\mathrm{T}}\mathbf{r}$, holds for $j = 1, 2, 3$.

We now prove that points on the plane A_i^* are mapped on the sphere S_i through the (inverse) inversion transformation, i.e. we prove $S_i = \{\mathbf{z} | \ \mathbf{z} = F^{-1}(\mathbf{z}^*), \mathbf{z}^* \in A_i^*\}$. From $F^{-1}(\mathbf{z}^*) = \mathbf{z}^*/\|\mathbf{z}^*\|^2$ and equation (2.4.7), we obtain:

$$F^{-1}(\mathbf{z}^*)^{\mathrm{T}}(F^{-1}(\mathbf{z}^*) - 2\mathbf{r}) = \frac{\mathbf{z}^{*\mathrm{T}}}{\|\mathbf{z}^*\|^2}\left(\frac{\mathbf{z}^*}{\|\mathbf{z}^*\|^2} - 2\mathbf{r}\right)$$

$$= \frac{1}{\|\mathbf{z}^*\|^2}\left(1 - \sum_{j=1}^{3} \frac{2\lambda_j \mathbf{x}_{ij}^{\mathrm{T}}\mathbf{r}}{\|\mathbf{x}_{ij}\|^2}\right). \qquad (2.4.9)$$

Upon substituting $\mathbf{x}_{ij}^{\mathrm{T}}\mathbf{x}_{ij} = 2\mathbf{x}_{ij}^{\mathrm{T}}\mathbf{r}$ into this equation, we obtain:

$$F^{-1}(\mathbf{z}^*)^{\mathrm{T}}(F^{-1}(\mathbf{z}^*) - 2\mathbf{r}) = \frac{1}{\|\mathbf{z}^*\|^2}(1 - \lambda_1 - \lambda_2 - \lambda_3) = 0\,. \quad (2.4.10)$$

Thus $F^{-1}(\mathbf{z}^*)$ satisfies equation (2.4.8), showing that the plane A_i^* is mapped onto S_i, and that the closed half space, H^*, produced by A_i^* which contains the origin is mapped onto the space outside the sphere S_i (recall that the distance is reversed by the inversion transformation of equation (2.4.6)). The intersection of S_i with the plane parallel to the x_1–x_2 plane passing through $(0, 0, -1)$ is the circle C_i passing through points p_{i1}, p_{i2}, p_{i3}. Obviously the circle C_i is the circumcircle of the triangle T_i. Since all points in P^* are in H^*, all points in $P = F^{-1}(P^*)$ are on or outside C_i, indicating that circumcircle C_i is an empty circle. Thus the set of triangles $\{T_1, \ldots,$

$T_{n_v}\}$ forms a triangulation in which circumcircles of the triangles are all empty circles. From the empty circumcircle theorem (Property D6), we notice that this triangulation is the Delaunay triangulation spanning P. □

In Definition D2 we remarked that a Delaunay triangulation was not necessarily defined with a Voronoi diagram. In fact, Property D7 shows that once a set of points P is given, we can define a Delaunay triangulation as the triangulation obtained from the above procedure (which does not require a Voronoi diagram). We can also use Property D7 to give a definition of a Voronoi diagram alternative to Definitions V1 or V2. When a Delaunay triangulation is given, it is straightforward to construct a Voronoi diagram by inversing the procedure mentioned in Definition D1. Thus we may define a Voronoi diagram via the Delaunay triangulation obtained from the inversion transformation procedure. We also note that we can construct a Voronoi diagram directly through the inversion transformation. We leave it as an exercise.

In Property D7 we obtained the Delaunay triangulation $\mathcal{D}(P)$ from the convex hull of $F(P)$ in $\mathbb{R}^3$ through the inversion transformation F. This method can be extended to $\mathbb{R}^m$. We can construct an m-dimensional Delaunay tessellation $\mathcal{D}(P)$ or an m-dimensional Voronoi diagram $\mathcal{V}(P)$ from the convex hull of $F(P)$ in $\mathbb{R}^{m+1}$ through the inversion transformation (Brown, 1979, 1980).

We next show another correspondence between a Delaunay triangulation and a convex hull in $\mathbb{R}^3$ (Edelsbrunner and Seidel, 1986; O'Rourke *et al.*, 1986; Edelsbrunner, 1987; Aggarwal *et al.*, 1989).

Let P^* be the lift-up of $P = \{p_1, \ldots, p_n\}$ (where $p_i \in \mathbb{R}^2$) and CH(P^*) be the convex hull of P^*. A facet of CH(P^*) is called a *lower facet* if the plane containing the facet passes through or below any point in P^*. The surface formed by all the lower facets of CH(P^*) is called the *lower boundary* of CH(P^*).

Suppose without loss of generality that three points p_i, p_j, p_k are placed on the x–y plane so that the cycle $p_ip_jp_kp_i$ has a counterclockwise direction when it is viewed from a point far above the x–y plane (see Figure 2.4.8). Let

$$H(p_i, p_j, p_k, p) = \begin{vmatrix} 1 & x_i & y_i & x_i^2 + y_i^2 \\ 1 & x_j & y_j & x_j^2 + y_j^2 \\ 1 & x_k & y_k & x_k^2 + y_k^2 \\ 1 & x & y & x^2 + y^2 \end{vmatrix} \qquad (2.4.11)$$

It is an exercise to see that the circle passing through p_i, p_j, p_k is represented by $H(p_i, p_j, p_k, p) = 0$, and that this circle is an empty circle if and only if $H(p_i, p_j, p_k, p_l) \geq 0$ for any point $p_l \in P \backslash \{p_i, p_j, p_k\}$. In this connection,

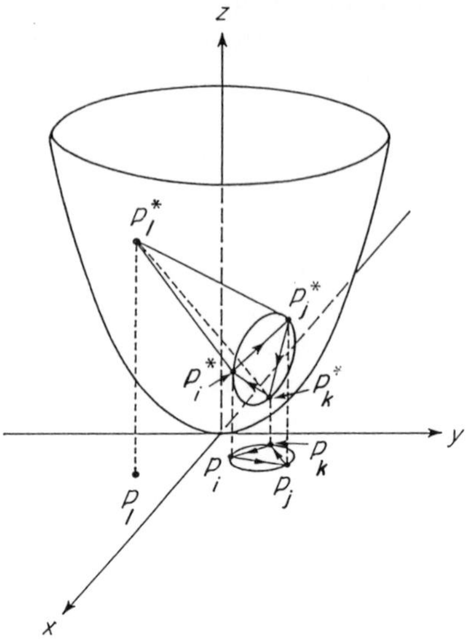

Figure 2.4.8 A Delaunay triangulation obtained through the lift-up transformation.

recall that the three points p_i, p_j, p_k form a Delaunay triangle if and only if the circle passing through these points is an empty circle (Property D6).

$H(p_i, p_j, p_k, p)$ gives another implication. The value of $H(p_i, p_j, p_k, p_l)/6$ can be interpreted as the signed volume of the tetrahedron with vertices p_i, p_j, p_k, p_l. Since we choose p_i, p_j, p_k so that the cycle $p_ip_jp_kp_i$ has a counterclockwise direction, the volume of the tetrahedron is positive if and only if p_l^* is above the plane containing p_i^*, p_j^*, p_k^* (see Figure 2.4.8). This means that p_i, p_j and p_k form a Delaunay triangle if and only if p_i^*, p_j^* and p_k^* give a facet of the lower boundary of CH(P^*). Thus we obtain the following property.

Property D8 The Delaunay triangulation $\mathcal{D}(P)$ is obtained as the orthographic projection of the lower boundary of CH(P^*) onto the x–y plane, where P^* is the lift up of P.

This property gives an interesting implication to the interpolation problem (Chapter 6). Suppose that we are given the height $z_i = x_i^2 + y_i^2$ at each site $p_i = (x_i, y_i)$, and that we want to interpolate these data by a collection of triangles whose vertices are the given points p_1^*, ..., p_n^*, where p_i^* is the lift-up of p_i. Every triangulation spanning P gives such an interpolation, but among them, the Delaunay triangulation is the only one that gives the surface of interpolation convex downward. Recall that the points in P^* lie on the paraboloid of revolution, which is convex downward. It is hence desirable that the interpolant is also convex downward. This requirement is satisfied only by the Delaunay triangulation. Since this property is a distinct property, we state it as Property D9.

Property D9 The Delaunay triangulation is the only triangulation $\mathcal{T}(P)$ spanning P such that the lift-up transformation of $\mathcal{T}(P)$ gives a surface convex downward.

Property D9 gives a justification for the use of the Delaunay triangulation as a tool for interpolation. Applications of Voronoi diagrams and Delaunay triangulations to interpolation will be discussed in detail in Chapter 6.

Unlike a Voronoi diagram, a Delaunay pretriangulation or a Delaunay triangulation has edges of finite length. Thus we can regard a Delaunay pretriangulation or a Delaunay triangulation as a geometric graph. This graph is a connected planar graph consisting of the set of Delaunay vertices, P, and the set of Delaunay edges, E_d. We call this graph the *Delaunay graph* of $\mathcal{D}(P)$ and denote it by $G(P, E_d)$. It is easy to see from Definition D2 that the Voronoi graph and the Delaunay graph have the following relation.

Property D10 The Delaunay graph $G(P, E_d)$ is the dual graph of the Voronoi graph $G(Q_{+1}, E_b)$.

Recalling Definitions V3 and D1, we notice that a Voronoi edge and its corresponding Delaunay edge are orthogonal under the non-degeneracy assumption. When two geometric figures form the primal–dual graphs, and their corresponding edges are orthogonal, these two figures are called *reciprocal figures* (Maxwell, 1864). A Voronoi diagram generated by P and the Delaunay triangulation spanning P are hence reciprocal figures. This property is strongly related to the realizability of polyhedra from the Delaunay triangulation and the Voronoi diagram (Properties D8 and V17). The relation between realizability and reciprocity is discussed by Sugihara (1986).

From Properties D4 and D10 we can obtain a few relationships among the numbers of Delaunay triangles, Delaunay edges, external Delaunay edges and Delaunay vertices, denoted by n_v, n_e, n_c and n, respectively. First, since the Delaunay graph is a planar graph, Euler's formula (equation (1.3.61)) holds, i.e. $n - n_e + (n_v + 1) = 2$, which is equivalent to Property V9. Second, more generally, for an m-dimensional Delaunay tessellation, Euler–Shaefli's formula (equation (1.3.62)) holds, which corresponds to Property V10. Third, for a Delaunay triangulation $\mathcal{D}(P)$ spanning P with the non-cocircularity assumption, let n_v be the number of Delaunay triangles and n_c be that of Delaunay triangles that have external edges (equivalently n_c is the number of infinite Voronoi polygons in $V(P)$). Since every internal edge is shared by two Delaunay triangles and every external edge belongs to only one Delaunay triangle, the number of Delaunay edges is given by

$$n_e = \tfrac{1}{2}(3n_v + n_c) \ . \tag{2.4.12}$$

Upon substituting equation (2.4.12) into $n - n_e + n_v = 1$, we have the following property.

Property D11 For the Delaunay triangulation $\mathcal{D}(P)$ spanning P, which satisfies the non-cocircularity assumption, let n_e be the number of Voronoi or Delaunay edges, n_v be the number of the triangles in $\mathcal{D}(P)$ and n_c be the number of the vertices on the boundary of CH(P). Then, the following equations hold:

$$n_v = 2n - n_c - 2 \ , \tag{2.4.13}$$

$$n_e = 3n - n_c - 3 \ . \tag{2.4.14}$$

Recalling Property D4, we notice that the above equations hold for a non-degenerate Voronoi diagram $\mathcal{V}(P)$, where n_e, n_v and n_c are the numbers of Voronoi edges, Voronoi vertices and infinite Voronoi polygons, respectively.

In the same way that a Delaunay triangulation can be regarded as a geometric graph, any triangulation can be regarded as a geometric graph with the vertices and edges of the triangles. Figure 2.4.9(a) shows an example in which the geometric graph $G(P, L)$ consists of a set of nine points denoted by $P = \{p_0, p_1, \ldots, p_4, p_1', \ldots, p_4'\}$ and a set of 16 line segments denoted by $L = \{L_1, \ldots, L_8, L_1', \ldots, L_8'\}$ (they are labelled as in Figure 2.4.9(a)). We now ask whether or not there exists a non-degenerate Delaunay triangulation whose geometric graph is isomorphic to the geometric

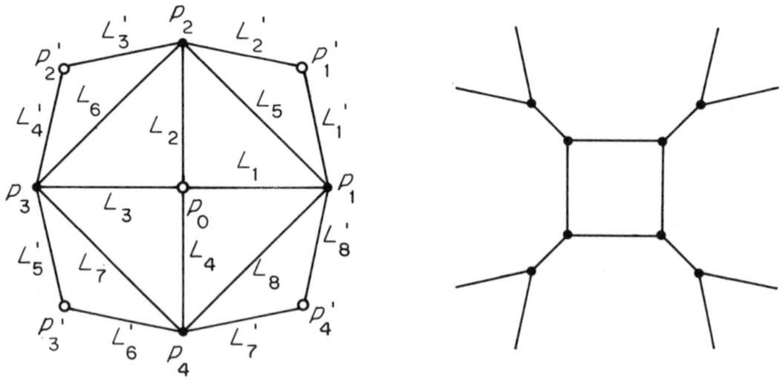

(a) (b)

Figure 2.4.9 (a) A triangulation whose geometric graph is not isomorphic to any non-degenerate Delaunay triangulation, and (b) its dual diagram whose geometric graph is not isomorphic to any non-degenerate Voronoi diagram.

graph $G(P, L)$ in Figure 2.4.9(a). More generally, we ask whether or not there exists a non-degenerate Delaunay triangulation whose geometric graph is isomorphic to the geometric graph of a given triangulation.

To answer the first question we construct a geometric subgraph, $G_s(P_s, L_s)$, consisting of $P_s = \{p_0, p_1, \ldots, p_4\}$ and $L_s = \{L_1, \ldots, L_8\}$ in such a way that $L_5, \ldots, L_8$ and $p_1, \ldots, p_4$ form an arbitrary convex quadrangle and the point p_0 is placed in its interior (Figure 2.4.10). Obviously subgraph $G_s(P_s, L_s)$ is isomorphic to the corresponding subgraph in Figure 2.4.9(a). We next add four points $p'_1, \ldots, p'_4$ and eight line segments $L'_1, \ldots, L'_8$ to $G_s(P_s, L_s)$ so that the resulting geometric graph is isomorphic to the geometric graph $G(P, L)$ in Figure 2.4.9(a). The number of such geometric graphs is infinitely many. To examine which geometric graph is a Delaunay triangulation, we construct the circle C_i which passes through p_0, p_i, p_{i+1} for $i = 1, \ldots, 4$ (p_5 is read as p_1), and consider the following three possible cases relative to the circle C_i.

First, suppose that we place p'_i in the region encircled by $\overline{p_i p_{i+1}}$ and the arc from p_i to p_{i+1} of C_i that does not contain p_0, and let $L'_{2i-1} = \overline{p_i p'_i}$, $L'_{2i} = \overline{p'_i p_{i+1}}$, $i = 1, \ldots, 4$ (the broken lines in Figure 2.4.10). Then $\triangle p_i p'_i p_{i+1}$ does not satisfy the empty circle criterion. Thus, the resulting triangulation is not a Delaunay triangulation.

Second, suppose that p'_i is on the arc of C_i from p_i to p_{i+1} that does not contain p_0 (the dash-dot lines in Figure 2.4.10). Then the triangles in the

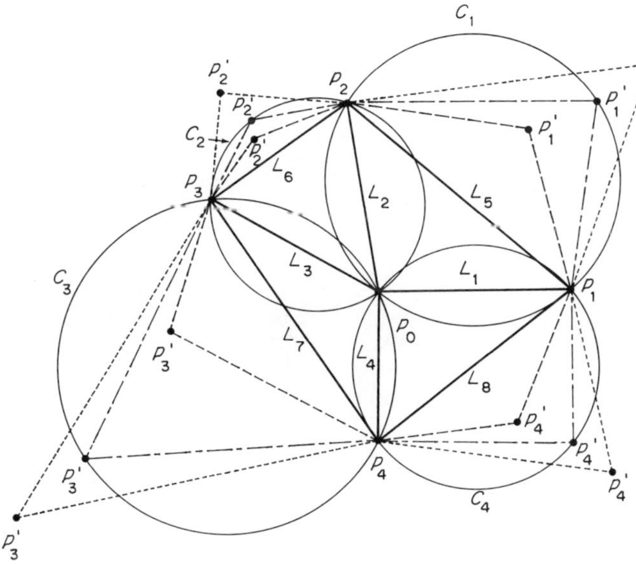

Figure 2.4.10 Three triangulations (the solid lines with the broken lines, the dash-dot lines, or the dot lines).

resulting triangulation satisfy the empty circle criterion. If the octagon with vertices $p_1, p_1', \ldots, p_4, p_4'$ is convex, this triangulation is a Delaunay triangulation. At first glance the convexity is not satisfied in general, but it may be satisfied in a special case. To examine the convexity, we construct a triangulation by adding $p_1', \ldots, p_4'$, $L_1', \ldots, L_4'$ in the following way.

Choose a point p_1' on C_1 such that p_0 and p_1' are on mutually opposite sides of the line $\overline{p_1p_2}$ (Figure 2.4.11). For $i = 2, \ldots, 4$, let p_i' be the point (other than p_i) of intersection between C_i and the line $\overline{p_{i-1}'p_i}$ (the broken lines), $i = 2, \ldots, 4$. We now show that p_1 is on the line $\overline{p_1'p_4'}$. Let $\alpha_i = \angle p_ip_i'p_{i+1}$ and $\beta = \angle p_ip_0p_{i+1}$ for $i = 1, \ldots, 4$, where $p_5 = p_1$. Since p_0, p_i, p_i', p_{i+1} are on C_i, we get $\alpha_i + \beta_i = \pi$ for $i = 1, \ldots, 4$. Substituting these equations into $\beta_1 + \cdots + \beta_4 = 2\pi$, we obtain $\alpha_1 + \cdots + \alpha_4 = 2\pi$. On the other hand, the sum of the inner angles of the pentagon with vertices $p_1', p_2', \ldots, p_5', p_1$ is 3π, i.e. $\alpha_1 + \cdots + \alpha_4 + \angle p_4'p_1p_1' = 3\pi$. From these equations we have the equation $\angle p_4'p_1p_1' = \pi$. Hence p_1, p_1' and p_4' are on the same line (note that this property can be easily generalized for m-gon $p_1, \ldots, p_m$ and $p_1', \ldots, p_m'$ obtained in the same manner). Thus the polygon with vertices $p_1, p_1', p_2, p_2', \ldots, p_4, p_4'$ is a convex quadrangle. Since the triangulation in Figure 2.4.11 is a triangulation spanning the convex quadrangle $p_1', \ldots, p_4'$ and every triangle satisfies the empty circle criterion, the triangulation in Figure 2.4.11 is the Delaunay triangulation. The graph $G(P, L)$ in Figure 2.4.9(a) is hence isomorphic to the Delaunay triangulation in Figure 2.4.11. However, the Delaunay triangulation is degenerate because every circle C_i has four points in P on it.

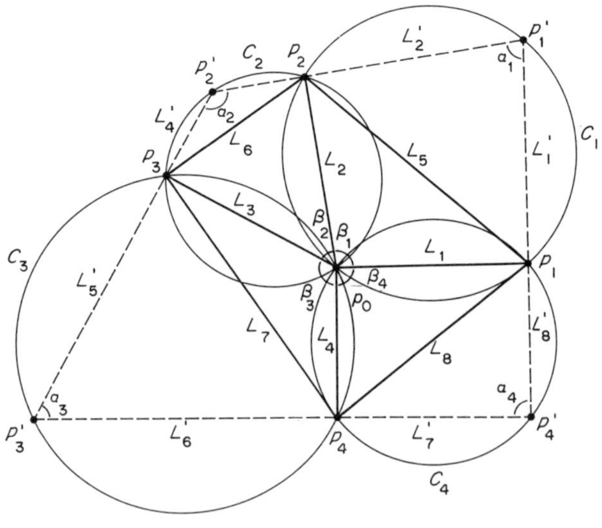

Figure 2.4.11 The degenerate Delaunay triangulation isomorphic to the geometric graph in Figure 2.4.9(a).

Last, suppose that p_i' is placed outside C_i in such a way that the geometric graph of the resulting triangulation is isomorphic to the geometric graph $G(P, L)$ (the dot lines in Figure 2.4.10). In this case every triangle in the triangulation satisfies the empty circle criterion, but using the triangulation in Figure 2.4.11 we can show that the octagon with vertices $p_1, p_1', \ldots, p_4, p_4'$ cannot be convex.

To sum up, we come to the conclusion that there does not exist a non-degenerate Delaunay triangulation whose geometric graph is isomorphic to the geometric graph of the triangulation in Figure 2.4.9(a). This conclusion in turn implies that the diagram shown in Figure 2.4.9(b), which is the dual of the diagram in Figure 2.4.9(a), is not isomorphic to any non-degenerate Voronoi diagram. Similarly, using the triangulation shown in Figure 2.4.11, we can show that there does not exist a Delaunay triangulation whose geometric graph is isomorphic to the geometric graph shown in Figure 2.4.12(a). Consequently, the diagram shown in Figure 2.4.11(b), which is the dual of the diagram shown in Figure 2.4.12(a), is not isomorphic to any Voronoi diagram.

Dillencourt (1990a, 1990b) investigated these unrealizability properties more generally, and showed the following two properties.

Property D12 Triangulation $\mathcal{T}$ is not isomorphic to a non-degenerate Delaunay triangulation if there exists a subset $P' \subset P$ such that the subgraph of $\mathcal{T}$ induced by the vertex set $P \backslash P'$ contains more than $|P|$ connected components.

Property D13 Triangulation $\mathcal{T}$ is not isomorphic to a Delaunay triangulation if there exists a subset $P' \subset P$ such that the subgraph of $\mathcal{T}$ induced by the

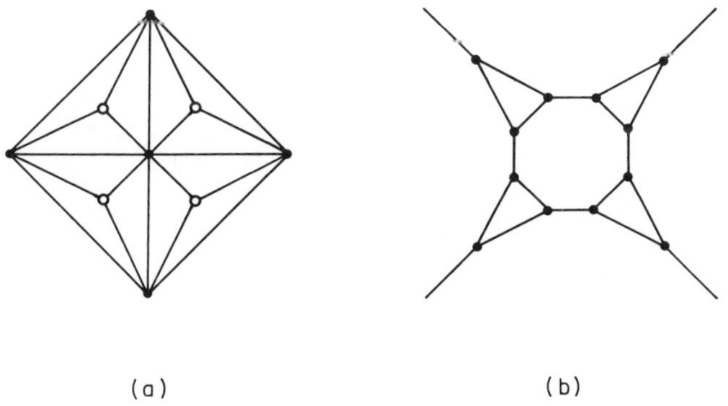

Figure 2.4.12 (a) A triangulation whose geometric graph is not isomorphic to any Delaunay triangulation, and (b) its dual diagram whose geometric graph is not isomorphic to any Voronoi diagram.

vertex set $P \backslash P'$ contains more than $|P| - 2$ connected components that do not contain any vertex on the boundary of $\mathcal{T}$.

The proofs are provided by Dillencourt (1990a).

We can recognize the unrealizability of the triangulation in Figure 2.4.9(a) by Property D12. Indeed, if we remove the four vertices represented by the filled circles, we get five connected components, each consisting of an isolated vertex represented by an unfilled circle. We can also recognize the unrealizability of the triangulation in Figure 2.4.12(a) by Property D13. If we remove the five vertices represented by the filled circles, we get four connected components each of which consists of an isolated vertex (represented by an unfilled circle) that is not on the boundary of the triangulation.

Properties D12 and D13 present only the sufficient conditions for a triangulation not to be isomorphic to a Delaunay triangulation. Characterization of the necessary and sufficient conditions in combinatorial terms, however, is still an open problem (Dillencourt, 1990a, 1990b).

For a given set of points P we have many possible triangulations of CH(P) spanning P. In some applications we want to choose a triangulation in which triangles are as closely equiangular as possible (practical examples will be shown in Chapter 6). One of the criteria is to choose a triangulation in which the minimum angle in each triangle is as large as possible. To state this criterion more explicitly, let us consider an internal edge $\overline{p_{i1}p_{i2}}$ in a triangulation $\mathcal{T}$, and let $\triangle p_{i1}p_{i2}p_{i3}$ and $\triangle p_{i1}p_{i2}p_{i4}$ be triangles sharing the edge $\overline{p_{i1}p_{i2}}$ (Figure 2.4.13(a), (b)). The quadrangle $p_{i1}p_{i3}p_{i2}p_{i4}$ may be non-convex (Figure 2.4.13(a)) or convex (Figure 2.4.13(b)). If it is convex and it does not degenerate into a triangle (p_{i1} is on $\overline{p_{i3}p_{i4}}$ or p_{i2} is on $\overline{p_{i3}p_{i4}}$), we have another possible triangulation, i.e. $\triangle p_{i1}p_{i3}p_{i4}$ and $\triangle p_{i2}p_{i3}p_{i4}$ (Figure 2.4.13(c)). We are concerned with which triangulation is locally better ('locally' in the sense that a triangulation is made in a local area, i.e.

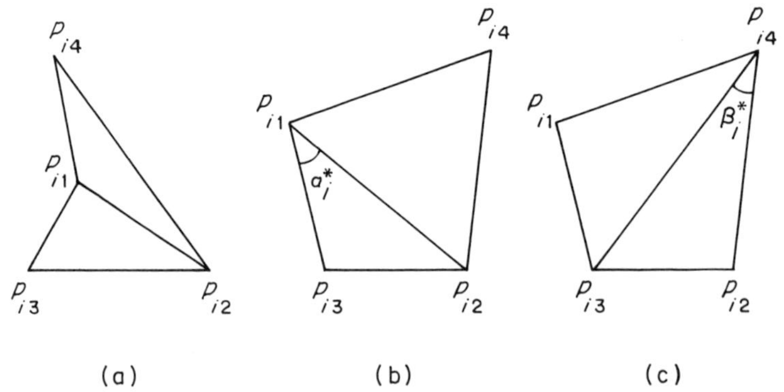

Figure 2.4.13 The local max-min angle criterion.

quadrangle $p_{i1}p_{i3}p_{i2}p_{i4}$). In the triangulation in panel (b), the minimum angle among six angles in $\triangle p_{i1}p_{i2}p_{i3}$ and $\triangle p_{i1}p_{i2}p_{i4}$ is $\angle p_{i2}p_{i1}p_{i3} = \alpha_i^*$. In the triangulation in panel (c), the minimum angle among the six angles in $\triangle p_{i1}p_{i3}p_{i4}$ and $\triangle p_{i2}p_{i3}p_{i4}$ is $\angle p_{i2}p_{i4}p_{i3} = \beta_i^*$. Comparing α_i^* and β_i^* in Figures 2.4.13(b) and 2.4.13(c), we notice that $\alpha_i^* > \beta_i^*$, or $\alpha_i^* = \max\{\alpha_i^*, \beta_i^*\}$. We may thus conclude that the triangulation in panel (b) is locally better than that in panel (c), because the minimum angle is maximized in the triangulation in panel (b). This criterion may be written generally as follows.

The local max-min angle criterion For a triangulation $\mathcal{T}$ of CH(P) spanning P, let $\overline{p_{i1}p_{i2}}$ be an internal edge in CH(P), and $\triangle p_{i1}p_{i2}p_{i3}$ and $\triangle p_{i1}p_{i2}p_{i4}$ be two triangles sharing the edge $\overline{p_{i1}p_{i2}}$. For a convex quadrangle $p_{i1}p_{i3}p_{i2}p_{i4}$ which does not degenerate into a triangle, let α_{ij}, $j \in I_6$, be six angles in $\triangle p_{i1}p_{i2}p_{i3}$ and $\triangle p_{i1}p_{i2}p_{i4}$; β_{ij}, $j \in I_6$, be six angles in $\triangle p_{i1}p_{i3}p_{i4}$ and $\triangle p_{i2}p_{i3}p_{4}$. If quadrangle $p_{i1}p_{i3}p_{i2}p_{i4}$ is non-convex or it degenerates into a triangle, or if it is a convex quadrangle which does not degenerate into a triangle and the relation

$$\min_j\{\alpha_{ij}, j \in I_6\} \geq \min_j\{\beta_{ij}, j \in I_6\} \tag{2.4.15}$$

holds, then we say that the edge $\overline{p_{i1}p_{i2}}$ satisfies the *local max-min angle criterion*.

In Figure 2.4.13 the edge $\overline{p_{i1}p_{i2}}$ in panels (a) and (b) satisfies the local max-min angle criterion, but the edge $\overline{p_{i3}p_{i4}}$ in panel (c) does not.

At first glance the practical operation in equation (2.4.15) (measuring angles and finding the minimum angles among them) appears a little complicated. In practice, we do not carry out such an operation but use the following relation.

Let $\alpha_i^* = \min_j\{\alpha_{ij}, j \in I_6\}$, and suppose, without loss of generality, that α_i^* is one of the angles of $\triangle p_{i1}p_{i2}p_{i3}$. Let C_i be the circumcircle of $\triangle p_{i1}p_{i2}p_{i3}$; H be the open half plane made by the line containing $\overline{p_{i1}p_{i2}}$ that does not contain $\triangle p_{i1}p_{i2}p_{i3}$; and B be the region indicated by the shaded region (including the boundary) in Figure 2.4.14(a). Obviously, p_{i4} is in H. Quadrangle $p_{i1}p_{i3}p_{i2}p_{i4}$ may be convex or may not. If p_{i4} is in B, quadrangle $p_{i1}p_{i3}p_{i2}p_{i4}$ is non-convex or it degenerates into a triangle ($\triangle p_{i4}p_{i2}p_{i3}$ or $\triangle p_{i4}p_{i1}p_{i3}$). Quadrangle $p_{i1}p_{i3}p_{i2}p_{i4}$ is a convex quadrangle which does not degenerate into a triangle if $p_{i4} \in H \backslash B$. Now suppose that p_{i4} is in $H \backslash [B \cup \mathrm{CH}(C_i)]$ and the minimum angle α_i^* is either $\angle p_{i3}p_{i1}p_{i2}$ (Figure 2.4.1(a)) or $\angle p_{i3}p_{i2}p_{i1}$. Let $\triangle p_{i1}p_{i3}p_{i4}$ and $\triangle p_{i2}p_{i3}p_{i4}$ be triangles constituting another triangulation of the quadrangle $p_{i1}p_{i3}p_{i2}p_{i4}$, and β_{ij}, $j \in I_6$, be angles of those triangles indicated in Figure 2.4.14(b). Using the theorem of equi-angles on a circle (see the two α_i^*'s in Figure 2.4.14(a)), we notice that

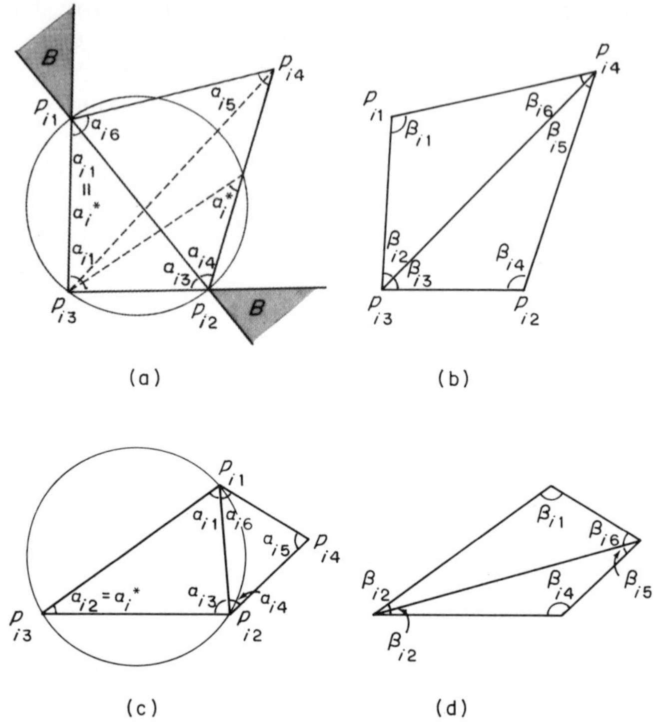

Figure 2.4.14 Conditions for the local max-min angle criterion.

$\alpha_i^* = \min_j\{\alpha_{ij}, j \in I_6\} = \alpha_{i1}$ (or α_{i3}) $> \beta_{i5}$ (or β_{i6}) $\geq \min_j\{\beta_{ij}, j \in I_6\} = \beta_i^*$. If the minimum angle α_i^* is $\angle p_{i1}p_{i3}p_{i2}$ as in Figure 2.4.14(c), $\alpha_j^* = \underline{\min_j}\{\alpha_{ij}, j \in I_6\} = \alpha_{i2} > \beta_{i2} \geq \min_j\{\beta_{ij}, j \in I_6\} = \beta_i^*$. Therefore the edge $\overline{p_{i1}p_{i2}}$ satisfies the local max-min angle criterion. Almost in the same manner, we can show that if $p_{i4} \in H \cap C_i$, then $\alpha_i^* = \beta_i^*$ holds, and if $p_{i4} \in H \cap [\mathrm{CH}(C_i) \backslash C_i]$, then $\alpha_i^* < \beta_i^*$ holds. Therefore we obtain the following relation:

$$\begin{aligned} &\min_j\{\alpha_{ij}, j \in I_6\} > \min_j\{\beta_{ij}, j \in I_6\} \quad \text{if } p_{i4} \text{ is outside } C_i\,, \\ &\min_j\{\alpha_{ij}, j \in I_6\} = \min_j\{\beta_{ij}, j \in I_6\} \quad \text{if } p_{i4} \text{ is on } C_i\,, \qquad (2.4.16) \\ &\min_j\{\alpha_{ij}, j \in I_6\} < \min_j\{\beta_{ij}, j \in I_6\} \quad \text{if } p_{i4} \text{ is inside } C_i \end{aligned}$$

(Lawson, 1977; Kishimoto, 1978; Sibson, 1978; Lee and Schachter, 1980). Note that the choice of a triangle from the two triangles $\triangle p_{i1}p_{i2}p_{i3}$ and $\triangle p_{i1}p_{i2}p_{i4}$ does not affect the result (it is an exercise of elementary geometry to prove that the circumcircle of $\triangle p_{i1}p_{i2}p_{i3}$ does not contain p_{i4} if and only if the circumcircle of $\triangle p_{i1}p_{i2}p_{i4}$ does not contain p_{i3}).

Using relation (2.4.16), we can prove the following property.

Property D14 (local max-min angle theorem) Let P be a set of distinct points $P = \{p_1, \ldots, p_n\} \subset \mathbb{R}^2$ $(3 \leq n < \infty)$ satisfying the non-cocircularity assumption, and $\mathcal{T}(P)$ be a triangulation of CH(P) spanning P. Every internal edge in $\mathcal{T}(P)$ satisfies the local max-min criterion if and only if $\mathcal{T}(P)$ is the Delaunay triangulation spanning P.

Proof From relation (2.4.16) it is obvious that if $\mathcal{T}(P)$ is the Delaunay triangulation spanning P, every internal edge satisfies the local max-min criterion, because if the circumcircle of $\triangle p_{i1}p_{i2}p_{i3}$ is an empty circle, p_{i4} is outside of the circumcircle. We shall prove that if every internal edge in $\mathcal{T}(P)$ satisfies the local max-min angle criterion, $\mathcal{T}(P)$ is the Delaunay triangulation spanning P. Since the edge $\overline{p_{i1}p_{i2}}$ satisfies the local max-min angle criterion, p_{i4} is outside C_i. What is left to prove is that all other vertices of the triangles are outside C_i (Figure 2.4.15). Since the edges $\overline{p_{i2}p_{i4}}$ and $\overline{p_{i1}p_{i4}}$ satisfy the local max-min angle criterion, there are no points in the horizontally and vertically hatched regions in Figure 2.4.15. Similarly, since the edges $\overline{p_{i1}p_{i3}}$ and $\overline{p_{i2}p_{i3}}$ satisfy the local max-min angle criterion, there are no points in the diagonally hatched regions. Obviously there are no points in $\triangle p_{i1}p_{i2}p_{i3}$ and $\triangle p_{i1}p_{i2}p_{i4}$. Therefore there are no points in the circumcircle of $\triangle p_{i1}p_{i2}p_{i3}$. Applying the same procedure to every triangle, we can prove that every circumcircle is an empty circle. From the empty circumcircle theorem (Property D6), we see that the triangulation spanning P is the Delaunay triangulation spanning P. □

Having understood Property D14 in theory, let us now consider a practical method for finding a triangulation satisfying the condition in Property D14 with an example shown in Figure 2.4.16.

We first construct an arbitrary triangulation of CH(P) spanning P, which is indicated by the continuous lines in Figure 2.4.16(a). Let E be the set of edges in this triangulation. We choose an edge, say $\overline{p_3p_5}$ (the heavy continuous lines in Figure 2.4.16(a)), from $E \backslash \partial\text{CH}(P)$ and examine if this edge satisfies the local max-min angle criterion. Since $\alpha^* > \beta^*$ in Figure

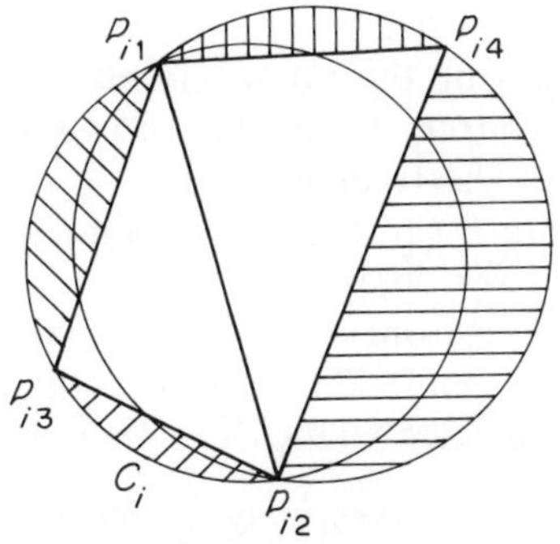

Figure 2.4.15 Illustration of the proof of Property D14.

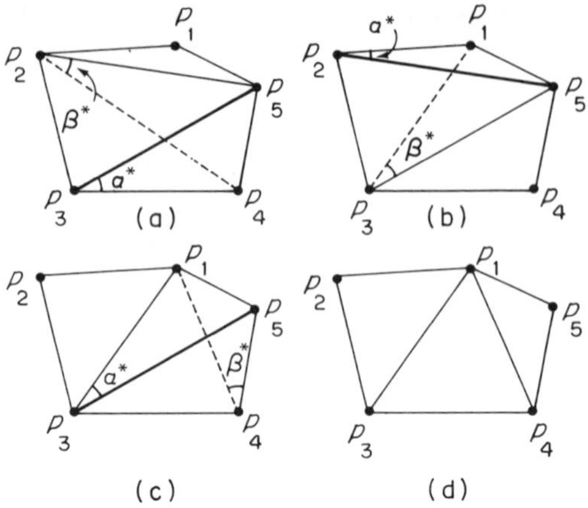

Figure 2.4.16 The swapping procedure.

2.4.16(a), the edge satisfies it. We next choose another edge, $\overline{p_2p_5}$ (the heavy continuous line segment in Figure 2.4.16(b)), from $E\backslash\mathrm{CH}(P)$, and examine if this edge satisfies the local max-min angle criterion. Since $\alpha^* < \beta^*$ in Figure 2.4.16(b), the edge does not satisfy it. In this case we swap $\overline{p_2p_5}$ (the heavy continuous line in Figure 2.4.16(b)) for $\overline{p_1p_3}$ (the broken line in Figure 2.4.16(b)), and replace E with $(E\backslash\{\overline{p_2p_5}\}) \cup \{\overline{p_1p_3}\}$. We next choose an edge, say $\overline{p_3p_5}$ (the heavy continuous line in Figure 2.4.16(c)), from $E\backslash\partial\mathrm{CH}(P)$. Since $\alpha^* < \beta^*$ in Figure 2.2.6(c), we swap $\overline{p_3p_5}$ (the heavy line in Figure 2.4.16(c)) for $\overline{p_1p_4}$ (the broken line in Figure 2.4.16(c)), and replace E with $(E\backslash\{\overline{p_3p_5}\}) \cup \{\overline{p_1p_4}\}$. We continue this procedure until we do not have to swap any edge in E. In the example of Figure 2.4.16(d), both internal edges satisfy the local max-min angle criterion, and so we stop. From Property D14, we notice that this triangulation is the Delaunay triangulation spanning P.

We now state the above procedure in a general form as follows.

The swapping procedure Let $P = \{p_1, \ldots, p_n\} \subset \mathbb{R}^2$ $(3 \leq n < \infty)$ be a set of n distinct points satisfying the non-cocircularity assumption.
Step 0. Construct an arbitrary triangulation of CH(P) spanning P, and let E be the set of edges in the triangulation.
Step 1. If all the edges in E satisfy the local max-min angle criterion, report E and stop; otherwise go to Step 2.
Step 2. Choose and delete from E an edge e that does not satisfy the local max-min angle criterion, and add to E the other diagonal edge of the quadrangle formed by the two triangles containing e, and go to Step 1.

We call this procedure the *swapping procedure.* Note that the above procedure is alternatively called the *local(ly) optimal procedure* in Lee and

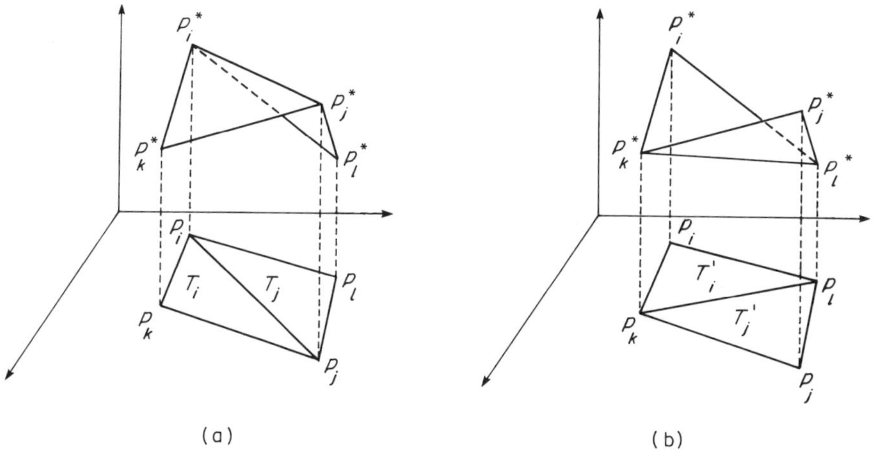

Figure 2.4.17 Two triangles (a) bending downward along the common edge and (b) those bending upward along the common edge.

Schachter (1980) or the *locally equiangular triangulation* in Sibson (1978).

The process of the swapping procedure can be easily understood by the lift-up transformation (Edelsbrunner, 1988). As shown in Figure 2.4.17, let $\mathcal{T} = \{T_1, \ldots, T_n\}$ be an arbitrary triangulation spanning P, and let $\mathcal{T}^* = \{T_1^*, \ldots, T_n^*\}$ be the surface formed by the collection of triangles obtained from $\mathcal{T}$ by the lift-up transformation (T_i^* is the lift-up of T_i). As we saw in Property D9, if $\mathcal{T}$ is not the Delaunay triangulation, the surface $\mathcal{T}^*$ is not convex downward. Let T_i^* and T_j^* be two triangles in $\mathcal{T}^*$ such that the triangles share the edge $\overline{p_i^* p_j^*}$ and they bend downward along the edge $\overline{p_i^* p_j^*}$ as in Figure 2.4.17(a). Such triangles make the surface $\mathcal{T}^*$ non-convex.

Let p_i, p_j, p_k be the vertices of T_i and p_i, p_j, p_l be the vertices of the triangle T_j. Let T_i' and T_j' be two triangles created by swapping the edge $\overline{p_i p_j}$ with the other diagonal edge $\overline{p_k p_l}$. By this swapping, we get a new pair of triangles $T_i'^*$ and $T_j'^*$ in $\mathbb{R}^3$ that bend upward along the edge $\overline{p_k^* p_l^*}$ as in Figure 2.4.17(b). Such triangles contribute to the convex surface $\mathcal{T}^*$. It is easy to see that by this swapping the edge shared by the resulting triangles satisfies the local max-min angle criterion. Hence swapping an edge according to the local max-min angle criterion corresponds to the procedure for changing the surface from $\mathcal{T}^*$ to $\mathcal{T}'^*$ by replacing two upper facets of the tetrahedron having the vertices $p_i^*, p_j^*, p_k^*, p_l^*$ with two lower facets. Intuitively, this procedure corresponds to gluing the tetrahedron on the surface $\mathcal{T}$ from below; the new lower surface gives $\mathcal{T}'^*$. We apply this replacement to any edge on the surface $\mathcal{T}^*$. Consequently, the swapping procedure terminates only when the lift-up surface $\mathcal{T}^*$ becomes convex downward. From this property and Property D9 we obtain the following property.

Property D15 The swapping procedure based upon the local max-min angle criterion always terminates with the Delaunay triangulation.

We know that the number of tetrahedra in the three-dimensional tessellation spanning n points is of order n^2 (Edelsbrunner, 1987). Consequently, the Delaunay triangulation is obtained from any initial triangulation by at most of order n^2 repetitions of the swappings.

A Delaunay triangulation satisfies a little stronger property than Property D15. Let $\alpha_i^*(\mathcal{T}_j)$ be the minimum angle in the triangle T_i, $i \in I_{n_v}$, in a triangulation $\mathcal{T}_j$ of CH(P) spanning P. We reindex $\{\alpha_1^*(\mathcal{T}_j), \ldots, \alpha_{n_v}(\mathcal{T}_j)\}$ as $\{\alpha_{(1)}^*(\mathcal{T}_j), \ldots, \alpha_{(n_v)}^*(\mathcal{T}_j)\}$, where $\alpha_{(1)}^*(\mathcal{T}_j) \leq \cdots \leq \alpha_{(n_v)}^*(\mathcal{T}_j)$, and call $\mathbf{a}(\mathcal{T}_j)^{\mathrm{T}} = (\alpha_{(1)}^*(\mathcal{T}_j), \ldots, \alpha_{(n_v)}^*(\mathcal{T}_j))$ the *ordered minimum angle vector* of $\mathcal{T}_j$, or briefly the *angle vector* of $\mathcal{T}_j$. For two angle vectors $\mathbf{a}(\mathcal{T}_1)$ and $\mathbf{a}(\mathcal{T}_2)$, if there exists $k \leq n_v$ such that

$$\begin{aligned} &\alpha_{(i)}^*(\mathcal{T}_1) = \alpha_{(i)}^*(\mathcal{T}_2), \quad i = 1, \ldots, k-1 , \\ &\alpha_{(k)}^*(\mathcal{T}_1) < \alpha_{(k)}^*(\mathcal{T}_2) , \end{aligned} \tag{2.4.17}$$

we say that $\mathcal{T}_2$ is *lexicographically greater* than $\mathcal{T}_1$, and denote this relation by $\mathcal{T}_1 \lessdot \mathcal{T}_2$. In terms of this relation, we can now state the following property.

Property D16 (lexicographically maximum triangulation theorem) A triangulation spanning P is a lexicographically maximum with respect to the ordered minimum angle vector if and only if the triangulation is the Delaunay triangulation spanning P.

Proof Suppose that an edge is swapped according to the local max-min angle procedure, and the triangulation changes from $\mathcal{T}_1$ to $\mathcal{T}_2$. By this swapping, suppose that two triangles sharing the edge, say T_1 and T_2 in $\mathcal{T}_1$, change into T_1' and T_2' in $\mathcal{T}_2$ (the other triangles remain the same), and that $\alpha_1^*(\mathcal{T}_1)$ and $\alpha_2^*(\mathcal{T}_1)$ are $\alpha_{(k)}^*(\mathcal{T}_1)$ and $\alpha_{(h)}^*(\mathcal{T}_1)$ in the ordered minimum angle vector $\mathbf{a}(\mathcal{T}_1)$, $k < h$. Since the edge is swapped according to the local max-min angle procedure, the relation

$$\min\{\alpha_1^*(\mathcal{T}_1), \alpha_2^*(\mathcal{T}_1)\} < \min\{\alpha_1^*(\mathcal{T}_2), \alpha_2^*(\mathcal{T}_2)\} \tag{2.4.18}$$

holds (recall relation (2.4.16)), where $\alpha_i^*(\mathcal{T}_2)$ is the minimum angle in the triangle T_i', $i = 1, 2$. Hence $\alpha_{(i)}^*(\mathcal{T}_2) = \alpha_{(i)}^*(\mathcal{T}_1)$, $i = 1, \ldots, k-1$, and $\alpha_{(k)}^*(\mathcal{T}_1) < \alpha_{(k)}^*(\mathcal{T}_2)$, from which we obtain $\mathcal{T}_1 \lessdot \mathcal{T}_2$. Whenever an edge is swapped, the new triangulation becomes lexicographically greater than the old triangulation. Moreover, Property D15 shows that the swapping procedure always terminates with the Delaunay triangulation for any initial triangulation. Therefore the Delaunay triangulation is lexicographically maximum.

As is shown in Property D16, the Delaunay triangulation is the triangulation with the lexicographically largest angle vector. This intuitively means that the Delaunay triangulation avoids thin and elongated triangles as much as possible. Noticing this property, one may expect that the length of the shortest path on the Delaunay triangulation $\mathcal{D}(P)$ between any two points in P is a good approximation to the Euclidean distance. This expectation is indeed true. To be precise, let $D(p_i, p_j)$ be the length of the shortest chain connecting p_i and p_j composed of Delaunay edges in $\mathcal{D}(P)$, where the length of a Delaunay edge is the Euclidean distance between the two endpoints of the edge and the length of the chain is the sum of the length of the edges on the chain. Then the following property holds.

Property D17 For any p_i and p_j in P,

$$D(p_i, p_j) \leq c \cdot d(p_i, p_j) \qquad (2.4.19)$$

holds, where $d(p_i, p_j)$ is the Euclidean distance between p_i and p_j, and

$$c = \frac{2\pi}{3\cos(\pi/6)} \approx 2.42\ . \qquad (2.4.20)$$

This was proved by Keil and Gutwin (1989); a little weaker result, where $c = (1 + \sqrt{5})\pi/2 \approx 5.08$, was proved by Dobkin *et al.* (1990).

Recall that for n points there are $O(n^2)$ pairs of points whereas there are only $O(n)$ edges in the Delaunay triangulation. Hence, the network composed of the $O(n^2)$ direct routes between points and the points of their intersection can be efficiently approximated by the Delaunay triangulation, which requires only $O(n)$ storage.

The Delaunay triangulation has an interesting monotonicity property with respect to ray shooting. To illustrate it, let us consider a triangulation $\mathcal{T}$ spanning P, and let p be a point in $\mathbb{R}^2$. For any triangles T_i and T_j in $\mathcal{T}$, we write $T_i \lessdot_p T_j$ if there exists a ray starting at p such that a traveller traversing along the ray from p passes through T_i and T_j in this order (see Figure 2.4.18). If there is no cyclic sequence $(T_1, T_2, \ldots, T_k, T_1)$ of triangles

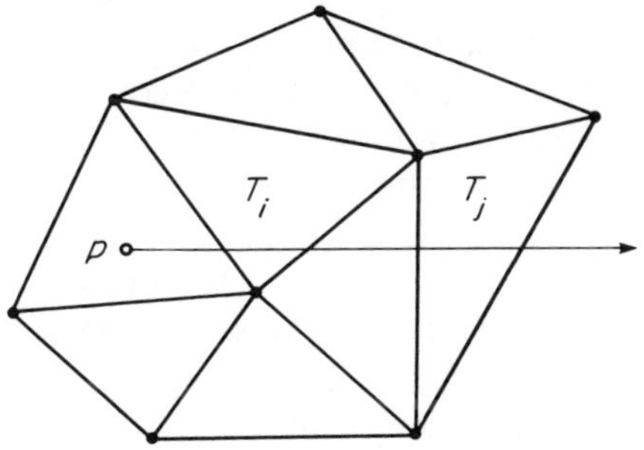

Figure 2.4.18 Monotonicity with respect to ray shooting.

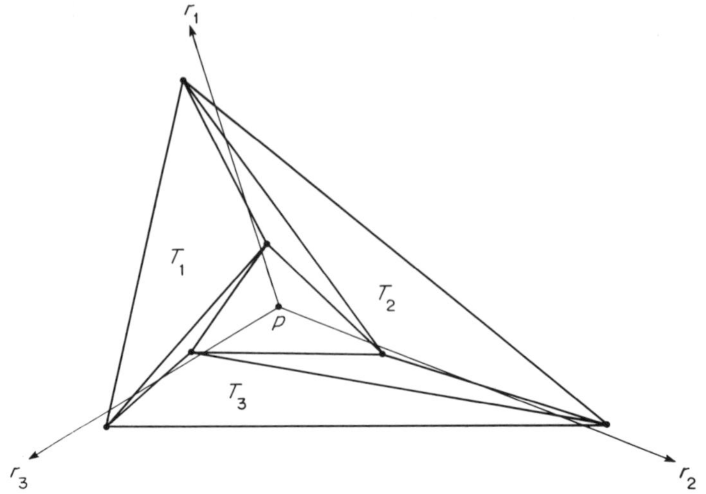

Figure 2.4.19 A triangulation not satisfying ray-shoot monotonicity.

satisfying $T_1 \lessdot_p T_2 \lessdot_p \cdots \lessdot_p T_k \lessdot_p T_1$, $k \geq 2$, we say that $\mathcal{T}$ is *monotone with respect to ray shooting at p*. If $\mathcal{T}$ is monotone with respect to ray shooting at any point in $\mathbb{R}^2$, we say that $\mathcal{T}$ is *ray-shoot monotone*.

The ray-shoot monotonicity is not necessarily satisfied by any triangulation. A counter example is shown in Figure 2.4.19. For the ray r_1, $T_1 \lessdot_p T_2$ holds; for the ray r_2, $T_2 \lessdot_p T_3$ holds; for the ray r_3, $T_3 \lessdot_p T_1$ holds. Thus $T_1 \lessdot_p T_2 \lessdot_p T_3 \lessdot_p T_1$, showing that the triangulation in Figure 2.4.19 is not monotone with respect to ray shooting at p. The Delaunay triangulation, however, has the following property.

Property D18 The Delaunay triangulation is ray-shoot monotone.

To show the outline of the proof, let $\mathcal{T}(P)$ be the Delaunay triangulation spanning P and p be any point in $\mathbb{R}^2$. For convenience, we transform the (x, y) coordinate system in such a way that p is at the origin, and construct the lower boundary of the convex hull of the lift-up of P, i.e. $\mathrm{CH}(P^*)$. For a Delaunay triangle T_i in $\mathbb{R}^2$, let T_i^* be the corresponding triangular facet on the boundary of $\mathrm{CH}(P^*)$, and $-\alpha_i$ be the z coordinate of the intersection point between the z-axis and the plane containing T_i^* (see Figure 2.4.20). Using the value of α_i, we define a binary relation, $<_p$: $T_i <_p T_j$ if and only if $\alpha_i \leq \alpha_j$. The binary relation $<_p$ gives a total order in the set of Delaunay triangles in $\mathcal{T}(P)$. Next consider a ray in $\mathbb{R}^2$ stating at the origin and passing through T_i and T_j in this order. Since the lower boundary of $\mathrm{CH}(P^*)$ lies on the paraboloid of revolution with the bottom at (0, 0, 0), the plane containing T_i^* cuts the z-axis at the higher point than the plane containing

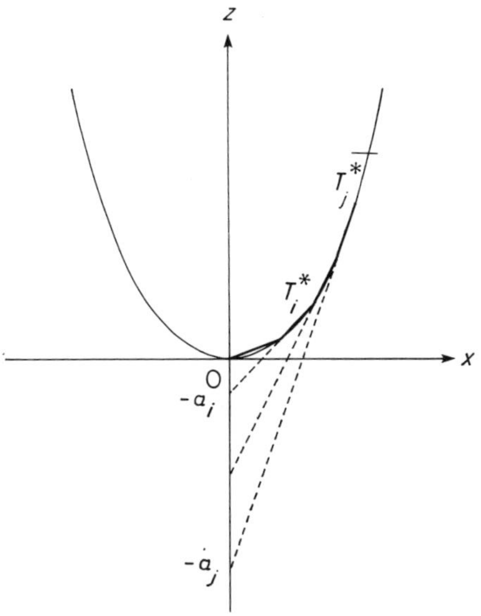

Figure 2.4.20 Monotonicity and order.

T_j^*, as shown in Figure 2.4.20. Thus, $T_i \leqslant_p T_j$ implies $T_i <_p T_j$. Since $<_p$ is a total order, we notice that $\mathcal{T}(P)$ is monotone with respect to ray shooting at p. This property holds for an arbitrary point p, and hence we obtain Property D18.

The ray-shoot property is a useful property when we consider problems related to visibility (De Floriani, 1989).

2.5 GRAPHS RELATED TO THE DELAUNAY TRIANGULATION

The Delaunay triangulation $\mathcal{D}(P)$ can be regarded as the connected geometric graph, called the Delaunay graph, consisting of the set of nodes given by P and the set of links given by the edges of Delaunay triangles. This Delaunay graph contains many subgraphs. Among them, notable subgraphs are the Gabriel graph, the relative neighbourhood graph, the Euclidean minimum spanning tree and the nearest neighbour graph. In this section we show the definitions of these graphs together with their geometric properties. Note that we assume throughout this section that P satisfies the non-cocircularity assumption, so that the Delaunay triangulation $\mathcal{D}(P)$ is defined uniquely for the set P of points.

First of all we notice from Property D2 that the boundary $\partial\mathrm{CH}(P)$ of the convex hull $\mathrm{CH}(P)$ is a subgraph of $\mathcal{D}(P)$. Figure 2.5.1 represents an example, where (a) shows a Delaunay triangulation for twelve points and (b) shows $\partial\mathrm{CH}(P)$ of the same set of points.

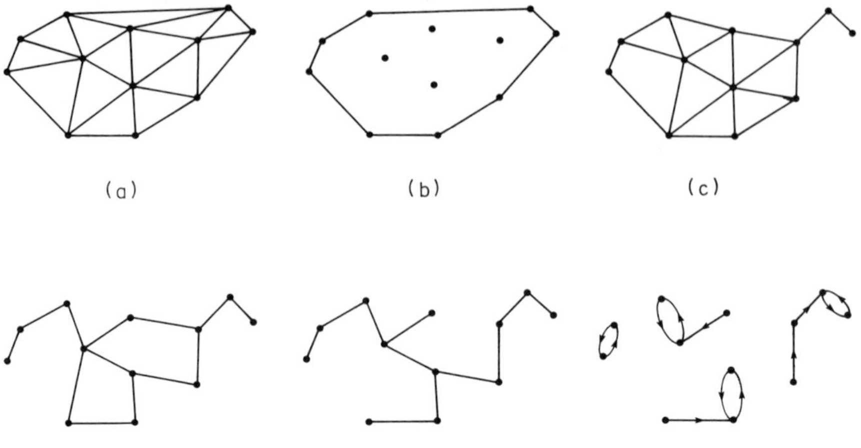

(a) (b) (c)

(d) (e) (f)

Figure 2.5.1 Graphs related to the Delaunay triangulation: (a) Delaunay triangulation; (b) convex hull; (c) Gabriel graph; (d) relative neighbourhood graph; (e) Euclidean minimum spanning tree; (f) nearest neighbourhood graph.

In Property D6 we characterized the Delaunay graph $\mathscr{D}(P)$ in terms of an empty circle. Similarly, we can characterize a subgraph of $\mathscr{D}(P)$, called the Gabriel graph, in terms of another type of empty circle. The *Gabriel graph*, denoted by GG(P), of P is defined by the graph in which $\overline{p_ip_j}$ is an edge of GG(P) if and only if the circle having $\overline{p_ip_j}$ as a diameter is an empty circle, that is, if and only if it contains no point of P in its interior (Gabriel and Sokal, 1969). Figure 2.5.1(c) presents the Gabriel graph for the same set of points as in (a). If $\overline{p_ip_j}$ is an edge of the Gabriel graph, p_i and p_j are said to be *Gabriel neighbours* of each other. Obviously, an edge of GG(P) is a Delaunay edge in $\mathscr{D}(P)$. It should be noted, however, that a Delaunay edge in $\mathscr{D}(P)$ is not necessarily an edge of GG(P).

To derive the necessary and sufficient conditions for an edge of GG(P) to be a Delaunay edge, let $\overline{p_ip_j}$ be an edge of GG(P), and C be the circle with diameter $\overline{p_ip_j}$, as shown in Figure 2.5.2. Let C' be the empty circle that passes through p_i, p_j and one more point, say p_k, in P. Since C' contains no point of P in its interior, the centre of C' is a Voronoi vertex shared by the Voronoi polygons of p_i, p_j and p_k (Property V7). Let C'' be the empty circle passing through p_i, p_j and one more point other than p_k, say p_l, such that the centre of C'' is on the other side of $\overline{p_ip_j}$ (see Figure 2.5.2). Then, the centre of C'' is also a Voronoi vertex, and the line segments connecting the centre of C' and that of C'' is a Voronoi edge shared by the two Voronoi polygons $V(p_i)$ and $V(p_j)$. Since the centres of C' and C'' are both on the perpendicular bisector of $\overline{p_ip_j}$, the line segment $\overline{p_ip_j}$ crosses the Voronoi edge shared by $V(p_i)$ and $V(p_j)$. In the above argument we implicitly assume

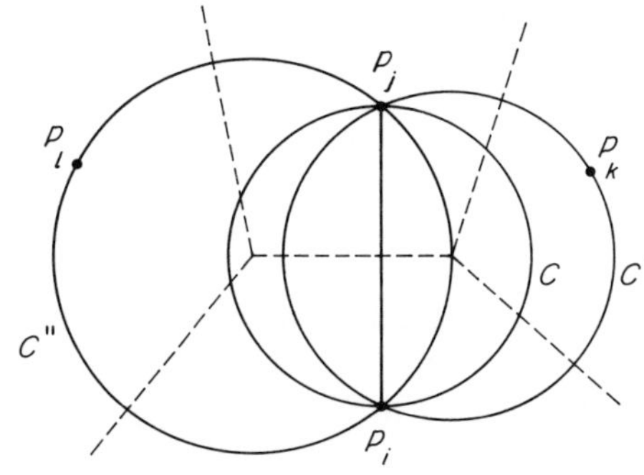

Figure 2.5.2 An edge of a Gabriel graph.

that both C' and C'' exist. If C' or C'' does not exist (which happens when p_i and p_j are consecutive vertices on the boundary of the convex hull of P), we can consider that the centre of C' or C'' is at infinity, and hence the above property still holds. Thus we get the next property.

Property D19 $\overline{p_ip_j}$ is an edge of GG(P) if and only if $\overline{p_ip_j}$ is a Delaunay edge in $\mathscr{D}(P)$ and $\overline{p_ip_j}$ crosses the corresponding Voronoi edge.

We can use GG(P) in defining adjacency in point patterns (Gabriel and Sokal, 1969; Matula and Sokal, 1980; Urquhart, 1982).

As is illustrated in Figure 2.5.3, the Gabriel graph is defined with the empty circle indicated by the broken lines. Modifying this empty circle, we may define a subgraph of $\mathscr{D}(P)$, called the relative neighbourhood graph, with an 'empty spindle', indicated by the shaded area in Figure 2.5.3. To be precise, the *relative neighbourhood graph*, denoted by RNG(P), is defined

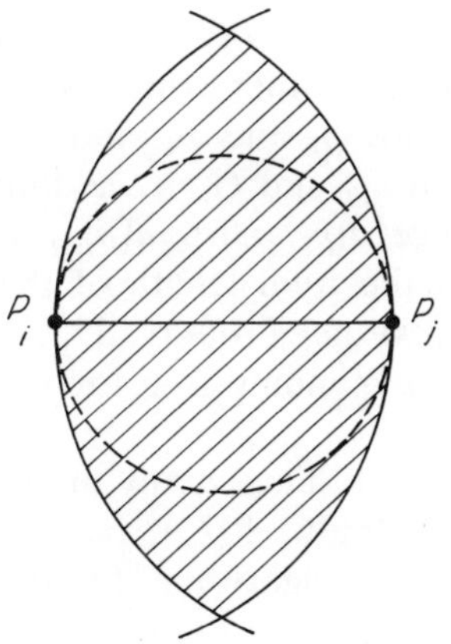

Figure 2.5.3 An edge of a relative neighbourhood graph.

as a geometric graph in which RNG(P) has an edge between p_i and p_j if and only if

$$d(p_i, p_j) \leq \min_{k(\neq i,j)} \max\{d(p_i, p_k), d(p_j, p_k)\} \tag{2.5.1}$$

(Toussaint, 1980a; Supowit, 1983). In other words, $\overline{p_ip_j}$ is an edge of RNG(P) if and only if there is no other point of P in the interior of the intersection of the disk with the centre at p_i and the radius $d(p_i, p_j)$ and the disk with the centre at p_j and the same radius (i.e. the open region represented by the shaded area in Figure 2.5.3). This region contains the interior of the circle (represented by the dashed line in Figure 2.5.3) having $\overline{p_ip_j}$ as a diameter, and hence we get the next property.

Property D20 An edge of RNG(P) is an edge of GG(P).

Figure 2.5.1(d) represents the relative neighbourhood graph for the same set of points as in (a). We can use RNG(P) as well as GG(P) for the analysis of point patterns (Toussaint, 1980a; Urquhart, 1982).

The subgraphs of graph $\mathcal{D}(P)$ include tree graphs. One of the most frequently referred to tree graphs is the Euclidean minimum spanning tree. The *Euclidean minimum spanning tree*, denoted by EMST(P), is defined as the tree having a vertex set P in which the sum of the Euclidean length of all the edges attains the minimum over all trees having the vertex set P. Regarding the relationship between an edge of EMST(P) and that of RNG(P), we can show the following property.

Property D21 An edge of EMST(P) is an edge of RNG(P).

To prove this property, assume, contrary to Property D21, that $\overline{p_ip_j}$ is an edge of EMST(P) but that there is a point, say p_k, in the intersection of the two disks with radius $d(p_i, p_j)$ centred at p_i and p_j, as shown in Figure 2.5.4(a). If we remove the edge $\overline{p_ip_j}$ from EMST(P), the tree is divided into two subtrees, G_1 and G_2; one contains p_i and the other contains p_j. Without loss of generality, suppose that G_1 contains p_i and p_k and G_2 contains p_j. Then we can obtain a new tree G' from G_1 and G_2 by adding a new edge $\overline{p_jp_k}$, as shown in Figure 2.5.4(b). Since $d(p_j, p_k) < d(p_i, p_j)$, the total length of the edges in G' is smaller than the total length of the edges in EMST(P), which is a contradiction. Thus, we get Property D21.

Figure 2.5.1(e) represents the Euclidean minimum spanning tree for the same set of points as in (a).

Minimum-cost spanning trees, EMST(P) in particular, have many applications such as clustering (Gower and Ross, 1969; Rohlb, 1973), pattern recognition (Osteen and Lin, 1974) and approximation algorithms for the travelling salesman problem (Rosenkrantz *et al.*, 1977).

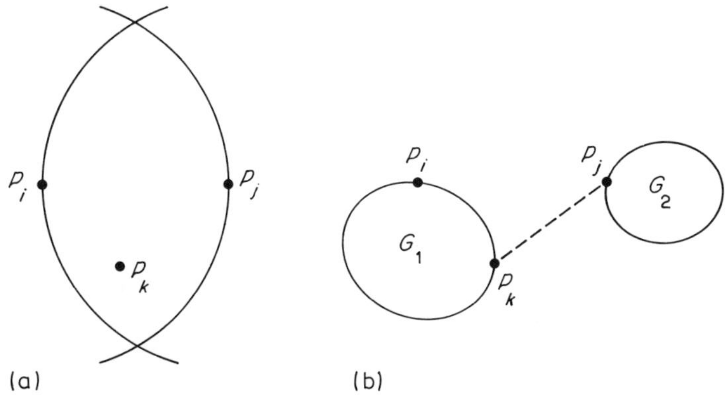

Figure 2.5.4 An edge in the EMST(P).

The geometric subgraphs of $\mathcal{D}(P)$ discussed above are all non-directed graphs. If we introduce a direction in a subgraph of $\mathcal{D}(P)$, we can define a directed subgraph, called the nearest neighbour graph. The *nearest neighbour graph*, denoted by NNG(P), is defined as the directed graph having a vertex set P such that $\overrightarrow{p_ip_j}$ is an edge of NNG(P) if and only if

$$d(p_i, p_j) = \min_{k(\neq i)} d(p_i, p_k) .$$

In other words, $\overrightarrow{p_ip_j}$ is an edge of NNG(P) if and only if p_j is the nearest neighbour of p_i. Note that the nearest neigbour relation is not symmetric; p_j being the nearest neighbour of p_i does not necessarily imply that p_i is the nearest neighbour of p_j. Every point has its nearest neighbour unless P consists of only one point. However, the nearest neighbour is not necessarily unique; there may be two or more points whose distances from p_i attain the minimum. Hence, in general, every vertex in NNG(P) has one or more edges going out from that vertex. The next property follows directly from the definition.

Property D22 If $\overrightarrow{p_ip_j}$ is an edge of NNG(P), $\overline{p_ip_j}$ is an edge of RNG(P).

Furthermore, we may derive a little stronger property than Property D22.

Property D23 If $\overrightarrow{p_ip_j}$ is an edge of NNG(P) and there is no other edge in NNG(P) that has the same initial vertex p_i, then $\overline{p_ip_j}$ is an edge of EMST(P).

To prove this property, assume, contrary to Property D23, that (p_i, p_j) is an edge of NNG(P) but $\overline{p_ip_j}$ is not an edge of EMST(P). If we add an edge $\overline{p_ip_j}$ to EMST(P), the resulting graph has exactly one cycle, and $\overline{p_ip_j}$ belongs to this cycle. Let $\overline{p_ip_k}$ be the other edge that is on this cycle and that has p_i as one of the terminal points. Let G be the graph obtained from

EMST(P) by replacing the edge $\overline{p_ip_k}$ with the edge $\overline{p_ip_j}$. Then G is a spanning tree. Moreover, since $\overrightarrow{p_ip_j}$ is the only edge starting at p_i in NNG(P), $\overline{p_ip_j}$ is shorter than $\overline{p_ip_k}$. Hence, the total length of the edges in G is smaller than that in EMST(P), which is a contradiction. Thus we get Property D23.

Property D23 implies that if no two pairs of points in P have the same distance, NNG(P) is a subgraph of the directed graph that is obtained from EMST(P) by replacing each edge with two parallel edges having opposite directions. Figure 2.5.1(f) represents the nearest neighbour graph for the same set of points as in (a). NNG(P) can be applied to geographic analysis (Kolars and Nystuen, 1974). The properties of NNG(P) are studied by Saaty (1970) and Pielou (1977).

For a set P of points in the plane, we have considered six graphs $\mathcal{D}(P)$, ∂CH(P), GG(P), RNG(P), EMST(P) and NNG(P). The relations of the first five undirected graphs are shown in Figure 2.5.5, where the lower graph is a subgraph of the upper graph. These relations come from Properties D2 and D19–D23. ∂CH(P) is a subgraph of $\mathcal{D}(P)$ (Property D2). EMST(P) is a subgraph of RNG(P) (Property D21). RNG(P) is a subgraph of GG(P) (Property D20). GG(P) is a subgraph of $\mathcal{D}(P)$ (Property D19). NNG(P) is a subgraph of the directed graph obtained from RNG(P) by replacing each edge with two parallel edges having opposite directions (Property D22). If no two pairs of points in P have the same distance, NNG(P) is a subgraph of the directed graph obtained from EMST(P) by replacing each edge with two parallel edges having opposite directions (Property D23). In particular, $\overline{p_ip_j}$ or $\overrightarrow{p_ip_j}$ can be a candidate for an edge of these graphs only when $\overline{p_ip_j}$ is a Delaunay edge. As we will see in Chapter 4, $\mathcal{D}(P)$ can be constructed efficiently. Hence, the above relationships give us methods for constructing these graphs through $\mathcal{D}(P)$; we first construct $\mathcal{D}(P)$ and next delete superfluous edges from $\mathcal{D}(P)$.

The construction of ∂CH(P) from $\mathcal{D}(P)$ is straightforward; indeed, what we have to do is simply to gather the outermost Delaunay edges in $\mathcal{D}(P)$.

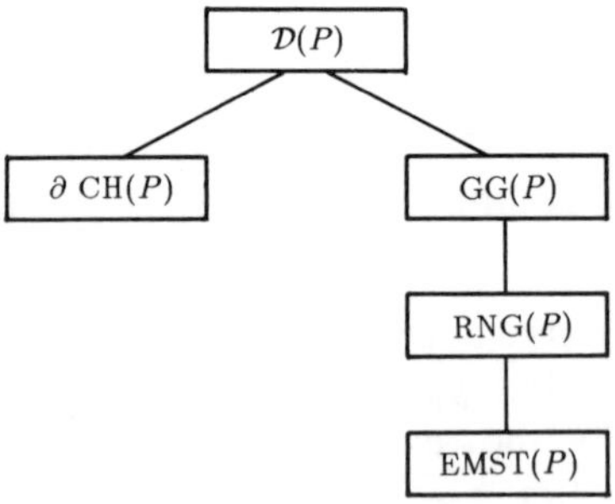

Figure 2.5.5 Relations among the Delaunay triangulation and the related graphs.

The construction of GG(P) from $\mathcal{D}(P)$ is likewise straightforward, because whether or not a Delaunay edge is an edge of GG(P) can be checked locally according to Property D19. Similarly, the construction of NNG(P) from $\mathcal{D}(P)$ is straightforward; all we have to do is to select, for each point in P, the shortest Delaunay edge incident to the point. The construction of RNG(P) and EMST(P) requires a more global check. See Toussaint (1980a) and Supowit (1983) for the construction of RNG(P). Prim (1957) showed some principles for finding the minimum-cost spanning tree, and those principles were used in many algorithms proposed thereafter. An efficient algorithm for finding the minimum-cost spanning tree for a planar graph is known (Cheriton and Tarjan, 1976). Since the Delaunay triangulation is a planar graph, this algorithm can be used for finding EMST(P) from $\mathcal{D}(P)$.

CHAPTER 3

Generalizations of the Voronoi Diagram

Since the early 1970s the ordinary Voronoi diagram has been extended or generalized in many directions, and those generalized Voronoi diagrams facilitate many practical applications in various fields. In this chapter we show the definitions of those generalized Voronoi diagrams together with their major geometric properties. First, to give a perspective of this section, we fix a general framework for possible generalizations and make general comments. (If the reader is interested in only a specific generalized Voronoi diagram, he or she may skip this introduction and directly go to the section on that Voronoi diagram.)

We notice from the definition of the ordinary Voronoi diagram (Definition V1) that an abstract idea for defining a Voronoi diagram is that every point in a space is assigned to at least one of the generators according to a certain assignment rule, and that the resulting sets of points associated with the generators are collectively exhaustive and mutually exclusive except for boundaries. To formalize this abstract idea, we consider a space S and a set of n distinct subsets of S, $A = \{A_1, \ldots, A_n\}$, which we shall call a *generator set*. In the ordinary Voronoi diagram the space S is the Euclidean space, and the generator set A is a set of points, $P = \{p_1, \ldots, p_n\}$ ($A = P$, $A_i = p_i$). In general, the space S may be the Euclidean space or non-Euclidean space, and an element A_i in A may be a point, a set of points, a line, an area, a polyhedron, etc.

Given a space S and a set $A = \{A_1, \ldots, A_n\}$, $2 \leq n < \infty$, we consider an assignment of a point p in S to at least one element in A. If we assign p to A_i, we put 1 for (p, A_i); if we do not assign p to A_i, we put 0 for (p, A_i), $p \in S$, $A_i \in A$. Then this assignment can be regarded as a mapping from $S \times A$ to $\{1, 0\}$. Mathematically, we consider a mapping δ from $S \times A$ to $\{0, 1\}$ such that

$$\delta(p, A_i) = \begin{cases} 1, & \text{if } p \text{ is assigned to } A_i, \\ 0, & \text{otherwise.} \end{cases} \tag{3.0.1}$$

We call δ an *assignment rule*. In the ordinary Voronoi diagram, for example, the assignment rule δ is given by

$$\delta(p, A_i) = \begin{cases} 1, & \text{if } \|\mathbf{x}-\mathbf{x}_i\| \le \|\mathbf{x}-\mathbf{x}_j\| \quad \text{for } j \in I_n \backslash \{i\}, \\ 0, & \text{otherwise,} \end{cases} \tag{3.0.2}$$

where $\mathbf{x}$ and $\mathbf{x}_i$ are the location vectors of p and p_i in $\mathbb{R}^m$, respectively. Under an assignment rule δ, we consider the set of points assigned to A_i, i.e.

$$V(A_i) = \{p | \delta(p, A_i) = 1, p \in S\}, \tag{3.0.3}$$

and the set of points assigned to both A_i and A_j, i.e.

$$e(A_i, A_j) = \{p | \delta(p, A_i) = \delta(p, A_j) = 1, p \in S\}, \quad i \neq j. \tag{3.0.4}$$

With $V(A_i)$ of equation (3.0.3), we define the set

$$\mathcal{V}(A, \delta, S) = \{V(A_1), \ldots, V(A_n)\}. \tag{3.0.5}$$

The mapping δ of equation (3.0.1) allows many assignment rules. Among them, we are concerned with the family of assignment rules that satisfy the following two conditions. First, every point p in S is assigned to at least one element of A, i.e.

$$\sum_{i=1}^{n} \delta(p, A_i) \ge 1, \quad p \in S. \tag{3.0.6}$$

Second, the set $e(A_i, A_j)$ is the boundary of $V(A_i)$, i.e. for any positive real ϵ, the open ball $N_\epsilon(p)$ with radius ϵ centred at point p in $e(A_i, A_j)$ contains not only interior points of $V(A_i)$ but also points in $S \backslash V(A_i)$. In mathematical terms, for $p \in e(A_i, A_j)$,

$$N_\epsilon(p) \cap [V(A_i) \backslash e(A_i, A_j)] \neq \emptyset, \quad N_\epsilon(p) \cap [S \backslash V(A_i)] \neq \emptyset \quad \text{for any } \epsilon > 0. \tag{3.0.7}$$

This condition intuitively means that if the space is a plane, the set $e(A_i, A_j)$ does not have a positive area; if the space is a three-dimensional space, it does not have a positive volume.

It is obvious from equation (3.0.6) that

$$\bigcup_{i=1}^{n} V(A_i) = S. \tag{3.0.8}$$

Moreover, from equations (3.0.3) and (3.0.4), the equation

$$[V(A_i) \cap V(A_j)] \backslash e(A_i, A_j) = \emptyset, \quad i \neq j, \tag{3.0.9}$$

holds. Therefore, under the assignment rule δ satisfying equations (3.0.6) and (3.0.7), elements in $\mathcal{V}(A, \delta, S)$ are collectively exhaustive and mutually exclusive except for boundaries, that is, $\mathcal{V}(A, \delta, S)$ forms a tessellation of S. We designate this tessellation the *generalized Voronoi diagram* generated by the generator set A with assignment rule δ in space S, and $V(A_i)$ the *generalized Voronoi region* associated with A_i. We call the assignment rule δ that generates a generalized Voronoi diagram the *Voronoi generation assignment rule*, or briefly the *V-assignment rule*.

In the literature, the generalized Voronoi diagram sometimes refers to more specific Voronoi diagrams than the above generalized Voronoi diagram. For example, the generalized Voronoi diagrams referred to by Drysdale and Lee (1978), Miles and Maillarelet (1982) and Ó'Dúnlaing *et al.* (1986) are what we call the higher-order Voronoi diagram, the line Voronoi diagram and the area Voronoi diagram, respectively, in what follows. The generalized Voronoi diagram used in this text is much wider than their generalized Voronoi diagrams.

The definition of the generalized Voronoi diagram by equation (3.0.3) is a generalization of Definition V2 of the ordinary Voronoi diagram. In Section 2.1 we defined the ordinary Voronoi diagram with half planes (Definition V3) as an alternative to Definition V2. When we examine the properties of Voronoi diagrams, we often find this definition more convenient and more operational. Motivated by this advantage, we next consider a generalization of Definition V3. In the ordinary Voronoi diagram, we alternatively call the half plane $H(p_i, p_j)$ the dominance region of p_i over p_j (recall equation (2.1.6)). To emphasize this notion, we may write $\mathrm{Dom}(p_i, p_j)$ for $H(p_i, p_j)$. For any $(p_i, p_j) \in P \times P$, $\mathrm{Dom}(p_i, p_j)$ determines a subregion in a plane. This implies that Dom can be regarded as a mapping from $P \times P$ to the family of all subregions in a plane. We now generalize this mapping.

Given a space S and a set of n distinct subsets of S, $A = \{A_1, \ldots, A_n\}$, $2 \leq n < \infty$, we consider a mapping Dom from $A \times A$ to the family of all subsets of S. For (A_i, A_j), we call $\mathrm{Dom}(A_i, A_j)$ (which is a subset of S) the *dominance region* of A_i over A_j. For a mapping Dom, we consider the set of points that belong to both $\mathrm{Dom}(A_i, A_j)$ and $\mathrm{Dom}(A_j, A_i)$, i.e.

$$b(A_i, A_j) = \mathrm{Dom}(A_i, A_j) \cap \mathrm{Dom}(A_j, A_i), \quad i \neq j. \tag{3.0.10}$$

We call the set $b(A_i, A_j)$ the *bisector* between A_i and A_j. Taking the intersection of $\mathrm{Dom}(A_i, A_j)$ for $j \in I_n \backslash \{i\}$, we define

$$V(A_i) = \bigcap_{j \in I_n \backslash \{i\}} \mathrm{Dom}(A_i, A_j), \tag{3.0.11}$$

and let

$$\mathcal{V}(A, \mathrm{Dom}, S) = \{V(A_i), \ldots, V(A_n)\}. \tag{3.0.12}$$

The dominance region defined above does not always generate a tessellation. First, elements in $\mathcal{V}(A, \mathrm{Dom}, S)$ may not satisfy equation (3.0.8). Second, equation (3.0.9) may not hold, because the set $b(A_i, A_j)$ may not be the boundary of $\mathrm{Dom}(A_i, A_j)$ (Figure 3.0.1(a); an actual example will be shown in Figures 3.7.2(c) and 3.7.4(b)). Since we are concerned with a tessellation,

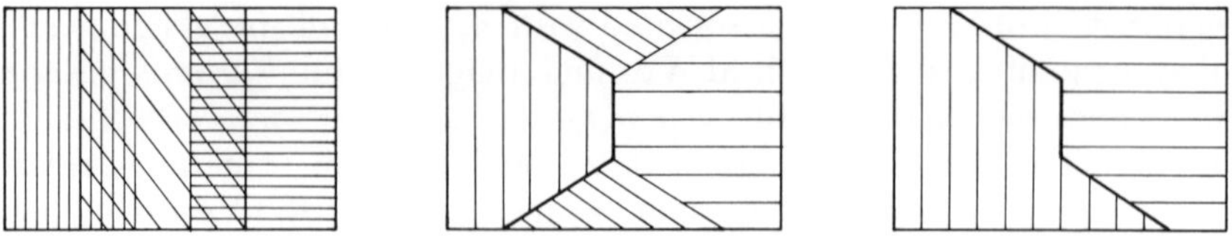

Figure 3.0.1 Dominance regions of A_i over A_j: (a) not well-behaving; (b) well-behaving; (c) regular. The vertically hatched region indicates $\text{Dom}(A_i, A_j)$; the horizontally hatched region indicates $\text{Dom}(A_j, A_i)$; the diagonally hatched region indicates $\text{Dom}(A_i, A_j) \cap \text{Dom}(A_j, A_i)$; the heavy lines show $b(A_i, A_j)$.

we confine ourselves to the family of dominance regions that satisfy the following two conditions. First, if $\text{Dom}(A_i, A_j) \neq \varnothing$, the bisector $b(A_i, A_j)$ is the boundary of $\text{Dom}(A_i, A_j)$, i.e. for any $p \in b(A_i, A_j)$,

$$N_\epsilon(p) \cap [V(A_i) \backslash b(A_i, A_j)] \neq \varnothing \quad N_\epsilon(p) \cap [S \backslash V(A_i)] \neq \varnothing \text{ for any } \epsilon > \varnothing. \tag{3.0.13}$$

Second, elements in $\mathcal{V}(A, \text{Dom}, S)$ satisfy equation (3.0.8). If the mapping Dom satisfies these two conditions, we say that $\text{Dom}(A_i, A_j)$ is a *well-behaving dominance region* (Figure 3.4.1(b)). Similarly, we call the bisector $b(A_i, A_j)$ of a well-behaving dominance region a *well-behaving bisector*. If $\text{Dom}(A_i, A_j)$ is a well-behaving dominance region, elements in $\mathcal{V}(A, \text{Dom}, S)$ are collectively exhaustive (because of equation (3.0.8)), and they are mutually exclusive except for boundaries (because of equation (3.0.11)). Consequently, $\mathcal{V}(A, \text{Dom}, S)$ forms a tessellation of S.

We should add one more important remark on $\text{Dom}(A_i, A_j)$. Unlike the case of $\text{Dom}(A_i, A_j) = H(p_i, p_j)$, the equation

$$\text{Dom}(A_i, A_j) \cup \text{Dom}(A_j, A_i) = S \tag{3.0.14}$$

may not hold even if $\text{Dom}(A_i, A_j)$ is a well-behaving dominance region (Figure 3.0.1(b); an actual example will be shown in Figure 3.2.4). If $\text{Dom}(A_i, A_j)$ is a well-behaving dominance region and equation (3.0.14) holds, we say that the assignment rule is *regular* (Figure 3.0.1(c)). If $\text{Dom}(A_i, A_j)$ is regular, then the bisector splits the space S into two subspaces which are mutually exclusive and collectively exhaustive (Figure 3.0.1(c)).

In the above, we defined a generalized Voronoi diagram in two ways: first, in terms of a mapping δ; second, in terms of a mapping Dom. Obviously, these two mappings are closely related. Namely, the mapping δ is written in terms of the mapping Dom as

$$\delta(p, A_i) = \begin{cases} 1, & \text{if } p \in \text{Dom}(A_i, A_j) \quad \text{for any } j \in I_n \backslash \{i\}, \\ 0, & \text{otherwise.} \end{cases} \tag{3.0.15}$$

To construct a generalized Voronoi diagram in practice, we have to specify the assignment rule δ in terms of algebraic relations such as equation (3.0.2) or, more broadly, logical relations. In the literature, V-assignment rules are often specified in terms of a distance relation or a set of distance relations. One of the simplest and most frequently adopted V-assignment rules is that a point p is assigned to the closest generator(s), i.e.

$$\delta(p, A_i) = \begin{cases} 1, & \text{if } d(p, A_i) \leq d(p, A_j) \quad \text{for any } j \in I_n \backslash \{i\}, \\ 0, & \text{otherwise,} \end{cases} \tag{3.0.16}$$

where $d(p, A_i)$ is the 'distance' from a point p to an element A_i in A. The assignment rule of equation (3.0.2) is a special case of the assignment rule of equation (3.0.16). When a generator is a point (i.e. $A_i = p_i$), the distance $d(p, p_i)$ usually implies a distance metric. The *distance metric* or *metric* for points in S is defined as a mapping, $d(p_i, p_j)$, of $S \times S$ into non-negative reals satisfying the following four axioms: for points $p_i, p_j, p_k \in S$, (i) $d(p_i, p_i) = 0$; (ii) $d(p_i, p_j) \leq d(p_i, p_k)+d(p_k, p_j)$; (iii) $d(p_i, p_j) = d(p_j, p_i)$; (iv) if $p_i \neq p_j$, then $d(p_i, p_j) > 0$. In the ordinary Voronoi diagram, the distance $d(p, A_i) = d(p, p_i)$ is given by the Euclidean distance, $d(p, p_i) = \|\mathbf{x}-\mathbf{x}_i\|$. Obviously the Euclidean distance is a distance metric. In general, however, a 'distance' $d(p, p_i)$ in the assignment rule of equation (3.0.16) for $A = P$ is not necessarily a distance metric. A distance metric $d(p_i, p_j)$ is defined for any point p_i in S and any point p_j in S, but when we define the assignment rule of equation (3.0.16) for $A = P$, it is sufficient to define $d(p, p_i)$ for any point p in S and any point p_i in P (not S). More generally, $d(p, A_i)$ in equation (3.0.16) is defined as a mapping of $S \times A$ into $\mathbb{R}$ (not $S \times S$ into $\mathbb{R}$). If the assignment rule δ of equation (3.0.16) with this mapping is a V-assignment rule, we call the mapping $d(p, A_i)$ a *Voronoi generation distance* or a *V-distance* (or just a *distance*) in short. The assignment rule of equation (3.0.16) with a V-distance produces a generalized Voronoi diagram, and we write it symbolically as $\mathcal{V}(A, d, S)$. The ordinary Voronoi diagram is written as $\mathcal{V}(P, d, \mathbb{R}^m)$, where $d(p, p_i) = \|\mathbf{x}-\mathbf{x}_i\|$.

A V-distance is not necessarily a distance metric. For example, consider a function given by

$$d(p_i, p_j) = \|\mathbf{x}_i-\mathbf{x}_j\|-w, \quad w \neq 0. \tag{3.0.17}$$

This function is not a distance metric because $d(p_i, p_i) = w \neq 0$, i.e. Axiom (i) is not satisfied. We notice, however, that the set

$$\text{Dom}(p_i, p_j) = \{\mathbf{x} | \, \|\mathbf{x}-\mathbf{x}_i\|-w \leq \|\mathbf{x}-\mathbf{x}_j\|-w\} \tag{3.0.18}$$

produces exactly the same dominance region defined with the Euclidean distance; consequently, the function given by equation (3.0.17) is a V-distance. This example shows that a V-distance allows many more various functions than a distance metric. We should note, however, that some of them are trivial. To show what generalizations are trivial, let F be a strictly increasing function of a V-distance, i.e.

$$F(d(p, A_i)) \leq F(d(p, A_j)), \quad \text{if and only if } d(p, A_i) \leq d(p, A_j). \tag{3.0.19}$$

It follows from this relation that $\text{Dom}(A_i, A_j)$ defined with $F(d(p, A_i))$ is the same as $\text{Dom}(A_i, A_j)$ defined with $d(p, A_i)$, and hence $\mathcal{V}(A, F(d), \mathbb{R}^m) = \mathcal{V}(A, d, \mathbb{R}^m)$. In this case we say that the distance $F(d(p, A_i))$ is *transformable* to the distance $d(p, A_i)$, and that the Voronoi diagram with the distance $F(d(p, A_i))$ is a *trivial generalization* of the Voronoi diagram with the distance $d(p, A_i)$. In applications, we often use the *exponential distance*, $d(p, p_i) = \exp(\|\mathbf{x}-\mathbf{x}_i\|)$, the *logarithmic distance*, $d(p, p_i) = \log(\|\mathbf{x}-\mathbf{x}_i\|)$ and the *power distance*, $d(p, p_i) = \|\mathbf{x}-\mathbf{x}_i\|^a$ $(a > 0)$. Apparently, these distances are the distances transformable to the Euclidean distance. Thus, Voronoi diagrams with those distances are trivial generalizations of the ordinary Voronoi diagram. In the following sections we show several generalized Voronoi diagrams defined with V-distances that are not transformable to the Euclidean distance. Stated a little more specifically, we show $\mathcal{V}(P, d, \mathbb{R}^m)$, where $d(p, p_i)$ is the weighted distance (Section 3.1), the shortest-path distance, the visibility-shortest-path distance (Section 3.4), the (modified) Minkowski metric, the (modified) Karlsruhe metric, and the Hausdorff distance (Section 3.7).

In the above generalized Voronoi diagrams, a generator set A is restricted to a set of points and a space S is restricted to the Cartesian space. Recalling the definition of equation (3.0.1), we can of course relax this restrictions. Regarding the generalizations with respect to a generator set A, we show $\mathcal{V}(A, d, \mathbb{R}^m)$, where the set A is a set of sets of points (Sections 3.2 and 3.3), a set of lines (Section 3.5), and a set of areas (Section 3.6). Regarding the generalized Voronoi diagrams with respect to a space S, we show $\mathcal{V}(P$, the length of the shortest path on S, $S)$, where S is a sphere, a cylinder, a cone, a polyhedral surface (Section 3.7), and a network (Section 3.8).

As is noticed from $\mathcal{V}(A, \delta, S)$, we can generalize the ordinary Voronoi diagram with respect to a generator set A, an assignment rule δ and a space S. This implies that the number of generalizations is potentially as many as the product of the numbers of generalizations of each factor. In the following sections, however, to avoid complicated exposition, we mainly deal with a single factor keeping the other factors the same as those of the ordinary Voronoi diagram. The reader may attempt to formulate a new generalized Voronoi diagram by combining those generalized Voronoi diagrams. We also note that the reader who wishes to understand generalized Voronoi diagrams from a more computational view point should consult Edelsbrunner and Seidel (1986).

3.1 WEIGHTED VORONOI DIAGRAMS

In the ordinary Voronoi diagram we implicitly assume that generator points are indifferent (except their locations) or that each generator point has the same weight. In some practical applications, however, this assumption may not be appropriate. Rather, it is more appropriate to assume that generator points have different weights reflecting the variable property of the generator points; for example, the population size of a settlement, the number of functions in a shopping centre, the amount of emissions from a polluter, the size of an atom in a crystal structure, and so forth. In this section we show the family of generalized Voronoi diagrams that take these different weights into account in terms of the 'weighted distance' $d_w(p, p_i)$.

We consider a set of distinct points, $P = \{p_i, \ldots, p_n\} \subset \mathbb{R}^m$ $(2 \leq n < \infty)$ $(A = P,\ S = \mathbb{R}^m)$ and assign a weight to each p_i which relates to some variable property of the phenomenon, such as those mentioned above. We represent this weight by a set of parameters, $W_i = \{w_{i1}, \ldots, w_{in_w}\}$ (if $n_w = 1$, we write w_i for W_i). With this weight we define a distance, $d_w(p, p_i)$ from p to p_i, called the *weighted distance*, which will be specified in the following subsections. The dominance region of p_i over p_j with the weighted distance is written as

$$\mathrm{Dom}(p_i, p_j) = \{p | d_w(p, p_i) \leq d_w(p, p_j)\}, \quad j \neq i. \tag{3.1.1}$$

Let

$$V(p_i) = \bigcap_{i=1}^{n} \mathrm{Dom}(p_i, p_j), \tag{3.1.2}$$

and $\mathcal{V}(P, d_w, \mathbb{R}^m) = \mathcal{V}_w = \{V(p_1), \ldots, V(p_n)\}$. If the dominance region given by equation (3.1.1) is well-behaving, the set $\mathcal{V}_w$ gives a generalized Voronoi diagram. We call this generalized Voronoi diagram the *weighted Voronoi diagram* generated by P with weight $\{W_1, \ldots, W_n\}$, and the set $V(p_i)$ the *weighted Voronoi region* associated with p_i.

Since the weighted distance allows many functional forms, we possibly have many weighted Voronoi diagrams. In the following subsections we show four types of weighted Voronoi diagrams that often appear in the related literature. We mainly deal with $\mathbb{R}^2$, but conceptually the extension to $\mathbb{R}^m$ is straightforward.

3.1.1 Multiplicatively weighted Voronoi diagram

The first type of weighted Voronoi diagram is characterized by the weighted distance given by

$$d_{mw}(p, p_i) = \frac{1}{w_i} \|\mathbf{x} - \mathbf{x}_i\|, \quad w_i > 0. \tag{3.1.3}$$

We call this distance the *multiplicatively weighted distance* or the *MW-distance* for short. The dominance region with the MW-distance is written as

$$\mathrm{Dom}(p_i, p_j) = \left\{\mathbf{x} \,\middle|\, \frac{1}{w_i}\|\mathbf{x} - \mathbf{x}_i\| \leq \frac{1}{w_j}\|\mathbf{x} - \mathbf{x}_j\|\right\}, \quad i \neq j. \tag{3.1.4}$$

After a few steps of calculation from equation (3.1.4), we obtain the bisector as

$$b(p_i, p_j) = \left\{ \mathbf{x} \left| \left\| \mathbf{x} - \frac{w_i^2}{w_i^2 - w_j^2} \mathbf{x}_j + \frac{w_j^2}{w_i^2 - w_j^2} \mathbf{x}_i \right\| = \frac{w_i w_j}{w_i^2 - w_j^2} \|\mathbf{x}_j - \mathbf{x}_i\| \right. \right\}, \quad (3.1.5)$$
$$w_i \neq w_j,\ i \neq j.$$

This bisector is the locus of a point p satisfying that the distance from p to the fixed points, $w_i^2\mathbf{x}_j/(w_i^2-w_j^2)-w_j^2\mathbf{x}_i/(w_i^2-w_j^2)$, is constant. Obviously, this locus is a circle in $\mathbb{R}^2$. This circle passes through the interior and exterior division points (denoted by p_{ij1} and p_{ij2}) of $\overline{p_ip_j}$ with ratio w_i,w_j, and its diameter is given by $\overline{p_{ij1}p_{ij2}}$ (Figure 3.1.1). In classic geometry this circle is known as the *Apollonius circle*. Figure 3.1.1 shows the Apollonius circles representing the bisector defined with the MW-distance for several ratios $\alpha = w_i/w_j$ ($w_i/w_j \geq 1$ is assumed without loss of generality). In the special case of $\alpha = 1$ ($w_i = w_j$), the bisector becomes a straight line, or a circle with an infinite radius.

In equation (3.1.3) we use $1/w_i$, but mathematically w_i is also acceptable. In equation (3.1.3) we adopt the former, because, as can be seen from Figure 3.1.1, the dominance region of p_i over p_j becomes larger as the weight w_i (or α) increases.

Since the Apollonius circle divides $\mathbb{R}^2$ into two disjoint regions, the bisector with MW-distance is well-behaving and regular. Thus the set $\mathcal{V}(P, d_{\mathrm{mw}}, \mathbb{R}^2) = \mathcal{V}_{\mathrm{mw}} = \{V(p_1), \ldots, V(p_n)\}$ gives a generalized Voronoi diagram, where $V(p_i)$ is given by equation (3.1.2) with equation (3.1.4). We call this generalized Voronoi diagram the *multiplicatively weighted Voronoi diagram*

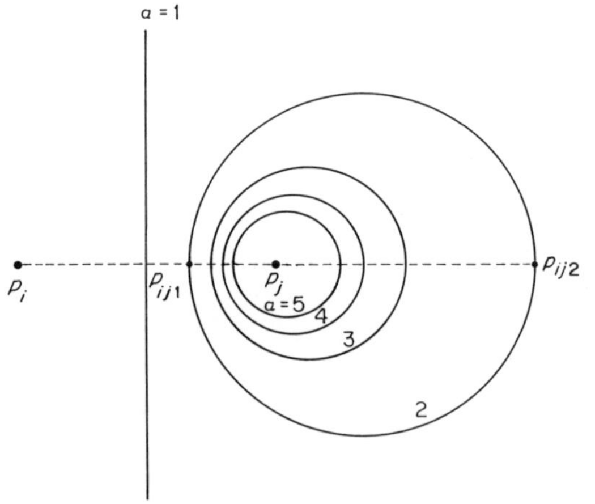

Figure 3.1.1 The bisectors with a multiplicatively weighted distance for several ratios $\alpha = w_i/w_j = 1,2,3,4,5$, or multiplicatively weighted Voronoi diagrams for $n = 2$.

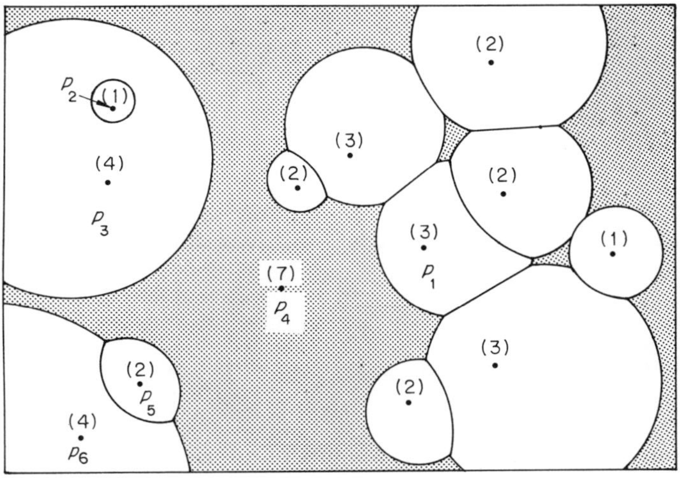

Figure 3.1.2 A multiplicatively weighted Voronoi diagram in $\mathbb{R}^2$ (the numbers in parentheses represent weights associated with the generators).

generated by P with the MW-distance or the *MW-Voronoi diagram* of P with $d_{mw}(p, p_i)$ (Aurenhammer and Edelsbrunner, 1984; Aurenhammer, 1988; Ash and Bolker, 1986, call it a *circular Dirichlet tessellation*). We call the set $V(p_i)$ the *multiplicatively weighted Voronoi region* associated with p_i or briefly the *MW-Voronoi region* of p_i. Figures 3.1.1 and 3.1.2 show planar MW-Voronoi diagrams. The MW-Voronoi diagram reduces to the ordinary Voronoi diagram when w_i is the same constant for all $i \in I_n$. Thus we can regard the MW-Voronoi diagram as a generalization of the ordinary Voronoi diagram.

To observe the geometric properties of an MW-Voronoi diagram, let us inspect Figure 3.1.2, where the numbers in the parentheses represent weights associated with the generators. First, $V(p_1)$ is not convex. Second, $V(p_2)$ is contained in $V(p_3)$ or $V(p_3)$ has a hole. Third, $V(p_4)$, indicated by the shaded area, is not connected. Fourth, the weights of the MW-Voronoi regions adjacent to a convex MW-Voronoi region, say $V(p_5)$, are not smaller than the weight of the convex MW-Voronoi region ($w_5 < w_4, w_6$). We can understand this convexity property from the dominance region in Figure 3.1.1 and the fact that $V(p_i)$ is the intersection of the dominance regions. Fifth, from the same figure and the same equation, we notice that $V(p_i)$ is not empty because the weights are finite. To sum up, we have the following property.

Property MW1 An MW-Voronoi region is a non-empty set; it need not be convex, or connected; and it may have a hole(s). An MW-Voronoi region

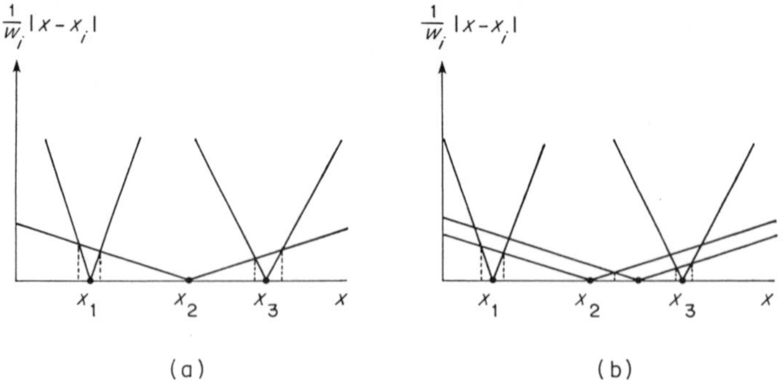

Figure 3.1.3 MW-Voronoi diagrams in $\mathbb{R}$.

$V(p_i)$ is convex if and only if the weights of adjacent MW-Voronoi regions are not smaller than w_i.

To observe another geometric property, we depict a one-dimensional MW-Voronoi diagram in Figure 3.1.3 where the value of $|x-x_i|/w_i$ is shown on the vertical axis. Obviously, the slope of $|x-x_i|/w_i$ becomes flatter as the weight w_i increases. From Figure 3.1.3(a) we readily notice that the generator whose weight is the largest eventually dominates the places far from the locations of the generators; consequently, the MW-Voronoi region of the largest weight is infinite. When two or more generators have the same largest weight as in Figure 3.1.3(b), however, the rightmost and leftmost Voronoi regions are infinite. Generalizing these properties in $\mathbb{R}^2$, we obtain the following property.

Property MW2 Let $w_{\max} = \max_j\{w_j, j \in I_n\}$ and $P_{\max}$ be the subset of P given by $P_{\max} = \{p_j | w_j = w_{\max}\}$. An MW-Voronoi region $V(p_i)$ is infinite if and only if $p_i \in P_{\max}$ and p_i is on the boundary of $\mathrm{CH}(P_{\max})$.

If the generator with the maximum weight is unique, we have only one infinite MW-Voronoi region. An example is shown by the shaded MW-Voronoi region in Figure 3.1.3.

We observe in Figure 3.1.1 that the bisector $b(p_i,p_j)$ is either a circle or a straight line. In Figure 3.1.2 we find that the Voronoi edge shared by $V(p_4)$ and $V(p_6)$ is disconnected. From these findings we have the following property with respect to edges.

Property MW3 Two MW-Voronoi regions may share disconnected edges. An edge is a circular arc if and only if the weights of the MW-Voronoi regions sharing the edge are different; an edge is a straight line if and only if the weights of the MW-Voronoi regions sharing the edge are the same.

3.1.2 Additively weighted Voronoi diagram

The second type of weighted Voronoi diagram is characterized by the weighted distance

$$d_{aw}(p, p_i) = \|\mathbf{x}-\mathbf{x}_i\|-w_i. \tag{3.1.6}$$

We call this distance the *additively weighted distance* or briefly the *AW-distance*. The dominance region of p_i over p_j is written as

$$\mathrm{Dom}(p_i, p_j) = \{\mathbf{x}|\ \|\mathbf{x}-\mathbf{x}_i\|-w_i \leq \|\mathbf{x}-\mathbf{x}_j\|-w_j\},\quad i \neq j, \tag{3.1.7}$$

and the bisector between p_i and p_j is written as

$$b(p_i, p_j) = \{\mathbf{x}|\ \|\mathbf{x}-\mathbf{x}_i\|-\|\mathbf{x}-\mathbf{x}_j\| = w_i-w_j\},\quad j \neq i. \tag{3.1.8}$$

The shape of the bisector or the dominance region varies according to the parameter values $\alpha = \|\mathbf{x}_i-\mathbf{x}_j\|$ and $\beta = w_i-w_j$ (where $w_i-w_j \geq 0$ is assumed without loss of generality). First,

$$\mathrm{Dom}(p_i, p_j) = \mathbb{R}^2,\quad \text{if } 0 < \alpha < \beta, i \neq j. \tag{3.1.9}$$

The generator p_i dominates the whole plane or the dominance region of p_j disappears. Such a case never happens for the dominance region defined with the MW-distance. Second,

$$\mathrm{Dom}(p_i, p_j) = \mathbb{R}^2 \backslash \{\mathbf{x}|\mathbf{x}_j+\lambda(\mathbf{x}_j-\mathbf{x}_i), \lambda \geq 0\},\quad \text{if } \alpha = \beta, i \neq j. \tag{3.1.10}$$

This dominance region is the whole plane except for the half life radiating from p_j in the direction from p_i to p_j (Figure 3.1.4, $\beta = 10$). Third,

$$b(p_i, p_j) = \{\mathbf{x}|\ \|\mathbf{x}-\mathbf{x}_i\|-\|\mathbf{x}-\mathbf{x}_j\| = \beta\},\quad \text{if } \alpha > \beta, i \neq j. \tag{3.1.11}$$

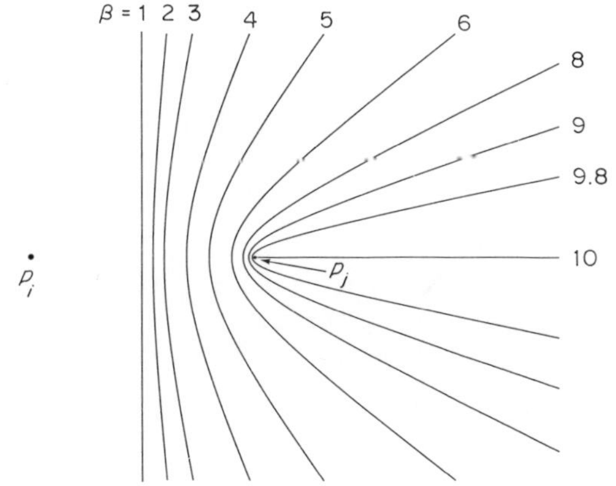

Figure 3.1.4 The bisectors with the additively weighted distance for parameter values $\alpha = \|\mathbf{x}_i-\mathbf{x}_j\| = 10$ and $\beta = w_i-w_j = 0, 1, 2, 3, 4, 5, 6, 8, 9, 9.8, 10$, or additively weighted Voronoi diagrams for $n = 2$.

This bisector is the locus of a point p satisfying that the difference between the distance from p to p_i and that from p to p_j is constant. This locus is known as a branch of the hyperbolic curve with foci p_i and p_j (Figure 3.1.4, β = 1–9.8). Last,

$$b(p_i, p_j) = \{\mathbf{x} | \|\mathbf{x}-\mathbf{x}_i\| = \|\mathbf{x}-\mathbf{x}_j\|\}, \quad \text{if } \beta = 0, i \neq j, \tag{3.1.12}$$

which is the straight line perpendicularly bisecting $\overline{p_ip_j}$ (Figure 3.1.4, $\beta = 0$). Figure 3.1.4 shows the bisector for several parameter values.

In equation (3.1.6) we do not restrict the sign of w_i. Usually we assume a positive w_i because, as can be seen in Figure 3.1.4, the dominance region of p_i over p_j becomes larger as the weight w_i increases. Obviously, both $-w_i$ and $+w_i$ are acceptable in equation (3.1.6); in some applications (for example, in the market area analysis in Section 7.3), the latter form is used.

Since the bisector defined with the AD-distance is well-behaving and regular, we can obtain a generalized Voronoi diagram $\mathcal{V}(P, d_{\text{aw}}, \mathbb{R}^m) = \mathcal{V}_{\text{aw}} = \{V(p_1), \ldots, V(p_n)\}$, where $V(p_i)$ is given by equation (3.1.2) with equation (3.1.7). We call the set $V(p_i)$ the *additively weighted Voronoi region* associated with p_i, or briefly the *AW-Voronoi region* of p_i, and the set $\mathcal{V}_{\text{aw}}$ the *additively weighted Voronoi diagram* generated by P with the AW-distance, or briefly the *AW-Voronoi diagram* of P with $d_{\text{aw}}(p, p_i)$ (Aurenhammer, 1988; Ash and Bolker, 1986, call it a *hyperbolic Dirichlet tessellation*). Figures 3.1.4 and 3.1.5 show planar AW-Voronoi diagrams. The AW-Voronoi diagram reduces to the ordinary Voronoi diagram when w_i is the same constant for all $i \in I_n$. Thus, we can regard the AW-Voronoi diagram as a generalization of the ordinary Voronoi diagram. We also note that the AW-Voronoi diagram can be regarded as a Voronoi growth model,

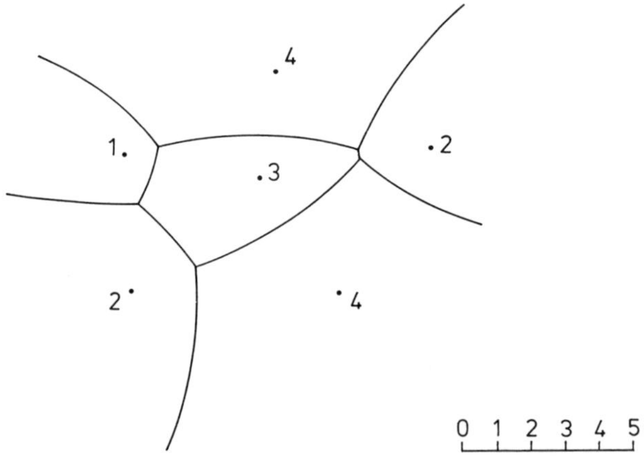

Figure 3.1.5 An additively weighted Voronoi diagram (the numbers indicate weights, w_i).

in particular the ordinary Johnson–Mehl model (see Chapter 7). Suppose that a finite number of points or a disk with zero radius are placed on a plane. Each disk grows with the same constant growth rate (i.e. the radius of every disk increases at the same constant rate) keeping its centre at the initial point but the growth of each disk starts at a different time. In the early stages disks do not overlap, but after a certain time a disk may touch another disk (see Figure 7.2.1). The growth of a disk ceases whenever two or more disks impinge. Eventually, every disk is deformed into a certain shape in the region, which is an AW-Voronoi region.

The properties of the AW-Voronoi diagram are quite different from those of the MW-Voronoi diagram. From equations (3.1.9) and (3.1.10), we obtain the following property.

Property AW1 The set $V(p_i)$ is empty if and only if

$$\min_{j,j\neq i} \{\|\mathbf{x}_j-\mathbf{x}_i\|-w_j, j\in I_n\} < -w_i. \quad (3.1.13)$$

The set $V(p_i)$ is a half line or a line segment if and only if

$$\min_{j,j\neq i} \{\|\mathbf{x}_j-\mathbf{x}_i\|-w_j, j\in I_n\} = -w_i. \quad (3.1.14)$$

The set $V(p_i)$ has a positive area if and only if

$$\min_{j,j\neq i} \{\|\mathbf{x}_j-\mathbf{x}_i\|-w_j, j\in I_n\} > -w_i. \quad (3.1.15)$$

The properties of equations (3.1.13) and (3.1.14) contrast with those of the ordinary Voronoi diagram (Property V1) and the MW-Voronoi diagram (Property MW1). In those Voronoi diagrams a Voronoi region is neither empty nor degenerates into a line.

From equations (3.1.11) and (3.1.12) we understand the following property.

Property AW2 An edge of an AW-Voronoi region is either a hyperbolic arc or a straight line segment.

As a consequence of Property AW2, we notice the following property.

Property AW3 If at least one weight, w_i, is different from the others and relation (3.1.15) holds, then there exists at least one non-convex AW-Voronoi region. Every non-convex AW-Voronoi region is star-shaped with respect to its generator.

Besides these properties, Ash and Bolker (1986) show the properties related to the generator recognition problem of the AW-Voronoi diagram (as well as the MW-Voronoi diagram).

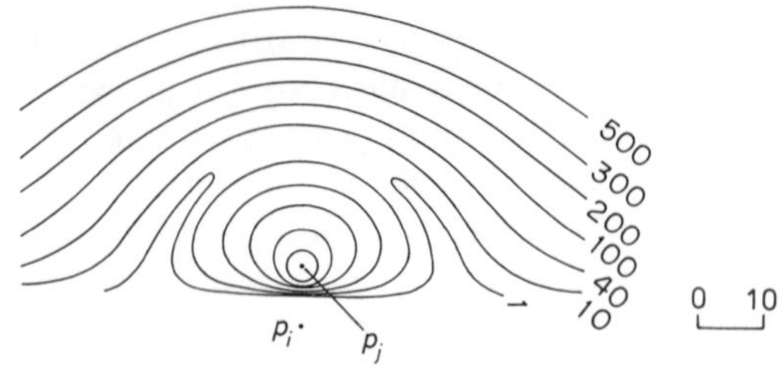

Figure 3.1.6 The bisectors with the generalized AW-distance given by $\phi(\|\mathbf{x}-\mathbf{x}_i\|) = (\|\mathbf{x}-\mathbf{x}_i\|-3)^3-w_i$, $\alpha = w_i-w_j = 1, 10, 40, 100, 200, 300, 500$, or additively weighted Voronoi diagrams with the generalized AW-distance for $n = 2$.

We may slightly generalize the AW-Voronoi diagram by replacing equation (3.1.6) with

$$d_{\text{g·aw}}(p, p_i) = \phi(\|\mathbf{x}-\mathbf{x}_i\|) - w_i, \tag{3.1.16}$$

where ϕ is usually a strictly increasing function (Ash and Bolker, 1986; Hanjoul *et al.*, 1989). We call this distance the *generalized AW-distance*. The generalized AW-distance is general in the sense that it includes not only the AW-Euclidean distance but also the MW-distance (the former is obtained by setting $\phi(t) = t$ and the latter is obtained by setting $\phi(t) = \log t$). Figure 3.1.6 shows planar AW-Voronoi diagrams with generalized AW-distances for $n = 2$.

3.1.3 Compoundly weighted Voronoi diagram

The third type of weighted Voronoi diagram is obtained by compounding the MW-distance and the AW-distance, i.e.

$$d_{\text{cw}}(p, p_i) = \frac{1}{w_{i1}}\|\mathbf{x}-\mathbf{x}_i\|-w_{i2} \quad w_{i1} > 0, \tag{3.1.17}$$

which we call the *compoundly weighted distance* or simply the *CW-distance*. The CW-distance becomes the AD-distance if $w_{i1} = 1$, $i \in I_n$; it becomes the MW-distance if $w_{i2} = 0$, $i \in I_n$. With this distance, the dominance region of p_i over p_j is written as

$$\text{Dom}(p_i, p_j) = \left\{\mathbf{x} \,\middle|\, \frac{1}{w_{i1}}\|\mathbf{x}-\mathbf{x}_i\|-w_{i2} \le \frac{1}{w_{j1}}\|\mathbf{x}-\mathbf{x}_j\|-w_{j2}\right\}, \quad i \neq j. \tag{3.1.18}$$

The boundary of this dominance region (the bisector) is the fourth-order polynomial function, and its shape is fairly complex, as is shown in Figure 3.1.7.

Although the shape of the bisector is fairly complex, it is easy to see from equation (3.1.18) that the bisector is well-behaving and regular. Thus we can obtain a generalized Voronoi diagram, $\mathcal{V}(P, d_{\text{cw}}, \mathbb{R}^m) = \mathcal{V}_{\text{cw}} = \{V(p_i),$

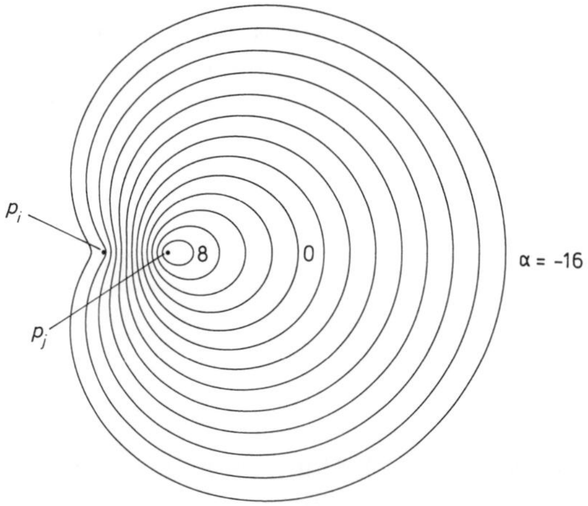

Figure 3.1.7 The bisectors with the compoundly weighted distance for several parameter values, $w_{i1}/w_{j1} = 1.5$, $\alpha = w_{i1}(w_{i2} - w_{j2}) = -16, \ldots, -2, 0, 2, \ldots, 8$ or compoundly weighted Voronoi diagrams for $n = 2$.

$\ldots, V(p_n)\}$, where $V(p_i)$ is given by equation (3.1.2) with equation (3.1.18). We call this generalized Voronoi diagram the *compoundly weighted Voronoi diagram* generated by P with the CW-distance, or the *CW-Voronoi diagram* of P with $d_{cw}(p, p_i)$. We call the set $V(p_i)$ the *compoundly weighted Voronoi region* associated with p_i, or briefly the *CW-Voronoi region* of p_i with $d_{cw}(p, p_i)$. Figure 3.1.8 shows a planar CW-Voronoi diagram. The CW-Voronoi diagram reduces to the ordinary Voronoi diagram when $w_{i1} = 1$, $w_{i2} = 0$, $i \in I_n$. Thus, the CW-Voronoi diagram can be regarded as a generalization of the ordinary Voronoi diagram.

Since the CW-Voronoi diagram becomes the MW-Voronoi diagram if $w_{i1} = 0$ for $i \in I_n$ and it becomes the MW-Voronoi diagram if $w_{i2} = 1$ for

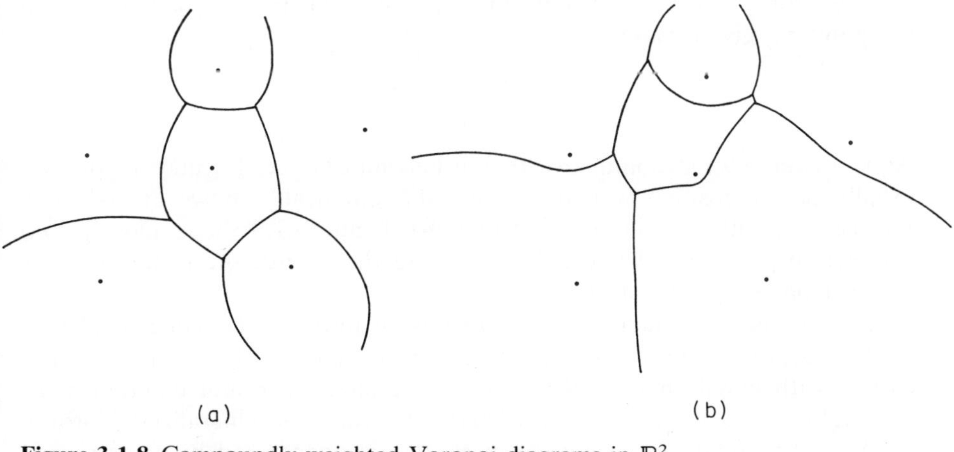

Figure 3.1.8 Compoundly weighted Voronoi diagrams in $\mathbb{R}^2$.

$i \in I_n$, the CW-Voronoi diagram shows the properties of the AW-Voronoi diagram and the MW-Voronoi diagram for these specific cases. In other cases its proper properties appear. We show one example.

Property CW1 An edge of a CW-Voronoi region is part of the fourth-order polynomial curve (a proper property), a hyperbolic arc (Property AW2), a circular arc (Property MW3) or a straight line (Properties AW2 and MW3).

3.1.4 Power diagram

The fourth type of weighted Voronoi diagram is characterized by the weighted distance

$$d_{\mathrm{pw}}(p, p_i; w_i) = \|\mathbf{x} - \mathbf{x}_i\|^2 - w_i, \tag{3.1.19}$$

which we call the *additively weighted power distance* or, following Aurenhammer (1988), the *power distance*. For simplicity, we usually employ the latter term, but the power distance should be distinguished from the ordinary power distance, $\|\mathbf{x} - \mathbf{x}_i\|^2$. Recalling the MW-distance, one may consider the multiplicatively weighted power distance, $\|\mathbf{x} - \mathbf{x}_i\|^2 / w_i$. This distance, however, is transformable to the MW-distance, and hence it is a trivial extension.

The dominance region with the power distance is written as

$$\mathrm{Dom}(p_i, p_j) = \{\mathbf{x} |\ \|\mathbf{x} - \mathbf{x}_i\|^2 - w_i \leq \|\mathbf{x} - \mathbf{x}_j\|^2 - w_j\}, \quad j \neq i. \tag{3.1.20}$$

After a few steps of calculation from equation (3.1.20), we obtain the bisector as

$$b(p_i, p_j) = \{\mathbf{x} | (\mathbf{x}_j - \mathbf{x}_i)^{\mathrm{T}} \mathbf{x} = \tfrac{1}{2}(\|\mathbf{x}_j\|^2 - \|\mathbf{x}_i\|^2 + w_i - w_j)\}, \quad j \neq i. \tag{3.1.21}$$

The bisector is hence the straight line perpendicular to $\overline{p_i p_j}$ passing through the point $\mathbf{x}^*_{ij}$ given by

$$\mathbf{x}^*_{ij} = \frac{\|\mathbf{x}_j\|^2 - \|\mathbf{x}_i\|^2 + w_i - w_j}{2\|\mathbf{x}_j - \mathbf{x}_i\|^2} (\mathbf{x}_j - \mathbf{x}_i). \tag{3.1.22}$$

As w_i increases, the dominance region becomes larger. Because of this we usually use a positive weight for w_i. Mathematically, however, we may replace $-w_i$ with $+w_i$ in equation (3.1.19). Figure 3.1.9 shows $\mathrm{Dom}(p_i, p_j)$ for several parameter values. As is seen in this figure, the generator p_j is not in $\mathrm{Dom}(p_j, p_i)$ for $w_i = 31, 41, 51$.

Since the line of equation (3.1.22) splits $\mathbb{R}^2$ into two disjoint half planes, the bisector of equation (3.1.21) is well-behaving and regular. Thus, equation (3.1.2) with equation (3.1.20) gives a generalized Voronoi diagram $\mathcal{V}(P, d_{\mathrm{pw}}, \mathbb{R}^m) = \mathcal{V}_{\mathrm{pw}} = \{V(p_1), \ldots, V(p_n)\}$. We call this generalized Voronoi diagram the *additively weighted power diagram* generated by P or simply

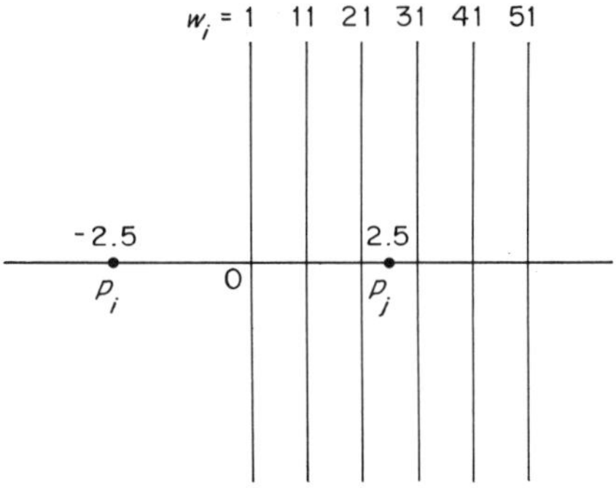

Figure 3.1.9 The bisectors with the additively weighted power distance for several parameter values, $\mathbf{x}_i^{\mathrm{T}} = (-2.5, 0.0)$, $\mathbf{x}_j^{\mathrm{T}} = (2.5, 0.0)$, $w_j = 1$, $w_i = 1, 11, 21, 31, 41$, or a Power diagram for $n = 2$.

the *power diagram* of P (Aurenhammer, 1988), and call the set $V(p_i)$ the *power Voronoi polygon* associated with p_i. As we shall show in Section 3.6, the power diagram can be regarded as a Voronoi diagram of circles or a Voronoi diagram with Laguerre geometry. Also, as we shall discuss in the next subsection, the power diagram can be regarded as a sectional Voronoi diagram.

We find in Figure 3.1.10 that some power polygons are infinite. Imai *et al*. (1985) derive the conditions for a finite or an infinite region as follows.

Property PW1 (i) The set $V(p_i)$ defined by equation (3.1.2) with equation (3.1.20) is non-empty and infinite if p_i is a vertex of CH(P).

(ii) The set $V(p_i)$ is either infinite or empty if p_i is on the boundary of CH(P), except the vertices.

(iii) The set $V(p_i)$ is either finite or empty if p_i is an interior point of CH(P).

The proof is provided by Imai *et al*. (1985, Lemma 3).

From the fact that the dominance region is a half plane and that $V(p_i)$ is the intersection of half planes, we note the following property.

Property PW2 If the set $V(p_i)$ is not empty, it is a convex polygon.

In Figure 3.1.9, we observe that the dominance region Dom(p_i, p_j) with the parameter values $w_i = 31, 41, 51$ includes not only p_i but also p_j. From this property, we notice the following property.

Property PW3 The generator of $V(p_i)$ may not be in $V(p_i)$.

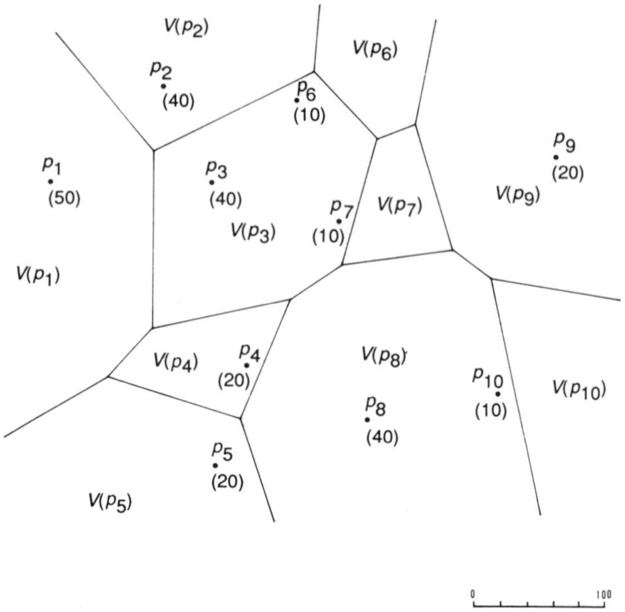

Figure 3.1.10 A(n) (additively weighted) power diagram.

3.1.5 Sectional Voronoi diagram

In the literature the sectional Voronoi diagram to be shown here is usually not subsumed under the family of weighted Voronoi diagrams. We, however, refer to it here because it can be regarded as a weighted Voronoi diagram.

Let $\mathcal{V} = \{V(p_1), \ldots, V(p_n)\}$ be the three-dimensional ordinary Voronoi diagram generated by P, and K be a plane in $\mathbb{R}^3$. Let

$$V_{\mathrm{sec}}(p_i') = V(p_i) \cap K \neq \varnothing, \tag{3.1.23}$$

$$\mathcal{V}_{\mathrm{sec}} = \{V_{\mathrm{sec}}(p_i'), i \in J_s\}, \tag{3.1.24}$$

where p'_i is the orthogonal projection of p_i on the plane K, and let J_s be the set of indices that satisfy $V(p_i) \cap K \neq \varnothing$. The set $V_{\mathrm{sec}}(p'_i)$ is the intersection of polyhedron $V(p_i)$ with the plane K. We call $V_{\mathrm{sec}}(p_i)$ a *sectional Voronoi polygon*. Obviously, the collection of resulting sectional Voronoi polygons covers the plane K and they are not overlapped, except at the boundaries. Thus the set $\mathcal{V}_{\mathrm{sec}}$ forms a tessellation. We call this tessellation the *sectional Voronoi diagram* obtained as the intersection of $\mathcal{V}$ with the plane K (Ash and Bolker, 1986; Aurenhammer, 1988; Imai *et al.*, 1985; Sibson, 1980). An example is shown in Figure 3.1.11 where generators are placed at (1, 0, 1), (−1, 0, 0), (0, −1, 0) in $\mathbb{R}^3$. The ordinary Voronoi diagram of those generator points are indicated by the heavy lines (to make visual perception

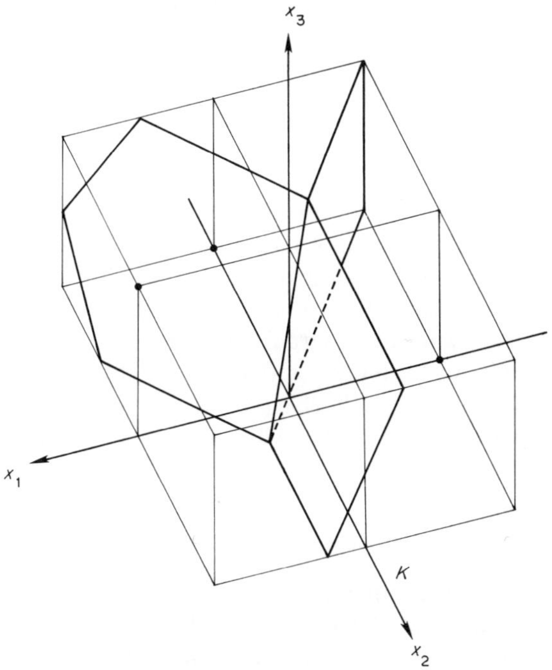

Figure 3.1.11 A sectional Voronoi diagram obtained from the three-dimensional ordinary Voronoi diagram with the x_1–x_2 plane, $n = 3$.

easy, the intersection of $\mathcal{V}$ with the rectangle parallopipedron are shown). The intersection of $\mathcal{V}$ with the x_1–x_2 plane (the plane K) is shown by the diagram on the x_1–x_2 plane.

Generalizing the concept of the sectional Voronoi diagram in $\mathbb{R}^3$, we can define a sectional Voronoi diagram of an m-dimensional Voronoi diagram with a hyperplane in $\mathbb{R}^m$.

Suppose, for simplicity, that the plane K is the x_1–x_2 plane (we can place the plane K on the x_1–x_2 plane by rotating the plane K with P). Then $V_{\text{sec}}(p'_i)$ is written as

$$V_{\text{sec}}((x_i,x_j)) = \{(x_1,x_2)|\sqrt{(x_1-x_{i1})^2+(x_2-x_{i2})^2+x_{i3}^2} \leq \sqrt{(x_1-x_{j1})^2+(x_2-x_{j2})^2+x_{j3}^2}, j\neq i, j\in J_s\}. \quad (3.1.25)$$

If we write $\|\mathbf{x}-\mathbf{x}_i\|^2$ for $(x_1-x_{i1})^2+(x_2-x_{i2})^2$, then equation (3.1.25) is written as

$$V_{\text{sec}}(\mathbf{x}_i, \mathbf{x}_j) = \{\mathbf{x}_i|\ \|\mathbf{x}-\mathbf{x}_i\|^2+x_{i3}^2 \leq \|\mathbf{x}-\mathbf{x}_j\|^2+x_{j3}^2, j\neq i, j\in J_s\}. \quad (3.1.26)$$

Recalling the power distance of equation (3.1.19), we note that this sectional Voronoi diagram is indeed the power diagram with the weight $\{-x_{i3}^2, i\in J_s\}$.

3.2 HIGHER-ORDER VORONOI DIAGRAMS

In the ordinary Voronoi diagram a generator is a point p_i, or a generator set is a set of points $P = \{p_i, \ldots, p_n\}$. In this section, which extends a point to a set of points, we consider the family of generalized Voronoi diagrams generated by a set of size-k subsets of P, $A^{(k)}(P) = \{\{p_{11}, \ldots, p_{1k}\}, \ldots, \{p_{l1}, \ldots, p_{lk}\}\}$, $p_{ij} \in P$, $l = {}_n\mathrm{C}_k$. We shall call this family the 'higher-order' Voronoi diagram. Here the 'order' means the number of points constituting a generator and 'higher' means more than one point. Note that 'higher' does not refer to the dimension of a space. In this section we deal with only $\mathbb{R}^2$, but conceptually the extension from $\mathbb{R}^2$ to $\mathbb{R}^m$ is straightforward.

The higher order Voronoi diagram has been studied by many researchers, including Miles (1970a), Shamos and Hoey (1975), Shamos (1978), Bentley and Maurer (1979), Lee (1982b), Miles and Maillardet (1982), Sibson (1980a), Chazelle *et al.* (1986). This section is based upon the results obtained by those researchers.

3.2.1 Order-k Voronoi diagram

Suppose, as is illustrated in Figure 3.2.1, that a set of distinct points (the filled circles) is placed in the Euclidean plane. We consider the Euclidean distances from a location, say the location indicated by the unfilled circle in the figure, to all points in the point set. As can be seen from the dashed lines in Figure 3.2.1, the first and the second nearest points from the location

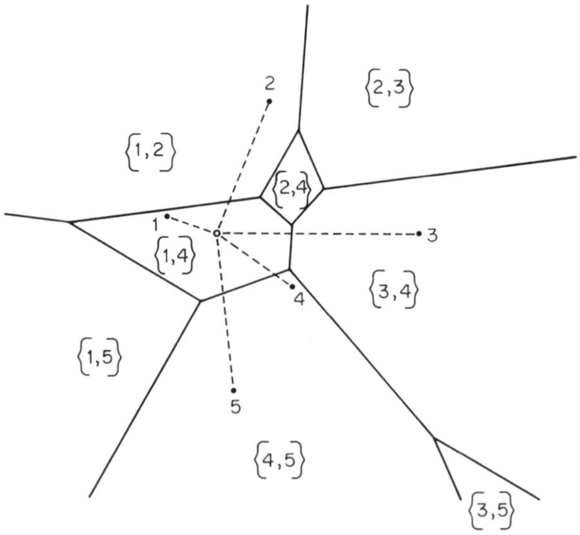

Figure 3.2.1 An order-2 Voronoi diagram (n = 5).

are p_1 and p_4. In this case we assign the location to the set $\{p_1, p_4\}$. In general, if a set of the first and the second nearest points from a location is $\{p_i, p_j\}$, we assign the location to $\{p_i, p_j\}$. If the set is not only $\{p_i, p_j\}$ but also another sets of two points, we assign the location to those sets. Note that we are not concerned with which point is the first nearest point in $\{p_i, p_j\}$; p_i may be the first or the second nearest point. Following this assignment rule, we assign all locations in the plane to at least one of the sets of two points. As a result, locations assigned to $\{p_i, p_j\}$ form a region, which we call the *order-2 Voronoi polygon* associated with $\{p_i, p_j\}$. A collection of the resulting order-2 Voronoi polygons is depicted in Figure 3.2.1, where the region with symbol $\{i, j\}$ indicates the order-2 Voronoi polygon associated with $\{p_i, p_j\}$ and i associated with a filled circle indicates the point p_i. This collection forms a tessellation, which we call the *order-2 Voronoi diagram* generated by the set of points.

Having understood the order-2 Voronoi diagram, the reader may be curious to know what the configurational relationship is between this diagram and the ordinary Voronoi diagram. Figure 3.2.2 shows this relationship. The dashed lines indicate the ordinary Voronoi diagram and the solid lines indicate the order-2 Voronoi diagram generated by the same generator set. We notice from this figure that the two Voronoi diagrams do not share edges but share some vertices (the unfilled circles). These vertices have an

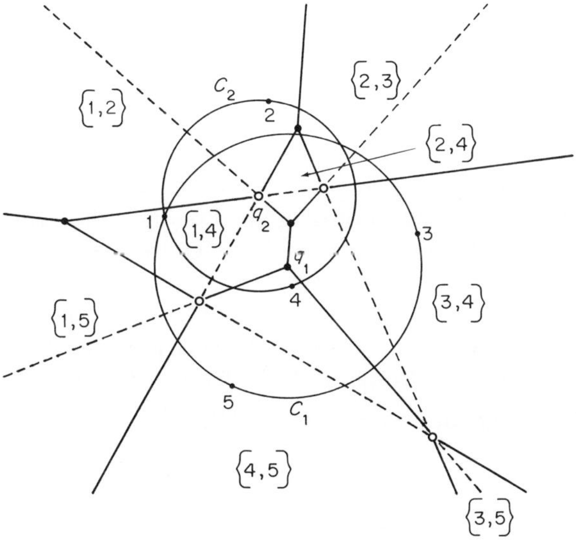

Figure 3.2.2 The ordinary Voronoi diagram (the dashed lines) and the order-2 Voronoi diagram (the solid lines) generated by the same set of generator points (n = 5, the locations of generator points are the same as those in Figure 3.2.1).

interesting property. To show it precisely, we first give the mathematical definition of the order-2 Voronoi diagram.

Let $P = \{p_1, \ldots, p_n\}$ be a set of distinct points in $\mathbb{R}^2$, where $2 \leq n < \infty$; $A^{(2)}(P) = \{P_1^{(2)}, \ldots, P_i^{(2)}, \ldots, P_l^{(2)}\}$, where $P_i^{(2)} = \{p_{i1}, p_{i2}\}$, $p_{i1}, p_{i2} \in P$, and $l = {}_nC_2$; $d(p, p_{ij})$ be the Euclidean distance from p to p_{ij}. The points p_{i1} and p_{i2} are the first and the second nearest points from p if and only if the distances from p to p_{i1} and p_{i2} are both shorter than or equal to the distances from p to the other points (i.e. points in $P \backslash \{p_{i1}, p_{i2}\}$). Thus the set of points $V(P_i^{(2)})$ assigned to $\{p_{i1}, p_{i2}\}$ is written as

$$V(P_i^{(2)}) = \{p | d(p, p_{i1}) \leq d(p, p_j) \text{ and } d(p, p_{i2}) \leq d(p, p_j) \text{ for } p_j \in P \backslash P_i^{(2)}\}. \quad (3.2.1)$$

The assignment rule in equation (3.2.1) is equivalent to the assignment rule that the longest distance among the distances from p to p_{i1} and p_{i2} is shorter than or equal to the shortest distance among the distances from p to the other points (i.e. points in $P \backslash \{p_{i1}, p_{i2}\}$). Thus $V(P_i^{(2)})$ is alternatively written as

$$V(P_i^{(2)}) = \{p | \max_{p_h}\{d(p, p_h) | p_h \in P_i^{(2)}\} \leq \min_{p_j}\{d(p, p_j) | p_j \in P \backslash P_i^{(2)}\}\}. \quad (3.2.2)$$

The extension of the order-2 Voronoi diagram to the order-k Voronoi diagram is straightforward. Let $A^{(k)}(P)$ be the set of all possible subsets consisting of k elements out of P, i.e. $A^{(k)}(P) = \{P_1^{(k)}, \ldots, P_i^{(k)}, \ldots, P_l^{(k)}\}$, where $P_i^{(k)} = \{p_{i1}, \ldots, p_{ik}\}$, $p_{ij} \in P$, and $l = {}_nC_k$. In these terms, we define

$$V(P_i^{(k)}) = \{p | \max_{p_h}\{d(p, p_h) | p_h \in P_i^{(k)}\} \leq \min_{p_j}\{d(p, p_j) | p_j \in P \backslash P_i^{(k)}\}. \quad (3.2.3)$$

We call the set $V(P_i^{(k)})$ the *order-k Voronoi polygon* associated with $P_i^{(k)}$ and the set of order-k Voronoi polygons, $\mathcal{V}(A^{(k)}(P), d, \mathbb{R}^m) = \mathcal{V}^{(k)} = \{V(P_1^{(k)}), \ldots, V(P_l^{(k)})\}$, the *order-k Voronoi diagram* generated by P. In the literature, $\mathcal{V}^{(k)}$ is sometimes called the generalized Voronoi diagram, the kth order Voronoi diagram and the Voronoi diagram of order k. Figure 3.2.3 shows an order-3 Voronoi diagram in which the region with symbol $\{i, j, k\}$ indicates the order-3 Voronoi polygon of $\{p_i, p_j, p_k\}$.

Alternatively, as we did with the ordinary as well as the weighted Voronoi diagrams, we can define an order-k Voronoi polygon by the intersection of the dominance regions. We should, however, pay careful attention to this definition. The dominance region of $P_i^{(2)}$ over $P_j^{(2)}$ is written as

$$\text{Dom}(P_i^{(2)}, P_j^{(2)}) = \{p | \max_{p_h}\{d(p, p_h) | p_h \in P_i^{(2)}\} \leq \min_{p_j}\{d(p, p_j) | p_j \in P_j^{(2)} \backslash P_i^{(2)}\}\}. \quad (3.2.4)$$

In Figure 3.2.4 we show $\text{Dom}(P_i^{(2)}, P_j^{(2)})$ and $\text{Dom}(P_j^{(2)}, P_i^{(2)})$ by the vertically and horizontally hatched regions, respectively. Since the hatched regions do not cover $\mathbb{R}^2$, $\text{Dom}(P_i^{(2)}, P_j^{(2)})$ is not regular. Although $\text{Dom}(P_i^{(2)}, P_j^{(2)})$ is not regular, it is well-behaving, and hence the set $\{V(P_1^{(2)}), \ldots, V(P_l^{(2)})\}$ produces a generalized Voronoi diagram, where $V(P_i^{(2)})$ is given by

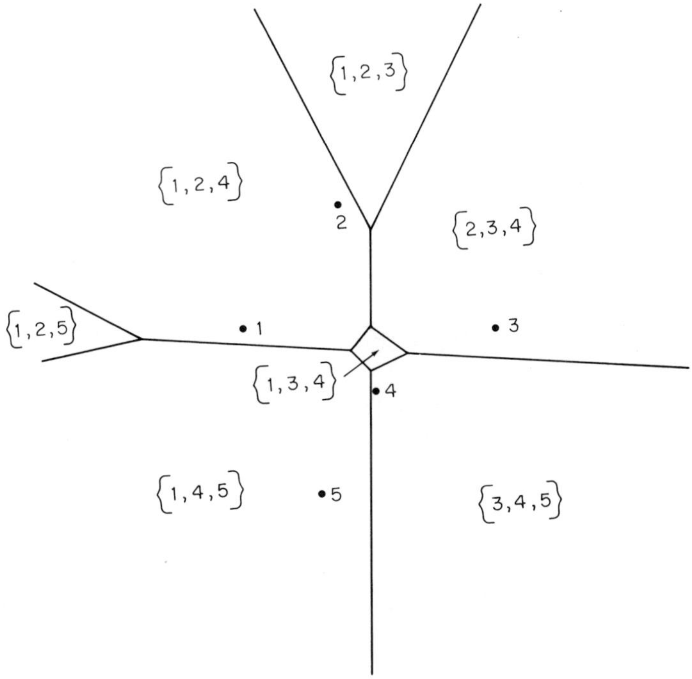

Figure 3.2.3 An order-3 Voronoi diagram (n = 5, the locations of points are the same as those in Figure 3.2.1).

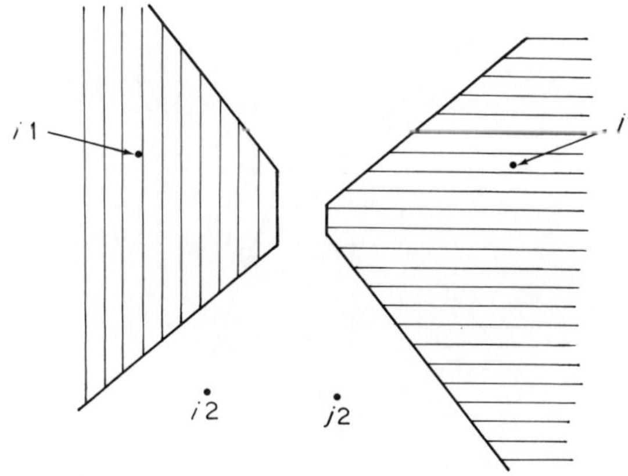

Figure 3.2.4 The dominance regions $\mathrm{Dom}(P_i^{(2)}, P_j^{(2)})$ (the vertically hatched region) and $\mathrm{Dom}(P_j^{(2)}, P_i^{(2)})$ (the horizontally hatched region), $P_i^{(2)} = \{p_{i1}, p_{i2}\}$, $P_j^{(2)} = \{p_{j1}, p_{j2}\}$.

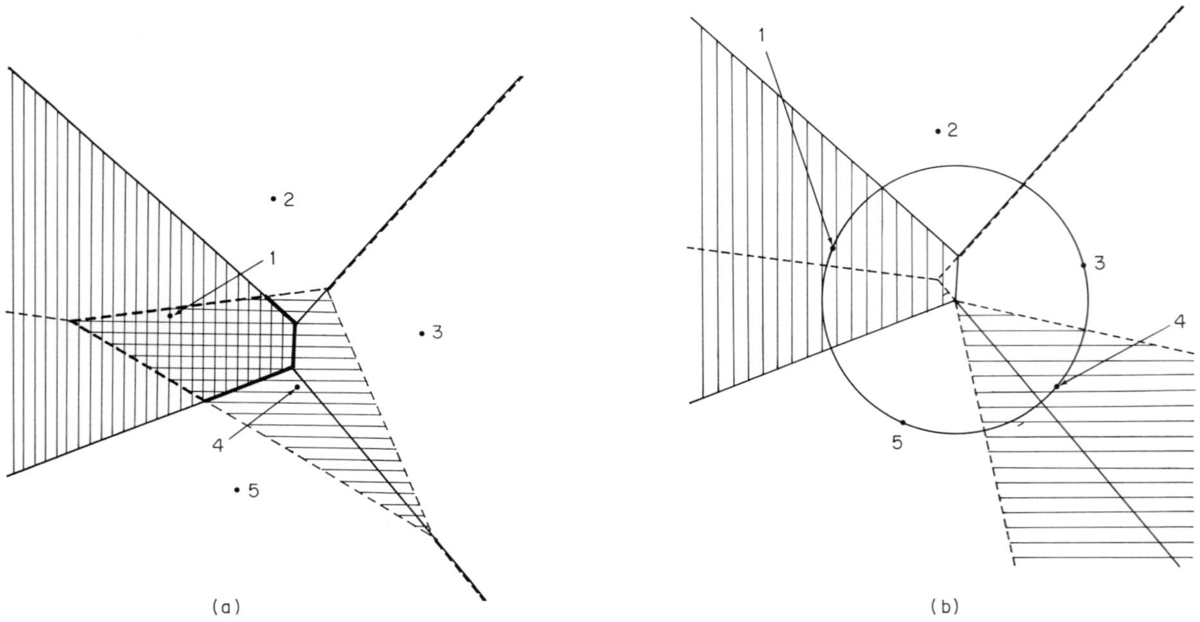

Figure 3.2.5 An order-2 Voronoi polygon obtained from the intersection of two ordinary Voronoi polygons: (a) the intersection has a non-zero area; (b) the intersection is a point.

$$V(P_i^{(2)}) = \bigcap_{j \in I_i \setminus \{i\}} \mathrm{Dom}(P_i^{(2)}, P_j^{(2)}). \tag{3.2.5}$$

The extension of this definition to a general k is immediate.

In addition to the definitions of $V(P_i^{(2)})$, using equations (3.2.1) and (3.2.5) we give one more definition. Since the set of points satisfying $d(p, p_{i1}) \leq d(p, p_j)$ and $d(p, p_{i2}) \leq d(p, p_j)$ is given by $H(p_{i1}, p_j) \cap H(p_{i2}, p_j)$, equation (3.2.1) is alternatively written as

$$V(P_i^{(2)}) = \bigcap_{p_j \in P \setminus \{p_{i1}, p_{i2}\}} [H(p_{i1}, p_j) \cap H(p_{i2}, p_j)], \tag{3.2.6}$$

where $H(p_{i1}, p_j)$ is the half plane used in defining the ordinary Voronoi diagram. Generalizing this definition for a general k, we obtain:

$$V(P_i^{(k)}) = \bigcap_{p_j \in P \setminus P_i^{(k)}} [H(p_{i1}, p_j) \cap \ldots \cap H(p_{ik}, p_j)]. \tag{3.2.7}$$

This definition is useful in examining the properties of the order-k Voronoi diagram.

The order, k, varies from one to n. An order-1 Voronoi diagram is the ordinary Voronoi diagram; thus an order-k Voronoi diagram can be regarded as a generalization of the ordinary Voronoi diagram. If $k = n$, then $A^{(n)}(P) = \{P\}$, and $\mathcal{V}^{(k)} = \{\mathbb{R}^2\}$. Since this diagram is too trivial, we exclude it from our discussion. If $k = n-1$, an order-$(n-1)$ Voronoi diagram has a special name, which we shall discuss in the next subsection.

As we remarked in the introduction of Chapter 3, we may generalize the order-k Voronoi diagram with respect to a distance. Among many possible distances, the order-k Voronoi diagram with the additively weighted power distance (equation (3.1.19)) is notable. In the literature this Voronoi diagram is called the *order-k power diagram* (Aurenhammer, 1987). The reader who wishes to understand this Voronoi diagram in depth should consult Aurenhammer (1988).

Having defined $\mathcal{V}^{(k)}$, we now wish to examine its geometric properties. From the definition of equation (3.2.6) we obtain:

$$V(P_i^{(2)}) = \left[\bigcap_{p_h \in P \setminus P_i^{(2)}} H(p_{i1}, p_h)\right] \cap \left[\bigcap_{p_h \in P \setminus P_i^{(2)}} H(p_{i2}, p_h)\right]. \tag{3.2.8}$$

Let $V(p_i|P\setminus\{p_j\})$ denote the Voronoi polygon of p_i in the ordinary Voronoi diagram generated by $P\setminus\{p_j\}$, i.e. $\mathcal{V}(P\setminus\{p_j\})$. Then, equation (3.2.8) is written as

$$V(P_i^{(2)}) = V(p_{i1}|P\setminus\{p_{i2}\}) \cap V(p_{i2}|P\setminus\{p_{i1}\}). \tag{3.2.9}$$

Figure 3.2.5(a) illustrates equation (3.2.9). The solid lines show $\mathcal{V}(P\setminus\{p_4\})$ and the vertically hatched region indicates $V(p_1|P\setminus\{p_4\})$. The hatched region consists of the region in which the first nearest point is p_1 and the region in which the first and second nearest points are p_4 and p_1, respectively. The dashed lines show $\mathcal{V}(P\backslash\{p_1\})$ and the horizontally hatched region indicates $V(p_4|P\backslash\{p_1\})$. This region consists of the region in which the first nearest point is p_4 and the region in which the first and second nearest points are p_1 and p_4 respectively. Thus the intersection of the vertically and horizontally hatched regions is the region in which the first and the second nearest points are p_1 and p_4, respectively, or they are p_4 and p_1, respectively. Thus the intersection gives $V(\{p_1, p_4\})$. In panel (a), the intersection is non-empty, but in some cases it may be empty. In a very special case, the intersection may be a point. In fact, panel (b) shows this example. This special case occurs when points in P are cocircular.

We can generalize equation (3.2.9) for a general k (Miles and Maillardet, 1982, Lemma on p. 102). From equation (3.2.7), we obtain the following property.

Property OK1

$$V(P_i^{(k)}) = \bigcap_{h=1}^{k} V(p_{ih}|[P\backslash P_i^{(k)}]\cup\{p_{ih}\}). \tag{3.2.10}$$

From this property we can derive a few important properties. First, as we mentioned above, the set $V(P_i^{(k)})$ may be empty. If $V(P_i^{(k)})$ is not empty, it is a point or an area. If $V(P_i^{(k)})$ is an area, it is a polygon and, moreover, it is convex, because $V(p_{ih}|[P\backslash P_i^{(k)}]\cup\{p_{ih}\})$ is, from Property V1, a convex polygon and the intersection of convex polygons is a convex polygon. To sum up, we obtain the following property.

Property OK2 The set $V(P_i^{(k)})$ given by equation (3.2.3) may be empty, a point or an area. Under the non-cocircularity assumption, a non-empty $V(P_i^{(k)})$ is a convex polygon.

In observing Figure 3.2.2 we notice that every order-2 Voronoi polygon is split by one Voronoi edge of the ordinary Voronoi polygon. We can intuitively understand this property from the fact that the first nearest point from a point in $V(\{p_i, p_j\})$ is p_i or p_j, i.e. $V(\{p_i, p_j\})$ consists of part of $V(p_i)$ and part of $V(p_j)$ which share one Voronoi edge. From this property and Property D7, the number, $n_{\rm f}^{(2)}$, of non-empty order-2 Voronoi polygons is equal to the number of Voronoi edges in the ordinary Voronoi diagram, i.e. $n_{\rm f}^{(2)} = n_{\rm e} = 3n-3-n_{\rm u}$. Generalizing this equation for k, Lee (1982b, Theorem 2 and Corollary 3) proves the following property.

Property OK3 If P satisfies the non-cocircularity assumption, the number, $n_{\rm f}^{(k)}$, of non-empty Voronoi polygons in $\mathcal{V}^{(k)}$ is given by

$$n_{\rm f}^{(k)} = (2k-1)n-(k^2-1) - \sum_{i=1}^{k} n_{\rm u}^{(i-1)}, \tag{3.2.11}$$

where $n_u^{(i-1)}$ is the number of infinite order-$(i-1)$ Voronoi polygons ($n_u^{(0)} = 0$). Consequently, $n_f^{(k)}$ is $O(k(n-k))$.

When we use equation (3.2.11), we have to count $n_u^{(i-1)}$. To this end, let us obtain a necessary and sufficient condition for an order-$(i-1)$ Voronoi polygon to be infinite. Let P_1 and P_2 be two disjoint sets of points in the plane. We say that the line L *separates* P_1 and P_2 if the points in P_1 and the points in P_2 lie in mutually opposite open half planes defined by L. In these terms the necessary and sufficient condition is written as follows.

Property OK4 The order-k Voronoi polygon $V(P_i^{(k)})$ is infinite if and only if either (i) there is a line that separates $P_i^{(k)}$ (the filled circles in Figure 3.2.6(a), (b)) and $P \backslash P_i^{(k)}$ (the unfilled circles in Figure 3.2.6(a), (b)) or (ii) there are two consecutive points p_j, p_l ($\in P \backslash P_i^{(k)}$) on $\partial CH(P \backslash P_i^{(k)})$ such that the points in $P_i^{(k)} \backslash \overline{p_j p_l}$ are in the open half plane defined by the line passing through p_j and p_l opposite to $CH(P \backslash P_i^{(k)})$ (Figure 3.2.6(c)).

Proof First, suppose that either (i) or (ii) holds. If (i) holds, we transform the coordinate system in such a way that the separating line L is vertical and $CH(P \backslash P_i^{(k)})$ is to the left of L (Figure 3.2.6(a)). If (ii) holds, we transform the coordinate system in such a way that $\overline{p_j p_l}$ is vertical, the midpoint of $\overline{p_j p_l}$ lies on the x-axis, and $CH(P \backslash P_i^{(k)})$ is to the left of $\overline{p_j p_l}$ (Figure 3.2.6(c)). In both cases the point at infinity on the x-axis in the positive x direction belongs to the region $V(p_{ih} | [P \backslash P_i^{(k)}] \cup \{p_{ih}\})$ for any $p_{ih} \in P_i^{(k)}$. Hence we see from Property OK1 that if either (i) or (ii) holds, the Voronoi polygon $V(P_i^{(k)})$ is infinite.

Next, suppose that $V(P_i^{(k)})$ is infinite. Then, there exists at least one ray X such that the point at infinity on X belongs to $V(P_i^{(k)})$. We transform the coordinate system in such a way that X lies on the x-axis facing toward the positive x direction. Let x_α be the maximum value of the x coordinates of the points in $P \backslash P_i^{(k)}$, and x_β be the minimum value of the x coordinates of

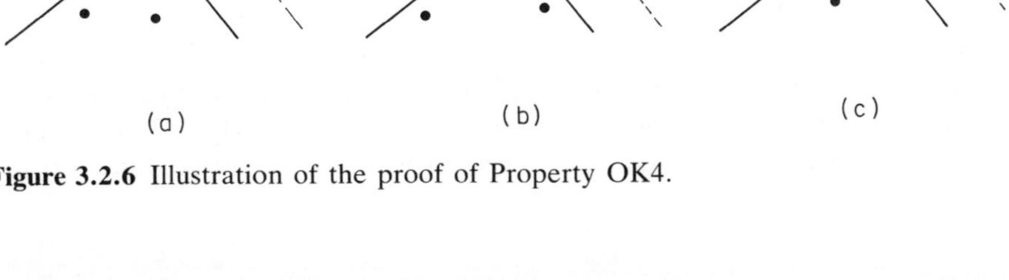

(a) (b) (c)

Figure 3.2.6 Illustration of the proof of Property OK4.

the points in $P_i^{(k)}$ (Figure 3.2.6(a), (b)). Then we get $x_\alpha \leq x_\beta$; this is because if $x_\alpha > x_\beta$, then for point p_{ih} with x coordinate x_β, the polygon $V(p_{ih}|[P\backslash P_i^{(k)}]\cup\{p_{ih}\})$ does not contain the point at infinity on the x-axis in the positive x direction.

Case 1. Suppose that $x_\alpha < x_\beta$ (Figure 3.2.6(a)). Then the vertical line defined by $x = (x_\alpha + x_\beta)/2$ separates $P_i^{(k)}$ and $P\backslash P_i^{(k)}$; hence (i) holds.

Case 2. Suppose that $x_\alpha = x_\beta$ (Figure 3.2.6(b), (c)). Let P_α be the set of points in $P\backslash P_i^{(k)}$ whose x coordinates are equal to x_α (the large filled circles in Figure 3.2.6(b), (c)), and let P_β be the set of points in $P_i^{(k)}$ whose x coordinates are equal to x_β $(= x_\alpha)$ (the large unfilled circles in Figure 3.2.6(c)).

Case 2.1. Suppose that P_α contains exactly one point: $P_\alpha = \{p_j\}$. Then either all the points in P_β are above p_j or all the points in P_β are below p_j, because if $p_{ih} \in P_i^{(k)}$ is above p_j and $p_{it} \in P_i^{(k)}$ is below p_j, the region $V(p_{ih}|[P\backslash P_i^{(k)}]\cup\{p_{ih}\})$ and $V(p_{it}|[P\backslash P_i^{(k)}]\cup\{p_{it}\})$ have no point in common. If the points in P_β are above p_j, we rotate the coordinate axes counterclockwise slightly, and get the situation where $x_\alpha < x_\beta$ and the point at infinity on the new x-axis in the positive x direction still belongs to $V(P_i^{(k)})$; thus we get (i). If the points in P_β are below p_j, we rotate the coordinate axes clockwise and get (i) similarly.

Case 2.2. Suppose that P_α contains two or more points. Then for any $p_j \in P_\alpha$ either all the points in P_β are above p_j or all the points in P_β are below p_j, because otherwise $V(P_i^{(k)})$ becomes empty, as we saw in Case 2.1. If all the points in P_β are above the highest point in P_α or below the lowest, we reduce the situation to Case 1 by slightly rotating the coordinate axes counterclockwise or clockwise, respectively, as we did in Case 2.1. Hence, the only remaining case is that there are two points p_j and p_l in P_α such that $\overline{p_j p_l}$ contains no other points in P_α and $\overline{p_j p_l}$ contains all the points in P_β, which implies (ii). Thus, in all cases either (i) or (ii) holds. □

In inspecting Figure 3.2.1, we notice that $V(\{p_1, p_5\})$ contains no points of P; $V(\{p_1, p_2\})$ contains one point of P; $V(\{p_4, p_5\})$ contains two points of P. In general, we notice the following property.

Property OK5 A non-empty order-k Voronoi polygon contains 0, ..., or k points of P.

In an ordinary Voronoi diagram, every Voronoi vertex has the empty circle (Property V7). To find the corresponding property in an order-k Voronoi diagram, let us first examine a vertex, say q_1, shared by $V(\{p_1, p_4\})$, $V(\{p_3, p_4\})$ and $V(\{p_4, p_5\})$ in Figure 3.2.2. We notice that p_1, p_3 and p_5 are equally distant from the vertex q_1, and hence we can draw the circle C_1 centred at q_1 which passes through points p_1, p_3 and p_5. This circle contains p_4. From this inspection and recalling Property V7, one might expect that for every vertex in an order-2 Voronoi diagram, there exists a circle which passes through three points of P and contains exactly one point of P in its interior. This expectation is, however, false. A counter example is given by the circle C_2 centred at the vertex q_2 shared by $V(\{p_1, p_2\})$, $V(\{p_2, p_4\})$ and $V(\{p_1, p_4\})$ (Figure 3.2.2). The circle C_2 does not contain any point of P in its interior. If we draw circles at all vertices in Figure 3.2.2, we can see that the circles centred at the unfilled circles (which are called 'old vertices' in Lee, 1982b) have no points of P in their interiors, and the circles centred at the large filled circles (which are called 'new vertices') have exactly one point of P in their interiors. This result can be theoretically proved and is generalized by Lee (1982b, Lemma 6) as follows.

Property OK6 Under the non-cocircularity assumption, for every vertex q_i of an order-k Voronoi polygon, there exists a unique circle centred at q_i which passes through three points of P and contains $k-2$ or $k-1$ points of P in its interior.

The condition for $k-2$ or $k-1$ is explicitly stated in Lee (1982b, Lemma 6).

Applying almost the same derivation used for Property D7, we obtain the following property.

Property OK7 Let $n_v^{(k)}, n_e^{(k)}, n_f^{(k)}, n_u^{(k)}$ be the numbers of vertices, edges, finite and infinite order-k Voronoi polygons, respectively. Then under the non-cocircularity assumption, the following equations hold:

$$n_v^{(k)} = 2(n_f^{(k)}-1) - n_u^{(k)}, \tag{3.2.12}$$

$$n_e^{(k)} = 3(n_f^{(k)}-1) - n_u^{(k)}. \tag{3.2.13}$$

These equations with $k = 1$ are, of course, the same as equations (2.4.13) and (2.4.14) in Property D11. The proof is shown by Lee (1982b, Lemma 11).

In addition to these properties, statistical properties of the order-k Voronoi diagram are investigated in spatial statistics. We shall show these properties in Chapter 5. Regarding the properties of the order-k power diagram, see Aurenhammer (1988).

3.2.2 Ordered order-*k* Voronoi diagram

The order-k Voronoi diagram has a closely related tessellation which we call the 'ordered' order-k Voronoi diagram. In the order-k Voronoi diagram, points in a generator are not ordered (recall that in $\mathcal{V}^{(k)}$ we were not concerned with which point was the first nearest point in a generator), whereas in the ordered order-k Voronoi diagram, they are ordered. To

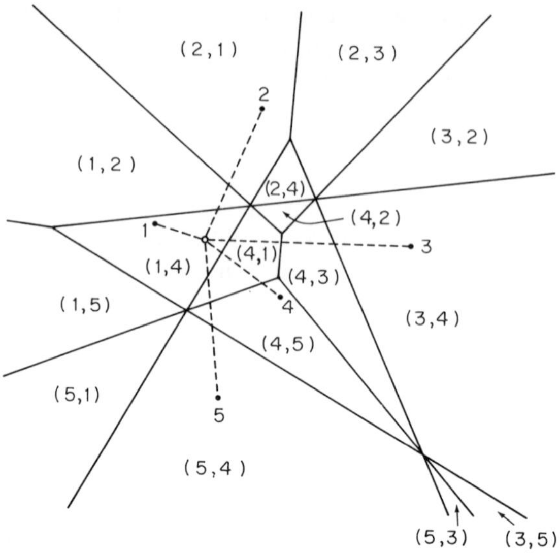

Figure 3.2.7 An ordered order-2 Voronoi diagram (n = 5, the locations of the generator points are the same as those in Figure 3.2.1).

show this contrast, let us begin our discussion with the same example used in Figure 3.2.1.

Suppose that a set of distinctive points (the filled circles in Figure 3.2.7) is placed in the Euclidean plane. We consider the Euclidean distances from a location, say the location indicated by the unfilled circle in Figure 3.2.7, to all points in the point set. The dashed lines in Figure 3.2.7 show that the first nearest point from the location is p_1 and the second nearest point from the location is p_4. In this case we assign this location to the ordered pair (p_1, p_4) (note that $(p_1, p_4) \neq (p_4, p_1)$). In general, if an ordered pair of the first and the second nearest points from a location is (p_i, p_j), we assign the location to (p_i, p_j). If the ordered pairs are not only (p_i, p_j) but also other ordered pairs, then we assign the location to those ordered pairs. In this manner we assign all locations in the plane to at least one ordered pair. As a result, locations assigned to each pair form a region, which we call the *ordered order-2 Voronoi polygon* associated with (p_i, p_j). The collection of ordered order-2 Voronoi polygons is shown in Figure 3.2.7 in which the region with symbol (i, j) indicates the ordered order-2 Voronoi polygon associated with (p_i, p_j). This collection forms a tessellation, which we call the *ordered order-2 Voronoi diagram* generated by the set of points.

To observe the geometric properties of an ordered order-k Voronoi diagram, it is instructive to compare this diagram with the ordinary Voronoi diagram, $\mathcal{V}$, generated by the same set of generator points. In Figure 3.2.8 the dashed lines show $\mathcal{V}$, and the solid lines show the ordered order-2

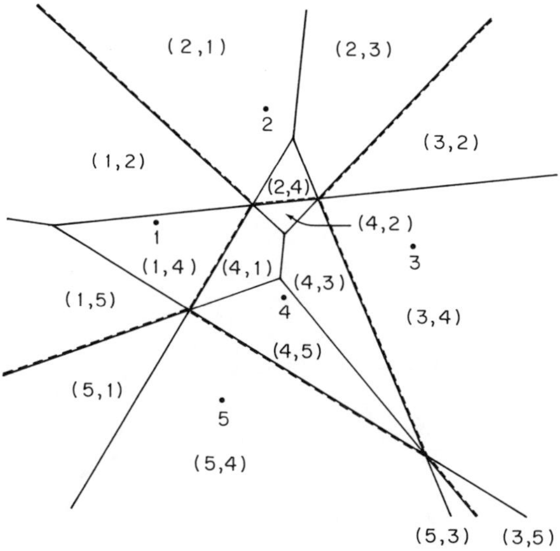

Figure 3.2.8 The ordered-2 Voronoi diagram (the solid lines) and the ordinary Voronoi diagram (the broken lines) generated by the same set of points (n = 5, the locations of the generator points are the same as those in Figure 3.2.1).

Voronoi diagram. In this figure, consider a Voronoi polygon, for example $V(p_4)$ in $\mathscr{V}$. Since the first nearest point of P from a location in $V(p_4)$ is p_4, the ordered order-2 Voronoi polygons associated with (p_4, p_j), $j = 1$, 2, 3, 5, are included in $V(p_4)$, and these polygons constitute $V(p_4)$. Thus the Voronoi edges in $\mathscr{V}$ are the edges of the order-2 Voronoi diagram. To deepen this analysis, let us restate the above verbal definition in mathematical terms.

Let $A^{\langle 2\rangle}(P)$ be the set of all ordered pairs of points obtained from $P = \{p_1, \ldots, p_n\}$, i.e. $A^{\langle 2\rangle}(P) = \{P_1^{\langle 2\rangle}, \ldots, P_i^{\langle 2\rangle}, \ldots, P_l^{\langle 2\rangle}\}$, where $P_i^{\langle 2\rangle} = (p_{i1}, p_{i2})$, p_{i1}, $p_{i2} \in P$, and $l = n(n-1)$. For a set $P_i^{\langle 2\rangle}$ in $A^{\langle 2\rangle}(P)$, we define

$$V(P_i^{\langle 2\rangle}) = \{p|d(p, p_{i1}) \leq d(p, p_{i2}) \leq d(p, p_j), p_j \in P \backslash \{p_{i1}, p_{i2}\}\}. \quad (3.2.14)$$

We call the set $V(P_i^{\langle 2\rangle})$ the *ordered order-2 Voronoi polygon* associated with $P_i^{\langle 2\rangle}$.

The extension to a general k is immediate. Let $A^{\langle k\rangle}(P)$ be the set of all ordered k-tuples of points obtained from $P = \{p_1, \ldots, p_n\}$, i.e. $A^{\langle k\rangle}(P) = \{P_1^{\langle k\rangle}, \ldots, P_i^{\langle k\rangle}, \ldots P_l^{\langle k\rangle}\}$, where $P_i^{\langle k\rangle} = (p_{i1}, \ldots, p_{ik})$, $p_{ij} \in P$, $j \in I_k$, and $l = n(n-1) \ldots (n-k+1)$. For a set $P_i^{\langle k\rangle}$ in $A^{\langle k\rangle}(P)$, we define

$$V(P_i^{\langle k\rangle}) = \{p|d(p, p_{i1}) \leq d(p, p_{i2}) \leq \ldots \leq d(p, p_{ik}) \leq d(p, p_j), \\ p_j \in P \backslash \{p_{i1}, \ldots, p_{ik}\}\}. \quad (3.2.15)$$

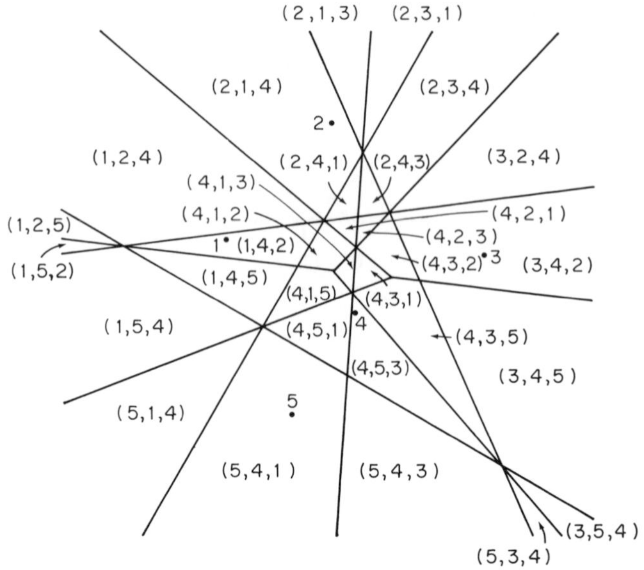

Figure 3.2.9 An ordered order-3 Voronoi diagram ($n = 3$, the locations of generator points are the same as those in Figure 3.2.1).

We call the set $V(P_i^{\langle k\rangle})$ the *ordered order-k Voronoi polygon* associated with $P_i^{\langle k\rangle}$, and $\mathcal{V}(A^{\langle k\rangle}(P), d, \mathbb{R}^m) = \mathcal{V}^{\langle k\rangle} = \{V(P_1^{\langle k\rangle}), \ldots, V(P_l^{\langle k\rangle})\}$ the *ordered order-k Voronoi diagram* generated by P. Figure 3.2.9 shows an ordered order-3 Voronoi diagram.

Alternatively, we can formally define $V(P_i^{\langle k\rangle})$ with the dominance region $\mathrm{Dom}(P_i^{\langle k\rangle}, P_j^{\langle k\rangle})$. This dominance region, however, is not regular (like Dom $(P_i^{\langle k\rangle}, P_j^{\langle k\rangle})$ shown in Figure 3.2.4 for $k = 2$). To analyse the geometric properties of $\mathcal{V}^{\langle k\rangle}$, the following alternative definition is more useful. The set of points satisfying $d(p, p_{i1}) \le d(p, p_{i2}) \le d(p, p_j)$ is given by $H(p_{i1}, p_{i2}) \cap H(p_{i2}, p_j)$. Thus, equation (3.2.14) is written as

$$V(P_i^{\langle 2\rangle}) = \bigcap_{p_j \in P \setminus \{p_{i1}, p_{i2}\}} H(p_{i1}, p_{i2}) \cap H(p_{i2}, p_j). \tag{3.2.16}$$

Similarly, for a general k, equation (3.2.15) is written as

$$V(P_i^{\langle k\rangle}) = \bigcap_{p_j \in P \setminus \{p_{i1}, \ldots, p_{ik}\}} H(p_{i1}, p_{i2}) \cap \ldots \cap H(p_{ik-1}, p_{ik}) \cap H(p_{ik}, p_j). \tag{3.2.17}$$

Having defined the ordered order-k Voronoi diagram $\mathcal{V}^{\langle k\rangle}$ mathematically, we are now ready to examine its geometric properties. From equation (3.2.16) we can write $V(P_i^{\langle 2\rangle})$ as

$$V(P_i^{\langle 2\rangle}) = \bigcap_{p_h \in P\backslash\{p_{i1}, p_{i2}\}} H(p_{i1}, p_{i2}) \cap H(p_{i1}, p_h) \cap H(p_{i2}, p_h)$$

$$= \left[\bigcap_{p_h \in P\backslash\{p_{i1}\}} H(p_{i1}, p_h) \right] \cap \left[\bigcap_{p_h \in P\backslash\{p_{i1}, p_{i2}\}} H(p_{i2}, p_h) \right]$$

$$= V(p_{i1} \cap V(p_{i2}|P\backslash\{p_{i1}\}). \tag{3.2.18}$$

Let us confirm this equation with $V((p_1, p_4))$ in Figure 3.2.8. Since the first nearest point in $V((p_1, p_4))$ is p_1, $V((p_1, p_4))$ is included in $V(p_1)$. The second nearest point in $V((p_1, p_4))$ becomes the first nearest point if p_1 is deleted from P. Thus $V((p_1, p_4))$ is given by $V(p_1) \cap V(p_4|P\backslash\{p_1\})$. Generalizing equation (3.2.18), we obtain the following equation.

Property OOK1 An ordered order-k Voronoi polygon is written in terms of ordinary Voronoi polygons as

$$V(P_i^{\langle k\rangle}) = \bigcap_{h=1}^{k} V(p_{ih}|P\backslash\{p_{i1}, \ldots, p_{ih-1}\}), \tag{3.2.19}$$

where $\{p_{i1}, \ldots, p_{ih-1}\} = \varnothing$ for $h = 1$.

In Figure 3.2.8 we observe that $V(p_4)$ consists of $V((p_4, p_2))$, $V((p_4, p_1))$, $V((p_4, p_5))$ and $V((p_4, p_3))$. We can generalize this property for a general k as follows.

Property OOK2 Let $A^{\langle k-1\rangle}(P\backslash\{p_{i1}\})$ be all possible ordered $(k-1)$-tuples obtained from $P\backslash\{p_{i1}\}$. Then

$$V(p_{i1}) = \bigcup_{(p_{i2}, \ldots, p_{ik}) \in A^{\langle k-1\rangle}(P\backslash\{p_{i1}\})} V((p_{i1}, p_{i2}, \ldots, p_{ik})). \tag{3.2.20}$$

This property implies that the ordered order-k Voronoi diagram $\mathcal{V}$ generated by $A^{\langle k\rangle}(P)$ is a refinement of the ordinary Voronoi diagram generated by P. Figure 3.2.8 shows the refinement of $\mathcal{V}(P)$ by $\mathcal{V}^{\langle 2\rangle}(A^{\langle k\rangle}(P))$.

An ordered order-k Voronoi diagram also refines the order-k Voronoi diagram. The region in which the first and the second nearest points are $\{p_i, p_j\}$ consists of the region in which the first and second nearest points are p_i and p_j, respectively, and the region in which the first and the second nearest points are p_j and p_i, respectively (observe this relation in Figure 3.2.8). Thus $V(\{p_i, p_j\}) = V((p_i, p_j)) \cup V((p_j, p_i))$. More generally, we have the following property.

Property OOK3

$$V(P_i^{(k)}) = \bigcup_{P_j^{\langle k\rangle} \in A^{\langle k\rangle}(P_i^{(k)})} V(P_j^{\langle k\rangle}), \tag{3.2.21}$$

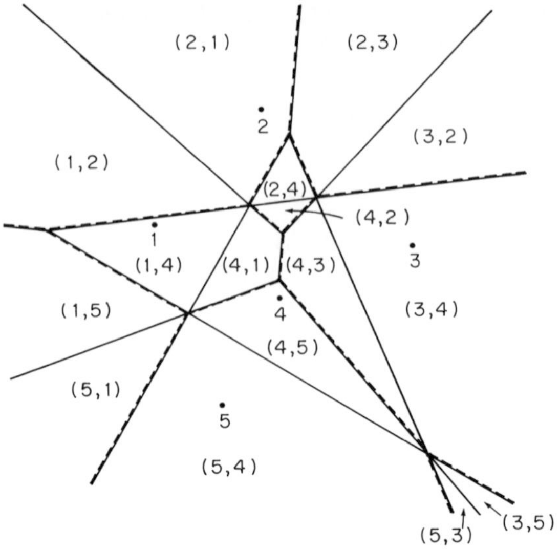

Figure 3.2.10 The order-2 Voronoi diagram (the dashed lines) and the ordered order-2 Voronoi diagram (the solid lines) generated by the same set of points; the ordered order-2 Voronoi diagram is a refinement of the order-2 Voronoi diagram.

where $A^{\langle k \rangle}(P_i^{\langle k \rangle})$ is the set of all possible k-tuples made of $p_{i1}, \ldots, p_{ik}$.

Equation (3.2.21) says that $\mathcal{V}^{\langle k \rangle}$ generated by P is a refinement of $\mathcal{V}^{(k)}$ generated by P. An example is shown in Figure 3.2.10.

From Property OOK1, we obtain the next property.

Property OOK4 The set $V(P_i^{\langle k \rangle})$ given by equation (3.2.15) may be empty. If $V(P_i^{\langle k \rangle})$ is not empty, it is a point or an area. Under the non-cocircularity assumption, a non-empty $V(P_i^{\langle k \rangle})$ is a convex polygon.

In Property OK4 we saw a necessary and sufficient condition for an order-k Voronoi polygon to be infinite. A similar condition is obtained for an ordered order-k Voronoi polygon. To avoid unnecessary complication, we next state the condition that is valid only when no two lines passing through two points in P are parallel to each other.

Property OOK5 Suppose that no two lines passing through two points in P are mutually parallel. Let $P_i^{\langle k \rangle} = (p_{i1}, p_{i2}, \ldots, p_{ik})$ be an ordered k-tuple of elements of P. Then the ordered order-k Voronoi polygon $V(P_i^{\langle k \rangle})$ is infinite if and only if there are k mutually parallel lines $L_1, L_2, \ldots, L_k$ such that L_h separates $\{p_{i1}, p_{i2}, \ldots, p_{ih}\}$ and $P \backslash \{p_{i1}, p_{i2}, \ldots, p_{ih}\}$ for $h = 1, 2, \ldots, k$ (Figure 3.2.11).

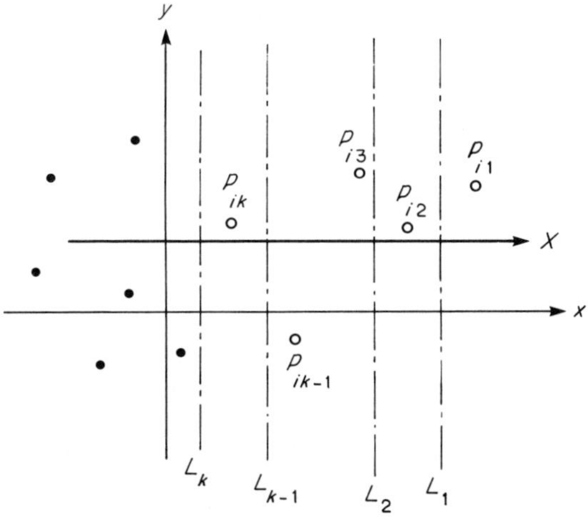

Figure 3.2.11 Illustration of the proof of Property OOK5.

Proof First, suppose that there exist k mutually parallel lines L_1, L_2, ..., L_k specified in Property OOK5 (Figure 3.2.11). We transform the coordinate system in such a way that L_1, L_2, ..., L_k are vertical and p_{i1} is the rightmost point in P. Then the point at infinity in the positive x direction is contained in the region $V(p_{ih}|P\backslash\{p_{i1}, p_{i2}, ..., p_{ih-1}\})$ for any $h = 1, 2, ..., k$. Hence it follows from Property OOK1 that $V(P_i^{\langle k \rangle})$ contains the point at infinity in the positive x direction; that is, $V(P_i^{\langle k \rangle})$ is infinite.

Next, suppose that $V(P_i^{\langle k \rangle})$ is infinite. Then there exists a ray X such that the point at infinity on X belongs to $V(P_i^{\langle k \rangle})$. Since no two lines passing through two points in P are mutually parallel, the perpendicular bisectors of two points in P are not parallel to each other. Hence in particular, no two edges on the boundary of $V(P_i^{\langle k \rangle})$ are parallel, and hence we can choose the ray X in such a way that X is not perpendicular to any line passing through two points in P. Now we transform the coordinate system in such a way that X is parallel to the x-axis (Figure 3.2.11). Let the x coordinate of p_{ih} be x_{ih} for $h = 1, 2, ..., k$. Property OOK1 implies that x_{ih} is greater than the x coordinate of any point in $P\backslash\{p_{i1}, p_{i2}, ..., p_{ih-1}\}$, and consequently we get $x_{i1} > x_{i2} > ... > x_{ik}$ and all the points in $P\backslash P_i^{\langle k \rangle}$ have smaller x coordinates than p_{ik}. Let L_h be the vertical line defined by $x = (x_{ih}+x_{ih+1})/2$ for $h = 1, 2, ..., k$, where x_{ik+1} is the x coordinate of the rightmost point in $P\backslash P_i^{\langle k \rangle}$. L_1, L_2, ..., L_k are mutually parallel (actually they are vertical) and L_{ih} separates $\{p_{i1}, p_{i2}, ..., p_{ih}\}$ and $P\backslash\{p_{i1}, p_{i2}, ..., p_{ih}\}$. □

Corresponding to Property D7 and Property OK7, the following equations hold for the numbers of vertices, edges and polygons.

Property OOK6 Let $n_v^{\langle 2 \rangle}$, $n_e^{\langle 2 \rangle}$, $n_f^{\langle 2 \rangle}$ and $n_u^{\langle 2 \rangle}$ be the numbers of vertices, edge, ordered order-2 Voronoi polygons and infinite ordered order-2 Voronoi polygons, respectively, in the ordered order-2 Voronoi diagram generated by P; $n_{\mathrm{v}}^{(2)}, n_{\mathrm{e}}^{(2)}, n_{\mathrm{f}}^{(2)}$ and $n_{\mathrm{u}}^{(2)}$ be those in the order-2 Voronoi diagram generated by P; and n_{v}, n_{e}, n_{f} and n_{u} be those in the ordinary Voronoi diagram generated by P which satisfies the non-cocircularity assumption. Then

$$n_{\mathrm{v}}^{\langle 2\rangle} = n_{\mathrm{v}}^{(2)}, \tag{3.2.22}$$

$$n_{\mathrm{e}}^{\langle 2\rangle} = n_{\mathrm{e}}^{(2)}+n_{\mathrm{e}}, \tag{3.2.23}$$

$$n_{\mathrm{f}}^{\langle 2\rangle} = 2n_{\mathrm{f}}^{(2)} = 2n_{\mathrm{e}} = 6n-6-2n_{\mathrm{u}}. \tag{3.2.24}$$

Equation (3.2.22) may be visually understood from Figure 3.2.10 in which vertices of $\mathcal{V}^{\langle 2\rangle}$ are coincident with vertices of $\mathcal{V}^{(2)}$. Also equation (3.2.23) is understandable from Figures 3.2.7 and 3.2.9 in which the edges of $\mathcal{V}^{\langle 2\rangle}$ consist of edges of $\mathcal{V}$ and $\mathcal{V}^{(2)}$. The first equation in equation (3.2.24) is obtained from the fact that $V(\{p_{i1}, p_{i2}\})$ consists of $V((p_{i1}, p_{i2}))$ and $V(p_{i2}, p_{i1}))$ (Figure 3.2.10). The second equation is obtained from $n_{\mathrm{f}}^{(2)} = n_{\mathrm{e}}$ (see the equation obtained below Property OK2), and the third equation is obtained from equation (2.3.20). Substituting equations (3.2.11), (3.2.12) and (3.2.13) into the above equations, we can write $n_{\mathrm{v}}^{\langle 2\rangle}, n_{\mathrm{e}}^{\langle 2\rangle}$ and $n_{\mathrm{f}}^{\langle 2\rangle}$ in different terms, some of which are shown in Shamos (1978).

Last, it is worth noting an interesting property found by Sibson (1980a), called the *local coordinates property*.

Property OOK7 The equation

$$\mathbf{x}_i = \frac{\sum_{pj\in P\setminus\{p_i\}} |V((p_i, p_j))|\mathbf{x}_j}{\sum_{pj\in P\setminus\{p_i\}} |V((p_i, p_j))|} \tag{3.2.25}$$

holds if the denominator is greater than zero and less than infinity, i.e. if p_i is the interior of the convex hull of $P\setminus\{p_i\}$.

This equation says that the location of the point p_i is the centroid of the locations of points of $p_j \in P\setminus\{p_i\}$ weighted with the area $|V((p_i, p_j))|$ of the ordered order-2 Voronoi polygon associated with (p_i, p_j). Sibson (1980a) proves this property in a little more general context. Since the proof requires a few pages, it is omitted here (see the proof on pp. 153–154 in Sibson, 1980a). This property is useful for interpolation (Sibson, 1980b). We shall discuss this subject in Chapter 6.

3.3 FARTHEST-POINT VORONOI DIAGRAM AND *k*TH NEAREST-POINT VORONOI DIAGRAM

The nearest generator point from a point in an ordinary Voronoi polygon is of course the generator point of the Voronoi polygon (Property V6). To emphasize this property, we alternatively call the ordinary Voronoi diagram the *nearest-point Voronoi diagram*. Opposite to the nearest point Voronoi diagram, we can consider the 'farthest-point' Voronoi diagram. More generally, we can consider the 'kth nearest-point' Voronoi diagram. In this section we show these generalized Voronoi diagrams, $\mathcal{V}(P$, a distance function to the kth nearest point, $\mathbb{R}^m)$.

3.3.1 Farthest-point Voronoi diagram

The definition of the farthest-point Voronoi diagram is almost the same as that of the nearest-point Voronoi diagram (Definition V1) except 'closest' is replaced with 'farthest'.

Given a set of distinct points in the Euclidean plane, we assign all locations in that plane to the farthest member(s) of the point set. The result is a tessellation of the plane into a set of the regions associated the members in the point set. We call this tessellation the *farthest-point Voronoi diagram* generated by the point set, and the regions constituting the Voronoi diagram *farthest-point Voronoi polygons*. Figure 3.3.1 shows a farthest-point Voronoi diagram.

To restate the above verbal definition mathematically, let $P = \{p_1, \ldots, p_n\}$ $(2 \leq n < \infty)$, $p_i \in \mathbb{R}^m$, $i \in I_n$. Then the farthest-point Voronoi polygon, $V_{\text{fp}}(p_i)$, associated with p_i is written as

$$V_{\text{fp}}(p_i) = \{p | d(p, p_i) \geq d(p, p_j), p_j \in P \backslash \{p_i\}\}, \tag{3.3.1}$$

$$= \{p | d(p, p_i) \geq \max_{p_j} \{d(p, p_j)\}, p_j \in P \backslash \{p_i\}\}. \tag{3.3.2}$$

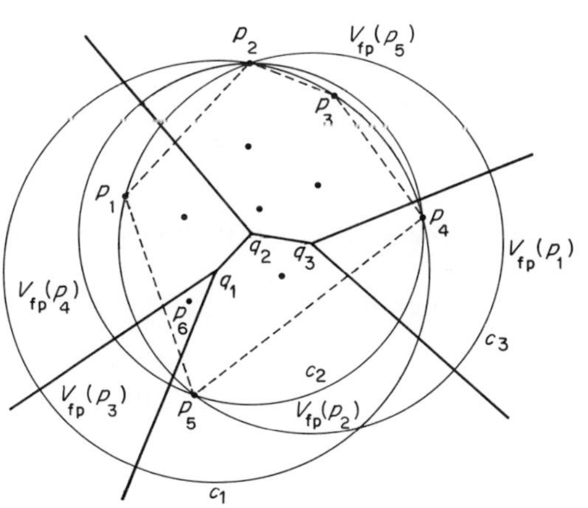

Figure 3.3.1 A farthest-point Voronoi diagram.

Alternatively, in terms of the dominance region

$$\begin{aligned}\mathrm{Dom}(p_i, p_j) &= \{p \mid d(p, p_i) \geq d(p, p_j)\} \\ &= H(p_j, p_i), i \neq j,\end{aligned} \tag{3.3.3}$$

the set $V_{\mathrm{fp}}(p_i)$ is written as

$$V_{\mathrm{fp}}(p_i) = \bigcap_{p_j \in P \setminus \{p_i\}} H(p_j, p_i). \tag{3.3.4}$$

For brevity, we call the region $V_{\mathrm{fp}}(p_i)$ the *FP-Voronoi polygon* of p_i. Since $\mathrm{Dom}(p_i, p_j)$ is a half plane, it is well-behaving and regular. Thus the set of FP-Voronoi polygons, $\mathscr{V}_{\mathrm{fp}} = \{V_{\mathrm{fp}}(p_1), \ldots, V_{\mathrm{fp}}(p_n)\}$, forms a tessellation of $\mathbb{R}^m$. We call this tessellation the *farthest-point Voronoi diagram* generated by P, or briefly the *FP-Voronoi diagram* of P.

The FP-Voronoi diagram is, in fact, an order-k Voronoi diagram. If p_i is the farthest point from p, then the set of points $P \setminus \{p_i\}$ is the set of the first, the second, ..., and the $(n-1)$th nearest point from p_i. Thus the $V_{\mathrm{fp}}(p_i)$ of $\mathscr{V}_{\mathrm{fp}}$ is the same as the order-$(n-1)$ Voronoi polygon associated with $P \setminus \{p_i\}$ of the order-$(n-1)$ Voronoi diagram generated by P. Following the notation of the order-k Voronoi diagram, we can write $V_{\mathrm{fp}}(p_i)$ as $V(P_i^{(n-1)})$. Although the FP-Voronoi diagram is subsumed under the order-k Voronoi diagram, we distinguish these two diagrams, because $V(P_i^{(k)})$ is associated with a set of points $P_i^{(k)}$, whereas $V_{\mathrm{fp}}(p_i)$ is associated with only one point, p_i.

Since the FP-Voronoi diagram is the order-$(n-1)$ Voronoi diagram, the properties of the FP-Voronoi diagram are implicitly stated in the properties of the order-k Voronoi diagram. We, however, explicitly state some of them here adding slightly stronger properties.

From Properties OK2 and OK4, we obtain the following property.

Property FP1 The set $V_{\mathrm{fp}}(p_i)$ given by equation (3.3.1) may be empty. If $V_{\mathrm{fp}}(p_i)$ is not empty, the set $V_{\mathrm{fp}}(p_i)$ is always an infinite convex polygon.

Observe that there is no FP-Voronoi polygon associated with p_6 in Figure 3.3.1. Note too that the above property holds even if points in P are cocircular.

The condition for a non-empty FP-Voronoi polygon is in Property FP2.

Property FP2 $V_{\mathrm{fp}}(p_i)$ is not empty if and only if p_i is at a vertex of the convex hull of P. A non-empty FP-Voronoi polygon need not contain generator points.

Observe the broken lines in Figure 3.3.1. The proof is given by Seidel (1982, pp. 25–26).

In Section 2.3.1 we solved the closest pair problem (Problem P1) with the ordinary Voronoi diagram (Property V6). Similarly, we can solve the farthest-pair problem corresponding to the closest-pair problem with the FP-Voronoi diagram.

Problem P7 (the reciprocal farthest-pair problem and the farthest-pair problem) For a given set P of distinct points in $\mathbb{R}^2$, if p_j is the farthest point from p_i and in addition p_i is the farthest point from p_j, we say that the pair $\{p_i, p_j\}$ is a *reciprocal farthest pair*. If the distance between p_i and p_j is the farthest among the distances between all possible pairs of points in P, we say that the pair $\{p_i, p_j\}$ is the *farthest pair*. Given P, find the reciprocal farthest pairs and the farthest pair.

It follows from the definition of the FP-Voronoi diagram that

Property FP3 $\{p_i, p_j\}$ is a reciprocal farthest pair if and only if $p_j \in V_{\text{fp}}(p_i)$ and $p_i \in V_{\text{fp}}(p_j)$.

Using this property and Property FP2, we can solve Problem P7. Let us solve, for example, the reciprocal farthest-pair problem in Figure 3.3.1. We notice from Property FP2 that reciprocal farthest pairs are found at points on the boundary of the convex hull of P, i.e. $\{p_1, \ldots, p_5\}$. First, consider p_1. This point is included in $V_{\text{fp}}(p_4)$, and p_4 is included in $V_{\text{fp}}(p_1)$. Thus $\{p_1, p_4\}$ is a reciprocal farthest pair. Second, consider p_2. This point is included in $V_{\text{fp}}(p_5)$, and p_5 is included in $V_{\text{fp}}(p_2)$. Thus $\{p_2, p_5\}$ also is a reciprocal farthest pair. Third, consider p_3. This point is included in $V_{\text{fp}}(p_5)$, but p_5 is not included in $V_{\text{fp}}(p_3)$. Thus p_3 does not give a reciprocal farthest pair. Consequently the reciprocal farthest pairs are $\{p_1, p_5\}$ and $\{p_2, p_5\}$. The farthest pair is given by the reciprocal pair whose distance is the greatest among the obtained reciprocal farthest pairs. Hence $\{p_2, p_5\}$ is the farthest pair. Note that the farthest pair gives the diameter of P. The *diameter* of P is defined by the greatest distance among the distances between any two points in P. Thus the diameter is given by the farthest pair. In the example of Figure 3.3.1, $\overline{p_2p_5}$ is the diameter. This method, however, is not the most efficient when we want to obtain only the diameter of P.

In Section 2.3 we considered the largest empty circle problem (Problem P4). Its counterpart problem is written as follows.

Problem P8 (the smallest enclosing circle problem) Given a set of distinct points P, find the smallest circle, called the *smallest enclosing circle*, which contains all points of P in its interior or on it.

We can solve this problem with the aid of the following property.

Property FP4 For every vertex q_i of FP-Voronoi polygons in the FP-Voronoi diagram generated by P, there exists a unique circle C_i centred at q_i which passes through three or more points of P, and in addition, which encloses all other points of P.

The smallest enclosing circle is determined either by the diameter of P or by three points of P. First, we examine if the circle whose diameter is given by the diameter of P encloses all points of P. If so, that circle is the smallest enclosing circle. If not, we draw the circles $\{C_i\}$ defined in Property FP4. Then the smallest enclosing circle is given by the smallest circle among $\{C_i\}$. In the example of Figure 3.3.1, since the circle with diameter $\overline{p_2p_5}$ does not include p_4, we draw circles C_1, C_2 and C_3, and notice that C_2 is the smallest enclosing circle.

In conjunction with Property FP4, it may be worth recalling the definition of the Delaunay triangulation with the inversion transformation (Property D7 in Chapter 2). We defined there 'near-side' and 'far-side' triangles and defined a Delaunay triangulation with near-side triangles. In the same manner, we can define a triangulation with far-side triangles. As we showed in Section 2.4 (in the derivation from equation (2.4.7) to equation (2.4.10)), the circumcircle of the triangle mapped from a near-sided triangle through the inversion transformation does not contain any points of P in its interior. In the same manner, we can show that the circumcircle of the triangle mapped from a far-side triangle through the inversion transformation contains all points of P in its interior or on the boundary. This circumcircle indeed corresponds to the circle referred to in Property FP4.

From Properties OK7 and FP1 (non-empty FP-Voronoi polygons are all infinite), we obtain the following property.

Property FP5 Let $n_{\mathrm{v}}^{(n-1)}$, $n_{\mathrm{e}}^{(n-1)}$ and $n_{\mathrm{f}}^{(n-1)}$ be the numbers of vertices, edges and FP-Voronoi polygons in $\mathcal{V}_{\mathrm{fp}}$. Then

$$n_{\mathrm{v}}^{(n-1)} = n_{\mathrm{f}}^{(n-1)} - 2 \leq n - 2, \qquad (3.3.5)$$

$$n_{\mathrm{e}}^{(n-1)} = 2n_{\mathrm{f}}^{(n-1)} - 3 \leq 2n - 3. \qquad (3.3.6)$$

3.3.2 *k*th nearest-point Voronoi diagram

The definition of the kth nearest-point Voronoi diagram is similar to that of the order-k Voronoi diagram.

Given a set of distinct points in the Euclidean plane, $P = \{p_1, \ldots, p_n\}$ $(2 \leq n < \infty)$, we assign a location p in the plane to a point p_i in the point set P if p_i is the kth nearest point from p; if the kth nearest points are not only p_i but also other points in P, we assign the location to those points. Following this assignment rule, we assign all locations in the plane to at least one point in P. As a result, we obtain a collection of regions associated

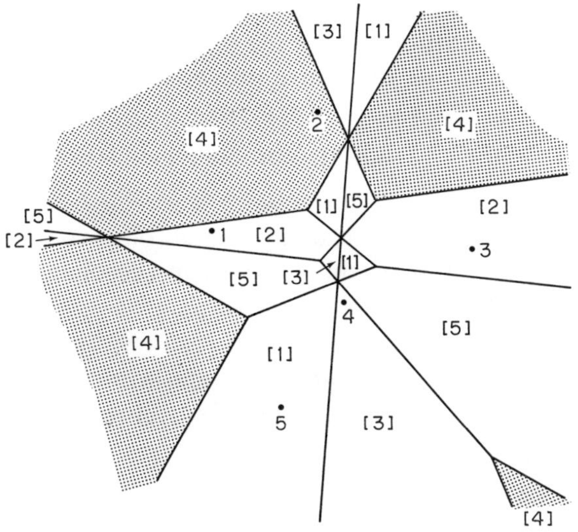

Figure 3.3.2 A third nearest-point Voronoi diagram ([i] indicates $V^{[3]}(p_i)$).

with points in P, denoted by $\mathcal{V}^{[k]} = \{V^{[k]}(p_1), \ldots, V^{[k]}(p_n)\}$, which forms a tessellation. We call the tessellation $\mathcal{V}^{[k]}$ the *kth nearest-point Voronoi diagram* generated by P, or briefly the *KNP-Voronoi diagram* of P, and the region $V^{[k]}(p_i)$ constituting $\mathcal{V}^{[k]}$ the *kth nearest-point Voronoi region* or the *KNP-Voronoi region* for short. An example is shown in Figure 3.3.2 where $k = 3$.

The definition of the KNP-Voronoi diagram with the dominance region $\text{Dom}(p_i,p_j)$ encounters a difficulty, because it is meaningless to consider the kth nearest point for only two points p_i and p_j. This difficulty is not so crucial because we usually define the KNP-Voronoi diagram through the ordered order-k Voronoi diagram. As a simple example, consider $V^{[2]}(p_1)$ in Figure 3.2.7. The region in which the second nearest point is p_1 is given by $V((p_2, p_1))$, $V((p_4, p_1))$ and $V(p_5, p_1))$, i.e. $V^{[2]}(p_1) = V((p_2, p_1))\cup V((p_4, p_1))\cup V((p_5, p_1))$. Generalizing this equation, we obtain the following property.

Property KNP1

$$V^{[k]}(p_j) = \bigcup_{\{p_{i1},\ldots,p_{ik-1}\}\in A^{(k-1)}(P\setminus\{p_j\})} V((p_{i1}, \ldots, p_{ik-1}, p_j)), \qquad (3.3.7)$$

where $A^{(k-1)}(P\setminus\{p_j\})$ is the set of all possible subsets consisting of $k-1$ elements out of $P\setminus\{p_j\}$.

Since this property holds, it is an exercise to drive the properties of the KNP-Voronoi diagram from those of the ordered order-K Voronoi diagram. Here we mention only one distinctive property.

Property KNP2 For $k \geq 2$, a KNP-Voronoi region is not necessarily convex, and may be disconnected.

In fact, $V^{[3]}(p_4)$ in Figure 3.3.2 is disconnected (observe the shaded region).

Generalizing the nearest neighbour problem (Problem P2) in Chapter 2, we have the following problem.

Problem P8 (the *k*th nearest neighbour problem) Given a set of distinct points P, find the kth nearest neighbour point of p_i for all $p_i \in P$.

Once the KNP-Voronoi diagram is given, we immediately obtain the solution. If $p_i \in V^{[k]}(p_j)$, then the kth nearest point is p_j.

3.4 VORONOI DIAGRAMS WITH OBSTACLES

The ordinary Voronoi diagram is defined with the Euclidean distance. The assumption underlying the Euclidean distance is that we can take a straight path between any two points in a region. In some applications, however, this assumption is not acceptable. For example, consider a region in which there exist obstacles, such as lakes, that prevent from traversing. If obstacles lie on the line between an origin and a destination, we cannot traverse straight; we have to make a detour around the obstacles. To deal with such obstacles in a Voronoi diagram, in this section we develop two types of generalized Voronoi diagrams, called the 'shortest-path' Voronoi diagram and the 'visibility-shortest-path' Voronoi diagram.

3.4.1 Shortest-path Voronoi diagram

We consider a generator set consisting of n distinct generator points, $P = \{p_1, \ldots, p_n\} \subset \mathbb{R}^2$, $(2 \leq n < \infty)$, and a set of n_0 closed regions, $O = \{O_1, \ldots, O_{n_0}\} \subset \mathbb{R}^2$, $(1 \leq n_0 < \infty)$. The set O represents a set of obstacles, which are neither transparent nor traversable. We assume that obstacles do not overlap each other, i.e. $O_i \cap O_j = \emptyset$, $i \neq j$, $i, j \in I_{n_0}$; points of P are not allowed to locate in the obstacles, i.e. $p_i \notin O_j$, $i \in I_n$, $j \in I_{n_0}$; each obstacle is connected and has no holes. In addition, for analytical as well as computational convenience, we assume that O_i is a polygon (not necessarily convex). Figure 3.4.1 shows an example of generator points (the filled circles) and obstacles (the shaded area) placed in the plane. The assumption of no holes can be easily relaxed. If a hole has no generator points, points in the hole are not assigned to any generator point. If there

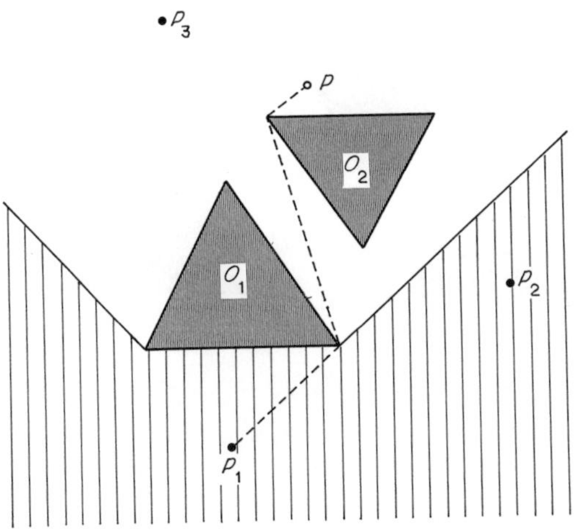

Figure 3.4.1 Generators (the filled circles), obstacles (the shaded areas), the visibility polygon with respect to p_1 (the hatched area) and the shortest path between p and p_1 (the dashed lines).

are some generator points in the hole, the Voronoi diagram in the hole can be treated as a Voronoi diagram in a simple polygon to be discussed in Section 3.4.2.

In the region $S = \mathbb{R}^2 \backslash O$, we define the distance between a point p and a point p_i in P by the length of the shortest path among all possible continuous paths connecting p and p_i that do not intersect obstacles $O_i \backslash \partial O_i$, $i \in I_{n_0}$ (note that a path can pass through points on the boundary of O_i). We call this distance the *shortest-path distance* between p and p_i, and denote it by $d_{sp}(p, p_i)$. In practice, we obtain the shortest-path distance with the aid of a 'visibility polygon' and a 'visibility graph'.

A *visibility polygon*, denoted by $\mathrm{VIS}(p_i)$, with respect to p_i is the set of points that are visible from p_i. Mathematically $\mathrm{VIS}(p_i)$ is defined by

$$\mathrm{VIS}(p_i) = \{p | \overline{p_i p} \cap [O_j \backslash \partial O_j] = \varnothing, p \in \mathbb{R}^2, j \in I_{n_0}\}. \tag{3.4.1}$$

An example is shown in Figure 3.4.1, where the visibility polygon with respect to p_1 is indicated by the hatched area. If p is visible from p_i, there are no obstacles between p and p_i, and hence the shortest-path distance between p and p_i is given by the Euclidean distance, i.e.

$$d_{sp}(p, p_i) = \|\mathbf{x} - \mathbf{x}_i\|, \quad \text{if } \mathbf{x} \in \mathrm{VIS}(p_i). \tag{3.4.2}$$

The computational methods for constructing a visibility polygon are developed by El Gindy and Avis (1981), Lee (1983), Lee and Chen (1985), Imai *et al.* (1985), Chazelle and Guibas (1985) and others.

A visibility graph is defined for a given geometric graph. Let $G(Q, L)$ be a given geometric graph, where Q is a set of n_q nodes and L is a set of

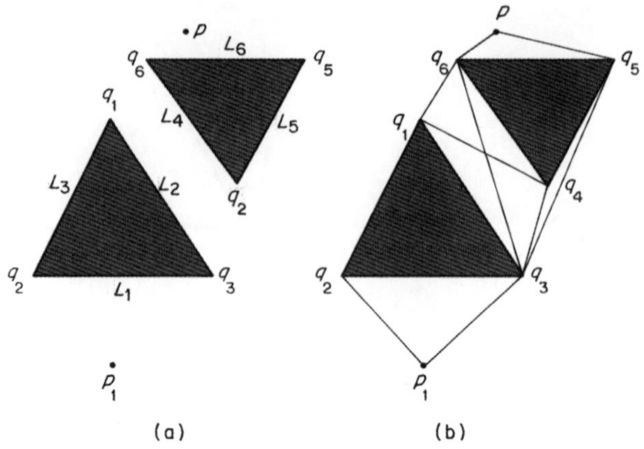

Figure 3.4.2 A geometric graph and its visibility graph.

links (Figure 3.4.2(a)). We assume that links are not transparent. Let L_{vis} be the set of line segments $\overline{q_iq_j}$ satisfying that q_j is visible from q_i, i.e.

$$L_{\text{vis}} = \{\overline{q_iq_j}|q_j \in \text{VIS}(q_i), i<j, i, j \in I_{n_q}\} \tag{3.4.3}$$

(all lines in Figure 3.4.2(b) where $Q = \{q_1, \ldots, q_6, p_1, p_2\}$). Note that line segments in L_{vis} may intersect. The *visibility graph* of $G(Q, L)$ is defined by the graph consisting of Q and L_{vis}, denoted by $G(Q, L_{\text{vis}})$.

With the aid of a visibility graph we can find the shortest path. As an example, let us obtain the shortest path between p and p_1 in Figure 3.4.1. We first consider the geometric graph $G(Q, L)$ in which Q is given by vertices of the obstacles O and P, and L is given by the edges of the obstacles O (Figure 3.4.2(a)). Second, we construct the visibility graph $G(Q, L_{\text{vis}})$ of the graph $G(Q, L)$ (Figure 3.4.2(b)). Third, we solve the shortest-path problem on this visibility graph (using, for example, the Dijkstra, 1959, method). Then, the shortest path between p and p_i on $G(Q, L\cup L_{\text{vis}})$ gives the shortest path between p and p_1 in Figure 3.4.1.

With the shortest path we define

$$V(p_i) = \{p|d_{\text{sp}}(p, p_i) \leq d_{\text{sp}}(p, p_j), j \neq 1, j \in I_n\}. \tag{3.4.4}$$

We call this set the *shortest-path Voronoi region* associated with p_i, or briefly the *SP-Voronoi region* of p_i, and the set of SP-Voronoi regions, $\mathcal{V}(P, d_{\text{sp}}, \mathbb{R}^2\backslash O) = \mathcal{V}_{\text{sp}} = \{V(p_1), \ldots, V(p_n)\}$, the *shortest-path Voronoi diagram* generated by P with obstacles O, or the *SP-Voronoi diagram* of P with O for short. Figure 3.4.3 shows the SP-Voronoi diagram generated by p_1, p_2 and p_3 with the obstacles O_1 and O_2 in Figure 3.4.1. The computational method for constructing an SP-Voronoi diagram is proposed by Asano and Asano (1987) and Seoung and Asano (1987).

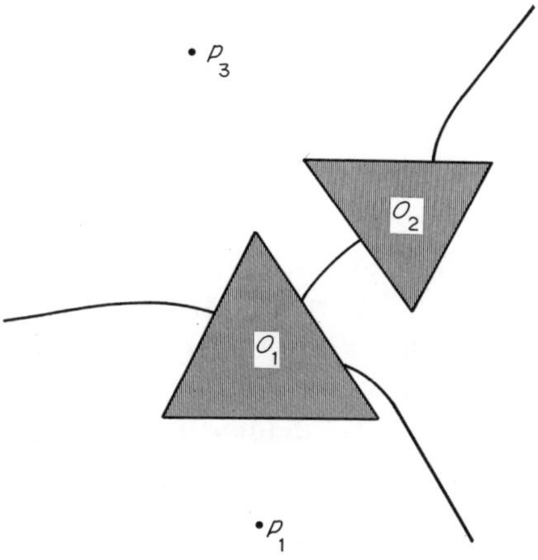

Figure 3.4.3 A shortest-path Voronoi diagram.

To observe the geometric properties of an SP-Voronoi diagram, consider the bisector defined by

$$b(p_i, p_j) = \{p | d_{\text{sp}}(p, p_i) = d_{\text{sp}}(p, p_j)\}, \quad i \neq j. \tag{3.4.5}$$

Figure 3.4.4 shows $b(p_1, p_3)$ with the obstacles O_1 and O_2 represented by the triangles $\triangle q_1q_2q_3$ and $\triangle q_4q_5q_6$. If a point p is in the region given by $\text{VIS}(p_1) \cap \text{VIS}(p_3)$ (the hatched region), then the shortest-path distances $d(p, p_1)$ and $d(p, p_3)$ are, from equation (3.4.2), given by the Euclidean

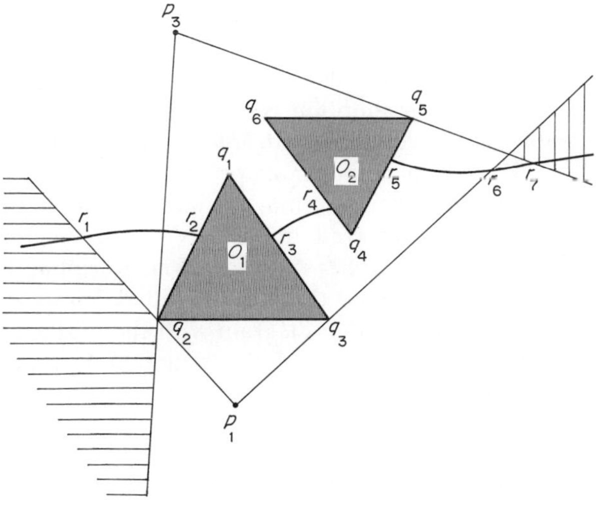

Figure 3.4.4 The bisector between p_1 and p_3, where the distance is given by the shortest-path distance.

distances. Thus the bisector between p_1 and p_3 in $\mathrm{VIS}(p_1) \cap \mathrm{VIS}(p_3)$ is given by

$$b(p_1, p_3) \cap [\mathrm{VIS}(p_1) \cap \mathrm{VIS}(p_3)] = \{\mathbf{x} \mid \|\mathbf{x} - \mathbf{x}_1\| = \|\mathbf{x} - \mathbf{x}_3\|\} \cap [\mathrm{VIS}(p_1) \cap \mathrm{VIS}(p_3)] \tag{3.4.6}$$

(the half lines radiating from r_1 and r_7 in Figure 3.4.4). Second, in the region $\mathrm{VIS}(q_2) \cap \mathrm{VIS}(p_3)$ the shortest-path distance $d_{sp}(p, p_3)$ is given by the Euclidean distance, but the shortest-path distance $d_{sp}(p, p_1)$ is not given by the Euclidean distance, but $\|\mathbf{x} - \mathbf{u}_2\| + \|\mathbf{u}_2 - \mathbf{x}_1\|$, where $\mathbf{u}_2$ is the location vector of q_2. This distance can be regarded as the additively weighted distance with the weight $-\|\mathbf{u}_2 - \mathbf{x}_1\|$ defined in Section 3.2. The bisector in $\mathrm{VIS}(q_2) \cap \mathrm{VIS}(p_3)$ is hence given by

$$b(p_1, p_3) \cap [\mathrm{VIS}(q_2) \cap \mathrm{VIS}(p_3)] = \{\mathbf{x} \mid \|\mathbf{x} - \mathbf{u}_2\| + \|\mathbf{u}_2 - \mathbf{x}_1\| = \|\mathbf{x} - \mathbf{x}_3\|\} \cap [\mathrm{VIS}(q_2) \cap \mathrm{VIS}(p_3)], \tag{3.4.7}$$

which is a hyperbolic arc (the curved line segment between r_1 and r_2 in Figure 3.4.4). Similarly, the bisector $b(p_1, p_3)$ in $\mathrm{VIS}(q_3) \cap \mathrm{VIS}(p_3)$, $\mathrm{VIS}(q_3) \cap \mathrm{VIS}(q_5)$ and $\mathrm{VIS}(p_1) \cap \mathrm{VIS}(q_5)$ is given by the hyperbolic arcs r_3 and r_4, r_5 and r_6, and r_6 and r_7, respectively, in Figure 3.4.4. From this examination we notice that the SP-Voronoi diagram is closely related to the additively weighted Voronoi diagram.

3.4.2 Voronoi diagram in a simple polygon

An SP-Voronoi diagram is useful to treat the bounded Voronoi diagram discussed in Section 2.1. We assume there that the boundary Voronoi polygons of a bounded Voronoi diagram, $\mathcal{V} \cap S$, are star-shaped with respect to their generators. This assumption is made because in many applications the shortest path that traverses outside of S is unrealistic (the dotted line in Figure 3.4.5). If we relax the star-shape assumption, we should consider the shortest path that is embedded in S. In the literature a Voronoi diagram with this modification is referred to as a Voronoi diagram in a simple polygon (Asano and Asano, 1987; Arnov, 1987). Recalling the SP-Voronoi diagram, the reader may notice that the bounded Voronoi diagram $\mathcal{V} \cap S$ can be regarded as an SP-Voronoi diagram with the obstacle given by $\mathbb{R}^2 \backslash S$. Figure 3.4.5 shows a modification of the bounded Voronoi diagram (the dashed lines) with its SP-Voronoi diagram.

3.4.3 Visibility-shortest-path Voronoi diagram

In the SP-Voronoi diagram we assumed that obstacles were polygons. Here we assume, for simplicity, that obstacles in O are all line segments that do

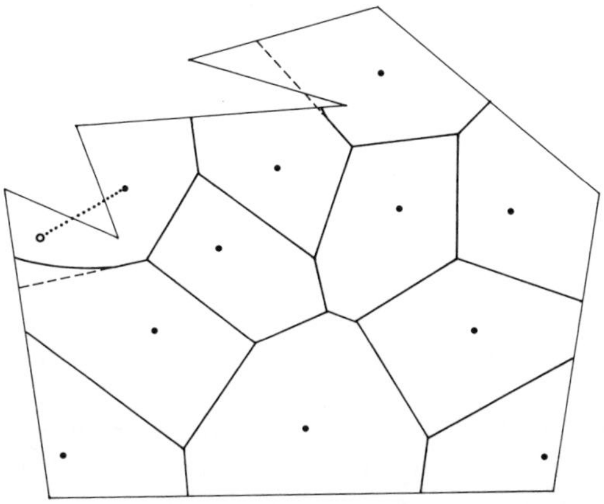

Figure 3.4.5 A Voronoi diagram in a simple polygon (a modification of the bounded Voronoi diagram (the dashed lines) with its SP-Voronoi diagram).

not intersect each other (note that the relaxation of the former assumption is conceptually easy). In addition, we assume that end points of the line segments in O are all generator points of P. Generator points in P are not necessarily end points of the line segments in O; some points may be isolated. An example is shown in Figure 3.4.6. where generator points are indicated by the filled circles.

Given a set of generator points $P = \{p_1, \ldots, p_n\}$, and a set of line obstacles $O = \{O_1, \ldots, O_n\}$, we define a distance by

$$d_{\mathrm{vsp}}(p, p_i) = \begin{cases} \|\mathbf{x} - \mathbf{x}_i\|, & \text{if } p_i \in \mathrm{VIS}(p), \\ \infty, & \text{otherwise.} \end{cases} \tag{3.4.8}$$

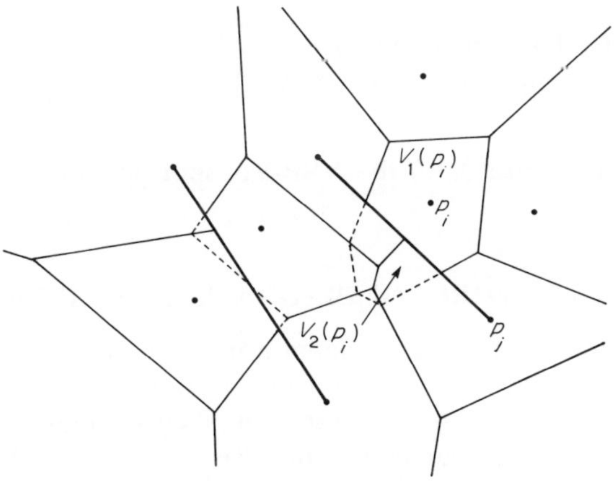

Figure 3.4.6 A visibility-shortest-path Voronoi diagram.

This distance implies that if a point p is visible from p_i, the distance is given by the Euclidean distance; if it is not visible, the distance is infinite. We call this distance the *visibility-shortest-path distance* (Aurenhammer, 1988, calls it the bounded distance) and abbreviate it to the *VSP-distance.*

With the VSP-distance, we define the set $V(p_i)$ by equation (3.4.4), where $d_{sp}(p, p_i)$ is replaced with $d_{vsp}(p, p_i)$. We call the set $V(p_i)$ the *visibility-shortest-path Voronoi polygon* associated with p_i, or briefly the *VSP-Voronoi polygon* of p_i, and the set of VSP-Voronoi polygons the *visibility-shortest-path Voronoi diagram* generated by P with obstacles O or the *VSP-Voronoi diagram* of P with O for short. Figure 3.4.6 illustrates a VSP-Voronoi diagram.

A VSP-Voronoi polygon is obtained from an ordinary Voronoi diagram. We first generate the ordinary Voronoi diagram by P (the dashed lines). For ease of exposition we assume that every Voronoi polygon is divided by at most one obstacle line segment. Let $V(p_i)$ be a Voronoi polygon divided by an obstacle line, and $V_1(p_i)$ and $V_2(p_i)$ be the resulting polygons, where $V_1(p_i)$ contains p_i. Then the VSP-Voronoi polygon of p_i is given by $V_1(p_i)$. Second, we assign points in $V_2(p_i)$ to the closest generator among the generators whose Voronoi polygon shares the edge of $V_2(p_i)$. Let $V_{2j}(p_i)$ be the set of points assigned to p_j (Figure 3.4.6). Then the VSP-Voronoi polygon of p_j is given by $V(p_j) \cup V_{2j}(p_i)$. We apply a similar modification to a Voronoi polygon whose generator is an end point of an obstacle line segment. We carry out this procedure for all obstacle line segments. The resulting tessellation is the VSP-Voronoi diagram of P with O.

We realize from the above procedure that a VSP-Voronoi diagram is a 'local' modification of an ordinary Voronoi diagram. We may thus draw the geometric properties of a VSP-Voronoi diagram from those of the ordinary Voronoi diagram paying attention to the above modification. Here we mention one distinct property.

Property VSP1 The set $V(p_i)$ given by equation (3.4.4) with equation (3.4.8) is a non-empty polygon. A VSP-Voronoi polygon may be non-convex.

For example, in Figure 3.4.6 the VSP-Voronoi polygon associated with p_j is non-convex.

3.5 VORONOI DIAGRAMS FOR LINES

In the preceding four sections, a generator is a point or a set of points. In this section, extending a generator from a point to a line, we consider a generalized Voronoi diagram generated by a set of lines. In theory we can consider a general line, but in practice we deal with points (as a special case of lines), straight lines, chains of straight line segments, circular arcs, chains of circular arcs and full circles. Curved lines may not be exactly represented by those lines, but this restriction is not crucial in practice because we can approximate a curved line by a chain of small straight line segments. In computational geometry a generalized Voronoi diagram for lines has been intensively studied since the late 1970s by Drysdale and Lee (1978), Drysdale (1979), Kirkpatrick (1979), Lee and Drysdale (1981), Imai *et al.* (1985), Kokubo (1985), Sharir (1985), Fortune (1986) and Yap (1987), among others. In this section we mainly following their results.

We consider a generator set $L = \{L_1, \ldots, L_n\} \subset \mathbb{R}^2$ $(1 \leq n < \infty)$, where L_i is a point, a line segment (which may be curved) or a geometric element consisting of line segments that are connected. We assume that elements in L do not intersect each other, i.e. $L_i \cap L_j = \emptyset$, $i \neq j$, $i, j \in I_n$. Under this assumption, we define the distance from p to L_i by the shorest distance between p and a point p_i on L_i, i.e.

$$d_s(p, L_i) = \min_{\mathbf{x}_i} \{\|\mathbf{x}-\mathbf{x}_i\| \,|\mathbf{x}_i \in L_i\}, \tag{3.5.1}$$

where $\mathbf{x}$ and $\mathbf{x}_i$ are the location vectors of p and p_i, respectively. With this distance, we define

$$V(L_i) = \{p|d_s(p, L_i) \leq d_s(p, L_j), j \neq i, j \in I_n\}. \tag{3.5.2}$$

Alternatively, let

$$\mathrm{Dom}(L_i, L_j) = \{p|d_s(p, L_i) \leq d_s(p, L_j)\}, \quad j \neq i, \tag{3.5.3}$$

$$b(L_i, L_j) = \{p|d_s(p, L_i) = d_s(p, L_j)\}, \quad j \neq i. \tag{3.5.4}$$

Then the set $V(L_i)$ is written as

$$V(L_i) = \bigcap_{j \in I_n \setminus \{i\}} \mathrm{Dom}(L_i, L_j). \tag{3.5.5}$$

Since $\mathrm{Dom}(L_i, L_j) \cup \mathrm{Dom}(L_j, L_i) = \mathbb{R}^2$, and the set $b(L_i, L_j)$ is the boundary of $\mathrm{Dom}(L_i, L_j)$, the dominance region $\mathrm{Dom}(L_i, L_j)$ is well-behaving and regular. Thus the resulting set $\mathcal{V}(L, d_s, \mathbb{R}^2) = \mathcal{V}(L) = \{V(L_1), \ldots, V(L_n)\}$ gives a generalized Voronoi diagram. We call this generalized Voronoi diagram the *line Voronoi diagram* generated by L, and the set $V(L_i)$ the *line Voronoi region* associated with L_i. If L_i degenerates into a point for all i, the line Voronoi diagram reduces to the ordinary Voronoi diagram. Thus the line Voronoi diagram is a generalization of the ordinary Voronoi diagram. Figure 3.5.1 shows the line Voronoi diagram generated by the set consisting of a point, a chain of straight lines, a circular arc and a full circle.

Figure 3.5.1 A line Voronoi diagram generated by the set of a point, a chain of straight line segments, a circular arc and a full circle.

3.5.1 Voronoi diagrams for a set of points and straight line segments

In the above definition we consider a general line, but in this subsection we assume that a generator L_i is a point or a straight line segment or a chain of straight line segments. We also assume that a straight line segment L_i contains both end points. Under these assumptions the shortest distance of equations (3.5.1) is explicitly written as

$$d_s(p,L_i) = \begin{cases} \|\mathbf{x}-\mathbf{x}_{i1}\|, & \text{if } p \in R_{i1}, \\ \|\mathbf{x}-\mathbf{x}_{i2}\|, & \text{if } p \in R_{i2}, \\ \left\| (\mathbf{x}-\mathbf{x}_{i1}) - \dfrac{(\mathbf{x}-\mathbf{x}_{i1})^{\mathrm{T}}(\mathbf{x}_{i2}-\mathbf{x}_{i1})}{\|\mathbf{x}_{i2}-\mathbf{x}_{i1}\|^2}(\mathbf{x}_{i2}-\mathbf{x}_{i1}) \right\|, & \text{if } p \in R_{i3} = \mathbb{R}^2 \backslash [R_{i1} \cup R_{i2}], \end{cases} \tag{3.5.6}$$

where $\mathbf{x}_{i1}$ and $\mathbf{x}_{i2}$ are the end points of L_i, and $R_{i1} = \{\mathbf{x} | (\mathbf{x}_{i2}-\mathbf{x}_{i1})^{\mathrm{T}}(\mathbf{x}-\mathbf{x}_{i1}) < 0\}$, $R_{i2} = \{\mathbf{x} | (\mathbf{x}_{i1}-\mathbf{x}_{i2})^{\mathrm{T}}(\mathbf{x}-\mathbf{x}_{i2}) < 0\}$ (see Figure 3.5.2(a)).

The bisectors between L_i and L_j with this distance are depicted in Figure 3.5.3, where L_i and L_j are: (a) a point and a point; (b) a point and a straight line segment; (c) a straight line segment and a straight line segment. In any case, the bisector splits $\mathbb{R}^2$ into two disjoint regions, and so the dominance region is well-behaving and regular. Thus the set $\mathcal{V}(L,d_s,\mathbb{R}^2)$ with the distance of equation (3.5.6) gives a line Voronoi diagram.

Figure 3.5.4 illustrates the line Voronoi diagram generated by the set L consisting of points, straight line segments and chains of straight line segments (the heavy lines). When we construct this Voronoi diagram, we add a modification to the generator set L to gain analytical and computational

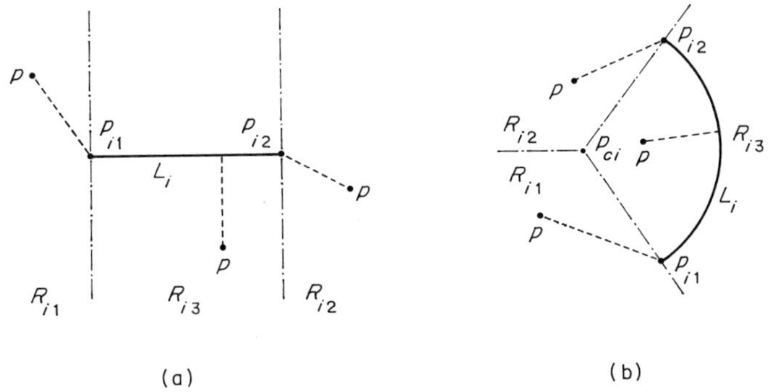

Figure 3.5.2 (a) The distance from a point to a straight line segment; (b) the distance from a point to a circular arc.

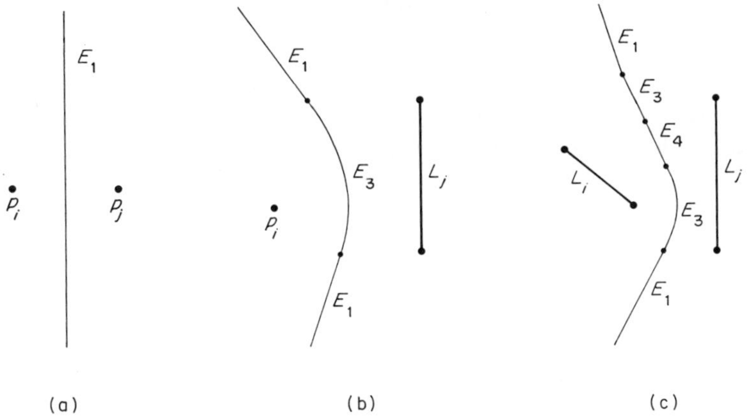

Figure 3.5.3 The bisectors between L_i and L_j: L_i and L_j are (a) a point and a point; (b) a point and a straight line segment; (c) a striaght line segment and a straight line segment.

tractability. To be explicit, suppose that the first l elements in L are points, i.e. $L_1 = p_1$, ..., $L_l = p_l$, and the rest of the elements are straight line segments. As we assume in the above, a straight line segment or a chain of straight line segments L_i includes both end points, but now we decompose this line segment into the end points, intermediate points and straight line segments without end points. The set of points is denoted by $\{p_{ij}, j\in I_{n_i+1}\}$ and the set of the lines is denoted by $\{L^o_{ij}, j\in I_{n_i}\}$ ($n_i = 1$ for a straight line segment). As a result, the generator L is decomposed into

$$L^{(d)} = \{p_i, i\in I_l\} \cup \left[\bigcup_{i=l+1}^{n} \{p_{ij}, j\in I_{n_i+1}\}\right] \cup \left[\bigcup_{i=l+1}^{n} \{L^o_i, j\in I_{n_i}\}\right]. \quad (3.5.7)$$

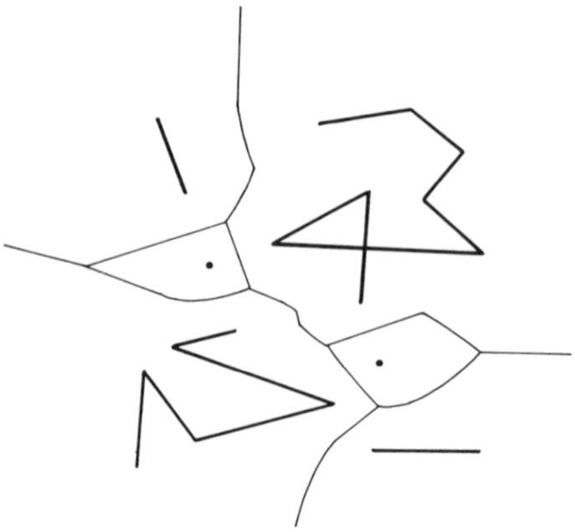

Figure 3.5.4 A line Voronoi diagram generated by the set consisting of points, two straight lines and two chains of straight line segments.

We call this set the *decomposed generator set* of L. If we decompose the generator set of L in Figure 3.5.4, we obtain the decomposed generator set depicted by the filled circles and the heavy lines in Figure 3.5.5 (where a line segment with small gaps at both ends symbolically indicates a line segment without end points). Since L_i^o does not include end points, the distance of equation (3.5.1) is alternatively defined by

$$d_s(p, L_i^o) = \sup_{\mathbf{x}_i} \{\|\mathbf{x}-\mathbf{x}_i\| \,|\, \mathbf{x}_i \in L_i^o\}, \tag{3.5.8}$$

where sup is a supremum operator (Section 1.3). Using this decomposed generator set with this distance, we generate the line Voronoi diagram $\mathcal{V}(L^{(d)})$, which is indicated by the light solid lines and the broken lines in Figure 3.5.5. Obviously, the edges of $\mathcal{V}(L)$ are included in $\mathcal{V}(L^{(d)})$. We can hence obtain $\mathcal{V}(L)$ from $\mathcal{V}(L^{(d)})$ by deleting the unnecessary edges indicated by the broken lines in Figure 3.5.5.

As is noticed from Figures 3.5.4 and 3.5.5, the line Voronoi diagram $\mathcal{V}(L^{(d)})$ of the decomposed set $L^{(d)}$ is a refinement of the line Voronoi diagram $\mathcal{V}(L)$ of L. This refinement is fundamental in the sense that every edge in $\mathcal{V}(L^{(d)})$ is part of a unique conic curve. For this reason, we examine the geometric properties of $\mathcal{V}(L)$ through $\mathcal{V}(L^{(d)})$.

From equation (3.5.5) with Figure 3.5.3, we obtain the following property.

Property L1 No line Voronoi region in $\mathcal{V}(L^{(d)})$ is empty. A line Voronoi region of a point to which three or more lines are incident is a point.

Substituting equation (3.5.6) into equation (3.5.4), we obtain the following property.

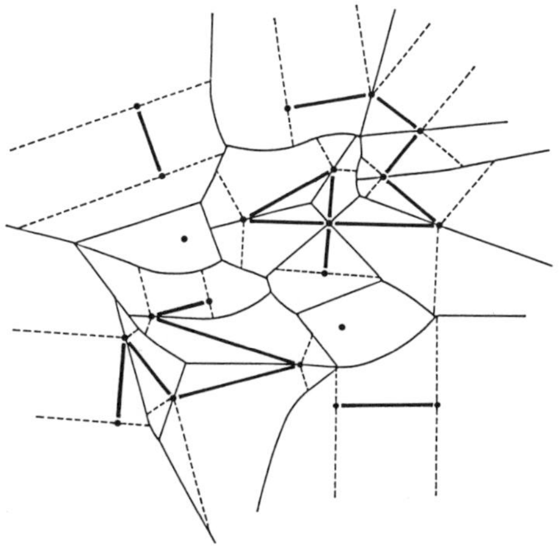

Figure 3.5.5 The line Voronoi diagram generated by the decomposed generator set obtained from the generator set in Figure 3.5.4.

Property L2 Edges in $\mathscr{V}(L^{(d)})$ are straight lines or parabolic arcs. These edges can be classified into the following four types according to the types of generators generating the edges.

Type E1. An edge generated by points p_i and p_j. The edge is a straight line perpendicularly bisecting $\overline{p_i p_j}$ (Figure 3.5.6(a)).

Type E2. An edge generated by a straight line segment L_i^o and its end point $p_{i1} \in \partial L_i$. The edge is a straight line perpendicular to L_i^o passing p_{i1} (Figure 3.5.6(b)).

Type E3. An edge generated by a point p_i and a straight line segment L_j^o $(p_i \notin \partial L_j)$. The edge is a locus of a point p satisfying that the distance from p to p_i is equal to the distance from p to the line L_j. This locus is known as a parabolic arc whose focus is p_i and whose directrix is L_j^o (Figure 3.5.6(c)).

Type E4. An edge generated by straight line segments L_i^o and L_j^o. The edge is a straight line dividing the angle made by L_i^o and L_j^o (Figure 3.5.6(d)).

The reader should not confuse Figures 3.5.6(c), (d) with Figures 3.5.3(b), (c). The bisectors in Figure 3.5.3 are generated by the generators that are not decomposed. Using Property L2, the reader should confirm that the bisectors in Figures 3.5.3(b), (c) consist of edges of types E1, E3 and E4. In Figure 3.5.3, we indicate the type of each edge by E1–E4.

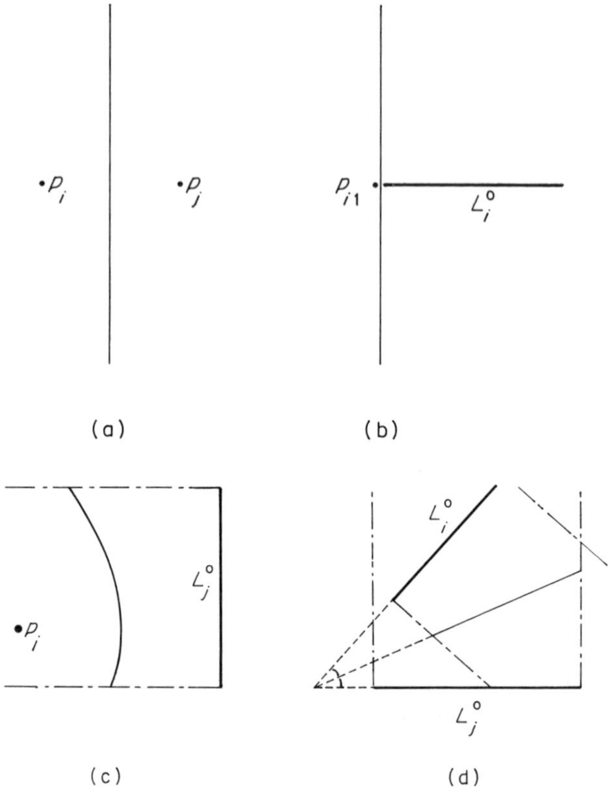

Figure 3.5.6 Four types of edges in a line Voronoi diagram: (a) type E1; (b) type E2; (c) type E3; (d) type E4.

As a consequence of Property L2, we obtain the following property.

Property L3 A line Voronoi region $V(L_i)$ is not necessarily convex.

When exactly three edges are incident at every vertex in the line Voronoi diagram, we say that the line Voronoi diagram is *non-degenerate*. Regarding a non-degenerate line Voronoi diagram $\mathcal{V}(L^{(d)})$, we obtain the following property from Property L2.

Property L4 Vertices $\{q_i\}$ in a non-degenerate line Voronoi diagram can be classified into the following six types according to the types of three edges, e_{i1}, e_{i2} and e_{i3}, which are generated by three generators.

Type V1. A vertex q_i generated by points p_i, p_j and p_k. The edges e_{i1}, e_{i2} and e_{i3} are all type E1 (Figure 3.5.7(a)).

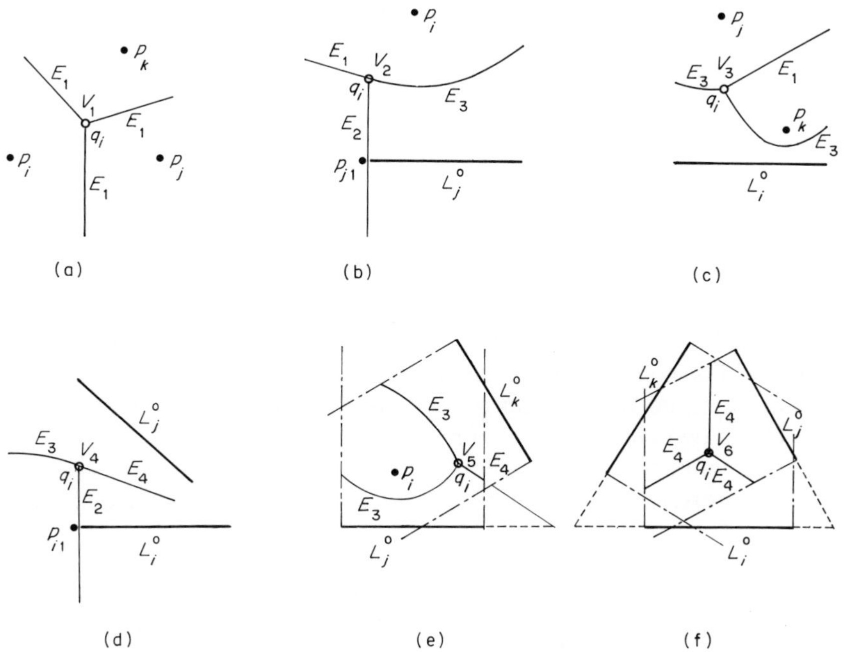

Figure 3.5.7 Types of vertices in a line Voronoi diagram: (a) type V1; (b) type V2; (c) type V3; (d) type V4; (e) type V5; (f) type V6.

Type V2. A vertex q_i generated by a point p_i, a straight line segment L_j^o and its end point $p_{j1} \in \partial L_j$. The edges are types E1, E2 and E3. The edge of type E1 and that of type E3 are tangent at the Voronoi vertex q_i (Figure 3.5.7(b)).

Type V3. A vertex q_i generated by a straight line segment L_i^o and two points p_j and p_k which are not the end points of L_i. The edges are types E1 and E3. The edges of type 3 have the common directrix L_i^o (Figure 3.5.7(c)).

Type V4. A vertex q_i generated by two straight line segments L_i^o and L_j^o, and its end point $p_{i1} \in L_i^o$. The edges are types E2, E3 and E4. The edges of type E3 and E4 are tangent at q_i (Figure 3.5.7(d)).

Type V5. A vertex q_i generated by a point p_i and two straight line segments L_j^o and L_k^o ($p_i \notin \partial L_j$, $p_i \notin \partial L_k$). The edges are types E3 and E4 (Figure 3.5.7(e)).

Type V6. A vertex q_i generated by straight line segments L_i^o, L_j^o and L_k^o. The edges are type E4. The extended lines of L_i^o, L_j^o and L_k^o form a triangle. If q_i is inside of this triangle, q_i is its inscribed centre. If q_i is outside of the triangle, q_i is one of its excentres (Figure 3.5.7(f)).

In Figure 3.5.7, types of vertices are indicated by V1–V6 with edge types. Finally, almost the same derivation used in Property V2 leads to the following property.

Property L5 A line Voronoi region $V(L_i^o)$ is infinite if and only if L_i^o is on the boundary of the convex hull of $L^{(d)}$.

3.5.2 Voronoi diagrams for a set of points, straight line segments and circular arcs

We now add circular arcs to the generator set examined in the preceding subsection. For computational convenience we assume that a circular arc is less than or equal to a semicircle. We treat a circular arc greater than a semicircle as two circular arcs connected at a point, and a full circle as two semicircles connected at two points. For a circular arc with radius r_i centred at $\mathbf{x}_{ci}$, the shortest distance of equation (3.5.1) is explicitly written as

$$d_s(p,L_i) = \begin{cases} \|\mathbf{x}-\mathbf{x}_{i1}\|, & \text{if } \mathbf{x} \in R_{i1}, \\ \|\mathbf{x}-\mathbf{x}_{i2}\|, & \text{if } \mathbf{x} \in R_{i2}, \\ |\|\mathbf{x}-\mathbf{x}_{ci}\| - r_i|, & \text{if } \mathbf{x} \in R_{i3} = \mathbb{R}^2 \backslash [R_{i1} \cup R_{i2}], \end{cases} \qquad (3.5.9)$$

where $\mathbf{x}$, $\mathbf{x}_{i1}$ and $\mathbf{x}_{i2}$ are the location vectors of p and the end points of L_i, respectively, and R_{i1}, R_{i2} and R_{i3} are the regions defined in Figure 3.5.2(b).

Since $\mathrm{Dom}(L_i,L_j) \cup \mathrm{Dom}(L_j,L_i) = \mathbb{R}^2$ and $b(L_i,L_j)$ is the boundary of $\mathrm{Dom}(L_i,L_j)$, the dominance region is well-behaving and regular. Thus a generator set consisting of points, straight line segments and circular arcs gives a line Voronoi diagram. An example is depicted in Figure 3.5.1. We construct this Voronoi diagram in the same manner as we employed in the preceding subsection. First, we decompose a given generator set L and next construct the line Voronoi diagram for the decomposed generator set $L^{(d)}$. The line Voronoi diagram $\mathcal{V}(L)$ is obtained from the line Voronoi diagram $\mathcal{V}(L^{(d)})$ by deleting superfluous edges.

To observe the geometric property of the line Voronoi diagram, let $L_{\mathrm{pl}}^{(d)}$ be the decomposed generator set consisting of points and straight lines, and $L_{\mathrm{plc}}^{(d)}$ be the decomposed generator set consisting of points, straight lines and circular arcs (the subscript 'plc' indicates points, lines and circles). Since the line Voronoi diagram $\mathcal{V}(L_{\mathrm{pl}}^{(d)})$ is a special case of the line Voronoi diagram $\mathcal{V}(L_{\mathrm{plc}}^{(d)})$, some properties of $\mathcal{V}(L_{\mathrm{pl}}^{(d)})$ mentioned in Section 3.5.1 hold in such a special case. In fact, Properties L1, L3 and L5 of $\mathcal{V}(L_{\mathrm{pl}}^{(d)})$ hold for $\mathcal{V}(L_{\mathrm{plc}}^{(d)})$. Also Property L2 of $\mathcal{V}(L_{\mathrm{pl}}^{(d)})$ holds for $\mathcal{V}(L_{\mathrm{plc}}^{(d)})$, but we should add the following seven types.

Property L7 Edges in $\mathcal{V}(L_{\mathrm{plc}}^{(d)})$ are straight line segments, parabolic arcs, elliptic arcs and hyperbolic arcs. Those edges can be classified into ten types according to the types of generators generating the edges. The first four types are given by E1–E4 in Property L2.

Type E5. An edge generated by a circular arc L_i^o and a point $p_j \notin \partial L_i$ which is inside the circle containing L_i^o centred at p_{ci}. The edge is the locus of a point p satisfying that the total distances from p to two fixed points p_j and p_{ci} is constant, which is known as an elliptic arc whose foci are p_{ci} and p_j (Figure 3.5.8(a)).

Type E6. An edge generated by a circular arc L_i^o and a point $p_j (\notin \partial L_i)$ which is outside the circle containing L_i^o centred at p_{ci}. The edge is the locus of a point p satisfying that the difference between the distances from p to p_{ci} and p_j is constant. This locus is known as a branch of a hyperbolic curve whose foci are p_{ci} and p_j (Figure 3.5.8(b)).

Type E7. An edge generated by a circular arc L_i^o and its end point $p_{i1} \in \partial L_i$. The edge generated by these generators is a straight line radiating from the centre of the circle containing L_i^o to p_{i1} (Figure 3.5.8(c)).

Type E8. An edge generated by a circular arc L_i^o and a straight line containing L_j^o which intersects the circle containing L_i^o centred at p_{cj} with radius r_i (note that $L_i^o \cap L_j^o = \emptyset$). The edge is the locus of a point p satisfying that the sum of the distances from p to the line L_i^o and to the fixed point p_{ci} is constant. This locus is known as a parabolic arc whose focus is p_{ci} and whose directrix is the line parallel to L_j at distance r_i (Figure 3.5.9(d)).

Type E9. An edge generated by two circular arcs L_i^o and L_j^o where the circle containing L_i^o centred at p_{ci} is contained in the circle containing L_j^o centred at p_{cj}. The edge is an elliptic arc whose foci are p_{ci} and p_{cj} (Figure 3.5.9(e)).

Type E10. An edge generated by two circular arcs L_i^o and L_j^o where the circle containing L_i^o centred at p_{ci} intersects the circle containing L_j^o centred at p_{cj} or the former circle is outside of the later circle. The edge is a branch of a hyperbolic curve whose foci are p_{ci} and p_{cj} (Figure 3.5.9(f)).

3.5.3 Voronoi diagrams for a set of circles

In the literature, special attention is paid to a line Voronoi diagram generated by a set L_c of full circles. To generate this special Voronoi diagram, we have to define a distance from a point to a circle. In the literature, a few distances are proposed. Probably the most natural distance is given by the shortest distance of equation (3.5.1), which is explicitly written as

$$d_{c1}(p, L_i) = | \, \|\mathbf{x} - \mathbf{x}_{ci}\| - r_i |, \tag{3.5.10}$$

where L_i is the circle centred at p_{ci} with radius r_i. A line Voronoi diagram $\mathcal{V}(L_c)$ generated by a set of circles with this distance is a special case of the line Voronoi diagram $\mathcal{V}(L_{plc})$ defined in the above subsection (the generator set is a set of paired semicircles). Hence the line Voronoi diagram $\mathcal{V}(L_c)$ shares the properties of $\mathcal{V}(L_{plc})$ mentioned in the above. In Figure 3.5.9 we

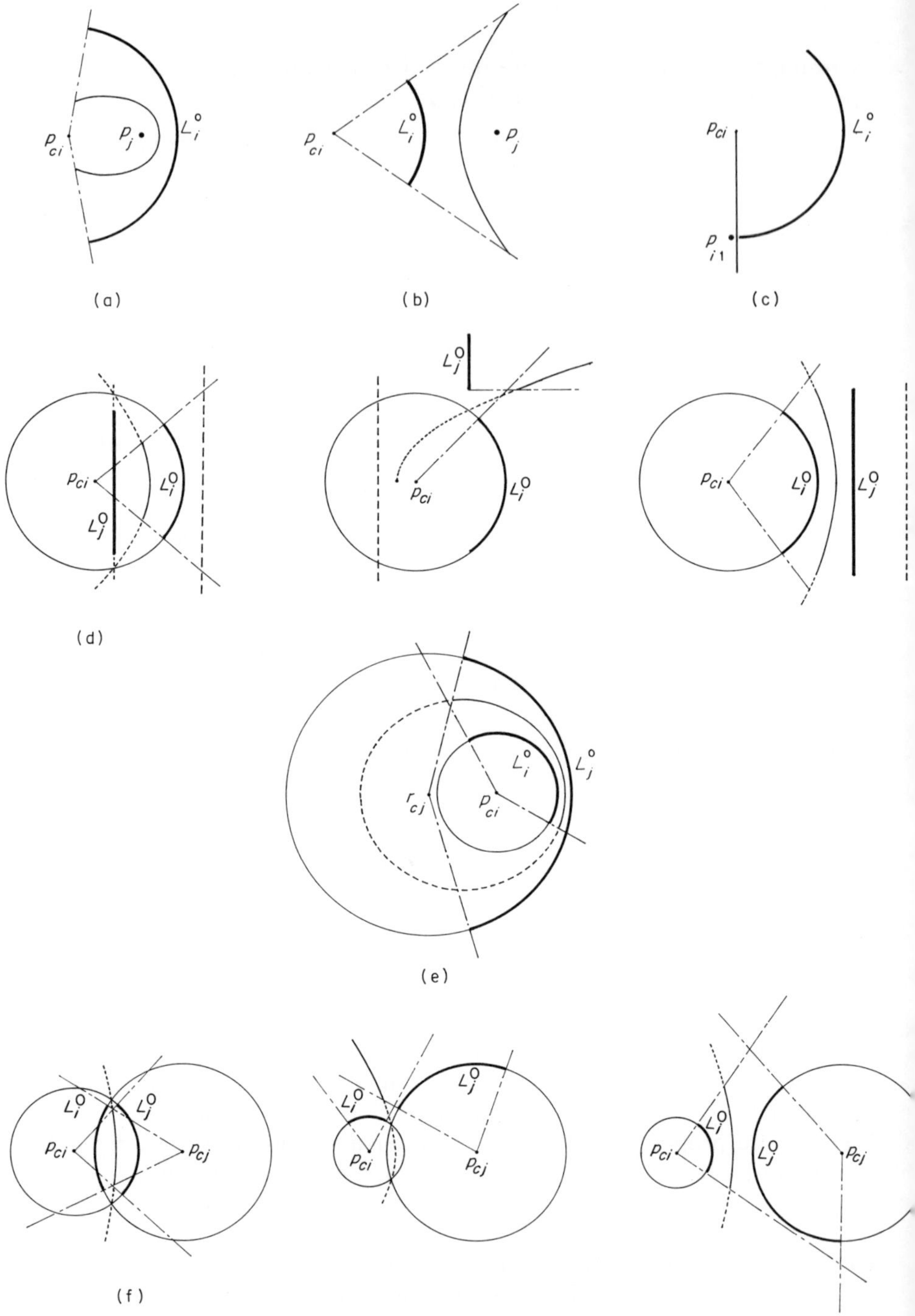

Figure 3.5.8 Types of edges generated by a circular arc and a point, a straight line or a circular arc: (a) type E5; (b) type E6; (c) type E7; (d) type E8; (e) type E9; (f) type E10.

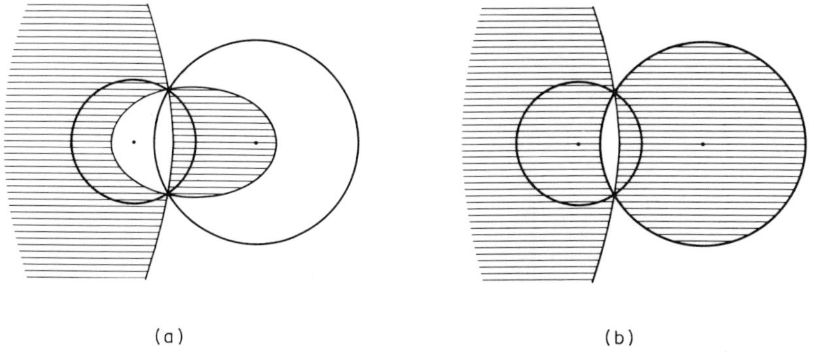

Figure 3.5.9 The line Voronoi diagrams generated by two intersecting circles with (a) the distance given by equation (3.5.10), and (b) the distance given by equation (3.5.11).

depict the line Voronoi diagram generated by two intersecting circles. The reader should confirm the types of edges using Property L7.

Lee and Drysdale (1981) and Sharir (1985) employ a similar but slightly different distance given by

$$d_{c2}(p, L_i) = \|\boldsymbol{x}-\boldsymbol{x}_{ci}\| - r_i. \tag{3.5.11}$$

This distance is a not a metric because it may take a negative value. The difference between $d_{c1}(p, L_i)$ and $d_{c2}(p, L_i)$ appears when $d_{c2}(p, L_i)$ and $d_{c2}(p, L_i)$ have different signs. We can observe this difference in Figure 3.5.9 where panel (a) uses the distance of equation (3.5.10) and panel (b) uses the distance of equation (3.5.11). The elliptic curve does not appear if we employ the distance of equation (3.5.11). The properties of this line Voronoi diagram are examined in detail by Sharir (1985). We also note that if the circles do not overlap, the distance $d_{c1}(p, L_i) = d_{c2}(p, L_i) = \|\boldsymbol{x}-\boldsymbol{x}_{ci}\|-r_i$ is the additively weighted distance (equation (3.1.6) in Section 3.1); thus $\mathcal{V}(L_c)$ reduces to the AW-Voronoi diagram.

A fairly different distance is employed by Imai *et al.* (1985), which is given by

$$d_{c3}(p, L_i) = \|\boldsymbol{x}-\boldsymbol{x}_{ci}\|^2 - r_i^2. \tag{3.5.12}$$

If p is outside the circle, $\sqrt{d_{c3}(p, L_i)}$ indicates the distance from point p to the point p_i on L_i, where $\overline{pp_i}$ is tangential to L_i at p_i (the broken line in Figure 3.5.10). We call the line Voronoi diagram with this distance the *Laguerre Voronoi diagram*. This term is named after Laguerre geometry in which a point $(x, y, z) \in \mathbb{R}^3$ is corresponded to a directed circle centred at (x, y) with radius $|z|$ on $\mathbb{R}^2$. Recalling Section 3.1.4, we notice that the distance of equation (3.5.12) is the power distance. Thus the Laguerre Voronoi diagram is exactly the same as the power Voronoi diagram.

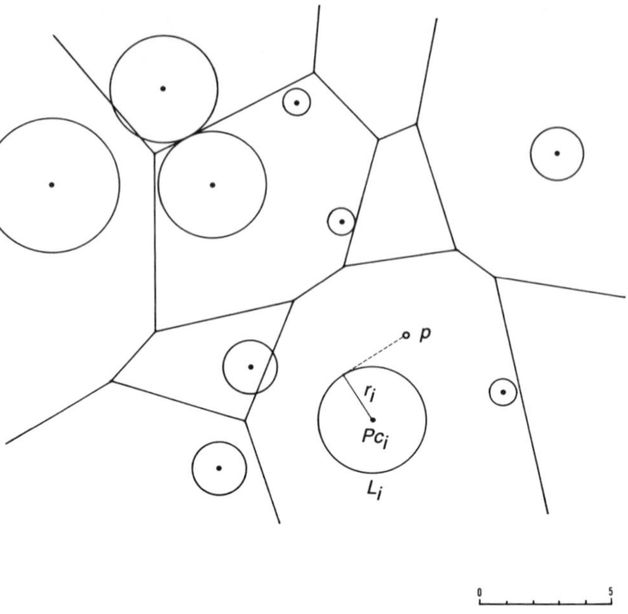

Figure 3.5.10 The Laguerre Voronoi diagram.

3.5.4 Medial axis

In conjunction with the line Voronoi diagram, we may refer to a medial axis which is often applied in pattern recognition. A medial axis can be defined for a connected geometric figure C (disconnected regions can be treated separately). The *medial axis* of such a geometric figure is a set of points $\mathbf{x} \in C$ for which there exist $\mathbf{x}_i$, $\mathbf{x}_j \in \partial C$ such that $\mathbf{x}_i \neq \mathbf{x}_j$ and $\|\mathbf{x}-\mathbf{x}_i\| = \|\mathbf{x}-\mathbf{x}_j\| = \min_{\mathbf{y}} \{\|\mathbf{x}-\mathbf{y}\| \,|\, \mathbf{y} \in \partial C\}$, that is, the medial axis of C, denoted by $M(C)$, is given by

$$M(C) = \{\mathbf{x} |\ \|\mathbf{x}-\mathbf{x}_i\| = \|\mathbf{x}-\mathbf{x}_j\| = \min_{\mathbf{y}} \{\|\mathbf{x}-\mathbf{y}\| \,|\, \mathbf{y} \in \partial C\},\ \mathbf{x}_i \neq \mathbf{x}_j,\ \mathbf{x}_i, \mathbf{x}_j \in \partial C, \mathbf{x} \in C\}. \tag{3.5.13}$$

An example is illustrated in Figure 3.5.11.

We can obtain a medial axis using the line Voronoi diagram. Suppose that a geometric figure C is represented by a polygon whose boundaries are given by a chain of line segments L (the dashed line in Figure 3.5.11). We first decompose L into $L^{(d)}$, and next construct the line Voronoi diagram $\mathcal{V}(L^{(d)})$. Recalling the property that points on the edges of line Voronoi regions are equally distant from generator lines (the boundaries of the geometric figure), we notice that the medial axis is included in the edges in $\mathcal{V}(L^{(d)})$. Deleting unnecessary edges in $\mathcal{V}(L^{(d)})$, we obtain the medial axis

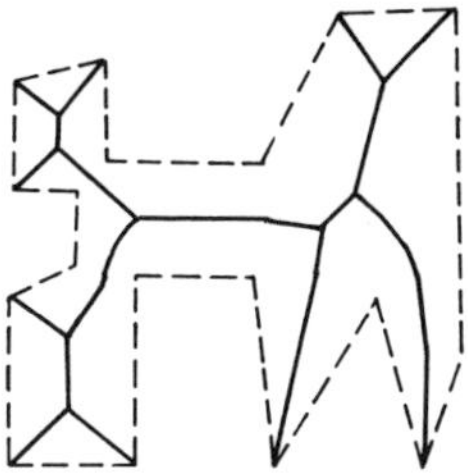

Figure 3.5.11 A medial axis.

of C (the solid line in Figure 3.5.11). The detailed computational method is shown in Preparata (1977) and Lee (1982a).

3.6 VORONOI DIAGRAMS FOR AREAS

Having now extended a generator from a point to a line, the reader may expect that the next extension is from a line to an area. This extension is possible and useful in many applications.

We consider a set of areas, $A = \{A_1, \ldots, A_n\}$ $(1 \leq n < \infty)$, in $\mathbb{R}^2$. We assume that A_i is a connected closed set and areas do not overlap each other, i.e. $A_i \cap A_j = \emptyset$, $i \neq j$, $i,j \in I_n$. An area A_i is not necessarily convex, and may have holes in which another area may exist. Under these assumptions, we define a distance from a point to an area A_i as the shortest distance from p to A_i, i.e.

$$d_s(p, A_i) = \min_{\mathbf{x}_i}\{\|\mathbf{x} - \mathbf{x}_i\| \, | \mathbf{x}_i \in A_i\}, \tag{3.6.1}$$

where $\mathbf{x}$ and $\mathbf{x}_i$ are the location vectors of p and p_i, respectively. With this distance, we define a set by

$$V(A_i) = \{p | d_s(p, A_i) \leq d_s(p, A_j), j \neq i, i, j \in I_n\}. \tag{3.6.2}$$

We call this set the *area Voronoi region* associated with A_i, and a set of the area Voronoi regions, $\mathcal{V}(A) = \{V(A_1), \ldots, V(A_n)\}$, the *area Voronoi diagram* generated by the generator set A. Mathematically, an area includes a line and a line includes a point. Thus the area Voronoi diagram includes the line Voronoi diagram, which includes the ordinary Voronoi diagram. We can thus regard the area Voronoi diagram as a generalization of the line Voronoi diagram as well as the ordinary Voronoi diagram. Figure 3.6.1(a) shows an area Voronoi diagram.

An area (which does not degenerate into a line) appears different from a line, but the area Voronoi diagram is quite similar to the line Voronoi diagram. Actually, we construct an area Voronoi diagram from its corresponding line Voronoi diagram. To be explicit, let $L(A)$ be the set of the boundaries of A, i.e. $L(A) = \{\partial A_1, \ldots, \partial A_n\}$. We first construct the line Voronoi diagram with $L(A)$ (Figure 3.6.1(b)). We next delete the

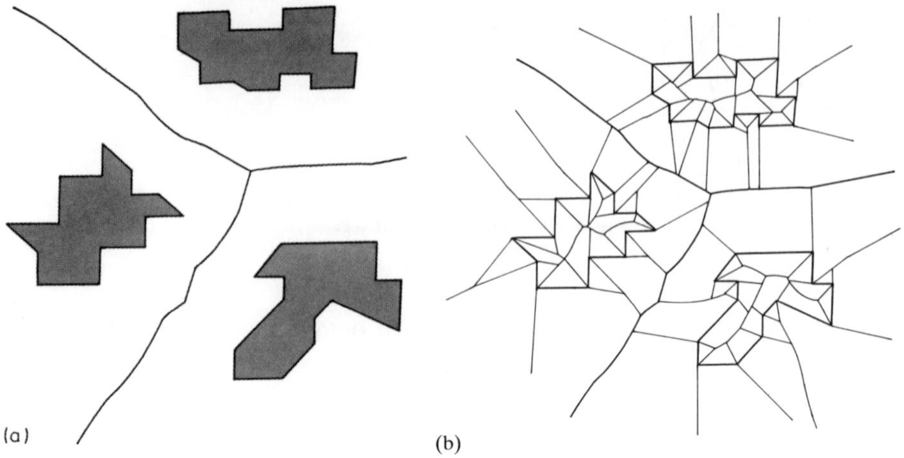

Figure 3.6.1 (a) An area Voronoi diagram and (b) its corresponding line Voronoi diagram.

superfluous edges. Then the remaining edges form the area Voronoi diagram (Figure 3.6.1(a)). As is noticed from this procedure, some properties of the area Voronoi diagram are similar to those of the line Voronoi diagram. It should be noted, however, that the properties of the line Voronoi diagram that are related to the inside of a circuit chain or a circle are not applicable to those of the area Voronoi diagram. For instance, an elliptic curve does not appear in the area Voronoi diagram because the curve appears only inside a circle.

Area Voronoi diagrams are useful for robot path planning. The problem is to find a path, if it exists, on which a polygon representing a robot can move without colliding with obstacles $A_1, \ldots, A_n$. To solve this problem we can use the area Voronoi diagram defined with the V-distance of equation (3.6.1), but in practice, to unburden algebraic complexity, we often use area Voronoi diagrams defined with alternative V-distances which give polyhedral (polygonal) Voronoi regions. The properties of such area Voronoi diagrams are discussed in detail by Ó'Dúnlaing *et al.* (1986), and Leven and Sharir (1987) and Canny and Donald (1988).

3.7 VORONOI DIAGRAMS WITH V-DISTANCES

As we mentioned in the introduction of Chapter 3, we can define a generalized Voronoi diagram with any distance as long as the distance is a V-distance. The distance is not necessarily the Euclidean distance. In fact, we showed in the preceding sections that the multiplicatively weighted distance, the additively weighted distance, the compoundly weighted distance, the power distance and the shortest-path distance, which are not the

Euclidean distance, produced generalized Voronoi diagrams. In this section we show other generalized Voronoi diagrams defined with various kinds of V-distances. First, we deal with V-distances defined in $\mathbb{R}^m$ (in particular, $m = 2$) and next V-distances defined on surfaces in $\mathbb{R}^3$.

3.7.1 Voronoi diagrams with the Minkowski metric L_p

We begin with a Voronoi diagram defined with the Minkowski metric in $\mathbb{R}^m$. The *Minkowski (power) metric* from a point p to a point p_i in $\mathbb{R}^m$ is defined by

$$d_{L_p}(p, p_i) = \left[\sum_{j=1}^{m} |x_j - x_{ij}|^p\right]^{1/p}, \tag{3.7.1}$$

where $(x_1, \ldots, x_m)$ and $(x_{i1}, \ldots, x_{im})$ are the Cartesian coordinates of p and p_i, respectively. Customarily, the symbol L_p is used for the Minkowski metric, where p refers to the degree of the power. The reader should not confuse p of a point with p of the degree of the power. The parameter p varies in the range of $1 \leq p < \infty$. If $p = 1$, equation (3.7.1) becomes

$$d_{L_1}(p, p_i) = \sum_{j=1}^{m} |x_j - x_{ij}|, \tag{3.7.2}$$

which is called the *Manhattan metric*, the *city-block distance* or the *taxi-cab distance*. If $p = 2$, the Minkowski metric is the Euclidean distance. If $p = \infty$, the Minkowski metric is called the *supremum metric* or *dominance metric*, which is alternatively written as

$$d_{L_\infty}(p, p_i) = \max_j\{|x_j - x_{ij}|, j \in I_m\}. \tag{3.7.3}$$

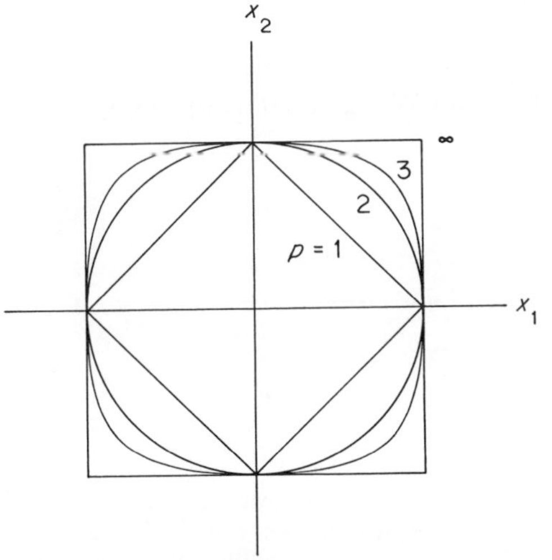

Figure 3.7.1 The contour lines of the Minkowski metric L_p for $p = 1,2,3,\infty$.

Figure 3.7.1 shows the contours of the Minkowski metric from the origin for p = 1, 2, 3, ∞. Since the Minkowski metric includes the Euclidean distance, we may say that Voronoi diagrams with the Minkowski metric are a generalization of the ordinary Voronoi diagram.

The geometric properties of Voronoi diagrams with the Minkowski matrices L_p drastically vary according to the value of p ($1 \leq p \leq \infty$). We first consider a planar Voronoi diagram defined with the Manhattan metric. To see its geometric properties, let us observe the bisector defined with the Manhattan metric. From equation (3.7.2) ($p = 1$), we notice that the shape of the bisector varies according to the parameter value $\alpha = (x_{j2}-x_{i2})/(x_{j1}-x_{i1})$, where $x_{j1} \geq x_{i1}$ is assumed without loss of generality. We have eight possible types: (i) $-\infty < \alpha < -1$; (ii) $-1 < \alpha < 0$; (iii) $0 < \alpha < 1$; (iv) $1 < \alpha < \infty$; (v) $\alpha = \pm\infty$; (vi) $\alpha = 0$; (vii) $\alpha = -1$; and (viii) $\alpha = 1$. However, those types are essentially grouped in three types: I = (i)–(iv); II = (v) and (vi); and III = (vii) and (viii), which are shown in Figure 3.7.2. The bisector of type I consists of three straight lines as depicted in Figure 3.7.2(a). The diagonal line segment has angle $\pi/4$ or $3\pi/4$ with the x_1-axis. This bisector is well-behaving and regular. The bisector of type II is a straight line (Figure 3.7.2(b)), which is also well-behaving and regular. The bisector of type III, however, is not well-behaving. In this case, as is indicated by the shaded area in Figure 3.7.2(c), the bisector consists of not only a straight line but also an area. To avoid this indeterminacy, we usually define the bisector as the straight line perpendicular bisecting $\overline{p_ip_j}$ (the solid and dashed lines in Figure 3.7.2(c)).

Once the above modification is made, the bisector becomes well-behaving, and hence the set $\mathcal{V}_M = \{V(p_1), \ldots, V(p_n)\}$ defined by equation (3.7.1) with equation (3.7.2) ($p = 1$) gives a generalized Voronoi diagram. We call the set $\mathcal{V}_M$ the *Manhattan-metric Voronoi diagram* generated by P or the *Manhattan Voronoi diagram* in brief, and the set $V(p_i)$ the *Manhattan(-metric) Voronoi polygon* associated with p_i (Carter *et al.*, 1972; Hwang, 1979; Lee, 1980; Lee and Wong, 1980). An example is illustrated in Figure 3.7.3.

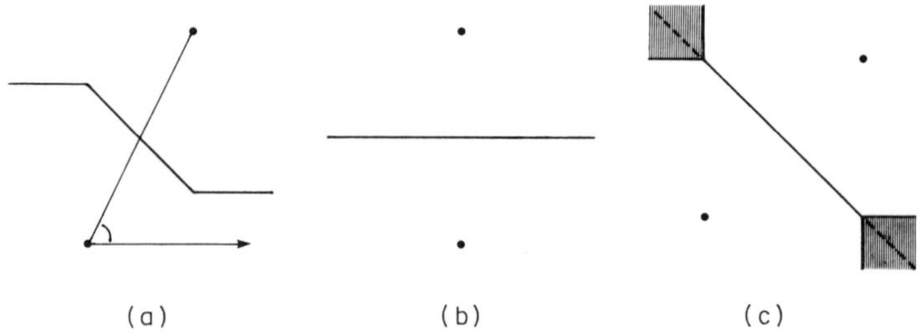

Figure 3.7.2 The bisector $b(p_i,p_j)$ defined with the Manhattan metric: (a) the bisector of type I ($\infty < \alpha < -1$, $-1 < \alpha < 0$, $0 < \alpha < 1$, $1 < \alpha < \infty$); (b) the bisector of type II ($\alpha = \pm\infty, 0$); (c) the bisector of type III ($\alpha = \pm 1$); where $\alpha = (x_{j2}-x_{i2})/(x_{j1}-x_{i1})$ and where $x_{j1} \geq x_{i1}$ is assumed without loss of generality.

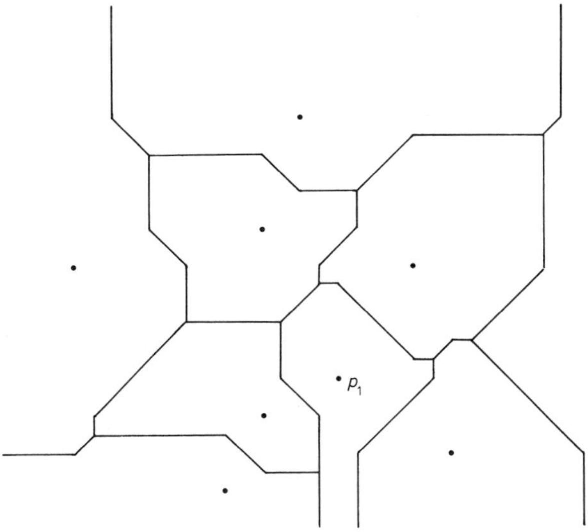

Figure 3.7.3 A Manhattan-metric Voronoi diagram.

From the properties of the bisector mentioned above with Figure 3.7.2, we note the following properties.

Property M1 The set $V(p_i)$ defined by equation (3.7.1) with the Manhattan metric (equation 3.7.2) is not empty, and forms a (Manhattan Voronoi) polygon. The Manhattan Voronoi polygon $V(p_i)$ is not necessarily convex, but it is always star-shaped with respect to the generator p_i.

Property M2 Every edge of a Manhattan Voronoi polygon consists of at most three straight lines which are parallel to the x_1-axis, x_2-axis, or diagonal lines with angle $\pi/4$ or $3\pi/4$.

Property M3 The Manhattan Voronoi polygon $V(p_i)$ is infinite if p_i is on the boundary of the convex hull of P, but not conversely.

Compared with Property V2 of the ordinary Voronoi diagram, the converse relation in Property M3 should be noted. An actual example is found in the Manhattan Voronoi polygon of p_1 in Figure 3.7.3.

In conjunction with the Manhattan Voronoi diagram, we may refer to the Voronoi diagram defined with the supremum metric, which we call the *supremum-metric Voronoi diagram*. The supremum-metric Voronoi diagram is quite similar to the Manhattan Voronoi diagram. To see this similarity, we depict the bisector of the supremum metric in Figure 3.7.4 in which the

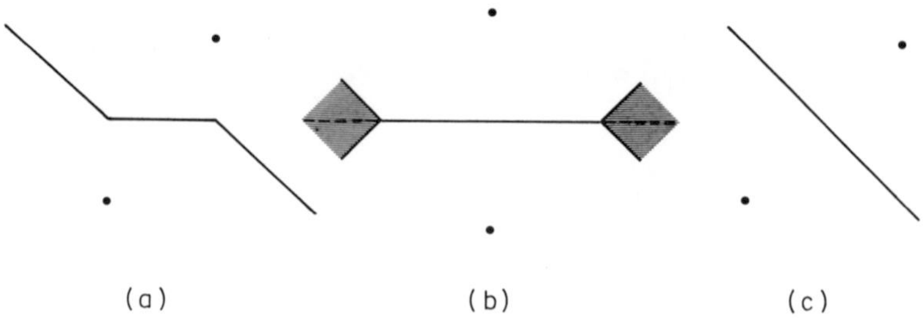

Figure 3.7.4 The bisector defined with the supremum metric: (a) the bisector of type I ($\infty < \alpha < -1$, $-1 < \alpha < 0$, $0 < \alpha < 1$, $1 < \alpha < \infty$); (b) the bisector of type II ($\alpha = \pm\infty, 0$); (c) the bisector of type III ($\alpha = \pm 1$); where $\alpha = (x_{j2} - x_{i2})/(x_{j1} - x_{i1})$ and where $x_{j1} \geq x_{i1}$ is assumed without loss of generality.

locations of the generators are the same as those of the bisector of the Manhattan metric in Figure 3.7.2.

Comparing Figure 3.7.4(b) with Figure 3.7.2(c), we notice that if we rotate the x_1–x_2 plane in Figure 3.7.2(c) with generators counterclockwise by $\pi/4$, we obtain Figure 3.7.4(b). Generally, if we rotate the bisector defined with the Manhattan metric counterclockwise by $\pi/4$, we obtain the bisector defined with the supremum metric. From this property, we obtain the following property.

Property SUP1 The supremum-metric Voronoi diagram of P is obtained by first rotating P counterclockwise by $\pi/4$, next constructing the Manhattan Voronoi diagram for the rotated set of points, and finally rotating the resultant diagram by $\pi/4$.

We should note that Property SUP1 does not say that the supremum-metric Voronoi diagram of P is the same as the Manhattan Voronoi diagram of P. As we notice from Figures 3.7.3 and 3.7.5, these two Voronoi diagrams are different.

3.7.2 Voronoi diagram with the Karlsruhe metric

In Manhattan, we can move along either North–South streets or East–West avenues. In Karlsruhe, we can move along either radiating streets from the centre or circular avenues around the centre. We call the shortest distance in Karlsruhe the Karlsruhe metric. To be precise, let (r, θ) and (r_i, θ_i) be the polar coordinates of p and p_i, respectively, where $0 \leq \theta, \theta_i < 2\pi$, r, $r_i > 0$, and

$$\delta(\theta, \theta_i) = \min\{|\theta - \theta_i|, 2\pi - |\theta - \theta_i|\}. \tag{3.7.4}$$

Then, the *Karlsruhe metric* or *Moscow metric* from p to p_i is defined by

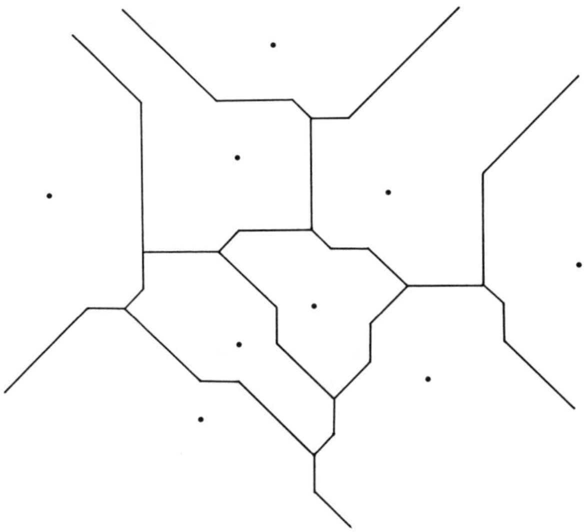

Figure 3.7.5 The supremum-metric Voronoi diagram generated by the generators whose locations are the same as those in Figure 3.7.3.

$$d_{\mathrm{K}}(p, p_i) = \begin{cases} \min\{r, r_i\}\delta(\theta, \theta_i) + |r - r_i| & \text{for } 0 \leq \delta(\theta, \theta_i) < 2, \\ r + r_i & \text{for } 2 \leq \delta(\theta, \theta_i) < \pi \end{cases} \quad (3.7.5)$$

(Koshizuka and Kurita, 1986; Klein, 1988). In Figure 3.7.6(a), the shortest path for $0 \leq \delta(\theta, \theta_i) < 2$ is indicated by the heavy solid line and that for $2 \leq \delta(\theta, \theta_i) \leq \pi$ is indicated by the broken line. In Figure 3.7.6(b) the contour lines of the Karlsruhe metric from the point indicated by the filled

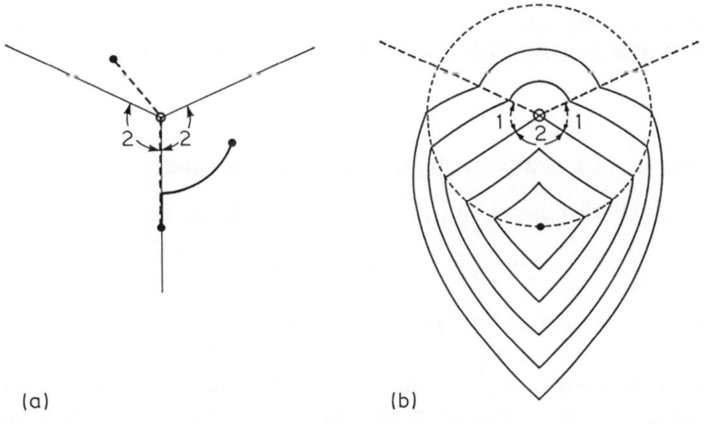

Figure 3.7.6 The contour lines of the Karlsruhe metric.

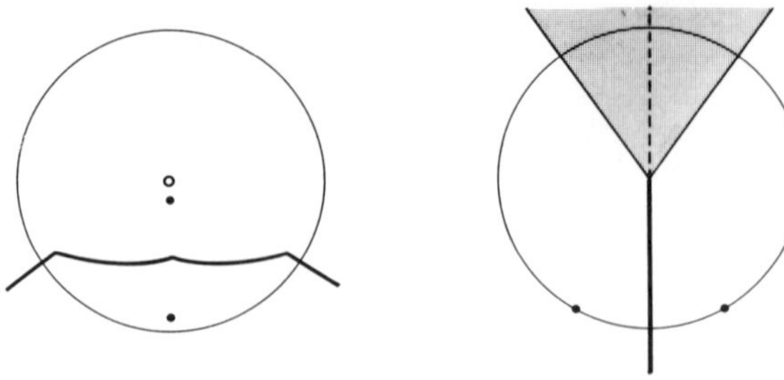

Figure 3.7.7 The bisectors defined with the Karlsruhe metric: (a) a well-behaving bisector; (b) not a well-behaving bisector and its modified bisector (the dashed line and heavy solid line).

circle (θ is measured from the line radiating from the unfilled circle (the origin) to the filled circle counterclockwise). In the fan-shaped region bounded by the lines radiating from the origin with 2 radians and $2\pi-2$ radians, the contour lines are circular arcs centred at the origin, and the contour lines radiating from the origin with 1 radian and $2\pi-1$ radians are straight lines in the dashed circle in Figure 3.7.6.

In most cases the bisector with the Karlsruhe metric is well-behaving, as is shown in Figure 3.7.7(a). In a special case (Figure 3.7.7(b)), however, the bisector is not well-behaving. The indeterminacy occurs when two points p_i and p_j have the same distance from the origin ($r_i = r_j$) and $\delta(p_i, p_j) < 2\pi-4$. In this case, as we did for the bisector of the Manhattan metric (Figure 3.7.2(c)), we define the bisector as the perpendicular line bisecting $\overline{p_ip_j}$ (the heavy solid and dashed lines in Figure 3.7.7(b)).

If we add the above modification to the bisector of the Karlsruhe metric, the set $\mathcal{V}_K = \{V(p_1), \ldots, V(p_n)\}$ gives a generalized Voronoi diagram. We call this the *Karlsruhe-metric Voronoi diagram* or simply the *Karlsruhe Voronoi diagram*, and the regions constituting the Karlsruhe Voronoi diagram *Karlsruhe(-metric) Voronoi regions*. An example is shown in Figure 3.7.8.

3.7.3 Voronoi diagrams with the Hausdorff distance

The *Hausdorff distance* from a point p to a closed set A is defined by

$$d_H(p,A_i) = \max_{\mathbf{x}_i}\{\|\mathbf{x}-\mathbf{x}_i\| \mid \mathbf{x}_i \in A_i\}. \tag{3.7.6}$$

This distance is useful in applications when we are concerned with the farthest distance. For instance, if we are concerned with the location of a fire station in a rural region where A_i represents the area of a village, the crucial house is the farthest house in each village from the fire station. Aurenhammer (1988) employs the Hausdorff distance to define a line

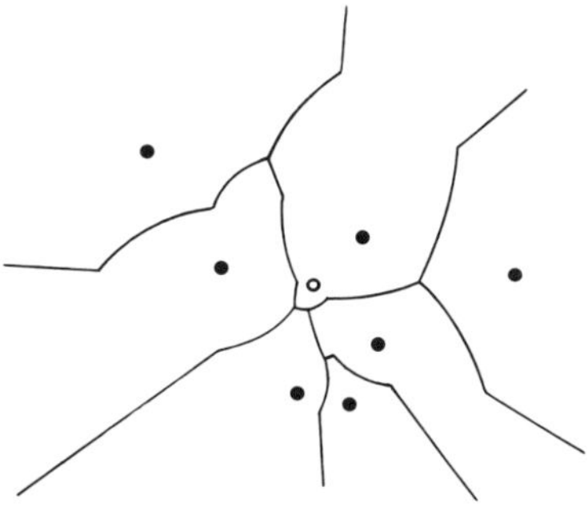

Figure 3.7.8 A Karlsruhe-metric Voronoi diagram. (*Source*: Klein (1988), Figure 12.)

Voronoi diagram. This line Voronoi diagram is quite different from the ordinary line Voronoi diagram in Section 3.4. To see this difference, let $b(L_i, L_j)$ be the bisector generated by two straight line segments L_i and L_j; p_{k1} and p_{k2} be the end points of $L_k, k = i, j$; and $H(p_i, p_j)$ be the half plane made by the perpendicular line bisecting $\overline{p_i p_j}$ that includes p_i. If a point p is in $H(p_{i2}, p_{i1}) \cap H(p_{j2}, p_{j1})$, then the farthest point in L_i from p is p_{i1}, and the farthest point in L_j from p is p_{j1}. Thus the bisector $b(L_i, L_j)$ in $H(p_{i2}, p_{i1}) \cap H(p_{j2}, p_{j1})$ is given by the line perpendicularly bisecting $\overline{p_{i1} p_{j1}}$. Generally, the bisector $b(L_i, L_j)$ in $H(p_{ik}, p_{ik}) \cap H(p_{jh}, p_{j\bar{h}})$ is written as

$$b(L_i, L_j) \cap [H(p_{ik}, p_{i\bar{k}}) \cap H(p_{jh}, p_{j\bar{h}})] =$$
$$\{\mathbf{x} | \, \|\mathbf{x} - \mathbf{x}_{i\bar{k}}\| = \|\mathbf{x} - \mathbf{x}_{j\bar{h}}\|\} \cap [H(p_{ik}, p_{i\bar{k}}) \cap H(p_{jh}, p_{j\bar{h}})], \quad k,h = 1,2, \qquad (3.7.7)$$

where $\bar{k} = 1$ if $k = 2$ and $\bar{k} = 2$ if $k = 1$, and the same for $\bar{h}$. An example is shown in Figure 3.7.9.

From Figure 3.7.9 we notice that the bisector consists of straight lines. A curved bisector observed in the orindary line Voronoi diagram never appears here. The bisector is well-behaving and hence we can obtain a line Voronoi

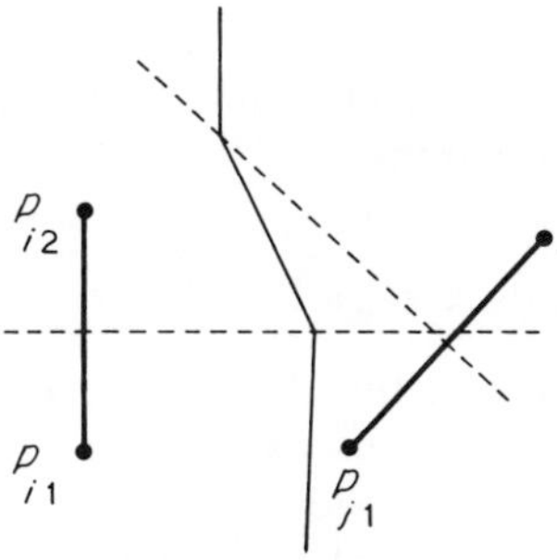

Figure 3.7.9 The bisector of two line segments defined with the Hausdorff distance.

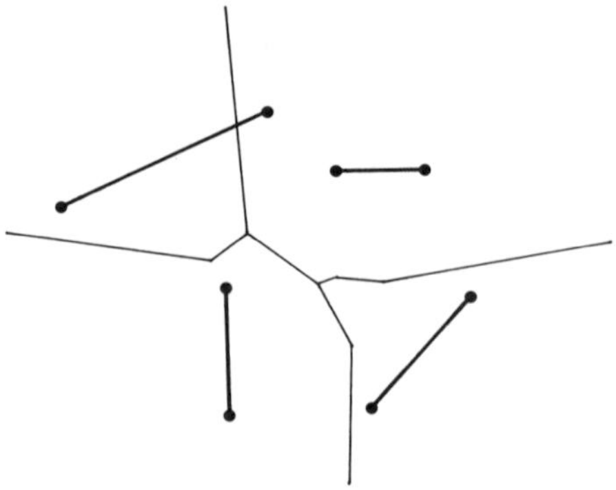

Figure 3.7.10 A Hausdorff line Voronoi diagram generated by a set of straight line segments.

diagram with the Hausdorff distance. We call this diagram the *Hausdorff line Voronoi diagram*. An example is shown in Figure 3.7.10.

From equation (3.7.6) or (3.7.7) the reader may notice that the Hausdorff Voronoi diagram can be constructed with the help of the farthest-point Voronoi diagram. The role of a farthest-point Voronoi diagram becomes more explicit when we construct a Hausdorff Voronoi diagram generated by a set of polygons. For instance, consider a set of polygons $A = \{A_1, \ldots, A_n\}$ and let $P_k = \{p_{il}, l \in I_{n_i}\}$ be the vertices of the polygon A_i, $i \in I_n$. To obtain the bisector $b(A_i, A_j)$ between polygons A_i and A_j, we first construct farthest-point Voronoi diagrams for P_k, $k = i, j$ (the dashed lines in Figure 3.7.11). Let $V_{\mathrm{fp}}(p_{kl}|P_k)$ be the farthest-point Voronoi polygon of p_{kl} in the farthest-point Voronoi diagram generated by P_k, $k = i, j$. Then part of the bisector is given by

$$\begin{aligned} & b(A_i, A_j) \cap \{V_{\mathrm{fp}}(p_{il}|P_i) \cap V_{\mathrm{fp}}(p_{jh}|P_j) \backslash [A_i \cup A_j]\} \\ & \quad = \{\mathbf{x} | \ \|\mathbf{x}-\mathbf{x}_{il}\| = \|\mathbf{x}-\mathbf{x}_{jh}\|\} \cap \{V_{\mathrm{fp}}(p_{il}|P_i) \cap V_{\mathrm{fp}}(p_{jh}|P_j) \backslash [A_i \cup A_j]\}, \quad (3.7.8) \\ & \quad l \in I_{n_i}, h \in I_{n_j} \end{aligned}$$

(the fine solid lines in Figure 3.7.11). Using this equation, we can construct a Hausdorff area Voronoi diagram.

3.7.4 Voronoi diagrams on a sphere

In the military district problem, the dominance regions or coverage areas of air bases in the world is often discussed. We may treat this problem by a Voronoi diagram defined on a sphere.

Let $P = \{p_1, \ldots, p_n\}$ $(2 \leq n < \infty)$ be distinct points on a sphere S with the unit radius centred at the origin, and $\mathbf{x}$ and $\mathbf{x}_i$ be the location vectors

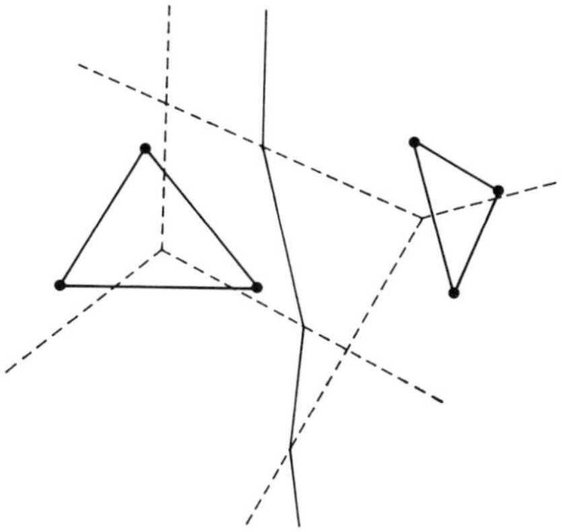

Figure 3.7.11 The bisector of two polygons with the Hausdorff distance, or a Hausdorff Voronoi diagram generated by a set of two polygons (the dashed lines show the farthest-point Voronoi diagrams generated by the vertices of the polygons).

of $p \in S$ and $p_i \in S$, respectively. The shortest distance from p to p_i on S is defined by the length of the lesser arc on the great circle (the circle whose centre is at the centre of S) passing through p and p_i. Mathematically, this distance is written as

$$d_{gc}(p, p_i) = \arccos(\mathbf{x}^T \mathbf{x}_i) \leq \pi. \tag{3.7.9}$$

We call this distance the *great circle distance*. The bisector defined with the great circle distance is given by the great circle that 'perpendicularly' passes through the mid-point of the great circular arc combining p_i and p_j ('perpendicularly' means that small segments of the two great circles around the mid-point are orthogonal). This bisector divides the sphere S into two disjoint hemispheres. Thus the bisector defined with the great circle distance is well-behaving, and

$$V(p_i) = \{p | d_{gc}(p, p_i) \leq d_{gc}(p, p_j), j \neq i, j \in I_n, p \in S\} \tag{3.7.10}$$

gives a non-empty set in S. We call this set the *spherical Voronoi polygon* associated with p_i. The set of resulting spherical Voronoi polygons gives a generalized Voronoi diagram, which we call the *spherical Voronoi diagram* generated by P on S (Miles, 1971; Brown, 1980; Paschinger, 1982; Ash and Bolker, 1985). Figure 3.7.12 shows a spherical Voronoi diagram.

The term 'spherical polygon' mentioned above is a 'polygon' on a sphere S. To be precise, let S_i be a hemisphere on S. Then, a *spherical polygon* is defined by the set of points in S which is obtained from the combination of intersections and/or unions of S_i, $i = 1,2,\ldots$ Like a convex polygon in $\mathbb{R}^2$, we can define a *convex spherical polygon* by the spherical polygon in which the lesser arc of a great circle passing through any two points in the

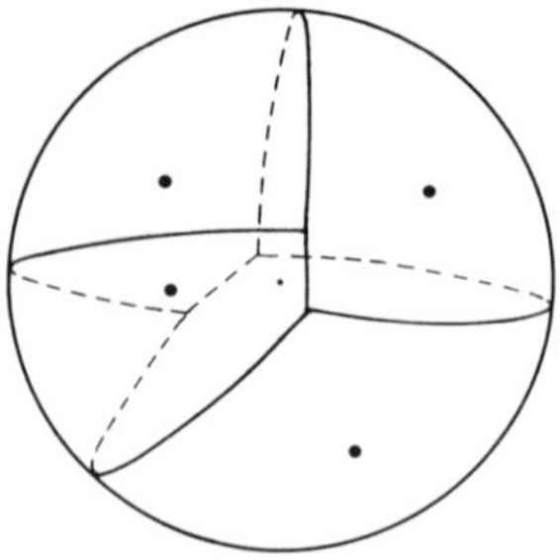

Figure 3.7.12 A spherical Voronoi diagram.

spherical polygon is embedded in the spherical polygon. From the definition of a spherical Voronoi polygon, we understand the following properties.

Property S1 The set $V(p_i)$ defined by equation (3.7.19) is a unique finite non-empty convex spherical polygon.

Property S2 An edge of a spherical Voronoi polygon is an arc of the great circle.

These properties correspond to Properties V1–V3 of the ordinary Voronoi diagram. Properties V4–V7 of the ordinary Voronoi diagram also hold for the spherical Voronoi diagram by replacing relevant terms. Property V9, however, becomes slightly different, because the ordinary Voronoi diagram is formulated in an unbounded space, whereas the spherical Voronoi diagram is formulated in a bounded space. From Euler's theorem, we obtain the following property.

Property S3 Let n_v and n_e be the number of vertices and edges in the spherical Voronoi diagram generated by $P = \{p_1, \ldots, p_n\}$. Then the following equation holds:

$$n_v - n_e + n = 2. \tag{3.7.11}$$

Replacing equation (2.3.4) with equation (3.7.11) in the derivations used for Properties V11–V13, the reader can obtain the properties corresponding to Properties V11–V13 of the ordinary Voronoi diagram.

3.7.5 Voronoi diagrams on a cylinder

When we meet phenomena in which the same configuration of generator points appears in cycles, we may analyse them with the aid of a Voronoi diagram on a cylinder.

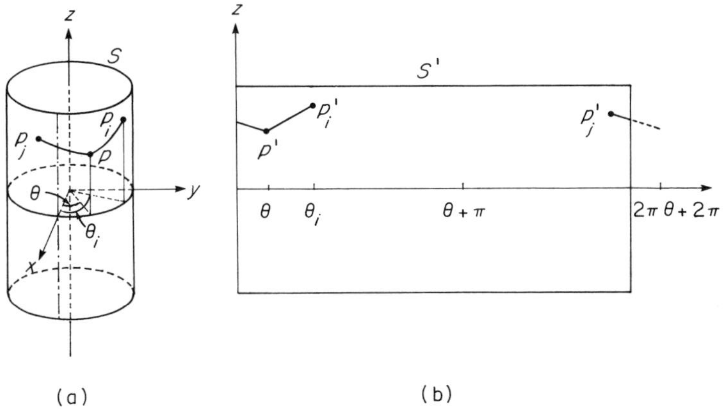

Figure 3.7.13 (a) A cylinder in $\mathbb{R}^3$, and (b) its developed region in $\mathbb{R}^2$.

Suppose that the axis of a cylinder S with unit radius is placed at the z-axis in $\mathbb{R}^3$, and that a set of distinct points, $P = \{p_1, \ldots, p_n\}$ $(2 \leq n < \infty)$, is placed on the cylinder S (Figure 3.7.13(a)). The location of p_i on S can be indicated by (x_i, y_i, z_i), where $x_i^2 + y_i^2 = 1$, and the distance between p and p_i is defined by the length of the shortest path on S connecting p and p_i. At first glance this measurement appears difficult in practice, but if we notice that a cylinder can be developed on a plane, this measurement becomes quite easy. If we cut the cylinder S along the line parallel to the z-axis passing through (1, 0, 0) (the dash-dotted line in Figure 3.7.13(a)), then S can be developed into S' on the plane, as is seen in Figure 3.7.13(b). The developed region S' is the region bounded by parallel lines at a distance 2π on $\mathbb{R}^2$. To treat the location of a point p_i' on S' which corresponds to p_i on S, let θ_i be the counterclockwise angle (viewed from above the x–y plane, i.e. $z > 0$) from the x-axis to the half line radiating from the origin to the point $(x_i, y_i, 0)$, then we can indicate the point p_i' by (θ_i, z_i) (Figure 3.7.12(b)). Since the shortest path on S is represented by the straight line on S', the distance between p and p_i on S is the same as the distance between $p' = (\theta, z)$ and $p_i' = (\theta_i, z_i)$ on S' (where $\theta \leq \theta_i$ is assumed without loss of generality), which is given by

$$d_{\text{cy}}(p, p_i) = \begin{cases} \sqrt{[(\theta - \theta_i)^2 + (z - z_i)^2]} & \text{for } \theta_i - \theta \leq \pi, \\ \sqrt{[(\theta + 2\pi - \theta_i)^2 + (z - z_i)^2]} & \text{for } \theta_i - \theta > \pi. \end{cases} \qquad (3.7.12)$$

Since the bisector $b(p_i, p_j)$ defined with the distance of equation (3.7.12) splits S into two disjoint regions (Figure 3.7.14(a)), the dominance region is well-behaving and regular. We can thus define a Voronoi diagram on S with this distance, which we call the *cylindrical Voronoi diagram* on S generated by P (Figure 3.7.14(b)).

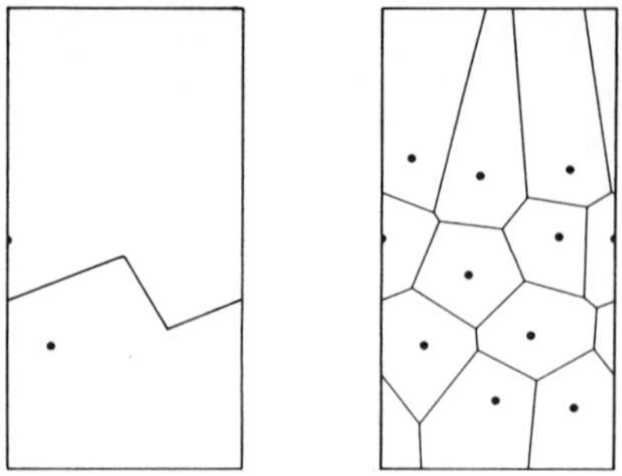

Figure 3.7.14 (a) A bisector on a cylinder, and (b) a cylindrical Voronoi diagram.

3.7.6 Voronoi diagrams on a cone

If a set of distinct points $P = \{p_1, \ldots, p_n\}$ is placed on a cone, we may consider a Voronoi diagram on a cone. Suppose that a cone S is placed in $\mathbb{R}^3$ in such a way that the axis of the cone S is the z-axis of $\mathbb{R}^3$, and the radius of the circle made by the intersection of the cone S with the x–y plane is unity (Figure 3.7.15(a)). Like a cylinder, a cone can be developed on a plane. If we cut the cone S along the line radiating from the top of the cone passing through (1, 0, 0) and develop it on the plane, we get a fan-shaped region S', as is shown in Figure 3.7.15(b). We can thus measure the distance between p and p_i on the cone S by the distance between p' and p'_i in the fan-shaped region S'. To write this distance mathematically, let (x_i, y_i, z_i) be the Cartesian coordinates of p_i in $\mathbb{R}^3$; θ_i be the counterclockwise angle (viewed from above the x–y plane, $z > 0$) from the x-axis to the half line radiating from the origin to the point $(x_i, y_i, 0)$; $\theta'_i = \theta_i \sin(\alpha/2)$, where α is the top angle of the cone (defined by the acute angle of the intersection lines made from the x–z plane and the cone);

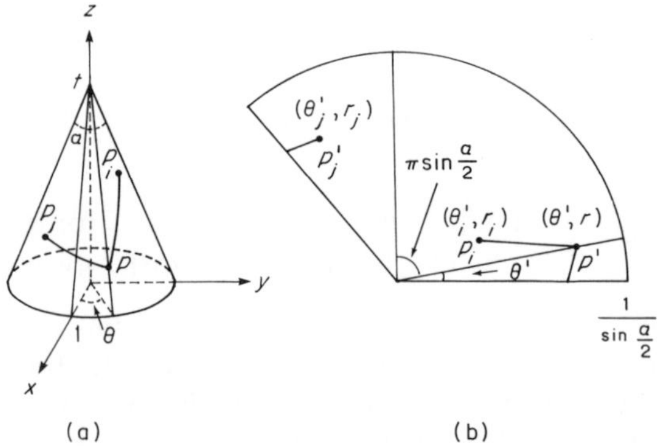

Figure 3.7.15 (a) A cone in $\mathbb{R}^3$, and (b) its developed region in $\mathbb{R}^2$.

$$r_i = \sqrt{x_i^2+y_i^2+(z_i-\cot(\alpha/2))^2}, \qquad (3.7.13)$$

which is the straight line distance from the top of the cone S to p_i. Then the point p_i on the cone S in $\mathbb{R}^3$ can be indicated by (θ'_i, r_i) in the fan-shaped region in $\mathbb{R}^2$. Thus the distance between p and p_i on the cone S is written as

$$d_{\text{cone}}(p,p_i) = \begin{cases} \sqrt{[r^2+r_i^2-2rr_i\cos(\theta'_i-\theta')]} \\ \qquad\qquad \text{for } \theta'_i \le \theta'+\pi\sin(\alpha/2), \\ \sqrt{[r^2+r_i^2-2rr_i\cos(\theta'+2\pi\sin(\alpha/2)-\theta'_1)]} \\ \qquad\qquad \text{for } \theta'_i > \theta'+\pi\sin(\alpha/2), \end{cases} \qquad (3.7.14)$$

where $\theta' \le \theta'_i$ is assumed without loss of generality (Figure 3.7.15(b)).

The bisector defined with this distance consists of a chain of straight line segments which divides the fan-shaped region S' into two disjoint regions, as is shown in Figure 3.7.16(a). The chain of the bisector may form a circuit on S. The bisector is a circuit on S if and only if the top angle is less than $\pi/3$ (Dehne and Klein, 1987, Lemma 2). Since a circuit bisector as well as a non-circuit bisector is well-behaving, we can define a Voronoi diagram with this bisector. We call the resulting diagram the *conic Voronoi diagram* generated by P on S (Dehne and Klein, 1987; Klein, 1988; Klein and Wood, 1988). An example is depicted in Figure 3.7.16(b).

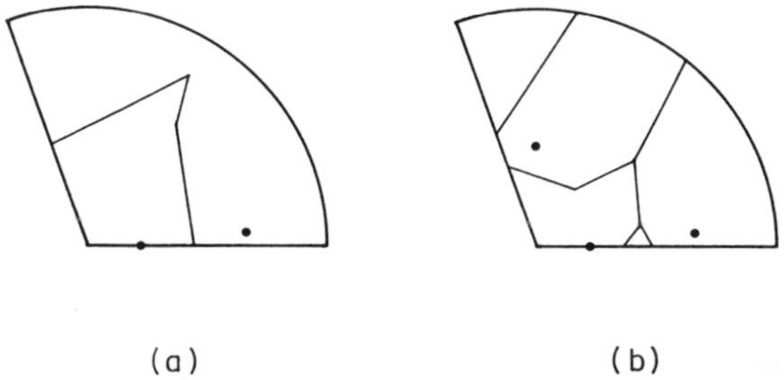

Figure 3.7.16 (a) A bisector in a cone, and (b) a conic Voronoi diagram.

3.7.7 Voronoi diagrams on a polyhedral surface

If the reader is interested in a terrain navigation or a Voronoi diagram defined on uneven ground, a Voronoi diagram on an uneven surface in $\mathbb{R}^3$ may be useful. In most practical applications a surface can be approximated by a polyhedral surface consisting of small polygonal faces. We can thus treat the above diagram as a Voronoi diagram on a polyhedral surface.

Let S be the surface of a polyhedron in $\mathbb{R}^3$, and $P = \{p_1, \ldots, p_n\}$ be a set of points on S. A simple example is shown in Figure 3.7.17(a), where S is a tetrahedron. We measure the distance $d(p, p_i)$ from a point p on S to the point p_i by the length of the shortest path on S from p to p_i. A

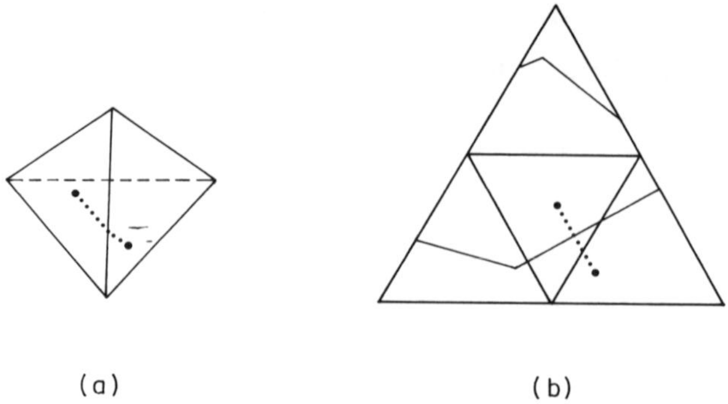

Figure 3.7.17 (a) A tetrahedron in $\mathbb{R}^3$, and (b) its developed triangle in $\mathbb{R}^2$ where the heavy solid line shows a bisector.

polyhedron can be developed into a region on a plane and we can measure this distance on the developed region. For instance, if we cut the tetrahedron in Figure 3.7.17(a) along three edges from the top, we obtain the triangle in Figure 3.7.17(b). The shortest path on the triangle is indicated by the dashed line in Figure 3.7.17. In general, a polyhedron developed on a plane is fairly complex. In particular, if a polyhedron is non-convex, facets developed on a plane may overlap, and it is not straightforward to find the shortest path. In computational geometry, Mount (1985), Sharir and Schorr (1984) and Sharir and Baltsan (1986) proposed a few methods for finding the shortest path. Using these methods, we can define the bisector between p_i and p_j (the heavy solid line in the example of Figure 3.7.17(b)), and define a Voronoi diagram on the surface of a polyhedron. We call this diagram the *polyhedral Voronoi diagram* (Figure 3.7.18).

3.7.8 Miscellany

In addition to the Voronoi diagrams with the above V-distances, we can find in the literature a few more Voronoi diagrams with other V-distances. We comment briefly on these diagrams.

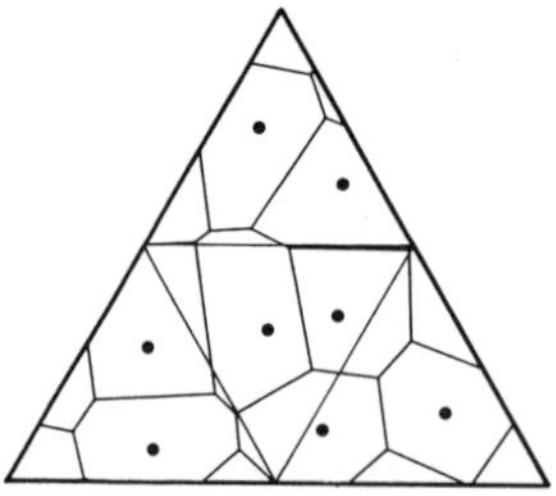

Figure 3.7.18 A polyhedral Voronoi diagram developed in $\mathbb{R}^2$.

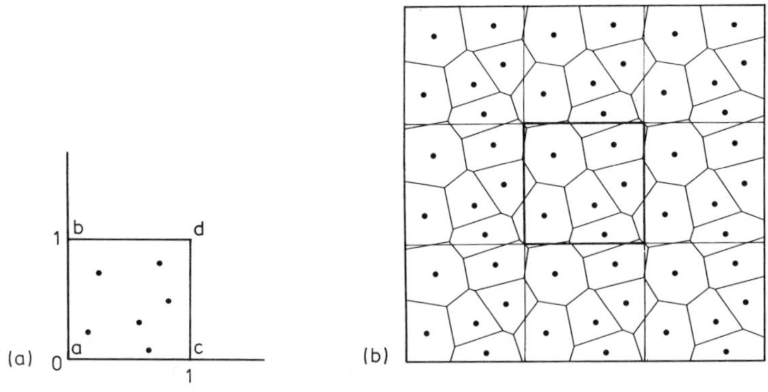

Figure 3.7.19 A Voronoi diagram on a quasi-torus.

A Voronoi diagram defined with the time–space distance is formulated by Seoung (1990), which will be shown in Chapter 8. A Voronoi diagram defined with the quadratic form distance is developed by Imai *et al.* (1985). A Voronoi diagram on a Riemannian manifold is investigated by Ehrlich and Im Hof (1979). A Voronoi diagram on a torus is referred to in conjunction with adjusting the boundary effect, which will be discussed in Chapter 5 (or Upton and Fingleton, 1985, p. 72). It should be noted that this torus is a 'quasi-torus' in the sense that a distance between two points on the torus is not measured by the shortest path distance on the torus but by the Euclidean distance on the 'developed' rectangle from the torus. An example is shown in Figure 3.7.19(a), where generators are distributed over a unit square whose vertices are given by (0,0), (1,0), (0,1) and (1,1). Conceptually, we connect the edge $\overline{ab}$ with the edge $\overline{cd}$ and the edge $\overline{ac}$ with the edge $\overline{bd}$ in Figure 3.7.19 by deforming the square. Although we deform the square, we measure a distance between two points (x_i, y_i) and (x_j, y_j) by

$$\begin{aligned} d_{\text{torus}}(p_i, p_j) = \min\{&\sqrt{(x_i-x_j)^2+(y_i-y_j)^2}, \sqrt{(x_i-x_j+1)^2+(y_i-y_j)^2}, \\ &\sqrt{(x_i-x_j)^2+(y_i-y_j+1)^2}, \sqrt{(x_i-x_j-1)^2+(y_i-y_j)^2}, \\ &\sqrt{(x_i-x_j)^2+(y_i-y_j-1)^2}, \sqrt{(x_i-x_j+1)^2+(y_i-y_j+1)^2}, \\ &\sqrt{(x_i-x_j+1)^2+(y_i-y_j-1)^2}, \sqrt{(x_i-x_j-1)^2+(y_i-y_j+1)^2}, \\ &\sqrt{(x_i-x_j-1)^2 + (y_i-y_j-1)^2}\}. \end{aligned} \quad (3.7.15)$$

This distance is readily obtained from the configuration which consists of nine squares with the same generators placed around the square, as shown in Figure 3.7.19(b). Then a Voronoi diagram on a quasi-torus is obtained as the solid lines in the central square in Figure 3.7.19(b).

3.8 NETWORK VORONOI DIAGRAMS

In Section 3.7 we showed that the Manhattan Voronoi diagram or the Karlsruhe Voronoi diagram were useful to investigate dominance regions in a grid street system or a radial–circular street system. Many actual streets are, however, not so regular. Even in Manhattan, the streets are not completely a grid. To deal with a Voronoi diagram in a more actual street system, it may be worth formulating a Voronoi diagram on a general network.

We consider a planar geometric graph $G(N,L)$ consisting of a set of nodes $N = \{p_1, \ldots, p_n, p_{n+1}, \ldots, p_l\}$ and a set of links $L = \{l_1, \ldots, l_k\}$ which form a connected component. For simplicity, we assume that $G(N, L)$ is a non-directed graph (the extension to a directed graph is not difficult). On $G(N, L)$ we define the distance from a point p on a link in L to a node p_i in N by the length of the shortest path from p to p_i. We call this distance the *network distance*, and denote it by $d_{\text{net}}(p, p_i)$. We call a geometric graph $G(N, L)$ with the network distance a *network*, and denote it by $\mathcal{N}(N, L)$. We assume without loss of generality that the generators are the first n elements of N, i.e. $P = \{p_1, \ldots, p_n\}$.

3.8.1 Network Voronoi node-diagram

We first consider that a space S is given by the node set N, called a node space (the unfilled circles, triangles and squares in Figure 3.8.1) of a network $\mathcal{N}(N, L)$ (the dashed lines with the nodes in Figure 3.8.1). Let

$$\text{Dom}(p_i, p_j) = \{p_k | p_k \in N; d_{\text{net}}(p_k, p_i) \leq d_{\text{net}}(p_k, p_j)\}, \quad i \neq j, \tag{3.8.1}$$

$$b(p_i, p_j) = \{p_k | p_k \in N; d_{\text{net}}(p_k, p_i) = d_{\text{net}}(p_k, p_j)\}, \quad i \neq j. \tag{3.8.2}$$

We call $\text{Dom}(p_i, p_j)$ the *dominance node set* of p_i over p_j and $b(p_i, p_j)$ the *bisector node set* between p_i and p_j. The bisector node set may be empty.

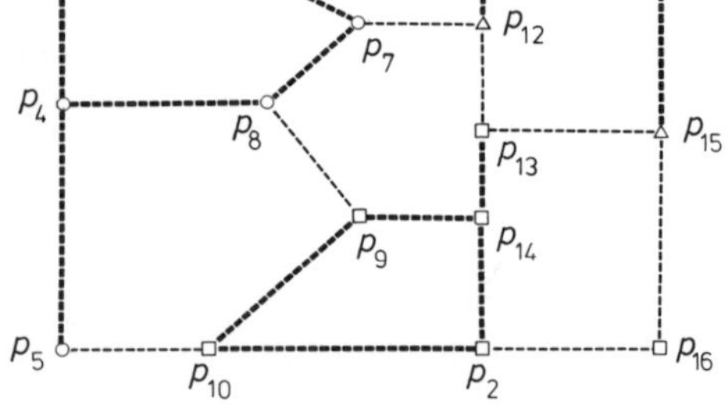

Figure 3.8.1 A network-node Voronoi diagram.

If fact, if no nodes are equally distant from any two nodes in N, the bisector node set is empty. Note that the bisector node set has quite different features from the bisector defined in the preceding sections, because the bisector node set is a set of discrete points, and so the concept of a boundary is meaningless. We can, however, develop a paralleling concept for the bisector node set. When the bisector node set is empty or the shortest path from $p_k \in b(p_i, p_j)$ to p_i and that from $p_l \in b(p_i, p_j)$ to p_j do not have common links for $p_k \neq p_l$, we say that the bisector node set is well-behaving, and $\mathrm{Dom}(p_i, p_j)$ is a well-behaving dominance node set.

For well-behaving dominance node sets, we define

$$V_{\mathrm{node}}(p_i) = \bigcap_{j=1}^{n} \mathrm{Dom}(p_i, p_j), \tag{3.8.3}$$

and let $\mathscr{V}(P, d_{\mathrm{net}}, N) = \mathscr{V}_{\mathrm{nn}} = \{V_{\mathrm{node}}(\mathrm{p}_1), \ldots, V_{\mathrm{node}}(\mathrm{p}_n)\}$. We call the set $\mathscr{V}_{\mathrm{nn}}$ the *network Voronoi node-diagram* generated by the node set P in the node space N, and call the set $V_{\mathrm{node}}(p_i)$ the *Voronoi node set* associated with the node p_i (Iri and Koshizuka, 1986, Chapter 4). In Figure 3.8.1, three Voronoi node sets generated by p_1, p_2 and p_3 are indicated by the sets of the unfilled circles, triangles and squares. It should be noted that to avoid lengthy treatments, we assume in the following analysis that the bisector node is empty.

The geometric properties of $\mathscr{V}_{\mathrm{nn}}$ appear quite different from those of the generalized Voronoi diagrams defined on a continuous space, because the node space N is discrete. We find, however, a few common properties. To discuss the 'shape' of the Voronoi node set, we consider the subgraph of the graph $G(N, L)$ in which the node set is given by $V_{\mathrm{node}}(p_i)$ and the link set $L(p_i)$ is given by the links in L both end points (nodes) of which are in $V_{\mathrm{node}}(p_i)$. We denote this subgraph by $G(V_{\mathrm{node}}(p_i), L(p_i))$. In Figure 3.8.1 the subgraphs $G(V_{\mathrm{node}}(p_i), L(p_i))$, $i \in I_n$, are indicated by the heavy dashed lines. If the shortest path between any two nodes in $V_{\mathrm{node}}(p_i)$ is embedded in $L(p_i)$ of $G(V_{\mathrm{node}}(p_i), L(p_i))$, we say that the Voronoi node set $V_{\mathrm{node}}(p_i)$ is *convex* (as an extension of convexity defined in Section 1.3). If the shortest path from any node in $V_{\mathrm{node}}(p_i)$ to the generator node p_i is embedded in $L(p_i)$ of $G(V_{\mathrm{node}}(p_i), L(p_i))$, we say that the Voronoi node set $V_{\mathrm{node}}(p_i)$ is *star-shaped* (as an extension of star-shape defined in Section 1.3). In observing the shortest path between p_{12} and p_{15} in Figure 3.8.1, we notice that the shortest path is not embedded in $L(p_3)$ of $G(V(p_3), L(p_3))$. Thus, in general a Voronoi node set is not necessarily convex. We notice, however, that every Voronoi node set in $\mathscr{V}_{\mathrm{nn}}$ is star-shaped. To sum up, we obtain the following property.

Property NN1 No Voronoi node set $V_{\mathrm{node}}(p_i)$ is empty. A Voronoi node set $V_{\mathrm{node}}(p_i)$ is not necessarily convex, but it is star-shaped with respect to the generator node p_i, $i \in I_n$.

3.8.2 Network Voronoi link-diagram

We next consider the space S given by points on links in L, i.e. $S = \cup_{i=1}^{k} l_i$, called a link space, and let

$$\mathrm{Dom}(p_i, p_j) = \left\{ p | p \in \bigcup_{i=1}^{k} l_i,\ d_{\mathrm{net}}(p, p_i) \le d_{\mathrm{net}}(p, p_j) \text{ for } p_j \in P \backslash \{p_i\} \right\}, \quad (3.8.4)$$

$$b(p_i, p_j) = \left\{ p | p \in \bigcup_{i=1}^{k} l_i,\ d_{\mathrm{net}}(p, p_i) = d_{\mathrm{net}}(p, p_j) \right\}, \quad i \neq j. \quad (3.8.5)$$

We call the set $\mathrm{Dom}(p_i, p_j)$ the dominance region of p_i over p_j on links in L, and the set $b(p_i, p_j)$ the bisector between p_i and p_j on links in L. If the bisector consists of finite points, we say that the bisector $b(p_i, p_j)$ is a well-behaving bisector and the dominance region $\mathrm{Dom}(p_i, p_j)$ is a well-behaving dominance region. A well-behaving bisector implies that the shortest path from $p \in b(p_i, p_j)$ to p_i and that from $p \in b(p_i, p_j)$ to p_j does not share a positive length line segment. For well-behaving dominance regions, we define

$$V_{\mathrm{link}}(p_i) = \bigcap_{j \in I_n \backslash \{i\}} \mathrm{Dom}(p_i, p_j), \quad (3.8.6)$$

and let $\mathcal{V}(P, d_{\mathrm{net}}, \cup_{i=1}^{k} l_i) = \mathcal{V}_{\mathrm{nl}} = \{V_{\mathrm{link}}(p_1), \ldots, V_{\mathrm{link}}(p_n)\}$. Since we define $\mathcal{V}_{\mathrm{nl}}$ for well-behaving dominance regions, elements in V_{nl} are collectively exhaustive and mutually exclusive, except the finite number of points. We call this tessellation the *network Voronoi link-diagram* generated by P on $\mathcal{N}(N, L)$ and the set $V_{\mathrm{link}}(p_i)$ the *Voronoi link set* associated with p_i.

Alternatively, we can define the set $V_{\mathrm{link}}(p_i)$ with the subgraph $G(V_{\mathrm{node}}(p_i), \mathrm{L}(p_i))$ defined in Section 3.8.1. From the definition of the Voronoi node set, we note that if $p \in \cup_{l_j \in L(p_i)} l_j$, then $p \in V_{\mathrm{link}}(p_i)$, that is, a point on the heavy lines in Figure 3.8.2 is assigned to its generator. What is left to assign is a point on the links indicated by the light lines in Figure 3.8.2. We denote the kth link connecting a node in $V_{\mathrm{node}}(p_i)$ and a node $V_{\mathrm{node}}(p_j)$ by l_{ijk}, $k \in I_{n_{ij}}$ (note that l_{ijk} may be empty for some pairs i and j). We divide the link l_{ijk} at the equally distant point from p_i and p_j and denote the resulting link connected to the node in $V_{\mathrm{node}}(p_i)$ by $l_{ijk/}$, i.e.

$$l_{ijk/} = \{ p \in l_{ijk} | d_{\mathrm{net}}(p, p_i) \le d_{\mathrm{net}}(p, p_j) \} \quad (3.8.7)$$

(Figure 3.8.2). Then the set $V_{\mathrm{link}}(p_i)$ given by equation (3.8.6) is written as

$$V_{\mathrm{link}}(p_i) = \left[\bigcup_{l_i \in L(p_i)} l_i \right] \cup \left[\bigcup_{j \in I_n \backslash \{i\}} \bigcup_{k \in I_{n_{ij}}} l_{ijk/} \right]. \quad (3.8.8)$$

An actual network Voronoi link-diagram of roads is shown in Figure 3.8.3.

For the network Voronoi link-diagram, almost the same property as in Property NN1 holds although a slight modification is necessary for the

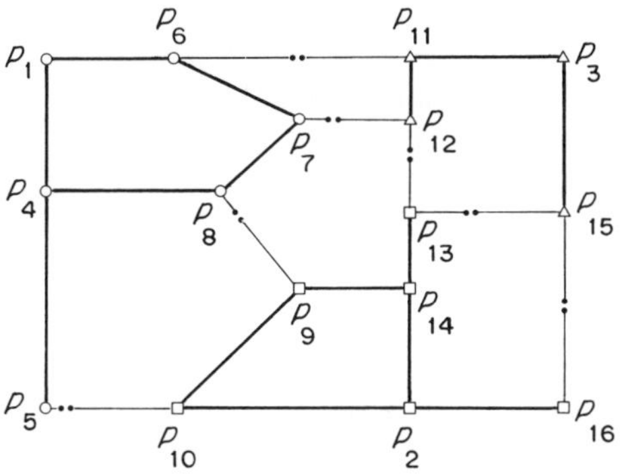

Figure 3.8.2 A network Voronoi link-diagram.

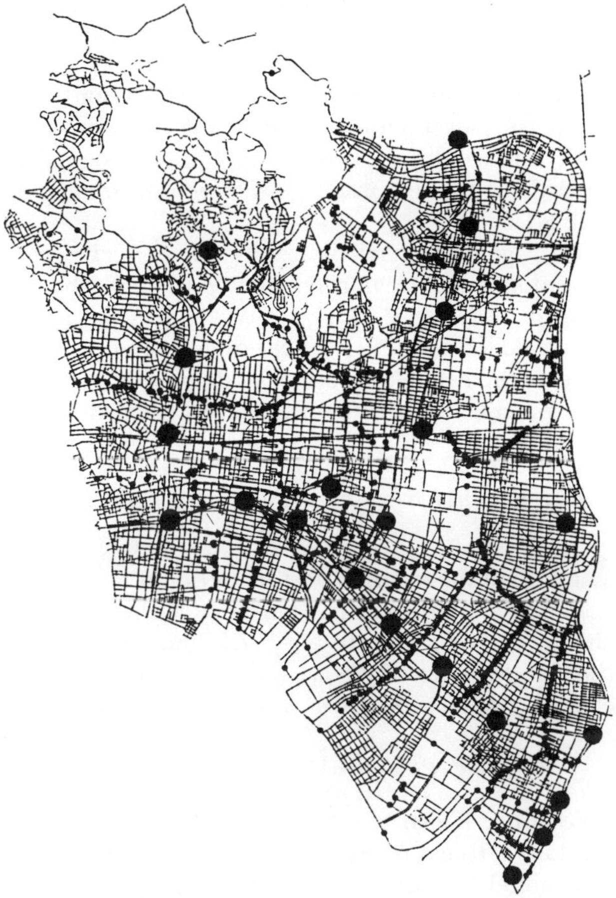

Figure 3.8.3 The network Voronoi link-diagram of roads in the Nishinomiya district of Osaka (the large filled circles indicate generators and the small filled circles indicate the boundary points). (*Source*: Yomono (1991), Figure 1.)

definition of convexity and star-shape. If the shortest path between any two points in $V_{\text{link}}(p_i)$ is embedded in $V_{\text{link}}(p_i)$, we say that the Voronoi link set is *convex*. If the shortest path from a point in $V_{\text{link}}(p_i)$ to the generator node p_i is embedded in $V_{\text{link}}(p_i)$, we say that the Voronoi link set $V_{\text{link}}(p_i)$ is *star-shaped*. Using these terms, we can state the following property.

Property NL1 No Voronoi node set $V_{\text{link}}(p_i)$ is empty. A Voronoi node set $V_{\text{link}}(p_i)$ is not necessarily convex, but it is star-shaped with respect to the generator node p_i, $i \in I_n$.

3.8.3 Network Voronoi area-diagram

In the network Voronoi node-diagram and link-diagram, a point to be assigned to a generator is restricted to a point on a node or a link. In some contexts, however, we find this restriction too strong. For example, if we can access from any point in a region to a network and take a path from the access point to a destination through links, every point (including a point on a node and a link) in a region should be assigned to a generator. To treat this assignment, we consider a Voronoi diagram formulated on $\mathbb{R}^2$ with a network $\mathcal{N}(N, L)$.

Let $P \in N$ be a generator set on a network $\mathcal{N}(N, L)$ placed in $\mathbb{R}^2$, and assume that the dominance region $\text{Dom}(p_i, p_j)$ defined by equation (3.8.4) is well-behaving for any $i, j \in I_l$, $i \neq j$. We first assign a point p in $\mathbb{R}^2$ to the nearest point, called the *access point*, on a link of L; second, the access point is assigned to the nearest generator node p_i with respect to the network distance. To be precise, let $a(p)$ be the nearest point on $\cup_{i=1}^{k} l_i$ from p, i.e.

$$d_{\text{acc}}(p, a(p)) = \|\boldsymbol{x} - \boldsymbol{x}_a\| = \min_{\boldsymbol{u}} \left\{ \|\boldsymbol{x} - \boldsymbol{u}\| \,|\, \boldsymbol{u} \in \bigcup_{i=1}^{k} l_i \right\}, \tag{3.8.9}$$

where $\boldsymbol{x}$, $\boldsymbol{x}_a$, and $\boldsymbol{u}$ are the location vectors of points p, $a(p)$ and an arbitrary point on $\cup_{i=1}^{k} l_i$ (the dotted lines in Figure 3.8.4), respectively. We call the point $a(p)$ the *access point* of p to $\mathcal{N}(N, L)$, and the distance $d_{\text{acc}}(p, a(p))$ the *access distance*. In terms of the access distance and the network distance, the distance from a point p in $\mathbb{R}^2$ to a node p_i in N is written as

$$d_{\text{acc}\cdot\text{net}}(p, p_i) = d_{\text{acc}}(p, a(p)) + d_{\text{net}}(a(p), p_i) \tag{3.8.10}$$

(Okabe and Yomono, 1988). We should note that this distance is not always the shortest distance from a point p to a node p_i, because the access distance is locally determined by equation (3.8.9). With the distance of equation (3.8.10), we define

$$V_{\text{area}}(p_i) = \{p | d_{\text{acc}\cdot\text{net}}(p, p_i) \leq d_{\text{acc}\cdot\text{net}}(p, p_j),\ j \neq i,\ j \in I_n\}. \tag{3.8.11}$$

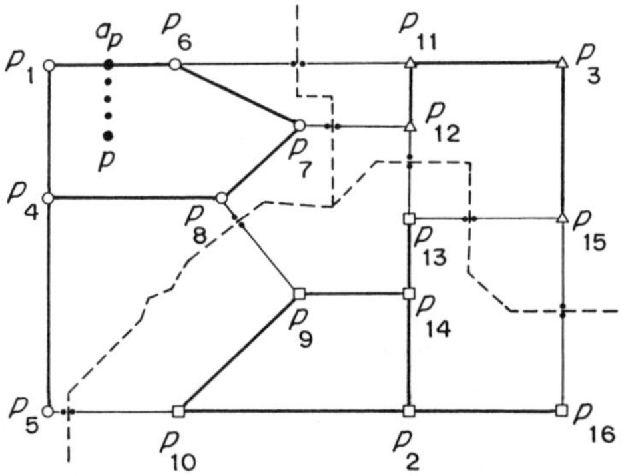

Figure 3.8.4 A network Voronoi area-diagram.

This set is alternatively defined with the line Voronoi diagram defined in Section 3.5. First, recall that we define $V_{\text{area}}(p_i)$ provided that the network Voronoi link-diagram can be defined for $\mathcal{N}(P, L)$. This means that $V_{\text{link}}(p_1)$, ..., $V_{\text{link}}(p_n)$ are mutually exclusive except at a finite number of points. If an access point $a(p)$ belongs to a link in $V_{\text{link}}(p_i)$, it follows from the definition of the network Voronoi link-diagram that the point p is assigned to the generator p_i. Thus, a point p being assigned to p_i is equivalent to a point p being assigned to the nearest generator among the generators given by $\{V_{\text{link}}(p_1), \ldots, V_{\text{link}}(p_n)\}$. This assignment is indeed the assignment of a point p in the line Voronoi diagram of $L = \{l_1, \ldots, l_n\}$, where l_i is given by $V_{\text{link}}(p_i)$. The set $V_{\text{area}}(p_i)$ is hence written as

$$V_{\text{area}}(p_i) = V_L(V_{\text{link}}(p_i)), \tag{3.8.12}$$

where $V_L(V_{\text{link}}(p_i))$ is the line Voronoi region associated with the generator consisting of lines in $V_{\text{link}}(p_i)$. When we refer to this line Voronoi diagram in a network study, we call the set $\mathcal{V}(P, d_{\text{acc-net}}, \mathbb{R}^2) = \mathcal{V}_{\text{na}} = \{V_{\text{area}}(p_i), \ldots, V_{\text{area}}(p_n)\}$ the *network Voronoi area-diagram*.

Since the network Voronoi area-diagram is exactly the same as the Voronoi line-diagram, the properties of the network Voronoi area-diagram are shared with those of the Voronoi line-diagram.

3.9 NOTES ON APPLICATIONS

Generalized Voronoi diagrams potentially provide many practical applications in various fields. Some of the applications are briefly reviewed here. Additional applications are discussed in Chapters 5–9.

The family of weighted Voronoi diagrams is often adopted in market area analysis to determine market areas of firms or stores, where weights w_{i1} and w_{i2} (in equation (3.1.17)) in the CW-distance correspond to a transportation cost and a mill price of goods or products, respectively. According to Shieh (1985), the earliest study dates back Rau (1841). Launhardt (1882) and Fetter (1924) are also forerunners, followed by Tuominen (1949), Hyson and Hyson (1950), Gambini *et al.* (1967), Beckmann (1971), Boots (1980), Aurenhammer (1987), Von Hohenbalken and West (1984a), Hanjoul and Thisse (1987), Sakamoto and Takagi (1987), Hanjoul *et al.* (1989), O'Kelly and Miller (1989) and many others.

The Johnson–Mehl model (Johnson and Mehl, 1939), which may be considered as the additively weighted Voronoi diagram (see Section 3.1.2), provides many diverse applications, which are reviewed in Section 5.6.

Gambini *et al.* (1967), Huff and Jenks (1968) and Huff (1973) studied national urban systems with the MW-Voronoi diagram. Huff and Lutz (1979) also used the same diagram to discuss a geographical delivery system for serving public interest in the Republic of Ireland. Illeris (1967) adopted the same diagram to determine functional regions of urban centres in Denmark; Hubbard (1970) and Wood (1974) carried out similar studies in Jamaica and Kenya, respectively. Cox and Agnew (1974) used the MW-Voronoi diagram to produce a partition of all Ireland into theoretical counties, which are compared with the actual counties (note that their map contains an error pointed by Weaire and Rivier (1984)). Archaeologists have used the MW-Voronoi diagram to define territories for various sites. Their work in this regard is reviewed by Hodder and Orton (1976, pp. 187–95). The most extensive of such studies is that of Hogg (1971) who used the MW-Voronoi diagram as a surrogate for territories of Neolithic hillforts south of the River Thames in England. Lesley (1976), Vaughan and Cousins (1977) and Wirasinghe and Ghoneim (1981) represented service areas of bus stops in terms of the AW-Voronoi diagram.

The sectional Voronoi diagram, which is regarded as the power diagram, may be utilized in stereology, which attempts to recover three-dimensional information from one- or two-dimensional samples. This diagram is also used in modelling the crystal structure of organic compounds (Fischer and Koch, 1979), metallic glass (Gellatly and Finney, 1982a) and globular proteins (Gellatly and Finney, 1982b). Aurenhammer (1987) refers to the power diagram in conjunction with the problem called the illumination of balls (Linhart, 1981): the union of n balls in a space can be illuminated by the vertices of the power diagram for the corresponding spheres.

The kth nearest point from a point can be readily obtained from the order-k Voronoi diagram, or more specifically the kth nearest-point Voronoi diagram. A natural application is found in a facility location problem in which the kth nearest facility is critical, for instance in the case of a single unit emergency depots when more than one unit is required or when the first nearest unit is in use (Keeney, 1972). The farthest-point Voronoi diagram, which is a special case of the order-k Voronoi diagram or the kth nearest-point Voronoi diagram, is also used to examine a facility location problem of the min-max criteria. We can find other applications in multivariate density estimation (Loftsgaarden and Quesenberry, 1965), classification (Cover and Hart, 1967) and information retrieval systems (Fukunaga and Narendra, 1975; Chazelle, 1985). In the study of waves and crystals, the kth nearest-point Voronoi diagram gives an important concept called the Brillouin zone (Brillouin, 1953). As we shall see in Chapter 6, the ordered order-k Voronoi diagram can be used for spatial interpolation (Sibson, 1980a), which is applied in statistical estimation (Ripley, 1981, Chapter 4) as well as in cartography (isoline generation).

Seoung and Asano (1987) suggested that high school districts in Seoul can be determined using the Voronoi diagram with obstacles, where the obstacles are the Han River and mountains. In robotics, a modified Voronoi diagram with obstacles is used to examine if a robot can pass between two given points without colliding with the obstacles (Lozano-Pérez and Wesley, 1979; Ó'Dúnlaing and Yap, 1985; Schwartz and Yap, 1986). Note that since a visibility polygon is used to construct the Voronoi diagram with obstacles (Section 3.4), this diagram is closely related to the problem of asking if one can see a given polygon from a given stand point on the plane with a set of disjoint obstacle polygons (Lee, 1978; Welzl, 1985; Asano *et al.*, 1986). The dual of the Voronoi diagram with line obstacles is the constrained Delaunay triangulation, which is used as a tool for mesh generation in finite-element methods where polygonal regions are triangulated preferably by not-so-thin triangles (Baker, 1989, Chew, 1989; Dey *et al.*, 1991).

Okabe and Fujii (1984) showed a possible application of the line Voronoi diagram for testing the effect of car emissions on withered trees along expressways. The same method was used by Okabe and Yoshikawa (1989) and Yoshikawa and Okabe (1991) to test the effect of arterial streets on the distribution of apartment buildings in Setagaya and Kohto-Sumida districts in Tokyo (see Section 8.3.2). In the same fashion, Yoshikawa *et al.* (1987) and Okabe *et al.* (1988) examined the effect of large parks on the distribution of high-class apartment buildings in Tokyo using the area Voronoi diagram (see Section 8.3.3). Fujii (1983), Fujii and Oikawa (1986) and Oikawa (1986) determine the sphere of parks with the weighted area Voronoi diagram, where the weights reflect the area of a park. Boggs (1951) and Ricketts (1986) dealt with the delimitation of seaward areas under national jurisdiction which implicitly uses the area Voronoi diagram, where areas correspond to national domains. The medial axis associated with the line Voronoi diagram in a simple polygon is applied to pattern recognition or image processing (Blum, 1967, 1973; Calabi and Hartnett, 1968; Fujii, 1976; Montarani, 1968; Philbrick, 1968; Lee, 1982a; Lantuejoul and Maisonneuve, 1984). The line Voronoi diagram can be also used to optimize service lines (Takeda, 1985) (see Chapter 9). The line and area Voronoi diagram is closely related to the buffer zones of lines and areas often used in geographical information systems (in fact, as is shown by Okabe and

Yoshikawa, 1989, the buffer zone is readily obtained once the line and area Voronoi diagrams are given; see Section 8.3.2).

Anjoumani (1981) analysed market areas in a rectangular network which corresponds to the Manhattan Voronoi diagram. Eaton and Lipsey (1980) and Von Hohenbalken and West (1984b) examined market areas with the Manhattan Voronoi diagram in comparison with the ordinary Voronoi diagram. Manhattan Voronoi diagrams or supremum Voronoi diagrams are effectively used in mass storage systems, where a record stored at a grid point is accessed by an electromechanical fetching mechanism, known as a read/write head, which moves with the Manhattan distance or the supremum metric (Lee and Wong, 1980). As in Sections 2.5 and 9.3, the Euclidean Steiner tree can be obtained from the ordinary Voronoi diagram. Similarly, the Steiner tree with the Manhattan distance can be obtained from the Manhattan Voronoi diagram. This tree has many applications in wire layout for printed circuit boards (Aho *et al.*, 1976; Hwang, 1976; Garey and Johnson, 1977).

The spherical Voronoi diagram is potentially useful to consider global problems, although few applications are known except the problem of dictators on a globe (Böröczy cited in Tóth, 1972). A related study to the spherical Voronoi diagram can be found in Ohsawa and Kijima (1991) who covered a dome with Delaunay triangles. The cylindrical Voronoi diagram is potentially useful to deal with cyclic problems over time. An example is shown by Takeda (1985) who studied the locational optimization of market places held with certain time intervals. This problem is also treated by Seoung (1990) using the time–space Voronoi diagram.

The network node-Voronoi diagram is implicitly used in multi-media problem (Hakimi, 1964), which is one of the central subjects in locational optimization on a network (Handler and Mirchandani, 1979). Few explicit applications are, however, known, except Yomono (1991) who showed the network Voronoi line-diagram with respect to stations in the Nishinomiya district of Osaka.

CHAPTER 4

Algorithms for Computing Voronoi Diagrams

In this chapter we consider practical methods for constructing Voronoi diagrams. Here we mainly concentrate our attention on the most basic case, namely the Voronoi diagram for points in a plane. First, in Section 4.1 we see a naive method which follows directly from the definition of the Voronoi diagram. The naive method is however inefficient, and many efforts have been made in the field of computational geometry to make efficient algorithms, which are also surveyed in this section. In Section 4.2 we present a typical data structure for representing the Voronoi diagram and thus make clear what is meant by the 'construction' of the Voronoi diagram. We also derive a lower bound of the time complexity for the construction of the Voronoi diagram. Then we write three typical algorithms in Sections 4.3, 4.4 and 4.5, respectively. These algorithms are designed on the assumption that no numerical error takes place in the course of computation, so that a direct translation of such an algorithm to a computer program does not necessarily give a numerically valid program; such a program may fail because of geometric inconsistency caused by numerical errors. To make a practically valid computer program, we have to consider numerical errors. Hence, in Section 4.6 we show two typical techniques for avoiding inconsistency due to numerical errors. The last three sections, Sections 4.7, 4.8 and 4.9, are for generalized Voronoi diagrams. In Section 4.7 we present a typical algorithm for constructing higher dimensional Voronoi diagrams, while in Section 4.8 we briefly touch upon algorithms for various kinds of generalized Voronoi diagrams in the plane. Finally, in Section 4.9 we see approximation algorithms for generalized Voronoi diagrams.

4.1 COMPUTATIONAL PRELIMINARIES

As in the previous chapter, let $P = \{p_1, \ldots, p_n\}$ be the set of generators, and $\mathcal{V} = \{V(p_1), \ldots, V(p_n)\}$ denote the Voronoi diagram for P. The construction of the Voronoi diagram is a procedure for generating $\mathcal{V}$ from P. In the previous chapter we saw several equivalent conditions for characterizing the Voronoi diagram. Among them the characterization based on half planes (Definition V3) directly gives us a naive method for constructing the diagram. The Voronoi polygon for the generator p_i is the intersection of all the half planes defined by the perpendicular bisectors of p_i and the other generators. According to this definition, we can construct the Voronoi polygons one by one. Thus we get the following method.

Naive method Input: n generators $p_1, p_2, \ldots, p_n$.
Output: Voronoi diagram $\mathcal{V} = \{V(p_1), V(p_2), \ldots, V(p_n)\}$.
Procedure:
Step 1. For each i such that $i = 1, 2, \ldots, n$, generate $n-1$ half planes $H(p_i, p_j)$, $1 \leq j \leq n$, $j \neq i$, and construct their common intersection $V(p_i)$.
Step 2. Report $\{V(p_1), V(p_2), \ldots, V(p_n)\}$ as the output and stop.

In this chapter we present a number of geometric algorithms and name them with sequential numbers. However, we do not want to call the above method a practical 'algorithm', because it is 'insufficient'. Let us see how 'insufficient' the above method is. To evaluate an algorithm, we have to consider at least correctness and efficiency, and if it involves numerical computation, we also have to consider robustness against numerical errors. The above method is obviously correct, because it is a restatement of the definition of the Voronoi diagram. So we consider the other two aspects.

A standard method for evaluating efficiency is to observe the asymptotic behaviour of the time required by an algorithm with respect to the size of input data (Knuth, 1968; Aho *et al.*, 1974). Let $T(n)$ denote the time required by an algorithm to process input data of size n. In general there are many different input data of the same size and they require different processing times. Hence we usually consider the worst case or the average case. In the worst case, $T(n)$ is the maximum time over all possible input of size n, whereas in the average case $T(n)$ is the average time over them. For some positive function $f(n)$ if there exists a constant C such that

$$\frac{T(n)}{f(n)} < C \quad \text{for all } n, \tag{4.1.1}$$

we write $T(n) = O(f(n))$ and say that $T(n)$ is of *order* $f(n)$. If $T(n) = O(f(n))$, we also say that the *time complexity* of the algorithm is of $O(f(n))$. Intuitively, $T(n)$ being of $O(f(n))$ implies that $T(n)$ does not increase more rapidly than $f(n)$ does as n grows. Hence the algorithm can be considered as being efficient if it admits a slowly increasing function $f(n)$ satisfying equation (4.1.1).

We may say, for example, an algorithm with $T(n) = O(1)$ is 'ideally fast', an algorithm with $T(n) = O(\log n)$ is 'very fast', and an algorithm with $T(n) = O(n)$ or with $T(n) = O(n \log n)$ is 'fast'. An algorithm with $T(n) = O(n)$ is sometimes called a *linear time* algorithm.

Let us return to the naive method, and consider its time complexity. To construct the half plane $H(p_i,p_j)$ for given two points p_i and p_j requires only constant time. Hence for each p_i the time required for constructing $n-1$ half planes is proportional to $n-1$, say $a(n-1)$, where a is a positive constant. To construct the intersection of the half planes, let us consider the following simple procedure. First we construct the intersection of two half planes, obtaining a polygon with two sides; next construct the intersection of this polygon with the third half plane, and so on. In the kth step of this procedure we have to find the intersection of a k-sided polygon (in the worst case) with another half plane, and this step requires time at least proportional to k if we check whether each of the k sides crosses the boundary line of the half plane. Thus, constructing the intersection of $n-1$ half planes requires time proportional to $1+2+\ldots+(n-2) = (n-2)(n-1)/2$, say $b(n-2)(n-1)$, where b is another positive constant. We have to repeat this process for all generators, and consequently the total time required by the naive method is

$$T(n) = n\{a(n-1)+b(n-2)(n-1)\} = \mathrm{O}(n^3).$$

This means that for a computer program based on the above procedure, the time for processing becomes eight ($=2^3$) times larger as the size of the input data becomes twice as large.

Employing a little more sophisticated technique, we can construct the intersection of $n-1$ half planes in $\mathrm{O}(n \log n)$ time (Preparata and Shamos, 1985). Hence the time complexity of the naive method can be decreased to $\mathrm{O}(n^2 \log n)$.

As we will see in the next section, any algorithm for constructing the Voronoi diagram requires at least $\mathrm{O}(n \log n)$ time in the worst case, and at least $\mathrm{O}(n)$ time on average. Moreover, there exist algorithms whose time complexities attain these lower bounds, and some of them actually run very fast even for small values of n (although the order itself represents the behaviour of the algorithm as n approaches infinity). The algorithm presented in Section 4.3 requires $\mathrm{O}(n)$ time on average and $\mathrm{O}(n^2)$ time in the worst case, while the algorithms presented in Sections 4.4 and 4.5 require $\mathrm{O}(n \log n)$ time in the worst case. Compared with these algorithms, the naive algorithm is not very satisfactory from the time-complexity point of view.

Another important point of view for evaluating the algorithm is robustness against numerical errors. From this point of view too the naive method is not satisfactory. This can be understood by the following example. Suppose that there are only four generators p_1, p_2, p_3 and p_4 and they come very near to a common circle. Then, as shown in Figure 4.1.1(a), the three perpendicular bisectors between p_1 and the other generators pass near to the centre of the circle; if numerical error takes place, it is difficult to decide the relative locations of the points of intersections of these bisectors. Hence

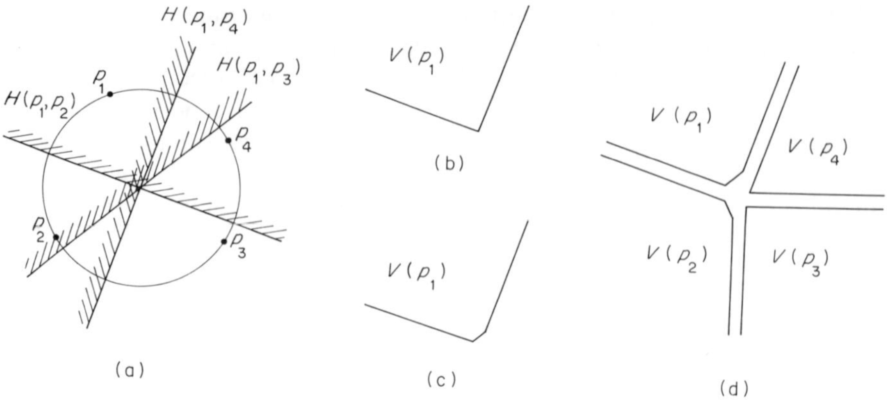

Figure 4.1.1 Inconsistency caused by numerical errors: (a) four generators on a common circle; (b) possible Voronoi polygon for p_1; (c) another possible Voronoi polygon for p_1; (d) inconsistent set of Voronoi polygons.

the result of intersecting the three half planes associated with p_1 may be a two-sided polygon as in (b) or a three-sided polygon as in (c); which result comes out depends on numerical errors. The other three Voronoi polygons have similar instability. In the naive method, each Voronoi polygon is computed independently, so that the result may become topologically inconsistent, as shown in Figure 4.1.1(d). This figure is contradictory because the Voronoi polygon $V(p_1)$ shows that there is a Voronoi edge between p_1 and p_3, but the Voronoi polygon $V(p_3)$ shows that such a Voronoi edge does not exist; we cannot glue the boundaries of the Voronoi polygons together.

From a theoretical point of view all the Voronoi polygons give a tessellation of the plane, but in actual computation the output of the naive method does not necessarily give a tessellation of the plane. Thus the naive method, if translated to a computer program in a straightforward manner, is not robust against numerical errors. Its output may be topologically inconsistent. We need careful consideration in order to make a numerically robust algorithm. Basic techniques for this purpose will be presented in Section 4.6.

Here we will briefly survey typical algorithms for constructing the ordinary Voronoi diagram in the plane. First of all, we should note that some algorithms are described in terms of the Voronoi diagram itself (e.g. Shamos and Hoey, 1975; Sibson, 1980a; and Ohya *et al.*, 1984a), while others are described in terms of the Delaunay diagram (e.g. Lee and Schachter, 1980; Guibas and Stolfi, 1985). As we saw in Section 2.2, however, these two diagrams are duals of each other, and we can transform one to the other and vice versa. Moreover, the description of the algorithm itself can be transformed from Voronoi diagram terminology to Delaunay diagram terminology, and vice versa. Hence the choice of terminology does not make any essential difference.

The naive method presented above is one of the easiest methods to understand (Rhynsburger, 1973) because it is a direct translation of the definition of the Voronoi diagram. Although this method is not recommended for general purposes, it may be used for some special types of simulation where the goal is not to construct the Voronoi diagram itself, but to generate many samples of Voronoi polygons to obtain statistical data such as the distribution of the number of edges per polygon (Crain, 1978; Boots and Murdoch, 1983; Quine and Watson, 1984). Bentley *et al.* (1980) combined this method with a bucketing technique and presented an algorithm that runs in $O(n)$ time on average, in which generators relevant to each Voronoi polygon are searched for from bucket to bucket in a spiral order (the bucketing technique will be described in Section 4.3). However, the output of the naive method is a simple collection of Voronoi polygons, and it does not include explicit information about the topological structure of the diagram. As we will see in more detail in the next section (Section 4.2), if we want to extract various kinds of information from the Voronoi diagram, the naive method is not suitable.

The second method, which is another naive method, can be categorized as a 'walking method', in which Voronoi vertices and Voronoi edges (or, equivalently, Delaunay polygons and Delaunay edges) are constructed one by one just in the order in which a traveller walks along the edges of the diagram. This method was described in Delaunay diagram terminology by Lawson (1977), and in Voronoi diagram terminology by Brassel and Reif (1979) and Cromley and Grogan (1985). While they did not consider the time complexity point of view, Maus (1984) utilized the bucketing technique to construct an $O(n)$ algorithm in the average sense, in which the algorithm is described in terms of the Delaunay triangulation.

The third method may be called a 'modification method', in which an initial diagram is construct first, and then it is modified step by step until it converges to the Voronoi diagram. Sibson (1978) showed that, starting with any triangulation of $CH(P)$, we can modify it to the Delaunay triangulation by flipping the diagonals of convex quadrilaterals according to the local max-min angle criterion, i.e. the diagonal of a convex quadrilateral is replaced by the other diagonal if the resulting pair of triangles gives a larger value for the minimum angle. As we saw in Section 2.4, Edelsbrunner (1988) gave an intuitive and simple interpretation of the procedure in the context of the lift-up transformation, and showed that $O(n^2)$ flipping of the diagonals is enough to obtain the Delaunay triangulation from any initial triangulation.

The fourth method, a simple but still very powerful method, is an incremental method, in which we start with a simple Voronoi diagram for two or three generators and modify it by adding generators one by one; the method was presented in Voronoi terminology by Green and Sibson (1978), Sibson (1980a, 1980b), Lee and Schachter (1980) and Shapiro (1981), and in Delaunay terminology by Watson (1981), Devijver and Dekesel (1982), Gowda *et al.* (1983) and Correc and Chapuis (1987). In the worst case, each addition of a generator requires time proportional to the number of generators added so far, and consequently the total time complexity is of $O(n^2)$. However, the average time complexity can be decreased to $O(n)$ by the use of special data structures such as buckets and a quaternary tree (Ohya, 1983; Iri *et al.*, 1984; Ohya *et al.*, 1984a). Also a numerically robust version of this method was proposed by Sugihara and Iri (1988, 1989b, 1989d). Thus, this method is one of most practical ones from both the time-complexity and the robustness points of view. This method will be presented in detail in Section 4.3 and the numerically robust version will be shown in Section 4.6.

The fifth method, another typical method for constructing the Voronoi diagram, is the use of the divide-and-conquer paradigm, in which the set of generators is recursively divided into smaller subsets and the Voronoi diagrams for those subsets of generators are merged into the final diagram. This method was first proposed in terms of the Voronoi diagram by Shamos and Hoey (1975); an early attempt to implement this method was made, for example, by Horspool (1979). Later this method was written in a simpler form using Delaunay triangulation terminology by Drysdale and Lee (1978), Lee and Schachter (1980) and Guibas and Stolfi (1985). The worst-case time complexity of this method is $O(n \log n)$, which is the best possible, as we will see in the next section. However, as was shown by Ohya *et al.* (1984a), the original divide-and-conquer method requires $O(n \log n)$ time also in the average sense. The average-case time complexity was lessened to $O(n \log \log n)$ by Dwyer (1987), and further to $O(n)$ by Katajainen and Koppinen (1988). Quite recently, a robust version of the method was also constructed (Ooishi, 1990; Sugihara *et al.*, 1990; Ooishi and Sugihara, 1991). The divide-and-conquer method will be shown in Section 4.4.

The sixth method is the plane sweep method (Fortune, 1986, 1987). A plane sweep is one of fundamental techniques in computational geometry, by which a two-dimensional problem can be reduced to an almost one-dimensional problem. A vertical line, called a sweep line, is moved over the plane from left to right, and the Voronoi diagram is constructed along this line. This method also attains the worst-case optimal time complexity, $O(n \log n)$. This method will be presented in Section 4.5.

The seventh method may be called a 'lift-up method', which utilizes the relationship between a two-dimensional Voronoi diagram and a three-dimensional convex hull (Properties D7, D8). In this method, first the generators in the plane are transformed to certain points in three-dimensional space, then their convex hull is generated, and finally the convex hull is inversely transformed to the original plane to obtain the Delaunay diagram.

Brown (1979, 1980), Aurenhammer and Edelsbrunner (1984) and Buckley (1988) used a transformation from the plane to a sphere (Property D7), and Edelsbrunner and Seidel (1986) and O'Rourke *et al.* (1986) used a transformation from the plane to a paraboloid of revolution (Property D8). This method can be extended to any dimension, and hence we will see this method in the context of the construction of higher-dimensional Voronoi diagrams (Section 4.7).

Other directions of algorithmic research include construction of the Voronoi diagram for convexly located generators (Aggarwal *et al.*, 1989), digital-image approximation of the Voronoi diagram (Toriwaki *et al.*, 1982; Mark, 1987; Yoshitake *et al.*, 1987), parallel computation of the Voronoi diagram (Saxena *et al.*, 1990) and constructing the time-varying Voronoi diagram (Tokuyama, 1988; Bajaj and Bouma, 1990).

For a while (i.e. from Section 4.2 to Section 4.5) we make the following assumptions.

Assumption A4.1 Numerical computation is carried out in precise arithmetic.
Assumption A4.2 No four generators align on a common circle.

The first assumption is necessary to consider algorithms in a 'theoretically closed world', and has been adopted, implicitly or explicitly, in almost all discussions for designing geometric algorithms. The second assumption is for avoiding degeneracy. Note that this assumption is not equivalent to the non-cocircularity assumption; the non-cocircularity assumption avoids an empty circle passing through four points, whereas Assumption A4.2 avoids any circle passing through four points; Assumption A4.2 is stronger than the non-cocircularity assumption. This assumption is necessary to avoid degeneracy in the course of constructing the Voronoi diagram, because we treat many Voronoi diagrams generated by subsets of the generator set. Whether a case is special or not sometimes depends on which algorithm we consider. Indeed, we will make other assumptions when we discuss other algorithms.

In Section 4.6, on the other hand, we remove all such assumptions and construct algorithms that are valid in the real world.

4.2 DATA STRUCTURE FOR REPRESENTING A VORONOI DIAGRAM

Before starting our consideration of algorithms, we have to make clear what is meant by 'construction of a Voronoi diagram'. Naively it might be thought that constructing a Voronoi diagram is equivalent to drawing it on a sheet of paper. Indeed, to draw the Voronoi diagram is a typical way to present it to the human eye. However, the drawn diagram itself, or a set of numerical data equivalent to the drawn diagram, does not convey enough information for many applications. Suppose that a computer program gives as output a list of line segments corresponding to the Voronoi edges. It is easy to draw the diagram from this list, but it is not easy to retrieve, for example, a sequence of Voronoi vertices on the boundary of one Voronoi polygon, or a set of Voronoi polygons contiguous to one Voronoi polygon. Thus, to 'draw' the Voronoi diagram is not enough in many applications. What we need to construct is a 'richer structure' from which we can extract a variety of information necessary for applications.

Another extremal way of representing the Voronoi diagram is to give all necessary data explicitly. For example, for each Voronoi polygon we may have the list of all Voronoi vertices on its boundary, the list of all Voronoi edges on its boundary, the list of all contiguous Voronoi polygons, and so on. However, this way of representation requires much space. Hence we have to choose some data structure in between which is concise but is still rich enough for us to retrieve various aspects of the diagram.

Here we choose one standard data structure, called a winged-edge data structure or a polygon data structure, which is common in geometric modelling and computer graphics (Baumgart, 1975; Ballard and Brown, 1982; Hoffmann, 1989). As will be shown in what follows, this data structure conveys explicit information about local incidence relations among Voronoi vertices, Voronoi edges and Voronoi polygons, so that a variety of information can be retrieved relatively easily.

The winged-edge data structure is a way of representing a geometric graph in the plane. However, the Voronoi diagram is slightly different from a geometric graph in that some of Voronoi edges extend infinitely. To consider the Voronoi diagram as a geometric graph, we introduce a closed curve which is large enough to surround all the Voronoi vertices and consider that the infinite Voronoi edges have their terminal points on this closed curve. These terminal points are intuitively considered as points at infinity. By this convention, the Voronoi diagram shown in Figure 4.2.1(a) is converted to

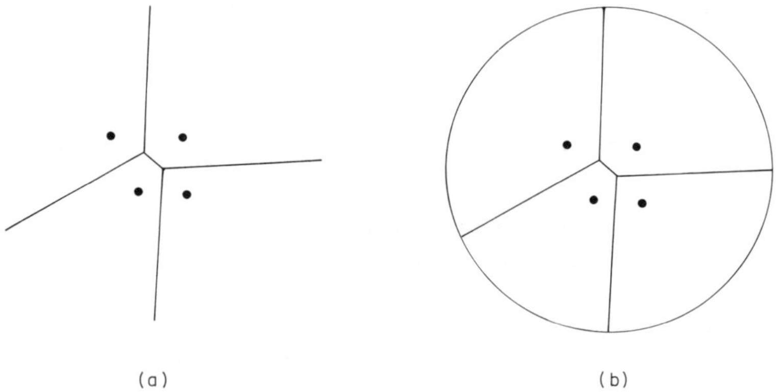

Figure 4.2.1 Voronoi diagram and the associated geometric graph: (a) Voronoi diagram; (b) geometric graph associated with the Voronoi diagram in (a).

the diagram in (b). At these terminal points the closed curve itself is divided into curved line segments; these line segments are also considered as Voronoi edges. Thus, by this convention the Voronoi diagram is converted into a geometric graph. We call this geometric graph the *augmented geometric graph* associated with the Voronoi diagram. (Note that the augmented geometric graph is a little different from the geometric graph of the Voronoi diagram introduced in Section 2.3; in the geometric graph all the infinite Voronoi edges are connected to a common additional vertex, whereas in the augmented geometric graph infinite Voronoi edges are connected to mutually different vertices 'at infinity'. We adopt this structure because we want to store the direction of the infinite edges as the 'coordinates' of these vertices, as we will see soon.)

Consider the Voronoi diagram for the set $P = \{p_1, \ldots, p_n\}$ of n generators. By the above convention the plane is divided into $n+1$ regions, n finite regions corresponding to the n Voronoi polygons and one outermost infinite region. We introduce a virtual generator, say p_∞, and consider that the outermost infinite region is the Voronoi region of p_∞. Thus the augmented geometric graph associated with the Voronoi diagram for P has $n+1$ regions, corresponding to $p_1, p_2, \ldots, p_n$ and p_∞. Let n_e and n_v be the number of edges and vertices, respectively, of the augmented geometric graph. Since at least three edges meet at a vertex, we get $3n_v \leq 2n_e$. From this inequality together with Euler's formula $(n+1)-n_e+n_v = 2$, we get

$$n_e \leq 3n-3 \quad \text{and} \quad n_v \leq 2n-2, \tag{4.2.1}$$

where the equality holds when no degeneracy takes place. (Note that these inequalities are a little different from those in Property V11 in Section 2.3; this is because the 'pseudo' Voronoi edges and the 'pseudo' Voronoi vertices on the outermost closed curve are also counted here.)

To represent the augmented geometric graph associated with the Voronoi diagram, we create the winged-edge data structure in the following way. First, for each edge we choose and fix the direction of the edge arbitrarily, and thus convert the geometric graph to a directed geometric graph. Next we arbitrarily choose and fix a linear order from 1 to n_v to vertices, and similarly choose and fix a linear order from 1 to n_e to edges. Furthermore, we call the Voronoi polygon associated with p_i *polygon* i ($i = 1, 2, \ldots, n, \infty$). The incidence relations among vertices, edges and polygons are represented by ten integer-value arrays, one for polygons, one for vertices and eight for edges, which are defined in the following way.

For polygon i ($i = 1, 2, \ldots, n, \infty$):

edge.around.polygon[i]: the ordinal number of an edge on the boundary of the polygon i.

For vertex j ($j = 1, 2, \ldots, n_v$):

edge.around.vertex[j]: the ordinal number of an edge incident to the vertex j.

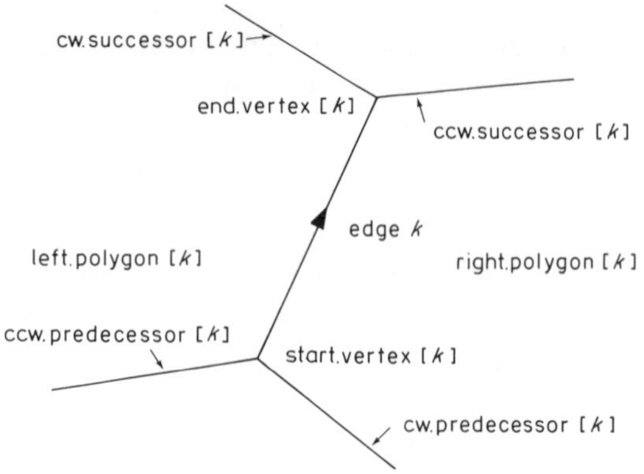

Figure 4.2.2 Geometric objects represented by the eight arrays associated with an edge.

For each edge k ($k = 1, 2, \ldots, n_e$) (Figure 4.2.2 shows the eight geometric objects represented by these arrays for edge k):

right.polygon[k]: the ordinal number of the polygon that is to the right of the edge k,
left.polygon[k]: the ordinal number of the polygon that is to the left of the edge k,
start.vertex[k]: the ordinal number of the start vertex of the edge k,
end.vertex[k]: the ordinal number of the end vertex of the edge k,
cw.predecessor[k]: the ordinal number of the edge next to edge k clockwise around the start vertex,
ccw.predecessor[k]: the ordinal number of the edge next to edge k counterclockwise around the start vertex,
cw.successor[k]: the ordinal number of the edge next to edge k clockwise around the end vertex,
ccw.successor[k]: the ordinal number of the edge next to edge k counterclockwise around the end vertex,

where cw and ccw indicate '*c*lock*w*ise' and '*c*ounter*c*lock*w*ise', respectively, and 'predecessor' and 'successor' represent edges around the start vertex and the end vertex, respectively. For each polygon i, we simply store in edge.around.polygon[i] one of the edges incident to the polygon, and similarly for each vertex j we store in edge.around.vertex[j] one of the edges incident to the vertex. Most information is stored in the eight arrays for edges (Figure 4.2.2).

Figure 4.2.3 shows a directed geometric graph obtained from the graph in Figure 4.2.1(b) by choosing directions arbitrarily and by ordering the vertices and the edges arbitrarily, and Table 4.2.1 shows the contents of the ten arrays representing this graph. Here we use the symbol ∞ to represent

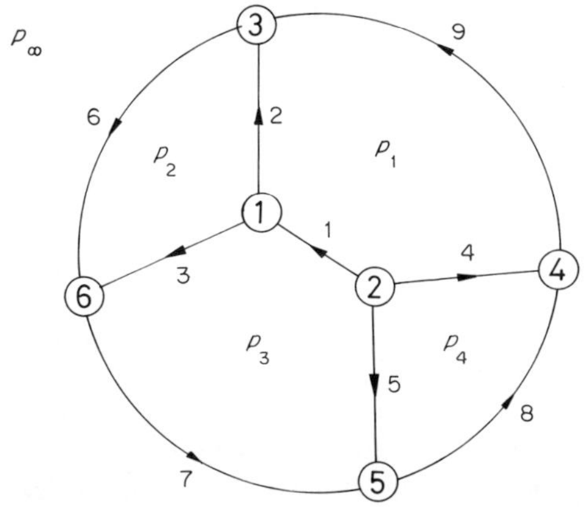

Figure 4.2.3 Directed augmented geometric graph associated with the Voronoi diagram in Figure 4.2.1.

the outermost infinite region; this is simply because we want to emphasize that this region does not correspond to an actual Voronoi polygon, and in practical implementations the symbol ∞ may be replaced by the integer $n+1$.

The above ten arrays together represent the topological structure of the Voronoi diagram, but the metric aspect of the Voronoi diagram should also be represented. For this purpose, we use three more arrays w.vertex[j],

Table 4.2.1 Entries of the ten integer arrays in the winged-edge data structure associated with the augmented geometric graph shown in Figure 4.2.3

k (edge number)	1	2	3	4	5	6	7	8	9
right.polygon[k]	1	1	2	4	3	∞	∞	∞	∞
left.polygon[k]	3	2	3	1	4	2	3	4	1
start.vertex[k]	2	1	1	2	2	3	6	5	4
end.vertex[k]	1	3	6	4	5	6	5	4	3
cw.predecessor[k]	4	1	2	5	1	9	6	7	8
ccw.predecessor[k]	5	3	1	1	4	2	3	5	4
cw.successor[k]	3	6	7	9	8	3	5	4	2
ccw.successor[k]	2	9	6	8	7	7	8	9	6

i (polygon number)	1	2	3	4	∞
edge.around.polygon[k]	1	2	1	4	6

j (vertex number)	1	2	3	4	5	6
edge.around.vertex[j]	1	1	2	4	5	3

x.vertex[j] and y.vertex[j] attached to the vertices. The array w.vertex[j] represents one-bit information in such a way that

$$\text{w.vertex}[j] = \begin{cases} 1 & \text{if vertex } j \text{ is an ordinary point,} \\ 0 & \text{if vertex } j \text{ is a point at infinity.} \end{cases}$$

If vertex j is an ordinary point, x.vertex[j] and y.vertex[j] represent the x and the y coordinates of this vertex, while if vertex j is a point at infinity, x.vertex[j] and y.vertex[j] represent the x and the y components of the unit vector designating the direction in which the associated (infinite) Voronoi edge runs. Thus, each entry of the array w.vertex requires one bit of memory space, while each entry of x.vertex or y.vertex requires one word to represent a floating-point number. The triple (w.vertex[j], x.vertex[j], y.vertex[j]) is a special case of what is usually called 'homogeneous coordinates', which are useful for representing ordinary points and points at infinity in a unified manner (e.g. Pedoe, 1970).

The collection of the above thirteen arrays is called the *winged-edge data structure*. In what follows by the construction of the Voronoi diagram $\mathcal{V}(P)$ we mean the construction of the winged-edge data structure for the augmented geometric graph associated with $\mathcal{V}(P)$.

In the augmented geometric graph associated with the Voronoi diagram, there are exactly $n+1$ polygons, and at most $3n-3$ edges and $2n-2$ vertices. Hence in total the winged-edge data structure requires memory space for $n+1+(2n-2)+8(3n-3) = 27n-25$ integers, $2n-2$ bits and $2(2n-2) = 4n-4$ floating-point numbers. To represent the Voronoi diagram completely, we also need $2n$ more floating-point numbers for the x and y coordinates of the generators in P, which is not the output from the algorithm but the input to the algorithm.

Once we obtain the winged-edge data structure, we can easily retrieve a variety of fundamental information about the Voronoi diagram. For example, if we want to obtain the list of edges and polygons that are incident to vertex j, we can use the next algorithm. In the algorithm, a 'list' means data containing a sequence of items, and 'to add an item to the tail of a list' means to augment the list by adding the item as the last element of the sequence. For example, if we add b to the tail of list $(b_1, b_2, \ldots, b_k)$, we get $(b_1, b_2, \ldots, b_k, b)$.

Algorithm 4.1 (Retrieve the edges and polygons incident to a vertex)
Input: Winged-edge data structure of $\mathcal{V}(P)$ and vertex j.
Output: List L_e of edges and list L_r of polygons that surround vertex j counterclockwise.
Procedure:
Step 1. $L_e \leftarrow$ empty list and $L_r \leftarrow$ empty list.
Step 2. $k \leftarrow$ edge.around.vertex[j], and *kstart* $\leftarrow k$.

Step 3. Add k to the tail of the list L_e.
Step 4. If j = start.vertex[k]
then add left.polygon[k] to the tail of L_r, and $k \leftarrow$ ccw.predecessor[k],
else add right.polygon[k] to the tail of L_r and $k \leftarrow$ ccw.successor[k].
Step 5. If k = *kstart* then return L_e and L_r, else go to Step 3.

In this description of the algorithm, we use several conventions common in computer science (Aho *et al.*, 1974). '$x \leftarrow y$' represents 'replace the value of variable x by the value of y', 'if C then X else Y' represents 'do X if the condition C is true, and do Y otherwise', and 'return x' represents 'report x as the output and exit from the procedure'.

In this algorithm, Steps 1 and 2 are done once, and Steps 3, 4 and 5 are repeated as many times as the number of edges (or, equivalently, the number of polygons) incident to vertex j. Hence, the total running time of this algorithm is proportional to the degree of vertex j.

If we want to get the list of edges and vertices that surround polygon i counterclockwise, we can use the next algorithm.

Algorithm 4.2 (Retrieve the edges and vertices surrounding a polygon)
Input: Winged-edge data structure of $\mathcal{V}(P)$ and polygon i.
Output: List L_e of edges and list L_v of vertices that surround polygon i counterclockwise.
Procedure:
Step 1. $L_e \leftarrow$ empty list, and $L_v \leftarrow$ empty list.
Step 2. $k \leftarrow$ edge.around.polygon[i], and *kstart* $\leftarrow k$.
Step 3. Add k to the tail of list L_e.
Step 4. If i = left.polygon[k]
then add end.vertex[k] to the tail of L_v and $k \leftarrow$ cw.successor[k],
else add start.vertex[k] to the tail of L_v and $k \leftarrow$ cw.predecessor[k].
Step 5. If k = *kstart* then return L_e and L_v, else go to Step 3.

This algorithm runs in time proportional to the number of edges on the boundary of polygon j, because Steps 1 and 2 are done once, and Steps 3, 4 and 5 are repeated as many times as the number of edges on the boundary of polygon j.

In particular, if polygon ∞ is chosen as the input polygon, this algorithm gives the list of edges and vertices on the boundary of the outermost infinite region, that is, the list of edges and vertices on the boundary of the convex hull CH(P) (recall Property V2).

Sometimes we want to retrieve the structure of the Delaunay diagram, the dual of the Voronoi diagram. The winged-edge data structure is also convenient for this purpose. Indeed, if we exchange the roles of the vertices and polygons, the winged-edge data structure for a geometric graph itself can be considered as the winged-edge data structure for the dual graph. For example, edge k corresponds to the Delaunay edge connecting two Delaunay vertices (i.e. generators) represented by right.polygon[k] and left.polygon[k], and the Delaunay polygon corresponding to the jth Voronoi vertex can be obtained as the output of Algorithm 4.1. Thus, what we have to do to retrieve information about the Delaunay diagram is simply to rename the geometric concepts by their dual concepts. That is, we consider that the edge k is a Delaunay edge connecting two generators right.polygon[k] and left.polygon[k], and that the Voronoi vertex i is a Delaunay triangle one of whose edges is edge.around.vertex[j], etc.

Before closing this section let us see the lower bound of the time complexity for constructing the Voronoi diagram. We can see that the lower bound of the worst-case time complexity for constructing the Voronoi diagram for n points is $O(n \log n)$. This lower bound can be derived from the well-known fact that any algorithm for sorting n real numbers in increasing order using comparisons requires at least $O(n \log n)$ time in the worst case (Aho *et al.*, 1974).

Let $X = \{x_1, x_2, \ldots, x_n\}$ be a set of n real numbers. We create the set $P = \{(x_1, x_1^2), (x_2, x_2^2), \ldots, (x_n, x_n^2)\}$ of n points in the plane and construct the Voronoi diagram for P. From this Voronoi diagram we obtain the cyclic sequence of points on the boundary of the convex hull $CH(P)$ by Algorithm 4.2 in $O(n)$ time. Since all the points in P are on the boundary of $CH(P)$, as shown in Figure 4.2.4, we obtain the increasing order of the n numbers in X. Thus, if the Voronoi diagram for P could be constructed faster than in $O(n \log n)$ time, then we could sort numbers in X faster than in $O(n \log n)$ time, which is a contradiction. Hence, any algorithm for constructing the

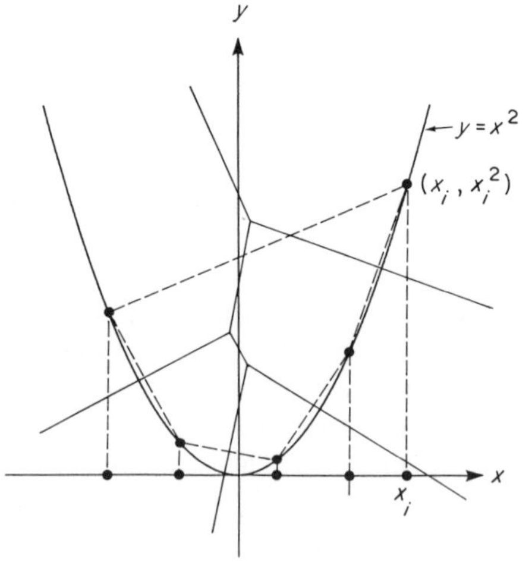

Figure 4.2.4 Relation between the sorting problem and the convex hull problem.

Voronoi diagram for n points takes at least $O(n \log n)$ time in the worst case.

This proof also contains the proof of the property that any algorithm for constructing the cyclic sequence of points on the boundary of the convex hull of n points takes at least $O(n \log n)$ time in the worst case.

On the other hand, the lower bound of the average-case time complexity for constructing the Voronoi diagram is obviously $O(n)$; this is because any algorithm should create the winged-edge data structure whose size is $O(n)$.

As we will see, both of the lower bounds are tight in the sense that we can actually construct an algorithm whose worst-case time complexity is $O(n \log n)$ and an algorithm whose average-case time complexity is $O(n)$.

4.3 INCREMENTAL METHOD

An incremental method is one of the most important methods because it is conceptually simple and its average time complexity can be decreased to $O(n)$ by some algorithmic techniques. This method starts with a simple Voronoi diagram for a few generators, say two or three generators, and modifies the diagram by adding other generators one by one. For $l = 1, 2, \ldots, n$, let $\mathcal{V}_l$ denote the Voronoi diagram for the first l generators $p_1, p_2, \ldots, p_l$. The main part of the incremental method is to convert $\mathcal{V}_{l-1}$ to $\mathcal{V}_l$ for each l.

Figure 4.3.1 shows an example of the addition of a generator. Suppose that we have already constructed the Voronoi diagram $\mathcal{V}_{l-1}$ as shown by the solid lines, and that we now want to add a new generator p_l. First, we find the generator, say p_i, whose Voronoi polygon contains p_l, and draw the perpendicular bisector between p_l and p_i. The bisector crosses the

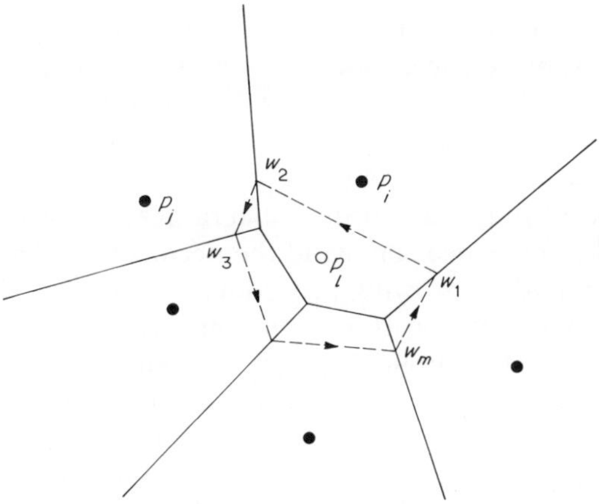

Figure 4.3.1 Addition of a new generator p_l.

boundary of $V(p_i)$ at two points; let the two points be w_1 and w_2 in such a way that p_l is to the left of the directed line segment $\overrightarrow{w_1w_2}$. The line segment $\overrightarrow{w_1w_2}$ divides the Voronoi polygon $V(p_i)$ into two portions, the one on the left belonging to the Voronoi polygon of p_l. Thus, we get a Voronoi edge on the boundary of the Voronoi polygon of p_l.

Starting with the edge $\overline{w_1w_2}$, we 'grow' the boundary of the Voronoi polygon of p_l by the following procedure, which we call the *boundary growing procedure*. The bisector between p_i and p_l crosses the boundary of $V(p_i)$ at w_2, entering the adjacent Voronoi polygon, say $V(p_j)$. So, we next draw the perpendicular bisector between p_l and p_j, and find the point (other than w_2) at which the bisector crosses the boundary of $V(p_j)$; let this point be w_3. In a similar way, we find the sequence of segments of perpendicular bisectors of p_l and the neighbouring generators until we reach the starting point w_1. Let this sequence be $(\overline{w_1w_2}, \overline{w_2w_3}, \ldots, \overline{w_{m-1}w_m}, \overline{w_mw_1})$. This sequence forms a counterclockwise boundary of the Voronoi polygon of the new generator p_l. Finally, we delete from $\mathcal{V}_{l-1}$ the substructure inside the new Voronoi polygon, and thus obtain $\mathcal{V}_l$.

The above description is a rough sketch of the incremental method. To make an exact and efficient algorithm we must be careful at several points.

First, we have to take care that the boundary-growing procedure works well only when the Voronoi polygon of the new generator p_l is finite, although a Voronoi polygon is not necessarily finite. An infinite region, if it happens, requires an exceptional branch of processing. However, such an exception can be avoided by a simple trick.

Recall that, in the Voronoi diagram $\mathcal{V}$ for the generator set P, the Voronoi polygon $V(p_i)$ is infinite if and only if p_i is on the boundary of the convex hull of P (Property V2). Keeping this in mind, let us renumber the generators as $p_4, p_5, \ldots, p_n$ (now n is the number of the generators plus three), and introduce three additional generators p_1, p_2 and p_3 in such a way that the triangle $p_1p_2p_3$ contains all the original generators $p_4, p_5, \ldots, p_n$ in its interior. The Voronoi diagram for p_1, p_2 and p_3 is a simple one, as shown in Figure 4.3.2(a). We start with this Voronoi diagram, and add other generators $p_4, p_5, \ldots, p_n$ one by one. The convex hull of $p_1, p_2, \ldots, p_l$ is the triangle $p_1p_2p_3$ itself for any $l = 4, 5, \ldots, n$, and consequently the Voronoi polygon of p_l in the Voronoi diagram $\mathcal{V}_l$ is always finite, as shown in Figure 4.3.2(b). Thus, we can avoid the occurrence of an infinite Voronoi polygon in the boundary-growing procedure.

The Voronoi diagram constructed in this way is not the same as the Voronoi diagram for the original generators. However, the difference due to the additional three generators can be eliminated substantially if a sufficiently large triangle is chosen. For example, if we magnify the central portion of Figure 4.3.2(b), we get (c), where the effect of the additional generators does not appear. Without loss of generality, let us assume that all the original generators $p_4, p_5, \ldots, p_n$ are located in the unit square

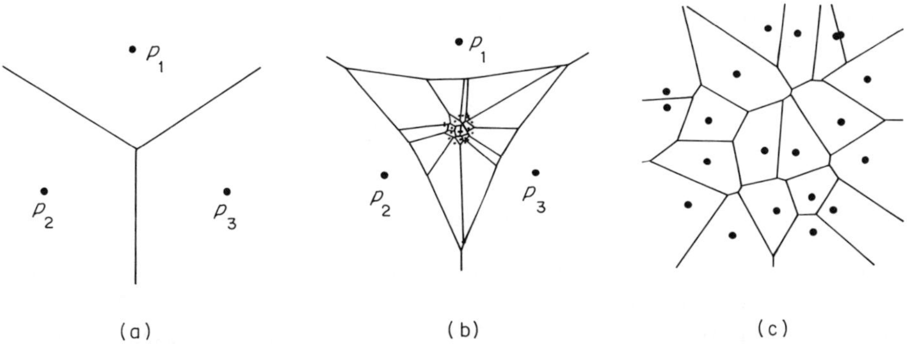

Figure 4.3.2 Avoidance of infinite Voronoi polygons: (a) Voronoi diagram for three additional generators; (b) Voronoi diagram for all the generators; (c) Voronoi diagram for the original generators.

$S = \{(x, y)|0 \leq x, y \leq 1\}$, and that we are interested in the Voronoi diagram in this square area. Then, we can choose the three additional generators, for example, by

$$\begin{aligned} p_1 &= (0.5, 3\sqrt{2}/2 + 0.5), \\ p_2 &= (-3\sqrt{6}/4 + 0.5, -3\sqrt{2}/4 + 0.5), \\ p_3 &= (3\sqrt{6}/4 + 0.5, -3\sqrt{2}/4 + 0.5). \end{aligned} \tag{4.3.1}$$

These three points form a sufficiently large regular triangle which contains the square area $\{(x,y)|0 \leq x,y \leq 1\}$; see Figure 4.3.2(b). We can see that no boundary of the Voronoi polygon of the additional generators goes through the interior of the square S unless the set of original generators is empty (actually, we can prove that no such boundary goes through the interior of the circle circumscribing S; the proof is left to the reader).

The second point we should take care over is the time complexity. The first task in the addition of a new generator p_l is to find the old generator p_i whose Voronoi polygon contains p_l. Obviously p_i satisfies $d(p_l, p_i) \leq d(p_l, p_j)$ for any $j = 1, 2, \ldots, l-1$. Hence, p_i can be found if the distances from p_l to all the other generators are computed. However, this requires $O(l)$ time, so that the total time complexity will become $O(1+2+\ldots+(n-1)) = O(n^2)$.

A little more intelligent way is to start with an initial guess and to search its neighbours for a closer generator. This can be done using the next algorithm.

Algorithm 4.3 (Nearest neighbour search) Input: l generators $p_1, p_2, \ldots, p_l$, Voronoi diagram $\mathcal{V}_{l-1}$, and initial guess p_i $(1 \leq i \leq l-1)$.
Output: the generator (other than p_l) that is closest to p_l.
Procedure:

Step 1. Among the generators adjacent to p_i, find the one, say p_j, with the minimum distance to p_l:

$$d(p_j, p_l) = \min_k d(p_k, p_l),$$

where the minimum is taken over all generators p_k whose Voronoi polygons are adjacent to $V(p_i)$.

Step 2. If $d(p_i, p_l) \leq d(p_j, p_l)$, return p_i, else $p_i \leftarrow p_j$ and go to Step 1.

The algorithm always terminates in a finite number of steps, because each replacement of p_i at Step 2 gives a closer generator than before. Moreover, the algorithm gives the closest generator as the output, because, if p_i is not closest to p_l, there always exists an adjacent generator closer to p_l (indeed, if one moves from p_i in the direction to p_l, one crosses the boundary of $V(p_i)$ and visits the neighbouring polygon, say the Voronoi polygon of p_j, which is closer to p_l; the proof is left to the reader).

Note that the time required by the algorithm depends greatly on the initial guess p_i; if it is near to the closest generator, we reach the closest generator quickly. Hence it is important to make a good initial guess. For this purpose we employ a bucketing technique. Before describing this technique, however, we describe another important point.

The final point we should be careful of is the size of the substructure to be deleted, because, if the substructure is large and complicated, we cannot find and delete it in constant time. As we have seen, the average number of edges on the boundary of a Voronoi polygon is six or less (recall Property V13 in Section 2.3). Consequently, we can expect that the average size of the substructure to be deleted is almost constant. However, we should not be too optimistic, because the average number of edges of a Voronoi polygon being six does not necessarily imply that the average number of edges of a 'new' Voronoi polygon will be six.

A counterexample is shown in Figure 4.3.3, where the generators align along a spiral and are numbered from outmost inward. The Voronoi polygon of the new generator, the innermost one, is adjacent to all the other polygons, so that the size of the substructure to be deleted is proportional to the number of generators added so far.

To restrict the average size of the substructure to be constant, the generators should be numbered in such a way that for any $l = 4, 5, \ldots, n$, the generators $p_4, p_5, \ldots, p_l$ distribute as uniformly as possible. Hence before starting the incremental method, we have to reorder the generators to meet this property as nearly as possible. For this purpose also, we use the bucketing technique.

A *rooted tree* is a tree in which one node is specified as the *root* (see Figure 4.3.4). When we draw a rooted tree, we put the root at the top, and draw other nodes downward in such a way that nodes at the same distance from the root align horizontally (where the distance of two nodes is the

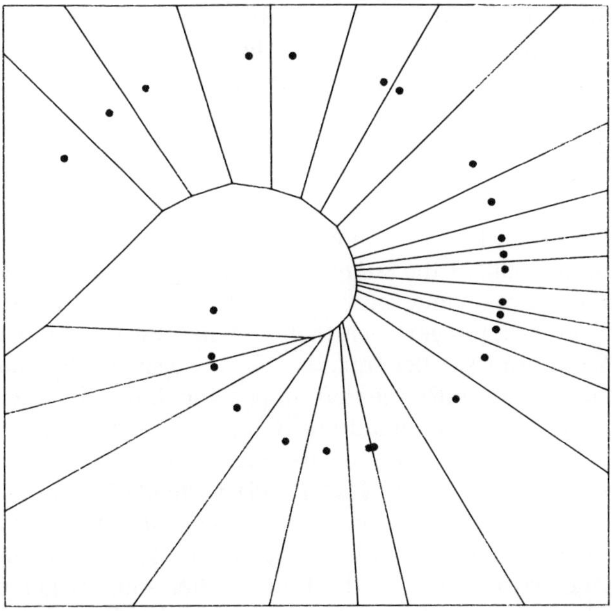

Figure 4.3.3 Voronoi diagram for points on a spiral.

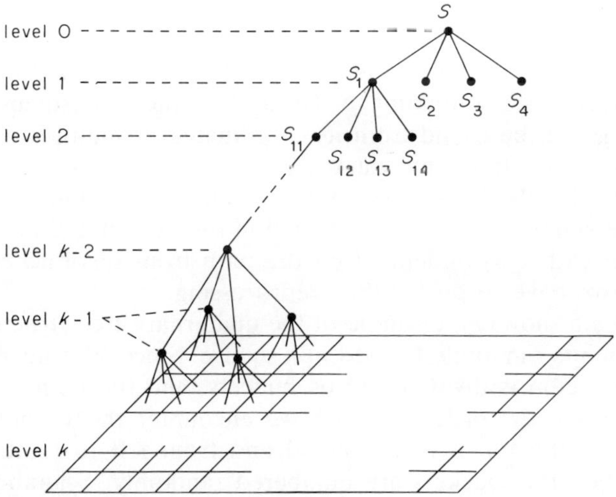

Figure 4.3.4 Quaternary tree with buckets as leaves.

number of links on the path connecting them). For each node u, the unique one-level upper node adjacent to u is called the *parent* of u, and one-level lower nodes adjacent to u (if any) are called *children* of u. The root is the only node that does not have a parent. A node without children is called a *leaf*. A node which is neither the root nor a leaf is called an *intermediate node*. If node u is on the path from node v to the root, u is called an *ancestor* of v and v a *descendant* of u. For any node u, the subgraph consisting of u and all its descendants is itself a rooted tree with the root u; this subgraph is called the *subtree* rooted at u. A rooted tree is called a *quaternary tree* if every node other than a leaf has exactly four children.

Let k be a positive integer such that 4^k is nearest to n. Let us divide the unit square S into four smaller squares, which we shall call quadrants; hence we divide the square into the left upper quadrant S_1, the right upper quadrant S_2, the left lower quadrant S_3 and the right lower quadrant S_4. We repeat this process k times recursively. That is, first S is divided into S_1, S_2, S_3 and S_4, and next S_i $(1 \leq i \leq 4)$ is divided into S_{i1}, S_{i2}, S_{i3} and S_{i4}, and so on. The division procedure can be represented by a quaternary tree; the root corresponds to the entire region S, and for each node, its four child nodes correspond to the four smaller squares generated by the partition of the corresponding square, as shown in Figure 4.3.4. The first division process generates 4 level-one nodes, the second division process generates 4^2 level-two nodes, and so on; finally the kth process generates 4^k level-k nodes. The smallest squares, the squares corresponding to the level-k nodes, are called *buckets*.

This quaternary tree structure is useful both for finding a good initial guess for Algorithm 4.3 and for reordering the generators. For these purposes, we need a little more preprocessing.

With each node u of the quaternary tree we associate a generator, say $g[u]$, according to the following procedure. First, for all $l = 4, 5, \ldots, n$, we find the bucket containing p_l (since S is the unit square, the bucket containing p_l can be found from its coordinates by multiplying 2^k and then truncating off the fractional parts). Next, we put $g[u] \leftarrow p_1$ for the root Thirdly, we visit the buckets one by one, and at each non-empty bucket we choose one generator, say p_i, contained in the bucket and put $g[u] \leftarrow p_i$ for all nodes u with $g[u]$ undefined on the path from this bucket to the root. This is all we have to do for the preprocessing.

Figure 4.3.5 shows an example of the quaternary tree structure. Panel (a) shows 60 points randomly located in a square. Since 60 is near to 64 $(= 4^3)$, the square region is divided into 64 buckets, and the points are numbered from 4 to 63 in the order in which we encounter them when we visit the buckets from the top row downward and from left to right at each row; points in the same buckets are numbered randomly. Scanning the buckets in the above order, we construct the quaternary tree, as shown in (b), where for each node the four child nodes correspond, from left to right, to the

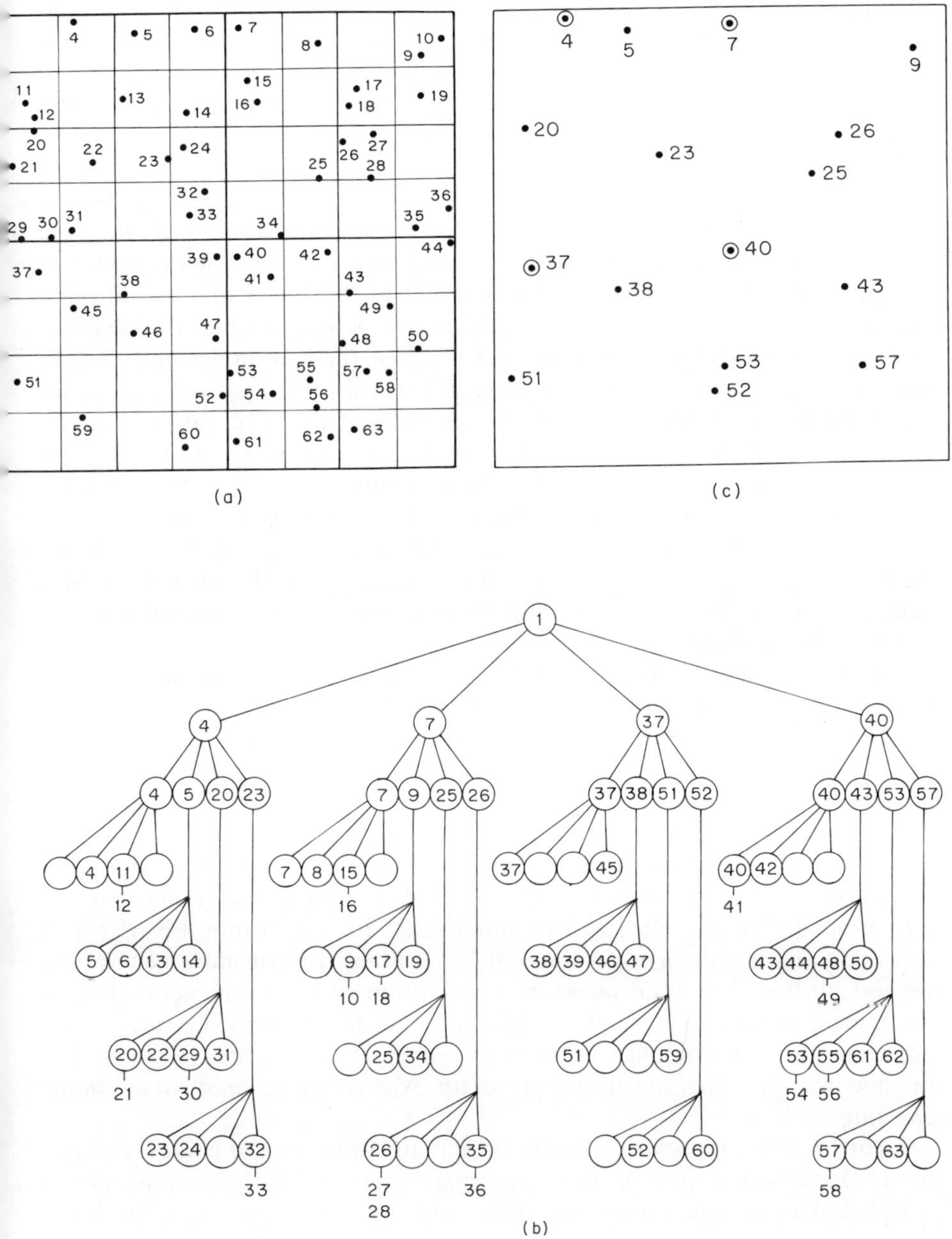

Figure 4.3.5 Randomly distributed points and the associated quaternary tree: (a) 60 points distributed randomly in a square and 64 buckets covering the square; (b) quaternary tree associated with the points and the buckets in (a); (c) points added when the nodes with level two or less are scanned.

left upper quadrant, the right upper quadrant, the left lower quadrant and the right lower quadrant, respectively. The number attached to each node represents the generator number $g[u]$. For a bucket containing two or more generators, the corresponding leaf node contains one of them and the other generators are stored in an additional list attached to the node.

Note that some buckets may contain two or more generators and some buckets may contain no generator, and hence some generators may not be assigned to any node and $g[u]$ may not be defined for some node u.

In the main processing of the incremental method, we visit the nodes of the quaternary tree from the level-one nodes to the leaves in a breadth-first manner, that is, we visit all the level-i nodes before visiting a level-$(i+1)$ node. Remember that we defined $g[u] = p_1$ for the root node, and that we start with the initial Voronoi diagram generated by p_1, p_2 and p_3. Hence we skip the root node, and start the addition at the level-one nodes. At each node u, if the associated generator $g[u]$ is new in the sense that we come across the generator $g[u]$ for the first time in the breadth-first visit, we add it to the current Voronoi diagram. Let u' be the parent node of u. Then the generator $g[u]$ is new if and only if $g[u] \neq g[u']$. If the visited node u is a bucket (i.e. a leaf node of the tree), and if the bucket contains generators other than $g[u]$, we add them to the current Voronoi diagram in an arbitrary order.

Adding generators in this order will keep the uniformness of the distribution of the generators relatively well, because at the end of visiting all the level-i nodes, exactly one generator is chosen from each non-empty square corresponding to that level. We call the above method for reordering *quaternary reordering*.

The quaternary tree is also useful for finding a good initial guess for Algorithm 4.3. Suppose that we are visiting node u. Let node u' be the parent of u. The generator $g[u]$ is added at this point if and only if $g[u] \neq g[u']$. We use $g[u']$ as the initial guess for Algorithm 4.3. If u is a leaf node and if the corresponding bucket contains two or more generators, we use $g[u]$ as the initial guess in the addition of the other generators in the bucket. This choice seems to give a good initial guess, because it lies either in the same square or in a one-level larger square. We call this method for choosing the initial guess for Algorithm 4.3 *quaternary initial guessing*.

Figure 4.3.5(c) represents the points that are added by the end of scanning up to level-two nodes of the quaternary tree in (b). The four points designated by double circles are those added by the time when the level-zero and level-one nodes are scanned, while the other points, represented by small filled circles, correspond to those added in the scan of the level-two nodes. We can see that the 16 points shown in (c) are distributed relatively uniformly in the square.

Point 4 is used as the initial guess for the nearest neighbour search in the addition of points 5, 20 and 23, and similarly point 7 is used as the initial guess in the addition of points 9, 25 and 26, and so on. We see that the initial guess is either the nearest point itself or a point very near to the nearest point.

Summing up the above considerations, we obtain the next algorithm.

Algorithm 4.4 (Quaternary incremental method) Input: Set $\{p_4, p_5, \ldots, p_n\}$ of $n-3$ generators located in the unit square $S = \{(x, y)|0 \le x, y \le 1\}$.

Output: Voronoi diagram $\mathcal{V}$ for n generators $\{p_1, p_2, \ldots, p_n\}$, where p_1, p_2 and p_3 are the additional generators defined by equation (4.3.1).

Procedure:

Step 1. Find positive integer k such that 4^k is closest to n, divide S into 4^k square buckets, and construct the quaternary tree having the buckets as leaves.

Step 2. Renumber the generators by quaternary reordering, and let the resultant order be $p_4, p_5, \ldots, p_n$.

Step 3. Construct the Voronoi diagram $\mathcal{V}_3$ for the three additional generators p_1, p_2 and p_3 defined by equation (4.3.1), as shown in Figure 4.3.2(a).

Step 4. For $l = 4, 5, \ldots, n$, do 4.1, ..., 4.4.

- 4.1. By Algorithm 4.3 with the initial guess given by quaternary initial guessing, find the generator p_i $(1 \le i \le l-1)$ closest to p_l.
- 4.2. Find the points w_1 and w_2 of intersections of the perpendicular bisector of p_i and p_l with the boundary of $V(p_i)$.
- 4.3. By the boundary-growing procedure, construct the closed sequence $(\overline{w_1w_2}, \overline{w_2w_3}, \ldots, \overline{w_{m-1}w_m}, \overline{w_mw_1})$ of segments of perpendicular bisectors forming the boundary of the Voronoi polygon of p_l.
- 4.4. Delete from $\mathcal{V}_{l-1}$ the substructure inside the closed sequence, and name the resultant diagram $\mathcal{V}_l$.

Step 5. Return $\mathcal{V} = \mathcal{V}_n$.

This algorithm does not refer to the associated winged-edge data structure explicitly, but it is not difficult to understand how to create and modify the winged-edge data structure accordingly; the initial winged-edge data structure for $\mathcal{V}_3$ is created in Step 3 and is modified in Step 4.4. To describe Steps 3 and 4.4 in terms of the winged-edge data structure is left as an exercise. Obviously, Steps 1 and 2 are done in O(n) time, and Steps 3 and 5 in O(1) time. Because of the quaternary reordering and the quaternary initial guessing, we can expect that Steps 4.1, ..., 4.4 can be done in O(1) time on average for each l. Thus, in total, Algorithm 4.4 runs in O(n) time on average if the generators are distributed at random in a square.

The use of the buckets with the quaternary tree was proposed by Ohya, Iri and Murota (Ohya, 1983; Ohya *et al.*, 1984a, 1984b), by which the average time complexity of the incremental method decreases from $O(n^2)$ to $O(n)$. They also recommended a certain specific order of visiting the buckets in the preprocessing and of visiting tree nodes in the main processing (see Ohya *et al.*, 1984a, for the details of their reordering).

4.4 DIVIDE-AND-CONQUER METHOD

In this and the next sections we give two $O(n \log n)$ algorithms for constructing the Voronoi diagram, which are optimal in a worst-case sense. They are indeed important from a theoretical point of view. However, since we have in practice already obtained a more efficient method in the previous section, we describe only the basic ideas of these algorithms, skipping details for implementation.

'Divide-and-conquer' is one of the fundamental paradigms for designing efficient algorithms. In this paradigm the original problem is recursively divided into several simpler subproblems of almost equal size, and the solution of the original problem is obtained by merging the solutions of the subproblems.

For our particular problem, the generators are sorted in increasing order of the x coordinates, and are represented by the 'list' $P = (p_1, p_2, \ldots, p_n)$ (unlike the incremental method, we do not use any additional generators; $p_1, \ldots, p_n$ are all original generators). This can be done in $O(n \log n)$ time by any optimal sorting algorithm such as a heap sort or a merge sort (cf. Knuth, 1973, or Aho *et al.*, 1974). In addition to Assumptions A4.1 and A4.2, here we assume the following.

Assumption A4.3 No two generators align vertically.

Hence the order $(p_1, p_2, \ldots, p_n)$ is defined uniquely. Next, the divide-and-conquer paradigm is applied in the following way (we first state the algorithm formally and then give an example).

Algorithm 4.5 (Divide-and-conquer method) Input: Number n of generators and list $P = (p_1, p_2, \ldots, p_n)$ of the generators arranged in increasing order of the x coordinates.
Output: Voronoi diagram $\mathcal{V}$ for P.
Procedure:
Step 1. If $n \leq 3$, then construct the Voronoi diagram $\mathcal{V}$ for P directly and go to Step 3.
Step 2. Otherwise do the following.
 2.1. Let t be the integral part of $n/2$, and divide P into $P_L = (p_1, \ldots, p_t)$ and $P_R = (p_{t+1}, \ldots, p_n)$.
 2.2. Construct the Voronoi diagram $\mathcal{V}_L$ for P_L by Algorithm 4.5.

2.3. Construct the Voronoi diagram $\mathcal{V}_{\mathrm{R}}$ for P_{R} by Algorithm 4.5.
2.4. Merge $\mathcal{V}_{\mathrm{L}}$ and $\mathcal{V}_{\mathrm{R}}$ into the Voronoi diagram $\mathcal{V}$ for P (this can be done by Algorithm 4.7).

Step 3. Return $\mathcal{V}$.

Note that the generators in P_{L} are to the left of the generators in P_{R}. We call elements of P_{L} *left generators* and those of P_{R} *right generators*. $\mathcal{V}_{\mathrm{L}}$ is called the *left Voronoi diagram* and $\mathcal{V}_{\mathrm{R}}$ the *right Voronoi diagram*.

The essential portion of this algorithm is Step 2.4, the step for merging two Voronoi diagrams. In what follows we consider how to do this step.

Suppose that we have obtained the left Voronoi diagram $\mathcal{V}_{\mathrm{L}}$ and the right Voronoi diagram $\mathcal{V}_{\mathrm{R}}$, as shown in Figure 4.4.1, where the left generators are represented by filled circles, the right generators by unfilled circles, $\mathcal{V}_{\mathrm{L}}$ by dot-and-dash lines and $\mathcal{V}_{\mathrm{R}}$ by broken lines. Now we have to merge them by generating new Voronoi edges and new Voronoi vertices and by deleting superfluous portions from $\mathcal{V}_{\mathrm{L}}$ and $\mathcal{V}_{\mathrm{R}}$. The new edges and the new vertices come from the interaction between left generators and right generators; that

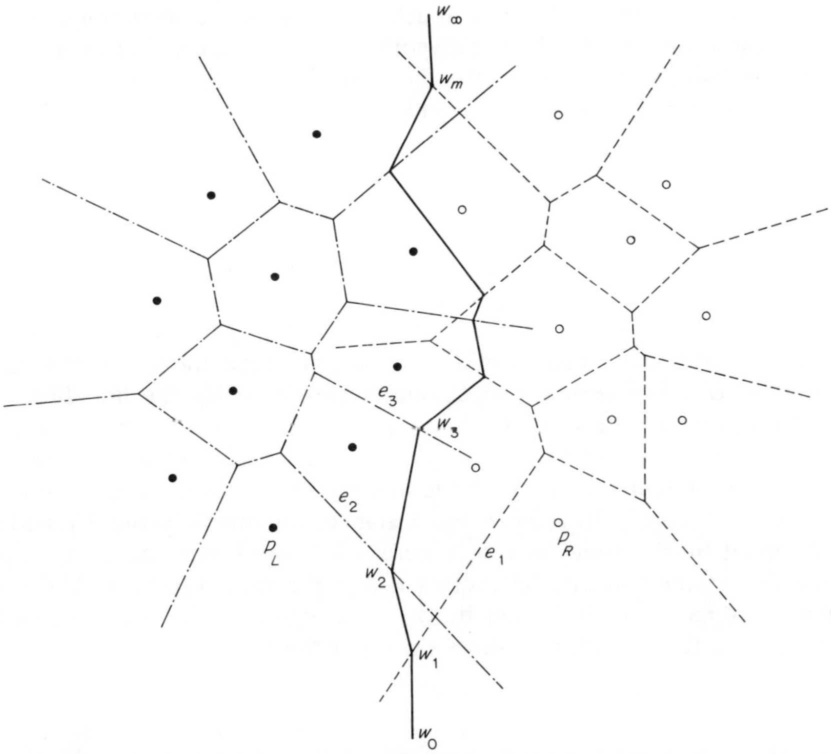

Figure 4.4.1 Merger of left and right Voronoi diagrams.

is, each new edge is part of the perpendicular bisector between a left generator and a right generator, and each new vertex is in an equidistance either from two left generators and one right generator or from one left generator and two right generators.

For any point p in the plane, let $d(p, P_L)$ denote the minimum distance from p to the generators in P_L, and $d(p, P_R)$ the minimum distance from p to the generators in P_R. Then any point p on a new Voronoi edge satisfies $d(p, P_L) = d(p, P_R)$. Consider a point p that moves from left to right along an arbitrary horizontal line. Since P_L is to the left of P_R, there is a unique point p^* such that $d(p^*, P_L) = d(p^*, P_R)$ and that $d(p, P_L) < d(p, P_R)$ before p reaches p^* and $d(p, P_L) > d(p, P_R)$ after p passes through p^*. As the horizontal line moves from the bottom upward, p^* moves continuously, tracing a polygonal line running between the left generators and the right generators. Our task is to construct this polygonal line and to remove from $\mathcal{V}_L$ and $\mathcal{V}_R$ the portions that lie to the right and to the left, respectively, of the polygonal line.

We construct the polygonal line from the bottom upward. To find the bottommost edge, we need one more preparation. We present an algorithmic tool for finding a line such that it touches two disjoint convex polygons and that all the other points of the polygons are on the same side of the line.

For any two points p and q, the directed line passing through p and q in this order is denoted by $L(p, q)$, and the directed line segment starting at p and ending at q is denoted by $\overrightarrow{pq}$.

Let U be a convex polygon defined by the cyclic list $(u_1, u_2, \ldots, u_s, u_1)$ of vertices which surround the polygon counterclockwise. For vertex u in U, let cnext[u] and ccnext[u] be the immediate predecessor and the immediate successor, respectively, of u in the list (cnext[u] is the clockwise *next* vertex of u and ccnext[u] is the counterclockwise *next* vertex of u).

Let U_L and U_R be two convex polygons, U_L being to the left of U_R, i.e. the x coordinates of vertices in U_L are smaller than the x coordinates of vertices in U_R. For vertex u in U_L and vertex w in U_R the line $L(u,w)$ is called a *common support* of U_L and U_R if all the nodes in U_L and U_R are on the same side of the line or on the line. U_L and U_R admit two common supports; the upper one is called the *upper common support* and the lower one the *lower common support*. For instance, for two polygons U_L and U_R represented by the solid lines in Figure 4.4.2, the lower common support is the line connecting the left vertex d and the right vertex g. The lower common support can be found by the next algorithm (an example of the behaviour of the algorithm follows the algorithm).

Algorithm 4.6 (Lower common support) Input: Two convex polygons U_L and U_R such that the maximum x coordinate over all vertices in U_L is smaller than the minimum x coordinate over all vertices in U_R.

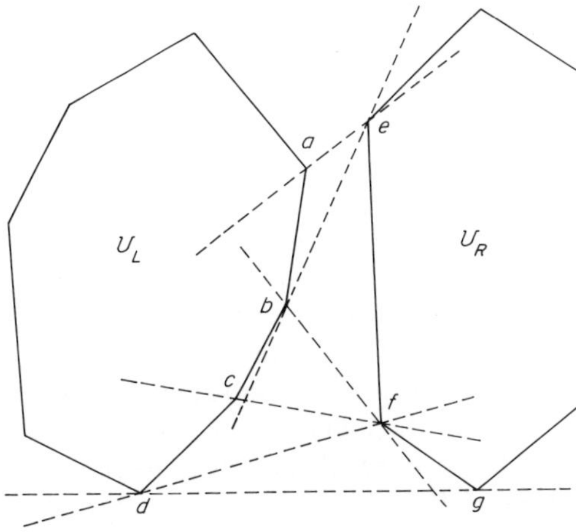

Figure 4.4.2 Search for the lower common support.

Output: Pair consisting of vertex u in U_L and vertex w in U_R such that $L(u,w)$ forms the lower common support of U_L and U_R.

Procedure:

Step 1. Find the vertex u in U_L with the largest x coordinate, and the vertex w in U_R with the smallest x coordinate.

Step 2. Do 2.1 and 2.2 alternately until u and w are not changed any more.

2.1. While vertex cnext[u] is lower than $L(u,w)$, repeat $u \leftarrow$ cnext[u].

2.2. While vertex ccnext[w] is lower than $L(u,w)$, repeat $w \leftarrow$ ccnext[w].

Step 3. Return $L(u,w)$.

An example of the behaviour of the algorithm is shown in Figure 4.4.2. First, the pair (u, v) is set to be (a, e) in Step 1. Next, the pair is changed to (b, e) in Step 2.1, to (b, f) in Step 2.2, through (c, f) to (d, f) in Step 2.1, and finally to (d, g) in Step 2.2.

This algorithm terminates in a finite number of steps and gives the lower common support as the output. This can be understood in the following way. Let $(u_1, u_2, \ldots, u_s, u_1)$ be the list of vertices in U_L appearing counterclockwise on the boundary. Without loss of generality, suppose that u_1 is the vertex on the lower common support, and that u_i is the vertex with the maximum x coordinate. Then, for any vertex w in U_R such that w is below the line $L\ (u_1, u_2)$, the vertex cnext[u_1] is not lower than the line $L(u_1, w)$. Thus, once u reaches u_1 in Step 2.1, it will not be updated any more. Similarly, once w reaches the vertex on the common lower support in Step 2.2, it will not be updated. Therefore Step 2 terminates in a finite number of repetitions. Moreover, the termination condition for Step 2 actually implies $L(u, w)$ being the lower common support.

Let n denote the total number of vertices in U_L and U_R. In Step 1, the initial pair (u, w) can be found by scanning the lists of vertices. In Step 2, the total number of updates of the pair (u, w) does not exceed n. Hence, Algorithm 4.6 runs in $O(n)$ time. Actually there is an $O(\log n)$ algorithm for finding the common support (Overmars and Leeuwen, 1981), but an $O(n)$ algorithm is enough for our purpose because the subproblem is not a bottleneck of the total time complexity.

Now we are ready to consider our main problem. We want to construct the polygonal line consisting of new Voronoi edges and new Voronoi vertices. As shown in Figure 4.4.1, let p_L and p_R be the pair of the left generator and the right generator forming the lower common support $L(p_L, p_R)$ of the convex hull of P_L and that of P_R, then $b(p_L, p_R)$, the perpendicular bisector of $\overline{p_L, p_R}$, gives the bottommost edge of the polygonal line, because $\overline{p_L p_R}$ is an edge of the convex hull of $P_L \cup P_R$ but is not an edge of the convex hull of P_L or of P_R. Hence, we start with this pair (p_L, p_R).

We consider a point w moving along the perpendicular bisector between p_L and p_R, from the bottom upward. If the y coordinate of the point w is sufficiently small, w belongs to both $V(p_L)$ and $V(p_R)$. Let w_0 be the point at infinity in the negative y direction along the bisector, and let w_1 be the point at which w crosses a Voronoi edge, say e_1, of $\mathcal{V}_L$ or $\mathcal{V}_R$ for the first time (as in Figure 4.4.1). Then the portion of the bisector along which w has moved so far, i.e. the half line $\overline{w_0 w_1}$, is the first new Voronoi edge, and w_1 is the first new Voronoi vertex.

At w_1 the moving point w changes the region either in $\mathcal{V}_L$ or in $\mathcal{V}_R$, depending on whether e_1 is an edge of $\mathcal{V}_L$ or of $\mathcal{V}_R$. If e_1 is an edge of $\mathcal{V}_L$, we replace p_L with the left generator that is on the other side of e_1. If e_1 is an edge of $\mathcal{V}_R$ (as in Figure 4.4.1), we replace p_R with the other right generator generating e_1. Next, we draw the perpendicular bisector between the updated pair of p_L and p_R. This bisector has w_1 on it. Suppose that w moves from w_1 along the new bisector upward until it crosses a Voronoi edge of $\mathcal{V}_L$ or $\mathcal{V}_R$ again. Let the crossing point be w_2. Thus we get the second new Voronoi vertex w_2 and the second new Voronoi edge $\overline{w_1 w_2}$.

We repeat similar procedures until we reach the pair (p_L, p_R) forming the upper common support. Thus, we obtain the next algorithm.

Algorithm 4.7 (Merger of two Voronoi diagrams) Input: Two Voronoi diagrams $\mathcal{V}_L$ and $\mathcal{V}_R$ for generator sets P_L and P_R, respectively, such that the generators in $\mathcal{V}_L$ have smaller x coordinates than those in $\mathcal{V}_R$.
Output: Voronoi diagram for $P_L \cup P_R$.
Procedure:
Step 1. Construct the convex hull of P_L and that of P_R.
Step 2. Find the lower common support $L(P_L, P_R)$ of them by Algorithm 4.6.

Step 3. $w_0 \leftarrow$ the point at infinity downward on $b(p_L, p_R)$, and $i \leftarrow 0$.

Step 4. While $L(p_L, p_R)$ is not the upper common support, repeat 4.1,..., 4.4.

4.1. $i \leftarrow i + 1$.

4.2. Find the point a_L (other than w_{i-1}) of the intersection of $b(p_L, p_R)$ with the boundary of $V(p_L)$.

4.3. Find the point a_R (other than w_{i-1}) of the intersection of $b(p_L, p_R)$ with the boundary of $V(p_R)$.

4.4. If a_L has a smaller y coordinate than a_R,
$w_i \leftarrow a_L$, and
$p_L \leftarrow$ the generator on the other side of the Voronoi edge containing a_L.
Otherwise
$w_i \leftarrow a_R$, and
$p_R \leftarrow$ the generator on the other side of the Voronoi edge containing a_R.

Step 5. $m \leftarrow i$.
$w_{m+1} \leftarrow$ the point at infinity upward on $b(p_L, p_R)$.

Step 6. Add the polygonal line $(w_0w_1, w_1w_2, \ldots, w_mw_{m+1})$, and delete from $\mathcal{V}_L$ the part to the right of the polygonal line and delete from $\mathcal{V}_R$ the part to the left of the polygonal line. Return the resultant diagram.

Step 1 is done in $O(n)$ time because we already obtained the Voronoi diagrams $\mathcal{V}_L$ and $\mathcal{V}_R$ (recall Algorithm 4.2). Step 2 is done in $O(n)$ time by Algorithm 4.6, and Step 3 in constant time. Step 4 is repeated at most $O(n)$ times, because there are at most $O(n)$ new Voronoi edges. Steps 4.1 and 4.4 require only constant time.

Steps 4.2 and 4.3 require time proportional to the number of edges on the boundary of $V(p_L)$ and to that of $V(p_R)$, respectively, which are not bounded by any fixed constant. However, we can decrease the total number of edges to be examined in Steps 4.2 and 4.3 over all repetitions of Step 4. For this purpose, we examine edges on the boundary of $V(p_L)$ counterclockwise, and those of $V(p_R)$ clockwise. Suppose that, in a certain repetition of Step 4, we have found the intersections a_L and a_R, as shown in Figure 4.4.3. Since a_L is lower than a_R in this example, p_L is replaced with p_L'; as a result of this, the polygonal line bends to the right. Therefore, when we search for the intersection a_R' of $b(p_L', p_R)$ with the boundary of $V(p_R)$, we need not examine the edges to the left of a_R; we can start searching at a_R clockwise. Thus we can avoid the repeated check of irrelevant edges. Similarly, if a_R is lower than a_L, p_R is replaced and the polygonal line bends to the left, so that we can avoid unnecessary check of edges by examining them counterclockwise. By doing so, we can execute Step 4 in $O(n)$ time.

Step 5 requires constant time. Step 6 requires $O(n)$ time. Hence, in total Algorithm 4.7 runs in $O(n)$ time.

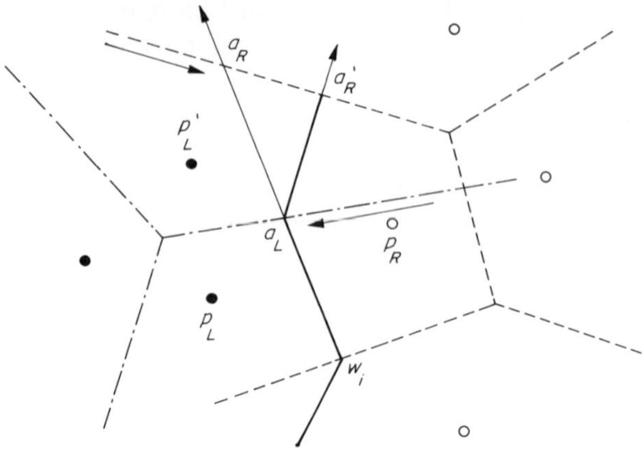

Figure 4.4.3 Search for the next Voronoi vertex on the boundary of the left and the right Voronoi diagrams.

Next, let us consider the time complexity of our main algorithm, Algorithm 4.5. The algorithm is recursive in the sense that the procedure calls itself at Steps 2.2 and 2.3. To evaluate the time complexity of such an algorithm, we first consider the other steps. Obviously, Steps 1, 2.1 and 3 can be done in constant time. Step 2.4 can be done in O(n) time by Algorithm 4.7. Thus, all steps other than the recursive calls require O(n) time.

Now, let $T(n)$ denote the time required by Algorithm 4.5 for n generators. Then, we get

$$T(n) = 2T(n/2) + \mathrm{O}(n),$$

because the algorithm calls itself twice using a half of the original input as the new input. With the consideration that $T(1)$ and $T(2)$ are constant, this equation implies that $T(n) = \mathrm{O}(n \log n)$ (Aho *et al.*, 1974).

Recall that any algorithm for constructing a Voronoi diagram requires O(n log n) time in the worst case (Section 4.2). Thus, Algorithm 4.5 is optimal from the viewpoint of the worst-case time complexity.

The divide-and-conquer technique was first applied to this problem by Shamos and Hoey (1975), and then many variants have been proposed (Lee and Schachter, 1980; Guibas and Stolfi, 1985). It is also known that a divide-and-conquer method of this type actually requires O(n log n) time not only in the worst case but also on average (Ohya *et al.*, 1984a). Thus, in the average sense, this method is not optimal. Recently, quite different types of divide-and-conquer methods have been proposed; for uniformly distributed generators, the algorithm proposed by Dwyer (1987) runs in O(n log log n) time on average, and the algorithm proposed by Katajainen and Koppinen (1988) runs in O(n) time on average.

4.5 PLANE SWEEP METHOD

'Plane sweep' is one of the fundamental techniques used to solve two-dimensional geometric problems. In this technique we use a special line called a *sweepline*, which is conventionally vertical, and sweep the plane with this line, say from left to right. As the sweepline moves, it hits geometric objects one by one. Each time an event happens, a portion of the problem is solved on the sweepline. Thus, this technique enables us to solve two-dimensional problems by a sequence of almost one-dimensional processing. Hence if successfully applied, the plane sweep technique gives a conceptually simple algorithm.

However, the plane sweep technique cannot be applied to our problem directly, because to construct Voronoi edges and Voronoi vertices on the sweepline we have to 'foresee' the locations of generators in the half plane that have not yet been swept; the sweepline reaches a Voronoi polygon before hitting the corresponding generator. If each generator p_i were to exist at the leftmost point of the Voronoi polygon $V(p_i)$, which is of course not the case, the plane sweep method could be applied. Indeed, there exists a transformation by which this 'dream' is realized (Fortune, 1986, 1987).

Suppose that the set $P = \{p_1, p_2, \ldots, p_n\}$ of n generators is given. In addition to Assumptions A4.1, A4.2 and A4.3, here we give the following assumption.

Assumption A4.4 Any generator and any Voronoi vertex on its boundary do not align horizontally.

This assumption is not critical, but makes the description of the algorithm much simpler.

For any point p, we define $r(p) = \min\{d(p, p_i)|1 \leq i \leq n\}$, i.e. $r(p)$ is the radius of the largest empty circle centred at p. Let $x(p)$ and $y(p)$ denote the x coordinate and the y coordinate, respectively, of p.

We define the map φ from $\mathbb{R}^2$ to itself by

$$\varphi(p) = (x(p) + r(p), y(p)) \tag{4.5.1}$$

for point p (see Figure 4.5.1). If $p \in V(p_i)$, we get $\varphi(p) = (x(p)+d(p, p_i), y(p))$ because p_i is the nearest generator to p. $\varphi(p) = p$ if and only if p is a generator. For the Voronoi vertex q, the Voronoi edge e, the Voronoi polygon $V(p_i)$ and the Voronoi diagram $\mathcal{V}$, their images using φ are denoted by $\varphi(q)$, $\varphi(e)$, $\varphi(V(p_i))$ and $\varphi(\mathcal{V})$, respectively.

As shown in Figure 4.5.1, let p_i and p_j be two generators such that $x(p_i) > x(p_j)$, and let e be the Voronoi edge between p_i and p_j. Then, φ maps e to part of a curved line whose leftmost point is p_i. Indeed, it is an easy exercise to see that the image $\varphi(e)$ of e is part of a hyperbola open rightward with the leftmost point p_i. The image of each Voronoi polygon is a connected region surrounded by segments of hyperbolas.

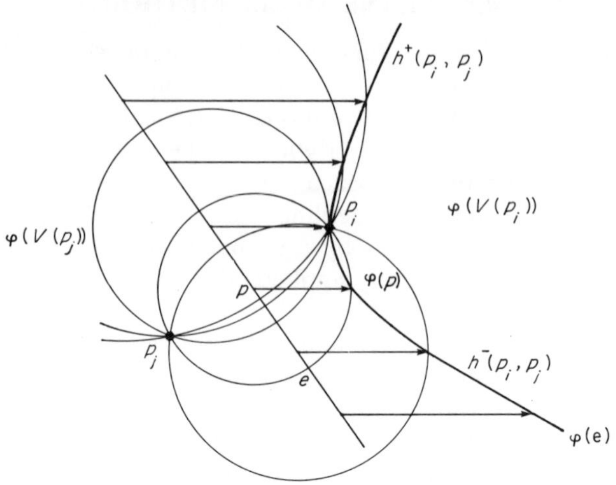

Figure 4.5.1 Perpendicular bisector and its image.

The mapping φ can be interpreted in the context of 'Voronoi diagrams in a river' (Sugihara, 1992a). Suppose that the generators are point-like islands in a river, and each island has a boat of the same speed. Suppose that $p \in V(p_i)$. If there is no flow of water, the boat starting at island p_i reaches p faster than any other boat. Now assume that the water in the river flows from left to right (i.e. in the positive x direction) at the same speed as the boat, as shown in Figure 4.5.2. Then, during the sailing of the boat from p_i to p, the boat is carried to the right by the same distance as it sails; hence the boat reaches the point $\varphi(p) = (x(p)+r(p), y(p))$ instead of the point p. Thus, $\varphi(p)$ represents the point reached by the boat that leaves the island nearest to p facing toward p. Hence $\varphi(V)$ can be interpreted as the tessellation on the surface of the river into 'territories' of the islands

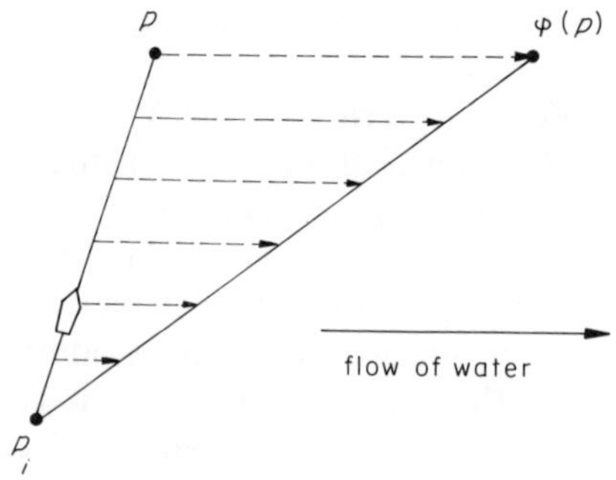

Figure 4.5.2 Path of a boat on the surface of a river.

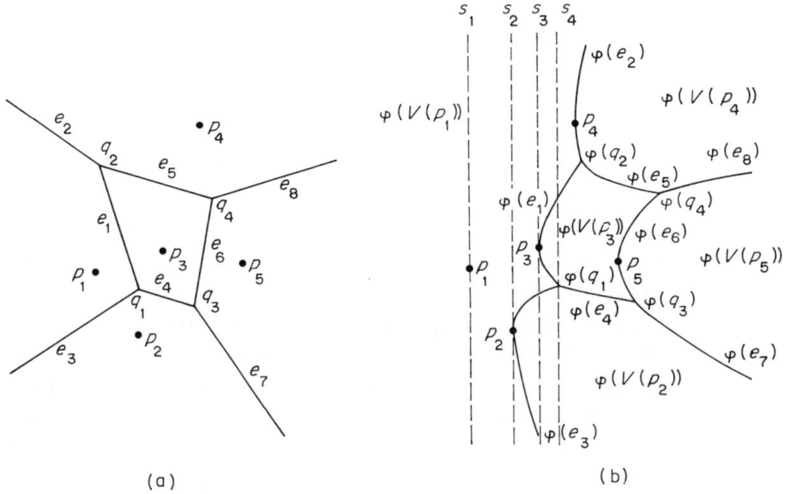

Figure 4.5.3 Voronoi diagram and its image: (a) Voronoi diagram $\mathcal{V}$ for five generators; (b) its image $\varphi(\mathcal{V})$).

such that any point in a territory can be reached by the boat starting at the corresponding island faster than any other boat.

An example of the map φ applied to the Voronoi diagram is shown in Figure 4.5.3, where (a) is the Voronoi diagram for five points and (b) is its image by φ. (If we want to interpret the diagram (b) as the tessellation on the surface of the river, the vertical line passing through p_1, the broken line s_1, is also necessary, because no boat can reach the region to the left of s_1; recall that the speed of the boat is equal to the speed of the flow of the water. The vertical line s_1 may be interpreted as the image of the line at infinity.)

A similar tessellation on the surface of the river can be defined for any speed of the flow of water. If the flow speed is 0, the associated tessellation is the ordinary Voronoi diagram, and as the flow speed becomes larger, the deformation of the associated tessellation from the ordinary Voronoi diagram becomes larger. However, we can expect that if the flow speed is not too large, the topological structure of the ordinary Voronoi diagram is preserved in the tessellation. Indeed, it can be shown that if the speed of the water flow is not greater than the speed of the boat, the geometric graph defined by the tessellation on the surface of the river is isomorphic to the geometric graph of the ordinary Voronoi diagram (Sugihara, 1992a). Hence, determining the topological structure of $\mathcal{V}$ is equivalent to determining that of $\varphi(\mathcal{V})$. Moreover, for any generator p_i other than the leftmost one, the transformed region $\varphi(V(p_i))$ has the leftmost point p_i; the images of all the Voronoi edges and Voronoi vertices related to p_i appear to the right of p_i. Hence we can apply the plane sweep technique for constructing $\varphi(\mathcal{V})$.

In what follows we use the terminology for Voronoi diagrams to designate objects in $\varphi(\mathcal{V})$. For instance, the image of a Voronoi vertex is called a Voronoi vertex of $\varphi(\mathcal{V})$, and the image of a Voronoi edge is called a Voronoi edge of $\varphi(\mathcal{V})$.

Let us note at this point that, because of Assumption A4.4, the image $\varphi(q)$ of any Voronoi vertex q is not the leftmost point of any Voronoi region in $\varphi(\mathcal{V})$, and consequently at $\varphi(q)$ exactly one Voronoi edge extends to the right and the other two Voronoi edges extend to the left. For instance, at the vertex $\varphi(q_1)$ in Figure 4.5.3(b), the edge $\varphi(e_4)$ extends to the right and the edges $\varphi(e_1)$ and $\varphi(e_3)$ extend to the left.

The plane sweep method moves a vertical sweepline across the plane from left to right, and each time an event happens, the structure of the Voronoi diagram $\varphi(\mathcal{V})$ is updated on the sweepline. There are two types of events. An event of the first type occurs when the sweepline hits a generator, as the lines s_1, s_2 and s_3 in Figure 4.5.3(b). An event of the second type occurs when the sweepline hits a Voronoi vertex of $\varphi(\mathcal{V})$, as the line s_4 in Figure 4.5.3(b).

Let p_i and p_j be two generators. The perpendicular bisector of p_i and p_j is transformed by φ to a hyperbola. Let us cut this hyperbola at the leftmost point, and let $h^+(p_i, p_j)$ and $h^-(p_i, p_j)$ be the upper part and the lower part, respectively, of the hyperbola; we call them the upper and lower *half hyperbolas* (see Figure 4.5.1). Note that $h^+(p_i, p_j) = h^+(p_j, p_i)$ and $h^-(p_i, p_j) = h^-(p_j, p_i)$.

To represent the structure of $\varphi(\mathcal{V})$ along the sweepline, we use an alternating list L of regions and boundary edges which appear on the sweepline from the bottom to the top. We also use Q, which is a set of points in the plane. Q represents the potential points at which events may happen in future. At the beginning, Q is the set of all generators, and candidates of Voronoi vertices are added to Q whenever they are found. The algorithm is as follows (the example that follows the algorithm may help the reader to understand the algorithm).

Algorithm 4.8 (Plane sweep method) Input: Set $P = \{p_1, p_2, \ldots, p_n\}$ of n generators.
Output: Voronoi diagram $\varphi(\mathcal{V})$.
Procedure:
Step 1. $Q \leftarrow P$.
Step 2. Choose and delete the leftmost point, say p_i, from Q.
Step 3. $L \leftarrow$ the list consisting of a single region $\varphi(V(p_i))$.
Step 4. While Q is not empty, repeat 4.1, 4.2 and 4.3.
  4.1. Choose and delete the leftmost point w from Q.
  4.2. If w is a generator, say $w = p_i$, do 4.2.1, 4.2.2 and 4.2.3.
    4.2.1. Find the region $\varphi(V(p_j))$ on L containing p_i.

4.2.2. Replace $\varphi(V(p_j))$ on L by the subsequence $(\varphi(V(p_j)), h^-(p_j, p_i), \varphi(V(p_i)), h^+(p_i, p_j), \varphi(V(p_j)))$.

4.2.3. Add to Q the intersection of $h^-(p_j,p_i)$ with the immediate lower half hyperbola on L, and the intersection of $h^+(p_i, p_j)$ with the immediate upper half hyperbola on L.

4.3. If w is an intersection, say $w = \varphi(q_t)$, do 4.3.1, ..., 4.3.4 (where we suppose that $\varphi(q_t)$ is the intersection of $h^{\pm}(p_i, p_j)$ and $h^{\pm}(p_j, p_k)$).

4.3.1. Replace subsequence $(h^{\pm}(p_i, p_j), \varphi(V(p_j)), h^{\pm}(p_j, p_k))$ on L by $h = h^-(p_i, p_k)$ or $h = h^+(p_i, p_k)$ appropriately.

4.3.2. Delete from Q any intersections of $h^{\pm}(p_i, p_j)$ or $h^{\pm}(p_j, p_k)$ with others.

4.3.3. Add to Q any intersections of h with its immediate upper half hyperbola and its immediate lower half hyperbola on L.

4.3.4. Mark $\varphi(q_t)$ as a Voronoi vertex incident to $h^{\pm}(p_i, p_j)$, $h^{\pm}(p_j, p_k)$ and h.

Step 5. Report all the half hyperbolas ever listed on L, all the Voronoi vertices marked in Step 4.3.4 and the incidence relations among them.

Steps 1, 2 and 3 are for initialization. In the example in Figure 4.5.3, Q is set to be $\{p_1, p_2, ..., p_5\}$ in Step 1, $p_i = p_1$ is chosen and Q is changed to $\{p_2, ..., p_5\}$ in Step 2, and L is set to be

$$L = (\varphi(V(p_1)))$$

in Step 3. Steps 2 and 3 correspond to the event that the sweepline hits the leftmost generator (the vertical line s_1 in the figure). The list $L = (\varphi(V(p_1)))$ implies that, if one moves along the sweepline from the bottom upward, one always lies in the Voronoi region for p_1.

Step 4 is the main part of the algorithm. In Step 4.1 we choose a point at which the next event occurs. In the example shown in Figure 4.5.3, the algorithm goes in the following way. We have $Q = \{p_2, ..., p_5\}$ at the end of Step 3, and hence at Step 4.1 point $w = p_2$ is chosen from Q and Q is changed to $Q = \{p_3, p_4, p_5\}$. This corresponds to the event that the sweepline hits p_2, the sweepline being at s_2 in Figure 4.5.3(b).

According to the type of event, Step 4.2 or 4.3 is done; Step 4.2 is for an event of the first type and Step 4.3 is for an event of the second type. Since $w = p_2$ is a generator, Step 4.2 is done. First, we find the region $\varphi(V(p_j))$ containing the new generator p_i in Step 4.2.1; next we create a new Voronoi region $\varphi(V(p_i))$ and a new Voronoi edge consisting of $h^-(p_i, p_j)$ and $h^+(p_j, p_i)$ in Step 4.2.2; and finally we insert candidates of Voronoi vertices which may be end points of $h^-(p_i, p_j)$ and $h^+(p_i, p_j)$. In our example, in Step 4.2.1 $\varphi(V(p_j))$ is recognized as $\varphi(V(p_1))$, and in Step 4.2.2 L is changed to

$$L = (\varphi(V(p_1)), h^-(p_1, p_2), \varphi(V(p_2)), h^+(p_2, p_1), \varphi(V(p_1))).$$

The list L at this point implies that, if one moves along the sweepline from the bottom upward, one traverses the region $\varphi(V(p_1))$, crosses the edge $h^-(p_1, p_2)$, traverses the region $\varphi(V(p_2))$, crosses the edge $h^+(p_2, p_1)$ and traverses the region $\varphi(V(p_1))$ in this order. This is the situation we have immediately after the sweepline hits p_2. Since there is no half hyperbola between $h^-(p_1, p_2)$ or above $h^+(p_2, p_1)$, nothing is done in Step 4.2.3.

In the second repeat of Step 4.1, $w = p_3$ is chosen and Q is changed to $Q = \{p_4, p_5\}$. This corresponds to the event that the sweepline hits p_3 (the line s_3 in Figure 4.5.3(b)). Hence Step 4.2 is done again.

The generator p_3 lies above the upper hyperbola $h^+(p_1, p_2)$, and hence in Step 4.2.2 the second appearance of $\varphi(V(p_1))$ in L is replaced by the sublist $(\varphi(V(p_1)), h^-(p_1, p_3), \varphi(V(p_3)), h^+(p_3, p_1), \varphi(V(p_1)))$, which changes L to

$$L = (\varphi(V(p_1)), h^-(p_1, p_2), \varphi(V(p_2)), h^+(p_2, p_1), \varphi(V(p_1)), h^-(p_1, p_3), \varphi(V(p_3)), h^+(p_3, p_1), \varphi(V(p_1))).$$

The new list L represents the situation obtained immediately after the sweepline hits p_3. In L there is a half hyperbola $h^+(p_2, p_1)$, which is immediately below $h^-(p_1, p_3)$, and consequently in Step 4.2.3 the point $\varphi(q_1)$ of intersection of these two half hyperbolas is added to Q; thus Q is changed to $Q = \{\varphi(q_1), p_4, p_5\}$.

In the third repeat of Step 4.1, $w = \varphi(q_1)$ is chosen and Q is changed to $Q = \{p_4, p_5\}$. This corresponds to the event that the sweepline comes to s_4 in Figure 4.5.3(b). Since $w = \varphi(q_1)$ is a Voronoi vertex, Steps 4.3.1–4.3.4 are done; throughout this step we should take the same superscript + or − for all appearances of $h^{\pm}(p_i, p_j)$ as we should do for all appearances of $h^{\pm}(p_j, p_k)$. In Step 4.3.1 the old region $\varphi(V(p_j))$ and its two boundaries $h^{\pm}(p_i, p_j)$ and $h^{\pm}(p_j, p_k)$ are deleted from L, and a new Voronoi edge h is inserted into L, where h is either $h^-(p_i, p_k)$ or $h^+(p_i, p_k)$, depending on the configuration around. At this point, $\varphi(q_t)$ is recognized as the end point of the edges $h^{\pm}(p_i, p_j)$ and $h^{\pm}(p_j, p_k)$. All the other candidates for the end points of them are deleted from Q in Step 4.3.2, and the candidates for the end point of the new edge h are added to Q in Step 4.3.3. In Step 4.3.4 the incidence structure around $\varphi(q_t)$ is recorded.

In our example, in Step 4.3.1 the subsequence $(h^+(p_2, p_1), \varphi(V(p_1)), h^-(p_1, p_3))$ is replaced by $h^-(p_2, p_3)$, changing L to

$$L = (\varphi(V(p_1)), h^-(p_1, p_2), \varphi(V(p_2)), h^-(p_2, p_3), \varphi(V(p_3)), h^+(p_3, p_1), \varphi(V(p_1))).$$

Nothing is done in Step 4.3.2 because Q contains no point of intersection of two half hyperbolas. Nothing is done in Step 4.3.3, either, because the half hyperbola $h^-(p_2, p_3)$ has no point of intersection with $h^-(p_1, p_2)$ or with $h^+(p_3, p_1)$. In Step 4.3.4 the incidence relations among the Voronoi vertices, the Voronoi edges and the Voronoi regions around $\varphi(q_1)$ are registered.

The algorithm treats the other events one by one in a similar manner until Q becomes empty. The Voronoi diagram $\varphi(\mathcal{V}(P))$ is obtained by collecting all the half hyperbolas ever listed in L, all the Voronoi vertices marked in Step 4.3.4 and the incidence relations among them.

We can understand the correctness of the algorithm by noting that (i) the generator p_i, other than the leftmost one, lies at the leftmost point of the Voronoi region $\varphi(V(p_i))$, (ii) at each Voronoi vertex in $\varphi(\mathcal{V})$ exactly one edge emanates to the right and the other two edges to the left, and that (iii) for each Voronoi vertex in $\varphi(\mathcal{V})$, the two edges emanating to the left become neighbours of each other on L at some time during the sweep (and hence the points inserted in Steps 4.2.3 and 4.3.3 include all the Voronoi vertices). See Fortune (1986, 1987) for details of the proof.

To execute the procedure efficiently, we need some techniques in data structure. The basic operations on Q are the insertion of points and the deletion of the minimum point with respect to the x coordinate. A storage requiring these operations is called a *priority queue*, and can be implemented, for instance, by a data structure called a heap. Indeed, using a heap, we can carry out each of the above operations in $O(\log n)$ time, where n is the number of points stored in Q (Aho *et al.*, 1974). In Step 4.3.2 we also have to delete from Q some points that are not necessarily minimum in the x coordinate. However, these points can be accessed in constant time if we store and maintain pointers at the associated half hyperbolas, and can be deleted in $O(\log n)$ time by the same manner as the insertion and the minimum deletion.

The other important storage L is an alternating list of regions and half hyperbolas along the sweepline. The intersections of the regions with the sweepline give intervals on the sweepline, and the intersections of the half hyperbolas with the sweepline give boundary points of the intervals. The interval corresponding to each region $\varphi(V(p_i))$ changes as the sweepline moves. The basic operations in L are to search for a relevant interval in Step 4.2.1 and the deletion and insertion of substrings in Steps 4.2.2 and 4.3.1. A storage requiring these operations is called a *dictionary*, and is implemented efficiently by a data structure such as a 2-3 tree or an AVL tree (Aho *et al.*, 1974). With such a data structure, we can execute each operation in $O(\log n)$ time.

Employing these techniques in data structure, we can execute Algorithm 4.8 in $O(n \log n)$ time. Indeed, Step 1 requires $O(n \log n)$ time, Step 2 $O(\log n)$ time and Step 3 constant time. Note that the total number of elements ever listed on L is of $O(n)$; this is because the number of Voronoi edges and Voronoi vertices are both of $O(n)$. The total number of elements ever stored in Q is also of $O(n)$, because there are only $O(n)$ pairs of

Voronoi edges that ever become neighbours of each other on L and hence there are only O(n) candidates for Voronoi vertices. Therefore, Steps 4.1, 4.2 and 4.3 require O(log n) time, and they are repeated O(n) times in Step 4. Thus, the total time complexity is of O(n log n).

The plane sweep technique has been applied to many geometric problems, such as point location, the extraction of intersections between line segments or between polygons and the construction of an arrangement consisting of lines (Preparata and Shamos, 1985). Fortune (1986) was the first to apply this technique to the construction of Voronoi diagrams and obtained an algorithm equivalent to Algorithm 4.8. The plane sweep technique was also applied to a Voronoi diagram for line segments (Fortune, 1986, 1987), to an additively weighted Voronoi diagram (Fortune, 1986, 1987) and to a Voronoi diagram on a cone (Dehne and Klein, 1987).

4.6 PRACTICAL TECHNIQUES FOR IMPLEMENTING THE ALGORITHMS

4.6.1 Inconsistency caused by numerical errors

So far we have assumed that neither degeneracy nor numerical error takes place in computation. However, algorithms designed in such an artificial world are not necessarily valid in practice, because in computers numerical values are treated only in finite-precision arithmetic. In degenerate or nearly degenerate cases, numerical error sometimes causes inconsistency in topological structures, and thus makes algorithms invalid.

An example of topological inconsistency we often encounter in the incremental method is shown in Figure 4.6.1. The boundary-growing procedure employed in the incremental method is based on the property that the boundary of a Voronoi polygon forms a closed sequence of segments of perpendicular bisectors. However, if the bisectors pass near to a Voronoi vertex, the topological relations between the boundaries and the Voronoi vertex are sometimes misjudged, resulting in a sequence in which the end point does not coincide with the start point, as shown in Figure 4.6.1. If such an inconsistency happens, the incremental algorithm cannot accomplish the task.

In this section we remove Assumptions A4.1–A4.4 which we gave for the sake of simplicity, and consider how to translate the incremental method into computer programs that work for any input data, degenerate or non-degenerate, in finite-precision computation. We present two quite different principles. The first one, presented in Section 4.6.2, is for constructing an error-free closed world in which all the topological structures are judged correctly. In this world, degeneracy is also detected exactly, so that we have another problem, i.e. the problem of preparing complicated branches of exceptional procedure for treating degenerate cases. To circumvent this problem, we also present a method for avoiding degeneracy by symbolic

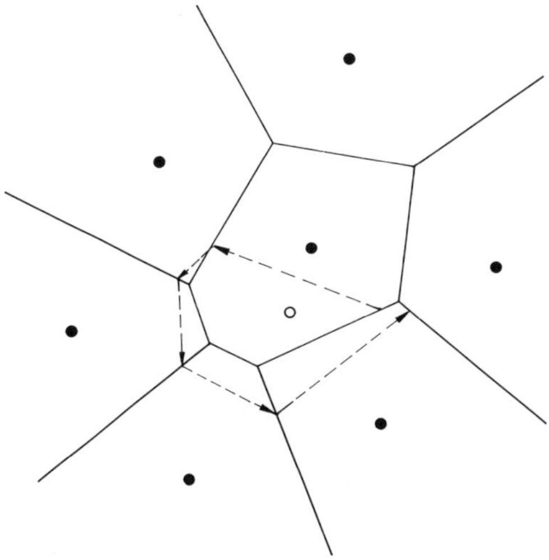

Figure 4.6.1 Unclosed boundary caused by numerical errors.

perturbation. In Section 4.6.3, on the other hand, we consider a world with numerical errors and present the second principle, in which we avoid inconsistency by placing a higher priority on topological structures than on numerical values. This principle automatically enables us to avoid degeneracy because it is impossible to detect degeneracy by computation with errors.

4.6.2 Construction of an error-free world

The first principle is to compute numerical values so precisely that they are good enough always to judge topological structures correctly.

Suppose that we fix the set $P = \{p_1, \ldots, p_n\}$ of generators. Let (x_i, y_i) denote the coordinates of the generator p_i. Let us assume that the coordinates are represented by integers whose absolute values are not greater than a certain fixed integer K:

$$-K \leq x_i, \quad y_i \leq K \quad \text{for } i = 1, 2, \ldots, n. \tag{4.6.1}$$

This assumption is not unrealistic, because in a digital computer the coordinates are usually represented by a fixed number of bits, so that they can be transformed into integers if a sufficiently large-scale factor is multiplied.

Suppose that we have constructed a Voronoi diagram $\mathcal{V}_{l-1}$ for the first $l-1$ generators $p_1, p_2, \ldots, p_{l-1}$ and now want to add a new generator p_l. For this purpose we need to find Voronoi vertices of $\mathcal{V}_{l-1}$ that should be contained in the Voronoi polygon of the new generator.

Let q_{ijk} denote the Voronoi vertex incident to three Voronoi polygons $V(p_i)$, $V(p_j)$ and $V(p_k)$ counterclockwise in this order. For three generators

p_i, p_j, p_k and a point $p = (x, y)$, let $H(p_i, p_j, p_k, p)$ be as defined by equation (2.4.11). The equation $H(p_i, p_j, p_k, p) = 0$ represents the circle that passes through p_i, p_j and p_k, and this circle contains p if and only if $H(p_i, p_j, p_k, p) < 0$. Recall that q_{ijk} belongs to the Voronoi polygon of the new generator p_l if and only if p_l is contained in this circle, which means that q_{ijk} belongs to $V(p_l)$ if and only if $H(p_i, p_j, p_k, p_l) < 0$.

Hence, what we have to do to find the points $w_1, w_2, \ldots, w_m$ in Step 4.3 of Algorithm 4.4 is to find Voronoi edges having $H(p_i, p_j, p_k, p_l)$ with opposite signs at their two end points.

Note that $H(p_i, p_j, p_k, p_l)$ is an integer, so that it can be computed correctly unless it overflows memory. From the Hadamard inequality (which says that the absolute value of the determinant of a matrix is not greater than the multiplication of the lengths of all the column vectors; see, for instance, Greub, 1975) and assumption (4.6.1), we obtain

$$
\begin{aligned}
&|H(p_i, p_j, p_k, p_l)| && (4.6.2)\\
&\leq \sqrt{1^2+1^2+1^2+1^2}\ \sqrt{x_i^2+x_j^2+x_k^2+x_l^2}\ \sqrt{y_i^2+y_j^2+y_k^2+y_l^2}\\
&\times \sqrt{(x_i^2+y_i^2)^2+(x_j^2+y_j^2)^2+(x_k^2+y_k^2)^2+(x_l^2+y_l^2)^2}\\
&\leq 2 \times 2K \times 2K \times 4K^2\\
&= 32K^4. && (4.6.2)
\end{aligned}
$$

If we compute the value of $H(p_i, p_j, p_k, p_l)$ according to the Laplace expansion of the right-hand side of (4.6.2), the absolute value of any number that appears in the course of computation does not exceed the bound $32K^4$, either. Hence the integer $H(p_i, p_j, p_k, p_l)$ can be computed without overflow if the computation is carried out so precisely that a number as large as $32K^4$ is representable.

Let $N = \lceil \log_2(K+1) \rceil$, where $\lceil x \rceil$ is the smallest integer not less than x. Then each coordinate can be represented by $N+1$ bits, N bits for the absolute value and 1 bit for the sign. Any integer between $-32K^4$ and $32K^4$ can be represented by $4N+6$ bits, $4N+5$ bits for the absolute value and 1 bit for the sign. Hence, using $4N+6$ bits for computation, we can always correctly determine the Voronoi vertices to be deleted in the addition of the new generator. Thus we can avoid misjudgment in the topological structure of the Voronoi diagram.

For example, if we use 64 bits (2 words in an ordinary computer) to represent each integer appearing in the computation (i.e. 1 bit for the sign and 63 bits for the absolute value), we can represent integers as large as $M = 2^{63}-1$. In this case, consequently, we can choose $K = 23\,170$, because

$$M = 2^{63}-1 = 2^5 \times 2^{58}-1 = 32(2^{14.5})^4-1 > 32\,(23\,170)^4,$$

which seems to give sufficient resolution for ordinary purposes in point pattern analysis.

By the above method we can indeed avoid misjudgment in topological structures, but we still have another problem, i.e. the treatment of degenerate cases. Degeneracy takes place when $H(p_i, p_j, p_k, p_l) = 0$. In this case the boundary of the Voronoi polygon of the new generator passes through the Voronoi vertex of $\mathscr{V}_{l-1}$, giving a Voronoi vertex with four or more edges. Such degeneracy usually requires exceptional branches of processing, which makes construction and maintenance of a computer program difficult.

Fortunately, however, we can avoid degeneracy by a very simple trick; intuitively, we 'shrink' the Voronoi polygon of the new generator infinitesimally and thus avoid creating degenerate Voronoi vertices.

From a topological point of view, the Voronoi diagram can be represented by an augmented geometric graph, a planar graph embedded in the plane; recall Section 4.2. Let G_i denote the augmented geometric graph associated with the Voronoi diagram $\mathscr{V}_i$ for the first i generators. G_i represents the topological structure of $\mathscr{V}_i$. If no degeneracy takes place, the degree of any vertex (i.e. the number of edges incident to the vertex) in G_i is three. In the case of degeneracy, on the other hand, G_i has vertices with degree four or more.

To represent G_i, we adopt the convention that a vertex with degree four or more, if any, is replaced with a microstructure consisting of degree-three vertices and 'length-zero' edges. For example, if four generators p_i, p_j, p_k and p_l are on a common circle and give a degenerate Voronoi vertex with degree four, as shown in Figure 4.6.2(a), we generate two degree-three vertices and one length-zero edge, as shown in (b) or (c); we can choose either of the two alternative structures. This convention implies that we need not generate a vertex of degree more than three even if degeneracy takes place.

Our idea to avoid exceptional processing for degeneracy is just to treat

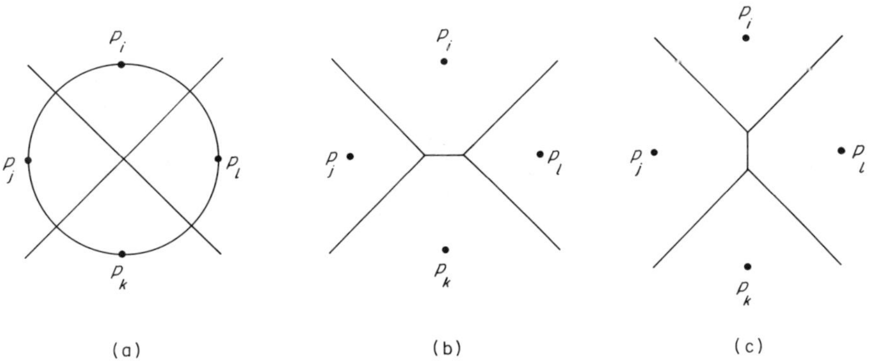

Figure 4.6.2 Degeneracy and perturbation: (a) four generators on a common circle; (b) introduction of a length-zero edge; (c) another way of introducing a length-zero edge.

the case where $H(p_i, p_j, p_k, p_l) = 0$ as if $H(p_i, p_j, p_k, p_l) > 0$. That is, we judge that q_{ijk} is inside $V(p_l)$ if $H(p_i, p_j, p_k, p_l) < 0$, and outside if $H(p_i, p_j, p_k, p_l) \geq 0$; note that the latter inequality allows the equality. This strategy intuitively means that each time degeneracy takes place, we shrink the Voronoi polygon of the new generator only slightly so as not to disturb other parts of the Voronoi diagram. Hence every Voronoi vertex of G_{l-1} is judged either inside $V(p_l)$ or outside; no vertex is 'on' the boundary of $V(p_l)$.

Figure 4.6.3 shows an example of how our method avoids degeneracy. Here we assume that eight generators $p_1, p_2, \ldots, p_8$ lie on a common circle, so that in an ordinary sense a Voronoi vertex of degree eight will appear, as shown in (a). Suppose that the generators $p_1, p_2, \ldots, p_8$ are added in this order. Then the resultant topological structure will be as shown in (b), where the degenerate vertex is replaced by a tree structure consisting of six vertices and five length-zero edges. This structure is the result of our strategy. This may be understood more easily when we consider the Delaunay triangulation, dual to the Voronoi diagram, shown in (c). Since the primal diagram (b) consists of degree-three vertices, the dual diagram (c) is composed of triangles. These triangles are generated in such a way that every time a new generator is added, it is connected to old generators to form a triangle. This is the interpretation of our shrink strategy in terms of the dual graph.

This strategy can avoid degeneracy because, if G_{l-1} is a geometric graph with degree three, then so is G_l. Moreover, the strategy does not disturb our original purpose in the sense that, when length-zero edges are contracted, the resulting diagram gives the correct topological structure of the Voronoi diagram.

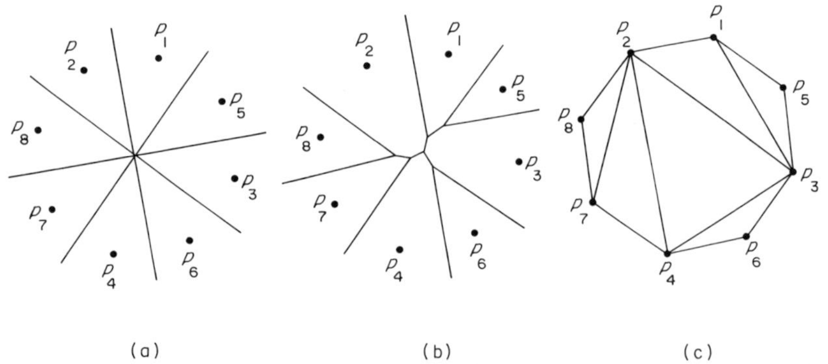

Figure 4.6.3 Degeneracy avoidance: (a) degenerate vertex of degree eight; (b) result of avoiding degeneracy; (c) the corresponding Delaunay triangulation.

Summing up the above consideration, we get the next principle for implementation.

Implementation Principle 4.1 (Error-free world) Choose and fix the upper bound K of the absolute values of the coordinates of the generators, and put $N = \lceil \log_2(K+1) \rceil$. Represent the x and y coordinates of the generators by $N+1$ bit integers (N bits for the absolute value and 1 bit for the sign), and compute the values of $H(p_i, p_j, p_k, p_l)$ in $4N+6$ bits. Judge that the Voronoi vertex q_{ijk} is inside the Voronoi polygon of the new generator p_l if $H(p_i, p_j, p_k, p_l) < 0$, and outside if $H(p_i, p_j, p_k, p_l) \geq 0$.

This principle is based on the recognition that 'if the input data are represented by a finite number of bits, the sign of the result of an algebraic computation over the data can be determined correctly in a certain finite-precision arithmetic'. Recently this principle has gradually been recognized and utilized in several fields of computer sciences such as linear programming (Khachian, 1980), computational algebra (Lenstra, 1984) and solid modelling (Sugihara and Iri, 1989a).

The shrink strategy presented here is a kind of 'symbolic perturbation'. Symbolic perturbation is a technique for avoiding degeneracy originally used in linear programming by Bland (1977). Recently the same idea was introduced in computational geometry, and general frameworks are discussed by Edelsbrunner and Mücke (1987) and Yap (1988), by which we can obtain a non-degenerate configuration that is topologically equivalent to a configuration resulting from some perturbation of the input numerical data. The strategy presented in this section can be considered as a symbolic perturbation in the context of the Voronoi diagram with the Laguerre metric (Sugihara, 1992b).

4.6.3 Topology-oriented approach

We next consider what we can do if we are not allowed to compute with sufficiently high precision. In poor precision arithmetic, numerical errors are inevitable, so that judgement based on numerical values cannot be relied upon; we can rely only on combinatorial computation. Hence it is natural to place higher priority on topological structures than on numerical values. In this subsection we treat topological structures as higher-priority information than numerical values and thus make the incremental method robust against numerical errors.

As we saw in Section 4.2, the topological structure of the Voronoi diagram $\mathcal{V}_l$ can be considered as the augmented geometric graph G_l. From a topological point of view, therefore, the main task in the incremental method is to convert G_{l-1} to G_l for $l = 4, 5, \ldots, n$. We first characterize this task in terms of combinatorial computation only.

For $i = 3, 4, \ldots, n$, let us define $\boldsymbol{A}_i$ as the set of all augmented geometric graphs G satisfying the following conditions.

(C4.1) The degree of any vertex in G is exactly three.

(C4.2) G divides the plane into $i+1$ regions; let us call these regions *cells*.

(C4.3) Every cell except for the outermost infinite one is simply connected, i.e. it is connected and does not have a hole.

(C4.4) Two cells share at most one common edge.

(C4.5) The outermost circuit of G (i.e. the boundary of the outermost infinite cell) consists of exactly three edges and exactly three vertices.

If the Voronoi diagram $\mathcal{V}_i$ is not degenerate, the augmented geometric graph G_i associated with $\mathcal{V}_i$ belongs to $\mathbf{A}_i$. Indeed, (C4.1) holds first because $\mathcal{V}_i$ is non-degenerate so that all the Voronoi vertices are of degree three, and secondly because the infinite Voronoi edges were connected to the outermost closed curve (recall Section 4.2). Condition (C4.2) holds because $\mathcal{V}_i$ divides the plane into i Voronoi polygons and we introduced the outermost closed curve to surround the diagram. (C4.3) and (C4.4) hold because Voronoi polygons are convex. To see that (C4.5) holds, recall that the additional three generators p_1, p_2 and p_3 were chosen in such a way that they formed a triangle containing all the original generators $p_4, p_5, \ldots, p_n$. Hence as shown in Figures 4.3.2(a) and (b), only p_1, p_2 and p_3 have infinite Voronoi polygons; the three cells corresponding to these three Voronoi polygons only are incident to the outermost circuit of G. Thus, G_i satisfies (C4.5).

Let us represent the augmented geometric graph G_i by the triple $G_i = (W_i, E_i, C_i)$, where W_i, E_i and C_i are the sets of vertices, edges and cells, respectively, of G_i. For any subset $T \subset W_i$, let $G_i(T)$ denote the subgraph of G_i consisting of vertices in T and those edges in E_i that connect two vertices in T. For any cell $c \in C_i$, let $W_i(c)$ denote the set of vertices on the boundary of the cell c. In the addition of a new generator p_l, the incremental method changes the topological structure of the Voronoi diagram from $G_{l-1} = (W_{l-1}, E_{l-1}, C_{l-1})$ to $G_l = (W_l, E_l, C_l)$. As shown by an example in Figure 4.6.4(a), this task can be done by first choosing a set T of vertices to be deleted (the vertices represented by filled circles in the figure), then creating new vertices (represented by unfilled circles) on all edges connecting a vertex in T and a vertex not in T, and new edges (represented by broken lines) forming a circuit enclosing T, and finally deleting the substructure inside the circuit.

Let $T(\subset W_{l-1})$ be the set of vertices that is deleted in the change from G_{l-1} to G_l. Then T should satisfy the following conditions.

(C4.6) T is non-empty.

(C4.7) T does not contain a vertex on the outermost circuit of G_{l-1}.

(C4.8) $G_{l-1}(T)$ is a tree, i.e. a connected acyclic graph.

(C4.9) For any $c \in C_{l-1}$, $G_{l-1}(T \cap W_{l-1}(c))$ is connected.

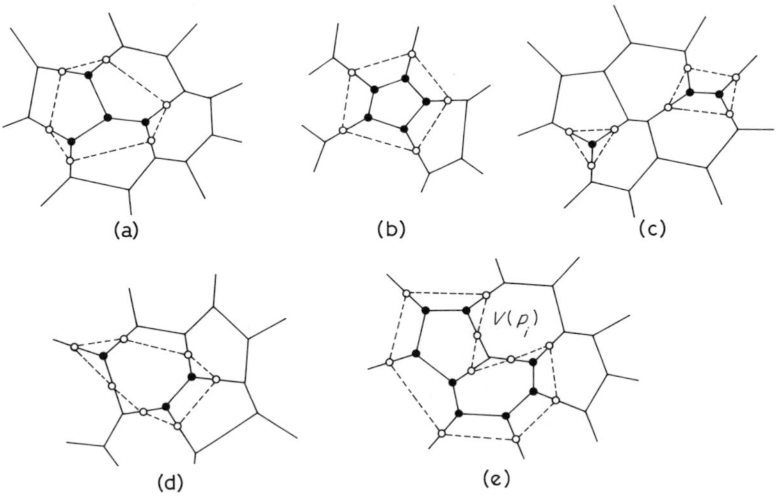

Figure 4.6.4 Topological aspect of the incremental method: (a) deletion of a tree; (b) deletion of a cycle; (c) deletion of two or more connected components; (d) another way of deleting two or more components; (e) deletion of a subgraph that violates condition (C4.9).

Condition (C4.6) is obvious because the new generator p_l should have its own non-empty Voronoi polygon. Condition (C4.7) holds because we introduced the three additional generators and hence the Voronoi polygon of p_l ($l = 4, 5, \ldots, n$) is finite. If $G_{l-1}(T)$ has a circuit, the Voronoi region of an old generator is removed entirely, as shown in Figure 4.6.4(b), which contradicts the basic property that every generator has its own non-empty region. If $G_{l-1}(T)$ is disconnected, then either the Voronoi region of p_l has two or more components, as shown in (c), or an old Voronoi region has two or more components, as shown in (d). Hence T satisfies (C4.8). If $G_{l-1}(T \cap W_{l-1}(c))$ is not connected as in (e) (where c corresponds to the boundary of Voronoi region $V(p_i)$), then the Voronoi regions $V(p_i)$ and $V(p_l)$ have two or more edges in common, which contradicts condition (C4.4). Thus, (C4.9) is satisfied.

Note that conditions (C4.6)–(C4.9) are written in terms of combinatorial properties only. Hence we can check the conditions by combinatorial computation; we need not employ numerical computation. Moreover, it is easy to see that, if G_{l-1} belongs to $\mathbf{A}_{l-1}$ and T satisfies (C4.6)–(C4.9), the augmented geometric graph G_l belongs to $\mathbf{A}_l$.

However, there is freedom in the choice of the subset T. Hence we employ numerical computation to choose the most promising subset T, and for this purpose only. Thus we obtain the next algorithm for changing G_{l-1} to G_l, which should be used in place of the boundary-growing procedure.

Algorithm 4.9 (Tree expansion and deletion) Input: Generators $p_1, p_2, \ldots, p_l$ and geometric graph $G_{l-1}(\in \mathbf{A}_{l-1})$.
Output: Geometric graph G_l $(\in \mathbf{A}_l)$.
Procedure:
Step 1. Find the generator p_i $(1 \leq i \leq l-1)$ such that $d(p_i, p_l)$ is minimum.
Step 2. Among the Voronoi vertices q_{ijk}'s on the boundary of $V(p_i)$, find the one that gives the smallest value of $H(p_i, p_j, p_k, p_l)$. Let T be the set consisting of this Voronoi vertex alone.
Step 3. Repeat 3.1 until T cannot be augmented any more.
3.1. For each Voronoi vertex q_{ijk} that is connected by a Voronoi edge to an element of T, add q_{ijk} to T both if $H(p_i, p_j, p_k, p_l) < 0$ and if the resultant T satisfies conditions (C4.7)–(C4.9).
Step 4. For every edge connecting a vertex in T with a vertex not in T, create a new vertex on the edge and thus divide the edge into two edges.
Step 5. Create new edges connecting the vertices created in Step 4 in such a way that the new edges form a circuit enclosing the vertices in T.
Step 6. Remove the vertices in T and the edges incident to them (and consider the interior of the circuit as the Voronoi polygon of p_l), and let the resultant geometric graph be G_l.

This algorithm corresponds to Steps 4.1–4.4 in Algorithm 4.4. More specifically, Step 1 in this algorithm is equivalent to Step 4.1 in Algorithm 4.4, Steps 2, 3 and 4 correspond to Steps 4.2 and 4.3, and Steps 5 and 6 correspond to Step 4.4. Hence if there is no numerical error, and if G_{l-1} corresponds to the Voronoi diagram $\mathcal{V}_{l-1}$, the output G_l correctly corresponds to the Voronoi diagram $\mathcal{V}_l$.

The main difference is that in this algorithm the initial vertex found in Step 2 is used as a 'seed' from which a tree is expanded, while in the boundary-growing procedure the boundary enclosing this tree is traced. This difference is essential, because even if computed values of $H(p_i, p_j, p_k, p_l)$ contain numerical errors, the subset T obtained in the algorithm satisfies (C4.6)–(C4.9). We need not worry abut inconsistency; the algorithm always carries out its task and the output G_l is consistent in the sense that it belongs to $\mathbf{A}_l$.

Note that degeneracy can be avoided automatically in the algorithm. We judge the Voronoi vertex q_{ijk} as being inside the Voronoi polygon of the new generator if and only if $H(p_i, p_j, p_k, p_l) < 0$. This is equivalent to the shrink strategy presented in the previous subsection.

The conditions (C4.8) and (C4.9) can be checked efficiently in the following manner. We assign three labels, 'in', 'out' and 'undecided', to the Voronoi vertices; 'in' is assigned to the Voronoi vertices that are added to T, 'out' to the Voronoi vertices that are judged not to be added to T, and 'undecided' to the Voronoi vertices that are not yet judged. Also, we assign two labels 'incident' and 'non-incident' to the Voronoi polygons; 'incident' to the Voronoi polygons whose boundary has a vertex in T, and 'non-incident' to the other Voronoi polygons. In Step 3.1, q_{ijk} is chosen among the Voronoi vertices that are labelled 'undecided' and that are adjacent to a vertex in T. Then, $G_{l-1}(T \cup \{q_{ijk}\})$ is a tree if and only if

(C4.10) q_{ijk} is not adjacent to two or more 'in' Voronoi vertices.

For any cell c of G_{l-1}, the subgraph $G_{l-1}((T \cup \{q_{ijk}\}) \cap W_{l-1}(c))$ is connected if and only if

(C4.11) for any 'incident' Voronoi polygon having q_{ijk} on its boundary, q_{ijk} is adjacent to an 'in' Voronoi vertex on this boundary.

Hence conditions (C4.8) and (C4.9) can be examined locally by checking the labels of the Voronoi polygons incident to q_{ijk} and those of the Voronoi vertices adjacent to q_{ijk}.

For the initial geometric graph G_3, we assign 'out' labels to the three degree-one vertices, and these labels are permanently fixed. This is because the Voronoi polygon of a new generator is always finite, so that these three vertices should not belong to T. At the start of Algorithm 4.9 all the other Voronoi vertices are labelled 'undecided', and all the Voronoi polygons are labelled 'non-incident'. Each time a Voronoi vertex is added to T, its label is changed from 'undecided' to 'in' and the labels of the Voronoi polygons incident to the Voronoi vertex, if they are 'non-incident', are changed to 'incident'. Similarly, each time a Voronoi vertex is judged not to be added to T, its label is changed from 'undecided' to 'out'. In Algorithm 4.9 the Voronoi vertices and the Voronoi polygons whose labels are changed are listed, and at the end of the algorithm these labels are cleared (that is, 'in' and 'out' are changed to 'undecided', and 'incident' is changed to 'non-incident'). In this way Algorithm 4.9 can be executed in time proportional to the size of the substructure that is deleted for the addition of the new generator.

Thus, we obtain the next principle for implementing the incremental method.

Implementation Principle 4.2 (Topology-oriented approach) We concentrate our attention on the construction of the topological structure G_n of the Voronoi diagram (that is, we replace $\mathcal{V}$, $\mathcal{V}_3$, $\mathcal{V}_{l-1}$, $\mathcal{V}_l$ and $\mathcal{V}_n$ in Algorithm 4.4 with G, G_3, G_{l-1}, G_l and G_n, respectively), and replace Steps 4.1–4.4 in Algorithm 4.4 with Algorithm 4.9.

Once G_n is obtained, it is easy to draw the Voronoi diagram itself, because a Voronoi vertex is the centre of the circle passing through the three surrounding generators, and a Voronoi edge is the perpendicular bisector of the two side generators.

An example of the behaviour of the computer program constructed on the basis of this principle is shown in Figure 4.6.5, where (a) is the output

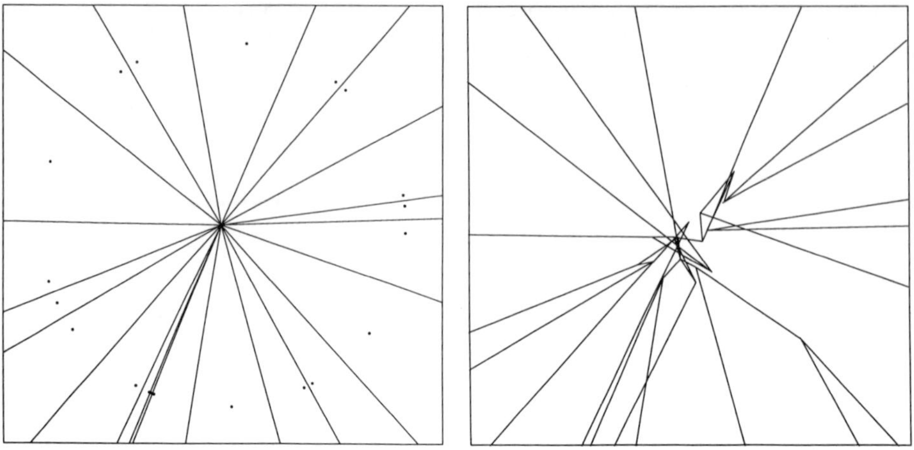

(a) (b)

Figure 4.6.5 Behaviour of a computer program based on the topology-oriented approach: (a) output of the program for 20 generators; (b) central portion of (a) magnified by 10^5.

constructed in single-precision floating point arithmetic for 20 generators placed at random on a common circle, and (b) represents the central portion of the same output magnified by 10^5. The diagram in (a) seems a correct Voronoi diagram but, as (b) shows, the central portion of the output is far from the correct Voronoi diagram. This set of generators gives high degeneracy, and the criss-cross microstructure of this kind seems natural in single precision arithmetic. It should be noted that even for such a degenerate set of generators, the program carried out its task and gave an output, and that the output is topologically consistent in the sense that it belongs to $\mathbf{A}_n$. A conventional computer program usually fails in processing such a degenerate set of generators.

We place higher priority on topological consistency than on numerical values, but this does not mean that numerical computation is less important. In our algorithm the most important numerical computation is that of $H(p_i, p_j, p_k, p_l)$ defined by (2.4.11). $H(p_i, p_j, p_k, p_l)$ can be rewritten in the following way:

$$H(p_i, p_j, p_k, p_l) = J^2_{ijk}(x_l - x_k) - J^3_{ijk}(y_l - y_k) + \tfrac{1}{2} J^4_{ijk}((x_l - x_k)^2 + (y_l - y_k)^2), \tag{4.6.3}$$

where

$$J^2_{ijk} = \begin{vmatrix} y_i - y_k & ((x_i - x_k)^2 + (y_i - y_k)^2)/2 \\ y_j - y_k & ((x_j - x_k)^2 + (y_j - y_k)^2)/2 \end{vmatrix}, \tag{4.6.4}$$

$$J^3_{ijk} = \begin{vmatrix} x_i - x_k & ((x_i - x_k)^2 + (y_i - y_k)^2)/2 \\ x_j - x_k & ((x_j - x_k)^2 + (y_j - y_k)^2)/2 \end{vmatrix}, \tag{4.6.5}$$

$$J^4_{ijk} = \begin{vmatrix} x_i - x_k & y_i - y_k \\ x_j - x_k & y_j - y_k \end{vmatrix}. \tag{4.6.6}$$

This expression shows that the value of $H(p_i, p_j, p_k, p_l)$ can be computed in terms of the coordinates of the generators with respect to the origin at p_k. The computation based on this expression is stable in a numerical sense, because the translation of the origin of the coordinate system to one of the relevant generators enables us to avoid dealing with unnecessarily large numbers. So this way of computation is recommended. Indeed, by this way of computation a Voronoi diagram for one million generators scattered at random in a unit square can be constructed in single precision arithmetic (Sugihara and Iri, 1989b).

The principle presented here was proposed by Sugihara and Iri (1988). A FORTRAN program called VORONOI2 was constructed on the basis of this principle, and is open for public use (Sugihara and Iri, 1989c).

4.7 ALGORITHMS FOR HIGHER-DIMENSIONAL VORONOI DIAGRAMS

A Voronoi diagram in a three- or higher-dimensional space is more complicated, at least in the sense that the size of the Voronoi diagram for n generators cannot be bounded by O(n). It is known that the number of k-faces of the Voronoi diagram for n generators in an m-dimensional space is of $O(n^{\min\{m+1-k, \lceil m/2 \rceil\}})$ for $0 \le k \le m$ (Klee, 1980).

In a three-dimensional space, for instance, the Voronoi diagram is a partition of the space into n convex polyhedra, but the total number of vertices of these polyhedra can be as many as $O(n^2)$. An example of such a situation is shown in Figure 4.7.1. Suppose that a half of n generators align on the x-axis and the other half align on the line passing through (0, 0,1) parallel to the y-axis. Recall that a Voronoi vertex, and hence its dual Delaunay tetrahedron, is created by each set of those four generators whose circumscribing sphere is an empty sphere. Hence, each set of two consecutive generators on one line and two consecutive generators on the other line create a Delaunay tetrahedron as shown in the figure, giving $O(n^2)$ tetrahedra in total. Thus, $O(n^2)$ is a lower bound of the worst-case time complexity of an algorithm for constructing a Voronoi diagram in three-dimensional space.

A simple and unifying approach to constructing an m-dimensional Voronoi diagram is to reduce the problem to a convex hull problem in an $(m+1)$-dimensional space.

As we saw in Chapter 2, we can establish a correspondence between the Voronoi diagram in the plane and a convex polyhedron in the three-

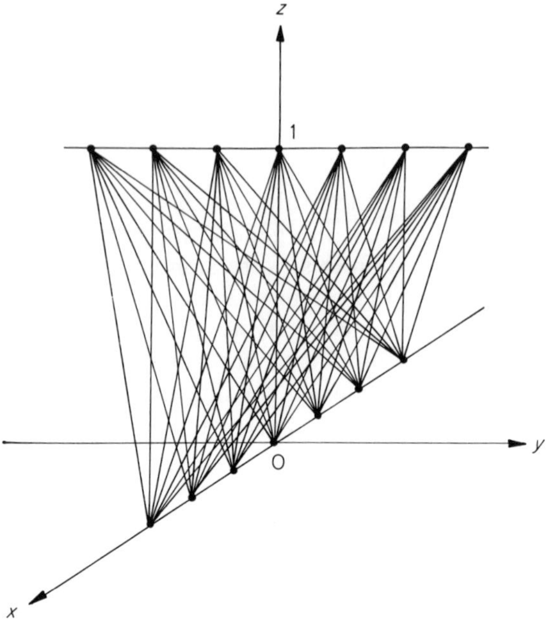

Figure 4.7.1 Three-dimensional Delaunay diagram with $O(n^2)$ tetrahedra.

dimensional space through the lift-up transformation. A similar correspondence is established in any dimension, and this fact can be used to construct the Voronoi diagram in the following way.

Let $P = \{p_1, p_2, ..., p_n\}$ be a set of n generators in an m-dimensional space, and $(x_{i1}, x_{i2}, ..., x_{im})$ be the Cartesian coordinates of p_i. We consider the configuration of $t = m+1$ generators. To avoid unnecessarily messy expressions, let us choose the first t generators $p_1, p_2, ..., p_t$ in P. These generators give a Voronoi vertex if and only if the hypersphere (a circle for $m = 2$, a sphere for $m = 3$, etc.) passing through them contains no other generators in its interior; actually the Voronoi vertex is the centre of this hypersphere.

The hypersphere S in an m-dimensional space that passes through $p_1, p_2, ..., p_t$ is represented by the equation

$$\begin{vmatrix} 1 & x_{11} & x_{12} \dots x_{1m} & x_{11}^2 + x_{12}^2 + \dots + x_{1m}^2 \\ 1 & x_{21} & x_{22} \dots x_{2m} & x_{21}^2 + x_{22}^2 + \dots + x_{2m}^2 \\ \vdots & \vdots & \vdots \ddots \vdots & \vdots \\ 1 & x_{t1} & x_{t2} \dots x_{tm} & x_{t1}^2 + x_{t2}^2 + \dots + x_{tm}^2 \\ 1 & x_1 & x_2 \dots x_m & x_1^2 + x_2^2 + \dots + x_m^2 \end{vmatrix} = 0, \qquad (4.7.1)$$

with the variable point $p = \{x_1, x_2, \ldots, x_m)$ (note that this equation is the extension of $H(p_i, p_j, p_k, p) = 0$, where $H(p_i, p_j, p_k, p)$ is defined in equation (2.4.11)). Hence the sign of the determinant in the left-hand side changes when the point p moves from inside S to outside.

Let p_i^* be the point in the t-dimensional space with coordinates

$$(x_{i1}, x_{i2}, \ldots, x_{im}, x_{i1}^2+x_{i2}^2+\ldots+x_{im}^2),$$

p^* be a variable point in the t-dimensional space with coordinates

$$(x_1, x_2, \ldots, x_m, x_1^2+x_2^2+\ldots+x_m^2),$$

and let $P^* = \{p_1^*, p_2^*, \ldots, p_n^*\}$. As illustrated in Figure 4.7.2, which is the case where $m = 2$, p_i^* is obtained by lifting up p_i in the positive direction of the tth coordinate axis until we reach the surface of the paraboloid of revolution

$$x_t = x_1^2+x_2^2+\ldots+x_m^2. \tag{4.7.2}$$

Let S^* denote the hyperplane in the t-dimensional space that passes through $p_1^*, p_2^*, \ldots, p_t^*$. Then, S coincides with the projection along the tth axis of the intersection of S^* with the surface defined by equation (4.7.2).

The determinant in the left-hand side of equation (4.7.1) can be considered as the signed volume of the t-dimensional parallelepiped defined by t edges

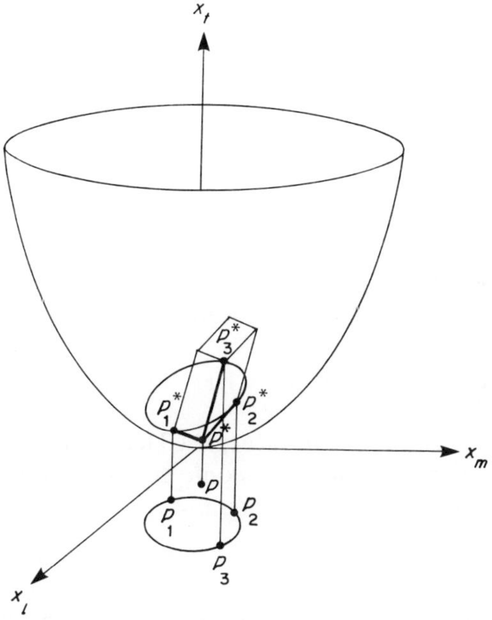

Figure 4.7.2 Lifting transformation that maps the plane to a hyperboloid of revolution.

$\overline{p^*p_1^*}, \overline{p^*p_2^*}, ..., \overline{p^*p_t^*}$. Since the t-dimensional region consisting of points above the surface defined by equation (4.7.2) is convex, we can see that the point p is in the interior of the sphere S in the m-dimensional space if and only if the point p^* is below the hyperplane S^*. This implies that the set of t generators p_1, p_2, ..., p_t gives a Voronoi vertex (that is, p_1, ..., p_t form the vertices of a Delaunay simplex) if and only if p_{t+1}^*, ..., p_n^* are all above the hyperplane S^*. Thus, all the Voronoi vertices (equivalently, all the Delaunay simplices) can be obtained by constructing the convex hull CH(P^*) of P^* in the t-dimensional space.

Suppose that p_1^*, p_2^*, ..., p_t^* form an m-dimensional face (m-face, in short) of the convex hull CH(P^*) in the (m+1)-dimensional space. There is a unique hyperplane passing through these points, and all the other points p_{t+1}^*, ..., p_n^* are on the same side of the hyperplane. We call the m-face a *lower* m-face if the other points are above the hyperplane, and an *upper* m-face otherwise. As we have seen, p_1^*, ..., p_t^* forming a lower m-face implies that the hypersphere passing through p_1, ..., p_t contains no other generators in its interior. Thus, the lower m-faces of CH(P^*) correspond to Voronoi vertices of $\mathcal{V}(P)$, or equivalently, the lower m-faces of CH(P^*) correspond to Delaunay simplices.

If p_1^*, ..., p_t^* form an upper m-face of CH(P^*), the corresponding hypersphere in the m-dimensional space contains all the other generators p_{t+1}, ..., p_n in its interior, which implies that p_1, ..., p_t form a Voronoi vertex of the farthest-point Voronoi diagram for P.

Algorithm 4.10 (m-dimensional Voronoi diagram) Input: set $P = \{p_1, p_2, ..., p_n\}$ of n generators in the m-dimensional space with the coordinates $(x_{im}, x_{i2}, ..., x_{im})$, $i = 1, 2, ..., n$.
Output: Delaunay tessellation spanning P.
Procedure:

Step 1. Create set $P^* = \{p_1^*, p_2^*, ..., p_n^*\}$ of points in a (m+1)-dimensional space, where the first m coordinate of p_i^* coincide with those of p_i and the (m+1)st coordinate is $x_{i1}^2+x_{i2}^2+...+x_{im}^2$.
Step 2. Construct the convex hull CH(P^*) of P^* in the (m+1)-dimensional space.
Step 3. Project all the lower m-faces of CH(P^*) in the direction parallel to the (m+1)st coordinate axis onto the original m-dimensional space, and return the resulting diagram.

From the time-complexity point of view, the critical step is Step 2. There are many algorithms for constructing convex hulls in higher dimensions. One popular method is gift wrapping (Chand and Kapur, 1970; Preparata and Shamos, 1985; Swart, 1985), where one supporting hyperplane is found initially and the neighbouring m-faces are found one by one just as we wrap the points using an elastic sheet. This method requires time proportional to $O(n^{\lfloor t/2 \rfloor +1})$ for n points in the t-dimensional space.

Another popular method is an incremental method (which is sometimes called a beneath–beyond method) (Preparata and Shamos, 1985; Seidel, 1986; Edelsbrunner, 1987), where a convex hull of a small number of points is constructed initially and then the other points are added one by one. When a new point is added, the m-faces of the previous convex hull are divided into those visible from the new point and the others, and the former m-faces are replaced with new m-faces containing the new point as a vertex. This method requires $O(n^{\lfloor (t+1)/2 \rfloor} + n \log n)$ time.

Other higher-dimensional convex hull algorithms include a shelling method, which runs in $O(n^{\lfloor t/2 \rfloor} \log n)$ time (Seidel, 1987), and a divide-and-conquer method, which runs in $O(n \log n + n^{\lfloor t/2 \rfloor})$ time (Buckley, 1988).

The farthest-point Voronoi diagram (actually the dual diagram of the farthest-point Voronoi diagram) can be obtained if we project all the upper m-faces, instead of the lower m-faces, in Step 3 of Algorithm 4.10.

The reduction of the Voronoi diagram construction in the m-dimensional space to convex hull construction in the $(m+1)$-dimensional space was pointed out by Brown (1979, 1980), where another transformation called inversion (which maps the m-dimensional plane to a hypersphere in the $(m+1)$-dimensional space) was used. That transformation was used by Aurenhammer and Edelsbrunner (1984), Edelsbrunner (1987) and Buckley (1988). The transformation presented in this section (i.e. lift-up to the paraboloid of revolution) was used by Edelsbrunner and Seidel (1986) and Edelsbrunner (1987). This transformation can be considered as an extremal case of the inversion in which the centre of inversion is taken as a point at infinity.

It should be noted that the lift-up transformation can be used also for two-dimensional points (O'Rourke *et al.*, 1986). In this case the problem is converted to the construction of a three-dimensional convex hull, and hence this method runs in $O(n \log n)$ time if the shelling method or the divide-and-conquer method is applied; this time complexity is optimal in the worst-case sense. The lift-up transformation gives also a linear-time algorithm for the Voronoi diagram for points forming a convex polygon (Aggarwal *et al.*, 1989).

Other algorithms for constructing higher-dimensional Voronoi diagrams include the naive method (Brostow *et al.*, 1978; Finney, 1979), a walking method (Avis and Bhattacharya, 1983) and incremental methods (Bowyer, 1981; Watson, 1981; Devijver and Dekesel, 1983; Tanemura *et al.*, 1983).

4.8 ALGORITHMS FOR GENERALIZED VORONOI DIAGRAMS

Let us return to two-dimensional Voronoi diagrams. As we saw in Chapter 3, there is a wide variety of generalization of the concept of the Voronoi diagram. However, we do not have space to treat them individually. Hence we try to present a general framework for constructing Voronoi diagrams of various types.

To represent a generalized Voronoi diagram in a unified manner, we introduce the following notations. Let $\Gamma = \{A_1, A_2, \ldots, A_n\}$ be the set of generators, and for any point z in the plane let $d(z, A_i)$ denote a 'distance' representing how far the point z is from the generator A_i (here we call d the 'distance' even if d does not satisfy the distance axiom). As in Chapter 3, we define the bisector of A_i and A_j by

$$b(A_i, A_j) = \{z | d(z, A_i) = d(z, A_j)\}, \tag{4.8.1}$$

and the dominance region of A_i over A_j by

$$\mathrm{Dom}(A_i, A_j) = \{z | d(z, A_i) \leq d(z, A_j)\}. \tag{4.8.2}$$

For any generator A_i, we define the Voronoi region for A_i with respect to distance d by

$$V(A_i) = \bigcap_{j \neq i} \mathrm{Dom}(A_i, A_j). \tag{4.8.3}$$

The collection of $V(A_1)$, $V(A_2)$, ..., $V(A_n)$ is called the *Voronoi diagram for Γ with respect to d*.

For an ordinary Voronoi diagram, generators are distinct points, and d is the Euclidean distance. For a weighted Voronoi diagram, generators are distinct points, and d is an additively weighted, a multiplicatively weighted or a compoundly weighted distance. For a farthest-point Voronoi diagram, generators are distinct points, and $d(z, A_i)$ is a monotone decreasing function of the Euclidean distance between z and A_i; for example, the inverse of the Euclidean distance between z and A_i can be used as $d(z, A_i)$. For a Voronoi diagram with obstacles, generators are distinct points, and $d(z, A_i)$ is the minimum length of the obstacle-avoiding path from A_i to z. For a Voronoi diagram of line segments or areas, generators are line segments or areas, respectively, and $d(z, A_i)$ is the minimum of the Euclidean distance from z to a point in A_i. For a Voronoi diagram with a non-Euclidean distance, generators are distinct points, and d is the associated non-Euclidean distance.

We restrict our consideration to the case where every bisector $b(A_i, A_j)$ is a single curve without self-intersection, closed or extended infinitely in both directions, dividing the plane into two connected regions $\mathrm{Dom}(A_i, A_j)$ and $\mathrm{Dom}(A_j, A_i)$. This case includes almost all of the generalized Voronoi diagrams listed in Chapter 3. An example of an exception is a bisector $b(A_i, A_j)$ with respect to an L_∞-distance of two generators aligning horizontally or vertically, where $b(A_i, A_j)$ includes a two-dimensional region.

For an ordinary Voronoi diagram, $b(A_i, A_j)$ is the straight line forming the perpendicular bisector of $\overline{A_iA_j}$. For a Voronoi diagram with a power distance (i.e. a power diagram) and for a farthest-point Voronoi diagram, the bisector $b(A_i, A_j)$, if it exists, is a straight line. In general, however, $b(A_i, A_j)$ is a complicated curve. For instance, the bisector of $b(A_i, A_j)$ of two line segments A_i and A_j with respect to the Euclidean distance can be a concatenation of up to seven segments of parabolas and straight lines, as shown in Figure 4.8.1, where segments 1 and 7 are perpendicular bisectors of two terminal points, segments 2, 4 and 6 are parabolas with one generator being the directorix and a terminal point of the other generator being the focus, and segments 3 and 5 are bisectors of the angles formed by the two generators. The bisector $b(A_i, A_j)$ with respect to a multiplicatively weighted distance is a circle containing the generator with the smaller weight. The bisector $b(A_i, A_j)$ with respect to an additively weighted distance is a hyperbola with A_i and A_j being the foci. Thus, the bisector changes its type of line according to the generators and the distance. In what follows we assume that we can construct $b(A_i, A_j)$ for any pair of generators in constant time, and that we can find the points of intersection of two bisectors also in constant time.

The concepts of the bisector and the dominance region for a pair of generators can be extended to those for a pair of disjoint subsets of generators. Let Γ' and Γ'' be mutually disjoint subsets of Γ. We define

$$b(\Gamma', \Gamma'') = \left\{ z \,\middle|\, \min_{A_i \in \Gamma'} d(Z, A_i) = \min_{A_j \in \Gamma''} d(z, A_j) \right\} \tag{4.8.4}$$

and call it the *bisector* of Γ' and Γ''. Similarly, we define

$$\mathrm{Dom}(\Gamma', \Gamma'') = \left\{ z \,\middle|\, \min_{A_i \in \Gamma'} d(z, A_i) \leq \min_{A_j \in \Gamma''} d(z, A_j) \right\} \tag{4.8.5}$$

and call it the *dominance region* of Γ' over Γ''. $b(\{A_i\}, \Gamma'')$ and $\mathrm{Dom}(\{A_i\}, \Gamma'')$ are abbreviated to $b(A_i, \Gamma'')$ and $\mathrm{Dom}(A_i, \Gamma'')$, respectively.

In the incremental method, the Voronoi diagram $\mathscr{V}(\{A_1, \ldots, A_l\})$ is constructed from the Voronoi diagram $\mathscr{V}(\{A_1, \ldots, A_{l-1}\})$ for each l, where

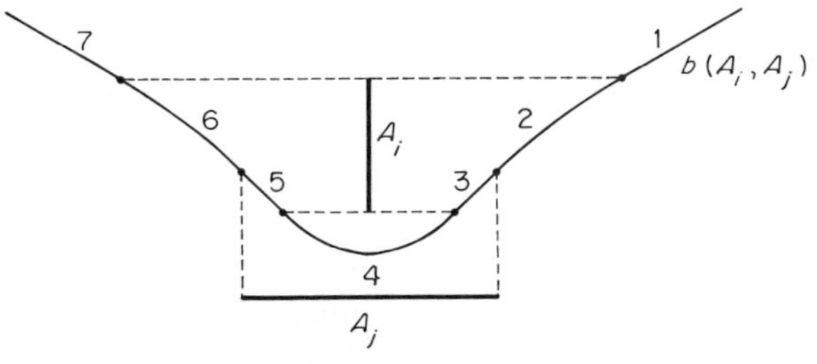

Figure 4.8.1 Bisector of two line segments.

the main task is to find the bisector $b(A_l, \{A_1, ..., A_{l-1}\})$. Let us assume that the Voronoi region $V(A_i)$ is simply connected. Then, for any $\Gamma' \subset \Gamma - \{A_i\}$, $b(A_i, \Gamma')$ consists of a single connected curve, a closed curve or a curve extended infinitely in both directions. This is the case for an additively weighted Voronoi diagram, a power diagram, a farthest-point Voronoi diagram, a Voronoi diagram of lines or areas, and a Voronoi diagram with an L_p-metric for $1 < p < \infty$. For those types of Voronoi diagrams the following incremental method can be used.

Algorithm 4.11 (Incremental method for a generalized Voronoi diagram)
Input: Set $\Gamma = \{A_1, A_2, ..., A_n\}$ of n generators and distance d.
Output: Voronoi diagram $\mathcal{V}(\Gamma) = \{V(A_1), V(A_2), ..., V(A_n)\}$ for Γ with respect to the distance d.
Procedure:
Comment. Let $\Gamma_l = \{A_1, A_2, ..., A_l\}$ for $1 \le l \le n$.
Step 1. Construct $\mathcal{V}(\Gamma_2)$.
Step 2. For $l = 3, 4, ..., n$, do 2.1, 2.2 and 2.3.
 2.1. Find a point on $b(A_l, \Gamma_{l-1})$.
 2.2. Starting with the point found in 2.2, trace the whole bisector $b(A_l, \Gamma_{l-1})$.
 2.3. Construct $\mathcal{V}(\Gamma_l)$ by inserting $b(A_l, \Gamma_{l-1})$ to $\mathcal{V}(\Gamma_{l-1})$ and then deleting the substructure inside $\mathrm{Dom}(A_l, \Gamma_{l-1})$.
Step 3. Return $\mathcal{V}(\Gamma) = \mathcal{V}(\Gamma_n)$.

An example of the addition of a generator is shown in Figure 4.8.2 for the case of a line Voronoi diagram. In this example each generator is partitioned into two terminal points and an open line segment, and they are treated as distinct generators. The Voronoi diagram for all the terminal points is generated first, and then open line segments are added one after another. The double line segment represents the new generator g_l and the broken lines represent $b(A_i, \Gamma_{l-1})$.

If the Voronoi region $V(A_i)$ contains the associated generator A_i, just as in the cases for a Voronoi diagram with a non-Euclidean metric and a Voronoi diagram for line segments or areas, the starter point in Step 2.2 can be found easily. All we have to do is first to find the Voronoi region $V(A_i)$ containing at least one point in the new generator A_l and then find the point of the intersection of $b(A_i, A_l)$ with the boundary of $V(A_i)$. If $V(A_i)$ does not necessarily contain A_i, as is the case for a power diagram and a farthest-point Voronoi diagram, some non-trivial method should be constructed to find a starter point.

An incremental method for Voronoi diagrams for line segments was proposed by Kokubo (1985), and a numerically stable version based on Principle 4.2 was constructed by Imai (1989) and Imai and Sugihara (1989). Other incremental-type algorithms are studied for farthest-point Voronoi

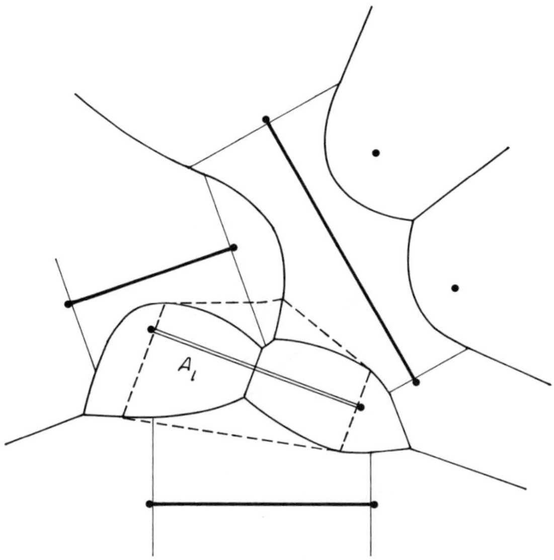

Figure 4.8.2 Incremental method for a Voronoi diagram for line segments.

diagrams (Suzuki, 1989) and Voronoi diagrams for polygons (Meshkat and Sakkas, 1987; Srinivasan and Nackman, 1987).

Next, let us consider the divide-and-conquer method. In this method two Voronoi diagrams $\mathcal{V}(\Gamma')$ and $\mathcal{V}(\Gamma'')$ are merged to $\mathcal{V}(\Gamma' \cup \Gamma'')$, where the main task is to find the bisector $b(\Gamma', \Gamma'')$.

For an ordinary Voronoi diagram the bisector $b(\Gamma', \Gamma'')$ is always a single connected curve (consisting of line segments). For a generalized Voronoi diagram, on the other hand, $b(\Gamma', \Gamma'')$ is not necessarily connected, but is a collection of a finite number of connected curves, some are closed and the others are extended infinitely in both directions.

In Figure 4.8.3 an example of the situation where $b(\Gamma', \Gamma'')$ is disconnected is shown for the case of the Voronoi diagram for line segments, where $\Gamma' = \{A_1, A_2, A_3\}$ and $\Gamma'' = \{A_4, A_5\}$. Dot-and-dash lines represent $\mathcal{V}(\Gamma')$, broken lines represent $\mathcal{V}(\Gamma'')$ and solid lines represent $b(\Gamma', \Gamma'')$. Note that the partition of Γ into Γ' and Γ'' is not very sophisticated; we obtain this partition if we choose the leftmost points of the line segments as the representatives and divide them into a left set Γ' and a right set Γ''.

To enumerate all the connected curves, we need a set of 'starters' from which we start tracing the curves. The set S of points is called a *starter set* of $b(\Gamma', \Gamma'')$ if any connected component of $b(\Gamma', \Gamma'')$ has at least one point in S. To trace the curves, S need contain only one point for each connected component. However, it is not always easy to tell the number of components of $b(\Gamma', \Gamma'')$ before we construct $b(\Gamma', \Gamma'')$. Hence, we put in S candidates

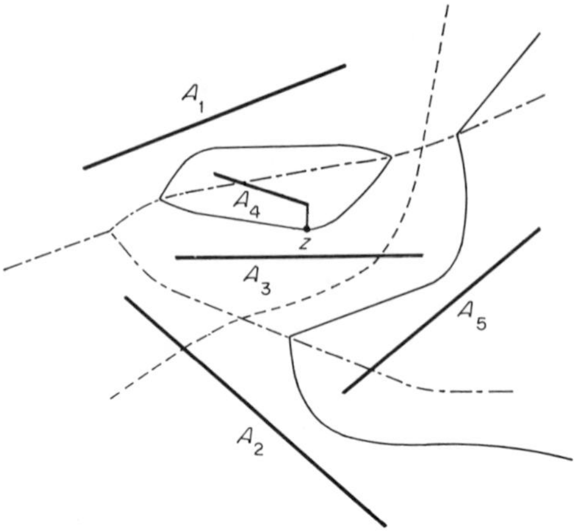

Figure 4.8.3 Merger of two Voronoi diagrams $\mathcal{V}(\{g_1, g_2, g_3\})$ and $\mathcal{V}(\{g_4, g_5\})$.

of starter points that are enough to trace all components. Thus, a divide-and-conquer algorithm can be described in the following way.

Algorithm 4.12 (Divide-and-conquer method for a generalized Voronoi diagram) Input: Set $\Gamma = \{A_1, A_2, \ldots, A_n\}$ of n generators and distance d.
Output: Voronoi diagram $\mathcal{V}(\Gamma) = \{V(A_1), V(A_2), \ldots, V(A_n)\}$ for Γ with respect to the distance d.
Procedure:
Step 1. If $n \leq 2$, construct $\mathcal{V}(\Gamma)$ directly and return it. Otherwise, do Steps 2, 3, ..., 8.
Step 2. Divide Γ into two disjoint subsets Γ' and Γ'' of almost the same size.
Step 3. Construct $\mathcal{V}(\Gamma')$ by Algorithm 4.12.
Step 4. Construct $\mathcal{V}(\Gamma'')$ by Algorithm 4.12.
Step 5. Find a starter set S of $b(\Gamma', \Gamma'')$, and put $B \leftarrow \emptyset$.
Step 6. While S is not empty, choose and delete an element s from S and if s is not on any curve in B, trace the connected components of $b(\Gamma', \Gamma'')$ containing s and put it in B.
Step 7. Merge $\mathcal{V}(\Gamma')$ and $\mathcal{V}(\Gamma'')$ into $\mathcal{V}(\Gamma)$ by inserting all the curves in B and deleting the substructure inside Dom(Γ', Γ'') from $\mathcal{V}(\Gamma'')$ and the substructure inside Dom(Γ'', Γ') from $\mathcal{V}(\Gamma')$.
Step 8. Return $\mathcal{V}(\Gamma)$.

The most difficult step in this algorithm is Step 5. We prefer a smaller starter set, but in general it is not easy to tell the number of components in $b(\Gamma', \Gamma'')$ unless we construct $b(\Gamma', \Gamma'')$ itself. For any $A_i \in \Gamma'$ and $A_j \in \Gamma''$, if point z with $d(z, A_i) = d(z, A_j)$ lies both in $V(A_i)$ of $\mathcal{V}(\Gamma')$ and in $V(A_j)$ of $\mathcal{V}(\Gamma'')$, then z is on $b(\Gamma', \Gamma'')$, and hence can be a starter point (Lee and

Drysdale, 1981). For instance, the point z in Figure 4.8.3 satisfies the above condition; this point is the midpoint of the perpendicular line segment dropped from the rightmost point of A_4 to A_3.

If the generators are points, we can partition Γ into a left generator set Γ' and a right generator set Γ'', and in this case we can sometimes find the starter set easily. For a Voronoi diagram with the Laguerre metric, $b(\Gamma', \Gamma'')$ consists of a single curve extended infinitely, and the starter point can be found from the common support of Γ' and Γ'', just as in the case of an ordinary Voronoi diagram (Imai *et al.*, 1985). For a Voronoi diagram with the L_p-metric for $1 < p < \infty$, $b(\Gamma', \Gamma'')$ is not necessarily connected, but every connected component in $b(\Gamma', \Gamma'')$ extends infinitely and the starter set can be found in a similar way from the convex hulls of Γ' and Γ'' (Chew and Drysdale, 1985).

The divide-and-conquer technique is used for order-k Voronoi diagrams (Lee, 1982b), for Voronoi diagrams with Laguerre distance (Imai *et al.*, 1985) and for line Voronoi diagrams (Lee and Drysdale, 1981). A slightly different type of the divide-and-conquer technique is used for Voronoi diagrams for line segments and circular arcs (Yap, 1987) and those for convex polygons (Leven and Sharir, 1987). Unifying studies of divide-and-conquer methods can be found in Chew and Drysdale (1985), Klein (1988, 1989), Klein and Wood (1988), and Klein, Melhorn and Meiser (1990).

There are a variety of other algorithms for individual types of generalized Voronoi diagrams. The plane sweep technique can be used for line Voronoi diagrams (Fortune, 1986, 1987), for additively weighted Voronoi diagrams (Fortune, 1986, 1987), for Voronoi diagrams on a cone (Dehne and Klein, 1987) and for Voronoi diagrams in a river (Sugihara, 1992a). Interpretation of a Voronoi diagram as an arrangement in one higher-dimensional space can be used for farthest-point Voronoi diagrams (Brown, 1979; Buckley, 1988), for multiplicatively weighted Voronoi diagrams (Aurenhammer and Edelsbrunner, 1984), for ordered order-k Voronoi diagrams (Edelsbrunner and Seidel, 1986) and for order-k Voronoi diagrams (Aurenhammer, 1990). Other algorithms include those for Voronoi diagrams with obstacles (Asano and Asano, 1987), for Voronoi diagrams on a polyhedral surface (Mount, 1985) and for Delaunay triangulation for points and line segments (Lee and Lin, 1986).

4.9 APPROXIMATION ALGORITHMS

Sometimes we want to construct Voronoi diagrams for only a few different sets of generators. In such a situation we cannot ignore the time required in developing a computer program; the time for writing a program is often more serious than the running time itself. This problem might be solved by approximation schemes.

If generators are mutually disjoint sets of points (such as curves and areas) and if we want to construct a Voronoi diagram for them with respect to distance d, the problem can be reduced to the construction of a Voronoi diagram for points with respect to d in the following way. Given a set of generators, we first replace each generator with a finite number of points that approximate the original generator, then construct the Voronoi diagram for these points, and finally remove superfluous Voronoi edges and superfluous Voronoi vertices. Thus, we obtain the next algorithm.

Algorithm 4.13 (Approximation by points) Input: Set $\Gamma = \{A_1, A_2, \ldots, A_n\}$ of n disjoint figures in the plane.
Output: Approximation of the Voronoi diagram for Γ.
Procedure:
Step 1. For each $i = 1, 2, \ldots, n$, create a finite set P_i of points that approximates the boundary of A_i.
Step 2. Construct the Voronoi diagram $\mathcal{V}$ for the point set $P_1 \cup P_2 \cup \ldots \cup P_n$.
Step 3. Delete from $\mathcal{V}$ those Voronoi edges whose generator points belong to the same original figure. Delete isolated Voronoi vertices, if any, from $\mathcal{V}$.
Step 4. Return $\mathcal{V}$.

Figure 4.9.1 shows an example of this scheme applied to a set of generators containing different types of figures such as line segments, circular arcs and polygons; (a) represents the Voronoi diagram for 882 points located on the boundaries of the original figures, and (b) shows the approximation of the Voronoi diagram for the original figures obtained from (a).

To obtain a good approximation, we have to use many points to approximate the original generators. Hence, in general, the number of generators (= points) becomes very large. Moreover, these points often give highly degenerate configurations. Therefore the approximation scheme is valid only when we can use a fast and robust algorithm, such as the one developed in Section 4.6. This approximation scheme can be extended to an m-dimensional generalized Voronoi diagram directly, if an algorithm for Voronoi diagrams for points in the m-dimensional space is available.

Another, more brute-force approximation scheme is the use of digital image techniques. A digital image is a two-dimensional array, say $I(i, j)$, $1 \le i, j \le N$, where the entry $I(i, j)$ represents some property of the image at the point with coordinate (i, j). Each point (i, j) is called a *pixel*, and $I(i, j)$ is called a *pixel value*. A pixel with a positive pixel value is called a *positive pixel*. The *four-neighbour set* of pixel (i, j) is defined by

$$\mathrm{NB}_4(i,j) = \{(i,j-1), (i,j+1), (i-1,j), (i+1,j)\},$$

and the *eight-neighbour set* of pixel (i, j) is defined by

$$\mathrm{NB}_8(i,j) = \mathrm{NB}_4(i,j) \cup \{(i-1,j-1), (i-1,j+1), (i+1,j-1), (i+1,j+1)\},$$

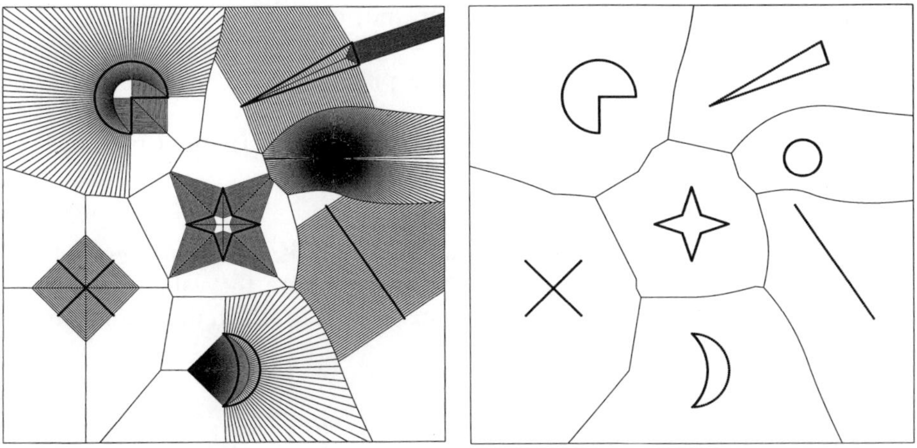

Figure 4.9.1 Approximation of generators by sets of points: (a) Voronoi diagram for 882 points that approximate seven generator figures; (b) approximation of the generalized Voronoi diagram obtained by deleting superfluous edges and superfluous vertices from (a).

where pixels outside the array are ignored. The four-neighbour set or the eight-neighbour set of (i, j) is simply called a neighbour set of (i, j) and is denoted by $\mathrm{NB}(i, j)$.

In the next scheme, an approximation of a Voronoi diagram is constructed in the form of a digital image. First, each generator is replaced with a set of pixels, and then from these pixels the 'territories' of the generators are expanded simultaneously at the same speed until the territories collide with one another. This can be done by repeating local parallel operations over the neighbours of pixels in the following way (an example of the behaviour of the algorithm immediately follows the algorithm).

Algorithm 4.14 (Approximation by a digital image) Input: Integer N and set $\Gamma = \{A_1, A_2, \ldots, A_n\}$ of n disjoint figures in the plane.

Output: Digital image $I(i, j)$ $(1 \leq i, j \leq N)$ approximating the Voronoi diagram for Γ, where $I(i, j) = k > 0$ if pixel (i, j) belongs to the Voronoi region of A_k, and $I(i, j) = -1$ if pixel (i, j) is incident to two or more different Voronoi regions.

Procedure:

Step 1. For all $1 \leq i, j \leq N$,
$I(i, j) \leftarrow k$ if there exists k such that $(i, j) \in A_k$,
$I(i, j) \leftarrow 0$ otherwise.

Step 2. Repeat 2.1 until pixel values are not changed any more.

2.1. For all (i, j) such that $I(i, j) = 0$, do the following simultaneously. $I(i, j) \leftarrow k$ if all the positive pixels in NB(i, j) have the same value k,
$I(i, j) \leftarrow -1$ if the positive pixels in NB(i, j) admit two or more different values.

Step 3. Return the array $I(i, j)$.

An example of the behaviour of the algorithm is shown in Figure 4.9.2. Panel (a) shows the initial value of the array, where the value i in a circle represents the ordinal number of the generator A_i to which the pixel belongs,

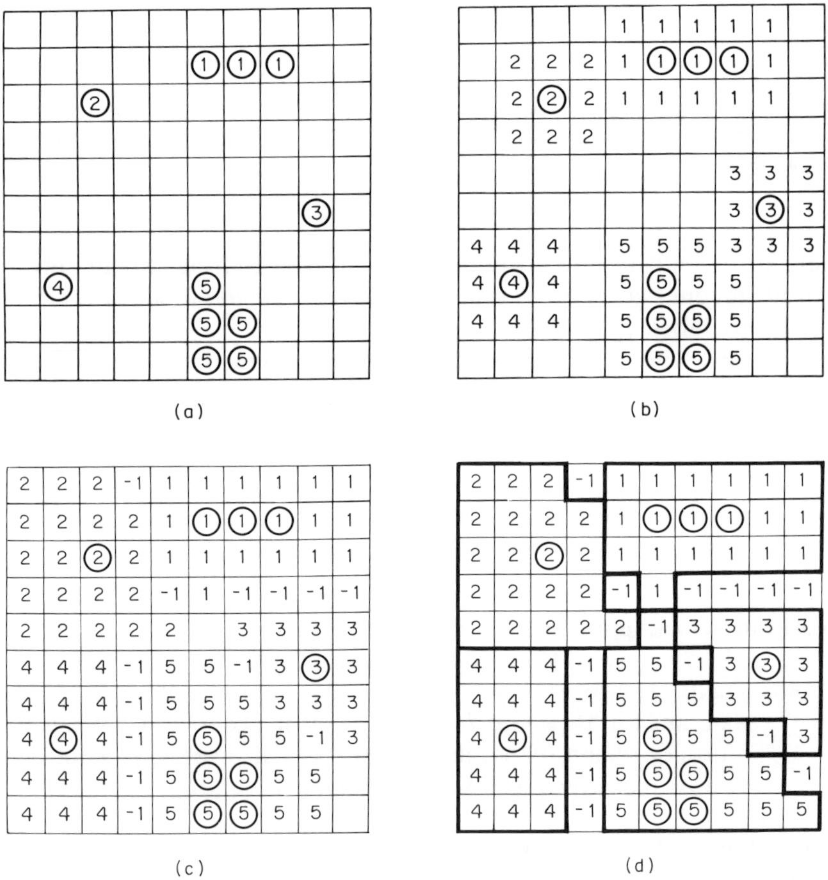

Figure 4.9.2 Approximation of the Voronoi diagram by a digital image: (a) digital image of five generators; (b) first execution of Step 2.1; (c) second execution of Step 2.1; (d) final result.

and blank pixels represent those with value 0. The generators A_2, A_3 and A_4 are represented by one pixel, whereas the generators A_1 and A_5 occupy more than one pixel. The array after the first, the second and the third execution of Step 2.1 are shown in (b), (c) and (d), respectively, where the eight-neighbour set $NB_8(i, j)$ is used as the neighbour set in the algorithm. After the third execution of Step 2.1 all the pixels have non-zero values, and hence the algorithm terminates. In (d), the boundaries of the Voronoi regions are represented by bold lines.

In Step 2.1 the pixel values are changed simultaneously; this is a kind of parallel procedure. To implement this procedure in the usual sequential computer, we use one more array, say $J(i, j)$, of the same size; we write the resulting pixel values in $J(i, j)$ in the first execution of Step 2.1, and switch the roles of these two arrays $I(i, j)$ and $J(i, j)$ every time we execute Step 2.1.

This approximation scheme is conceptually simple and does not require any exact algorithm for Voronoi diagrams for points, but is very time and space consuming. Moreover, the output of this algorithm is not a winged-edge data structure; the output merely gives us approximations of Voronoi regions and we need further processing to extract the incidence relations among the regions, the edges and the points. Furthermore, the distance is not Euclidean. If the four-neighbour set is used in Step 2.1, the 'territories' expand according to the L_1-metric, while if the eight-neighbour set is used, they expand according to the L_∞-metric. If these two kinds of the neighbour sets are used alternately in the repetition of Step 2.1, the output gives a little bit better approximation of a Euclidean-metric Voronoi diagram. Therefore this approximation scheme should be used only for special purposes such as for Voronoi diagrams with the L_1- or L_∞-metric or for very rough approximations of Voronoi diagrams.

On the other hand, if the original generators themselves are given in the form of a digital image, as is the case in digital pattern analysis, Algorithm 4.4 can be applied directly to various kinds of proximity analysis (Toriwaka *et al.*, 1982; Mark, 1987; Yoshitake *et al.*, 1987).

CHAPTER 5

Poisson Voronoi Diagrams

In Section 2.1 we introduced the idea of a generator set $P = \{p_1, \ldots, p_n\}$ as a set of n points which generates the ordinary Voronoi diagram $\mathcal{V}$. The only conditions imposed on P were that n was finite and greater than one and that all points were distinct (i.e. no two or more points were spatially coincident). Other than this, no other aspects of the locations of the points in m-dimensional Euclidean space, $\mathbb{R}^m$, were specified. In this chapter we consider $\mathcal{V}$, its dual Delaunay tessellation $\mathcal{D}$, and some generalizations of $\mathcal{V}$ when the members of P are located in $\mathbb{R}^m$ according to the homogeneous Poisson point process Θ_P discussed in Section 1.3.8. We refer to the resulting tessellation as the *Poisson Voronoi diagram* $\mathcal{V}_P$, and the *Poisson Delaunay tessellation* $\mathcal{D}_P$. We use the terms *Poisson Voronoi cell* (PVC) and *Poisson Delaunay cell* (PDC) to refer to the individual m-dimensional polytopes ($m \geq 2$) of the respective tessellations. Figure 5.0.1 shows a portion of $\mathcal{V}_P$ in $\mathbb{R}^2$, while a typical PVC in $\mathbb{R}^3$ is illustrated in Figure 5.0.2.

We begin our treatment of $\mathcal{V}_P$ by considering its properties in Section 5.1. Some of these properties are inherited from properties of Θ_P by which the members of P are generated. Other properties of $\mathcal{V}_P$ are common to the set of all 'random' tessellations of which $\mathcal{V}_P$ is a member. The properties of $\mathcal{V}_P$ make it very attractive for use in a number of ways in an extensive range of empirical circumstances which are described in Section 5.2. In view of this widespread use of $\mathcal{V}_P$ (and $\mathcal{D}_P$), there is great interest in obtaining information on both PVCs and PDCs.

Such information includes moments, distributions and correlations of various characteristics such as the number of sides, area, perimeter (in $\mathbb{R}^2$), number of faces, volume and total edge length (in $\mathbb{R}^3$) of a typical cell. Considerable success has been achieved in deriving analytically the first-order moments of variety of PVC and PDC characteristics and these are reported in Sections 5.4 and 5.9, respectively. In principle it is also possible to derive integral formulae to obtain other moments of cell characteristics but often these are notoriously complicated, especially for PVCs. Consequently, many researchers have resorted to Monte Carlo approaches where the

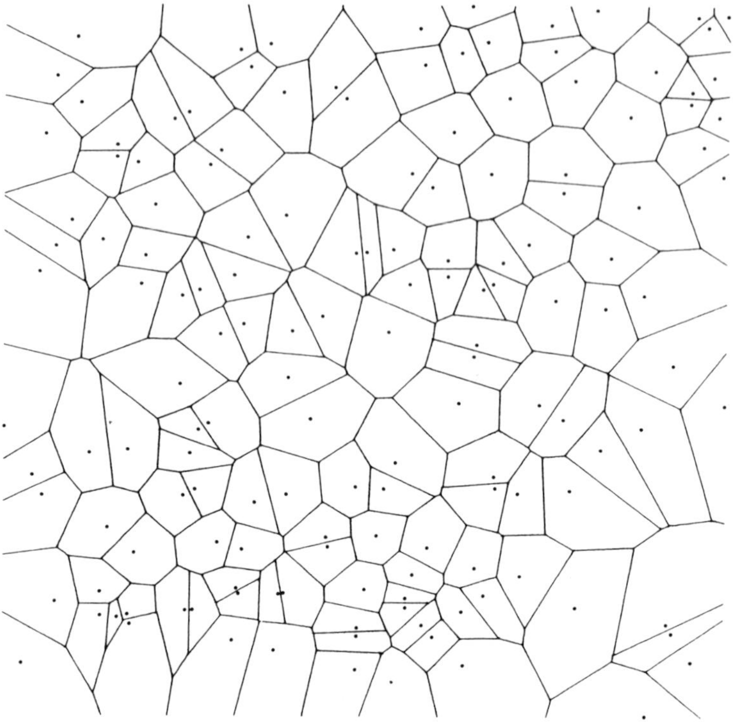

Figure 5.0.1 A portion of a planar Poisson Voronoi diagram.

characteristics of cells are estimated by simulation. Several different simulation approaches are possible for both $\mathcal{V}_P$ and $\mathcal{D}_P$ and these are outlined in Section 5.3. Estimates obtained from such simulations of PVCs and PDCs are reported in Sections 5.4 and 5.9, respectively.

In many applications involving $\mathcal{V}_P$ in $\mathbb{R}^3$, information concerning $\mathcal{V}_P$ is not directly available but instead is acquired from planar sections or linear probes. Thus, such applications require information on sections of PVCs rather than the PVCs themselves. This information is reported in Section 5.5.

In Chapter 3 we described various ways of generalizing $\mathcal{V}$. Many of these generalizations can also be pursued when the generator set is a realization of Θ_P. In fact, as noted in Section 3.1.5, the sectional Voronoi diagram of Section 5.5 can be considered as a type of weighted Voronoi diagram. Section 5.6 considers a particular type of additively weighted Poisson Voronoi diagram more commonly known as the Johnson–Mehl model. This creates a dynamic version of the Poisson Voronoi model by associating with each generator point a weight which reflects the arrival time of the point

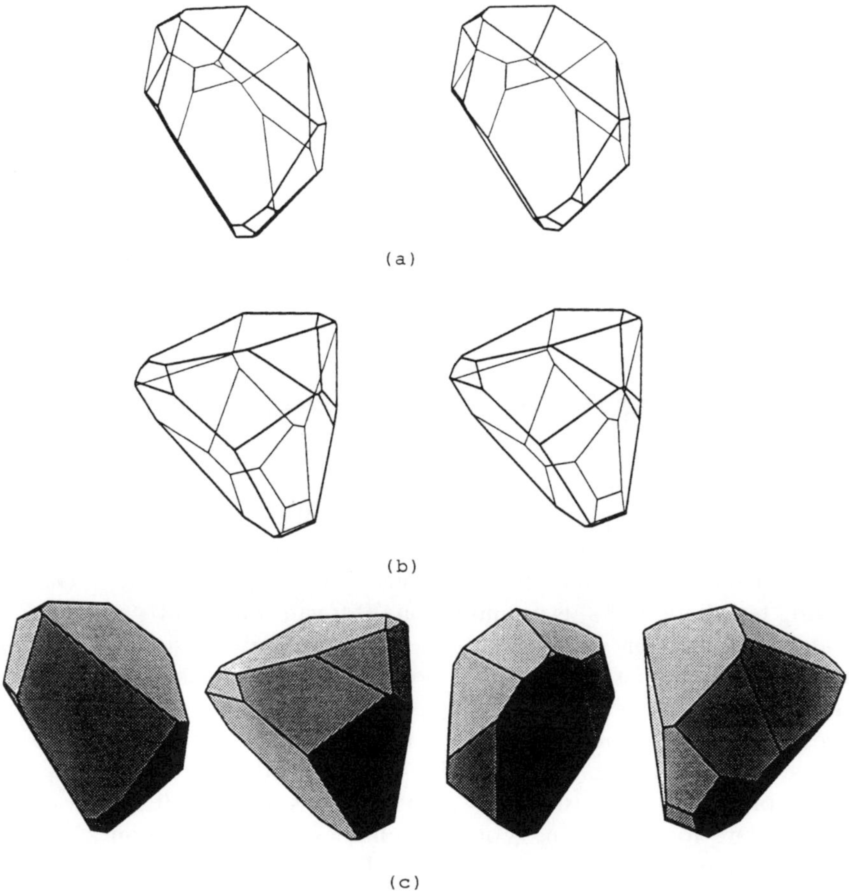

Figure 5.0.2 (a) Stereo pair of typical Poisson Voronoi cell in $\mathbb{R}^3$. (b) The stereo pair after rotation of 90° about a fixed axis. (c) Cell shown after rotation of 0°, 90°, 180°, 270° about the fixed axis. (*Source*: Møller *et al.* (1989).)

in $\mathbb{R}^m$. Order-k and ordered order-k Voronoi diagrams and Voronoi diagrams on the surface of a sphere are discussed in Sections 5.7 and 5.8, respectively.

In general our knowledge of PDCs, particularly that derived by analytical methods, is greater than that of PVCs. This information, together with a brief treatment of sections of $\mathcal{D}_P$ and $\mathcal{D}_P$ on the surface of a sphere, is reported in Section 5.9.

For some applications the conditions inolved in generating $\mathcal{V}_P$ are inappropriate, perhaps because there are additional conditions influencing the locations of the members of P or because there are constraints on the sizes of the cells. In such circumstances modifications may be undertaken which produce other random tessellations. A selection of such tessellations is briefly surveyed in Section 5.10.

In view of the widespread interest in $\mathcal{V}_P$ in empirical circumstances, considerable effort has been extended in many different disciplines in the pursuit of information on PVCs and PDCs and consequently the results of such endeavours are widely dispersed over a far-flung range of publications. In light of this it is not surprising that much duplication or near duplication has occurred because individual researchers (and reviewers and editors!) have been ignorant of similar, previous work of others. To reduce further duplication and in an attempt to synthesize previous activity, Sections 5.4–5.9 emphasize the reporting and reconciliation of the results of many previously disparate studies.

5.1 PROPERTIES OF POISSON VORONOI DIAGRAMS

As indicated in Section 2.1, an m-dimensional Voronoi diagram $\mathcal{V}$ constitutes a tessellation of $\mathbb{R}^m$ ($m = 1, 2,\ldots$) since the interiors of the individual cells of $\mathcal{V}$ are non-overlapping and together with their edges are collectively space-filling. In particular, when the generator set P consists of points which are the realization of the homogeneous Poisson point process Θ_P, the resulting tessellation, $\mathcal{V}_P$, is a random one (Stoyan *et al.*, 1987, pp. 258–259; Møller, 1989, Section 4). In $\mathbb{R}^m$ the cells of the m-dimensional $\mathcal{V}_P$ are m-dimensional polytopes. In turn, the boundary of each polytope consists of an hierarchy of s-dimensional polytopes ($s = 0, 1, \ldots, m-1$) which we term s-faces. Thus, for example in $\mathbb{R}^3$ the 0-faces, 1-faces and 2-faces are the vertices, edges and faces, respectively, of a polyhedron.

Some of the characteristics of $\mathcal{V}_P$ and its constituent PVCs derive from the procedure for generating Voronoi cells described in Section 2.1, while others are inherited from Θ_P (see Section 1.3.8). The former characteristics include:

Property PV1 The cells of $\mathcal{V}_P$ are convex (see Property V1 in Section 2.3).

Property PV2 $\mathcal{V}_P$ is a regular tessellation.

A regular tessellation is one in which the s-faces of the individual cells coincide with those of the tessellation. A Voronoi diagram is regular because each s-face ($s < m$) lies in the set of s-dimensional points which are equidistant from the generators of the cells which contain the s-face.

Properties which are inhereited from Θ_P include:

Property PV3 $\mathcal{V}_P$ is a stationary and isotropic tessellation.

This means that its characteristics and those of its constituent PVCs are invariant under both translation and rotation about the origin in $\mathbb{R}^m$. Thus, $\mathcal{V}_P$ can be considered a homogeneous (or motion-invariant) tessellation.

Like Θ_P, $\mathcal{V}_P$ is ergodic and satisfies the mixing condition by which spatially distant features of the tessellation are independent.

Property PV4 $\mathcal{V}_P$ is a normal tessellation.

A normal tessellation is one in which every s-face of the tessellation lies in the boundaries of $(m - s + 1)$ cells $(0 \le s \le m - 1)$. Møller (1989, Proposition 7.1) provides a proof of this property. Equivalent terms to normal are ordinary or natural or in equilibrium state, all of which are used because empirical tessellations (for $1 \le m \le 3$) frequently possess this property. In terms of Section 2.1, $\mathcal{V}_P$ is non-degenerate. Because of Property PV2 we may also express Property PV4 in terms of the constituent PVCs. Each s-face of a PVC lies in the s-dimensional set of points which are equidistant from a set of $(m - s + 1)$ points of P. Thus, in $\mathbb{R}^2$, each vertex is shared by three cells and each edge is common to two cells. In $\mathbb{R}^3$ four cells meet at each vertex, each edge is shared by three cells and each face is common to two cells.

Once $\mathcal{V}_P$ is generated using Θ_P, other point process can be constructed from $\mathcal{V}_P$. One set of such point processes, Θ_{Ps} $(s = 0, 1, \ldots, m)$, of intensities λ_s $(s = 0, 1, \ldots, m)$, consists of the centroids of the s-faces of the cells of $\mathcal{V}_P$. For example in $\mathbb{R}^2$ the appropriate processes are Θ_{P0} for the vertices, Θ_{P1} for the edge midpoints and θ_{P2} for the cell centroids. Møller (1989, p. 52) proves that an iterative relationship exists between the intensities λ_s (which also holds for all normal, regular random tessellations). This is given by

$$[1 - (-1)^t]\lambda_t = \sum_{s=0}^{t-1} (-1)^s \binom{m-s+1}{t-s} \lambda_s \quad (t=1, \ldots, m) . \tag{5.1.1}$$

For $m = 2$, equation (5.1.1) yields

$$3\lambda_0 = 2\lambda_1 = 6\lambda_2 ,$$

and for $m = 3$

$$2\lambda_0 = \lambda_1 = 2(\lambda_2 - \lambda_3) .$$

Since Θ_P is stationary, Θ_{Ps} $(s = 0, 1, \ldots, m)$ also have this property. Stationarity also holds for any process with intensity Ψ_s, generated by the s-faces $(s = 0, 1, \ldots, m)$ of $\mathcal{V}_P$ (e.g. in $\mathbb{R}^2$ the fibre process, intensity Ψ_1 consisting of the set of edges of $\mathcal{V}_P$).

Other relationships which hold for all stationary tessellations and which may be derived by either ergodic arguments (Miles, 1972a; Cowan, 1978, 1980) or Palm distributions (Radecke, 1980; Mecke, 1980, 1984; Møller, 1989), relate mean values of cell characteristics. For $\mathbb{R}^m$ general expressions are given by Møller (1989, Section 5), while specific relationships for $m = 2$ and $m = 3$ are given by Stoyan *et al.* (1987, p. 265 and p. 269, respectively).

Furthermore, Mecke (1984) shows that for $\mathbb{R}^2$ these relationships can be expressed in terms of three parameters, λ_0, λ_2 and Ψ_1, while for $\mathbb{R}^3$ seven parameters are required (λ_0, λ_3, $(\lambda_1 + \lambda_2)$, Ψ_1, Ψ_2, Ψ_0' and Ψ_1'), where Ψ_s' $(s = 0,1)$ is the intensity of the s-face process in which the s-faces are weighted by the numbers of adjacent cells. Both Miles (1972a) and Møller (1989) exploit these relationships in deriving many of the properties of PVC characteristics reported in Section 5.4. Møller also derives a series of relationships between mean values of face and tessellation characteristics.

Finally, since the intersection between an arbitrary p-dimensional hyperplane $(p = 1, \ldots, m)$ and $\mathcal{V}_P$ in $\mathbb{R}^m$ yields a random tessellation, the properties of such sections correspond to those of $\mathcal{V}_P$ reported above. This permits the generation of characteristics of their constituent cells from those of $\mathcal{V}_P$ (Miles, 1972a, Section 2.5; 1972b; Møller, 1989, Section 6) and these are reported in Section 5.5.

5.2. USES OF POISSON VORONOI DIAGRAMS

Given the way in which the Poisson Voronoi diagram $\mathcal{V}_P$ is generated and its resulting properties, there has been widespread interest in its use in a variety of empirical situations. Such uses of $\mathcal{V}_P$ take two main forms. The most straightforward use is as a model of a given empirical structure, while the other use is as a normative model against which other tessellations can be evaluated. Sometimes the distinction between these two uses becomes blurred because, once established as a structural model, $\mathcal{V}_P$ is often subsequently used in a normative way.

An example of the direct use of $\mathcal{V}_P$ is provided by Christ *et al.* (1982) and Drouffe and Itzykson (1984) in the investigation of quantum field theory. More recent work here has proposed replacing the traditional concept of a space–time continuum by a discrete lattice. Initially, regular lattices were used. However, although the links between individual nuclei (points) are clearly defined for such lattices, they are not motion invariant and thus do not possess properties assumed in the traditional concept. As we have seen, the homogeneous Poisson point process Θ_P does possess these properties (see Section 1.3.8) and so Christ *et al.* (1982) proposed using a random lattice which is defined as one in which the nuclei are generated by Θ_p and the links between the nuclei correspond to the adjacencies between the Voronoi regions of the Voronoi diagram of nuclei. In this way the random lattice is equivalent to $\mathcal{V}_P$.

In studies of conductivity and percolation in granular composites, the set of links defined by contiguous cells in $\mathcal{V}_P$ (equivalent to the Delaunay triangulation of Θ_P) has been proposed as a model of a topologically random network in both $\mathbb{R}^2$ (Jerauld *et al.*, 1984a) and $\mathbb{R}^3$ (Jerauld *et al.*, 1984b; Rivier *et al.*, 1985). In such circumstances, once the links are identified each may be assigned an associated level of conductance according to a selected prescription. In this way random percolative networks can be generated by randomly assigning zero conductance to a specified proportion of links.

DiCenzo and Wertheim (1989) have also used $\mathcal{V}_p$ in modelling the growth of clusters formed by metal vapour deposited on amorphous substrates. The model is a two-step one in which first nucleation sites are distributed in $\mathbb{R}^2$ ($\mathbb{R}^3$) according to Θ_p and then atoms landing on the surface of the substrate are assigned to the nearest nucleation site. The number of atoms in a cluster formed at a nucleation site will be proportional to the area (volume) of the surface closest to the nucleation site which is equal to the area (volume) of the Voronoi cell of the nucleation site. Thus, the distribution of cell areas (volumes) can be used to derive the distribution of cluster sizes.

$\mathcal{V}_P$ also forms the basis of another frequently used model of structure. This is a specific form of a random mosaic which we label the *Poisson Voronoi Random Mosaic* (PVRM). The creation of a random mosaic involves two steps (Schachter and Ahuja, 1979):

Step 1: for some bounded region B of $\mathbb{R}^m$, tessellate B,
Step 2: independently assign one of k phases to each cell of the tessellation according to a fixed set of probabilities, $p_1, \ldots, p_k$; $\Sigma_{i=1}^{k} p_i = 1$.

In this way B is partitioned into subregions $B_1, \ldots, B_k$, each consisting of the union of all cells of phase k, so that $\cup_{i=1}^{k} B_k = B$. If the tessellation of Step 1 is the Poisson Voronoi diagram, we produce the PVRM. In ecology this model is referred to as the *S-Mosaic* (Pielou, 1977, p. 185).

The simplest PVRM is two-phase. Winterfield *et al.* (1981) used such a mosaic as a model of a disordered composite in $\mathbb{R}^2$ to study conduction and percolation in continuous media. The two phases are conducting and non-conducting (insulating). Using this approach they were able to study the effects on various conduction and percolation properties of the medium by varying the probabilities associated with each phase. Once established, they also used the two-phase PVRM as a normative model against which other mosaics involving tessellations composed of regular hexagonal and square cells were judged. The two-phase PVRM has also been used to model microemulsions in $\mathbb{R}^3$ (Talmon and Prager, 1977, 1978; Kaler and Prager, 1982). Microemulsions are thermodynamically stable three-component systems consisting of an aqueous component, an oil component and a surfactant. The aqueous and oil components are the two phases of the mosaic (with probabilities p_a, p_o, respectively) and the surfactant is distributed over the resulting aqueous–oil interface (the boundaries of the subregions defined by the two phases). Such a model is capable of generating all three classes of microemulsion geometry; oil in water (small values of p_o/p_a), bicontinuous ($p_o/p_a \simeq 1$) and water in oil (large values of p_o/p_a).

A three-phase PVRM in $\mathbb{R}^3$ is used by Brumberger and Goodisman (1983) in the interpretation of small angle X-ray scattering observed for heterogeneous catalysts such as porous oxides (e.g. Al_2O_3, S_iO_2) and oxide supported metals (e.g. Pt/Al_2O_3, Rh/S_iO_2). Here the three phases are void, support and metal, each of which is considered to contain an internally uniform electron density and whose associated probabilities are equal to the proportion of the total volume of the sample catalyst occupied by that phase.

A variant of the PVRM is described by Ahuja *et al.* (1985) and An *et al.* (1983) who use it in the representation of an image with k colours. A Poisson Voronoi tessellation is superimposed on the image and the individual cells of the tessellation are assigned the colour of the underlying image. In those cases where more than one colour occurs in a cell, the cell is assigned the colour which has the largest representation. The image can then be represented by the set of coordinates of the generators of the individual cells of the tessellation together with their associated colours. These authors suggest that such a representation is particularly useful for the secure transmission of images since, if the coordinates of the cell generators are produced by a pseudo-random number generator, only the seed for this generator and the colour of each cell need to be transmitted.

Although we have illustrated a number of empirical situations where our knowledge of the phenomenon suggests that $\mathcal{V}_P$ may be an appropriate model there are many other occurrences where such direct use may be unjustified. For instance, we may have little or no information on how a particular tessellation was generated or we may have sufficient information to suggest that it is unlikely that it was created under conditions analogous to those generating $\mathcal{V}_P$. In the former case our first concern might be to determine if the empirical tessellation differs from $\mathcal{V}_p$, while in the latter we may be interested in knowing in what ways the tessellation differs from $\mathcal{V}_p$. In both instances the results of our investigations may provide additional insight into the conditions under which the empirical tessellation was generated. In this way $\mathcal{V}_P$ is used as a normative model against which other tessellations are compared. Such comparisons may be informal in nature or may involve the formal testing of hypotheses concerning the nature of the difference between the empirical tessellation and $\mathcal{V}_P$.

Numerous examples of the use of $\mathcal{V}_p$ in this normative role occur for phenomena in both $\mathbb{R}^2$ and $\mathbb{R}^3$ in both the natural and the social sciences. In the former $\mathcal{V}_P$ has been used to evaluate actual galaxy distributions in $\mathbb{R}^3$ (Yoshioka and Ikeuchi, 1989; Zaninetti, 1990) and in $\mathbb{R}^2$ when these are considered as either slices of the universe or as projections on the sky (Icke and Van der Weygaert, 1987). Another example is provided by sections through various geological materials (Crain, 1976). $\mathcal{V}_p$ in $\mathbb{R}^3$ has been used to evaluate the structure of several monatomic liquids and solids which may be modelled as sphere packings (Rahman, 1966; Finney, 1970a). In animal ecology tessellations created by territorial units have been compared with $\mathcal{V}_P$ in $\mathbb{R}^2$ (Hamilton, 1971; Hasegawa and Tanemura, 1976; Buckley and Buckley, 1977; Tanemura and Hasegawa, 1980). In the social sciences Boots

(1973, 1975a) has used $\mathcal{V}_P$ in $\mathbb{R}^2$ to evaluate service areas associated with public bus services in parts of England and Wales, while Singh and Singh (1978) and Singh (1979) have used them in the evaluation of territories of villages in India.

All the above examples illustrate the use of $\mathcal{V}_P$ as a normative model in the examination of empirical tessellations. However, Θ_P has also been used extensively as a normative model in the analysis of point patterns. This is because we can compare characteristics of the Voronoi diagram generated by a given set of points (or more frequently, the Delaunay tessellation) with those of $\mathcal{V}_P$ (or $\mathcal{D}_P$). This provides an indirect way of comparing the empirical point pattern with one generated by Θ_P which provides the generators for $\mathcal{V}_P$. Such endeavours have received considerable attention and as a result are dealt with in more detail in Sections 8.1 and 8.2.

5.3 SIMULATING POISSON VORONOI CELLS AND POISSON DELAUNAY CELLS

Simulations have frequently been used in studying the characteristics of PVCs, particularly for estimating the distribution of those characteristics. Table 5.3.1 summarizes those published studies known to the authors, although there are doubtless some others. As noted in Section 1.3.8, the homogeneous Poisson point process Θ_P is ergodic so that limiting distributions of the characteristics of the cells in a single, large Poisson Voronoi diagram $\mathcal{V}_P$ will coincide with those obtained from a sequence of individual, 'typical' cells sampled from separate $\mathcal{V}_P$ values. Consequently, two main simulation approaches are possible.

The first of these (referred to as method A in Table 5.3.1) involves simply generating a large number of points inside a bounded region B according to Θ_P, constructing $\mathcal{V}_P$ and measuring the characteristics of all its cells. There are essentially two complications with this approach. The first is that, no matter how large $\mathcal{V}_P$ is, allowance must be made for edge effects introduced by those cells close to the boundary of B. One way to treat such edge effects is to exclude from consideration any cells of $\mathcal{V}_P$ in B for which a circle, centred at any vertex v of the cell and passing through the three points of Θ_P which are equidistant from v, intersects the boundary of B (Kenkel *et al.*, 1989) (cell A in Figure 5.3.1 is such a cell). If B is a rectangle, an alternative procedure involves creating additional points outside of each edge of B which are translates of points of Θ_P inside the opposite edge of B (the points in the shaded polygons in Figure 5.3.2). The translated points are then used to complete the polygons of those points of $\mathcal{V}_P$ whose polygons intersect the boundary of B. This procedure, which is usually referred to as a periodic boundary condition, is equivalent to converting B into a torus (see Section 3.7.8). The other complication arises because the characteristics of neighbouring cells are not independent (see Section 5.4). Thus, the

Table 5.3.1 Simulation studies of Poisson Voronoi cells

Study	Method	Number of cells	Characteristics
Two-dimensions			
Boots (1987)	R	30 000	L
Boots and Murdoch (1983)	R	50 000	A N P
Crain (1972, 1976)	T	5–11 000	A N P
Crain (1978)	T	25–59 000	A L N P
DiCenzo and Wertheim (1989)	A	~193 500	A
Drouffe and Itzykson (1984)	R	27 000	A N
Hamilton (1971)	A	376	N
Hasegawa and Tanemura (1976)	A	500	A N α
Hermann *et al.* (1989)	A	~2 000	A
Hinde and Miles (1980)	T	2 000 000	A N P α
Hutchings and Discombe (1986)	A	42 318	A N P
Kiang (1966)	A	8 960	A
Le Caer and Ho (1990)	A	600 696	A L N P
		1 001 500	
		1 020 800	
Quine and Watson (1984)	R	50 000	A N P
Stoyan and Hermann (1986)	A	1 290	A
Stoyan and Stoyan (1990)	A	~3 000	L
Tanemura and Hasegawa (1980)	A	500	A L N P α
Vincent *et al.* (1976, 1983)	A	10 000	N
Weaire *et al.* (1986)	A	79 400	A
		734 639	A
		929 070	A
Wilkinson (1988)	A	2 048	A
Winterfield *et al.* (1981)	A	~96 500	N
Three dimensions			
Hanson (1983)	T	3 000	F M N S V
Jerauld *et al.* (1984b)	A	10 184	F
Kiang (1966)	A	12 800	V
Lorz (1990b, 1991)	R,A	82 870	A B F L N M P S V b α_1 α_2 α_3
Quine and Watson (1984)	R	2 500	B C F N S V
Rahman (1966)	A	528	F N V
Tanemura (1988)	A	100 000	F M S V
Yoshioka and Ikeuchi (1989)	A	64 000	V

Characteristics:

Two dimensions

A = Area
L = Edge length
N = Vertices (edges)
P = Perimeter
α = Interior angle

Three-dimensions

A = Area/face
B = Total edge length/polyhedron
C = Full neighbours
F = Faces/polyhedron
L = Edge length
M = Vertices/polyhedron
N = Edges/face
P = Edge length/face
S = Surface area/polyhedron
V = Volume
b = Mean breadth/polyhedron
α_1 = Face angle of an edge
α_2 = Interior angle of a face
α_3 = Face angle of a cell

For definitions of methods, see text.

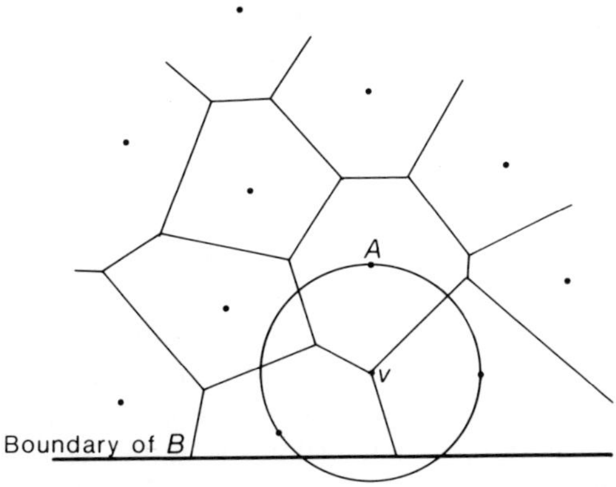

Figure 5.3.1 A procedure for identifying cells subject to edge effects.

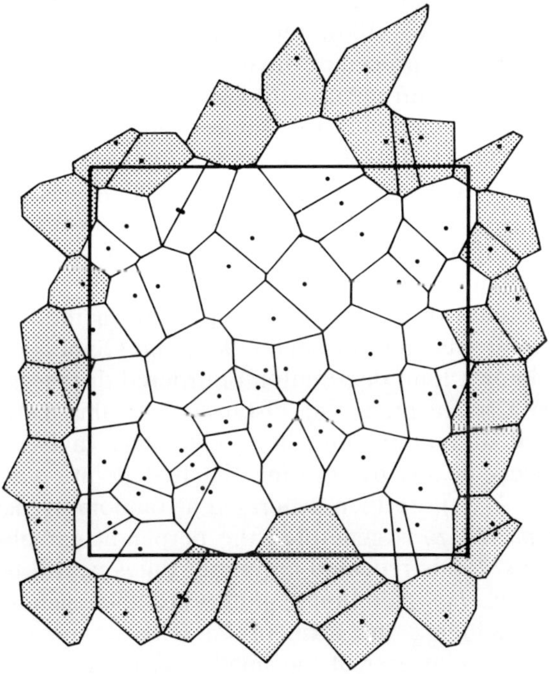

Figure 5.3.2 Periodic boundary condition to avoid edge effects.

number of cells in the aggregate must be sufficiently large to avoid any bias introduced by this source.

The second simulation approach involves generating a sequence of independent, typical PVCs, measuring the characteristics of each and then aggregating them to obtain the required distributions. The primary advantage of this approach over the previous one is that it avoids problems of edge effects. There are two variants of this approach. The first, method T in Table 5.3.1, involves repeatedly simulating Θ_P within a bounded region B, constructing $\mathcal{V}_P$ and then identifying and measuring only the central cell. This approach involves two considerations: the choice of a single parameter λ (the intensity of Θ_P) and the number of replications n of the procedure. The selection of n is determined primarily by the degree of accuracy at a specified confidence level required of the simulation estimates. The choice of λ represents a compromise between two concerns. If λ is too small the central polygon may intersect the boundary of B or be subject to edge effects (see above), but if it is too large many points will be generated that have no effect on the construction of the central polygon. Estimates of the appropriate value of λ vary considerably even in the case when B is a unit square (for example, Crain 1978, suggests $\lambda = 35$ is sufficient, while Hinde and Miles, 1980, use $\lambda = 100$). Once the values of λ and n have been selected this simulation procedure involves two steps: generating a Poisson distributed variable k with mean $\lambda\,|B|$ (where $|B|$ is the size of B) and then generating k independent uniform random points in B. However, Hinde and Miles (1980) have suggested that, rather than selecting k independently, it is more efficient to choose the values deterministically as the $(i-1/2)/n$ quantiles (i=1, ..., n) of the Poisson distribution. While they indicate that this approach is superior to selecting the $i/(n + 1)$ quantiles ($i = 1, \ldots, n$), they make no claims for its optimality.

The second variant of the sequence approach, method R in Table 5.3.1, was proposed by Boots and Murdoch (1983) and Quine and Watson (1984). It exploits the known joint distribution of ordered distances from the origin, 0, in Θ_P, since a set of increasing distances with this distribution together with a sequence of random directions gives Θ_P. In two dimensions this simulation process begins by locating the first point, p_0, at 0, additional points p_i ($i = 1, 2, \ldots$) are then generated at random in increasing order of distance from p_0. As p_i is generated the perpendicular bisector of $\overline{p_0 p_i}$ is drawn. After this step is repeated a number times p_0 will be surrounded by a convex polygon, P_0. Additional points are then generated until these points cease to influence the form of P_0. This occurs when the distance from p_i exceeds the diameter of the smallest circle centred at p_0 containing P_0. At that time the required characteristics of the polygon P_0 are measured and the entire procedure repeated for the next polygon. Unlike method T this procedure is essentially scale invariant and so the choice of λ is no longer a consideration. Also, since the number of points required to produce the final P_0 is usually about 15 to 20, this procedure is more efficient computationally than method T.

Quine and Watson (1984, p. 551) suggest that a modification of this procedure may be used to gain information about the tails of the distributions of the characteristics of the PVCs. This involves generating the polygon P_1 around p_1, ignoring p_0 in the construction of P_1. Thus, P_1 contains p_0. Since it is known that in any random tessellation such as $\mathcal{V}_P$ the cell containing the origin is larger on average than other cells (Matheron, 1975, pp. 168–169; Stoyan, 1986), polygons which result from this modified procedure will be larger than those generated by the basic procedure. Thus, knowledge of their characteristics will help improve the estimates of the tails of the distributions. Drouffe and Itzykson (1984) use a refinement of this approach to ensure that only polygons with a specified number of sides are created.

In $\mathbb{R}^3$ Quine and Watson suggest it is easier to work with $\mathcal{D}_P$ rather than $\mathcal{V}_P$. An initial very large simplex is established with its circumcentre at p_0. This is subdivided to create a list of Delaunay simplices. As a new point is established this list is updated to include those simplices with the new point as a vertex. This process continues until the distance from p_0 to a new point exceeds twice the maximum circumradius of those simplices with a vertex at p_0.

Relative to simulating $\mathcal{V}_P$, less effort has been expended in the simulation of $\mathcal{D}_P$. In part this is because analytical approaches (particularly the work of Miles, 1970a) have been successful in deriving more thorough knowledge of typical cells than is the case for $\mathcal{V}_P$. In addition, for $m \leq 3$ the methods of simulating $\mathcal{V}_P$ are also efficient for constructing $\mathcal{D}_P$. However, this is not the case for $m > 3$ and Kendall (1989) has developed a procedure for simulating individual cells which is particularly efficient in such circumstances. The approach is based on a theorem derived by Miles (1970a) (see equation (5.9.2)) which shows that the size and shape of a PDC are independent if the size is measured in terms of the radius r of its circumsphere and the shape by the angles at its vertices. Furthermore, since the distribution of r is known in terms of m (see Section 5.9), the problem of simulation reduces to that of constructing the shape of the PDC. Kendall's approach involves creating a sequence of simplices, each of which consists of $(m + 1)$ independent points in $\mathbb{R}^m$, from a standard Gaussian distribution. For each simplex the values of r (the circumradius), and l, the square root for the sum of the squares of the distances of the $(m + 1)$ points from their centroid, are used to create a shape variable $\rho = r/l$. The jth such simplex in the sequence, with shape measure ρ_j, is retained and its shape recorded if and only if

$$U_j \leq 1/(\alpha_m \rho_j^{m^2}) ,$$

where $\alpha_m = (m + 1)^{m^2/2}$ and U_j is the jth of an independently identically distributed (i.i.d.) sequence of random variables uniformly distributed on the unit interval [0,1]. Kendall (1987) proves that the sequence of recorded shapes will be an i.i.d. sequence of shapes of PDCs. The probability of accepting a simplex in this procedure is known exactly as a function of m (Kendall, 1989). As $m \to \infty$ this probability tends to zero, thus making the method unpractical for values of $m > 6$. Kendall (1983) has used such an approach to simulate the shapes of 5000 PDCs for $m = 2$ (see Section 5.9).

For larger values of m, Kendall (1988) has developed another simulation method which generates values of two regularity indices for the cells rather than the cells themselves. These indices are F, the volume of a PDC relative to the volume of a regular simplex having the same circumradius, and $\gamma = \log(1/F)$. For $m \geq 3$, Kendall obtains an expression for the expectation of the k moments of F as the product of $2m$ beta functions. Since each of these beta functions is the $(k/2)$th moment of a random variable on [0,1], F can be expressed as the square root of a product of $2m$ independent beta variables. Simulation procedures for such beta variables are available (Ripley, 1987) and so Kendall's method provides a means of simulating an i.i.d. sequence of F values. Initial results suggest that the approach produces useful results for values of m up to at least $m = 10\,000$.

5.4 PROPERTIES OF POISSON VORONOI CELLS

For the Poisson Voronoi diagram $\mathcal{V}_P$ to be used in evaluating empirical tessellations, extensive knowledge of the characteristics of its constituent Poisson Voronoi cells (PVCs) is required. This information includes moments and distributions of individual characteristics of PVCs and correlations between pairs of characteristics. Although in principle it should be possible to derive this information analytically, the success of this approach has been limited to the derivation of first-order moments. With only a few exceptions, higher order moments, distributions and correlations have been estimated using Monte Carlo methods described in Section 5.3.

Here we begin by considering first-order moment measures. Many of these can be obtained from general expressions for $\mathcal{V}_P$ in $\mathbb{R}^m$ which can be derived by either ergodic approaches (Miles, 1970b, 1974) or Palm measures (Møller, 1989). If λ is the intensity of the homogeneous Poisson point process, Θ_P in $\mathbb{R}^m$, a main result (Møller, 1989, Theorem 7.2, p. 63) states that the expected value $\lambda_t \mathrm{E}_t V_s$ of the total s-content of the s-faces contained in a typical t-face of $\mathcal{V}_P$ is

$$\lambda_t \mathrm{E}_t V_s = \frac{\lambda^{[(m-s)/m]} 2^{(m-s+1)} \pi^{[(m-s)/2]} \Gamma[(m^2-sm+s+1)/2] \, \Gamma(m/2+1)^{(m-s+s/m)} \Gamma(m-s+s/m)}{(t-s)!(m-t+1)!\Gamma[(m^2-sm+s)/2]\Gamma[(m+1)/2]^{(m-s)}\Gamma[(s+1)/2]m} \quad (0 \leq s \leq t \leq m) . \quad (5.4.1)$$

If we limit our attention to the m-dimensional polytopes of $\mathcal{V}_P$, $t = m$ and

Θ_{Pm} is the set of centroids of the PVCs, so that $\lambda_m = \lambda$ and (5.4.1) reduces to

$$E_m V_s = \frac{\dfrac{2^{(m-s+1)}\pi^{[(m-s)/2]}\Gamma[(m^2-sm+s+1)/2]}{\Gamma(m/2+1)^{(m-s+s/m)}\Gamma(m-s+s/m)}}{(m-s)!\Gamma[(m^2-sm+s)/2]\Gamma[(m+1)/2]^{(m-s)}\Gamma[(s+1)/2]m\lambda^{s/m}} \quad (0 \leq s \leq m) \qquad (5.4.2)$$

(see also Miles, 1970b, 1974, equation (75)). Also for $s = 0$ the total s-content of the s-faces, V_0, is the number of s-faces, N_0, so that

$$\lambda_t E_t N_0 = \frac{\lambda 2^{(m+1)}\pi^{[(m-1)/2]}\,\Gamma[(m^2+1)/2](m-1)!}{t!(m-t+1)!\Gamma(m^2/2)m}\left\{\frac{\Gamma(m/2+1)}{\Gamma[(m+1)/2]}\right\}^m \qquad (5.4.3)$$

(see Møller, 1989, p. 66, equation 7.6). Furthermore, setting $t=0$ in equation (5.4.3) yields

$$\lambda_0 = \frac{\lambda 2^{(m+1)}\pi^{[(m-1)/2]}\Gamma[(m^2+1)/2]}{(m+1)\, m^2\Gamma(m^2/2)}\left\{\frac{\Gamma(m/2+1)}{\Gamma[(m+1)/2]}\right\}^m, \qquad (5.4.4)$$

so that using equation (5.1.1) from Section 5.1:

$$\lambda_1 = \frac{\lambda 2^m \pi^{[(m-1)/2]}\Gamma[(\mathrm{m}^2+1)/2](\mathrm{m}-1)!}{\mathrm{m}^2\Gamma\,(\mathrm{m}^2/2)}\left\{\frac{\Gamma(\mathrm{m}/2+1)}{\Gamma[(\mathrm{m}+1)/2]}\right\}^m. \qquad (5.4.5)$$

In accordance with our concern for empirical applications we present the values of equations (5.4.1)–(5.4.5) for $m = 2$ and $m = 3$ in Tables 5.4.1 and 5.4.2, respectively. To minimize the incidence of subscripts in the text,

Table 5.4.1 Expected values of characteristics of a typical Poisson Voronoi cell in $\mathbb{R}^2$

Characteristic	Symbol	Numerical value
Intensity of cell vertices	λ_0	2λ
Intensity of mid-points of cell edges	λ_1	3λ
Intensity of cell centroids	λ_2	λ
E_2N_0 (E_2N_1) Number of vertices (edges)	$E(N)$	6
Number of full neighbours	$E(N^2)$	37.781
E_2V_2 Area	$E(C)$	4
	$E(A)$	λ^{-1}
E_2V_1 Perimeter	$E(A^2)$	$1.280\lambda^{-2}$
	$E(P)$	$4\lambda^{-1/2}$
E_1V_1 Length of a cell edge	$E(P^2)$	$16.945\lambda^{-1}$
	$E(L)$	$0.667\lambda^{-1/2}$
	$E(L^2)$	$0.630\lambda^{-1}$

λ = intensity of Θ_P.
Sources: Meijering (1953), Gilbert (1962), Hanson (1983) and Brakke (1986).

Table 5.4.2 Expected values of characteristics of a typical Poisson Voronoi cell in $\mathbb{R}^3$

Characteristic	Symbol	Exact value	Numerical value
Intensity of cell vertices	λ_0	$(24/35)\pi^2\lambda$	6.768λ
Intensity of mid-points of cell edges	λ_1	$(48/35)\pi^2\lambda$	13.535λ
Intensity of centroids of cell faces	λ_2	$(24\pi^2/35+1)\lambda$	7.768λ
Intensity of cell centroids	λ_3	λ	λ
E_3N_0 Number of vertices/cell	$E(M)$	$(96/35)\pi^2$	27.071
E_3N_1 Number of edges/cell	$E(E)$	$(144/35)\pi^2$	40.606
E_3N_2 Number of faces/cell	$E(F)$	$(48\pi^2/35)+2$	15.535
	$E(F^2)$		252.392*
Number of full neighbours	$E(C)$	8	8
	$E(C^2)$		68.115†
E_2N_0 (E_2N_1) Number of vertices (edges)/face	$E(N)$	$(144\pi^2)/(24\pi^2+35)$	5.228
	$E(N^2)$		29.822*
E_3V_3 Volume	$E(V)$	λ^{-1}	λ^{-1}
	$E(V^2)$		$1.180\lambda^{-2}$
E_3V_2 Surface area	$E(S)$	$(256\pi/3)^{1/3}\Gamma(5/3)\lambda^{-2/3}$	$5.821\lambda^{-2/3}$
	$E(S^2)$		$36.076\lambda^{-4/3}$*
E_3V_1 Total edge length/cell	$E(B)$	$[(4\pi)^{5/3}\Gamma(1/3)]/[5(9\lambda)^{1/3}]$	$17.486\lambda^{-1/3}$
	$E(B^2)$		$319.453\lambda^{-2/3}$*
E_2V_2 Area of a face	$E(A)$	$[(35)2^{8/3}\pi^{1/3}\Gamma(2/3)]/[(9\lambda)^{2/3}(24\pi^2+35)]$	$0.375\lambda^{-2/3}$
	$E(A^2)$		$0.283\lambda^{-4/3}$*
E_2V_1 Total edge length/face	$E(P)$	$[(7)2^{10/3}\pi^{\prime/3}\Gamma(1/3)]/[(9\lambda)^{1/3}(24\pi^2+35)]$	$2.252\lambda^{-1/3}$
	$E(P^2)$		$6.453\lambda^{-2/3}$*
E_1V_1 Edge length	$E(L)$	$[7\Gamma(1/3)]/(36\pi\lambda)^{1/3}$	$0.431\lambda^{-1/2}$
	$E(L^2)$		$0.291\lambda^{-2/3}$*
Average breadth‡	$E(b)$	$[(16\pi^5)/243\lambda]^{1/3}[\Gamma(1/3)/5]$	$1.458\lambda^{-1/3}$
	$E(b^2)$		$2.156\lambda^{-2/3}$*
Randomly selected dihedral angle§	$E(\alpha_3)$		2.070*
	$E(\alpha_3^2)$		4.464*
Randomly selected interior angle of a typical face	$E(\alpha_2)$		1.826*
	$E(\alpha_2^2)$		3.762*
Dihedral angle on a typical edge	$E(\alpha_1)$	$2/3\pi$	2.094
	$E(\alpha_1^2)$		4.558*

λ = intensity of Θ_P.
*Estimates from Monte Carlo simulations by Lorz (1990b, 1991).
†Estimates from Monte Carlo simulations by Quine and Watson (1984).
‡Mean length of the cell's protection onto an isotropic random line (also known as the mean caliper diameter).

each of these values, $E_m V_s$, is given a simpler symbol, which is identified in Tables 5.4.1 and 5.4.2. Many of these values were first obtained individually by Meijering (1953), including the expected number of full neighbours $E(C)$. Two cells are full (or Gabriel) neighbours in $\mathbb{R}^2$ ($\mathbb{R}^3$) if the line joining their generator points intersects their common edge (polygonal face) (see Section 2.5). Also included in Table 5.4.1 are second-order moments derived by numerical integration for the number of cell vertices (edges) N, the cell perimeter P, the length of a cell edge L (Brakke, 1986) and cell area A (Gilbert, 1962; Hanson, 1983; Brakke, 1986). Similarly, Table 5.4.2 includes the second-order moment for cell volume V derived by numerical integration (Gilbert, 1962) and higher order moments from Monte Carlo simulations by Quine and Watson (1984) and Lorz (1990b, 1991). In Table 5.4.2 emphasis is given to the results of Lorz since his study involves 82 870 cells compared with 3000 and 2500 cells in the studies of Hanson (1983) and Quine and Watson, respectively. Note that the value of $E(N) = 6$ in Table 5.4.1 is a specific instance of a general result

$$E(N) = 2\chi/(\chi - 2) \,, \qquad (5.4.6)$$

where χ is the mean number of sides meeting at each vertex of the tessellation, which holds for all homogeneous random planar tessellations (Matschinski, 1954).

For $m = 2$, in addition to the unconditional moments reported in Table 5.4.1, Quine and Watson (1984, Tables 1 and 2) also estimate the first four conditional moments for the area A and perimeter P, given the number of vertices (sides) N of a PVC. Conditional moments of A are also estimated by Crain (1978, Table 3), Drouffe and Itzykson (1984) and Le Caer and Ho (1990). Crain (1978) and Boots (1987) have also estimated conditional moments of P and the length of a cell edge L, respectively. These results are summarized in Table 5.4.3. On the basis of the results for A, Quine and Watson conjectured that

$$E(A|N) = (2N-3)/9\lambda \,, \qquad (5.4.7)$$

where $(A|N)$ is the area A conditional on N, which is consistent with $E(A) = \lambda^{-1}$, but they are unable to offer a proof. The values for $E(A|N)$ in Table 5.4.3 are also consistent with 'Lewis's law' which describes a linear relationship between $E(A|N)$ and N of the form

$$E(A|N) = b(A_0/c)(N-6) + (A_0/c) \qquad (5.4.8)$$

where c is the number of cells in the tessellation and A_0 is the total area of the tessellation. This relationship was originally observed for a variety of biological tessellations (including cucumber epidermis and pigmented epithelium of the retina) (Lewis, 1928, 1930, 1931, 1943, 1944) and has more recently been derived (Rivier and Lissowski, 1982; Rivier, 1985) using an approach which maximizes the entropy of the distribution of N, subject to the following three constraints:

Table 5.4.3 Conditional moments of characteristics of a typical Poisson Voronoi cell in $\mathbb{R}^2$

Number of edges N	Area $E(A\|N)\lambda$ a	b	c	d	Perimeter $E(P\|N)\lambda^{1/2}$ c	d	Edge length $E(L\|N)\lambda^{1/2}$ e
3	0.342	0.342	0.359	0.360	2.803	2.828	0.910
4	0.560	0.558	0.558	0.567	3.217	3.260	0.805
5	0.777	0.774	0.778	0.773	3.650	3.660	0.733
6	0.996	0.996	0.995	0.994	4.030	4.040	0.672
7	1.228	1.222	1.223	1.214	4.382	4.396	0.625
8	1.463	1.451	1.447	1.447	4.701	4.728	0.590
9	1.693	1.688	1.661	1.710	4.981	5.068	0.553
10	1.930	1.938	1.868	1.850	5.244	5.240	0.519
11	2.155	2.16	2.063		5.475		0.512
12	2.400	2.37	1.990		5.415		0.473
13	2.691	2.6					
14	2.930						
15	3.200						
16	3.46						
18	4.00						
20	4.53						
25	5.84						
30	7.17						
35	8.42						
40	9.73						
45	11.02						
50	12.28						

λ = intensity of Θ_P.
Sources: a Drouffe and Itzykson (1984, Table 1).
b Le Caer and Ho (1990, Table 6).
c Quine and Watson (1984, Tables 1 and 2).
d Crain (1978, Table III).
e Boots (1987, Table 1).

$$\sum_N p(N) = 1 \ , \tag{5.4.9}$$

$$\sum_N p(N)\mathrm{E}(A|N) = A_0/c \ , \tag{5.4.10}$$

$$\sum_N p(N)N = 6 \ . \tag{5.4.11}$$

Equation (5.4.9) is simply the normalization of the probability distribution of N. Equation (5.4.10) ensures that the tessellation is space exhaustive, while equation (5.4.11) is a derivative of Euler's theorem which relates the number of edges, vertices and cells for a tessellation in $\mathbb{R}^2$. Least squares fits of the data in Table 5.4.3 to equation (5.4.8) yield values of $b = 0.200$ (Quine and Watson, 1984), 0.219 (Crain, 1978), 0.229 (Le Caer and Ho, 1990) and 0.256 (Drouffe and Itzykson, 1984).

Using numerical integration Brakke (1986) derives the following correlations:

$$r(A,P) = 0.953 ,$$

$$r(A,N) = 0.568 ,$$

$$r(P,N) = 0.501 .$$

Almost identical values are obtained by Hinde and Miles (1980, Table V) from their simulation study. Estimations of the correlations $r(A,P|N)$ between A and P, controlling for N, are given by Crain (1978, Table 3). His results show that in general $r(A,P|N)$ increases with N, indicating increasing 'circularity'.

For $m = 3$, Quine and Watson (1984, Tables 4, 5 and 6) have estimated the first four conditional moments of the volume V, surface area S, and total edge length B, given the number of faces F of a PVC, while Lorz (1990b) estimates the same conditional moments for V, and also for the mean breadth b, and a randomly selected dihedral angle α_3 of a PVC. Again we report only the estimates for the first conditional moment of these characteristics (see Table 5.4.4). Lorz (1990b, 1991) also estimates various unconditional and conditional correlation coefficients for these characteristics (see Table 5.4.5). The values of $\mathrm{E}(V|F)$, where $V|F$ is the volume V conditional on F, in Table 5.4.4 show a reasonable fit with the three-dimensional version of Lewis's law (equation (5.4.7)) suggested by Rivier (1982, 1983a, 1983b, 1986a) and given by

$$\mathrm{E}(V|F) = b(V_0/c)[F - \mathrm{E}(F)] + (V_0/c) , \qquad (5.4.12)$$

where V_0 is the total volume of the tessellation and c is the total number of cells in the tessellation. In the derivation of equation (5.4.12) the two-dimensional space-filling constraint given by equation (5.4.10) is replaced by

$$\sum_F p(F)\mathrm{E}(V|F) = V_0/c , \qquad (5.4.13)$$

while the topological constraint

$$\sum_F p(F)F = 12/[6 - \mathrm{E}(N)] \qquad (5.4.14)$$

replaces the constraint in equation (5.4.11). Using the values of Lorz and Quine and Watson in Table 5.4.4 gives least squares estimates of $b = 0.095$ and $b = 0.088$, respectively, for equation (5.4.12).

Conditional moments and unconditional and conditional correlation coefficients for selected characteristics of the faces and edges of a PVC have

Table 5.4.4 Conditional moments of characteristics of a typical Poisson Voronoi cell in $\mathbb{R}^3$

Number of faces/cell F	Volume V $E(V\|F)\lambda$ a	b	Surface area S $E(S\|F)\lambda^{2/3}$ a	Total edge length B $E(B\|F)\lambda^{1/3}$ a	Average breadth $\bar{b}$ $E(\bar{b}\|F)\lambda^{1/3}$ a	Face angle α_3 $E(\alpha_3\|F)$ a
5	0.160	—	2.192	7.096	0.939	1.420
6	0.225	0.179	2.628	8.046	1.042	1.571
7	0.277	0.218	2.858	8.967	1.082	1.630
8	0.342	0.356	3.183	9.861	1.127	1.770
9	0.423	0.400	3.616	10.955	1.192	1.774
10	0.511	0.507	4.018	12.074	1.246	1.864
11	0.590	0.583	4.359	13.081	1.290	1.916
12	0.670	0.697	4.686	14.049	1.331	1.952
13	0.759	0.754	5.029	15.045	1.372	2.001
14	0.844	0.834	5.340	15.999	1.408	2.032
15	0.942	0.940	5.691	17.018	1.449	2.066
16	1.031	1.040	5.996	17.970	1.483	2.093
17	1.127	1.141	6.308	18.943	1.516	2.132
18	1.225	1.218	6.620	19.898	1.549	2.154
19	1.328	1.273	6.934	20.860	1.581	2.180
20	1.429	1.463	7.234	21.860	1.611	2.209
21	1.518	1.582	7.494	22.691	1.637	2.227
22	1.636	1.603	7.836	23.725	1.671	2.235
23	1.715	1.794	8.058	24.542	1.693	2.258
24	1.822	1.861	8.334	25.519	1.717	2.283
25	1.944	1.926	8.693	26.474	1.753	2.354
26	1.988	1.885	8.787	27.063	1.761	2.321
27	2.149	—	9.226	28.161	1.802	2.312
28	2.285	—	9.575	29.390	1.833	2.366
29	2.553	1.878	10.150	30.373	1.876	2.226
30	2.106	—	9.010	29.881	1.773	2.229
31	2.838	2.072	11.192	31.026	1.984	2.378

λ = intensity of Θ_P.
Sources: a Lorz (1990b).
b Quine and Watson (1984).

been estimated by Lorz (1990b) and are summarized in Tables 5.4.6 and 5.4.7, respectively.

Since many of the moment measures for the PVCs given in Tables 5.4.1–5.4.4 and 5.4.6 may not differ greatly from the corresponding measures for cells in tessellations where the generating set is located in $\mathbb{R}^m$ according to point processes other than Θ_P (Hermann *et al.*, 1989), for empirical applications it is desirable to know the nature of the distributions of the PVC characteristics. In particular most attention has been focused on the distributions of N, A in $\mathbb{R}^2$ and V in $\mathbb{R}^3$, since the first of these is easy to

Table 5.4.5 Correlation coefficients for characteristics of a typical Poisson Voronoi cell in $\mathbb{R}^3$

	V	S	B	b	α_3
F	0.7356	0.7116	0.8743	0.6713	0.2503
V	—	0.9819	0.9374	0.9448	0.1921
S	—	—	0.9247	0.9874	0.1852
B	—	—	—	0.8980	0.2261
b	—	—	—	—	0.1749

F = Number of faces/cell.
V = Volume.
S = Surface area.
B = Total edge length/cell.
b = Average breadth.
α_3 = Angle of a uniformly random selected face.
Source: Lorz (1991).

Table 5.4.6 Conditional moments of characteristics of the faces of a typical Poisson Voronoi cell in $\mathbb{R}^3$

Number of edges/face N	Area of a face A $E(A\|N)$	Total edge length/face P $E(P\|N)$	Interior angle of a face α_2 $E(\alpha_2\|N)$
3	0.035	0.754	1.048
4	0.136	1.528	1.571
5	0.303	2.237	1.884
6	0.504	2.831	2.095
7	0.709	3.309	2.245
8	0.908	3.699	2.358
9	1.095	4.024	2.448
10	1.260	4.278	2.512
11	1.432	4.530	2.573
12	1.537	4.665	2.632
13	1.724	4.933	2.706
14	1.807	5.152	2.639
15	2.166	5.532	2.778

Source: Lorz (1990b).

measure manually and the second can simply be determined by image processing systems.

We begin by considering $\mathcal{V}_P$ in $\mathbb{R}^2$. Although Miles and Maillardet (1982, p. 108) have been able to derive an expression for the probability $p(N)$ that a PVC has N vertices (edges), this is computationally complex and the only distributions that have not been obtained from simulation studies are those

Table 5.4.7 Estimates of correlation coefficients for characteristics of the faces and edges of a typical Poisson Voronoi Cell in $\mathbb{R}^3$

	A	P	α_2	α_1
N	0.7510	0.7468	0.5774	—
A	—	0.9280	0.4183	—
P	—	—	0.4564	—
L	—	—	—	−0.0000

N = Number of vertices (edges)/face.
A = Area of a face.
P = Total edge length/face.
L = Edge length.
α_2 = Interior angle of a face.
α_1 = Face (dihedral) angle.
Source: Lorz (1990b).

relating to the interior angles of a PVC. These distributions can be derived because of the duality between $\mathcal{V}_P$ and $\mathcal{D}_P$ (see Section 2.2). In the course of an extensive study of $\mathcal{D}_P$ (see Section 5.9), Miles (1970a) obtained the probability density function (pdf) of a pair of angles selected at random from an arbitrary triangle of $\mathcal{D}_P$ (see equation (5.9.7)). Since there is a complementarity in π between the interior angles of the polygons at a vertex in $\mathcal{V}$ and the angles of the Delaunay triangle whose circumcentre is that vertex, equation (5.9.8) can be used to derive the pdf $f(\alpha)$ of a randomly selected angle α of an arbitrary PVC of $\mathcal{V}_P$. From Miles (1970a, equation 9.3) (see also Icke and Van de Weygaert, 1987) we obtain

$$f(\alpha) = [(4\sin \alpha/3\,\pi)](\sin \alpha - \alpha \cos \alpha) \quad (0 \leq \alpha \leq \pi)\ , \qquad (5.4.15)$$

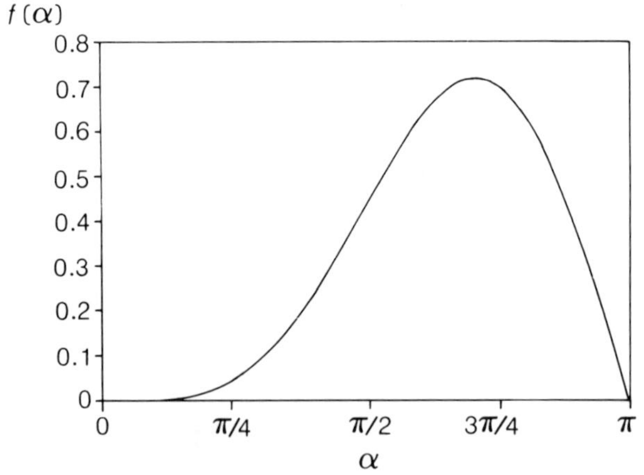

Figure 5.4.1 Probability density function of randomly selected angle α of a typical Poisson Voronoi cell in $\mathbb{R}^2$.

which gives $E(\alpha) = 2\pi/3$ and $E(\alpha^2) = (5\pi^2/2 - 5/6)$. The mode of equation (5.4.15) is given by the root of $\tan 2\alpha = 2\alpha$ ($\alpha<\pi$) which is approximately equal to 2.2467 radians (128° 43′ 36″). In addition, $p(\alpha < \pi/2) = 1/6$. The distribution of $f(\alpha)$ is shown in Figure 5.4.1. We may also condition α in terms of N and examine the distribution of α_N, a randomly selected angle of an arbitrary N-sided PVC. Such angles will have a continuous distribution $f(\alpha_N)$ on $(0, \pi)$ with an expected mean value

$$E(\alpha_N) = (N-2)\,\pi/N\,. \tag{5.4.16}$$

Hinde and Miles (1980, Figure 5) have estimated $f(\alpha_N)$ for $N = 3, 4, \ldots, 14$.

Table 5.4.8 (see also Figure 5.4.2) gives the estimated values $\hat{p}(N)$ of $p(N)$ obtained by Hinde and Miles (1980), Drouffe and Itzykson (1984) and Le Caer and Ho (1990). Note that the largest value of N encountered amongst the two million polygons generated by Hinde and Miles is 14. On

Table 5.4.8 Estimates, $\hat{p}(N)$, of the probability that a typical Poisson Voronoi cell in $\mathbb{R}^2$ has N sides

	(1)		(2)		(3)	
N	$\hat{p}(N)$	$\hat{\sigma}(N)$	$\hat{p}(N)$	$\hat{\sigma}(N)$	$\hat{p}(N)$	$\hat{\sigma}(N)$
3	0.01131	0.00007	0.01127	0.00894	0.0113	0.0002
4	0.1071	0.00022	0.1077	0.03316	0.1068	0.0002
5	0.2591	0.00031	0.258	0.04472	0.2595	0.0005
6	0.2944	0.00032	0.294	0.05477	0.2946	0.0003
7	0.1991	0.00028	0.198	0.05477	0.1986	0.0001
8	0.0902	0.00020	0.090	0.14142	0.0905	0.0003
9	0.0295	0.00012	0.0288	0.02646	0.0295	0.0003
10	0.00743	0.00006	0.00695	0.01414	0.0074	0.0001
11	0.00149	0.00003	0.00153	0.00949	0.00144	0.00003
12	0.00025	0.00001	0.00024	0.00548	—	—
13	3.4×10^{-5}	0.4×10^{-5}	2.9×10^{-5}	0.00224	—	—
14	5.0×10^{-6}	1.6×10^{-6}	2.8×10^{-6}	0.00063	—	—
15	—	—	2.6×10^{-7}	0.00022	—	—
16	—	—	2.3×10^{-8}	0.00005	—	—
18	—	—	1.3×10^{-10}	7.1×10^{-6}	—	—
20	—	—	1.5×10^{-13}	2.8×10^{-7}	—	—
25	—	—	9.6×10^{-21}	7.7×10^{-11}	—	—
30	—	—	1.3×10^{-29}	3.3×10^{-15}	—	—
35	—	—	3.6×10^{-40}	—	—	—
40	—	—	2.4×10^{-50}	—	—	—
45	—	—	2.0×10^{-63}	—	—	—
50	—	—	1.5×10^{-75}	—	—	—

$\hat{\sigma}(N)$ = standard deviation of $\hat{p}(N)$.
Sources: (1) Hinde and Miles (1980, Table III).
(2) Drouffe and Itzykson (1984, Table 1).
(3) Le Caer and Ho (1990, Table 1).

the basis of their estimates, Drouffe and Itzykson suggest that asymptotically $p(N) \sim N^{-2N}$.

Numerous researchers, including Crain (1972, 1978), Di Cenzo and Wertheim (1989), Hermann *et al.* (1989), Kiang (1966) and Weaire *et al.* (1986), have produced estimates of the distribution $f(A)$ of cell area, while Quine and Watson (1984) have examined the distribution of A conditioned by N. Figure 5.4.3 shows the distribution of a, a standardized measure of A (where $a = \lambda A$) (Hinde and Miles, 1980, Figure 4).

There has been considerable conjecture on the theoretical form of $f(A)$. Hinde and Miles (1980, p. 219) fit a three-parameter generalized gamma function of the form

$$f(x) = rb^{q/r}x^{q-1}\exp(-bx^r)/\Gamma(q/r) \quad (r>0, b>0, q>0)\ . \qquad (5.4.17)$$

Values obtained from equation (5.4.17) with $r = 1.0787$, $b = 3.0328$ and $q = 3.3095$ are plotted in Figure 5.4.3. Visually this provides a good fit although it is rejected by a chi-square test. Hermann *et al.* (1989, p. 194) claim that Maxwell, lognormal and gamma distributions fit equally well the values of A derived from their simulations. The general form of the last distribution is that of equation (5.4.17) with $r = 1$,

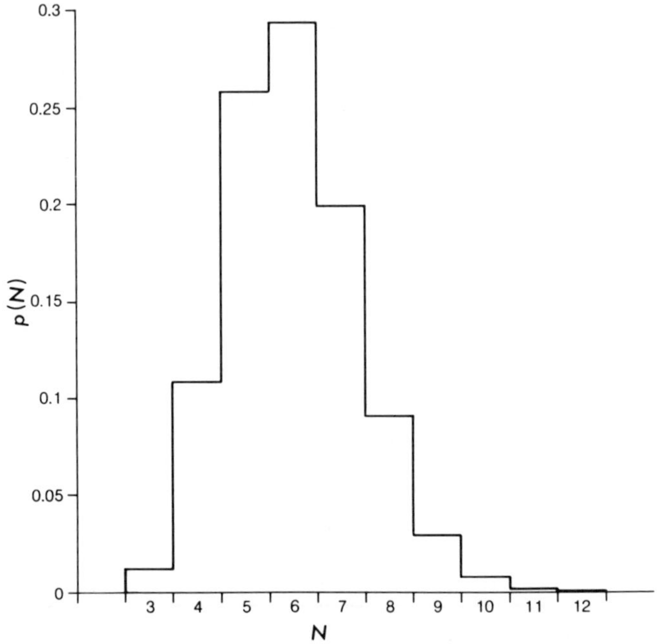

Figure 5.4.2 Normalized histogram of the number of sides N of a typical Poisson Voronoi cell in $\mathbb{R}^2$ (values from Drouffe and Itzykson, 1984).

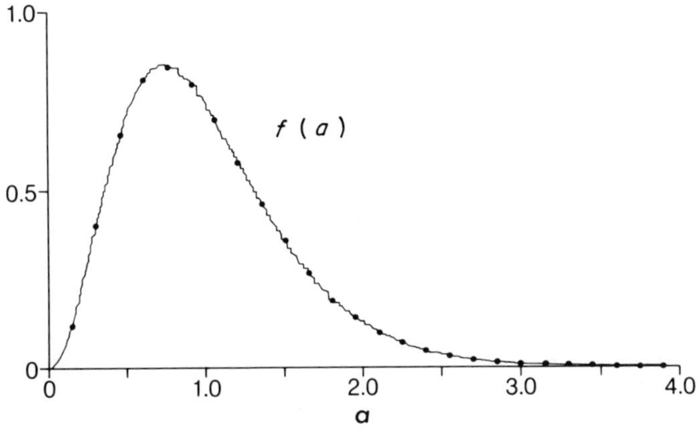

Figure 5.4.3 Normalized histogram of standardized area $a(=\lambda A)$ of a typical Poisson Voronoi cell in $\mathbb{R}^2$ showing point values of the generalized gamma fit (5.4.17). (*Source*: Hinde and Miles (1980).)

$$f(x) = b^q x^{q-1} \exp(-bx)/\Gamma(q) \ . \tag{5.4.18}$$

The moments of (5.4.18) are given by

$$\mu_k = [\Gamma(q+k)/\Gamma(q)]b^{-k} \ . \tag{5.4.19}$$

On the basis of his simulations, Kiang (1966) suggests that $f(A)$ follows equation (5.4.18) with $q = 4$ and $b = 4\lambda$. Both Rivier (1986b) and Weaire *et al.* (1986) offer some limited analytical support for $f(A)$ having a gamma distribution, but the latter authors suggest values of $q = 3.61$ and $b = 3.61\lambda$. Using equation (5.4.19) these values yield $E(A^2) = 1.277$, which is not consistent with Gilbert's value reported in Table 5.4.1. However, the fit to their observed values is excellent (Weaire *et al.*, 1986, p. L104). Di Cenzo and Wertheim (1989) also report an excellent fit to equation (5.4.18) with $q = 3.61$ and $b = 3.57\lambda$. Furthermore, Miles and Maillardet (1982, p. 101) suggest that when the number of sides N of a cell is large, the values of $A|N$ also approximate equation (5.4.18) with values $q = N$ and $b = 4\lambda$.

Hinde and Miles (1980, Figure 3) have estimated the distribution of p, a standardized measure of cell perimeter P (where $p = (\lambda^{1/2}/4)P$) (see Figure 5.4.4), while Quine and Watson (1984, Figure 3) have studied the distribution of P conditioned on N. Hinde and Miles again fit equation (5.4.17) to their data with $r = 2.3389$, $b = 2.9563$ and $q = 7.5579$. This fit is shown on Figure 5.4.4. As with a, while the visual impression is one of a good fit, it is rejected by a chi-square test (at the 0.05 significance level). They also note that, since their estimates of the coefficients of skewness and kurtosis of p are 0.193 and 2.983, respectively, p is approximately normally distributed with mean 1 and variance 0.0592. Values for this normal distribution are also plotted in Figure 5.4.4.

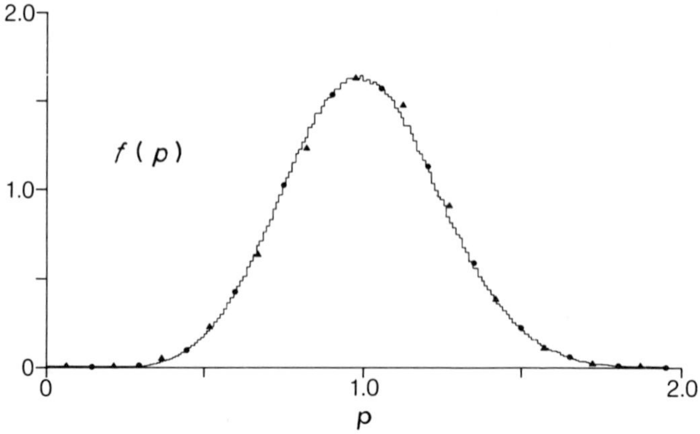

Figure 5.4.4 Normalized histogram of standardized perimeter $p(\lambda^{1/2}P/4)$ of a typical Poisson Voronoi cell in $\mathbb{R}^2$ showing point values for two fits: ● generalized gamma fit (5.4.17); ▲ normal fit. (*Source*: Hinde and Miles (1980).)

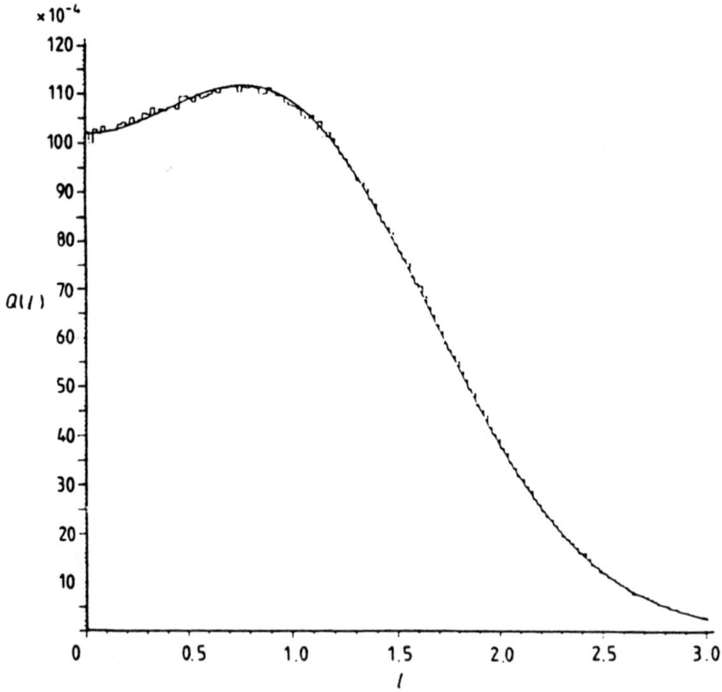

Figure 5.4.5 Normalized histogram of standardized length $l(=3\lambda^{1/2}L/2)$ of an edge of a typical Poisson Voronoi cell in $\mathbb{R}^2$ showing the fit (full curve) of (5.4.20). (*Source*: Le Caer and Ho (1990).)

Le Caer and Ho (1990) show that the distribution of l, a standardized measure of L, the length of an individual edge of a PVC (where $l = 3\lambda^{1/2} L/2$) is well approximated by the distribution of the absolute value of the Gaussian variable with mean μ and variance σ^2:

$$f(l) = \{1/[\sigma(2\pi^{1/2}]\} \{\exp[-(l-\mu)^2/2\sigma^2] + \exp[-(l+\mu)^2/2\sigma^2]\} \quad (l \geq 0) , \tag{5.4.20}$$

with $\mu = 0.9144$ and $\sigma = 0.7623$. This distribution, together with the distribution obtained from their simulation studies, is shown in Figure 5.4.5. This result is consistent with that from the simulations of Boots (1987).

Turning now to $\mathbb{R}^3$, Hanson (1983), Quine and Watson (1984) and Lorz (1990b, 1991) have estimated the distributions of the number of faces per cell F, the number of vertices (edges) per face N and cell volume V. The distribution of V was also estimated by Kiang (1966) and Tanemura (1988). Lorz's estimates for F and N are given in Tables 5.4.9 and 5.4.10 and illustrated in Figures 5.4.6 and 5.4.7, respectively. Both Quine and Watson and Lorz have also estimated the distributions of the surface area S and the total edge length B of a PVC, while Quine and Watson estimated the distribution of the number of full neighbours C and Lorz estimated the distributions of the mean cell breadth b and a uniformly selected face angle α_3. These distributions are illustrated in Figure 5.4.8. In addition, Lorz has also estimated the distributions of the perimeter P, the area A and a uniformly random selected interior angle α_2 of a face of a PVC and the

Table 5.4.9 Estimates, $\hat{p}(F)$, of the probability that a typical Poisson Voronoi cell in $\mathbb{R}^3$ has F faces

F	$\hat{p}(F)$	F	$\hat{p}(F)$
5	0.00006	20	0.04500
6	0.00019	21	0.02993
7	0.00193	22	0.01895
8	0.00578	23	0.01144
9	0.01496	24	0.00691
10	0.03102	25	0.00373
11	0.05164	26	0.00215
12	0.07806	27	0.00091
13	0.09865	28	0.00042
14	0.11492	29	0.00017
15	0.12088	30	0.000048
16	0.11418	31	0.000024
17	0.10297	32	0.000012
18	0.08219	33	—
19	0.06287	34	—
		35	0.000012

Source: Lorz (1991).

Table 5.4.10 Estimates, $\hat{p}(N)$, of the probability that a face of a typical Poisson Voronoi cell in $\mathbb{R}^3$ has N edges (vertices)

N	$\hat{p}(N)$	N	$\hat{p}(N)$
3	0.1355	10	0.0068
4	0.2301	11	0.0017
5	0.2404	12	0.00038
6	0.1900	13	0.000066
7	0.1162	14	0.0000047
8	0.0569	15	0.0000039
9	0.0220		

Source: Lorz (1990b).

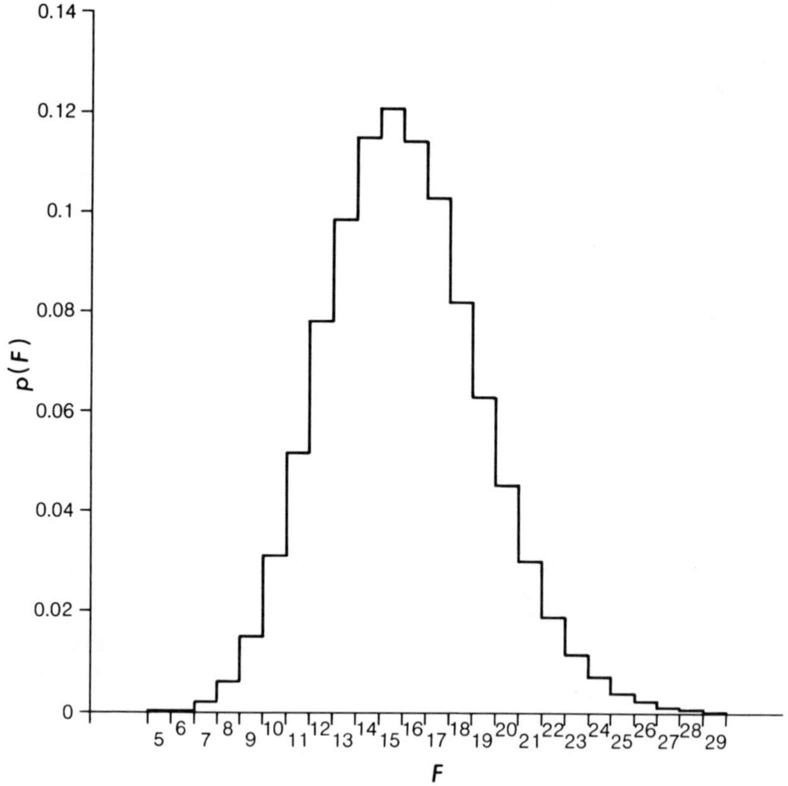

Figure 5.4.6 Normalized histogram of the number of faces F of a typical Poisson Voronoi cell in $\mathbb{R}^3$ (values from Lorz, 1991).

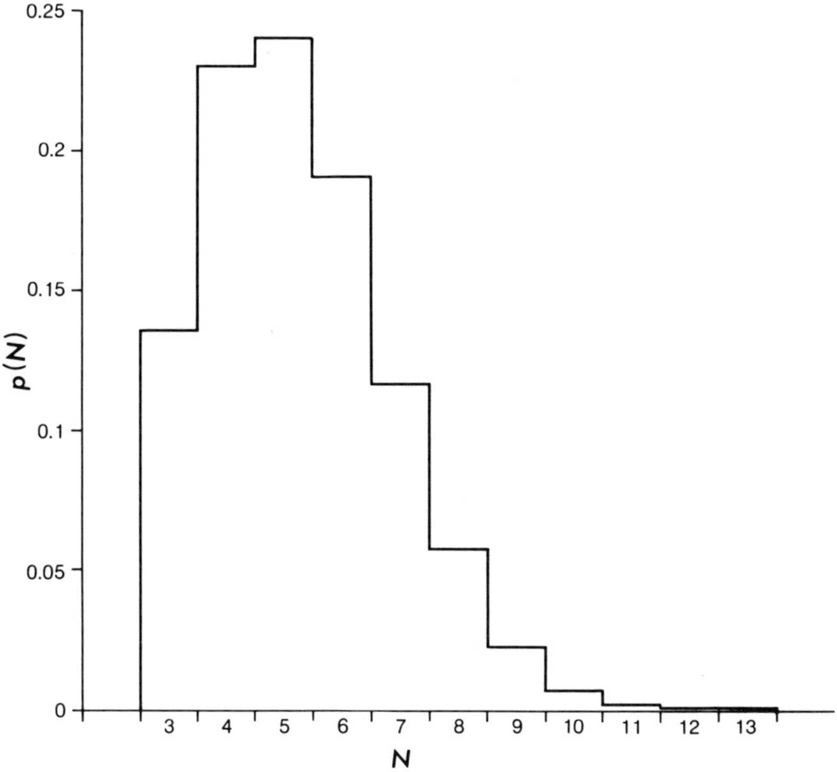

Figure 5.4.7 Normalized histogram of the number of vertices (edges) N of a randomly selected face of a typical Poisson Voronoi cell in $\mathbb{R}^3$ (values from Lorz, 1990b).

length L and the angle α_1 of an edge of a PVC, all of which are illustrated in Figure 5.4.9. On the basis of his estimates, Kiang (1966) conjectures that the distribution of V is of the form of equation (5.4.18) with $q = 6$ and $b = 6\lambda$ although this is not supported by the estimates of Tanemura (1988).

In some areas of application, particularly materials science and geography, there is considerable interest in how the individual cells are arranged within $\mathcal{V}_P$. Although $\mathcal{V}_P$ is an ergodic tessellation, we know that the characteristics of an individual cell cannot be independent of those of contiguous cells. So far these dependences have been explored by examining the relationships between the value of a given characteristic for an individual cell and the mean value of that characteristic for contiguous cells. Using a simulation approach (see Section 5.3) Boots and Murdoch (1983) estimate such relationships for the number of vertices (sides) and standardized measures of the area and perimeter of a PVC in $\mathbb{R}^2$. Figure 5.4.10(a) shows the values

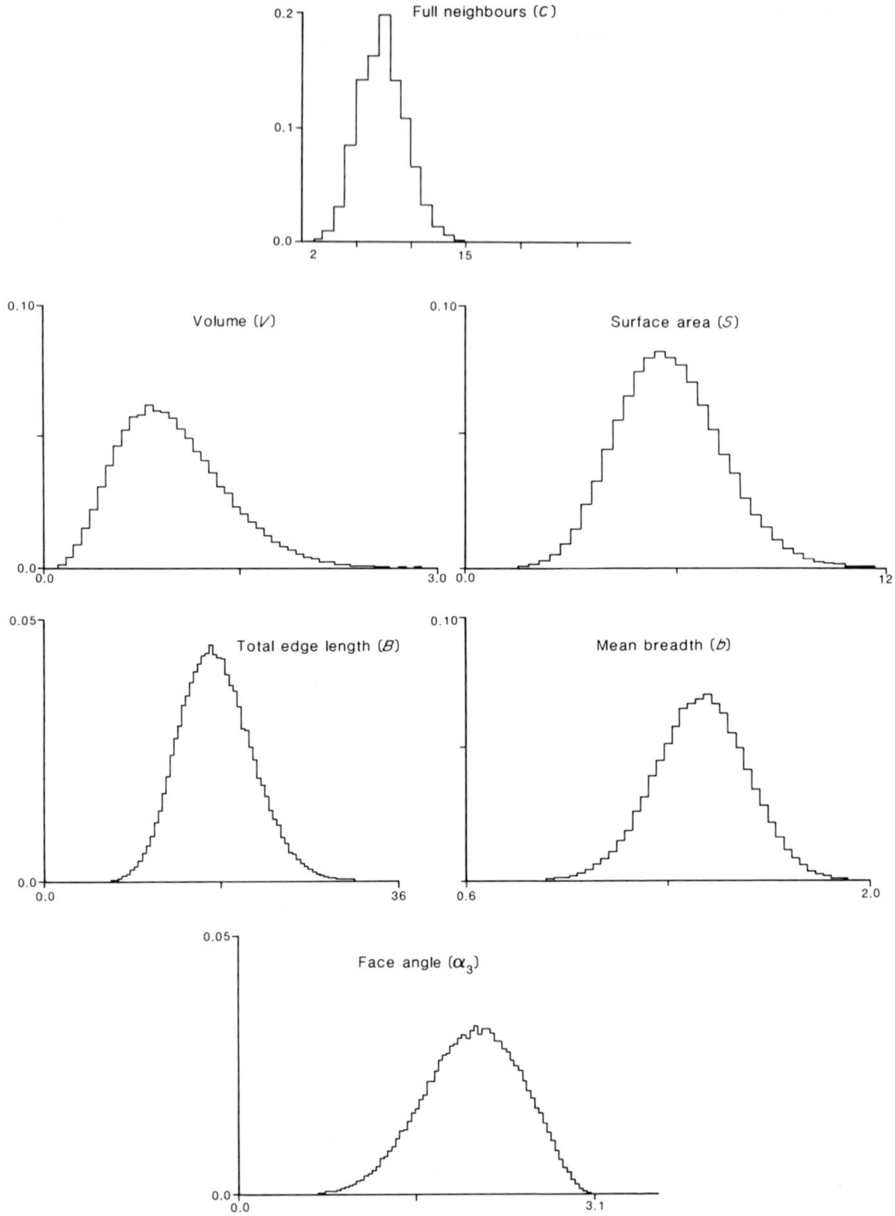

Figure 5.4.8 Normalized histograms of the number of full neighbours C, volume V, surface area S, total edge length b, mean breadth b and face angle α_3 of a typical Poisson Voronoi cell in $\mathbb{R}^3$. (*Source*: Quine and Watson (1984); Lorz (1990b).)

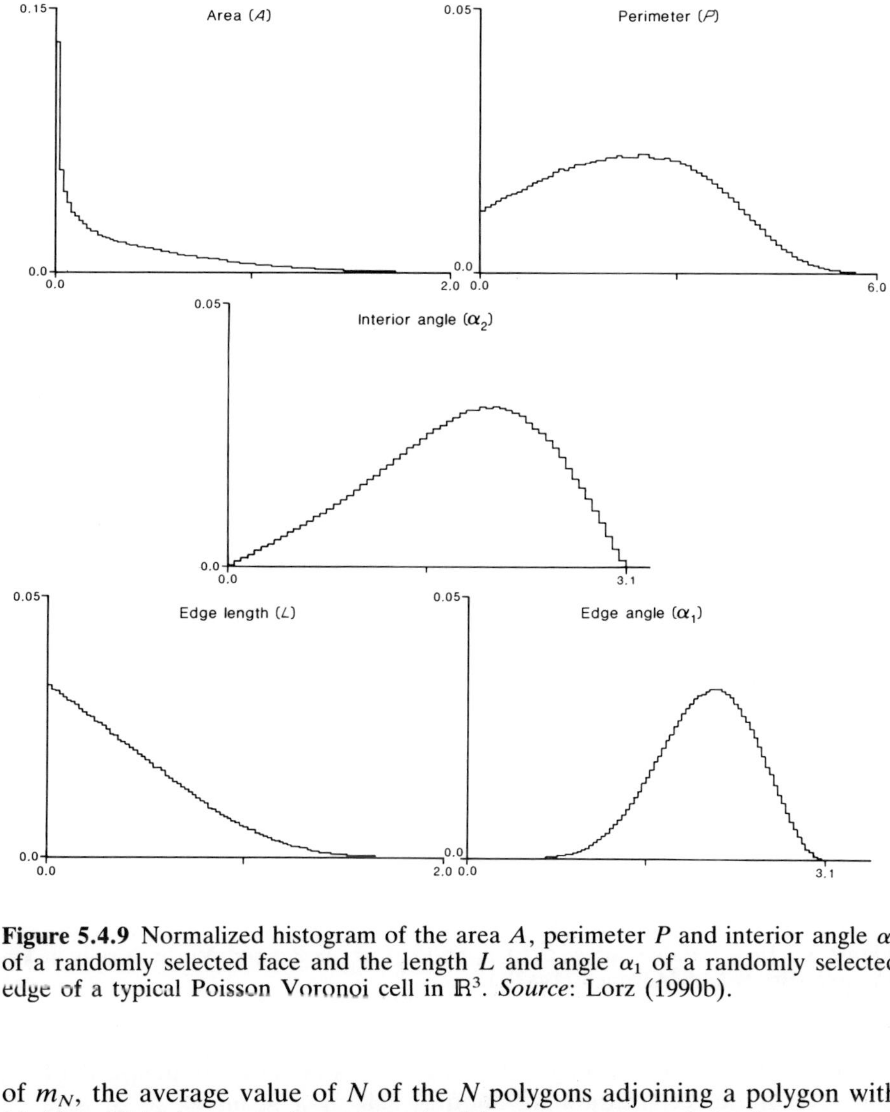

Figure 5.4.9 Normalized histogram of the area A, perimeter P and interior angle α_2 of a randomly selected face and the length L and angle α_1 of a randomly selected edge of a typical Poisson Voronoi cell in $\mathbb{R}^3$. *Source*: Lorz (1990b).

of m_N, the average value of N of the N polygons adjoining a polygon with N sides. This data may be summarized by an equation of the form

$$m_N = q + b/N + cN^2 , \tag{5.4.21}$$

where $q = 5.01$, $b = 9.68$ and $c = -11.18$. Aboav (1987) has also suggested that this data fits equally well the following equation, which is devoid of any arbitrary constants:

$$m_N = \mathrm{E}(N) + 2/N + (\mathrm{E}(N^{1/2}) - N^{1/2})/2 , \tag{5.4.22}$$

although he offers no theoretical support for this form.

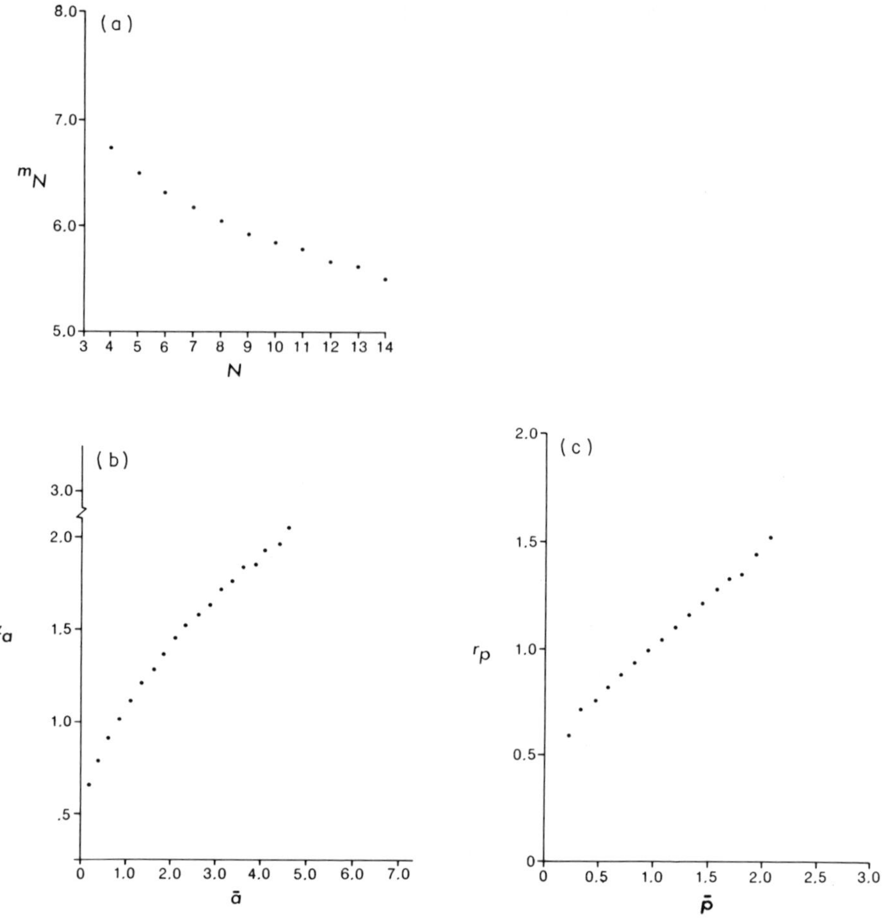

Figure 5.4.10 (a) Values of m_n; (b) z_a and (c) r_p for Poisson Voronoi cells in $\mathbb{R}^2$. (*Source*: Boots and Murdoch (1983).)

Values of z_a, the average value of a (a standardized measure of A; $a = \lambda A$) of the cells adjoining another cell of a given value of a in an interval δa are illustrated in Figure 5.4.10(b). The relationship may be summarized by

$$z_a = q + b\bar{a}^{1/2} + c\bar{a} \ , \tag{5.4.23}$$

where $q = 0.364$, $b = 0.629$, $c = 0.080$ and $\bar{a}$ is the mean value of observations in the interval δa. Finally, the values of r_p, the average value of p (a standardized measure of P; $p = \lambda^{1/2}P/4$), for cells adjoining a cell with a given value of p in the interval δp, are shown in Figure 5.4.10(c). These values indicate a linear relationship described by

$$r_p = q + b\bar{p} \,, \tag{5.4.24}$$

where $q = 0.564$, and $b = 0.456$ and $\bar{p}$ is the mean value of the observations in the interval δp.

Note that equation (5.4.21) and similar results of Le Caer and Ho (1990) are not consistent with 'Aboav's law' given by

$$m_N = 5 + b/N \,, \tag{5.4.25}$$

which describes the relationship originally observed in a number of empirical tessellations (Aboav, 1970, 1980, 1983, 1984, 1985). This relationship was later given theoretical justification by Weaire (1974) and Blanc and Mocellin (1979). Using the sum rule

$$\mathrm{E}(Nm_N) = \mathrm{E}(N^2) = \mathrm{Var}(N) + \mathrm{E}(N)^2 \,, \tag{5.4.26}$$

where Var(N) is the variance of N. Lambert and Weaire (1983) suggested that b in equation (5.4.25) is equal to $6 + \mathrm{Var}(N)$. Rivier (1985a) speculated that the lack of fit of $\mathcal{V}_P$ to equation (5.4.25) may be due to the fact that its theoretical derivation requires that there is no correlation beyond immediately contiguous cells and that such correlation may be present in $\mathcal{V}_P$. Rivier (1985a) has also derived a three-dimensional version of equation (5.4.25) which relates the average number of sides, m_N, of the contiguous faces of a N-sided face to N, given by

$$m_N = 5F - 11 - b(F - 1 - N) \,. \tag{5.4.27}$$

Paralleling (5.4.26) Rivier suggested that

$$\mathrm{Var}(N|F) = (F - 4)\,(F - 3)\,(5 - b - 12/F)/F \,. \tag{5.4.28}$$

However, since no information exists regarding m_N for $\mathcal{V}_P$ in $\mathbb{R}^3$, the applicability of equation (5.4.27) remains untested.

As indicated in Section 5.1 it is possible to generate other point processes from $\mathcal{V}_P$ which share properties of the point process Θ_P which generated $\mathcal{V}_P$. Two such processes with motion invariance are those of the vertices of $\mathcal{V}_P$ and the mid-points of the edges (edge centres) of $\mathcal{V}_P$. Concentrating on $\mathbb{R}^2$, if ρ is the intensity of the point process, the probability that an infinitesimally small disc of area $|A|$ contains a point of the process is $\rho|A|$. For two such discs with a distance r between their centres, the probability $p(r)$ that both discs contain a point of the process is

$$p(r) = (\rho|A|)^2 g(r) \,, \tag{5.4.29}$$

where $g(r)$ is known as the correlation function (Stoyan *et al.*, 1987, p. 120). A second quantity, $\delta(r)$, known as the second-order product density, is defined as

$$\delta(r) = \rho^2 g(r) \,. \tag{5.4.30}$$

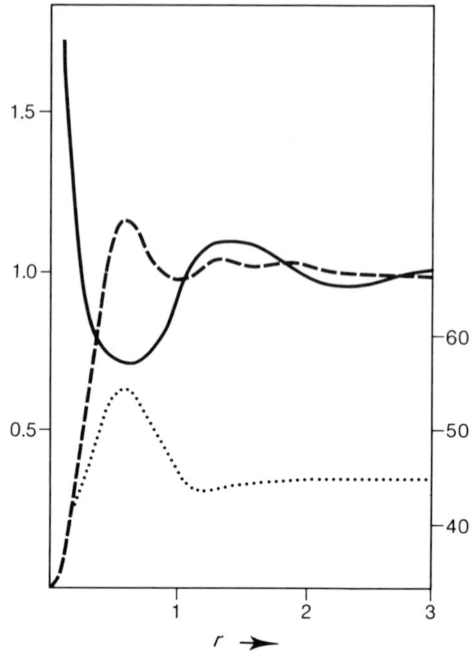

Figure 5.4.11 Pair correlation and angle correlation functions for the Poisson Voronoi diagram in $\mathbb{R}^2$: —— pair correlation function for vertices; – – – – pair correlation function for edge centres; –·–·–·–. angle correlation function $g_\alpha(r)$. *Source*: Stoyan and Stoyan (1990).

For Θ_P, $g(r) = 1$ and $\delta(r) = \lambda^2$. Stoyan and Stoyan (1990) provided methods for estimating ρ, $g(r)$ and $\delta(r)$ from point patterns observed within a bounded region in $\mathbb{R}^2$. Using Monte Carlo simulation, they estimate $g(r)$ for both the vertices and edge centres of $\mathcal{V}_P$. These functions are shown in Figure 5.4.11. Note that for both distributions, for $r > 1$, $g(r)$ approximates that for Θ_P.

This approach may be developed further by considering marked point processes in which one or more values (or marks) are associated with each of the points. Stoyan and Stoyan (1990) considered the situation where each edge centre c_i is marked by the angle $\alpha(c_i)$ between the line through the end points of the edge and a given fixed line ($0 \leq \alpha(c_i) \leq 2\pi$). They defined an angle correlation function $g_\alpha(r)$ as

$$g_\alpha(r) = \delta_\alpha(r)/\delta(r) \, , \tag{5.4.31}$$

where $\delta_\alpha(r)$ is a further product density. To interpret $\delta_\alpha(r)$, again consider two infinitesimally small discs of area $|A|$ with distance r between their centres and let $\delta(r)\ |A|^2$ be the probability that there is an edge centre in each of the discs. Define

$$f(u,\upsilon) = \min[|u - \upsilon|, 2\pi - |u - \upsilon|] \qquad (5.4.32)$$

as the minimum angle between the two lines, of directions u and v, containing the edge centres. Then $\delta_\alpha(r) = f(u,v)\ |A|^2$ and $g_\alpha(r)$ can be interpreted as the conditional mean of the angles between the two edges. Stoyan and Stoyan (1990) estimated $g_\alpha(r)$ for the edge centres of $\mathcal{V}_P$ and this is illustrated in Figure 5.4.11. Stoyan and Hermann (1986) considered marking c_i using the areas A_1 and A_2 of the two cells on each side of the edge containing c_i, where A_1 is the area of the cell lying in the upper half of $\mathbb{R}^2$ defined by a line through the end points of the edge. Thus, A_1 and A_2 have the same distribution. Possible marks for c_i include $A_1 \cdot A_2$ and $(A_1 + A_2)$. They calculate the means and variances for both these marks for $\mathcal{V}_P$.

It is also possible to mark the original generator points of $\mathcal{V}_P$. Stoyan and Hermann (1986) did this in terms of the area of the cell associated with the generator to define an area correlation function $g_A(r)$ analogous to $g_\alpha(r)$ defined above, where $g_A(r)$ may be interpreted as the mean of the product of the areas of two cells with generators a distance r apart. They estimated $g_A(r)$ for $\mathcal{V}_P$ using a Monte Carlo simulation approach.

5.5 SECTIONAL POISSON VORONOI DIAGRAMS

In many empirical situations involving tessellations in $\mathbb{R}^3$ it is not possible to observe the constituent cells directly. Instead the cells are examined by means of linear probes or planar sections. Indeed, the prevalence of such occurrences led to the development of the subject of stereology which is concerned with drawing inferences about the geometric properties of three-dimensional material on the basis of observed, lower dimensional information.

We label the structure produced by the intersection between an s-dimensional hyperplane H_s $(1 \le s \le m)$ and $\mathcal{V}_P$ in $\mathbb{R}^m$ as the s-dimensional sectional Poisson Voronoi tessellation denoted by $\mathcal{V}_P\ (s,m)$. Thus, $\mathcal{V}_P$ in $\mathbb{R}^3$ is equivalent to $\mathcal{V}_P\ (3,3)$. Also the intersection of H_t with $\mathcal{V}_P\ (s,m)$ is equal to $\mathcal{V}_P\ (t,m)$, $t<s$ (Miles, 1986a, p. 153). Figure 5.5.1 shows a portion of $\mathcal{V}_P$ (2,3). Since $\mathcal{V}_P$ in $\mathbb{R}^3$ is homogeneous, isotropic and ergodic, $\mathcal{V}_P\ (s,m)$ has the same properties. Similarly, since $\mathcal{V}_P$ is normal so too is $\mathcal{V}_P\ (s,m)$. However, as the results below indicate and a comparison of Figures 5.0.1 and 5.5.1 suggest, $\mathcal{V}_P\ (m,m+1)$ is not equal to $\mathcal{V}_P$ in $\mathbb{R}^m$. For example, a planar section of $\mathcal{V}_P$ in $\mathbb{R}^3$ is not equivalent to $\mathcal{V}_P$ in $\mathbb{R}^2$. Indeed, it is not known if there exists a planar point process which produces $\mathcal{V}$ identical to $\mathcal{V}_P$ (2,3) (Stoyan *et al.*, 1987, p. 106).

As with $\mathcal{V}_P$ only limited information about the characteristics of the individual cells of $\mathcal{V}_P\ (s,m)$ has been obtained analytically and researchers have again resorted to Monte Carlo simulations as a source of supplementary

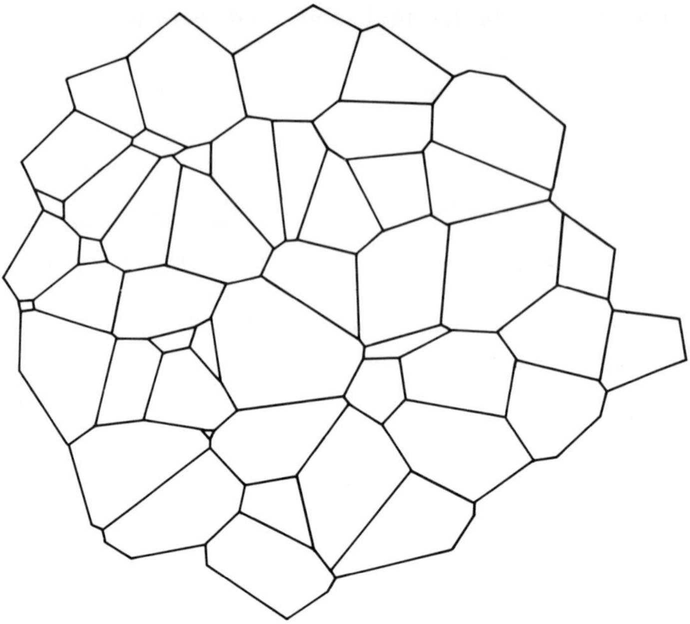

Figure 5.5.1 Portion of a two-dimensional sectional Poisson Voronoi tessellation $\mathcal{V}_P$ (2,3). (*Source*: Lorz (1991).)

information. General expressions paralleling equations (5.4.1)–(5.4.5) for $\mathcal{V}_P$ may be derived by either ergodic approaches (Miles, 1972a, 1984, 1986a) or Palm measures (Møller, 1989). We use the same symbols employed in Section 5.4 and in addition use the superscript (*s*) to refer to the dimensionality of the section. Møller (1989) showed that the intensities of the vertices $\lambda_0^{(s)}$ and the midpoints of the edges of the cells $\lambda_1^{(s)}$ in $\mathcal{V}_P$ (*s*, *m*) are given by

$$\lambda_0^{(s)} = [2/(s+1)]\lambda_1^{(s)}\ , \tag{5.5.1}$$

$$\lambda_1^{(s)} = \frac{\dfrac{\lambda^{s/m}\pi^{s/2}\Gamma[(ms+m-s+1)/2]}{\Gamma(m/2+1)^{[s+1-(s/m)]}\Gamma(s+1-s/m)}}{m\Gamma(s/2+1)\Gamma[(ms+m-s)/2]\Gamma[(m+1)/2]^{s+1}}\ , \tag{5.5.2}$$

and the expected length of a line segment $E_1^{(s)}V_1$ in $\mathcal{V}_P$ (*s*,*m*) is

$$E_1^{(s)}\,V_1 = \frac{\dfrac{\Gamma[(ms-s)/2+1]\Gamma[(ms+m-s)/2]}{\Gamma[(m+1)/2]\Gamma(s-s/m+1/m)}}{\dfrac{\lambda^{1/m}\Gamma[(ms-s+1)/2]\Gamma[(ms+m-s+1)/2]}{\Gamma(s-s/m+1)\Gamma(m/2+1)^{(1-1/m)}}}\ . \tag{5.5.3}$$

For $s = 1$ and $m \geq 2$,

$$\lambda_0^{(1)} = \lambda_1^{(1)} = (\mathrm{E}_1^{(1)} V_1)^{-1} = \frac{(m-1)!\lambda^{1/m}2\Gamma(m/2+1)^{[2-(1/m)]}\Gamma(2-1/m)}{m\Gamma(m-1/2)\Gamma[(m+1)/2]^2} . \tag{5.5.4}$$

For $s = 2$ and $m \geq 3$,

$$\lambda_1^{(2)} = \frac{\lambda^{2/m}\pi\Gamma[(3m-1)/2]\Gamma(m/2+1)^{[3-(2/m)]}\Gamma(3-2/m)}{m\Gamma(3m/2-1)\Gamma[(m+1)/2]^3} , \tag{5.5.5}$$

$$\lambda_0^{(2)} = 2\lambda_1^{(2)}/3 , \tag{5.5.6}$$

$$\lambda_2^{(2)} = \lambda_1^{(2)}/3 , \tag{5.5.7}$$

$$\mathrm{E}_2^{(2)}V_0 = \mathrm{E}_2^{(2)}N_0 = \mathrm{E}_2^{(2)}N_1 = 6 , \tag{5.5.8}$$

$$\mathrm{E}_2^{(2)}V_1 = \frac{(m-1)!6\Gamma(2-1/m)\Gamma(3m/2-1)\Gamma[(m+1)/2]}{\lambda^{1/m}\Gamma(m/2+1)^{(1-1/m)} \Gamma(m-1/2)\Gamma[(3m-1)/2]\Gamma(3-2/m)} , \tag{5.5.9}$$

$$\mathrm{E}_1^{(2)}V_1 = \frac{(m-1)!\Gamma(3m/2-1)\Gamma(m+1)/2]\Gamma(2-1/m)}{\lambda^{1/m}\Gamma(m-1/2)\Gamma[(3m-1)/2] \Gamma(3-2/m)\Gamma(m/2+1)^{[1-(1/m)]}} , \tag{5.5.10}$$

$$\mathrm{E}_0^{(2)}V_1 = 3\mathrm{E}_1^{(2)}V_1 . \tag{5.5.11}$$

As our primary concern is with applications, the values of the above expressions for s=1, 2 and m=2,3 are given in Table 5.5.1. As in Section 5.4, to avoid excessive use of subscripts in the text, these values are given simpler symbols which are identified in the table. This table also includes a value for $\mathrm{E}^{(1)}(L2^2)$ derived by numerical integration (Gilbert, 1962) and other moments estimated by Lorz (1990b, 1991) using Monte Carlo simulations. Lorz (1990b) has also estimated moments and correlation coefficients for the area, perimeter and an arbitrary interior angle conditional on the number of edges for a typical cell of $\mathcal{V}_\mathrm{P}(2, 3)$.

In addition to the exact mean sectional values given in Table 5.5.1, Miles (1972a, 1984) has derived asymptotic ones as $m \to \infty$. These are reported in Table 5.5.2. Recall that the values for $\mathcal{V}_\mathrm{P}(1, m)$ and $\mathcal{V}_\mathrm{P}(2, m)$ are equivalent to the line sections of both $\mathcal{V}_\mathrm{P}(3, m)$ and $\mathcal{V}_\mathrm{P}(2, m)$ and the plane section of $\mathcal{V}(3, m)$, respectively.

A lemma due to Miles (1972a, p. 260) shows that an s-dimensional section of $\mathcal{V}_\mathrm{P}$ is stochastically equivalent to an s-dimensional section of a corresponding inhomogeneous $(s + 1)$-dimensional structure. In particular, let $\Theta_\mathrm{P}{}^*$ be an inhomogeneous Poisson point process in $\mathbb{R}^m$ with intensity

Table 5.5.1 Expected values of characteristics of a typical cell in an s-dimensional sectional Poisson Voronoi diagram

Characteristic	Symbol	Exact value	Numerical value
$s=1$, $m=2$			
Intensity of vertices (mid-pints of line segments)	$\lambda_0^{(1)}$	$4\lambda^{1/2}/\pi$	$1.273\lambda^{1/2}$
$E_1^{(1)}V_1$ Length of a line segment	$E^{(1)}(L2)$	$\pi/4\lambda^{1/2}$	$0.785\lambda^{-1/2}$
	$E^{(1)}(L2^2)$		$0.804\lambda^{-1}$
$s=1$, $m=3$			
Intensity of vertices (mid-points of line segments)	$\lambda_0^{(1)}$ $(\lambda_1^{(1)})$	$\Gamma(2/3)(32\pi\lambda/81)^{1/3}$	$1.455\lambda^{1/3}$
$E_1^{(1)}V_1$ Length of a line segment	$E^{(1)}(L3)$	$(32\pi\lambda/81)^{-1/3}/\Gamma(2/3)$	$0.687\lambda^{-1/3}$
	$E^{(1)}(L3^2)$		$0.632\lambda^{-2/3}$
$s=2$, $m=3$			
Intensity of vertices	$\lambda_0^{(2)}$	$2\Gamma(1/3)[16\pi^5\ \lambda^2/9]^{1/3}/15$	$2.916\lambda^{2/3}$
Intensity of mid-points of polygon sides	$\lambda_1^{(2)}$	$\Gamma(1/3)\ [16\pi^5\lambda^2/9]^{1/3}/5$	$4.374\lambda^{2/3}$
Intensity of polygon centroids	$\lambda_2^{(2)}$	$\Gamma(1/3)[16\pi^5\ \lambda^2/9]^{1/3}/15$	$1.458\lambda^{2/3}$
$E_2^{(2)}N_0(E_2^{(2)}N_1)$ Number of vertices edges/polygon	$E^{(2)}(N)$	6	6
	$E^{(2)}(N^2)$		38.827*
$E_2^{(2)}V_2$ Area of a polygon	$E^{(2)}(A)$	$15/\Gamma(1/3)\ [16\pi^5\lambda^2/9]$	$0.686\lambda^{-2/3}$
	$E^{(2)}(A^2)$		$0.699\lambda^{-4/3}$
$E_2^{(2)}V_1$ Perimeter of a polygon	$E^{(2)}(P)$	$(6/\pi)^{1/3}$ $5\Gamma(2/3)\lambda^{-1/3}/\Gamma(1/3)$	$3.136\lambda^{-1/3}$
	$E^{(2)}(P^2)$		$11.308\lambda^{-2/3}$*
$E_1^{(2)}V_1$ Length of an edge	$E^{(2)}(L)$	$5\Gamma(2/3)$ $(36\pi\lambda)^{-1/3}/\Gamma(1/3)$	$0.523\lambda^{-1/3}$
	$E^{(2)}(L^2)$		$0.404\lambda^{-2/3}$
Uniformly random selected interior angle of a polygon	$E^{(2)}(\alpha_2)$		2.000*
	$E^{(2)}(\alpha_2^2)$		4.349*
Angle of an edge at a typical vertex	$E^{(2)}(\alpha_1)$	$2\pi/3$	2.094
	$E^{(2)}(\alpha_1^2)$		4.689*

λ = intensity of Θ_P.
*Estimates from Monte Carlo simulations by Lorz (1990b, 1991).

$\lambda_m = \lambda_m\ (x_4, \ldots, x_m)$ and $\mathcal{V}_P{}^*$ be its associated Voronoi diagram. If $^0[3]$ is the three-dimensional subspace of $\mathbb{R}^m$ defined by $x_4 = \ldots x_m = 0$, then since λ_m is functionally independent of (x_1, x_2, x_3), the three-dimensional section of $\mathcal{V}_P{}^*$ by $^0[3]$ is a homogeneous normal tessellation, denoted by $\mathcal{V}^*(3,m)$. Thus, $\mathcal{V}_P(3,m)$ is stochastically equivalent to $\mathcal{V}^*(3, 4)$ with respect to

Table 5.5.2 Asymptotic moments of characteristics of a typical cell in an s-dimensional sectional Poisson Voronoi diagram

Characteristic	Symbol	Exact value	Numerical value
$s=1$			
$E_1^{(1)}V_1$ Length of a line segment	$E^{(1)}(L1)$	$(2e)^{1/2}\lambda^{-1/2}$	$0.429\lambda^{-1/2}$
$s=2$			
$E_2^{(2)}N_0$ ($E_2^{(2)}N_1$) Number of vertices (edges)/polygon	$E^{(2)}(N)$	6	6
$E_2^{(2)}V_2$ Area of a polygon	$E^{(2)}(A)$	$3^{1/2}\lambda^{-2/3}/\pi e$	$0.203\lambda^{-2/3}$
$E_2^{(2)}V_1$ Perimeter of a polygon	$E^{(2)}(P)$	$(6/e)^{1/2}\lambda^{-1/3}$	$1.486\lambda^{-1/3}$
$s=3$			
$E_3^{(3)}N_0$ Number of vertices/cell	$E^{(3)}(M)$	$1/\varphi$	22.79
$E_3^{(3)}N_1$ Number of edges/cell	$E^{(3)}(E)$	$3/2\varphi$	34.19
$E_3^{(3)}N_2$ Number of faces/cell	$E^{(3)}(F)$	$2 + (1/2\varphi)$	13.39
$E_2^{(3)}N_0(E_2^{(3)}N_1)$ Number of vertices (edges)/face	$E^{(3)}(N)$		5.106
$E_3^{(3)}V_3$ Volume	$E^{(3)}(V)$	$1/16\pi e^{3/2}\varphi\lambda$	$0.101\lambda^{-1}$
$E_3^{(3)}V_2$ Surface area	$E^{(3)}(S)$	$1/2^{3/2}\pi e\varphi\lambda^{2/3}$	$0.944\lambda^{-2/3}$
$E_3^{(3)}V_1$ Total edge length/cell	$E^{(3)}(B)$	$3/4(3e)^{1/2}\varphi\lambda^{1/3}$	$5.987\lambda^{-1/3}$
$E_2^{(3)}V_2$ Area of a face	$E^{(3)}(A)$		$0.0705\lambda^{-2/3}$
$E_2^{(3)}V_1$ Total edge length/face	$E^{(3)}(P)$		$0.894\lambda^{-1/3}$
$E_1^{(3)}V_1$ Edge length	$E^{(3)}(L)$		$0.175\lambda^{-1/3}$

λ = intensity of Θ_P.
$\varphi = 1/8–3/4\pi \sin^{-1}(1/3)$.
Source: Miles (1972a, 1984).

$$\lambda_4(x_4) = \{2\pi^{(m-3)/2}/\Gamma[(m-3)/2]\}\ \lambda x_4^{m-4} \tag{5.5.12}$$

(Miles, 1972a, pp. 260–261). This lemma extends from three-dimensional sections to general s-dimensional sections and Miles (1984) exploited it to derive a four-fold integral expression for the distribution of line segments when $s = 1$ but this is hardly tractable. However, Muche and Stoyan (1990) have derived numerically tractable double integral formulae for the chord length ($s = 1$) probability distribution functions $f(L2)$ and $f(L3)$ in $\mathbb{R}^2$ and $\mathbb{R}^3$, respectively. These distributions are derived from linear distributions functions (Gilbert, 1962) which give the probabilities that a fixed line segment of length $L2$ (in $\mathbb{R}^2$), $L3$ (in $\mathbb{R}^3$), with one end point in the origin, lies completely in the cell P_o containing the origin. Muche and Stoyan (1990) also derived the spherical distribution function which gives the probabilities that a fixed disc in $\mathbb{R}^2$ or sphere in $\mathbb{R}^3$ of radius r, centred at the origin, lies completely within P_o. The cumulative frequency distributions of $L2$ and $L3$ as derived by Muche and Stoyan using numerical integration are given in Table 5.5.3.

Table 5.5.3 Cumulative frequency distributions of line segments $L2$ for $\mathcal{V}_P(1,2)$ and $L3$ for $\mathcal{V}_P(1,3)$

$L2,L3$	$F(L2)$	$F(L3)$	$L2,L3$	$F(L2)$	$F(L3)$
0.00	0.000	0.000	1.25	0.851	0.911
0.05	0.022	0.032	1.30	0.874	0.931
0.10	0.045	0.065	1.35	0.894	0.947
0.15	0.069	0.099	1.40	0.912	0.960
0.20	0.094	0.133	1.45	0.928	0.971
0.25	0.120	0.168	1.50	0.941	0.979
0.30	0.149	0.204	1.55	0.952	0.985
0.35	0.179	0.240	1.60	0.962	0.990
0.40	0.212	0.278	1.65	0.970	0.993
0.45	0.246	0.316	1.70	0.976	0.996
0.50	0.282	0.355	1.75	0.981	0.997
0.55	0.321	0.396	1.80	0.985	0.998
0.60	0.361	0.437	1.85	0.989	0.999
0.65	0.402	0.478	1.90	0.991	0.999
0.70	0.444	0.520	1.95	0.993	1.000
0.75	0.487	0.563	2.00	0.995	1.000
0.80	0.530	0.605	2.05	0.996	1.000
0.85	0.573	0.647	2.10	0.997	1.000
0.90	0.615	0.688	2.15	0.998	1.000
0.95	0.655	0.727	2.20	0.999	1.000
1.00	0.694	0.765	2.25	0.999	1.000
1.05	0.730	0.800	2.30	0.999	1.000
1.10	0.764	0.833	2.35	0.999	1.000
1.15	0.796	0.862	2.40	1.000	1.000
1.20	0.825	0.888	2.45	1.000	1.000

Source: Muche and Stoyan (1990).

As with $\mathcal{V}_P$ knowledge of such distributions is valuable in empirical work because existing evidence indicates that moment measures of many characteristics of individual cells of $\mathcal{V}_P(s,m)$ vary little from those cells in sectional tessellations where the set of generator points is not a realization of Θ_P (Lorz, 1990a). However, there have been far fewer reported simulation studies of $\mathcal{V}_P(s,m)$ than $\mathcal{V}_P$, all of which emphasize $\mathcal{V}_P(2,3)$ (Mahin *et al.*, 1976, 1980; Saetre *et al.*, 1986; Lorz, 1990a, 1990b, 1991). Of these the most extensive are those of Lorz (1990b, 1991) whose study involved 1 050 036 cells. Both he and Mahin *et al.* (1980) estimated the distribution of the number of vertices (edges) per cell N and cell area A for $\mathcal{V}_P(2,3)$ (see also Saetre *et al.*, 1986; Lorz, 1990a), while Lorz (1990b, 1991) also estimated the distributions of the perimeter of a cell P, individual edge length L, a uniformly random selected internal cell angle α_2 and the angle of an edge at a vertex α_1 for $\mathcal{V}_P(2,3)$ and Mahin *et al.* (1980) estimated the distribution of the length of a line segment $L3$ for $\mathcal{V}_P(1,3)$ based on 1241 observations. All of these distributions are illustrated in Figure 5.5.2. Note that in the case of $L3$, it is the standardized value, $L3/E^{(1)}(L3)$, which is reported. The distribution of N for $\mathcal{V}_P(2,3)$ is also reported in Table 5.5.4. Comparison of Figure 5.5.2 with Figures 5.4.2–5.4.5 and Table 5.5.4 with Table 5.4.1 clearly reinforces the observation that $\mathcal{V}_P(2,3)$ is not the same as $\mathcal{V}_P$ in $\mathbb{R}^2$.

Distributions of area, perimeter and interior angle, conditional on the number of cell vertices (edges), have also been estimated by Lorz (1990b) who also derives estimates of both the unconditional and conditional correlations between most of these properties (see Table 5.5.5).

Table 5.5.4 Estimates, $\hat{p}(N)$, of the probability that a typical polygon in a two-dimensional section of a Poisson Voronoi Diagram in $\mathbb{R}^3$ has N edges (vertices)

N	$\hat{p}(N)$	N	$\hat{p}(N)$
3	0.0631	10	0.0178
4	0.1358	11	0.0051
5	0.2047	12	0.0012
6	0.2273	13	0.000263
7	0.1837	14	0.000042
8	0.1104	15	0.0000048
9	0.0506	16	0.00000095

Source: Lorz (1990b, 1991).

Table 5.5.5 Estimates of correlation coefficients for characteristics of a typical polygon and a typical edge in a two-dimensional section of a Poisson Voronoi diagram in $\mathbb{R}^3$

	A	P	α_2	α_1
N	0.7530	0.7276	0.5653	—
A	—	0.9407	0.4331	—
P	—	—	0.4581	—
L	—	—	—	−0.2006

N = Number of edges (vertices)/polygon.
A = Area of a polygon.
P = Perimeter of a polygon.
L = Length of an edge.
α_2 = Interior angle of a polygon.
α_1 = Angle of an edge at a vertex.
Source: Lorz (1990b, 1991).

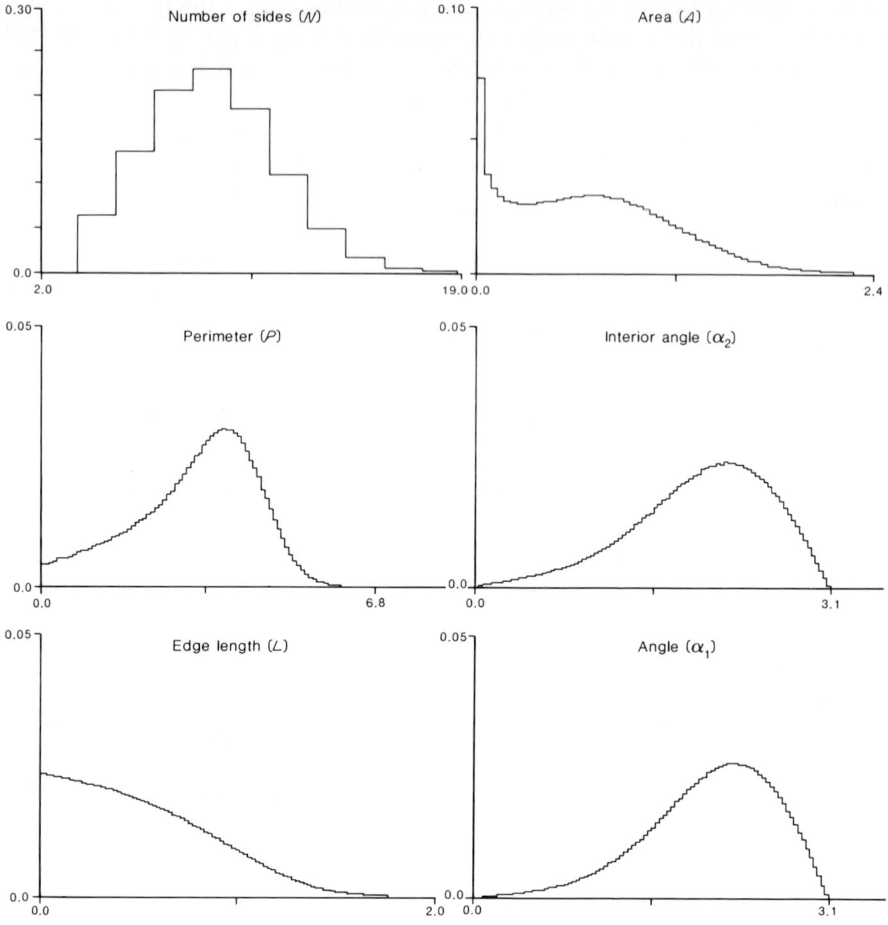

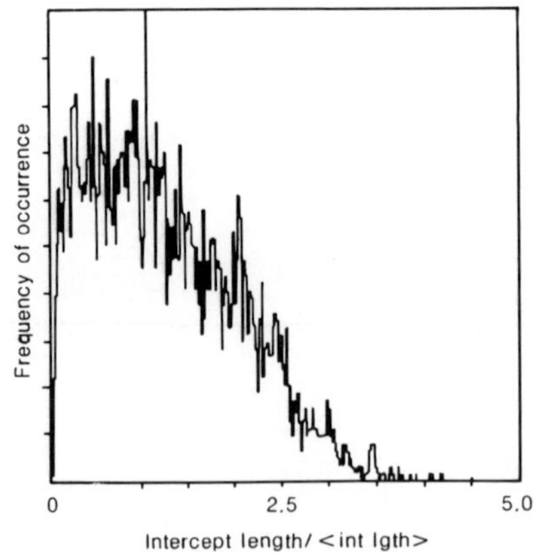

Figure 5.5.2 Normalized histograms of the number of vertices (edges) N, area A, perimeter P, individual edge length L, internal cell angle α_2 and edge angle α_1 for $\mathcal{V}_P$ (2, 3) and standardized line segment length l_3 $(=L_3/\mathrm{E}^{(1)}(L3))$ for $\mathcal{V}_P$ (1, 3).

5.6 ADDITIVELY WEIGHTED POISSON VORONOI DIAGRAMS: THE JOHNSON–MEHL MODEL

In Section 5.2 we discussed a variety of circumstances in which the Poisson Voronoi model could be used in a normative way in the examination of empirical tessellations. There are, however, other empirical tessellations for which it is not an appropriate model. One group of such tessellations consists of those in which the members of the generator set P are not contemporaneous. In such instances we may assume that each generator p_i has both a location $\mathbf{x}_i$ $(x_{i1}, x_{i2}, \ldots, x_{im})$ in $\mathbb{R}^m$ and an associated birth time $t_i(t_i \geq 0)$. Once born, each point generates a cell which grows at a constant rate of radial growth, v. Interaction between growing cells is subject to the following conditions:

C1 If a point is born at a location that is already occupied by a growing cell, it disappears without trace.

C2 When radiating rays of two growing cells meet, both cease radiating so that the cell associated with a point is fully grown when growth ceases in all directions.

As we indicated in Section 3.1.2, this model can be considered equivalent to an additively weighted Voronoi diagram in which the weight associated with p_i corresponds to t_i, since for the cell $C(p_i)$ which is generated by growth from p_i,

$$C(p_i) = \{\mathbf{x} | (t_i + \|\mathbf{x} - \mathbf{x}_i\|/v) \leq (t_j + \|\mathbf{x} - \mathbf{x}_j\|/v), \text{for } i \neq j, j \in I_n\} \qquad (5.6.1)$$

where $\|\mathbf{x} - \mathbf{x}_i\|$ and $\|\mathbf{x} - \mathbf{x}_j\|$ are the Euclidean distances from $\mathbf{x}$ to p_i, p_j, respectively.

If we assume that both $\mathbf{x}_i$ and t_i are generated by the homogeneous Poisson point process, so that the generators appear at a constant rate α per unit of $\mathbb{R}^m$ per time interval, we have the Johnson–Mehl (J–M) (or continuous nucleation) model. This model was originally developed to study the growth of crystal aggregates (Johnson and Mehl, 1939) and surface film on metal (Evans, 1945). It has been independently rediscovered at least once since then (Glass, 1973, 1974; Armstrong, 1974; Jackson, 1974). The development of a tessellation over time in $\mathbb{R}^2$ according to the J–M model is illustrated in Figure 5.6.1.

Moments of several characteristics of a typical cell in a tessellation generated by the J–M model in $\mathbb{R}^2$ and $\mathbb{R}^3$ have been derived and these are summarized in Table 5.6.1. To facilitate comparison with similar moments reported in Tables 5.4.1 and 5.4.5 for the Poisson Voronoi tessellation, the numerical values in Table 5.6.1 are given in terms of the density of generator points per unit space after all cell growth has ceased rather than in terms of the birth rate, α, and the growth rate, v. Note that inequalities arise in

Table 5.6.1 Expected values of characteristics of a typical cell for the Johnson–Mehl model

Characteristic	Symbol	Exact value	Numerical value
Two-dimensional			
Intensity of cell centroids	λ	$(3/\pi)^{1/3}\Gamma(4/3)(\alpha/v)^{2/3}$	$0.879(\alpha/v)^{2/3}$
Number of vertices (edges)	$E(N)$	—	<6
Number of full neighbours	$E(C)$	4	4
Area	$E(A)$	$[(\pi/3)^{1/3}/\Gamma(4/3)](v/\alpha)^{2/3}$	λ^{-1}
Perimeter	$E(P)$	$[4(3/\pi)^{1/3}\Gamma(5/3)/\Gamma(4/3)](v/\alpha)^{1/3}$	$3.734\lambda^{-1/2}$
Length of cell edge	$E(L)$	$<[2(3/\pi)^{1/3}\Gamma(5/3)/3\Gamma(4/3)](v/\alpha)^{1/3}$	$<0.623\lambda^{-1/2}$
Three-dimensional			
Intensity of cell centroids		$[4\Gamma(5/2)/\pi^{3/2}]^{1/4}\Gamma(5/4)(\alpha/v)^{3/4}$	$0.896(\alpha/v)^{3/4}$
Number of vertices/cell	$E(M)$	$16\pi^2/7$	22.559
Number of edges/cell	$E(E)$	$>24\pi^2/7$	>33.839
Number of faces/cell	$E(F)$	$>8\pi^2/7+2$	>13.280
Number of full neighbours	$E(C)$	7	7
Volume	$E(V)$	$[(\pi/3)^{1/4}/\Gamma(5/4)](v/\alpha)^{3/4}$	λ^{-1}
	$E(V^2)$		$2.236\lambda^{-2}$
Surface area	$E(S)$	$[16\pi^{1/2}\Gamma(7/4)]/[3^{3/2}\Gamma(5/4)](v/\alpha)^{1/2}$	$5.143\lambda^{-2/3}$
Total edge length/cell	$E(P)$	$[3^{1/4}4\pi^{9/4}]/[5\Gamma(5/4)](v/\alpha)^{1/4}$	$14.71\lambda^{-1/3}$
Sections			
Surface area	$E^{(2)}(A)$	$\{3^{1/2}5[\Gamma(1/4)]^{2/3}\}/\{2^{4/3}\pi^2\ [\Gamma(5/4)]^{2/3}\}(v/\alpha)^{1/2}$	$0.816\lambda^{-2/3}$
Length of a line segment	$E^{(1)}(L3)$	$\{3^{1/4}[\Gamma(1/4)]^{1/3}\}/\{2^{2/3}\pi^{1/4}\Gamma(3/4)\ [\Gamma(5/4)]^{1/3}\}(v/\alpha)^{1/4}$	$0.778\lambda^{-1/3}$

α = birth rate.
v = radial growth rate.
Sources: Evans (1945), Meijering (1953) and Gilbert (1962).

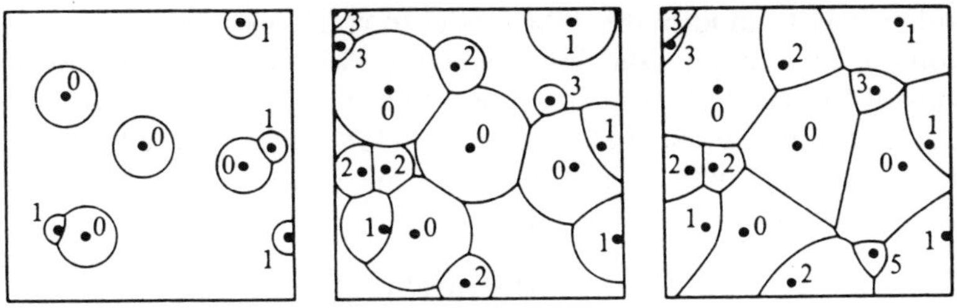

Figure 5.6.1 The Johnson–Mehl model for times (a) $t=1$, (b) $t=3$ and (c) $t=7$.

Table 5.6.1 because it is possible that the common edge (in $\mathbb{R}^2$) or the common face (in $\mathbb{R}^3$) between two cells may be interrupted by one or more smaller cells. Frost and Thompson (1987a) have estimated the distributions of area, edge length and number of sides (vertices) of a typical cell using information from 10 000 cells from a computer simulation.

The J–M model has been developed further by considering changes in either the spatial location process for the generators or their birth rate or both. So far most attention has been focused on considering the birth rate as a function of time while retaining the assumption that the locations are generated by the homogeneous Poisson point process. We refer to these as *time inhomogeneous J–M* (TIJ–M) models. Miles (1972a) has derived general integral expressions for the expected values of the volume, surface area, total edge length and number of vertices of a typical TIJ–M cell in $\mathbb{R}^3$ when the nucelation rate is an inhomogeneous Poisson process of intensity $\alpha(t)$ (see Section 1.3.8). Horalek (1988, 1989) evaluated Miles' integrals for the case where

$$\alpha(t) = \alpha t^{\beta} \quad (\beta > -1, \alpha > 0) .$$

Note that $\beta=0$ gives the J–M model while as $\beta \to -1$ the tessellation approaches $\mathcal{V}_P$. Frost and Thompson (1987a) used a simulation approach in studying a TIJ–M model in $\mathbb{R}^2$ in which the nucleation rate declines over time with

$$\alpha(t) = \alpha e^{-\beta t} \quad (\alpha > 0, \beta > 0) .$$

In this case as $\beta \to 0$ we obtain the J–M model, while $\beta \to \infty$ produces $\mathcal{V}_P$. Frost and Thompson (1987a) derived estimates of the distributions of the area, edge length and number of sides of a typical cell. They also estimated the same distributions for situations in which the nucelation rate remains constant but in which new generators are prohibited from being born within a distance δ from any edge of an existing cell (note that the Voronoi hard core models discussed in Section 5.10 may be considered as a limiting case of this modified J–M model).

Another way to modify the J–M model to conform with certain empirical situations is to maintain the homogeneous Poisson point process for the locations of the generators in space and time but to alter the growth rate v to vary over time. Frost and Thompson (1987b) considered situations in which the radial growth of a cell C_i at time t_i, $v_i(t)$, is proportional to a power of its radius $r_i(t)$ at that time. Thus,

$$v_i(t) \propto r_i(t)^{\beta} ,$$

which is equivalent to considering $r_i(t)$ as a function of the time elapsed $(t-t_i)=\Delta t$ since the birth of the cell's generator. In particular, they examined the distributions of the area and number of sides (vertices) of a typical cell in models with values of $\beta = -3, -2, -1, 0, 1$ (equivalent to Δt^{ρ} ; $\rho = 1/4$, $1/3$, $1/2$, 1) and $\exp(\Delta t)$, respectively. When $\beta = 0$ we have the basic J–M model, while $\beta = -1$ yields a Voronoi power diagram (see Section 3.1.4) and $\beta = 1$ produces a multiplicatively weighted Voronoi diagram (see Section 3.1.1).

Because the J–M model has been of particular interest in materials sciences applications, there has been interest in the nature of one- and two-dimensional sections of the model in $\mathbb{R}^3$. Horalek (1989, 1990) derived mean values for the length of the intercept through a typical cell $\mathrm{E}(L)$ and the area of a section of a typical cell $\mathrm{E}(A)$ for both the J–M model and the TIJ–M model with $\alpha(t) = \alpha t^{\beta}$ ($\alpha > 0$, $-1 < \beta < \infty$) (see Table 5.6.1 for these values for the J–M model). Note that an exact value cannot be obtained for $\mathrm{E}(A)$ because the star-shaped nature of the cells means that in two-dimensional sections it is possible to have subareas which are entirely surrounded by the planar section of an adjoining cell. Mahin *et al.* (1976, 1980) have used computer simulation to esimate the distribution of the cross-sectional area, intercept length and number of sides (vertices) of cells in sectional tessellations, while Saetre *et al.* (1986) have examined the distribution of the areas of cells in sections of TIJ–M models with declining birth rates (chi-square distribution for t_i), increasing birth rates (a Weibull distribution for t_j) and with growth rates declining over time.

Applications of the family of J–M models have been confined mainly to the natural sciences and examples can be found in most of the references cited in this section. Additional astrophysical applications are presented by Zaninetti (1989, 1990) who has used them in the study of supernovae explosions. In the social sciences Boots (1973, 1975a) has used the models to explore the development of service areas associated with the provision of public, intercity bus services in England and Wales.

5.7 HIGHER ORDER POISSON VORONOI DIAGRAMS

In Section 3.2 we generalized the basic concept of the ordinary Voronoi diagram by considering diagrams in which an individual cell is created with respect to a subset $P_i^{(k)}$ of k points ($1 \le k \le n$) of the n points of P rather than an individual point of P. In the case of the order-k Voronoi diagram $\mathcal{V}^{(k)}$, an individual cell consists of all locations in $\mathbb{R}^m$ for which the members of $P_i^{(k)}$ are the set of k nearest points, while in the ordered order-k Voronoi diagram $\mathcal{V}^{\langle k \rangle}$ we add the additional constraint that the k points are ordered so that all locations in a cell share the same first, second, ..., kth nearest points of P. When $\mathcal{V}^{(k)}$ and $\mathcal{V}^{\langle k \rangle}$ are defined with respect to Θ_P we label them $\mathcal{V}_P^{(k)}$ and $\mathcal{V}_P^{\langle k \rangle}$, respectively. Like $\mathcal{V}_P$, $\mathcal{V}_P^{(k)}$ is a normal tessellation while $\mathcal{V}_P^{\langle k \rangle}$ is not. Figure 5.7.1 shows $\mathcal{V}_P^{(k)}$ in $\mathbb{R}^2$ for several different values of k.

Although Miles (1984, p. 311) has been able to derive an expression for the expectation $E_m V_s$ of the total s-content of the s-faces of an arbitrary cell of $\mathcal{V}_P^{(k)}$ in $\mathbb{R}^m$, which is analogous to expression (5.4.2) for $\mathcal{V}_P$, it has only proved possible to evaluate it for m=1,2. Thus, the known properties of $\mathcal{V}_P^{(k)}$ are limited to some first-order moments for m=2. The same holds for $\mathcal{V}_P^{\langle k \rangle}$. These values are given in Table 5.7.1.

Note that for both tessellations E(N) is a consequence of the general result given in equation (5.4.6). Similarly, E(A) for $\mathcal{V}_P^{(k)}$ holds for any underlying point process which is homogeneous and ergodic (Miles, 1986b, p. 569). This value is also a specific instance of the more general relationship (Miles, 1984, 1986a)

$$E_m V_m^{(2)} = 2E_m V_m^{(1)} / E_m^{(1)} N_{m-1} , \tag{5.7.1}$$

where (k) refers to the order of $\mathcal{V}_P^{(k)}$, and $\mathcal{V}_P^{(1)} = \mathcal{V}_P$, which holds because of the one-to-one correspondence between the cells of $\mathcal{V}_P^{(2)}$ and the ($m-1$) faces of $\mathcal{V}$.

The values of E(A) for $\mathcal{V}_P^{(k)}$ and $\mathcal{V}_P^{\langle k \rangle}$ can be combined to define the expectation of the multiplicity m of a cell of $\mathcal{V}_P^{(k)}$ as the number of cells of $\mathcal{V}_P^{\langle k \rangle}$ it comprises so that

$$E(m) = k\{1 + [64(k-1)(k-2)]/[9\pi^2(2k-1)]\} , \tag{5.7.2}$$

which tends to $32k^2/9\pi^2$ as $k \to \infty$ (Miles, 1970a, p. 119; Santalo, 1976, p. 20).

For $\mathcal{V}_P^{\langle k \rangle}$ in $\mathbb{R}^2$, as $k \to \infty$ the proportion of cells with four vertices tends to unity (Miles, 1970a, p. 118) and $\mathcal{V}_P^{\langle k \rangle}$ tends to become equivalent to the tessellation $\mathcal{L}$ formed by an isotropic Poisson line process of intensity $\tau = (8/3)\,(\lambda k^3/\pi)^{1/2}$. Such a process involves defining lines in terms of their polar coordinates (r, Θ) representing the perpendicular distance and direction from the origin, o, respectively, where the range of (r, Θ) is taken as $\Pi = \{(r, \Theta)|\ -\infty<r<\infty,\ 0 \le \Theta < \pi\}$. Then for $r > 0$ the Poisson point process (see Section 1.3.8) of intensity τ/π in Π induces the Poisson line process of intensity τ in $\mathbb{R}^2$. Thus, any values for characteristics of an arbitrary cell of $\mathcal{L}$ carry over asymptotically for $\mathcal{V}^{\langle k \rangle}$. Knowledge of such characteristics is quite extensive with all the first- and second-order moments,

Table 5.7.1 Expected values of characteristics of a typical cell of order-*k* and ordered order-*k* Poisson Voronoi diagrams in $\mathbb{R}^2$

Characteristic	Symbol	Exact value	Limit ($k\to\infty$)
Order-k			
$E_2N_0(E_2N_1)$ Number of vertices (edges)	$E(N)$	6	
Number of full neighbours	$E(C)$	4	
E_2V_2 Area	$E(A)$	$1/(2k-1)\lambda$	
E_2V_1 Perimeter	$E(P)$	$(2k)!/[k!(k-1)!(2k-2)2^{(2k-3)}\lambda^{1/2}]$	$4/(\pi k\lambda)^{1/2}$
Ordered order-k			
$E_2N_0(E_2N_1)$ Number of vertices (edges)	$E(N)$	$[54\pi^2k+256(k-1)(k-2)]/[9\pi^2(2k-1)+64(k-1)(k-2)]$	4
E_2V_2 Area	$E(N^2)$		$(\pi^2+24)/2$
	$E(A)$	$1/\{k[2k-1+(64/9\pi^2)\ (k-1)(k-2)]\lambda\}$	$9\pi^2/64k^3\lambda$
E_2V_1 Perimeter	$E(A^2)$		$81\pi^6/2^{13}k^6\lambda^2$
	$E(P)$	$\{2\Sigma_1^k[(2m)!/m!(m-1)!2^{(2m-2)}\ ]\}/\{k[2k-1+(64/9\pi^2)(k-1)(k-2)]\lambda^{1/2}\}$	$(3/4)[\pi^3/k^3\lambda]^{1/2}$
	$E(P^2)$		$9\pi^3(\pi^2+4)/128k^3\lambda$

λ = intesnity of Θ_P.
Source: Miles (1964, p. 904; 1970, p. 114).

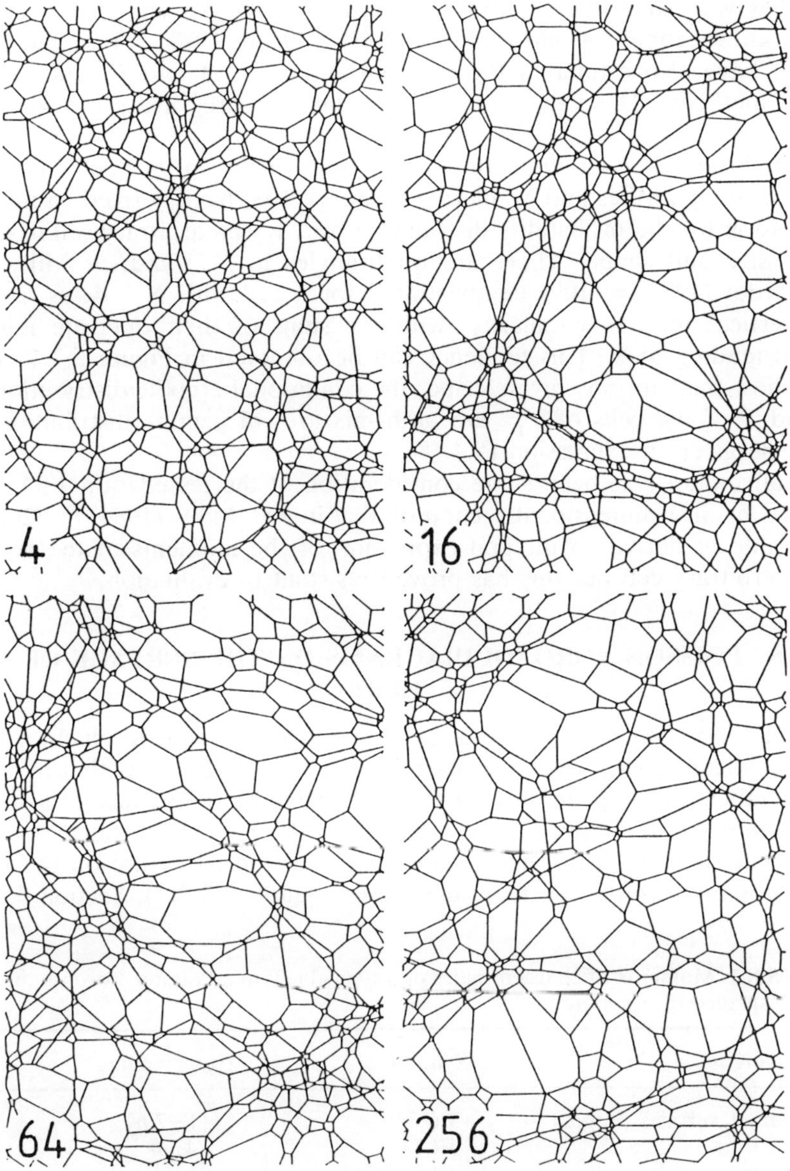

Figure 5.7.1 Portions of an order-k Poisson Voronoi diagram $\mathcal{V}_P^{(k)}$ for k=4, 16, 64, 256. (*Source*: Miles and Maillardet (1984, Figure 5).)

together with some higher order moments and the correlation coefficients being known for N, A and P (Miles, 1964a, 1964b, 1973). Crain and Miles (1976) have also estimated the distributions of N, A and P using Monte Carlo methods. The values of the moment measures are given in Table 5.7.1.

Very little is known about the distributions of the various cell characteristics, although Miles and Maillardet (1982, p. 108) have derived an integral expression for the probability distribution of N for an arbitrary cell in terms of its size and shape. There are no triangles in $\mathcal{V}_{\mathrm{P}}^{(k)}$ and, as Miles and Maillardet (1982, p. 109) demonstrated for a cell with $N = 4$, as $k \to \infty$, the vertices tend to circularity, with the sums of their opposite interior angles tending to π. This tendency can be observed in Figure 5.7.1. They also show that, for any given shape, the conditional ergodic distributions of A and P of the cells of $\mathcal{V}_{\mathrm{P}}^{(2)}$ are both mixtures of gamma distributions of the type given in expression (5.4.18).

Finally, there has been some consideration of the tessellation $\mathcal{V}_{\mathrm{P}}^{(k)}$ (s,m) created by an s-dimensional section of $\mathcal{V}_{\mathrm{P}}^{(k)}$ in $\mathbb{R}^m$. Miles (1984, p. 326) has been able to derive an integral expression of the moments of the volume of an arbitrary cell but this has proved resistant to evaluation.

5.8 POISSON VORONOI DIAGRAMS ON THE SURFACE OF A SPHERE

One of the instances of Voronoi diagrams involving spaces which are not the Euclidean space, which was examined in Section 3.7, was that in which the members of the generator set P were located on the surface of a sphere S with unit radius, centred at the origin o. In this case the distance between two points is defined in terms of the length of the lesser arc on the great circle passing through the two points (see equation (3.7.9), Section 3.7.4).

Table 5.8.1 Moments of characteristics of a typical cell in a Poisson Voronoi diagram on the surface of a sphere

Characteristic	Symbol	Exact value
Number of vertices (edges)	E(N)	$6(1-2/n)$
Area	E(A)	$4\pi/n$
Perimeter	E(P)	$\pi(n-1)! \sum_{i=0}^{(n/2-1)} [2^{(n+2i-3)} i!(i+1)!(n-2i-2)!]^{-1}$

n = number of points in P.
Source: Mills (1971).

If P consists of n points selected at random on S, the resulting Voronoi diagram with respect to P is a random Voronoi diagram which we label $\mathcal{V}_S$. Knowledge of $\mathcal{V}_S$ is limited to the first-order moments derived by Miles (1971) and given in Table 5.8.1. Note that as $n \to \infty$, E(N) tends to 6, which corresponds to E(N) = 6 for $\mathcal{V}_P$ in $\mathbb{R}^2$.

5.9 PROPERTIES OF POISSON DELAUNAY CELLS

As demonstrated in Section 2.3 the Poisson Delaunay tessellation $\mathcal{D}_P$ can be considered the dual tessellation of $\mathcal{V}_P$ in $\mathbb{R}^m$. Each Poisson Delaunay cell (PDC) of $\mathcal{D}_P$ is an $(m + 1)$-dimensional simplex whose vertices $\mathbf{x}_0,\ldots,\mathbf{x}_m$ are points of Θ_P. For $m = 2$ and $m = 3$ the PDCs are triangles and tetrahedrons, respectively. Any s-face of $\mathcal{D}_P$ is an $(s + 1)$-dimensional simplex with vertices $\mathbf{x}_0,\ldots,\mathbf{x}_s$, also points of Θ_P. Alternatively, an s-face can be considered the convex hull of $\mathbf{x}_0,\ldots,\mathbf{x}_s$. There are $\binom{d+1}{s+1}$ s-faces contained in a PDC. Because of the duality between $\mathcal{V}_P$ and $\mathcal{D}_P$, there is a simple relationship between the intensities of their s-faces (Møller, 1989, equation (7.3.1)). If $\lambda_s\,(\mathcal{D})$ and $\lambda_s\,(\mathcal{V})$ are the intensities of the s-faces of $\mathcal{D}$ and $\mathcal{V}$, respectively, then

$$\lambda_s\,(\mathcal{D}) = \lambda_{m-s}(\mathcal{V}), \quad s = 0,\ldots,m\ . \tag{5.9.1}$$

The major result relating to PDCs was derived by Miles (1970b, 1974) (see also Møller, 1989, Section 7). Let c and r be the circumcentre and circumradius, respectively, of an $(m + 1)$-dimensional PDC in $\mathbb{R}^m$. Then the $(m + 1)$ vertices of the cell are the points $\{c + r\mathbf{u}_i\}$, where $\{\mathbf{u}_i\}$ are the unit vectors (i=0, . . .,m). The ergodic probability distribution (pdf) of $\mathcal{D}_P$ is completely specified by the joint pdf

$$f(r,\mathbf{u}_0,\ldots,\mathbf{u}_m) = a(\lambda,m)\,\Delta_m r^{m^2-1}\exp(-\lambda\omega_m r^m)\ , \tag{5.9.2}$$

where Δ_m is the volume of the $(m + 1)$-simplex with vertices $\mathbf{u}_0,\ldots,\mathbf{u}_m$, $\omega_m = \pi^{m/2}/\Gamma\,(m/2 + 1)$ and

$$a(\lambda,m) = \frac{\pi^{(m^2+1)/2}\Gamma(m^2/2)\,\{2\lambda\Gamma[(m+1)/2]\}^m}{m^{(m-2)}\Gamma(m/2)^{2m+1}\Gamma[(m^2+1)/2]}\ . \tag{5.9.3}$$

Furthermore, r, which may be considered a measure of the size of the $(m + 1)$-simplex, is independent of $\{\mathbf{u}_0,\ldots,\mathbf{u}_m\}$ which represents the shape of the simplex. It was this independence which was exploited by Kendall (1989) in deriving a method of simulating individual $(m + 1)$-dimensional simplices (see Section 5.3). The pdf of r is a generalized gamma function (equation (5.4.17)) with $r = m$, $q = m^2$ and $b = \lambda_m$. Miles (1974) showed that equation (5.9.2) can be used to derive a general expression for the kth moment of the volume of a typical PDC,

$$E_m V_m^k = \frac{(m+k-1)!\Gamma(m^2/2)\Gamma(m^2+mk+k+1)/2] \; \Gamma[(m+1)/2]^{(m-k+1)} \prod_{i=2}^{(m+1)} \Gamma[(k+i)/2]/\Gamma(i/2)}{(m-1)!\Gamma[(m^2+1)/2]\Gamma(m^2+mk)/2) \; \Gamma[(m+k+1)/2]^{(m+1)}(2^m \pi^{(m-1)/2} \lambda)^k}$$

$$\text{for } k = 1,2,\ldots . \tag{5.9.4}$$

When $m = 2$, this reduces to

$$E_2 V_2^k = \frac{\Gamma[(3k+5)/2]\Gamma(k/2+1)}{3\Gamma[(k+3)/2]^2 2^k \pi^{(k-1/2)} \lambda^k}, \quad k = 1,2,\ldots, \tag{5.9.5}$$

Table 5.9.1 Expected values of characteristics of a typical Poisson Delaunay cell in $\mathbb{R}^2$

Characteristic	Symbol	Exact value	Numerical value
Intensity of cell vertices	λ_0	λ	
Intensity of midpoints of cell edges	λ_1	3λ	
Intensity of cell centroids	λ_2	2λ	
Number of edges/vertex	$E(E)$	6	
Number of cells/vertex	$E(M)$	6	
Number of cells/edge	$E(N)$	2	
Area	$E(A)$	$1/2\lambda$	$0.5\lambda^{-1}$
	$E(A^2)$	$35/8\pi^2\lambda^2$	$0.443\lambda^{-2}$
Perimeter	$E(P)$	$32/3\pi\lambda^{1/2}$	$3.395\lambda^{-1/2}$
	$E(P^2)$	$125/3\pi\lambda$	$13.263\lambda^{-1}$
Length of a cell edge	$E(L)$	$32/9\pi\lambda^{1/2}$	$1.132\lambda^{-1/2}$
	$E(L^2)$	$5/\pi\lambda$	$1.592\lambda^{-1}$
	$E(L^k)$	$2^{(k+3)}\Gamma[(k+5)/2]/[3(k+2)\pi^{(k+1)/2}\lambda^{1/2}]$	—
Inradius	$E(I)$	$1/4\lambda^{1/2}$	$0.250\lambda^{-1/2}$
	$E(I^2)$	$1/4\pi\lambda$	$0.080\lambda^{-1}$
Circumradius	$E(R)$	$3/4\lambda^{1/2}$	$0.750\lambda^{-1/2}$
	$E(R^2)$	$2/\pi\lambda$	$0.637\lambda^{-1}$
	$E(R^k)$	$\Gamma(k/2+2)/\pi\lambda^{k/2}$	—

λ = intensity of Θ_P.
Sources: Martin (1965) and Miles (1970a).

while for $m = 3$,

$$E_3V_3^k = \frac{35\pi^{1/2}(k+1)!(k+2)!(2k+4)!}{256\Gamma(k/2+2)^3\,\Gamma[(9+3k)/2]\,(16\pi\lambda)^k}, \quad k = 1,2,\ldots \,. \qquad (5.9.6)$$

Values of $E_2V_2^k$ and $E_3V_3^k$ for $k = 1,2$ are given in Tables 5.9.1 and 5.9.2, respectively.

For $m = 2$, Miles (1970a) has also derived first- and second-order moments of a number of characteristics of a typical PDC (see Table 5.9.1). In addition, the relationship described in equation (5.9.1) allows the derivation of other moment measures for $\mathscr{D}_P$ in both $\mathbb{R}^2$, $\mathbb{R}^3$ and $\mathbb{R}^4$ which are reported in Tables 5.9.1, 5.9.2 and 5.9.3, respectively. Correlation coefficients for pairs of selected characteristics of PDCs in $\mathbb{R}^2$ have also been obtained by Miles (1970a, Table 1) and are given in Table 5.9.4.

In terms of probability distributions of PDC characteristics, knowledge is limited to $m = 2$ and focuses on the internal angles of the triangles. Miles (1970a) derives the pdf of a pair of angles α, β selected at random from an arbitrary triangle of $\mathscr{D}_P$,

$$f(\alpha,\beta) = (8/3\,\pi)[\sin\alpha \sin\beta \sin(\alpha+\beta)] \quad (\alpha > 0,\ \beta > 0,\ (\alpha+\beta) < \pi)\,. \qquad (5.9.7)$$

Table 5.9.2 Expected values of characteristics of a typical Poisson Delaunay cell in $\mathbb{R}^3$

Characteristic	Symbol	Exact value	Numerical value
Intensity of cell vertices	λ_0	λ	λ
Intensity of midpoints of cell edges	λ_1	$(24\pi^2/35+1)\lambda$	7.768λ
Intensity of centroids of cell faces	λ_2	$(48/35/\pi^2\lambda$	13.535λ
Intensity of cell centroids	λ_3	$(24/35)\pi^2\lambda$	6.768λ
Number of edges/vertex	E(E)	$(48\pi^2/35)+2$	15.535
Number of faces/vertex	E(F)	$(144/35)\pi^2$	40.606
Number of cells/vertex	E(M)	$(96/35)\pi^2$	27.071
Number of faces/edge	E(G)	$(144\pi^2)/(24\pi^2+35)$	5.228
Number of edges/cell	E(N)	$(144\pi^2)/(24\pi^2+35)$	5.228
Volume	E(V)	$35/24\pi^2\lambda$	$0.148\lambda^{-1}$
	E(V^2)		$0.037\lambda^{-2}$
Area of a face	E(A)	$[(3/4\pi)^{2/3}(7)(5)^3\Gamma(2/3)]/3^5\pi\lambda^{2/3}$	$0.597\lambda^{-2/3}$
Length of an edge	E(L)	$[(3/4\pi)^{1/3}(5)(7)^3\Gamma(1/3)]/(3)^2\,(4)^4\lambda^{1/3}$	$1.237\lambda^{-1/3}$

λ = intensity of Θ_P.

Table 5.9.3 Expected values of characteristics of a typical Poisson Delaunay cell in $\mathbb{R}^4$

Characteristic	Exact value	Numerical value
Number of edges/vertex	340/9	37.778
Number of triangles/vertex	590/3	196.667
Number of tetrahedra/vertex	2860/9	317.778
Number of cells/vertex	1430/9	158.889
Number of triangles/edge	177/17	10.412
Number of tetrahedra/edge	429/17	25.235
Number of cells/edge	286/17	16.824
Number of tetrahedra/triangle	286/59	4.847
Number of cells/triangle	286/59	4.847
Number of tetrahedra/cell	2	2
Volume	$9/286\lambda$	$0.0315\lambda^{-1}$
Volume of a tetrahedron	$(2/\pi^2)^{3/4}(32/385)\ (14/13)^2(18!!/17!!)\ \Gamma(3/4)\lambda^{-3/4}$	$0.192\lambda^{-3/4}$
Area of a triangle	$(2)^{1/2}(2/13)\ (8/11)^2(17!!/16!!)\ \Gamma(1/2)\lambda^{-1/2}$	$0.681\lambda^{-1/2}$
Length of an edge	$(2/\pi^2)^{1/4}\ (27/77\pi)\ (16!!/15!!)\Gamma(1/4)\lambda^{-1/4}$	$1.3825\lambda^{-1/4}$

λ = Intensity of Θ_P.
Source: Christ *et al.* (1982, Tables 1 and 2).

Table 5.9.4 Correlation coefficients for characteristics of a typical Poisson Delaunay cell in $\mathbb{R}^2$

Characteristic		P	I	R
Area	A	0.916	0.921	0.783
Perimeter	P	—	0.878	0.936
Inradius	I	—	—	0.694
Circumradius	R	—	—	—

Source: Miles (1970a, p. 113).

The mode of this distribution is $\alpha = \beta = \pi/3$, which indicates that the most likely triangle in $\mathcal{D}_P$ is an equilateral one. By integrating equation (5.9.7) over all values of β, we obtain

$$f(\alpha) = 4[(\pi - \alpha)\cos\alpha + \sin\alpha]\sin\alpha/3\pi \quad (0 < \alpha < \pi) \qquad (5.9.8)$$

as the pdf of a randomly selected angle of an arbitrary triangle of $\mathcal{D}_P$. This distribution is illustrated in Figure 5.9.1. Equation (5.9.8) was also obtained

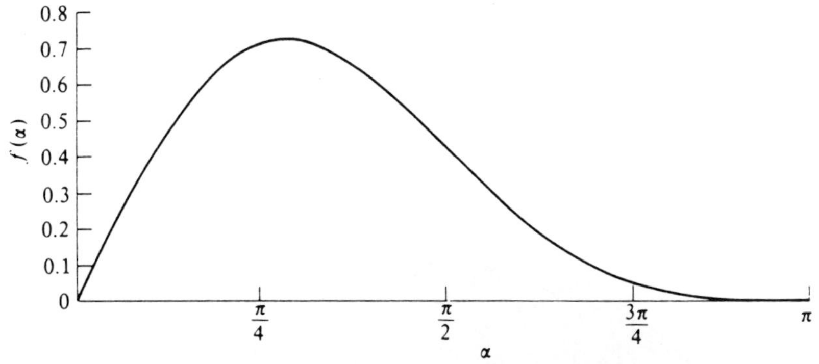

Figure 5.9.1 Probability density function of a randomly selected angle of a typical Poisson Delaunay cell in $\mathbb{R}^2$.

by Collins (1968). The mean of equation (5.9.8) is $E(\alpha) = \pi/3$ and the second moment is

$$E(\alpha^2) = [(2\pi^2/9 - (5/6)] ,$$

while the mode is given by the root of $\tan 2\alpha = 2(\alpha - \pi)$ which is approximately equal to 0.894 89 radians (51°16′24″). Problems in using equation (5.9.8) in empirical applications led to the use of Miles' approach to derive both the minimum and maximum angles, α_{min} and α_{max}, respectively of an arbitrary PDC. Mardia *et al.* (1977) derived

$$f(\alpha_{min}) = (2/\pi)[(\pi - 3\alpha_{min}) \sin 2\alpha_{min} + \cos 2\alpha_{min} - \cos 4\alpha_{min}] \quad (0 < \alpha_{min} < \pi/3) \qquad (5.9.9)$$

and Boots (1986) gave

$$f(\alpha_{max}) = (2/\pi)(3\alpha_{max} \sin 2\alpha_{max} - \cos 2\alpha_{max} + \cos 4\alpha_{max} - \pi \sin 2\,\alpha_{max}) \quad (\pi/3 < \alpha_{max} < \pi/2)$$

and

$$f(\alpha_{max}) = (1/\pi)(4\pi \cos \alpha_{max} \sin \alpha_{max} + 3 \sin(\alpha_{max})^2 - \cos(\alpha_{max})^2 - 4\alpha_{max} \cos \alpha_{max} \sin \alpha_{max} + 1) \qquad (5.9.10)$$
$$(\pi/2 < \alpha_{max} < \pi) .$$

Further understanding of the angular properties of $\mathcal{D}_P$ in $\mathbb{R}^2$ is provided by Kendall (1983) whose approach involved an examination of the shape of Delaunay triangles. Kendall's theory of shape involved associating a triad of labelled points from a pattern with a location in a shape space (for an overview of how this is achieved see Stoyan *et al.*, 1987, pp. 214–227, or Small, 1988). The particular shape space derived by Kendall is represented as a flat, curvilinear triangle referred to as the spherical blackboard. This

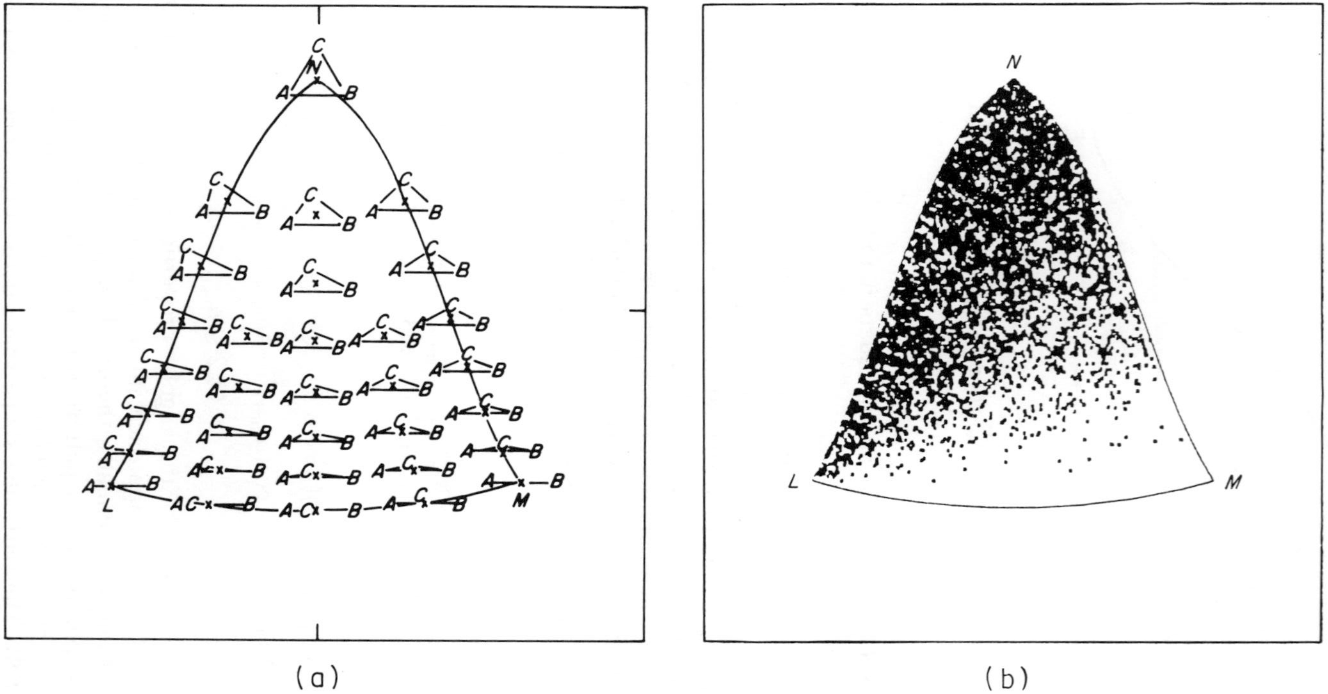

Figure 5.9.2 (a) The spherical blackboard showing 32 triangles located according to their shapes. (b) 5000 Poisson Delaunay cells in $\mathbb{R}^2$ located on the spherical blackboard. *Source*: (a) Kendall (1981); (b) Kendall (1983).

is illustrated in Figure 5.9.2(a), which also shows triangles of representative shapes located at the appropriate positions. Figure 5.9.2(b) shows the shapes of 5000 PDCs obtained by Kendall using the simulation procedure described in Section 5.3. This clearly illustrates the dominance of more equilateral shapes.

The distribution of the length of an arbitrary edge L of $\mathcal{D}_P$ in $\mathbb{R}^2$ has been derived by Collins (1968) and is given by

$$f(L) = (\pi\lambda L/3)[\lambda^{1/2}L\exp(-\pi\lambda L^2/4) + \operatorname{erfc}(\pi^{1/2}\lambda^{1/2}L/2)] \;, \qquad (5.9.11)$$

where erfc is the complimentary error function

$$\operatorname{erfc}(z) = (2/\pi^{1/2})\int_z^\infty e^{-t^2}\,dt = \pi^{-1/2}\,\Gamma(1/2,z^2) \;. \qquad (5.9.12)$$

Drouffe and Itzykson (1984) have also estimated $E(L|N)$, the average length of an edge emanating from a vertex as a function of the number of edges incident at the vertex, which they find grows approximately as $N^{1/2}$.

With regard to s-dimensional sections of $\mathcal{D}_P$ in $\mathbb{R}^m$, note that $\mathcal{D}_P$ (2,3) consists of triangles and quadrilaterals while $\mathcal{D}_P$ (3,4) consists of polyhedra with four and five faces and that the plane section of $\mathcal{D}_P$ (3,4) contains polygons with 3, 4 and 5 sides (Miles, 1986a, p. 153).

So far the only work on the nature of $\mathcal{D}$ from the various generalizations of $\mathcal{V}$ discussed in Section 3.7 has been that of Miles (1971) on the nature of the Delaunay tessellation $\mathcal{D}_S$ for p random points distributed on the surface of a sphere of unit radius and centre at the origin o in $\mathbb{R}^3$. In this case $\mathcal{D}_S$ has $(2p-4)$ triangular faces and $(3p-6)$ edges. The expected area of a face $E(A)$ is given by

$$E(A) = 2\pi/(p-2) \;. \qquad (5.9.13)$$

In adition, Miles (1971, p. 167, equation 7.6) derived an expression for the probability distribution of $\mathcal{D}_S$ but is not able to evaluate it.

5.10 OTHER RANDOM VORONOI DIAGRAMS

If, as the result of comparing characteristics of an empirical tessellation with those of the Poisson Voronoi diagram $\mathcal{V}_P$, we determine that the latter is inappropriate as a model of the former or if we have other evidence which supports this viewpoint, we may wish to consider other random tessellations as models. One approach is to modify the properties of the homogeneous Poisson point process Θ_p (see Section 1.3.8) which provides the generator set for $\mathcal{V}_P$. For example, the independence property of Θ_P means that there is no constraint on interpoint distances, while in empirical situations inhibitive or repulsive effects may limit how closely individual points can be located with respect to others. If such limits are imposed, the locations of the points in $\mathbb{R}^m$ are more dispersed than they would be in Θ_P. Following established convention in point pattern analysis (see Section 8.1), we label the resulting pattern of points as regular. In some areas of application in the natural sciences such patterns are also referred to as ordered. Alternatively, we may relax the homogeneity property of Θ_P and permit the intensity of points to become variable in $\mathbb{R}^m$. The resulting pattern of points is one in which there is a relative concentration of points in some regions of $\mathbb{R}^m$ and a relative absence in others. Again following convention, we refer to such patterns as clustered (or highly disordered).

In this section we briefly review different point processes which have been used to create generator sets for Voronoi diagrams. We do not report results for these diagrams although Table 5.10.1 provides a summary of the studies known to the authors together with a listing of the characteristics examined in them. We also note that, with the exception of those studies by Finney (1970b, 1975a) where the point locations are measured from laboratory models of sphere packings in $\mathbb{R}^3$, all studies involve estimates from Monte Carlo simulations.

We begin by considering processes that produce point patterns which are more regular than those generated by Θ_P. A common feature of many of these processes is that a minimum interpoint distance r is specified. In one group of processes, collectively known as *hard core processes*, this constraint is imposed on points generated by Θ_P. The process is equivalent to a random packing of equal discs or spheres (in $\mathbb{R}^2$, $\mathbb{R}^3$, respectively) if the resulting points are considered to be the centres of non-intersecting discs or spheres of diameter r. If we are concerned with a bounded region B of $\mathbb{R}^m$, hard core processes terminate when either a previously specified number of points have been located in B or it is no longer possible to add more points in B without violating the minimum interpoint distance. The latter case is referred to as *complete random packing*. In either case the process can be characterized by the *packing density*, which is the proportion of B occupied by discs or spheres. Hard core processes may take a variety of forms. In the *simple sequential inhibition (SSI) process* (Hasegawa and Tanemura, 1980; Tanemura and Hasegawa, 1980; Lotwick, 1982; Hanson, 1983) discs of fixed diameter r are located sequentially in B until the specified number of discs is obtained (or complete random packing is achieved). At each step the location of the centre of the next disc is chosen uniformly at random from those points which ensure that no two discs overlap. An alternative approach involves a *thinning process* in which first a specified number of points is generated in B using Θ_P and then points are removed in various ways until the minimum interpoint distance constraint is satisfied for all remaining points (Vincent and Howarth, 1982; Hermann *et al.*, 1989). Another hard core process involves a *re-arrangement process* (Finney, 1975a; Fraser, 1991). Initially, a set of points is generated by Θ_P. Then, for any pair of points separated by a distance less than r, the points are moved apart along a line passing

Table 5.10.1 Studies of other random Voronoi and Delaunay cells

Study	R^m	Model type	Characteristics
Boots (1977)	2	CP HC	N
Boots (1982)	2	CP	N
Boots (1984)	2	DL	N
Burger *et al.* (1988)	2	CA DL	$A\ L_D\ N$
Burger *et al.* (1989)	2	CA DL	$A\ L_D\ N$
Finney (1970a)	3	HC	$A\ F\ N\ V$
Finney (1975a)	3	HC	$F\ N$
Fraser (1991)	2	HC	$A\ L\ N$
Frost and Thompson (1987a)	2	HC	$A\ L\ N\ \alpha$
Frost and Thompson (1987b)	2	HC	$A\ N$
Hanson (1983)	3	HC	$F\ M\ N\ V$
Hasegawa and Tanemura (1980)	2	HC	L_D
Hermann *et al* (1989)	2	HC	A
Hutchings and Discombe (1986)	2	DL PC	$A\ L_D$
Icke and van de Weygaert (1987)	2	HC SC	$A\ L\ N\ P\ \alpha$
Le Caer and Ho (1990)	2	RM	$A\ L\ N\ P\ A_D\ L_D\ P_D$
Smalley (1966)	2	HC	N
Tanemura and Hasegawa (1980)	2	HC	$A\ L\ N\ P\ \alpha$
Vincent and Howarth (1982)	2	HC	$L_D\ N\ \alpha_D$
Wilkinson (1988a)	2	CA	A

Model type:
CA Cluster amplification
CP Compound Poisson
DL Displaced lattice
HC Hard core
PC Poisson cluster
RM Random matrix
SC Sequential clustering

Characteristics: Two dimensions

Voronoi cells:
A = area
L = Edge length
N = Vertices
P = Perimeter
α = Interior angle

Delaunay cells:
A_D Area
L_D Edge length
P_D Perimeter
α_D Interior angle

Three dimensions
A = Area/face
F = Faces/polyhedron
M = Vertices/polyhedron
N = Edges/face
V = Volume

through them until their interpoint distance is equal to r. This procedure is repeated until the minimum interpoint distance condition is met.

Hasegawa *et al.* (1981) proposed a variant of the SSI version of the hard core process which they labelled *Voronoi polygonal areal random packing*. This involved defining a minimum size a for the Voronoi polygon associated with a point. Initially, three points are located in B according to Θ_P. A new point is then selected at random and the Voronoi diagram of the points is generated. The new point is retained only if the sizes of each individual

Voronoi polygons exceed a. This procedure is continued until no additional points can be located in B without violating the minimum size constraint.

An alternative way of generating point patterns which are more regular than those resulting from Θ_P is to take a completely ordered situation and introduce a random component. Point lattices are instances of such ordered situations (see Section 8.1) and provide the basis of *displaced lattice processes*. If d is the distance separating neighbouring points on a lattice, a displaced lattice process involves moving each lattice point a distance $\alpha\beta d$ in a random direction ϕ ($0 < \phi < 2\pi$), where β is a displacement function and α is a random value ($0 < \alpha < 1$). As $\beta \rightarrow 1$ the points of the displaced lattice tend to Θ_P. So far the Voronoi diagrams for displaced lattice points have only been explored in $\mathbb{R}^2$ for triangular (Boots, 1984; Burger *et al.*, 1988, 1989; Wilkinson, 1988a) and square lattices (Hutchings and Discombe, 1986).

Le Caer and Ho (1990) and Le Caer (1990) provided another point process in $\mathbb{R}^2$ which is more regular than Θ_P and which is associated with the eigenvalues of random matrices. For an ensemble of asymmetric ($n \times n$) complex random matrices M_n in which the real and imaginary parts of the matrix elements are independently and identically distributed according to a Gaussian distribution with mean μ and standard deviation $\sigma/\sqrt{2}$, the eigenvalues of $M_n/n^{1/2}$ are distributed uniformly over a disc of radius σ. The eigenvalues also display an inhibitive effect since the probability of finding two identical eigenvalues in such matrices is zero. Furthermore, this process is unique in the statistical sense, isotropic whatever the value of n and homogeneous when $n \rightarrow \infty$ (see Section 1.3.8 for a discussion of isotropy and homogeneity in point processes). These properties led Le Caer and Ho (1990) to suggest that the resulting Voronoi diagram, which they label the *random matrix Voronoi diagram*, could be used in the same normative fashion as $\mathcal{V}_P$.

Less effort has been expended on the Voronoi diagrams associated with point patterns more clustered than those resulting from Θ_P. Clustered point patterns can be generated by relaxing the homogeneity property of Θ_P (see Section 1.3.8). Processes of this kind in which the intensity of the process varies spatially are usually referred to as *compound point processes*. One example is provided by the negative binomial process in which the intensity of Θ_P varies over a discrete set of subareas in B according to a gamma distribution. Boots (1977, 1982) reported on Voronoi diagrams for point processes generated in this way in $\mathbb{R}^2$.

An alternative way of destroying the homogeneity of Θ_P occurs in *Poisson cluster processes*. Such processes involve three steps. First, a set of locations for cluster centres in B is defined using Θ_P. Then a random variable describing the number of points in each cluster is generated from a specified probability distribution. Finally, the points in each cluster are distributed with respect to the cluster centre according to some spatial process. Hutchings and Discombe (1986) and Hermann *et al.* (1989) both described characteristics of Voronoi diagrams for points generated by a Poisson cluster process where the number of points in a cluster has a Poisson distribution and where the points are distributed according to Θ_P over a circle of fixed radius centred on the cluster centre.

Burger *et al.* (1988, 1989) and Wilkinson (1988a) used the opposite of the thinning process described above to generate points which are more clustered than those of Θ_P. We label this a *cluster amplification process*. First a set of points is generated by Θ_P with intensity λ. Then any point which is further than a critical distance $\alpha\lambda^{-1/2}$ $(0< \alpha < 1)$ from any other point is removed. The points removed in this way are replaced by an equivalent number of points located in B according to Θ_P. These steps are then repeated until the critical distance condition is satisfied. Note that as $\alpha \to 1$, the cluster amplification process tends to Θ_P.

Finally, we note that Icke and Van de Weygaert (1987) described a *sequential clustering process* in which an initial point is located at random in B and then each subsequent point is located at a distance $r = -\alpha \log d$ $(\alpha \geq 0)$, where d is a random number $(0 \leq d \leq 1)$, and a random direction ϕ $(0 \leq \phi \leq 2\pi)$ from the previous point. As $\alpha \to \infty$, the process approaches Θ_P.

CHAPTER 6

Spatial Interpolation

In Chapter 2 the ordinary Voronoi diagram was defined with respect to a generator set of n distinct points, $P = \{p_1,\ldots,p_n\}$, located in m-dimensional Euclidean space, $\mathbb{R}^m$. As with the bounded ordinary Voronoi diagram of Section 2.1, our concern here will be with a finite region S in $\mathbb{R}^m$. The location of each point p_i in S is indicated by $\mathbf{x}_i = (x_{i1},\ldots,x_{im})$ and for the purposes of this chapter we label these points as *data sites*. Unlike Chapter 5 we make no assumption regarding how the data sites are located in S, although in most of the applications discussed below the data sites are irregularly spaced in S. In addition, for each data site we have a numerical observation z_i which we call the *data value*. If we assume that the data values represent observations from a surface defined by $(\mathbf{x},z)$ *spatial interpolation* involves finding a function $f(\mathbf{x})$ which best represents the entire surface and which predicts values of z for locations other than P in S. Such a function is referred to as an *interpolant*. The particular form we select for $f(\mathbf{x})$ depends on the type of measurement scale of z and the use to which the interpolated surface is put. However, we can make an immediate distinction between *exact* and *approximate interpolants* in terms of whether or not they reproduce the data values; i.e. for an exact interpolant, $f(\mathbf{x}_i) = z_i$. Approximate interpolants may be considered a form of data smoothing which are more appropriate than exact interpolants if the data values are subject to inaccurate measurement or errors. This distinction is not of primary concern in this chapter although most of the interpolants discussed specifically are exact ones.

Various other characteristics may also be used to distinguish between interpolants. One which is relevant here is the difference between *global* and *local interpolants*. With the former the value of $f(\mathbf{x})$ at any given location in S depends on all the data values while the latter involves use of only those values at 'nearby' points. Various definitions of nearby are provided below. In general, local interpolants involve partitioning S into s subdomains S_i; $S = \cup_{i=1}^{s} S_i$ and then selecting a function of the form

$$f(\mathbf{x}) = f_i(\mathbf{x}), \quad \mathbf{x}\in S_i,\ i = 1,\ldots,s\ . \tag{6.0.1}$$

Thus, global interpolants are affected by the addition or deletion of data values or by changing the location of data sites. For local interpolants such changes only produce effects at locations in the vicinity of the changes. Consequently, fairly large systems of equations are needed to evaluate $f(\boldsymbol{x})$ for global interpolants, which means that exact versions are not feasible when n is large, say in the order of 100–200 data sites. In contrast, local interpolants typically involve only small systems of equations although there may be a large number of these. In this chapter we consider local interpolants exclusively since global interpolants do not involve Voronoi diagrams.

We focus our discussion on $\mathbb{R}^2$. Although the methods described can be extended in theory to $\mathbb{R}^3$, complications created by the bounded nature of S quickly dominate the interpolation. Also the majority of applications relate to $\mathbb{R}^2$. Typical examples of data values include soil and rock types, elevations of a geological strata, barometric pressure and thickness of an ore body or of a manufactured part. In such cases our choice of an interpolant (surface) will be influenced by the nature of the data values. Values measured on a nominal scale, such as rock and soil types, require a piecewise continuous surface which may be represented in two-dimensions as a k-phase mosaic (or in cartographic terms, a choropleth map) (see Figure 6.0.1 where k=3). Other values such as elevation which are continuous random variables require at least a continuous surface. These can be represented in a number of ways including contour (isarithmic) maps,

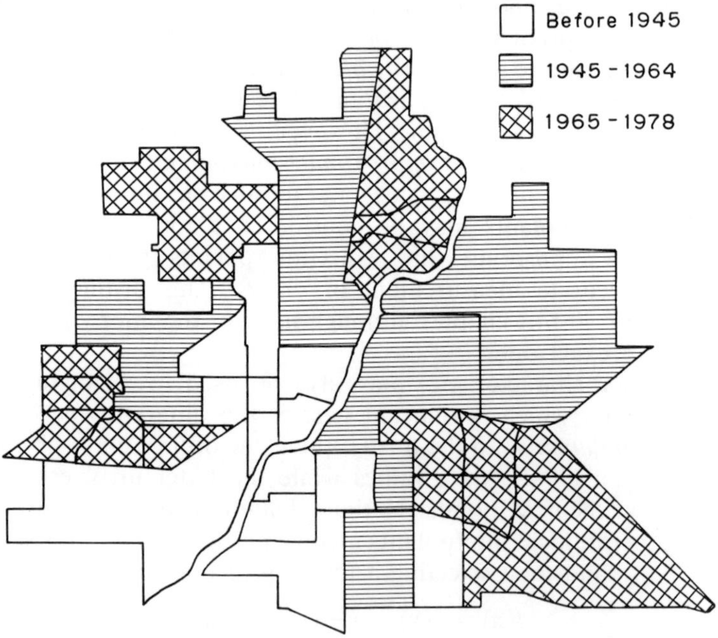

Figure 6.0.1. Choropleth map: Saskatoon Census Metropolitan Area, Canada. Time period of immigration of the majority of immigrants by census tract, 1981.

perspective views, hill shading and slope maps (see Figures 6.0.2(b), (c), (d) and (e), respectively). The nature of the continuity of the interpolant (surface) is designated C^k which indicates that the k-th or lower order derivatives of the interpolant are all continuous. Thus, for a C^0 interpolant the derivatives are in general discontinuous at the data sites so that a contour map produced from it will appear angular (see Figure 6.0.3). In contrast, a C^1 interpolant such as that used in Figure 6.0.2(b) produces smooth contours. For most practical applications, such as those listed below, C^1 interpolants are sufficient, particularly since, with the exception of special cases, higher order smoothness does not appear to be detectable by the human eye (Sibson, 1981, p. 22).

In terms of the uses of the interpolant, by far the most common is simply to produce a contour map or other visual representation of the surface. Other uses involve calculating some property of the surface at a specified location, calculating the volume under the surface, as in the estimation of ore bodies, and calculating global statistics for the surface, such as estimating the mean precipitation over a region.

This chapter considers only those interpolation procedures that involve Voronoi diagrams or Delaunay triangulations. We refer to these as polygonal and triangular methods, respectively, although in view of the duality between Voronoi diagrams and Delaunay triangulations, this distinction is one adopted primarily for the convenience of presentation. General reviews of spatial interpolants can be found in Rhind (1975), Schumaker (1976), Schut (1976), Barnhill (1977), Ripley (1981, Chapter 4), Franke (1982), Sabin (1982), Lam (1983) and Gold (1989). Section 6.1 deals with polygonal methods while triangular methods are the subject of Section 6.2. The methods described in Section 6.1 include both the use of the ordinary Voronoi diagram and the ordered order-k Voronoi diagram. The main thrust of Section 6.2 is a consideration of the merits of using triangular domains and the characteristics of different triangulations, including the Delaunay. Initially, this discussion assumes that S is convex, usually rectangular. Section 6.3 considers the modifications necessary to extend the Delaunay triangulation to situations in which S is neither convex nor simply connected. This section also considers ways of incorporating specific data features into the triangulation. In Section 6.4 we consider how the Delaunay triangulation can be used in approximating surfaces with a predefined level of accuracy. Finally, in Section 6.5 we note that a triangulation may be considered an ordering of the data sites and we examine how the Delaunay triangulation can be used in this regard for multivariate data.

6.1 POLYGONAL METHODS

In general all local interpolants represent the value of the surface $f(\mathbf{x})$ at an arbitrary point p located at $\mathbf{x}$ in S as a weighted, usually linear function of the values at a set of 'nearby' data sites, $D(p)$ ($D(p) \subset P$), so that

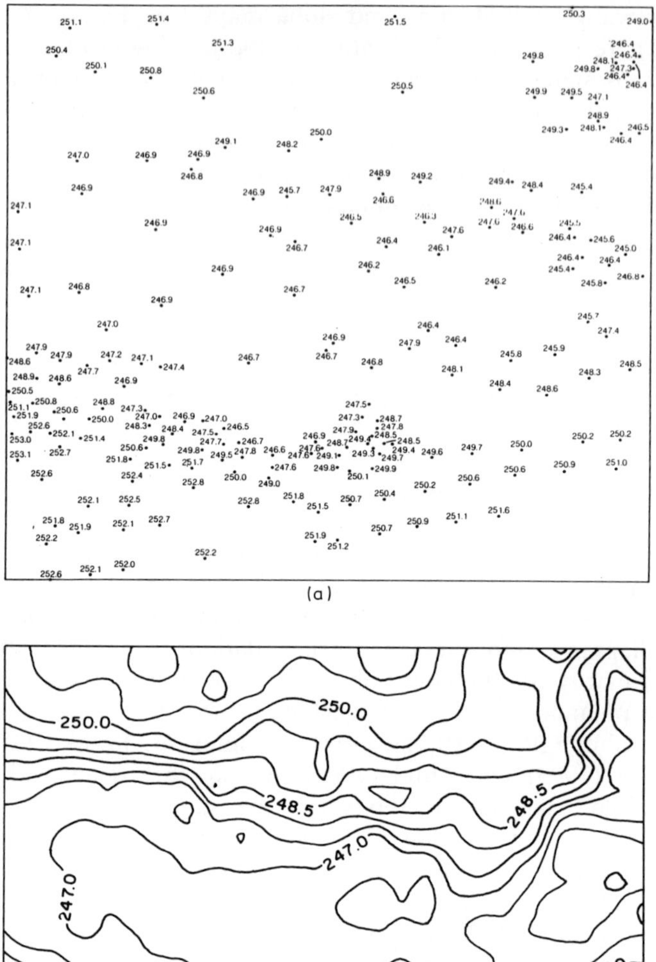

Figure 6.0.2 Different representations of a continuous surface. (a) Original data sites and data values. (b) Contour (isarithmic) map. (c) Perpsective view. View direction (azimuth) = 45°; viewing angle = 30°. (d) Hill shading. Sun azimuth = 35°; sun elevation angle = 145°. (e) Slope. Classes: 1, 0–10%; 2, 10–20%; 3, 20–30%; 4, 30–40%; 5, 40–50%; 6, 50–60%.

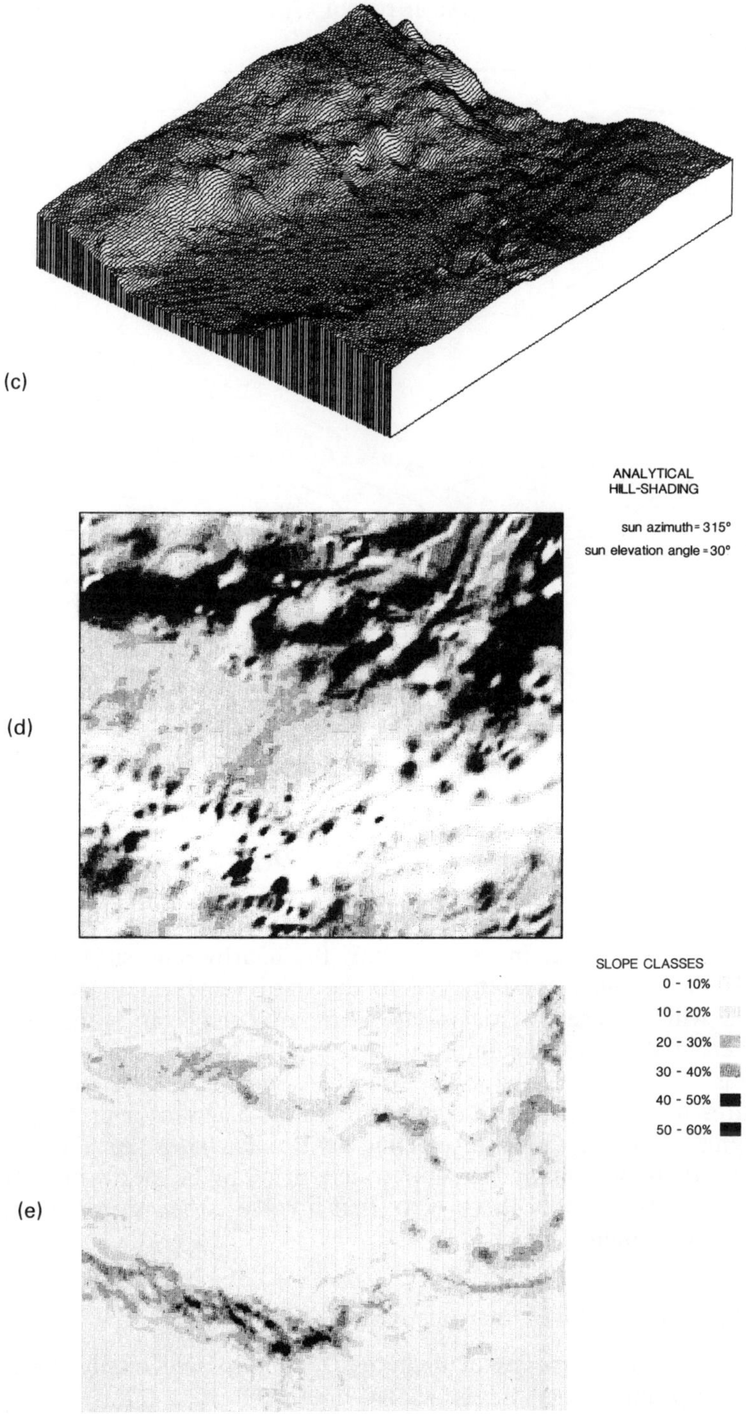

Figure 6.0.2. Continued.

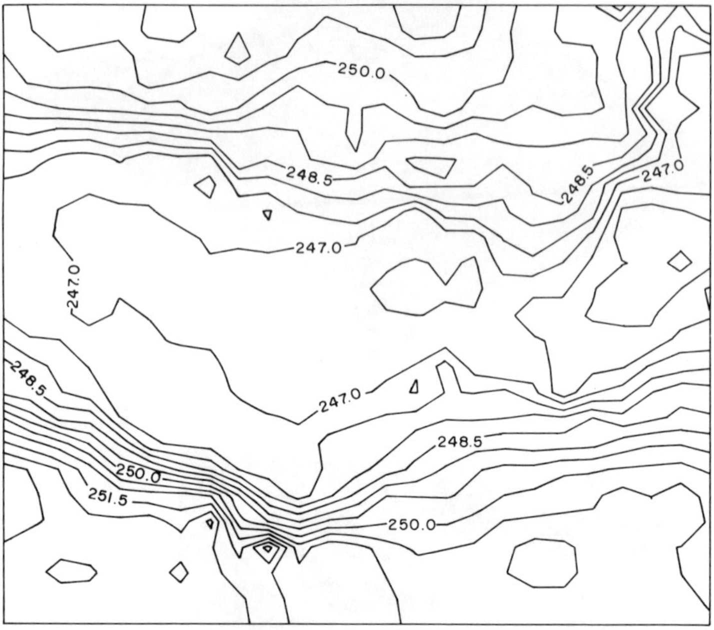

Figure 6.0.3. Contour map of data in Figure 6.0.2 (a) using a C^0 interpolant.

$$f(\mathbf{x}) = \sum_{i=1}^{n_D} w_i z_i, \quad p_i \in D(p) \ , \tag{6.1.1}$$

where n_D is the number of nearby sites and w_i is the weight attached to p_i. Local interpolants differ in terms of how the nearby data sites are selected and how the weights are attached to the data values at those sites. The extent to which these two decisions are related may vary considerably in that different weighting schemes may be applied to the same set of nearby data sites. Since, with one exception (see Section 6.1.2), the weighting procedures do not involve concepts of Voronoi diagrams, our presentation will emphasize the methods of selecting nearby data sites. As Tobler (1975) indicated, there are many ways in which this can be performed. In this chapter we consider only those ways that involve either Voronoi diagrams or Delaunay triangulations.

6.1.1 Nearest neighbour interpolation

Perhaps the simplest means of defining $D(p)$ is to select only the nearest data site to p and assign the associated data value to z. Thus,

$$f(\mathbf{x}) = z_i, \quad d(p, p_i) < d(p, p_j), \quad p_i \in P, i \neq j \ , \tag{6.1.2}$$

where $d(p, p_i)$ and $d(p, p_j)$ are the Euclidean distances between p and p_i and p and p_j, respectively. Note that this definition is essentially equivalent to that of an open Voronoi polygon given in Section 2.1. If $\boldsymbol{x}$ is equidistant from two or more members of P, $f(\boldsymbol{x})$ is undefined. Such locations correspond to the boundaries of the Voronoi polygons of $\mathcal{V}(P)$ so that the resulting surface is discontinuous. In this particular application the individual polygons are usually referred to as *proximal polygons* (Peucker and Chrisman, 1975).

Such an interpolant is most appropriate when z is measured on a nominal scale as with land use, soil, vegetation or rock type whose individual values form patches or phases over S. In such cases once the initial proximal polygons are formed, adjacent polygons with equal values can be merged to form larger patches (see Figure 6.1.1). Arnold and Milne (1984) used this approach to derive soil maps for Greece based on data values collected at data sites spaced every 50–200 m over the study area.

This application may be considered the equivalent of the nearest-point rule for pattern reconstruction in image analysis (Ahuja and Schachter, 1983, Section 1.4.3.3). Here the problem is one of reconstructing a continuous image from the colours observed at a number of fixed sample points. Under

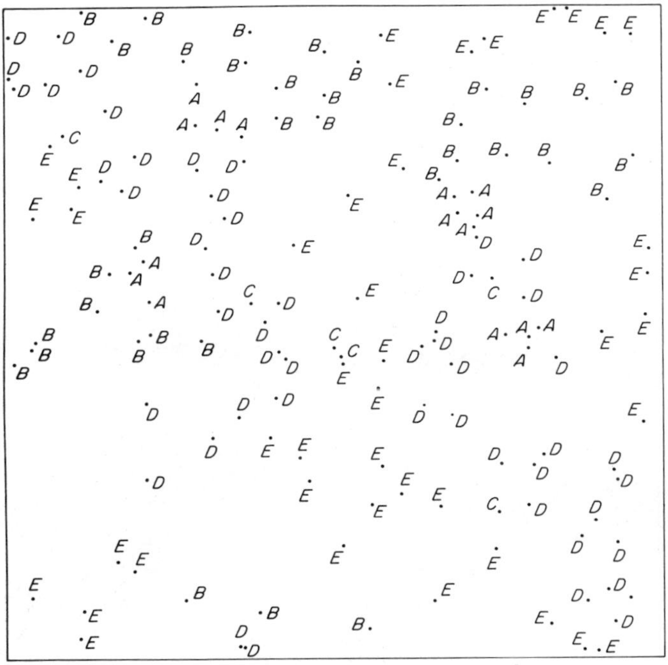

(a)

Figure 6.1.1. Creation of a discontinuous surface by nearest neighbour interpolation: (a) Data sites and data values. (b) Voronoi diagram (proximal polygons). (c) Final surface.

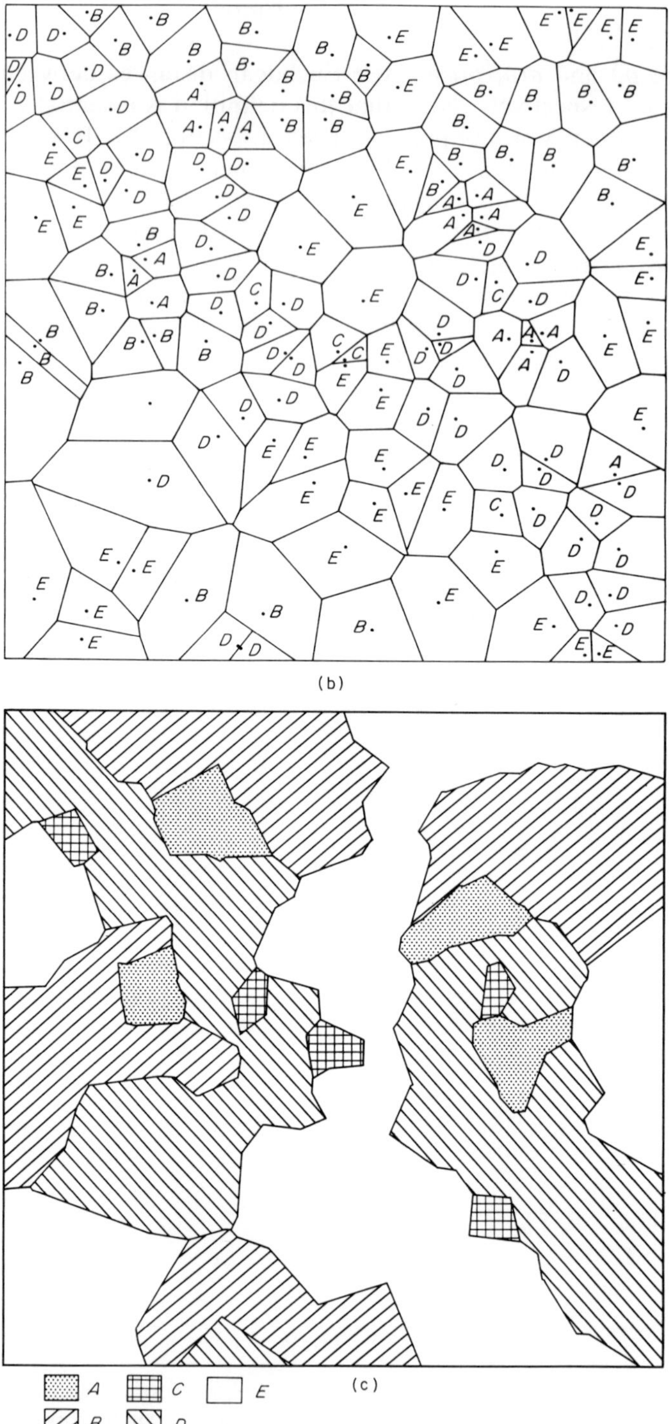

Figure 6.1.1. Continued.

the nearest-point rule the colour assigned to each image element is that of the nearest sample point. A parallel procedure is also used in quantization which involves analog to digital conversion (Conway and Sloane, 1988, pp. 56–62) and has applications in digital measuring instruments or recording devices and in digital communications systems. Here the value at any location $\boldsymbol{x}$ is 'rounded off' to that of the closest data site.

Proximal polygon interpolation can also be used when z is not nominally scaled although it will not reproduce the underlying continuous surface. The application which gives rise to the term 'Thiessen polygon' is just such an example (Thiessen, 1911; Horton, 1917). Thiessen's concern was to estimate the mean value $\bar{z}$ of a surface (in his case precipitation) over a bounded region for which data values z_i only existed for an irregularly spaced set of n data sites (precipitation gauges). Given the spatial distributions of gauges Thiessen felt it inappropriate to estimate $\bar{z}$ simply as

$$\bar{z} = \sum_{i=1}^{n} z_i/n \ . \tag{6.1.3}$$

Instead he used

$$\bar{z} = \sum_{i=1}^{n} w_i z_i / \sum_{i=1}^{n} w_i \ , \tag{6.1.4}$$

where w_i is the area of the proximal polygon associated with p_i. This example is also discussed in Chapter 8.

The same approach is sometimes used in estimating the average grade of an ore deposit, although as befits a different application with different origins, the proximal polygons are usually termed *area of influence polygons* (Popoff, 1966; Reedman, 1979; Hayes and Koch, 1984). The approach involves representing the surface as a set of polygonal prisms whose heights reflect the thicknesses of the ore body (see Figure 6.1.2). The average grade $\bar{z}$ is calculated using expression (6.5) with w_i equal to the volume of the prism associated with p_i and z_i equal to the ore grade.

A related procedure has been proposed by Dahlberg (1967) as a way of representing a pattern of points in S, where each point indicates the occurrence of a given phenomenon, as a surface. Voronoi polygons are generated about each point which are then represented as prisms the heights of which are inversely proportional to the area of the polygon. Dahlberg suggested that the prisms may provide a more easily interpreted representation of the variation in the intensity of the phenomenon over the region.

6.1.2 Natural neighbour interpolation

Another way of defining $D(p)$ for an arbitrary point p at location $\boldsymbol{x}$ in S is to identify those members of P which are Voronoi or, to use Sibson's term, 'natural' neighbours of p (Sibson, 1981; Watson and Philip, 1987). This

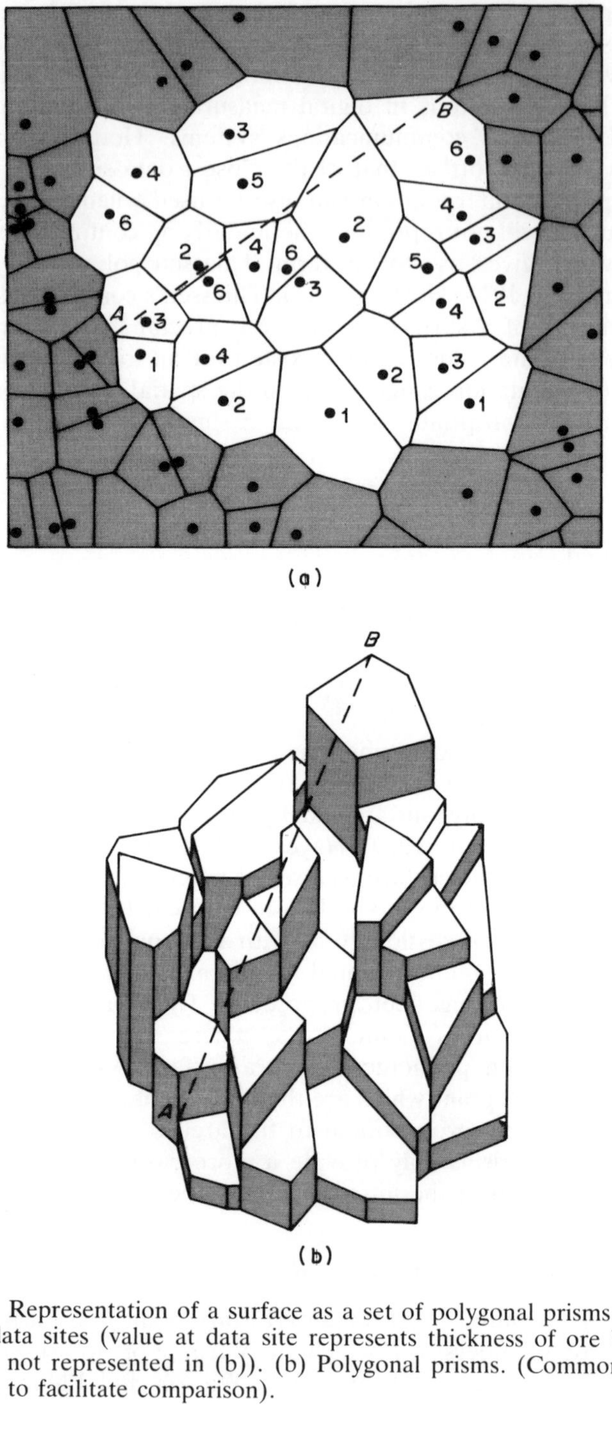

Figure 6.1.2. Representation of a surface as a set of polygonal prisms: (a) Voronoi diagram of data sites (value at data site represents thickness of ore body; shaded polygons are not represented in (b)). (b) Polygonal prisms. (Common dashed line AB is shown to facilitate comparison).

involves generating the Voronoi diagram of $(P \cup p)$. Like all the members of P, p has its associated Voronoi region, $V(p)$ (the shaded polygon in Figure 6.1.3). Those points of P whose Voronoi regions are contiguous to $V(p)$ form the natural neighbours of p (the filled circles in Figure 6.1.3). The data values at the natural neighbours are then used in equation (6.1.1).

A special case of spatial interpolation is the missing value problem where the unknown value of a variable at a given location must be estimated from the known values of the variable at other locations. In their study of rainfall data for Kansas and Nebraska, U.S.A., Haining *et al.* (1984) used natural neighbours to identify which weather stations to use in estimating missing values at other stations.

There are also other situations besides interpolation where it may be necessary to identify nearby points and in which natural neighbours have been proposed as the solution (Besag, 1974, 1975; Ord, 1975). A particular instance is the study of spatial autocorrelation (Goodchild, 1986; Griffith, 1987; Odland, 1988). This involves measuring the extent of dependence in the values of a variable observed at nearby locations. Spatial autocorrelation measures ρ often have the general form

$$\rho = \sum_{i=1}^{n} \sum_{i=1}^{n} w_{ij} z_i z_j \,, \tag{6.1.5}$$

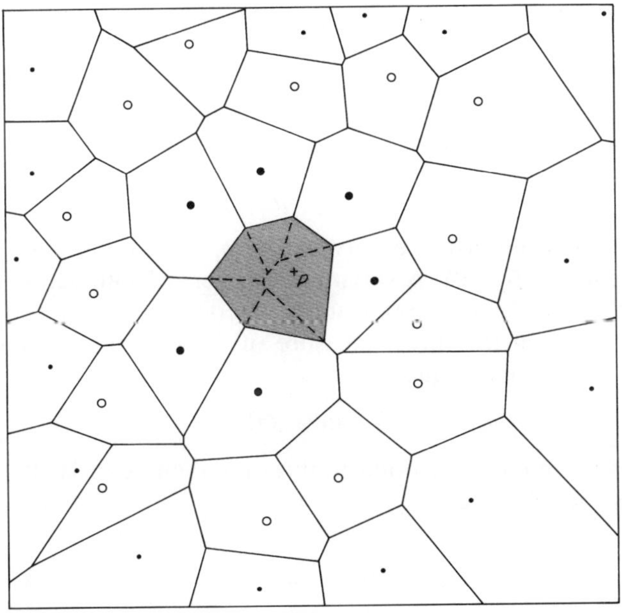

Figure 6.1.3 Voronoi diagram $(P \cup p)$. *Key*: ●, natural neighbours of p; points contributing to interpolated value at p using equation (6.1.9); ○, points contributing to interpolated values at p using equation (6.1.13); ·, other points of p.

where z_i and z_j are the values of the variable at data sites p_i and p_j, respectively, and w_{ij} is a weight reflecting the spatial relationship between p_i, p_j. In a study of the grain handling system of the province of Manitoba in Canada from 1943 to 1975, Griffith (1982) suggested setting $w_{ij}=1$ in equation (6.1.5) if p_i p_j are natural neighbours and $w_{ij}=0$ otherwise.

It is worth noting that the natural neighbour definition has also been frequently proposed as a way of identifying neighbours in a variety of other circumstances. In $\mathbb{R}^3$ these include identifying neighbouring atoms, molecules or particles in both crystalline and amorphous structures including monatomic liquids, solutions, metallic glasses, metals and alloys and for identifying neighbouring cracks in percolation structures. The definition of neighbouring capillaries in muscle tissue is an example involving two-dimensional sections of material in $\mathbb{R}^3$, while the recognition of neighbours for plants and animals and for stores in competitive situations in ecology and retailing, respectively, are examples of the use for phenomena represented in $\mathbb{R}^2$. Discussion of these uses together with the appropriate references is contained in Chapter 7.

If the members of P have associated weights (see Section 3.1) these may be combined with natural neighbour concepts to define other types of neighbour relationships. One example occurs in the field of spatial interaction modelling (Hayes and Fotheringham, 1984) which attempts to predict the movement of people, goods, information, etc. between places located in $\mathbb{R}^2$. Typically, the amount of interaction I_{ij} between two places p_i and p_j, with weights w_i and w_j, respectively (where the weights usually reflect some measure of the sizes of the places such as their populations), is considered to be directly proportional to the product of their weights and inversely proportional to the distance separating them so that

$$I_{ij} = kw_iw_jd(p_i, p_j)^{-\beta} , \tag{6.1.6}$$

where $d(p_i, p_j)$ is the distance between p_i, p_j, k is a constant, and β is the distance exponent ($\beta > 0$). A variant of this general interaction formulation, known as the intervening opportunities model, was proposed by Stouffer (1940). He replaced the direct measure of distance used in equation (6.7) by an indirect one and argued that

$$I_{ij} = kw_iw_jO_{ij}^{-1} , \tag{6.1.7}$$

where O_{ij} is the number of opportunities intervening in $\mathbb{R}^2$ between p_i and p_j which can be measured by

$$O_{ij} = \sum_{\substack{k=1 \\ k\neq i,j}}^{m} w_k , \tag{6.1.8}$$

where m is the number of points in P which intervene between p_i and p_j. The problem with this modification is the definition of the intervening places. One way is to generate the Voronoi diagram $\mathcal{V}(P)$ of P and to consider the places intervening between p_i and p_j as those whose Voronoi polygons are intersected by the line segment, $\overline{p_ip_j}$. For example, in Figure 6.1.4(a) the places intervening between p_{18} and p_{35} are p_6, p_{13} and p_{30}. In those circumstances where w_i is recorded as on ordinal rank r_i (the larger the rank the more important the place), it is often suggested that interaction between p_i and p_j will only occur if no points with higher ranks intervene between the two. Thus, to determine which places interact, we can use the same procedure as above. In this case p_i and p_j will interact if the line segment $\overline{p_ip_j}$ does not intersect a Voronoi polygon of a point with a higher rank than either p_i or p_j. Thus, in Figure 6.1.4(a) if the numbers associated with the points represent their ranks, the points which interact with p_{18} are shown by the line segments linking them to p_{18}.

Elliott (1981, 1982, 1983, 1985) has proposed a further variant of this method. He labelled the set of places for which interaction occurs with p_i as the surrounding larger neighbours (SLNs) of p_i. The SLNs of p_i may be defined by generating $\mathcal{V}(P)$ (see Figure 6.1.4(a)) which in turn may be represented by the dual Delaunay triangulation, $\mathcal{D}(P)$ (see Figure 6.1.4(b)). If we consider $\mathcal{D}(P)$ as a graph (see Section 1.3.6) we can measure how far each point p_j is from p_i in terms of the number of edges e_j of $\mathcal{D}(P)$ (see Figure 6.1.4(b)). Point p_j is an SLN of p_i if it has a higher rank than p_i and a path of length e_j can be traced from p_j to p_i which does not pass through any other vertex of $\mathcal{D}(P)$ with a higher rank than p_j. Figure 6.1.4(c) shows the SLNs for p_{18}. These may be compared with the higher ranked places defined by the previous method (the solid lines in Figure 6.1.4(a)). Elliott (1983, 1985) made use of SLNs derived in this way in modelling inter-city airline passenger traffic in the United States.

Sibson (1981) has also proposed using the natural neighbour concept in deriving the appropriate weights used in equation (6.1.1). We know from our discussion of ordered order-k Voronoi diagrams in Section 3.2 (Property OOK2) that in the Voronoi diagram of $(P \cup p)$ the Voronoi region of p, $V(p)$, can be exhaustively subdivided into ordered order-2 regions $V((p, p_j))$, such that the first and second nearest points of $(P \cup p)$ from any location in $V((p, p_j))$ are p and p_j, respectively (see Figure 6.1.3). Let $\kappa(\boldsymbol{x}, \boldsymbol{x}_j)$ be the area of $V((p, p_j))$. In the special case $p = p_j$, we set $\kappa(\boldsymbol{x}, \boldsymbol{x}_j) > 0$ and $\kappa(\boldsymbol{x}, \boldsymbol{x}_i) = 0$ for $i \neq j$. Sibson used the normalized values $\lambda(\boldsymbol{x}, \boldsymbol{x}_j) = \kappa(\boldsymbol{x}, \boldsymbol{x}_j)/\Sigma_i\, \kappa(\boldsymbol{x}, \boldsymbol{x}_i)$ in deriving the weights for z_j in expression (6.1.1). These values have two fundamental properties:

Property LCP1 $\lambda(\boldsymbol{x}, \boldsymbol{x}_j)$ is a continuous function of $\boldsymbol{x}$ which is continuously differentiable everywhere except for the data sites.

Property LCP2 Provided that $V(p)$ does not meet the boundary of S,

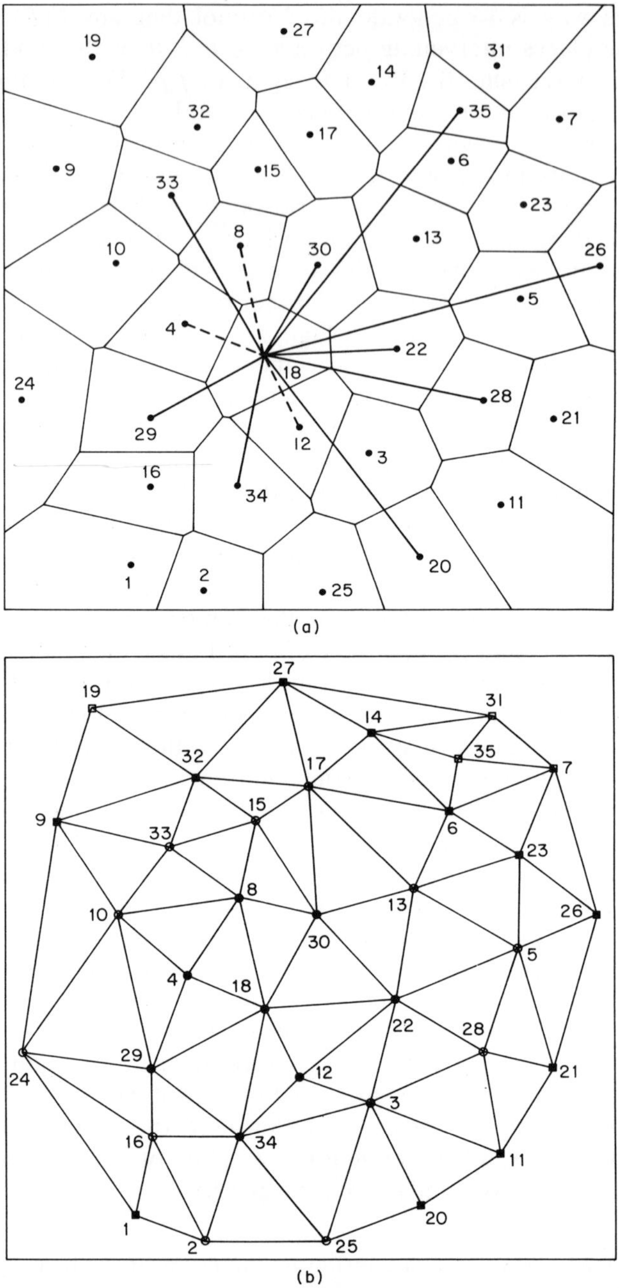

Figure 6.1.4 Definition of intervening opportunities. (a) Places interacting with p_{18}: —— higher ranked places; ---- lower ranked places. (b) Distance from p_{18} in terms of number of Delaunay edges: ●, 1; ○, 2; ■, 3; □, 4. (c) Surrounding larger neighbours of p_{18}.

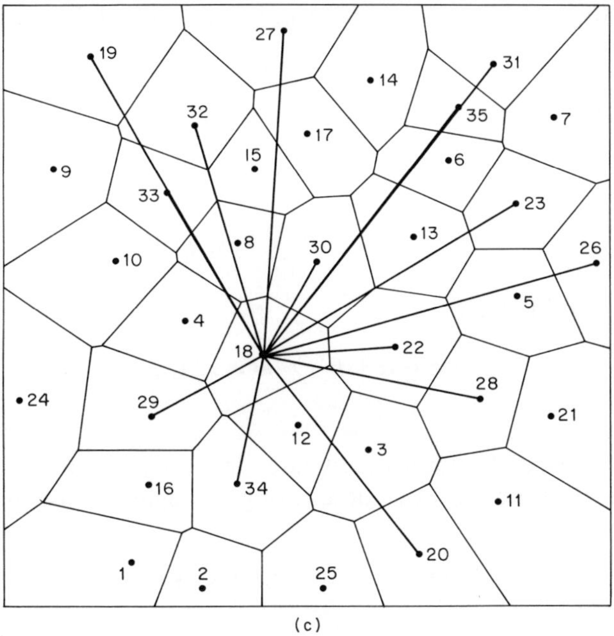

Figure 6.1.4. Continued.

$$\sum_j \lambda(\mathbf{x}, \mathbf{x}_j)\mathbf{x}_j = \mathbf{x} .$$

A proof of LCP2 is given by Sibson (1980a) (see Section 3.2.2, Property OOK7). In view of this representation of $\mathbf{x}$, Sibson labelled the $\lambda(\mathbf{x},\mathbf{x}_j)$'s the local coordinates of point p. If $V(p)$ meets the boundary of S, $\lambda(\mathbf{x}, \mathbf{x}_j)$ does not have any physically useful meaning because it is not independent of the shape of region S. Indeed, in what follows we use the value of $\lambda(\mathbf{x}, \mathbf{x}_j)$ only for points p such that $V(p)$ does not meet the boundary of S.

A C^0 interpolant, f_0 ($\mathbf{x}$), can be obtained by using the local coordinates directly so that

$$f_0(\mathbf{x}) = \sum_j \lambda(\mathbf{x}, \mathbf{x}_j)z_j , \tag{6.1.9}$$

Thus the data values that contribute to the interpolated values at p in Figure 6.1.3 are those for the points shown by filled circles. If the data values come from a linear function of the form

$$z = \alpha + \beta^{\mathrm{T}}\mathbf{x} ,$$

$f_0(\mathbf{x})$ correctly reproduces it, which can be shown as follows. Substituting

$$z_j = \alpha + \beta^{\mathrm{T}}\mathbf{x}_j$$

in equation (6.1.9) gives

$$f_0(\mathbf{x}) = \sum_j \lambda(\mathbf{x}, \mathbf{x}_j)\,(\alpha + \beta^{\mathrm{T}}\mathbf{x}_j)$$

$$= \alpha + \beta^{\mathrm{T}} \sum_j \lambda(\mathbf{x}, \mathbf{x}_j)\mathbf{x}_j\ ,$$

Using Property LCP2,

$$z = \alpha + \beta^{\mathrm{T}} \sum_j \lambda(\mathbf{x}, \mathbf{x}_j)\mathbf{x}_j\ ,$$

Thus, $f_0(\mathbf{x}) = z$.

Sibson also used the local coordinates in deriving a C^1 interpolant. He fitted planes $\zeta(\mathbf{x}_i)$ for the value at $\mathbf{x}_i$, using a weighted least squares fit with weights

$$w(\mathbf{x}_i, \mathbf{x}_j) = \lambda(\mathbf{x}_i, \mathbf{x}_j)/d(\mathbf{x}_i, \mathbf{x}_j)^2\ , \tag{6.1.10}$$

where

$$d(\mathbf{x}_i, \mathbf{x}_j)^2 = (\mathbf{x}_j - \mathbf{x}_i)^{\mathrm{T}}(\mathbf{x}_j - \mathbf{x}_i)\ .$$

Explicitly, $\boldsymbol{b}_i$, the gradient at $\mathbf{x}_i$, is given by

$$\boldsymbol{b}_i = H_i^{-1}\boldsymbol{q}_i \tag{6.1.11}$$

where

$$H_i = \sum_j w(\mathbf{x}_i, \mathbf{x}_j)\,(\mathbf{x}_j - \mathbf{x}_i)\,(\mathbf{x}_j - \mathbf{x}_i)^{\mathrm{T}}\ ,$$

$$\boldsymbol{q}_i = \sum_j w(\mathbf{x}_i, \mathbf{x}_j)\,(\mathbf{x}_j - \mathbf{x}_i)\,(z_j - z_i)\ .$$

Thus,

$$\zeta_i(\mathbf{x}) = z_i + \boldsymbol{b}_i^{\mathrm{T}}\,(\mathbf{x} - \mathbf{x}_i)$$

is the linear function through $(\mathbf{x}_i, z_i)$ with slope $\boldsymbol{b}_i$. Combining these functions with weights $W(\mathbf{x}, \mathbf{x}_i) = \lambda(\mathbf{x}, \mathbf{x}_i)/d(\mathbf{x}, \mathbf{x}_i)$, gives the interpolant

$$f^*(\mathbf{x}) = z_i, \quad \text{if } \mathbf{x} = \mathbf{x}_i\ ,$$

$$f^*(\mathbf{x}) = \sum_j [W(\mathbf{x}, \mathbf{x}_j)\,\zeta_j(\mathbf{x})] \sum_j W(\mathbf{x}, \mathbf{x}_j) \quad \text{if } \mathbf{x} \neq \mathbf{x}_i\ , \tag{6.1.12}$$

The final C^1 interpolant, $f_1(\mathbf{x})$, is a linear combination of $f_0(\mathbf{x})$ and $f^*(\mathbf{x})$ chosen to ensure that it fits exactly restricted quadratic surfaces of the form

$$z = \alpha + \beta^{\mathrm{T}}\,\mathbf{x} + \gamma\mathbf{x}^{\mathrm{T}}\mathbf{x}\ .$$

Specifically,

$$f_1(x) = \frac{\left\{\left[\sum_i \lambda(\mathbf{x}, \mathbf{x}_i)d(\mathbf{x}, \mathbf{x}_i)\right] \Big/ \sum_i W(\mathbf{x}, \mathbf{x}_i)\right\} f_0(\mathbf{x}) + \sum_i [\lambda(\mathbf{x}, \mathbf{x}_i)d(\mathbf{x}, \mathbf{x}_i)^2] f^*(\mathbf{x})}{\left[\sum_i \lambda(\mathbf{x}, \mathbf{x}_i)d(\mathbf{x}, \mathbf{x}_i)\right] \Big/ \sum_i W(\mathbf{x}, \mathbf{x}_i) + \sum \lambda(\mathbf{x}, \mathbf{x}_i)d(\mathbf{x}, \mathbf{x}_i)^2} \tag{6.1.13}$$

Note that while $f_0(\mathbf{x})$ in (6.1.9) is valid for all locations which satisfy the condition imposed in Property LCP2 above, $f(\mathbf{x})$ in (6.1.13) holds only for a more limited set of locations. This set consists of those locations p for which neither $V(p)$ nor any Voronoi polygon adjacent to $V(p)$ in the Voronoi diagram $\mathcal{V}(P \cup p)$ meets the boundary of S. The set of data values which contribute to the value of $f^*(\mathbf{x})$ at p in Figure 6.1.3 are those for the points shown by both filled and unfilled circles.

6.2 TRIANGULAR METHODS

In this approach interpolation of z at an arbitrary location p in S involves three steps. First, we construct a triangulation $\mathcal{T}(P)$ of the data sites in P using the data sites as vertices of the triangles. Next we identify the triangle $T(p)$ of $\mathcal{T}(P)$ which contains p and define the set of nearby points $D(p)$ as the vertices of $T(p)$. Finally, we compute z using the values of z_i (i = 1,2, 3) for the members of $D(p)$.

The same procedure is also used in two other applications. One is in finite element methods, where solutions of partial differential equations of planar regions are obtained by decomposing the regions into triangular elements (Schwarz, 1988). The other is the estimation of ore reserves where triangular prisms may be used in place of polygonal ones (see Section 6.1.1) (Harding, 1920–21, 1923; Popoff, 1966, pp. 78–84; Reedman, 1979, pp. 438–439).

In all of these application areas the methods used in the last two steps are independent of the nature of $\mathcal{T}(P)$ and since they do not involve Voronoi or Delaunay concepts, we confine our attention in this chapter to the construction of $\mathcal{T}(P)$. Reviews of different triangle-based interpolants can be found in Barnhill (1977), Gold (1980), McCullagh (1981) and Watson and Philip (1984a).

When the main purpose of interpolation is to create contours from data values at a set of data sites irregularly spaced over S the triangulation approach possesses several advantages over the other most frequently used approach, the regular grid. An instance of this is a digital terrain mode (DTM) where the data values are elevations of some region of a planetary surface. In this context $\mathcal{T}(P)$ is referred to as a triangulated irregular network (TIN). In the regular grid approach a regular, usually square grid is superimposed on S and values of z are estimated at the grid intersections on the basis of their spatial relationships to the data values for the members of P. There are a variety of ways that such an estimation can be performed (see Monmonier, 1982, pp. 58–65; Watson and Philip, 1985; Jones *et al.*, 1986, pp. 44–57 for reviews). The main advantage of the regular grid approach is that the set of grid intersections may be stored as a matrix in which both their locations and topological relationships are defined implicitly so that it is necessary only to store the z values assigned to them. In contrast, the topological relationships of the facets of a TIN must be stored explicitly, although this is not necessary if they are to be used only for hill shading. If the facets of the TIN are used as the units of storage, each facet requires six pointers, three for the vertices and three to identify the neighbouring triangles. Since the total number of triangles in a TIN is $2n-n_c-2$, where n_c is the number of data sites on the boundary of the convex hull of P (see Property D11, Section 2.4), this scheme requires a total of $12n-6n_c-12$ pointers. Alternatively, we can list in a consistent order (e.g. clockwise from the north) the neighbouring data sites of each data site. This requires a total of $6n-2n_c-6$ pointers, equal to twice the number of edges in the TIN (see Property D11, Section 2.4) for about a 50% savings in storage.

Topographical surfaces are not stationary (see Section 1.3.8) since the variation in terrain height changes from one landform to another (compare a jagged mountain range with a coastal plain). Thus, the spacing of points of any regular grid superimposed on S must be adjusted to capture height changes in areas of greatest variability. This means that many of the grid points will be redundant in areas of constant relief. TINs with their natural variable resolution can easily adapt to such variations. Mark (1975) suggests that in a typical DTM 14 regular data points are required for each irregular data point to produce the same level of representation. Also it is much easier to incorporate linear and other required features (such as ridge and valley lines) in a TIN structure (see Section 6.3). In such circumstances Peucker *et al.* (1976) suggest that the ratio of regular to irregular data points rises to about 287:1. Furthermore, since the data sites are the vertices of the TIN, the data values will be interpolated exactly. It is unlikely that this will happen in the regular grid approach which, as a result, should more properly be considered a smoothing procedure rather than an interpolation one.

The facets of a TIN also avoid the ‘saddle point’ problem which can arise with the regular grid. If the data values associated with four neighbouring grid intersections take any of the forms in Figure 6.2.1(a), then the path of the contour through the grid square defined by the intersections is unambiguous. However, if the data values are configured as in Figure 6.2.1(b), then several different paths for the contour are possible. Finally, TINs also appear to have advantages in terms of the computational effort and speed involved in calculating the final contours. Although there is little systematic information on this, at least one study (McCullagh and Ross,

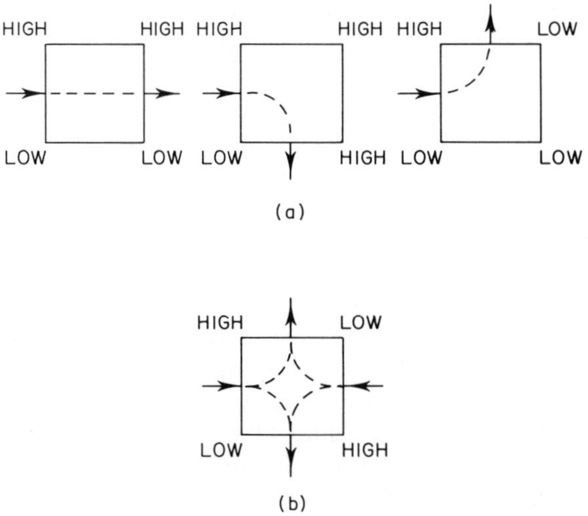

Figure 6.2.1 The saddle-point problem. (a) Unambiguous paths for the contour. (b) Ambiguous paths for the contour.

1980) suggests that the identification of the TIN structure takes on average only one-tenth of the time required to compute the values of z at the grid intersections. The computing time for C^0 contours is approximately equal for the regular grid and the TIN while a C^1 interpolation on the fomrer takes about two-thirds of the time taken on the latter. Nevertheless, this suggests that overall the TIN method is at least five times faster than the regular grid in this context. Computational speed for an actual contouring situation reported by Philip and Watson (1982) supports this suggestion.

If there are n data sites in P, let E be the set of $n(n-1)/2$ line segments (edges) joining all pairs of data sites. Then $\mathcal{T}(P)$ is a maximal subset of E in which no two edges intersect except at their end points. $\mathcal{T}(P)$ is not unique. The number of possible triangulations $N[\mathcal{T}(P)]$ of P depends in part on the location of the members of P in S and no general expression exists for evaluating $N[\mathcal{T}(P)]$. However, each triangulation has $3(n-1) - n_c$ edges and $2(n-1) - n_c$ triangles (see Section 2.4).

In the three areas of application mentioned above it is desirable that the triangulation selected avoids the occurrence of triangles with two highly acute interior vertex angles and that the triangles be as equiangular as possible. This is because elongated triangles involve greater separation of data sites and reduce the ability of the triangulation to reflect local variation in Z. Whitney (1929) appears to be the first to have recognized the importance of this consideration to spatial interpolation. In Section 2.4 (Property D14) we saw that the Delaunay triangulation $\mathcal{D}(P)$ of P satisfies a local max-min angle criterion (also referred to as an equiangular criterion) in that the diagonal of every strictly, convex quadrilateral Q in $\mathscr{D}(P)$ is chosen so that it maximizes the minimum interior angle of the two resulting triangles (see Figure 2.4.13 in Section 2.4). Sibson (1978) demonstrated that $\mathscr{D}(P)$ is the only triangulation with this property. Further, as reported in Property D16 in Section 2.4.2, $\mathscr{D}(P)$ is also globally equiangular in that it maximizes the minimum angle occurring in $\mathscr{T}(P)$.

Lawson (1977, pp. 178–181) demonstrates that the equiangular criterion is equivalent to another, the circle criterion. If k is a circle passing through any three vertices of Q, and the fourth vertex is interior to k, we select the diagonal linking this vertex to the opposite vertex (see Figure 6.2.2(a)). If the fourth vertex is exterior to k we select the other diagonal (Figure 6.2.2(b)). If the fourth vertex is on k we select either diagonal (Figure 6.2.2(c)). This criterion avoids the selection of triangles with relatively large circumcircles and Watson and Philip (1984b) conjecture that $\mathscr{D}(P)$ is the $\mathscr{T}(P)$ which minimizes the mean circumcircle diameter.

It has also been suggested (Lawson, 1977; De Floriani *et al.*, 1982, 1983, 1985) that the equiangular criterion of $\mathscr{D}(P)$ is equivalent to the 'Pitteway' criterion proposed by McLain (1976), which requires that any location within any triangle of $\mathscr{T}(P)$ should have one of the vertices of that triangle as its closest data point. However, Sabin (1976) and Ripley (1981, p. 40) show that there are some patterns of data sites for which no $\mathscr{T}(P)$ satisfies the Pitteway criterion. In Section 2.4 (Property D3) we give a necessary and sufficient condition for $\mathscr{D}(P)$ to satisfy the Pitteway criterion.

Additional speculation concerns the relationship of $\mathscr{D}(P)$ to other triangulations, in particular the *Minimum length triangulation* (also known as the minimum weight triangulation and the optimal triangulation), MLT(P), and the *'greedy' triangulation*, GT(P). Let $|\mathscr{T}(P)|$ denote the sum of the lengths of the edges in $\mathscr{T}(P)$. Then MLT(P) is the $\mathscr{T}(P)$ which minimizes $|\mathscr{T}(P)|$. MLT(P) is unique for general sites. This triangulation was originally proposed by Bengtsson and Nordbeck (1964) and Düeppe and Gottschalk (1970). Steele (1982) showed that for n points ($1 \leq n < \infty$) located in a unit square according to a homogeneous planar Poisson point process (see Section 1.3.8), $|\text{MLT}(P)|/n^{1/2}$ tends to a constant as n tends to infinity.

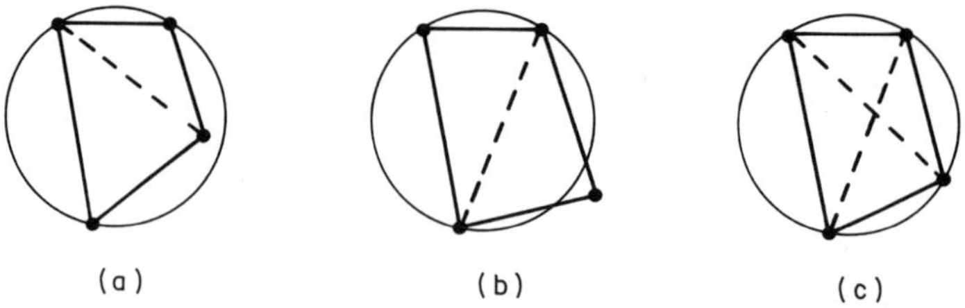

Figure 6.2.2. The circle criterion.

The criterion for the definition of GT(P) is that no edge should be included in $\mathcal{T}(P)$ if there is a shorter edge which would properly intersect it. GT(P) may be found by considering all possible edges in order of increasing length and inserting an edge in S provided that no edge already exists which properly intersects it. If no pair of edges has the same length, GT(P) is unique.

Often $\mathcal{D}(P)$, MLT(P) and GT(P) are identical (see Figure 6.2.3(a)). However, they need not be (see Figure 6.2.3(b)) as was first noted by Lawson (1977) and Lloyd (1977). In most situations where $\mathcal{D}(P)$, MLT(P) and GT(P) are not identical, GT(P) is a better approximator of MLT(P) than $\mathcal{D}(P)$. For arbitrarily large n, Manacher and Zobrist (1979) constructed sets of n data sites P_1 and P_2 for which

$$|\mathcal{D}(P_1)|/|\mathrm{MLT}(P_1)| \geq \mathrm{O}(n/\log n) \tag{6.2.1}$$

and

$$|\mathrm{GT}(P_2)|/|\mathrm{MLT}(P_2)| \geq \mathrm{O}(n^{1/3}) \ , \tag{6.2.2}$$

where $|\cdot|$ is the length of the triangulation as defined above and O(k) is the order of k (see Section 4.1). P_2 involves non-convex polygons. However, Lingas (1986a) constructed a positive real $\epsilon < 1$ for which

$$|\mathrm{GT}(P)|/|\mathrm{MLT}(P)| \leq \mathrm{O}(n^{\epsilon}) \tag{6.2.3}$$

holds for any convex, planar point set P with n data sites.

Kirkpatrick (1980) identified arbitrarily large convex sets of n data sites P_3 for which

$$|\mathcal{D}(P_3)|/|\mathrm{MLT}(P_3)| \geq \mathrm{O}(n) \ , \tag{6.2.4}$$

and so $\mathcal{D}((P_3)$ is asymptotically no better than an arbitrary triangulation in approximating MLT(P_3). However, Lingas (1986b) provded that when P consists of data sites located in a unit square according to a homogeneous planar Poisson point process (see Section 1.3.8),

$$|\mathcal{D}(P)|/|\mathrm{MLT}(P)| \leq \mathrm{O}(\log n) \tag{6.2.5}$$

with probability of at least $1 - cn^{\alpha}$, where c and α are constants satisfying $c > 0$ and $\alpha > 1$. Furthermore, Chang and Lee (1984) showed that when the data points are located in this way the average length of $\mathcal{D}(P)$, E($|\mathcal{D}(P)|$) is of the same order as that of MLT(P), E($|$MLT(P)$|$); more precisely they proved that

$$\mathrm{E}(|\mathcal{D}(P)|)/\mathrm{E}(|\mathrm{MLT}(P)|) \leq 64/9\pi \ . \tag{6.2.6}$$

Other triangulations have been suggested, but unlike $\mathcal{D}(P)$, MLT(P) and GT(P) these may be prone to problems such as lack of uniqueness and the creation of unnecessarily elongated triangles. Typically, these triangulations involve an iterative procedure in which first an initial triangulation is

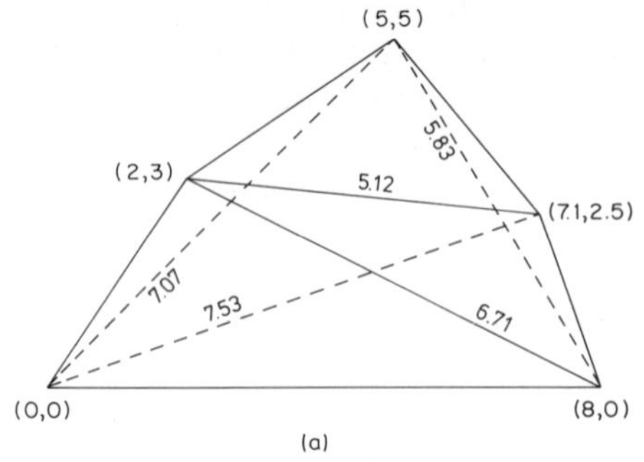

Figure 6.2.3 (a) A set of data sites in $\mathbb{R}^2$ for which $\mathcal{D}(P)$ = MLT(P) = GT(P) (solid line). (b) A set of data sites in $\mathbb{R}^2$ for which $\mathcal{D}(P) \neq$ MLT(P) $\neq$ GT(P) (redrawn from material in Watson and Philip (1980)).

constructed and then this is 'optimized' according to specific criteria. Examples of such optimization criteria include (a) requiring that any location in a triangle must be closer to one of the vertices of that triangle than to any other vertex of the triangulation (i.e. the 'Pitteway' criterion discussed above) (McLain, 1976; Kleinstreuer and Holderman, 1980); (b) minimizing the length of the diagonal of the convex quadrilateral Q (Zienkiewicz and Phillips, 1971; Fraser and Van den Driessche, 1972; Fraser, 1977; Mirante and Weingarten, 1982); and (c) maximizing the minimum triangle height perpendicular to the proposed diagonal of Q (Gold *et al.*, 1977; Gold, 1980).

Uniqueness and direct construction are particularly important to automated triangulation procedures. These characteristics of $\mathcal{D}(P)$ together with its equiangular property has led to it being proposed most frequently as a good triangulation for both spatial interpolation (Lewis and Robinson, 1977; Peucker *et al.*, 1978; Fowler and Little, 1979; McCullagh and Ross, 1980; De Floriani *et al.*, 1982, 1983, 1985; Philip and Watson, 1982; Watson and Philip, 1984b; Jones 1989) and finite element methods (Bramble and Zlamal, 1970; Cavendish, 1974; Lawson, 1977; Akima, 1978; Joe, 1986), although in the latter area of application Babuska and Aziz (1976) argued that what is essential is not a minimum angle condition but a maximum angle (min-max) condition in which no angle is too close to 180°. Edelsbrunner *et al.* (1990) have proposed an algorithm for generating the min-max triangulation.

6.3 MODIFYING DELAUNAY TRIANGULATIONS

One problem in using $\mathcal{T}(P)$ as the basis for interpolation over S is that $\mathcal{T}(P)$ is only defined in the interior of the convex hull CH(P) of P and therefore $\mathcal{T}(P)$ does not usually cover S. Furthermore, particularly for irregularly spaced data sites, $\mathcal{T}(P)$ invariably contains some undesirable

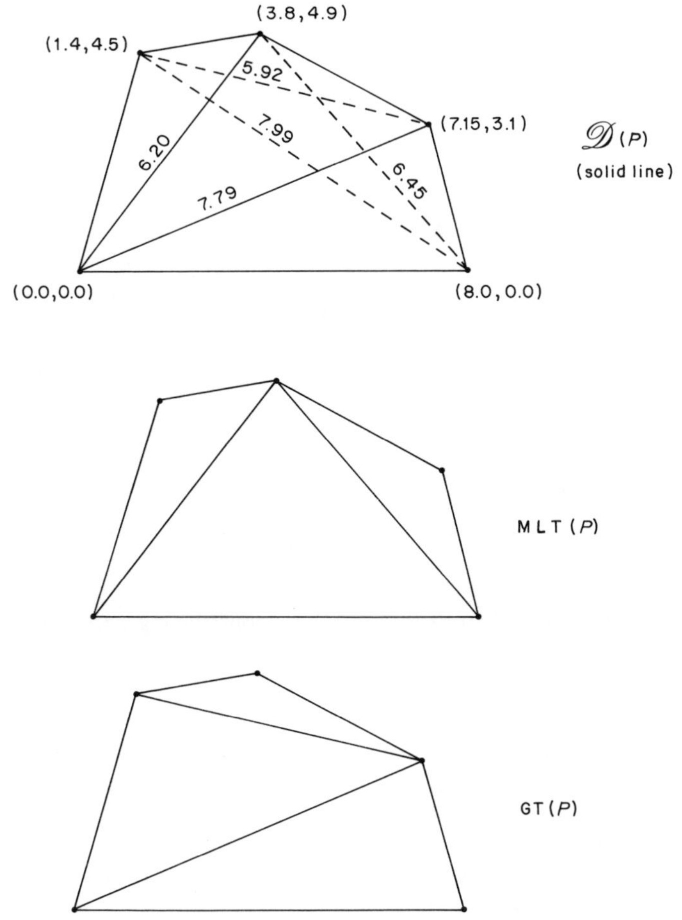

(b)

Figure 6.2.3. Continued.

long, thin triangles associated with points on the boundary of CH(P) regardless of the criterion used to define $\mathcal{T}(P)$. One way to extend a function defined over $\mathcal{T}(P)$ is to divide the area of S outside CH(P), $S\backslash$CH(P), into trapezoids and triangles by inserting perpendiculars to the exterior edges of CH(P) at each of the vertices, as shown in Figure 6.3.1 (Akima, 1978; Ripley, 1981, pp. 42–43; Franke, 1982, p. 788). The value of the interpolant for any location in S\CH(P) is then obtained from the data values at one (triangular area) or two (trapezoidal area) nearest data sites. Strictly speaking such values should be considered extrapolations rather than interpolations.

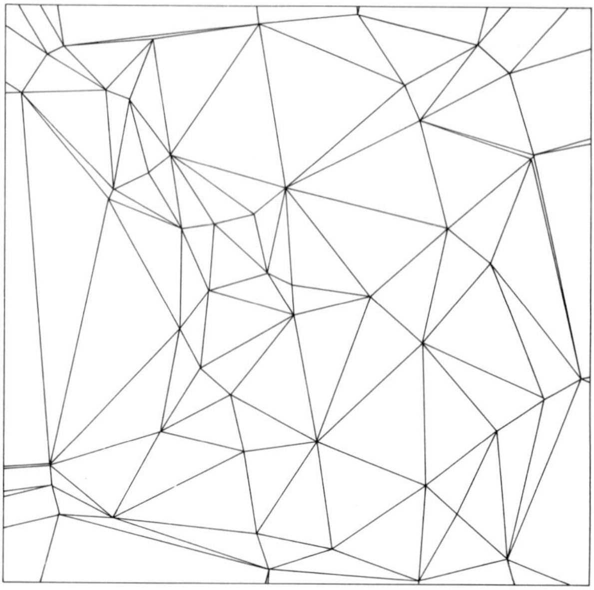

Figure 6.3.1. Extension of $\mathscr{D}(P)$ to cover S by constructing perpendiculars.

Another procedure which completes coverage of S and reduces the incidence of elongated triangles is to introduce imaginary data sites on the boundary of S (McCullagh and Ross, 1980; Franke, 1982) as indicated by the filled circles in Figure 6.3.2. The impact of this procedure on $\mathscr{T}(P)$ depends on the number and spacing of such imaginary data sites (compare Figures 6.3.2(a) and (6.3.2(b)).

So far in this chapter we have assumed that S is a simply connected, convex region. We now consider the construction of $\mathscr{D}(P)$ when S does not possess these properties as may be the case in some applications in digital terrain modelling (Yoeli, 1977;. De Floriani *et al.*, 1982; Mirante and Weingarten, 1982) and finite element methods (Lewis and Robinson, 1977; Kleinstreuer and Holderman, 1980; Sadek, 1980) where non-convex and multiply connected regions may be encountered. In such cases it is necessary to include edges of S as part of $\mathscr{T}(P)$. Another instance where specific edges must be incorporated into $\mathscr{T}(P)$ arises in digital terrain modelling when the data values refer to elevations. Although the data sites are usually irregular in terms of their location in the $(x–y)$ plane, often they are chosen to preserve both point features, such as pits and peaks, and linear features, such as ridges, valleys and fault lines, of special topographical significance (Peucker and Douglas, 1975; Yoeli, 1977; Fowler and Little, 1979; McCullagh, 1981; De Floriani *et al.*, 1985; Christensen, 1987).

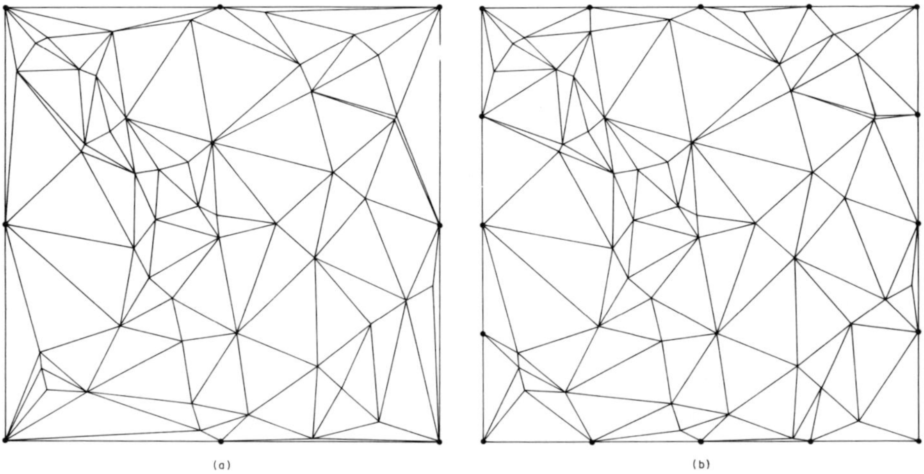

Figure 6.3.2. Extension of $\mathscr{D}(P)$ to cover S by introducing imaginary points (filled circles): (a) 8 imaginary points; (b) 16 imaginary points.

In all of the above instances we can consider the linear feature as a set of non-intersecting edges L with end points in P, which together with P can be considered as a planar, straight line graph $G(P,L)$. Thus, the problem becomes one of constructing a triangulation $\mathscr{T}(G)$ for $G(P,L)$. Our concern is extending the Delaunay criteria (see Section 6.2) to $\mathscr{T}(G)$ and we refer to the resulting triangulation as the *constrained Delaunay triangulation,* CDT(G), of $G(P,L)$.

The modifications of $\mathscr{D}(P)$ which must be undertaken to create CDT(G) arise because the presence of L in S affects the mutual visibility of members of P. Two sites in P, p_i and p_j, are said to be mutually visible if the line segment $\overline{p_i p_j}$ does not properly intersect any member of L. For example, in Figure 6.3.3 points p_1 and p_2 are mutually visible while points p_2 and p_4 are not. Let V be the set of edges joining those data sites which are mutually visible. Then CDT(G) consists of L together with a subset of V in which the circumcentre of each triangle with vertices p_i, p_j and p_k does not contain in its interior any other data site which is visible from p_i, p_j and p_k. Thus, points p_1, p_2 and p_3 in Figure 6.3.3 form a triangle of CDT(G) even though the circle passing through them contains p_4 because p_2 and p_4 are not mutually visible. If no four points of P are circular, CDT(G) is unique. CDT(G) is a subgraph of the visibility graph of $G(P,L)$ (see Section 3.4.1). For graphs $G(P,\text{O})$, CDT(G) is equal to $\mathscr{D}(P)$. Figure 6.3.4 shows examples of CDT(G) together with their unconstrained counterparts $\mathscr{D}(P)$.

Like $\mathscr{D}(P)$, CDT(G) satisfies the local equiangular criterion since it applies to any strictly convex quadrilateral of $\mathscr{T}(G)$ and only one diagonal can be

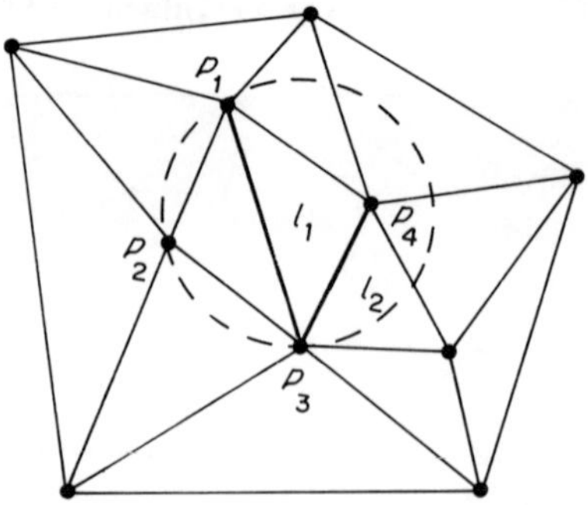

Figure 6.3.3. Illustration of criterion for defining edges in a constrained Delaunay triangulation.

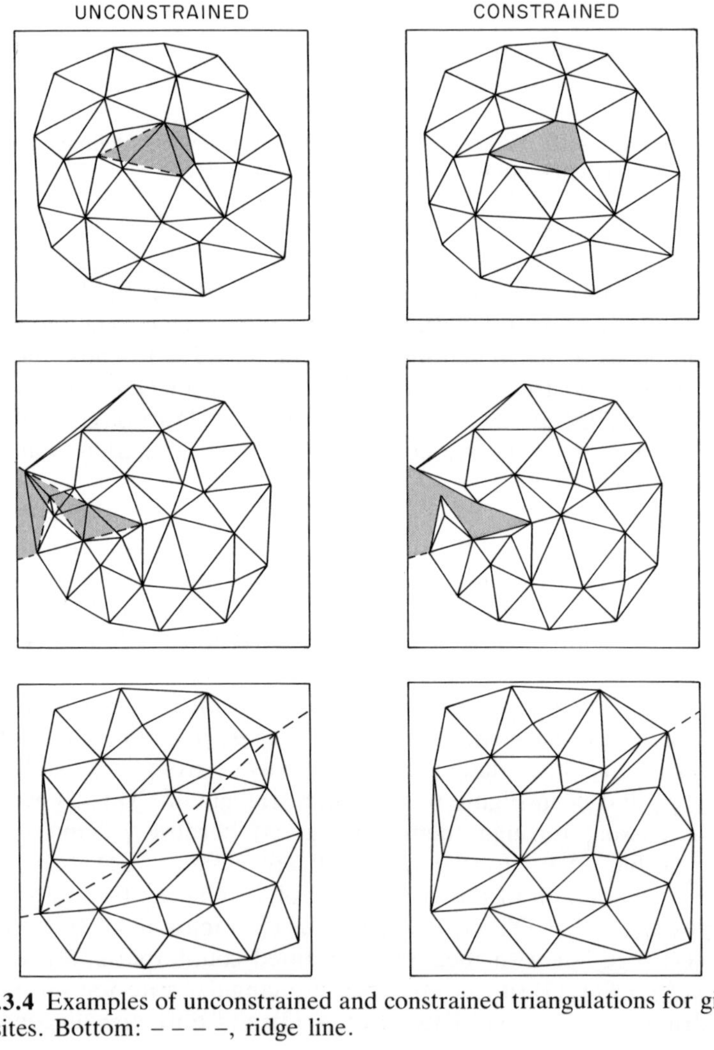

Figure 6.3.4 Examples of unconstrained and constrained triangulations for given sets of data sites. Bottom: – – – –, ridge line.

selected if a quadrilateral of $\mathcal{T}(G)$ is not convex or if one of the diagonals is a member of L. CDT(G) also satisfies global equiangularity (see Lee and Lin, 1986, for a proof). Lee and Lin (1986) provide an $O(n^2)$ algorithm for computing CDT(G) which reduces to $O(n \log n)$ when $G(P,L)$ is a simple polygon.

6.4 APPROXIMATING SURFACES

There are some applications where the number of data sites P may be extremely large. One instance is digital terrain modelling where the data sites may be in the form of a dense raster (grid) structure, particularly if they have been obtained by remote sensing (Fowler and Little, 1979; De Floriani *et al.*, 1983, 1985; Chen and Guevara, 1987). To capture the fundamental features of the surface being modelled while avoiding subsequent unnecessary time-consuming search and retrieval operations during its analysis and inefficiencies in its storage, it may be undesirable to use $\mathcal{T}(P)$ in deriving the function to represent the surface. Instead, we can select a representative subset P' of data sites in P which enables the surface to be represented with less than some predetermined level of error. P' should be generated so that only those members of P necessary to reduce the error level below the specified level are selected. Typically, this is achieved by an incremental procedure in which an initial set P' is defined and $\mathcal{T}(P')$ is generated. Further data sites are added to P' until the surface generated using $\mathcal{T}(P')$ no longer exceeds the specified error level. Such an incremental procedure is facilitated by using the Delaunay triangulation $\mathcal{D}(P')$ to define $\mathcal{T}(P')$ since the local optimization character of $\mathcal{D}(P')$ ensures that an existing triangle is only affected if the new data site is added within the circumcircle defined by the vertices of the existing triangle.

A typical procedure is that used by De Floriani *et al.* (1983, 1985). The initial set P' is chosen so that it contains only those data sites necessary to define the boundary of the region under consideration, which may be the boundary of the convex hull CH(P) of P or a modified form obtained using the procedures defined in Section 6.3. $\mathcal{D}(P')$ is then constructed. For each data site p''_i in the set P'' ($P'' = P \backslash P'$) of the data sites not in P', we can define an associated error value $e(p''_i)$ as

$$e(p''_i) = |z''_i - f(p''_i)/z''_i \;, \qquad (6.4.1)$$

where z''_i is the data value at p''_i and $f(p''_i)$ is the interpolated value (obtained using a specified interpolation procedure) at p''_i using $\mathcal{D}(P')$. The error associated with $\mathcal{D}(P')$, $e[\mathcal{D}(P')]$, is defined as

$$e[\mathcal{D}(P')] = \max_{p''_i \in P''} e(p''_i) \qquad (6.4.2)$$

and can be used as a measure of the error. If $e[\mathcal{D}(P')]$ exceeds the predetermined error level the data site p''_i ($p''_i \in P''$) for which $E(p''_i) = e\,[\mathcal{D}(P')]$

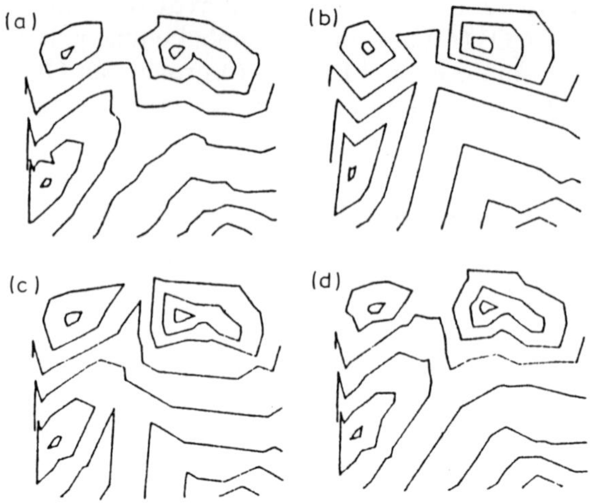

Figure 6.4.1 Estimating surfaces: (a) original surface; (b) 50% of error, obtained by selecting 20% of data sites; (c) 30% of error, obtained by selecting 30% of data sites; (d) 20% of error, obtained by selecting 50% of data sites. (*Source*: De Floriani *et al*. (1985).)

is added to P' and $\mathcal{D}(P')$ is updated. The process is repeated until $e[\mathcal{D}(P')]$ no longer exceeds the specified error level. Figure 6.4.1 shows a given surface at three different error levels when a C^0 interpolant is used in equation (6.2.5). Note that this procedure may be suboptimal since P' need not be the minimal subset of P necessary to produce a surface with a specified error level.

6.5 ORDERING MULTIVARIATE DATA

The creation of either the Voronoi diagram $\mathcal{V}(P)$ or the Delaunay tessellation $\mathcal{D}(P)$ of a set of points P imposes a spatial structure on P and so both $\mathcal{V}(P)$ and $\mathcal{D}(P)$ can be considered as a spatial ordering of the points of P (Shamos and Hoey, 1975; Gold, 1980; Watson and Philip, 1984b; Gold and Cormack, 1987). In Sections 6.1 and 6.2 we saw how the adjacencies obtained from such an ordering could be used to operationalize the definition of 'nearby' points required in procedures such as spatial interpolation, estimating missing values and spatial autocorrelation. Another application occurs in the ordering of a sample of multivariate data where the individual cases can be considered as members of a point set P in $\mathbb{R}^m$. Here data ordering is useful in such procedures as the identification and treatment of outliers in P, establishing significance levels for Monte Carlo tests, estimating correlation and deriving measures of location (Green, 1981).

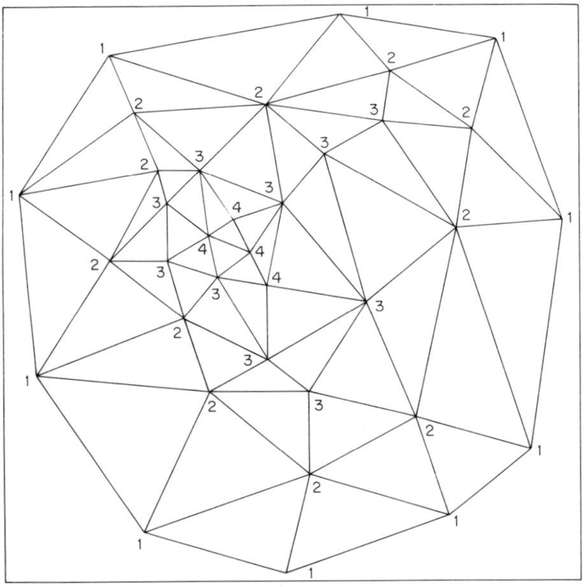

Figure 6.5.1. Peeling according to adjacencies defined by the Delaunay triangulation (numbers indicate the members of specific peels).

Unlike univariate data there is no unique way of ordering multivariate data. One approach which has received some attention is 'peeling' the data set. In general, peeling involves an iterative process which begins by finding the smallest region R of a given form which contains P. The points on the boundary of R are assigned the index 1. These points are then discarded and R is redefined for the remaining points. This process is repeated until no points are left. Those points which are discarded on the jth iteration form the jth peel. Various suggestions have been proposed for the form of R including, in the bivariate case, rectangles, circles and ellipses. The first two take no regard of the joint distribution of the sample and may thus fail to identify outliers in strongly correlated data. The last two forms create peels that may not be nested spatially. Consequently, other forms of R have been proposed. One is the boundary of the convex hull of P (Bartlett, 1976) which, as we saw in Section 2.3, is a subgraph of $\mathcal{D}(P)$. Another peel (Green, 1981) can be defined by the neighbour relationships in $\mathcal{V}(P)$ or $\mathcal{D}(P)$ as shown in Figure 6.5.1. Here those points on the boundary of CH(P) have the index 1 while their Delaunay neighbours have the index 2. The neighbours of these latter points are, in turn, indexed 3, and so on. So far there has been no analysis of the relative merits of the different peels in the various applications of data ordering mentioned above.

CHAPTER 7

Models of Spatial Processes

We have noted that the ordinary Voronoi diagram $\mathcal{V}(P)$ of a set of n points $P = \{p_1, p_2, \ldots, p_n\}$ in $\mathbb{R}^m$ is a tessellation of $\mathbb{R}^m$ since it consists of a set of regions which do not overlap, except possibly at their boundaries, and which completely cover $\mathbb{R}^m$ (see Section 2.3). Since there is a large number of empirical structures which also involve tessellations of $\mathbb{R}^m$, one of the most obvious direct applications of Voronoi concepts is in the modelling of such structures and the processes that generate them. Indeed, in Section 5.2 we have already shown how the Voronoi diagram associated with a set of points located in $\mathbb{R}^m$ according to the homogeneous Poisson point process (the Poisson Voronoi diagram) can be used as a normative model to evaluate and compare various empirical structures. In this chapter we take a more general approach which does not limit the location of the members of P to a particular spatial pattern and consider other processes which result in tessellations. We shall refer to these as models of spatial processes of which several different types are presented. In Section 7.1 we consider models involving spatial assignment processes, while in Section 7.2 we consider models involving concepts of growth.

The models presented in both these sections assume that once formed the regions remain unchanged and that the phenomenon associated with the regions is of a single type or belongs to a single group. In section 7.3 we relax the first assumption by permitting the regions to change in form after their initial creation. Since such situations involve an explicit time dimension, we refer to them as spatial–temporal processes. The second assumption is relaxed in Section 7.4 by considering situations in which regions are associated with different phenomena which compete for space (e.g. crops and weeds in a field, prey and predators in an ecosystem).

If we suspect that a given empirical structure is generated by spatial processes resulting in tessellations which are Voronoi diagrams, it is important to know if a particular instance of that structure is equivalent to a Voronoi diagram, and in Section 7.5.1 we consider how this may be done. Even if a given tessellation is shown to differ from a Voronoi diagram it still may be important to determine the extent of the difference and several techniques for doing this are discussed in Section 7.5.2.

For convenience, throughout this chapter we refer to the tessellation of $\mathbb{R}^m$ associated with a particular phenomenon as a spatial pattern of that phenomenon. Also, since our concern is with modelling empirical structures, our treatment is limited to two- and three-dimensional Euclidean space.

7.1 ASSIGNMENT MODELS

These models describe processes which produce spatial patterns by allocating individual locations in $\mathbb{R}^m$ (m = 2,3) or a bounded region in $\mathbb{R}^m$ (depending on the particular empirical circumstance) to a set of n points $P = \{p_1, p_2, ..., p_n\}$ at positions $\mathbf{x}_1$ $\mathbf{x}_2$, ..., $\mathbf{x}_n$, respectively, in the same space. If the following assumptions are made, the resulting set of regions is equivalent to the ordinary Voronoi diagram $V(P)$ of P:

Assumption A1 Each point p_i (i=1, ..., n) is located simultaneously before any assignment procedure occurs.

Assumption A2 Each point p_i remains fixed at $\mathbf{x}_i$ throughout the assignment process.

Assumption A3 Each point p_i is of equal importance.

Assumption A4 Each location which is closer in terms of Euclidean distance to p_i than to any other member of P is assigned to p_i; locations which are equidistant from two or more members of P are assigned to the boundaries of the regions of those points.

Collectively Assumptions A1–A4 define the *Voronoi Assignment Model.*

Given the simple assumptions of the Voronoi Assignment Model it is of no surprise that it has been evoked in a wide range of situations in an equally wide range of disciplines. These applications could be presented in a variety of ways but our treatment will simply make a general distinction between those in the natural sciences and those in the social sciences. We begin with the former.

One of the longer established areas of the use of the Voronoi Assignment Model is in the modelling of physical–chemical systems. Such systems are frequently considered to consist of a set of sites occupied by atoms, ions or molecules (depending on the specific application) which are represented as equal-sized spheres. When the sites are regularly arranged in $\mathbb{R}^m$, as they would be if they formed a lattice (see Figure 7.1.1(a)), the structure is said to be *crystalline*. Such structures are also referred to as *periodic* or *ordered* since the physical situation at a given site is reproduced at other sites

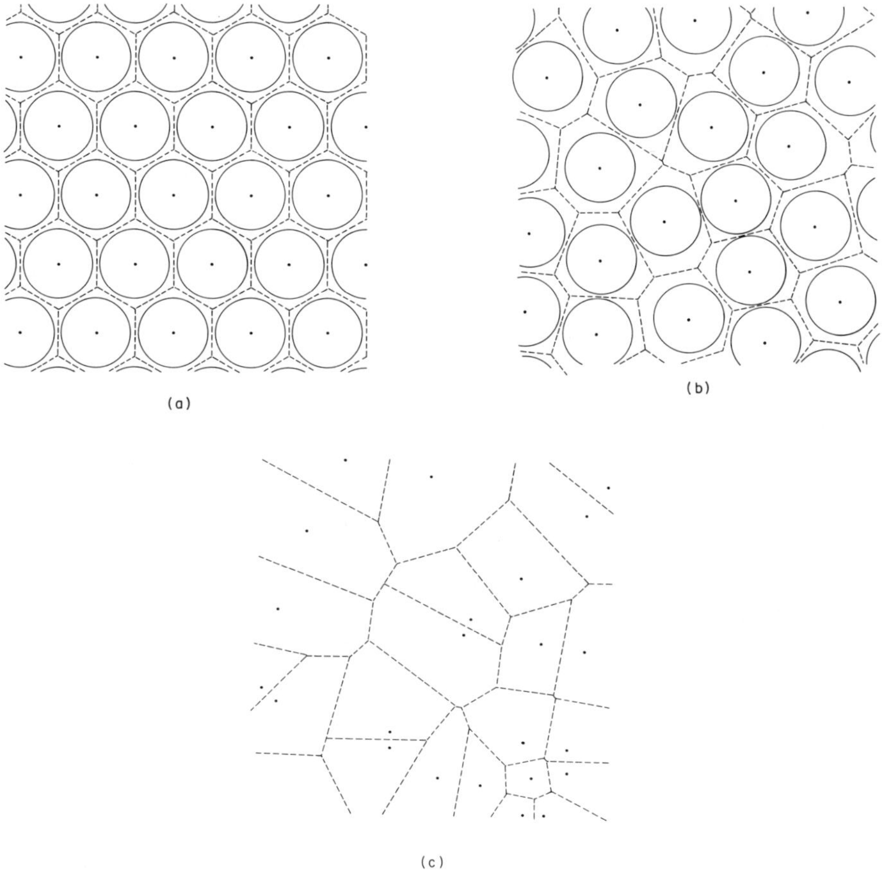

Figure 7.1.1 (a) Crystalline structure; sites and their associated spheres arranged on a lattice (dashed lines indicate Voronoi regions of sites). (b) Amorphous structure; sites and their associated spheres arranged irregularly (dashed lines indicate Voronoi regions of sites). (c) Ideal gas structure; sites located according to a homogeneous Poisson point process (dashed lines indicate Voronoi regions of sites).

throughout the structure. Examples of such structures in $\mathbb{R}^3$ are encountered frequently in inorganic chemistry (Wells, 1984). When the sites are no longer regularly arranged but the radii of the spheres are not negligible relative to the average distance between sites, we have a structure such as that illustrated in Figure 7.1.1(b) which is referred to as an *amorphous* (alternatively *non-crystalline, non-periodic* or *disordered*) structure (Yonezawa and Ninomiya, 1983). In such structures the size of the spheres has an inhibitive effect on how closely the sites may approach each other (no closer than the diameter of the spheres). Consequently, although amorphous structures lack the long-range order of crystalline structures, they may still exhibit some degree of short-range (local) order. Examples of such structures include monatomic liquids and metallic glasses (Zallen, 1979). Finally, if the sites are not regularly arranged and the radii of the spheres are considered negligible relative to the average inter-site distance, the inhibitive effect no longer occurs and the sites may occupy locations equivalent to those generated by a homogeneous Poisson point process (see Section 1.3.8) as shown in Figure 7.1.1(c). Such a structure is typical of many gases and that in Figure 7.1.1(c) is often referred to as an ideal gas. Both the crystalline and amorphous structures can thus be considered as consisting of closely spaced spherical atoms and it is in these situations that Voronoi concepts have been used extensively.

Regardless of how the sites are positioned in $\mathbb{R}^m$ it is possible to assign each location in $\mathbb{R}^m$ to the nearest site. The result is to produce the Voronoi diagram $\mathcal{V}(P)$ of the sites (see Figure 7.1.1). Once this is done, characteristic of $\mathcal{V}(P)$ be used to define various parameters of the atoms (ions, molecules) of the structure. One such property is the number of neighbours or *coordination number* c_i of the atom p_i which can be measured by the number of Voronoi regions adjacent to the Voronoi region $V(p_i)$ of p_i (Frank and Kasper, 1958; Fischer and Koch, 1979). The coordination number is considered a basic parameter on which many atomic properties, such as the atomic (ionic) radius, depend directly. The region with p_i as its centre and with vertices that are the sites whose Voronoi regions are adjacent to $V(p_i)$ is called the *coordination shell* (or *coordination polyhedron* in $\mathbb{R}^3$) of the site p_i (Frank and Kasper, 1958; Mackay, 1972; Loeb, 1976, p. 132). Examination of how the coordination polyhedra are connected together in space provides an alternative way of modelling the structure. In structures in $\mathbb{R}^3$ consisting of closely packed spheres the proportion of the volume of the structure contained within the spheres is referred to as the *atomic volume* or the *packing fraction* of the material, variations in which can be used to differentiate between materials. Thus, the ratio of the volume of the spherical atom p_i relative to the volume of $V(p_i)$ can be used as a measure of local atomic volume (Cape *et al.*, 1981).

When the sites are regularly arranged they may form a *lattice*. In this case each site has an identical environment in an identical orientation to every other site. More formally, in $\mathbb{R}^m$ a lattice consists of the set of points with position vectors $\boldsymbol{p}$ of the form

$$\boldsymbol{p} = l_1\boldsymbol{a}_1 + l_2\boldsymbol{a}_2 + l_3\boldsymbol{a}_3 + \ldots + l_m\boldsymbol{a}_m , \tag{7.1.1}$$

where $\boldsymbol{a}_1$, $\boldsymbol{a}_2$, $\boldsymbol{a}_3$, ..., $\boldsymbol{a}_\mathrm{m}$ are m linearly independent vectors and l_1, l_2, l_3, ..., l_m are integers. A lattice is infinite. In some applications it is necessary to consider only a finite portion of the lattice which is usually chosen so that it has the simplest possible form. For example, in $\mathbb{R}^3$ a finite lattice of n sites may be chosen as the set of sites of the form in equation (7.1.1) where $0 \leq l_1 < n_1$, $0 \leq l_2 < n_2$, $0 \leq l_3 < n_3$ and $n = n_1n_2n_3$. For lattices

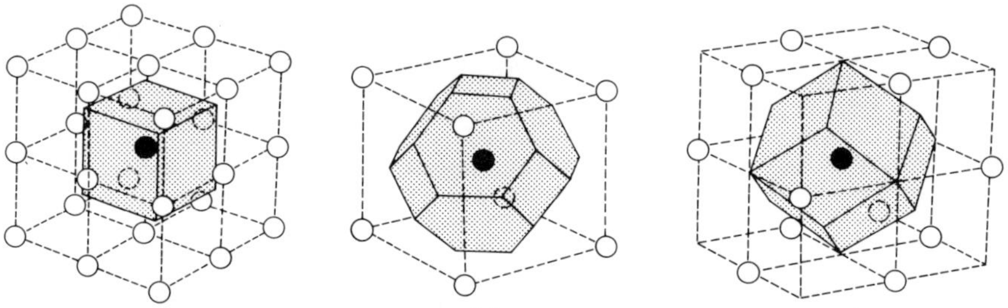

Figure 7.1.2 Some three-dimensional lattices and their associated Voronoi polyhedra: (a) simple cubic lattice, the Voronoi polyhedron is a cube: (b) body centred cubic lattice, the Voronoi polyhedron is a truncated octahedron; (c) face centred cubic lattice, the Voronoi polyhedron is a rhombic dodecahedron.

in $\mathbb{R}^3$ the Voronoi regions of the sites are identical polyhedra with identical orientations. Examples are shown in Figure 7.1.2. In some areas, particularly solid-state physics, Voronoi regions with these characteristics are called *Wigner–Seitz regions* (Wigner and Seitz, 1933) to distinguish them from the non-congruent Voronoi regions of amorphous structures (Rahman, 1966; Ashcroft and Mermin, 1976, pp. 73–75; Zinman, 1979). In inorganic chemistry the Voronoi region of an atom is more frequently referred to as *Wirkungsbereich* (domain of action) (Niggli, 1927) or as the *domain of the atom* (Frank and Kasper, 1958). The Voronoi region is also a primitive cell of the lattice. This is a bounded region which, when translated through all the vectors of the lattice, completely fills $\mathbb{R}^m$ without overlap, except possibly at boundaries. Although there is no unique way of defining the primitive cell of a lattice, the primitive cell defined by the Voronoi region of a site has the advantage that it has the same degree of symmetry as the lattice.

In some crystal structures the sites are not arranged on a lattice. One possibility is that all sites are equivalent with respect to other point group operations besides that of translation involved in a lattice, such as rotation about an axis or reflection. Such sites are said to form a *lattice complex*. In such cases the environments of the sites are identical, except for orientation, and the Voronoi regions are congruent but with more than one orientation. An example of a lattice complex in $\mathbb{R}^2$, where the sites are of two different orientations (indicated by dots and crosses), is shown in Figure 7.1.3. For lattices and lattice complexes in $\mathbb{R}^2$ all the different types of Voronoi polygons have been determined (Laves, 1930; Delone *et al.*, 1980; see Grunbaum and Shephard, 1987, Table 9.1.5 for a summary).

In $\mathbb{R}^3$ the Voronoi polyhedra are sometimes called *plesiohedra* (Grunbaum and Shephard, 1980). As such they are members of a larger set of regions called *stereohedra*, congruent copies of any one of which are capable of filling $\mathbb{R}^3$ without overlap except possibly for boundaries. There has been considerable interest in crystallography in determining the maximum number

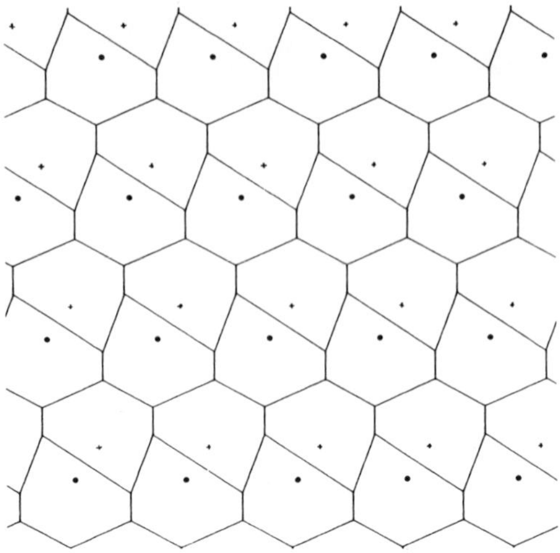

Figure 7.1.3 A lattice complex in $\mathbb{R}^2$ with sites of two different orientations indicated by $\cdot$ and $+$.

of $(m-1)$-dimensional faces F of a stereohedra. Delone (1961) has obtained the result that in m dimensions

$$F \leq 2^m(1+h) - 2\,, \tag{7.1.2}$$

where h is the number of sets of sites with different orientations of their environments. For a lattice $h = 1$, and so for $\mathbb{R}^3$, expression (7.1.2) gives an upper limit of 14, which is achieved by the Voronoi polyhedra of a body centred cubic lattice. The Voronoi polyhedra are truncated octahedrons with six square faces and eight regular hexagonal faces; all edges are of the same length (see Figure 7.1.2(b)). For $m = 3$ and $h = 2$ the upper limit of F given by expression (7.1.2) is 22. A stereohedron with this number of faces has not been identified but $F = 16$ for the lattice complex consisting of two interpenetrating face centred cubic lattices is usually referred to as the diamond structure (since the carbon atoms in a diamond crystal take on this arrangement) (see Figure 7.1.4(a)). A simplified version of this polyhedron has been used by Grigorovici and Manaila (1967/8, 1969) and Grigorovici (1969) as a model for germanium. Smith (1965) has identified stereohedra with 20 faces which are Voronoi polyhedra of a slightly distorted version of the diamond structure (see Figure 7.1.4(b)). Other examples of Voronoi polyhedra with 20 faces have been identified by Stogrin (1968, 1973). The upper limit of F for $m = 3$ is 390 since the maximum value for h in $\mathbb{R}^3$ is 48 (Henry and Lonsdale, 1965). So far only a number of Voronoi polyhedra with 38 faces have been identified (Engel, 1981). One of these is illustrated in Figure 7.1.4(c). Other polyhedra with numbers of faces ranging from 17 to 37 have also been reported by a number of other

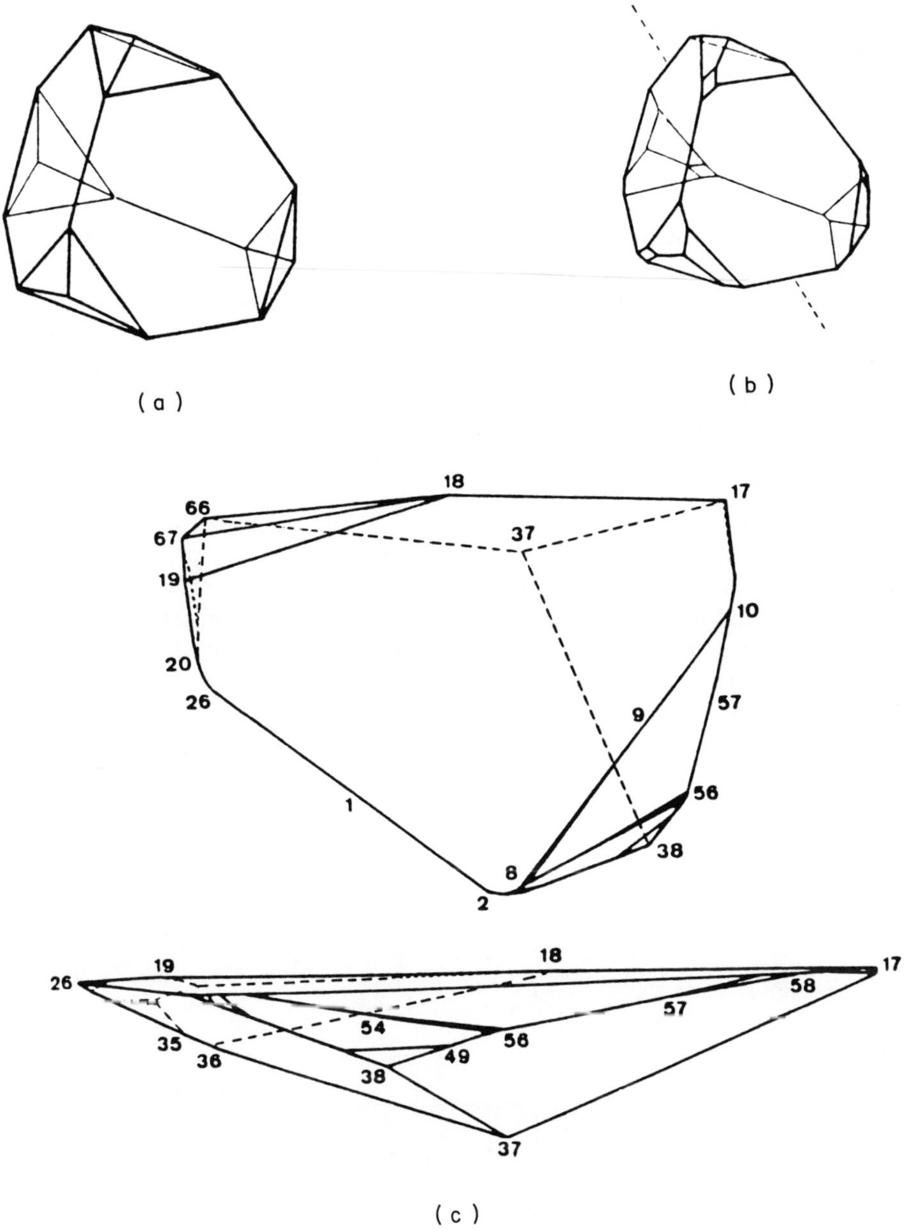

Figure 7.1.4 (a) 16-faced space-filling Voronoi region obtained from the 'diamond lattice'. *Source*: Grünbaum and Shephard (1980, Figure 8). (b) 20-faced space-filling Voronoi region obtained from a distorted 'diamond lattice'. (*Source*: Grünbaum and Shephard (1980, Figure 9(a)). (c) Two views of a 38-faced Voronoi region (some of the vertices are labelled to facilitate recognition. (*Source*: Engel (1981, Figure 4).)

researchers and these are reviewed in Grunbaum and Shephard (1980). Conway and Sloane (1988, Chapter 21) extend the treatment of Voronoi regions for lattices in $\mathbb{R}^m$ ($m \geq 4$).

Another alternative to a lattice which is sometimes encountered in crystallography occurs when the sites are not equivalent with respect to one or more point group operations but instead belong to k different sets of equivalent sites. In such instances the Voronoi regions form k different sets of mutually equivalent regions which together fill $\mathbb{R}^m$. An example in $\mathbb{R}^2$ for $k = 3$ is illustrated in Figure 7.1.5.

The lack of regularity and homogeneity in amorphous structures means that there is a much greater need to be able to characterize such materials at the individual atomic (ionic, molecular) level than for crystalline structures (Turnball and Polk, 1972; Jerauld *et al.*, 1984a). This is true of monatomic liquids which were conceptualized initially as equivalent to structures consisting of randomly packed spheres of equal size, as in Figure 7.1.1(b) (Bernal, 1937, 1959). Smith (1965), Collins (1967, 1968, 1972) and Finney (1970a, 1975a, 1975b) extended this to a packing of polyhedra by defining the Voronoi diagram of the set of sphere centres. This representation has been supported by others (e.g. Brostow and Sicotte, 1973, 1975; Rivier and Duffy, 1982; Rapoport, 1983), although some have suggested that it is preferable to represent the structure by means of the dual Delaunay tessellation rather than the Voronoi diagram (Smith, 1964; Ogawa and Tanemura, 1974). Either representation permits the definition of the coordination number of an atom and other measures using this information can be derived. For example, Finney (1970a) (see also Cape *et al.*, 1981)

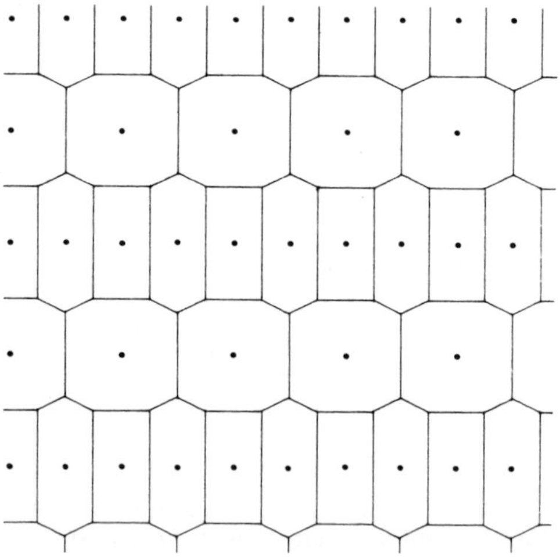

Figure 7.1.5 Voronoi diagram for sites belonging to three different sets of equivalent sites.

has proposed that the local configuration of an atom a_i, with associated Voronoi polyhedra $V(a_i)$ with f_i faces can be described by listing the number of faces $n_3, n_4, n_5, \ldots, n_j$ having 3, 4, 5, ..., j edges, where $\Sigma_{k=1}^{j} n_k = f_i$. Similarly, Rapoport (1983) suggested using the volume of $V(a_i)$ as a measure of the local density of the atoms since the former is inversely proportional to the latter. Regardless of which structural characteristics are measured, knowledge of the local structure of a liquid is essential for studying its behaviour under various conditions such as a crystallization (Tanemura *et al.*, 1977; Cape *et al.*, 1981) and instability induced by circular shear (Rivier *et al.*, 1984). This is because the arrangement of individual atoms, which changes with transitions of the liquid from one state to another (Collins, 1967), influences the macroscopic properties of the liquid (Bristow and Sicotte, 1975). Rivier and Duffy (1982) and Rivier (1985b) proposed that the arguments relating to the structure of liquids can be extended directly to models of the structure of metallic glasses.

The Voronoi concepts used above to study monatomic liquids have also been extended to non-monatomic ones. One instance is the modelling of solution structures consisting of a combination of a solvent and a solute, with particular attention being placed on aqueous solution. Initially, David and David (1982b) proposed constructing the Voronoi polyhedron of each atom of the solute using the oxide anions of the solvent water and the other atoms of the solute. Later they extended the construction of the Voronoi polyhedra to include the protons of the solvent as well (David and David, 1982c). Once defined, the Voronoi polyhedra of the atoms of the solute can be used in a variety of ways. First, the union of the polyhedra of the atoms of the solute can be considered the *solvent sheath* (i.e. the region which contains the solute as its interior region) (see Figure 7.1.7). The volume of the solvent sheath so defined can be considered the co-volume of the solvent as measured by water, while the area of its external faces measures the extent to which each solvent molecule coordinates each atom of the solute (David and David, 1983; Bellagamba *et al.*, 1985). Furthermore, Bellagamba *et al.* (1985) suggested that the properties of different solvents can be studied by examining the effect of each solvent in terms of the corresponding solvent sheath. Figure 7.1.6 shows the Voronoi polyhedra for the different types of atoms for two different solvation structures. David and David (1982a) also proposed using the Voronoi construction to determine if a given atom is a member of a set of solute atoms forming a single cluster in a solvate. An atom is considered to be a cluster member only if its Voronoi polyhedron, as defined using the other atoms of the set and the atoms of the solvent, is closed. David (1984) also suggested that the solvent accessible cleft regions of a solute can be identified using Voronoi concepts. The approach is illustrated in Figure 7.1.7, which for convenience shows a two-dimensional representation of a complex of solute molecules (shown by filled circles) solvated by water molecules (shown by unfilled circles). The

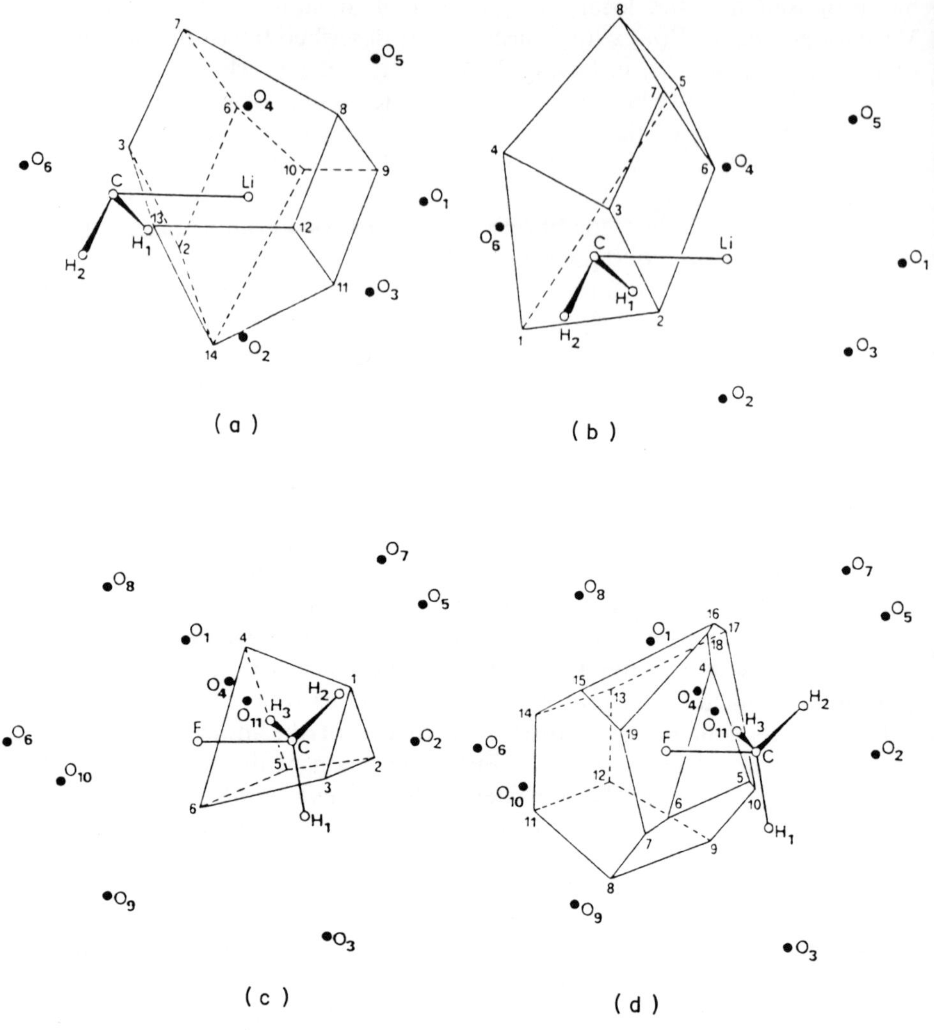

Figure 7.1.6 Voronoi polyhedra for different types of atoms in solvation structures: (a) lithium and (b) carbon atoms for $CH_2Li^+(H_2O)_6$; (c) carbon and (d) fluorine atoms for $CH_3F(H_2O)_{11}$. Filled circles represent the positions of oxygen atoms of water molecules. (*Source*: Bellagamba *et al.* (1985, Figures 1 and 4).)

union of the Voronoi polygons of the solute atoms constitutes the solvent sheath. If we consider the Voronoi polyhedra of the water molecules which are adjacent to the solvent sheath we observe that these may be either open or closed. The open Voronoi polygons can be considered as the surface solvating water molecules while the closed Voronoi polygons are of two types. Some (those shown stippled in Figure 7.1.7) exist because the molecule is, at least partially, on the dimple region of the solvent, while others (those

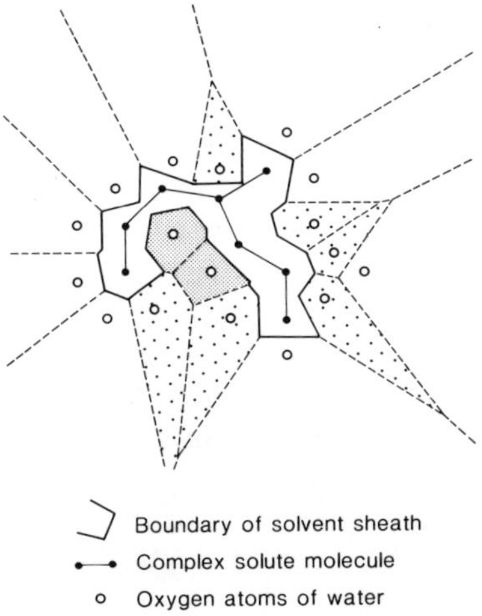

Figure 7.1.7 Identification of solvent accessible cleft regions of a solute. Stippled areas indicate molecules on the dimple region of the solvent; cross-hatched areas indicate molecules occupying a true cleft.

shown cross-hatched in Figure 7.1.7) occupy a true cleft of the solute molecules.

Voronoi concepts have also been used extensively in modelling metallurgical structures. This is because one prevalent approach used in quantitative metallography involves the identification of microstructural characteristics, their measurement and the deduction of empirical relationships linking these characteristics to the physical and mechanical properties of the material (Lantuejoul, 1980). Furthermore, many of the processes which affect such structures operate at the local level. Examples include competition between heterogeneous flow and decohesion and the development of recrystallization nuclei during plastic flow (Burger *et al.*, 1988). In particular, Voronoi assignment models have been employed to characterize the structure of second-order particles in metals by considering the particles as points and defining their Voronoi diagram (Wray *et al.*, 1983; Koken *et al.*, 1988). Examples include sulphide inclusions and carbides in steels and graphite fibres in aluminium (Spitzig *et al.*, 1985). Another metallurgical application is involved in modelling the accumulation of structural damage in materials subjected to stress at elevated temperatures. In such situations voids which nucleate on grain boundaries grow and subsequently coalesce causing fracture. One input parameter required by void growth models developed so far is the area A_i of the grain boundary affected by a given void v_i over which local stress occurs. Burger *et al.* (1988) and Wilkinson (1988a, 1988b) have used the area of the Voronoi polygon of v_i as a measure of A_i.

Another direct application of the Voronoi Assignment Model occurs in modelling capillary distribution in transverse sections of muscle tissue

(Egginton and Ross, 1989b). Voronoi polygons (in this application referred to as *capillary domains*) are defined with respect to capillaries represented as points (Hoofd *et al.*, 1985; Egginton *et al.*, 1988). Adjacencies of the Voronoi polygons in the resulting Voronoi diagram provide identification of neighbouring capillaries. A mean inter-capillary distance can then be defined for neighbouring capillaries which is a basic input parameter in models of oxygen transport in tissues (Egginton *et al.*, 1987; Egginton and Ross, 1989a). In this instance the assumptions involved in the Voronoi Assignment Model are equivalent to assuming that the capillaries possess equal transport capacity and that they have equal levels of partially pressurized oxygen (Egginton, 1990). Thus the models are most appropriate for muscle consisting of a single fibre type such as cardiac muscle (Turek *et al.*, 1986) and skeletal muscle for fish (Egginton *et al.*, 1987). A comprehensive review of this area of application is provided by Egginton (1990). More generally, Honda *et al.* (1979) and Saito (1982) also have used Voronoi Assignment Models for studying monolayer cells and epithelial cells in tissue.

We have already noted the use of Voronoi Assignment Models in studying periodic structures in chemistry. They also have been used in modelling a different type of periodicity which occurs in biology in phyllotaxis, the spiralling close packed arrangement of florets in composite plants such as daisies, asters and sunflowers and the scales of a pine cone or pineapple where successive florets are either identical or have a structure which can be obtained from the previous one. Rivier (1986c, 1988) and Rivier *et al.* (1984) have modelled the individual florets as Voronoi regions created with respect to points located on a parabolic spiral. Label the individual florets with positive integers ($i = 1, 2, \ldots$) with the magnitude of the integer reflecting the age of the floret. If the stem is the origin of the coordinate system, the polar coordinates (r_i, θ_i) of the generator point of a floret are

$$r_i = ai^{\frac{1}{2}} \quad \text{and} \quad \theta_i = 2\pi\alpha i,$$

where a is the average linear dimension of a floret (with area πa^2) and α ($0 < \alpha < 1$) is a parameter of the structure. Thus the florets are generated at regular intervals from the stem and at a given angle $2\pi\alpha$ from each other. Rivier *et al.* (1984) and Rivier (1986c, 1988) prove that, to create florets which are as uniform as possible in terms of size and shape (expressed by the number of edges of the floret), α must be an irrational number (rational α produces a spider-web-like structure) of the form

$$\alpha = 1/\{q_1 + 1/[q_2 + 1/(q_3 \ldots)]\},$$

where q_j are positive integers. The simplest solution occurs when $q_j = 1$ for all j and $\alpha \simeq 0.618$ which is also equal to the reciprocal of the golden mean ($[1 + \sqrt{5}]/2$). This structure is illustrated in Figure 7.1.8.

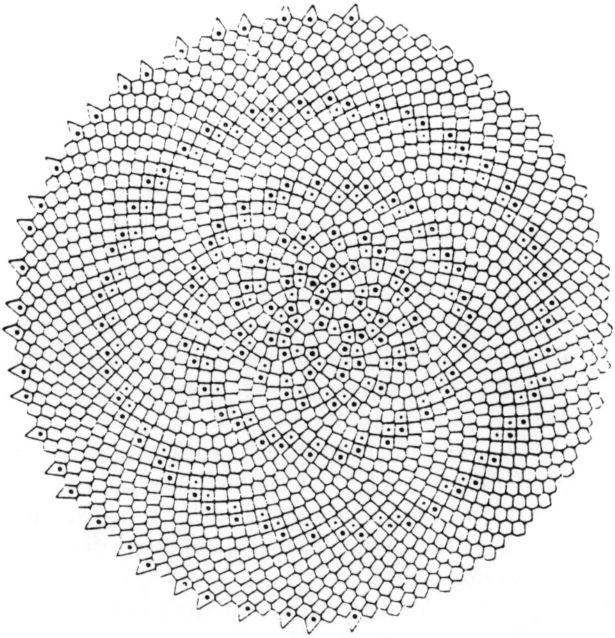

Figure 7.1.8 Simple daisy structure. (*Source*: Rivier *et al.* (1984, Figure 6).)

There is also extensive use of Voronoi Assignment Models in both plant and animal ecology. In the case of plants (and trees) the Voronoi polygon associated with a plant is considered an indicator of the portion of the environment available to that plant. Indeed, this application represents yet another area in which Voronoi polygons were discovered independently; not once but twice! Brown (1965) defined what he called the *area potentially available* (APA) to a plant, which is equivalent to the Voronoi polygon of the plant, as recognized by Jack (1967). Later Mead (1967, 1971) defined the same region labelling it the *plant polygon*, a name which has become reasonably current in plant ecology. Given the assumptions of the Voronoi Assignment Model the most appropriate application will be to monospecific stands of upright plants with simultaneous germination (Mithen *et al.*, 1984) especially in the first few months after germination when individual plant growth rates are most similar, thus ensuring that most plants are of approximately the same size (Matlack and Harper, 1986). However, Voronoi polygon models can also be useful when more than one species of plant occupy the same environment but since such models involve spatial competition they are discussed in Section 7.4.

By defining the Voronoi diagram of a set of plants the polygon adjacencies can be used to define the neighbours of a given plant, such information being useful as input to interplant competition models (Mead, 1966, 1971). Other characteristics of a plant polygon such as area are also used in studying aspects of a plant's behaviour including survival rate, growth rate, size and weight. Examples of such studies are provided by Hasegawa *et al.* (1981), Liddle *et al.* (1982), Watkinson *et al.* (1983), Bulow-Olsen *et al.* (1984), Mithen *et al.* (1984), Hutchings and Waite (1985), Matlack and Harper (1986) and Firbank and Watkinson (1987).

A variant of this use in plant ecology is provided by Hasegawa *et al.* (1981) in their study of the crown projection diagrams of trees in a forest. They propose that if identical trees develop uniformly in a uniform environment the crown projection diagram will be equivalent to a Voronoi diagram. To test this they construct the Voronoi diagram associated with the areal centres of gravity of the crowns of a stand of Japanese cedars (*Cryptomenia japonica*) and compare this with the actual crown projection diagram. A good measure of fit is found between the two diagrams but some differences do exist because some degree of overlap and small unfilled areas are observed in the actual crown projection diagram which are not present in the Voronoi diagram.

Voronoi Assignment Models are also applicable to animal territories (Tanemura and Hasegawa, 1980). However, despite being associated with specific point locations such as nests, roosts and food caches, animals, unlike plants, are able to move away from these locations and hence growth models of the kind described in the next section are more appropriate.

Social scientists studying various types of territorial structures associated with humans have also made extensive use of Voronoi Assignment Models. Unlike many of their physical counterparts, human spatial patterns rarely show a high degree of regularity although one celebrated instance occurs in geography in the context of central place theory (King, 1984). This is concerned with explaining the size and spacing of nucleated settlements (central places) which provide goods and services to rural customers distributed over a geographical region. One aspect of this theory relates to the definition of the trade areas (areas of influence) associated with the central places for a specified good or service. In classical central place theory (Christaller, 1933; Lösch, 1940) the locations of the central places are assumed to form a triangular lattice (see Figure 8.2.2) and the rural population is assumed to be uniformly distributed over the region. In addition, it is assumed that

(i) every central place provides the good or service under consideration;

(ii) consumers travel to a central place to acquire the good or service;

(iii) the market price m of the good or service is the same at all central places;

(iv) the cost c to the consumer of acquiring the good or service is equal to the market price plus the cost of transportation t to the central place ($c = m + t$);

(v) $t = rd$, where d is the Euclidean distance travelled by the consumer and r is the transportation rate per unit distance;

(vi) customers minimize the cost of purchasing the good or service.

Taken collectively, assumptions (ii)–(vi) imply that consumers will patronize the nearest central place. Thus, the trade areas are equivalent to the Voronoi polygons of the central places (Dacey, 1965) and in the case of classical central place theory they will be regular hexagons.

These concepts can be extended directly to situations involving any type of settlement spacing. Thus Bogue (1949) and Snyder (1962) use the Voronoi Assignment Model to define Voronoi polygons around US metropolitan centres and Uruguayan urban centres, respectively, which are considered proxies for the real market areas of these centres. Ohji (1986) used them to define market areas for periodic markets in Kanataka State in southern India. In each case the market areas so created correspond to areas over which the centres exercise a spatial monopoly. Identical concepts have been applied to a different geographic scale to model the trade areas of individual retail stores (West, 1981; Jones and Mock, 1984; Von Hohenbalken and West, 1984a; West and Von Hohenbalken, 1984; Jones and Simmons, 1987). Polygon adjacencies in the Voronoi diagram of the stores can be used to identify which stores can be considered neighbours and thus are likely to be in direct competition.

Such considerations are also relevant in modelling the service areas of facilities providing public services. Suppose that we consider a set of facilities at fixed locations in a region which provide a service to a set of individual users distributed over the same region. The service may either be dispensed to the users at their locations or offered at the facility to which the users must travel. Obvious examples of such facilities are those providing municipal services such as health and education. If we assume that the facilities are identical, that there are no constraints on the number of people who can use a facility and that, to minimize costs of acquiring the service, users are served by the nearest facility, the service areas of the facilities will be equivalent to their Voronoi polygons (Goodchild and Massam, 1969; Keeney, 1972). Note that the last of these assumptions means that the aggregate distance travelled between users and facilities is minimized. This situation is usually referred to as the *unconstrained transportation problem* (Scott, 1971, p. 118; Keeney, 1972; Massam, 1972, pp. 91–10; 1975, pp. 59–60). While such a formulation may be unrealistic in empirical situations since facility constraints almost always occur and that, if allowed to choose, users do not always go to the nearest facility, the Voronoi diagram is often used as a normative structure against which the empirical or other proposed structures are evaluated. Goodchild and Massam (1969) followed such a strategy in evaluating the efficiency of electricity supply service areas in southern Ontario, Canada. Similarly, Cox and Agnew (1976) used the Voronoi diagram of the existing county towns of Ireland to create a theoretical partition (Figure 7.1.9(a)) against which the existing set of counties (Figure 7.1.9(b)) is compared. Using an information theoretic measure (described in Section 7.5), they determine that there is an overall spatial correspondence of 72% between the two structures. However, the measures of agreement between individual counties is quite variable being highest for large coastal counties such as Clare (92%) and Kerry (90%) and lowest for small inland counties such as Carlow (40%) and Longford (44%).

Singh (1976, 1979) and Singh and Singh (1975, 1977, 1978), and Callen and Stephan (1975) use the edges of the Voronoi diagram of village sites in the Middle Ghanga Valley in India and Bougainville Island in the Solomons, respectively, to simulate village boundaries. In archaeology Voronoi Assignment Models have been used to identify possible chiefdom territories in Wessex, England (Renfrew, 1973a) and on the Island of Malta

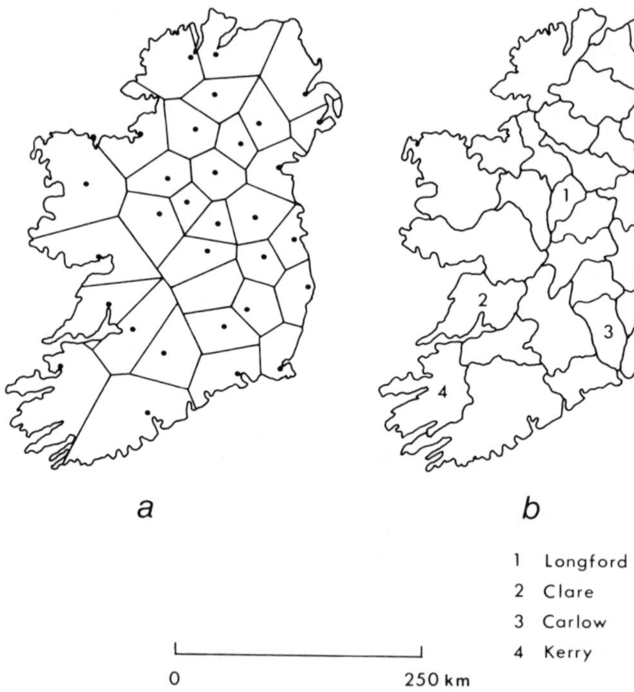

Figure 7.1.9 Counties of Ireland. (a) Theoretical partition created by Voronoi diagram of county towns. (b) Actual counties. (Redrawn from data in Cox and Agnew, 1976.)

(Renfrew, 1973b) by defining Voronoi polygons around the sites of neolithic long barrows and temples, respectively. Voronoi polygons have also been used as surrogates for territories of Mayan ceremonial centres (Hammond, 1972, 1974), Iron Age (First century BC) hill forts in Wiltshire, England (Cunliffe, 1971) and Romano–British walled centres (Hodder, 1972).

7.2 GROWTH MODELS

These models produce spatial patterns as the result of a simple growth process operating with respect to a set of n points (nucleation sites), $P = \{p_1, p_2, ..., p_n\}$, at positions $\mathbf{x}_1, \mathbf{x}_2, ..., \mathbf{x}_n$, respectively, in $\mathbb{R}^m$ or a bounded region of $\mathbb{R}^m$ ($m = 2, 3$). If we make the following assumptions, the resulting pattern will be equivalent to the ordinary Voronoi diagram $\mathcal{V}(P)$ of P:

Assumption G1 Each point p_i ($i = 1, 2, ..., n$) is located simultaneously.

Assumption G2 Each point p_i remains fixed at $\mathbf{x}_i$ throughout the growth process.

Assumption G3 Once p_i is established, growth commences immediately and at the same rate l_i in all directions from p_i.

Assumption G4 l_i is the same for all members of P.

Assumption G5 Growth ceases whenever and wherever the region growing from p_i comes into contact with that growing from p_j ($j \neq i$).

Collectively, Assumptions G1–G5 define the *Voronoi Growth Model*. Figure 7.2.1 shows a series of stages in such a growth process.

Although the Voronoi Growth Model has been applied in a variety of different circumstances, the range of applications is not as great as that for the Voronoi Assignment Model and most are found in the natural sciences. One obvious application is to model crystal growth about a set of nucleation sites (Avrami, 1939; Leonyuk and Mal'tseva, 1980). Here Assumptions G3–G5 are equivalent to assuming an omnidirectional, uniform supply of crystallizing material to all faces of the growing crystal in the absence of any adsorbable impurities. Assumption G3 also implies that the rate of growth of the volume of a crystal will be proportional to its surface area.

However, probably the most extensive use of growth models is in metallurgy. Here a major use is in modelling phase transitions involving transformation of an isotropic, one-component solid through nucleation, and isotropic growth of grains of a new or re-crystallized phase (Watson

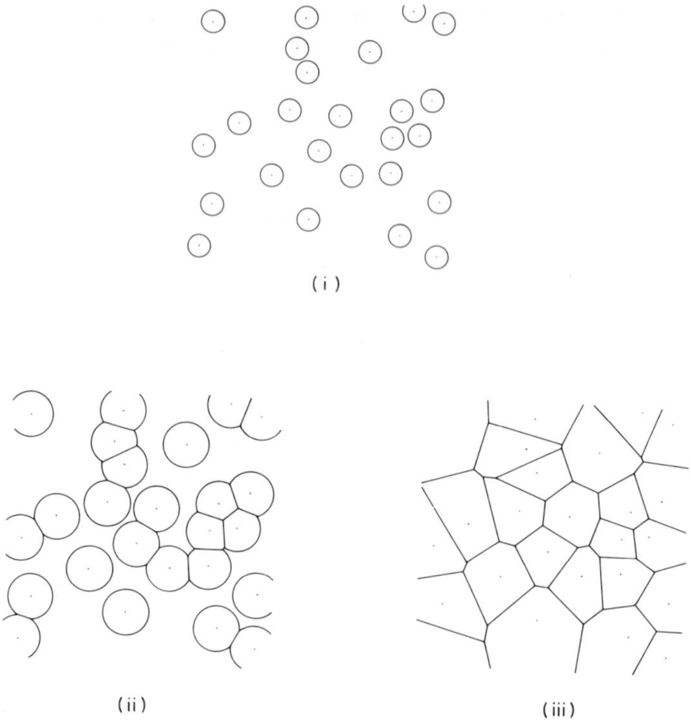

Figure 7.2.1 Stages (i), (ii) and (iii) of the Voronoi growth model.

and Smith, 1975; Mahin *et al.*, 1980; Burger *et al.*, 1988). In this context the Voronoi growth model is sometimes referred to as the *cell model* (Meijering, 1953) or the *site saturation model* (Saetre *et al.*, 1986). Specific examples include the covering of a metallic surface by films or layers of corrosion product (Evans, 1945) where the nucleation sites P might be surface imperfections such as impurities, points of intersection with bulk defects and surface pits (Frost and Thompson, 1987a). Another example is the growth of thin films of metal or semiconductors. In these examples if the thickness of the film is small relative to the spacing between the nucleation sites or if the grain boundaries are perpendicular to the plane of the film, a two-dimensional representation is appropriate.

An interesting application of the Voronoi Growth Model occurs in astronomy, where it has been used to describe the morphology of the large-scale structure of the universe (Matsuda and Shima, 1984; Icke and Van der Weygaert, 1987; Yoshioka and Ikeuchi, 1989). Icke and Van der Weygaert suggest that a collection of slightly underdense regions in a primordial density field will act as the seeds of voids or expansion centres ('superhubble bubbles') from which matter will flow outwards until it encounters similar material flowing from another void. If the voids are created simultaneously, and the excess expansion is the same for all voids, we have the equivalent assumptions to those specified above for the Voronoi Growth Model. Once defined, the Voronoi diagram associated with the voids forms a skeleton around which the galaxies assemble during the evolution of the universe. In $\mathbb{R}^3$ the faces of the Voronoi polyhedra can be interpreted as the contact surfaces ('Pancakes'), their edges as elongated 'superclusters' and their nodes as virialised 'Abel clusters'. Van de Weygaert and Icke (1989) propose that the density at the facets of the Voronoi polyhedra of material diffused from the voids is inversely proportional to the dimensionality of those facets so that over time galaxies tend to congregate at the highest density locations (the Abel clusters). Yoshioka and Ikeuchi's model involves seed objects, located in $\mathbb{R}^3$ by the homogeneous Poisson point process, which explode simultaneously and with the same energy thus forming shockwaves which expand into the intergalactic medium. The expanding shockwaves carry with them any mass particles which they encounter, thus concentrating the mass particles on the two-dimensional surfaces where the expanding shockwaves come into mutual contact. The result is again a Voronoi diagram in $\mathbb{R}^3$.

The Voronoi growth model has also been used in geology in several circumstances. One is in modelling crack patterns in general and cracking in basalt (lava flows) in particular, where the crack patterns result from the contraction of the material on cooling. An early application was provided by Stiny (1929) who proposed that, as a molten stream of basalt cools irregularly at its surface, the locations where hardening occurs first act as nuclei from which subsequent cooling extends in all directions throughout the material. Thus, tensile stress can be modelled as a set of expanding circles centred at the nuclei which, as they meet, develop lines along which cracking occurs. Smalley (1966) seems to have developed a similar model independently. He argues that if the basaltic material is of uniform temperature, the nuclei will appear simultaneously and form a regular spatial arrangement in the form of a triangular lattice (see Figure 8.2.2) so that growth centred on them produces hexagonal blocks of basalt. However, if uniformity of temperature does not occur, as is usually the case, the nuclei will not develop simultaneously and will not form a regular arrangement. Smalley (1966) suggested that in such circumstances the growth process should be applied to centres located according to a hard core process of the type described in Section 5.10, whereby centres are located at random subject to the constraint that centres which are closer than some specified distance to an already existing centre cannot develop. He ran a number of simulations of this model and examined the frequency distribution of N, the number of neighbours (edges, vertices) of an individual Voronoi polygon (block of basalt). These frequencies were compared with those obtained from real-world basalt patterns and found to differ in a statistically significant way. Smalley argued that such discrepancies could arise owing to the difficulty of identifying very short edges in field counts of N. Thus he ignored such small edges in his simulated patterns before recalculating the frequencies of N associated with them. These modified model frequencies provide a better fit to the empirical data than the original model frequencies but again the differences between the two sets are statistically significant. Further evidence for questioning the validity of the Stiny/Smalley model is provided by Stoyan and Stoyan (1980) who demonstrated that at least one empirical pattern of basalt blocks does not strongly approximate a Voronoi diagram (see also Section 7.5).

Also in geology, Williams (1972, p. 152) has proposed that the Voronoi Growth Model is appropriate for modelling the polygonal depressions which develop in karst topography. Here the nuclei around which growth develops are stream sinkholes.

An ecological application of the Voronoi Growth Model occurs in the study of animal territories where the Voronoi diagram defined with respect to a set of individual animals, each with an associated fixed location in $\mathbb{R}^2$, can be considered to represent ideal territories if the environment is uniform and the animals of equal strength (Hasegawa and Tanemura, 1976). An example is provided by the territories of male mouthbreeder fishes (*Tilapa mossambica*) observed by Barlow (1974). These fish were kept in an outdoor pool with an initially uniform sandy bottom. The males excavate breeding pits by spitting sand away from fixed pit centres towards the pit centres of their neighbours. Reciprocal spitting results in sand parapets which act as territorial boundaries. Figure 7.2.2 shows such boundaries traced by Hasegawa and Tanemura (1976) from a photograph taken by Barlow. Honda (1978) and Suzuki and Iri (1986a) (see Section 7.5) show that the pattern of such boundaries closely approximates the edges in the Voronoi diagram with the breeding pits as generators (see Figures 7.5.4 and 9.5.2).

Unlike the Voronoi Assignment Model, there have been far fewer applications of the Voronoi Growth Model to phenomena studied by social

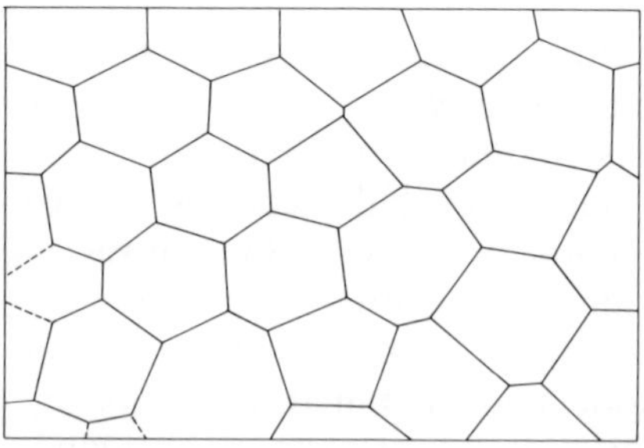

Figure 7.2.2 Schematic diagram of polygonal territories formed by male mouthbreeder fish replicated from a photograph in Barlow (1974). (*Source*: Hasegawa and Tanemura (1976, Figure 1).)

scientists. One example is provided by Boots (1973, 1975a) who used it to investigate patterns of service areas for bus companies as identified by Green (1955). In Britain bus services have their origins in the first few decades of the twentieth century when they first challenged the hegemony of the railways as the prime form of longer distance public transportation. The development of these services was not coordinated by any central agency and the process was a highly competitive one. Bus services tended to become established at the urban centres which already provided surrounding rural areas with other goods and services. In areas away from the major urban agglomerations, such centres tend to be relatively similar in size and so if the bus services were established at these places at approximately the same time, the Voronoi Growth Model could describe the service areas that arose under competition between centres. Although no formal statistical tests were undertaken, summary statistics for the characteristics of the bus service areas, such as number of neighbours, area and boundary length were found to be similar to those expected for Voronoi polygons.

Finally, it is important to note that both the Voronoi Assignment and Growth Models can be modified to produce a whole set of extended models by changing one or more of their underlying assumptions. For instance, by changing Assumption A1 or G1 the points can be introduced into $\mathbb{R}^m$ asynchronously which gives rise to the Johnson–Mehl type models described in Sections 3.1 and 5.6. Changing Assumption A2 or G2 produces dynamic models of the type discussed in Section 7.3. Assumption A3 can be modified by considering points to be of differing importance and assigning to each of them a weight which reflects their relative importance. Such differential weights might also be reflected in differences in the growth rates associated with individual points in the Voronoi Growth Model, thus revising Assumption G4. The weights could also be used in defining how the distance between a member of P and other locations in $\mathbb{R}^m$ is measured, thus leading to assignment procedures which differ from those in Assumption A4. Such models and their applications are discussed in Chapter 3.

7.3 SPATIAL–TEMPORAL PROCESSES

In the preceding sections we dealt with spatial processes leading to Voronoi diagrams in which the locations of generators do not change over time. In this section we show some models of spatial processes in which the location of generators change over time.

7.3.1 Spatial competition models: the Hotelling model

In economics, in particular in spatial economics (or urban economics), a considerable number of studies have been carried out to examine market area stability (a review is provided, for example, by Gabszewicz *et al.*,

1986). Economic units, say firms, are competing against each other to maximize their profits. In a spatial context, this competition takes the form of market area competition. Since consumers buy products from the firm that offers the lowest price, each firm has its own market area. The market areas are different in size, yielding different profits, and firms with small profits move to alternative locations to gain more profits. Through this relocation process the configuration of firms may vary over time, or it may eventually reach an equilibrium configuration. The pioneering work on this market area stability was done by Hotelling (1929). A variety of models has been developed from his model, which we call Hotelling models (processes) or, more broadly, spatial competition models (processes). In this subsection we investigate spatial competition processes with the simplest Hotelling model, i.e. the Alonso version (Alonso, 1964) of the original Hotelling (1929) model.

Suppose that there are n firms located at $\mathbf{x}_1, \ldots, \mathbf{x}_n$ in a region S, and these firms are selling the same products with the same mill price (the price not including a delivery cost). We assume that the delivery cost from a firm at $\mathbf{x}_i$ to a consumer at $\mathbf{x}$ is proportional to the Euclidean distance $\|\mathbf{x}_i - \mathbf{x}\|$, and that consumers buy the products from the firm that quotes the lowest delivered price (the mill price plus the delivery cost). Under these assumptions the configuration of n market areas is represented by the ordinary Voronoi diagram $\mathcal{V}(P) = \{V(\mathbf{x}_1), \ldots, V(\mathbf{x}_n)\}$, and the market area of firm i is represented by the Voronoi polygon $V(\mathbf{x}_i)$. We further assume that the demand for the products is uniformly distributed over region S; the marginal cost of the products is the same for all firms; relocation cost is negligibly small. Then the profit of firm i is proportional to the area $V(\mathbf{x}_i)$. The firms compete in terms of their locations to maximize their profits. As a result, we observe a spatial competition process of n firms over time. We specify this process by the following behavioral assumptions.

Assumption HP1 (relocation timing) Each firm considers relocation once in every period, and the interval is equal over time periods.

Assumption HP2 (zero conjectural variation) In considering possible relocation, every firm conjectures that all other firms will not change their locations.

Let $X_{-i} = \{\mathbf{x}_1, \ldots, \mathbf{x}_{i-1}, \mathbf{x}_{i+1}, \ldots, \mathbf{x}_n\}$ be the set of locations of all firms except firm i, and $\mathbf{x}_i^*$ be the location of firm i that yields the maximum profit provided that the locations of the other firms are fixed at the present locations, i.e.

$$|V(\mathbf{x}^*_i)| = \max_{\mathbf{x} \in S} \{|V(\mathbf{x})| \, |X_{-i} \text{ is fixed}, \mathbf{x} \neq \mathbf{x}_j, j \in I_n \setminus \{i\}\} , \qquad (7.3.1)$$

where $|V(\mathbf{x})|$ indicates the area of $V(\mathbf{x})$.

As is indicated by the condition in equation (7.3.1), firm i is not allowed to locate at the same location as any other firms.

We add one more behavioral assumption.

Assumption HP3 (relocation decision) Firm i moves to location $\mathbf{x}_i^*$ (given by equation (7.3.1)) if location $\mathbf{x}_i^*$ yields a greater profit than the present profit, i.e. $|V(\mathbf{x}_i^*)| > |V(\mathbf{x}_i)|$. Otherwise, firm i remains at the present location $\mathbf{x}_i$.

We call the process characterized by the above assumptions the *Hotelling process*. Stated abstractly, the Hotelling process is the process in which each generator independently moves to maximize the area of its Voronoi polygon. Through the Hotelling process, the Voronoi diagram may change over time, or it may eventually reach an equilibrium state in which all firms have no incentive to relocate. To be precise, the configuration of the n firms is in *global equilibrium* if an only if no firm can find a more profitable location than the present location, that is, the Voronoi diagram $\mathcal{V}^* = \{V(\mathbf{x}_1^*), \ldots, V(\mathbf{x}_n^*)\}$ is in the *global equilibrium* if and only if

$$|V(\mathbf{x}_i^*)| \geq \max_{\mathbf{x}_i \in S} \{|V(\mathbf{x}_i)| \, |X_{-i} = X_{-i}^*, \mathbf{x}_i \neq \mathbf{x}_j^*, j \in I_n \setminus \{i\}\}$$
$$\text{for any } i \in I_n \,, \tag{7.3.2}$$

where $X_{-i}^* = \{\mathbf{x}_1^*, \ldots, \mathbf{x}_n^*\}$.

When S is a bounded one-dimensional space, we can obtain the exact global equilibrium configurations $\mathcal{V}^*$ (Eaton and Lipsey, 1975), which are shown in Figure 7.3.1. For $n = 2$, 4 and 5, we have the unique global equilibrium configuration; for $n = 3$, we have no global equilibrium; for $n \geq 6$, we have an infinite number of equilibrium configurations. Two extreme configurations are shown in panels (d) and (e).

The Hotelling process in a two-dimensional space was first studied by Eaton and Lipsey (1975a, 1975b), followed by Shaked (1975), Okabe and Suzuki (1987) and Okabe and Aoyagi (1991). Compared with the one-

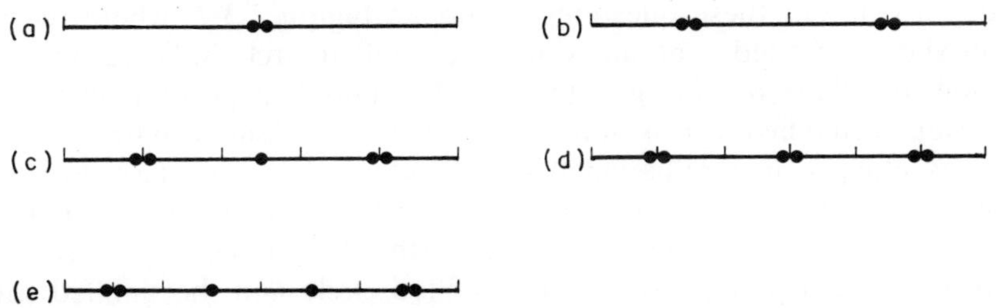

Figure 7.3.1 Global equilibrium configurations (Voronoi diagrams) of n firms on a bounded line market ($n = 2$, 4, 5, 6).

dimensional space, analytical examination becomes very difficult owing to complicated geometric calculations. We first show a few basic properties about the variation of a Voronoi polygon. These properties are useful not only for the following analysis but also for locational optimization which will be discussed in Chapter 9.

Consider a Voronoi diagram $\mathcal{V} = \{V(\mathbf{x}), V(\mathbf{x}_1), \ldots, V(\mathbf{x}_{n-1})\}$, where $\mathbf{x}^{\mathrm{T}} = (x,y)$ is the location of firm 0, and suppose that firms 1, ..., m share their market boundaries with firm 0 (firms 1,...,m are indexed counterclockwise around firm 0). Let $\mathbf{u}_i^{\mathrm{T}} = (u_i, v_i)$ be the Cartesian coordinates of the Voronoi vertex shared by $V(\mathbf{x})$, $V(\mathbf{x}_i)$ and $V(\mathbf{x}_{i+1})$. Recalling that a Voronoi vertex is the intersection of two bisectors (Chapter 2), we can write (u_i, v_i) in terms of $\mathbf{x}$, $\mathbf{x}_i$ and $\mathbf{x}_{i+1}$, from which we can obtain the area of a Voronoi polygon $V(\mathbf{x})$. Since this quantity is frequently used in many applications, we state it as a formula.

The area of a Voronoi polygon

$$|V(\mathbf{x})| = \frac{1}{2}\sum_{i=1}^{m} (u_i v_{i+1} - u_{i+1} v_i) , \tag{7.3.3}$$

where

$$u_i(x, y) = \frac{(x_i^2 + y_i^2 - x^2 - y^2)(y - y_{i+1}) - (x_{i+1}^2 + y_{i+1}^2 - x^2 - y^2)(y - y_i)}{2\{(x_i - x)(y_{i+1} - y) - (x_{i+1} - x)(y_i - y)\}},$$

$$v_i(x, y) = \frac{(x_i^2 + y_i^2 - x^2 - y^2)(x - x_{i+1}) - (x_{i+1}^2 + y_{i+1}^2 - x^2 - y^2)(x - x_i)}{2\{(x_i - x)(y_{i+1} - y) - (x_{i+1} - x)(y_i - y)\}}, \tag{7.3.4}$$

where $u_{m+1} = u_1$, $v_{m+1} = v_1$, $y_{m+1} = y_1$.

From equations (7.3.3) and (7.3.4) we readily notice that $|V(\mathbf{x})|$ is continuous except at $\mathbf{x} = \mathbf{x}_i$ unless m changes. We should next examine whether or not $|V(\mathbf{x})|$ is continuous even if m changes. This question is closely related to a more general problem, called the moving generator (particle) problem (Cruz Orive, 1979). The questions there are to ask how the area of $V(\mathbf{x})$ changes and how the shape of $V(\mathbf{x})$ (i.e. m) changes as the generator $\mathbf{x}$ moves in a plane provided that the other generators are fixed. To consider these questions we depict Figure 7.3.2, where the filled circles show the fixed generators and the unfilled circle indicates the initial location, o, of the moving generator $\mathbf{x}$. The initial shape of $V(\mathbf{x})$ at $\mathbf{x} = 0$ is a hexagon (the heavy broken lines in Figure 7.3.2). Suppose that generator $\mathbf{x}$ moves along a line radiating from o, indicated by the dash-dot line in Figure 7.3.2. When the generator $\mathbf{x}$ comes to the point indicated by the unfilled square, $V(\mathbf{x})$ becomes a septagon (the dotted lines in Figure 7.3.2). There must be a point between the unfilled circle and the unfilled square on the dash-dot line at which the shape of $V(\mathbf{x})$ changes from a hexagon to a septagon. The following property gives the answer to this question.

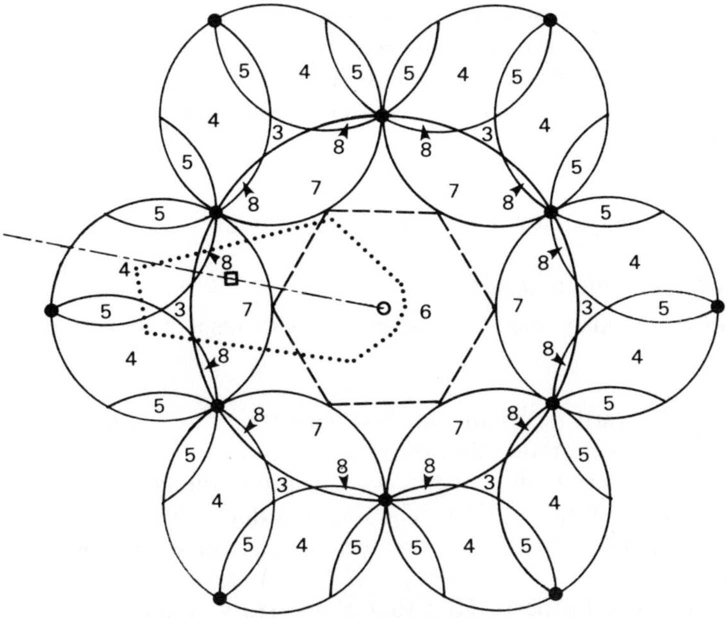

Figure 7.3.2 The shape of a Voronoi polygon $V(x)$ whose generator x moves on the plane where the other generators are fixed at regular triangular lattice points (the number m in each region enclosed by circular arcs indicates the number of vertices of $V(x)$ when x comes to that region).

Property HP1 Consider a Voronoi diagram $\mathscr{V} = \{V(\mathbf{x}), V(\mathbf{x}_1), \ldots, V(\mathbf{x}_n)\}$, where $V(\mathbf{x})$ is adjacent to $V(\mathbf{x}_1),\ldots,V(\mathbf{x}_m)$ when $\mathbf{x} = \mathbf{0}$. Suppose that as the generator $\mathbf{x}$ moves from $\mathbf{0}$ along a line L radiating from $\mathbf{0}$, a Voronoi vertex $\mathbf{u}_{m+1}$ appears between vertices $\mathbf{u}_i$ and $\mathbf{u}_{i+1}$ (or $\mathbf{u}_i$ and $\mathbf{u}_{i+1}$ are degenerated into one vertex $\mathbf{u}_m$). Let C be the circle that passes through points $\mathbf{x}_i$, $\mathbf{x}_{i+1}$ and $\mathbf{x}_m$, and $\mathbf{x}^*$ be a point at which the line L meets the circle C. Then $\mathbf{x}^*$ is the point at which $V(\mathbf{x})$ changes from m-gon to $(m + 1)$-gon (or $(m - 1)$-gon).

We can prove this property from the empty circle property (Property D5) in Chapter 2. The detailed derivation is shown by Cruz Orive (1979) for special lattice configurations and by Okabe and Aoyagi (1991) for general configurations.

Using Property HP1, we can determine regions for $\mathbf{x}$ in which $V(\mathbf{x})$ is an m-gon for a given configuration of generators. An example is shown in Figure 7.3.2, where generators are placed on regular triangular lattice points. (Note that generators are not necessarily spaced regularly.) The number in each region enclosed by circular arcs indicates m of $V(\mathbf{x})$. From equations (7.3.3) and (7.3.4) with Property HP1, we can show, after a few steps of derivation, that $|V(\mathbf{x})|$ is continuous with respect to $\mathbf{x}$ ($\neq \mathbf{x}_i$) not only when m does not change but also when m changes. In addition, we can show the following property.

Property HP2 For fixed $\mathbf{x}_1, \ldots, \mathbf{x}_{n-1}$ and $\mathbf{x} \neq \mathbf{x}_i$:

(i) $|V(\mathbf{x})|$ is continuous with respect to $\mathbf{x}$,
(ii) $\partial|V(\mathbf{x})|/\partial\mathbf{x}$ exists and it is continuous with respect to $\mathbf{x}$.

It is straightforward to obtain the first derivative of equation (7.3.3), from which we can easily notice that the derivative is continuous with respect to $\mathbf{x}$ unless m changes. If m changes, we can prove the continuity of the first derivatives using Property HP1. The detailed derivation is shown in Okabe and Aoyagi (1991). Note that the second derivatives are not necessarily continuous.

It follows from Property HP2 that a necessary condition for the global equilibrium in the Hotelling process is given by

$$\frac{\partial|V(\mathbf{x}_i)|}{\partial x_{ik}} = 0, \quad \kappa = 1, 2 \text{ for } i = 1, \ldots, n. \tag{7.3.5}$$

When this equation holds, we say that the configuration (Voronoi diagram) of the n firms is in *local equilibrium*.

When the region S is unbounded and firms are located on regular lattice points, it is an exercise to examine if the configuration of the firms is in local equilibrium. Using equations (7.3.3)–(7.3.5) we can show that the Voronoi diagram of regular triangular lattice points (Voronoi polygons are hexagons) and that of squares (Voronoi polygons are squares) are in local equilibrium, but the Voronoi diagram of regular hexagonal lattice points (Voronoi polygons are regular triangles) is not in local equilibrium. Furthermore, we can show a stronger property than this property.

Property HP3 The Voronoi diagram of regular triangular lattice points and that of square lattice points are in global equilibrium. The Voronoi diagram of regular hexagonal points is not in local equilibrium; consequently, it is not in global equilibrium.

We can prove this property by computing the Hessain, which is too lengthy to show here (see Okabe and Aoyagi, 1991, in which REDUCE III is used).

When the region S is bounded, analysis becomes extremely intractable except for $n = 2, 3$. Shaked (1975) shows the following property.

Property HP4 When the region S is a disk, two firms paired at the centre of the disk are in global equilibrium; there is no local equilibrium for three firms.

See Shaked (1975) for the proof.

The analytical difficulty encountered in a bounded two-dimensional space with $n \geq 4$ firms results from the fact that the optimization problem of equation (7.3.1) is a non-linear, non-convex programming problem. As is shown in the brief review of the non-linear, non-convex programming problem in Chapter 9, it is almost impossible to obtain the global maximum by an analytical method; only a local maximum is obtainable by the numerical method. This intractability prevents analytical examination of the Hotelling process over time, not only in a bounded two-dimensional space but also in an unbounded two-dimensional space. At present it is difficult to say if firms on an unbounded two-dimensional space reach the global equilibrium configurations stated in Property HP3 for any initial configuration. Giving up analytical examination, Eaton and Lipsey (1975b) employed a naive numerical method to examine the Hotelling process for 17 firms in a disk. They conjectured from their numerical results that the global equilibrium configuration would not be achieved. Alternatively, Okabe and Suzuki (1987) imposed the following weaker assumption than Assumption HP3 and examined the Hotelling process over time with numerical simulations.

Assumption HP4 Let q^* be the centre of the largest empty circle in the Voronoi diagram $\mathcal{V} = \{V(\mathbf{x}_1), \ldots, V(\mathbf{x}_n)\}$.

(i) If $|V(\mathbf{x}_i)| < |V(q^*)|$, firm i searches for a local maximum location $\mathbf{x}_i^*$ from q^* (for example, q^* in Figure 7.3.3). If $|V(\mathbf{x}_i^*)| > |V(q^*)|$, firm i moves to $\mathbf{x}_i^*$; otherwise, firm i moves to q^*.

(ii) If $|V(\mathbf{x}_i)| \geq |V(q^*)|$, firm i searches a local maximum location $\mathbf{x}_i^{**}$ from the present location (for example, $\mathbf{x}_2$ in Figure 7.3.3). If $|V(\mathbf{x}_i^{**})| > |V(\mathbf{x}_i)|$, firm i moves to $\mathbf{x}_i^{**}$; otherwise firm i remains at the present location $\mathbf{x}_i$.

Under Assumptions HP1, HP2 and HP4 with the steepest descent method (see Chapter 9), Okabe and Suzuki (1987) carried out the numerical simulation for 256 firms in a square over 100 periods (one period is defined by a period during which every firm makes the locational decision once). Part of the processes is shown in Figure 7.3.4, where the filled circles, squares, triangles and stars indicate the change in profit (i.e. a market area) being less than 1%, 1–3%, 3–5%, and more than 5%, respectively. From these numerical results, they noticed that although the configuration does

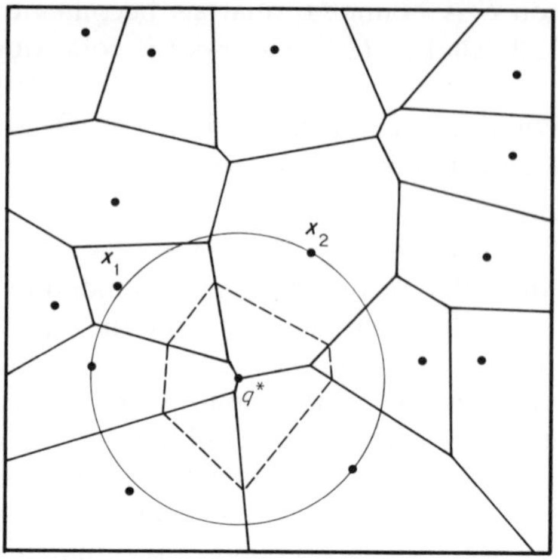

Figure 7.3.3 Searches for local maximum locations $|V(x_1)| < |V(q^*)|, |V(x_2)| > |V(q^*)|$

not achieve the global equilibrium by the 100th period (observe peripheral firms in Figure 7.3.4), the configuration of inner firms is stable after a certain period of time.

From the simulation of $n = 17$ in Eaton and Lipsey (1975b) and that of $n = 256$ in Okabe and Suzuki (1987), we notice that the stability of the Hotelling process depends upon the number of firms. Moreover, the shape of the region S crucially affects the stability. These effects are investigated by Aoyagi and Okabe (1992).

7.3.2 Adjustment models

Spatial–temporal processes are investigated not only in economic behaviour as we discussed in the above but also in animal behaviour. One well-known process is the adjustment process (Hasegawa and Tanemura, 1976, 1980, 1986). Consider individual animals settled down in a region. In the initial stage the individuals are fairly independent. As time goes by, however, the individuals interact with each other and each individual tends to occupy its territory in such a way that the individual is as distant from its neighbours as possible. Through this interaction process the territories of the individuals are adjusted together until a stable configuration is achieved.

To formulate this process mathematically, let $\mathbf{x}_1^{(t)}, \ldots, \mathbf{x}_n^{(t)}$ be the location of n individuals in a region S, $t = 0, 1, \ldots$. We assume that the initial locations, i.e. $\mathbf{x}_1^{(0)}, \ldots, \mathbf{x}_n^{(0)}$, are uniformly randomly distributed over S. We next assume that a location in S is dominated by the nearest individual. It follows from this assumption that the territory of individual i at time t is

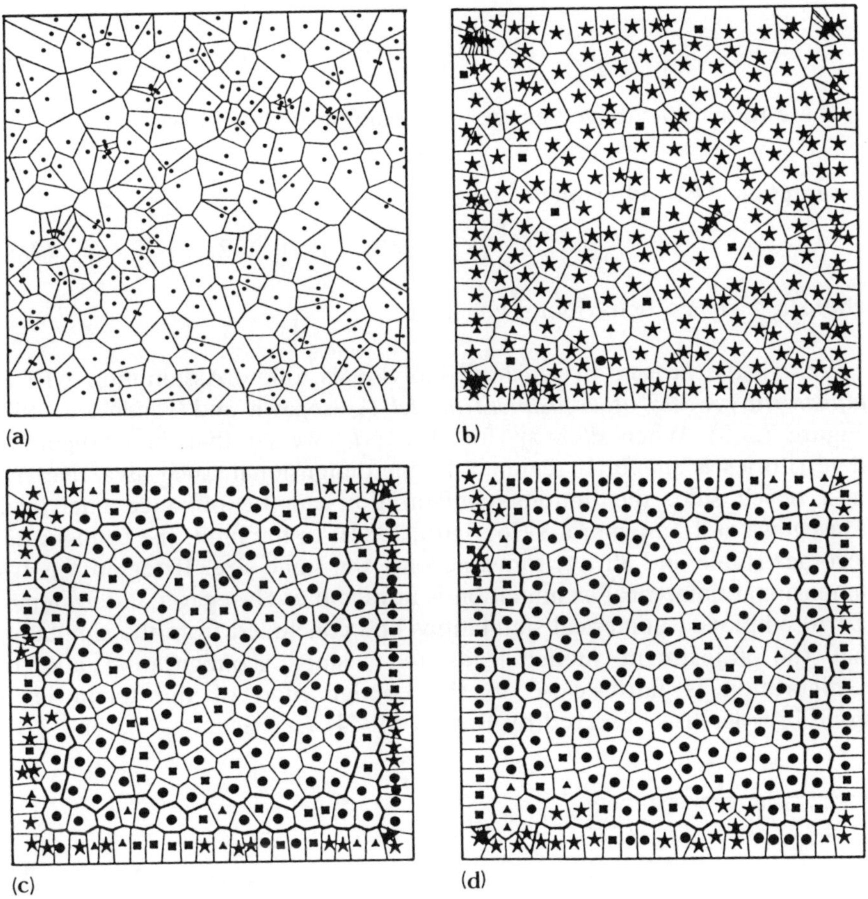

Figure 7.3.4 The Hotelling process of 256 firms in a square region. (a) The initial configuration (uniformly random), (b) the first period, (c) the tenth period, (d) the hundredth period. (The filled circles, squares, triangles and stars indicate the change in profit (i.e. a market area) being less than 1%, 1–3%, 3–5%, and more than 5%, respectively.) (*Source*: Okabe and Suzaki (1987, Figure 5).)

given by the Voronoi polygon $V(\mathbf{x}_i^{(t)})$ in the Voronoi diagram $\mathcal{V}^{(t)} = \{V(\mathbf{x}_1^{(t)}), \ldots, V(\mathbf{x}_n^{(t)})\}$. We further assume that individual i moves its location toward the centre of the gravitation of its territorial area. To be explicit, let $\mathbf{u}_{ij}^{(t)} = (u_{ij}^{(t)}, v_{ij}^{(t)})^{\mathrm{T}}$, $j \in I_{n_i}$, be Voronoi vertices of $V(\mathbf{x}_i^{(t)})$ which are indexed counterclockwise. Then the gravitation centre $\mathbf{x}_{gi}^{(t)} = (x_{gi}^{(t)}, y_{gi}^{(t)})^{\mathrm{T}}$ of $V(\mathbf{x}_i^{(t)})$ is given by

$$x_{gi}^{(t)} = \frac{1}{6|V(\mathbf{x}_i^{(t)})|} \sum_{j=1}^{n_i} (v_{ij}^{(t)} - v_{ij-1}^{(t)}) \{(u_{ij}^{(t)} + u_{ij-1}^{(t)})^2 - u_{ij}^{(t)} u_{ij-1}^{(t)}\} ,$$

$$y_{gi}^{(t)} = \frac{1}{6|V(\mathbf{x}_i^{(t)})|} \sum_{j=1}^{n_i} (u_{ij}^{(t)} - u_{ij-1}^{(t)}) \{(v_{ij}^{(t)} + v_{ij-1}^{(t)})^2 - v_{ij}^{(t)} v_{ij-1}^{(t)}\}. \quad (7.3.6)$$

In these terms, the above behavioural assumption may be written as follows.

Assumption AP1 (adjustment move)

$$\mathbf{x}_i^{(t+1)} = \mathbf{x}_i^{(t)} + \frac{1}{\alpha}(\mathbf{x}_{gi}^{(t)} - \mathbf{x}_i^{(t)}), \quad t = 0, 1, 2, \ldots, \quad (7.3.7)$$

where $\alpha \geq 1$ is a behavioural constant.

We call the process in which all individuals follow Assumption AP1 the *adjustment process* or *adjustment model* (Hasegawa and Tanemura, 1976) (Figure 7.3.5). When $\mathbf{x}_{gi}^{(t)} = \mathbf{x}_i^{(t)}$ for all $i \in I_n$, we say that the configuration of territories is *stable*. If region S is one-dimensional, such as a river side or a beach side, the limiting configuration can be obtained analytically (Hasegawa and Tanemura, 1976; Saito, 1982). In the limit, the configuration becomes stable and all territories become the same size. If region S is two-dimensional, analytical examination is intractable. Hasegawa and Tanemura (1976) employed numerical simulations and concluded from the simulations that the adjustment model leads to a stable configuration and this configuration is fairly close to the configuration of the territories of mouthbreeder fish observed by Barlow (1974) (see Sections 7.2 and 7.5).

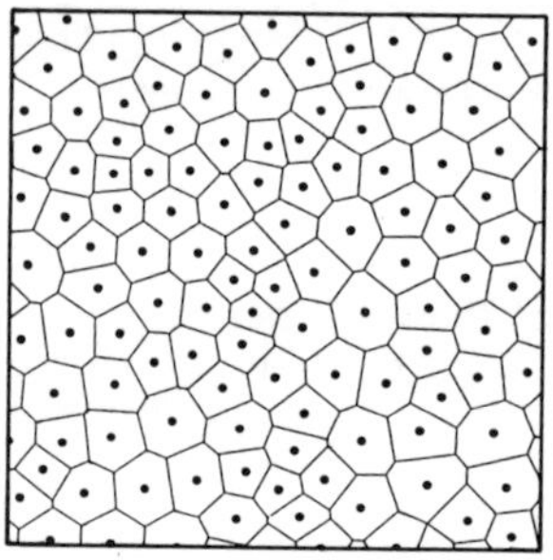

Figure 7.3.5 Adjustment process (t = 500). (*Source*: Hasegawa and Tanemura (1981, Figure 1).)

7.4 TWO-SPECIES MODELS

In Section 7.3 we showed the models of territories formed through the interaction within the same kind of behavioural units. In the real world we also find other types of territories which are formed not only through the interaction within the same kind of behavioural units but also through the interaction between different kinds of behavioural units. In this section we show a few Voronoi-type models that attempt to explain such phenomena.

In animal ecology many prey–predator relations are known, such as cows and lions, and it is reported that prey tend to form herds or flocks. To explain this tendency, Hamilton (1971) proposed a model which may be called the selfish herd model. Consider, as an example, cows and lions in a grassy land, and suppose that cows are grazing grass and lions are hiding in the grass to catch a cow. Lions tend to catch the cow nearest to their hiding site. Then a Voronoi polygon associated with a cow represents the 'domain of danger' (Hamilton, 1971) of that cow. To be explicit, let $\mathcal{V} = \{V(p_1), \ldots, V(p_n)\}$ be the Voronoi diagram generated by a set of sites of cows. If a lion happens to be in $V(p_i)$, the cow at p_i will be a prey of the lion. Thus the area $V(p_i)$ represents the magnitude of danger of the cow at p_i. When a cow senses the presence of a lion, the cow tends to move to the nearest cow. As a result, cows form a herd and each cow wants to be surrounded by the other cows. Actually, the domain of danger of a cow that is tightly surrounded by the other cows becomes small and the cow is less likely to be a prey of a lion. Since every cow wants to be tightly surrounded by the other cows, the configuration of cows within a herd is unstable, but as a whole, a herd guarantees more or less safety. With this model, Hamilton (1971) emphasizes that the selfish avoidance of a predator can lead to aggregation.

The area of a Voronoi polygon has a negative implication for a cow, but it may have a positive implication for certain kinds of insects. As an example Cannings and Cruz Orive (1975) referred to signalling females in relation to mobile males. Consider wingless females scattered over a region. Females send signals to males to be fertilized. Suppose that the intensity of the signal is inversely proportional to the distance from a female and the intensity of the signal is the same for all females. If males respond to the strongest signals and males are distributed fairly uniformly over the region, then the probability of a female being fertilized is proportional to the area of the Voronoi polygon of the female. Each female tries to maximize the probability of being fertilized by moving her location. As a result, the configuration of females changes over time and eventually it may reach an equilibrium configuration. Recalling the profit maximization behaviour in Section 7.3, we notice that this problem is similar to the problem considered in the Hotelling process. Canning and Cruz Orive (1975) examined whether or not a slight move of a female is profitable provided that the other females are placed on regular hexagonal, square or triangular lattice points. This problem is in fact the same as the problem of the Hotelling process in an unbounded plane, and the answer is shown in Property HP3. This problem is also theoretically investigated by Cruz Orive (1979).

In plant ecology, the prey–predator relation may correspond to the relationship between crop plants and weeds. This relationship was studied by Fischer and Miles (1973) with a Voronoi diagram. Usually crop plants are planted at rectangular lattice points (the filled circles with $a \times b$ intervals

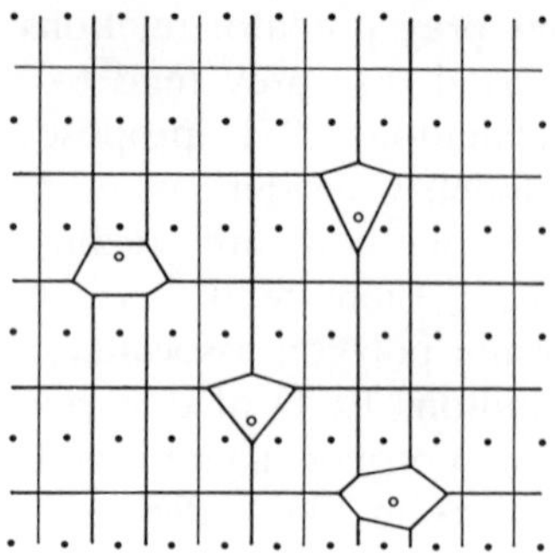

Figure 7.4.1 Two-species model (filled circles indicate crop plants and unfilled circles indicate weeds).

in Figure 7.4.1), whereas weeds are distributed randomly, i.e. the distribution of weeds follows a Poisson process (the empty circles in Figure 7.4.1). Suppose that crop plants and weeds emerge simultaneously and expand with equal rates. Ultimately, as we saw in Section 7.2, each plant establishes its individual zone of exploitation, and collectively form a Voronoi diagram. To estimate the yield of the crop, it is useful to know the expected area $\mu_c(\rho)$ of each crop plant domain (the mean area of each Voronoi polygon) when the density ρ of weeds and intervals a and b are given. Fischer and Miles (1973) obtained this value as

$$\mu_c(\rho) = \frac{4}{\rho}\,\Psi\left(\sqrt{\frac{\pi\rho}{2}}\,a\right)\Psi\left(\sqrt{\frac{\pi\rho}{2}}\,b\right), \tag{7.4.1}$$

where

$$\Psi(x) = \frac{1}{2\pi}\int_0^x e^{-z^2/2}\,dz\,. \tag{7.4.2}$$

Fischer and Miles (1973) also examined the cases of simultaneous emergence and different expansion rates, and different emergence and different expansion rates.

We may also find the prey–predator relation in economic behaviour. Von Hohenbalken and West (1984a) and West and Von Hohenbalken (1984) surveyed the entry and withdrawal of stores in Edomonton, Canada, during 1959–1973 to test Kreps and Wilson's (1982) economic predation theory. The predation process assumed by Kreps and Wilson may be summarized as follows.

Suppose that an established firm operates a set of stores in a region and a firm is entering this region with new stores.

(i) The new stores of the entrant are challenged by their neighbourhood stores of the established firm.

(ii) The established firm locates new stores to force losses on the entrant's new stores.

(iii) The entrant stops constructing new stores.

(iv) Eventually, the entrant withdraws from the region.

Von Hohenbalken and West (1984a) assume that a set of market areas is represented by a Voronoi diagram and the neighbourhood stores of a store, say store i, are given by stores whose Voronoi polygons are adjacent to the Voronoi polygon of store i. With these assumptions, they empirically investigated the above predation process in Edomonton. They concluded that generally the data support the Kreps and Wilson's predation process.

7.5 RECOGNIZING VORONOI DIAGRAMS

If the Voronoi assignment and growth models described in the preceding sections are appropriate for modelling a particular phenomenon, we would expect spatial patterns of the phenomenon to display characteristics of Voronoi diagrams. In this section we first consider whether an empirical pattern constitutes a Voronoi diagram. Recognizing that the recording of many types of empirical patterns often involve some measurement error, we may not be surprised if a given pattern does not correspond to a Voronoi diagram even when we suspect that the pattern was generated by processes such as those in the Voronoi models. In such instances we may wish to determine the degree of correspondence between the empirical pattern and a Voronoi diagram. Methods for doing this are considered in Section 7.5.2.

Throughout this section we focus on an empirical pattern S in a bounded region B of $\mathbb{R}^2$. We assume that S consists of a set, $\{s_1, s_2, \ldots, s_n\}$, of n regions such that $\cup_{i=1}^{n} s_i = B$ and $(s_i \backslash \delta s_i) \cap (s_j \backslash \delta s_j) = \emptyset$, $i \neq j$, $i, j \in I_n$, where δs_i and δs_j are the boundaries of s_i and s_j, respectively. For some S it may be possible to identify a single point g_i in the interior of s_i which may be considered the generator of s_i while for other S such points may be lacking. Examples of the former type of pattern encountered in the previous sections include market areas of retail outlets and tribal territories, while examples of the latter include basalt blocks and sections of metallic grains.

7.5.1 Reconstructing Voronoi diagrams

We begin by considering ways of determining if S is a Voronoi diagram. In the case where we have a set of generators $G = \{g_1, g_2, \ldots, g_n\}$ associated with S this simply involves generating the Voronoi diagram $\mathcal{V}(G)$ of G and observing if $\mathcal{V}(G) = S$. If G does not exist the question can be answered using Property V15 of Section 2.3. Recall that for each vertex q_j ($j = 1, \ldots, k$) of a given Voronoi polygon V_i with k sides (vertices), this property enables us to define a half line L_{ij} of a given direction radiating from q_j into the interior of V_i. The intersection of any two such half lines gives the

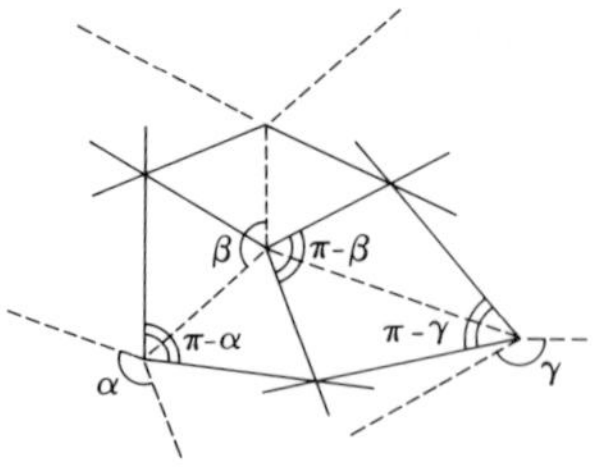

Figure 7.5.1 Illustration of a necessary and sufficient condition for a planar tessellation to be a Voronoi diagram. ---------, Voronoi edges; ———, half-lines, L_{ij}.

location of the generator of V_i. Ash and Bolker (1985) prove that if S is non-degenerate and all its members are convex, S is a Voronoi diagram if and only if all such half lines have a common point for each member of S (Property V16 in Section 2.3). This condition, illustrated in Figure 7.5.1, provides us with the required test. Stoyan and Stoyan (1980) used this procedure to demonstrate that a sample of basalt blocks is not a Voronoi polygon.

An alternative procedure has been proposed by Evans and Jones (1987). As we saw in Chapter 2, the line segment $\overline{p_k p_l}$ joining the generators p_k and p_l of two adjacent Voronoi polygons $V(p_k)$ and $V(p_l)$ is bisected perpendicularly by the common edge $e(p_k, p_l)$ of $V(p_k)$ and $V(p_l)$. This, in turn, means that p_k and p_l are subject to the following conditions:

C1 p_k and p_l lie on a line perpendicular to $e(p_k,p_l)$ (slope condition).

C2 p_k and p_l are equidistant from $e(p_k,p_l)$ (distance condition).

These conditions can be formulated algebraically to form a linear system of equations which can be solved to find the locations p_k and p_l, Let (u_p, v_p) and (u_q, v_q) be the locations of the end points of the common edge $e(s_k, s_l)$ of two adjacent members s_k and s_l, of S and (x_k, y_k) and (x_l, y_l) be the locations of the generators g_k and g_l, respectively. Let $e(s_k, s_l)$ be a segment of the line

$$y = t_i x + b_i\ ,$$

where $t_i = (v_p - v_q) / (u_p - u_q)$ and $b_i = v_p - t_i u_p$. If S is a Voronoi diagram, C1 gives

$$(x_k/t_i) + y_k - (x_l/t_i) - y_l = 0 \tag{7.5.1}$$

and C2 gives

$$t_i x_k - y_k + t_i x_l - y_l = -2b_i\ . \tag{7.5.2}$$

If S is non-degenerate and has m interior edges, equations (7.5.1) and (7.5.2) are systems of m equations with $2n$ unknowns. Equations (7.5.1) and (7.5.2) can be represented as two underdetermined matrix equations

$$B\mathbf{x} = 0 \qquad (7.5.3)$$

and

$$A\boldsymbol{x} = b\,, \qquad (7.5.4)$$

where $\boldsymbol{x} = (x_1, y_1, x_2, y_2, \ldots, x_n, y_n)^{\mathrm{T}}$, $\boldsymbol{b} = -2(b_1, b_2, \ldots, b_m)^{\mathrm{T}}$ and A and B are sparse matrices with non-zero elements given by equations (7.5.1) and (7.5.2). Taken jointly equations (7.5.3) and (7.5.4) have enough constraints to provide a least squares solution for **x**. Evans and Jones (1987) consider three different ways of solving equations (7.5.3) and (7.5.4). These are to simultaneously minimize C1 and C2,

$$\text{minimize}\left\|\begin{bmatrix} A \\ B \end{bmatrix}\boldsymbol{x} - \begin{bmatrix} \boldsymbol{b} \\ \boldsymbol{0} \end{bmatrix}\right\|^2, \qquad (7.5.5)$$

minimize C1 while constraining C2,

$$\text{minimize } \|B\mathbf{x}\|^2 \text{ subject to } A\boldsymbol{x} = \boldsymbol{b}\,, \qquad (7.5.6)$$

and to minimize C2 while constraining C1,

$$\text{minimize } \|A\boldsymbol{x} - \mathbf{b}\|^2 \text{ subject to } B\boldsymbol{x} = \boldsymbol{0}\,. \qquad (7.5.7)$$

Specific methods for solving equations (7.5.5), (7.5.6) and (7.5.7) are given by Evans and Jones (1987). If S is a Voronoi diagram, then equations (7.5.5), (7.5.6) and (7.5.7) will all yield the same solution for **x**.

7.5.2 Approximating Voronoi diagrams

Even if we find that S does not constitute a Voronoi diagram we may still wish to see the extent to which it approximates one. If G exists for S we can generate $\mathcal{V}(G)$, the Voronoi diagram of G, and compare this with S. When G is unknown we must first generate a set of Voronoi centres. This can be done using the methods described in Section 7.5.1 or by additional methods presented in this subsection.

An example of a situation where G is known is cited in Section 7.1, where we describe how Cox and Agnew (1976) proposed that Voronoi polygons defined around the existing country towns of Ireland can be considered efficient service areas for these towns. With this in mind Cox and Agnew (1974, 1976) developed an information theory statistic which measures the correspondence between the actual counties and the Voronoi polygons (see Figure 7.1.9).

Let $S = \{s_1, s_2, \ldots, s_n\}$ be the set of counties ($n = 32$), $\mathcal{V} = \{V(p_1), V(p_2), \ldots, V(p_n)\}$ be the set of Voronoi polygons and B the total region covered by S or $\mathcal{V}$ individually (in this case Ireland). Furthermore, let $L = \{l_1, l_2, ..., l_q\}$ be a set of q locations on a dense triangular lattice (see Figure 8.2.2) superimposed on B. Define f_{ij} as the number of members of L falling in $V(p_i) \cap s_j$ so that

$$\sum_{i=1}^{n} \sum_{j=1}^{n} f_{ij} = q \,. \tag{7.5.8}$$

Convert f_{ij} into an empirical probability, $p_{ij} = f_{ij}/q$, and define $p_{i\cdot}$ and $p_{\cdot j}$ as the probability of a member of L being in the ith region of $\mathcal{V}$ and the jth region of S, respectively. If we have no information other than $p_{i\cdot}$ and $p_{\cdot j}$ and we consider that $p_{i\cdot}$ and $p_{\cdot j}$ are independent (as would occur if there was no association between $\mathcal{V}$ and S), then our estimate $\hat{p}_{ij}$ of p_{ij} would be

$$\hat{p}_{ij} = p_{i\cdot}\, p_{\cdot j} \,, \tag{7.5.9}$$

The values of $\hat{p}_{ij}$ can be compared with the corresponding values of p_{ij} as defined above using Thiel's measure of mutual information (Thiel, 1967):

$$I_{V(p_i)s_j} = \log_2 (p_{ij}/\hat{p}_{ij}) \,. \tag{7.5.10}$$

If stochastic independence actually occurs, $p_{ij} = \hat{p}_{ij}$ and $I_{V(p_i)s_j} = 0$, implying that knowledge of p_{ij} provides us with no information additional to that provided by $p_{i\cdot}$ and $p_{\cdot j}$. Thus, the total expected mutual information for the two patterns is given by

$$I_{\mathcal{V}S} = \sum_{i=1}^{n} \sum_{j=1}^{n} p_{ij} \log_2 (p_{ij}/\hat{p}_{ij}) \,. \tag{7.5.11}$$

If there is a perfect correspondence between $\mathcal{V}$ and S, $p_{ij} > 0$ in equation (7.5.11) only if $i = j$ and for each $p_{ii} > 0$, $p_{ii} = p_{i\cdot} = p_{\cdot i}$. In this circumstance the expected mutual information $I^*_{\mathcal{V}S}$ for the two patterns is given by

$$I^*_{\mathcal{V}S} = \sum_{i=1}^{n} \sum_{j=1}^{n} p_{ij} \log_2(1/p_{ij}) = \sum_{j}^{n} p_{\cdot j} \log_2 (1/p_{\cdot j}) = \sum_{i=1}^{n} p_{i\cdot} \log_1 (1/p_{i\cdot}) \,, \tag{7.5.12}$$

which is equal to the entropy H_S of S (see Section 8.1.2). Thus, a percentage measure of the spatial correspondence C between $\mathcal{V}$ and S is provided by

$$C = (I_{\mathcal{V}S}/H_S)\, 100 \,. \tag{7.5.13}$$

Correspondence between individual members of $\mathcal{V}$ and S can be measured by a parallel procedure. For each member s_j of S, the expected mutual information $I_{\mathcal{V}s_j}$ is given by

$$I_{\mathcal{V}s_j} = \sum_{i=1}^{n} p_{ij} \log_2 (p_{ij}/\hat{p}_{ij}) \,, \tag{7.5.14}$$

while the expected mutual information $I^*_{\mathcal{V}s_j}$ under the condition of perfect correspondence of s_j to one member of $\mathcal{V}$ is given by

$$I^*_{\mathcal{V}S_j} = p_{\cdot j} \log_2 (1/p_{\cdot j}) , \tag{7.5.15}$$

so that a measure of correspondence c_j for the jth member of S is given by

$$c_j = (I_{\mathcal{V}S_j}/I^*_{\mathcal{V}S_j})\, 100 . \tag{7.5.16}$$

The results of applying this procedure to S consisting of Irish counties were reported in Section 7.1.

Several procedures are possible when the locations of the generators of S have to be estimated. One, proposed by Evans and Jones (1987), exploits the condition that every interior vertex in a non-degenerate Voronoi diagram is the centre of an empty circle passing through three generators (Property V7 in Section 2.3). Departures from this condition can be used to derive a measure of local deviation of members of S from Voronoi polygons. For each interior vertex q_p ($p = 1, \ldots, m$) located at (x_p, y_p), Evans and Jones defined the local deviation as the distance of q_p from the centre (x_p^*, y_p^*) of a circle passing through the generators g_k ($k = 1, 2, 3$) of the three polygons of S incident at q_p (see Figure 7.5.2). The average of such distances, divided by $\bar{e}$, the mean length of an interior edge of S, provides a normalized measure of fit τ, which is independent of scale. Specifically,

$$\tau = \sum_{p=1}^{m} [(x_p - x^*_p)^2 + (y_p - y^*_p)^2]^{\frac{1}{2}}/(m\bar{e}) . \tag{7.5.17}$$

τ approaches 0 as S approaches a Voronoi diagram. To help interpret values of τ for S, Evans and Jones (1987) examined the distribution of τ for a number of simulated patterns consisting of convex, non-Voronoi polygons. After estimating the generators using equation (7.5.5), they used this approach to show that the polygonal pattern formed by ice cracks (ice wedge

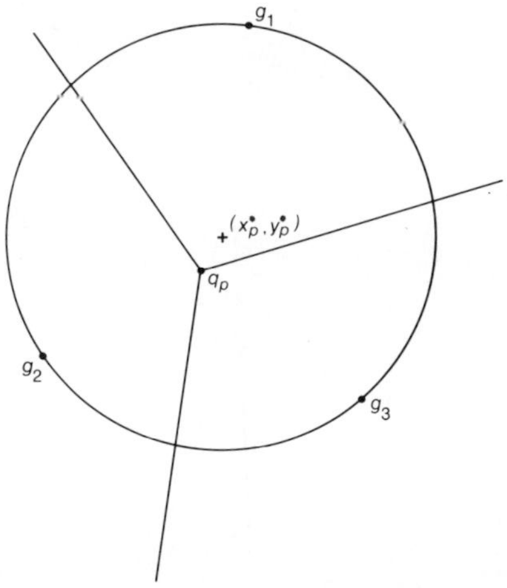

Figure 7.5.2 Illustration of a measure of a local deviation from a Voronoi diagram.

polygons) identified by Lachenbruch (1962, Plate 2, Figure 2) closely approximates a Voronoi diagram ($\tau = 0.15$).

Another approach, proposed by Honda (1978), involves the same property (Property V15 of Section 2.3.1) used in Section 7.5.1 to test if S is a Voronoi diagram. Recall that if s_i is a Voronoi polygon with k vertices, we may draw a half line L_{ij} in a specified direction from each vertex q_j ($j = 1, \ldots, k$) of s_i into the interior of s_i and that these half lines intersect at a single point p_i, which is the generator of s_i. If s_i is not a Voronoi polygon the half lines will not have a common point of intersection, as shown in Figure 7.5.3. In such cases Honda defined the generator g_i of s_i as the location which generates the minimum value D_i of

$$\sum_{j=1}^{k} d_j^2/k\,, \tag{7.5.18}$$

where d_j^2 is the Euclidean distance from g_i to L_{ij}. Once the set G of all such generators g_i is defined for S, we can construct the Voronoi diagram $\mathcal{V}(G)$

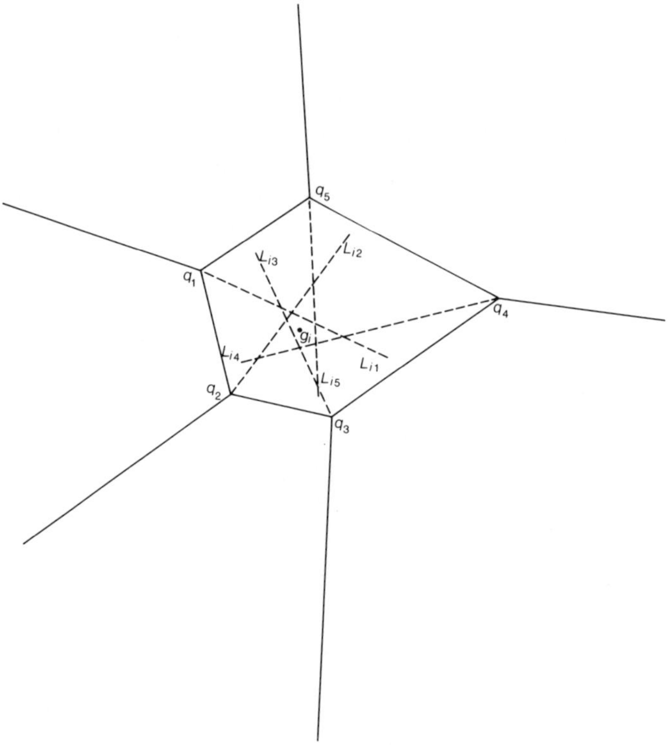

Figure 7.5.3 Procedure for estimating the generator of a polygonal region.

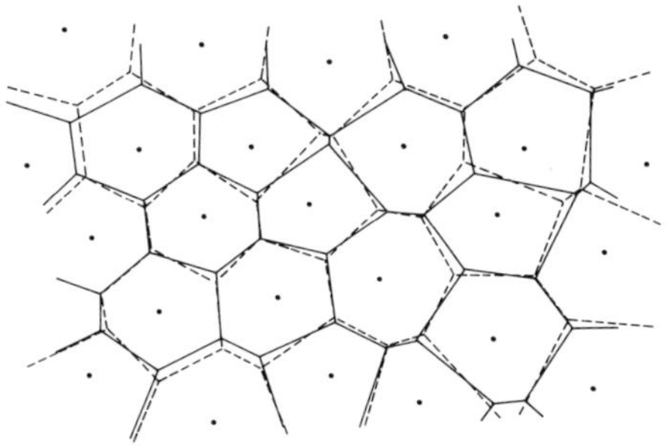

Figure 7.5.4 Comparison of Voronoi diagram for estimated Voronoi generators (dashed lines) and territories formed by male mouthbreeder fish (solid lines). (*Source*: Honda (1978, Figure 2(b)).)

of G. Figure 7.5.4 shows $\mathcal{V}(G)$ for the territories of male mouthbreeder fish (*Tilapia mossambica*) previously discussed in Section 7.2. A normalized measure Δ of the deviation of S from a Voronoi diagram can be defined as

$$\Delta = \left[\left(\sum_{i=1}^{n} D_i\right) \Big/ n\right] \Big/ \left[\left(\sum_{i=1}^{n} a_i\right) \Big/ n\right], \qquad (7.5.19)$$

where a_i is the area of s_i. Δ is 0 if S is a Voronoi diagram and increases as S diverges from a Voronoi diagram. Honda (1978) suggested that values of $\Delta \leq 1 \times 10^{-2}$ indicate a very close approximation of S to a Voronoi diagram, while values of 2–3 $\times 10^{-2}$ still indicate fairly good approximations. Using this approach Honda (1978, 1983) found that many patterns consisting of monolayer cells and epithelial cells in tissue closely approximate Voronoi diagrams. These findings are supported by observations from microscopic analyses (Prewitt, 1980).

Suzuki and Iri (1986a) also formulated the task of reconstructing the generators of S as a minimization problem. Their approach is discussed in detail in Section 9.5 and is only summarized here. It involves minimizing the discrepancy between S and a Voronoi diagram constructed for generators estimated from S. The discrepancy is defined as

$$\sum_{i=1}^{n} a_i - \sum_{i=1}^{n} \mu_2 (s_i \cap V(g_i)), \qquad (7.5.20)$$

where a_i is the area of s_i, $V(g_i)$ is the Voronoi polygon of the estimated generator of s_i and μ_2 is the planar Lebesgue measure. Suzuki and Iri apply their procedure to two empirical patterns. One consists of the districts of junior high schools in Tsukuba, Japan (see Figure 9.5.3), for which the discrepancy is 10% of the total area of S. The other pattern is that of the territories of the male mouthbreeder fish discussed in Section 7.2 and also examined by Honda (1978). Figure 9.5.2 shows the approximating Voronoi diagram for which the discrepancy is equal to 3.8% of the total area of S. Their solution may be compared with that obtained by Honda (1978) (see Figure 7.5.4). Stoyan and Stoyan (1990) also applied this procedure to examine a pattern of basalt columns for which they found a discrepancy equal to 4.4% of the total area of S.

CHAPTER 8

Point Pattern Analysis

A common concern which runs through a wide range of disciplines is the examination of the spatial occurrence of a particular phenomenon. Illustrative examples, which also reflect the extensive range of scales covered by such phenomena, include the location of sulphide inclusions in a steel (metallurgy), artefacts over a site (archaeology), nest sites of bird species in a habitat (ecology), towns in a state (geography), seismic events in a continent (geology) and galaxies of stars in the universe (astronomy). In each of these instances it is possible to represent individual incidences of the phenomenon as a set of n distinct points, $S = \{s_1, s_2, \ldots, s_n\}$, in a bounded region B in either $\mathbb{R}^2$ or $\mathbb{R}^3$. We call such representations *empirical point patterns*. Note that, while the individual objects themselves are not points, such representations are possible because the physical sizes of the objects are very small relative to both the distances between them and the extent of the region in which they occur.

Empirical point patterns are examined for a variety of reasons. A major one is the expectation that such patterns represent one source of evidence that may be useful in learning more about the phenomenon represented and the processes responsible for creating it. If our ideas concerning a phenomenon are sufficiently developed, we may be able to build an explanatory model of it. We can use such models to derive hypotheses concerning the locational behaviour of the phenomenon. For instance, our theory may suggest that the phenomenon in question may be regularly distributed over B. Support for such a hypothesis, and thus the model from which it is derived, can be obtained by analysing empirical point patterns showing the distribution of the phenomenon. Even when our knowledge of the phenomenon is more rudimentary, information gained from the analysis of empirical point patterns may enable us to acquire some initial insights into the phenomenon. For example, the discovery that objects are spaced differently towards the margins of B than they are at its centre may lead us to investigate the possibility of different forces operating at those locations or of the same forces operating but with different intensities.

In some instances our concern focuses on the location of the individual points of S with respect to other points of S or on their distribution over B. Such situations are dealt with in Sections 8.1 and 8.2. At other times we may be interested in how the members of S are located with respect to other objects, not members of S, also located in B. These objects may be points, lines or areas. We examine these circumstances in Section 8.3 which is divided in terms of the dimensionality of the other objects.

When our analysis is limited to a consideration of only the members of S, it is usual to adopt an approach which involves establishing a theoretical point pattern with respect to which other patterns are compared. The theoretical point pattern chosen is that which results from the operation of the homogeneous Poisson point process Θ_P described in Section 1.3.8. A pattern which results from such a process can be considered one that would arise by chance in a completely undifferentiated environment. Following Diggle (1983) we call such a point pattern *complete spatial randomness* (CSR). Figure 8.0.1(a) shows such an example of CSR. In view of the conditions involved in generating such a pattern, it is unlikely that true CSR occurs in any empirical situations. Instead, our concern with CSR is primarily for its role as an idealized standard. The two conditions that are assumed to exist when CSR results provide a simple model that can be useful in several circumstances. For instance, if we know little about the process responsible for a particular empirical point pattern, we can begin by testing a hypothesis that the pattern is CSR. This allows for exploration of the empirical point pattern. Whether or not the initial hypothesis relating to CSR is rejected, a description of the empirical point pattern remains. If CSR cannot be rejected, further analysis is unwarranted, otherwise the results of the test can be used as an aid in formulating new hypotheses regarding the empirical point pattern.

Classes of point patterns can be recognized using CSR as a benchmark. *Clustered* point patterns are those in which the points are significantly more grouped in B than they are in CSR (see the example in Figure 8.0.1(b)), whereas *regular* patterns (sometimes also called uniform or dispersed point patterns) are those in which the points are more spread out over B than they would be in CSR (see Figure 8.0.1(c)). Clustered and regular patterns can arise as the result of changing one or more of the postulates of Θ_P identified in Section 1.3.8.

Homogeneity may be relaxed by differentiating subareas of B in some way. Environmental inhomogeneity implies that some subareas of B are less likely to receive a point than other subareas, or might even be prohibited from receiving a point. In such circumstances we would expect to find more points in the favoured subareas of B than elsewhere, thereby producing a clustered pattern.

One way of relaxing independent scattering is to permit interaction among points. Points may either attract or repulse one another. Attraction may result from processes such as agglomeration, association and some types of diffusion and competition. In each instance the result is a clustered pattern. Such situations, where different processes result in similar types of point

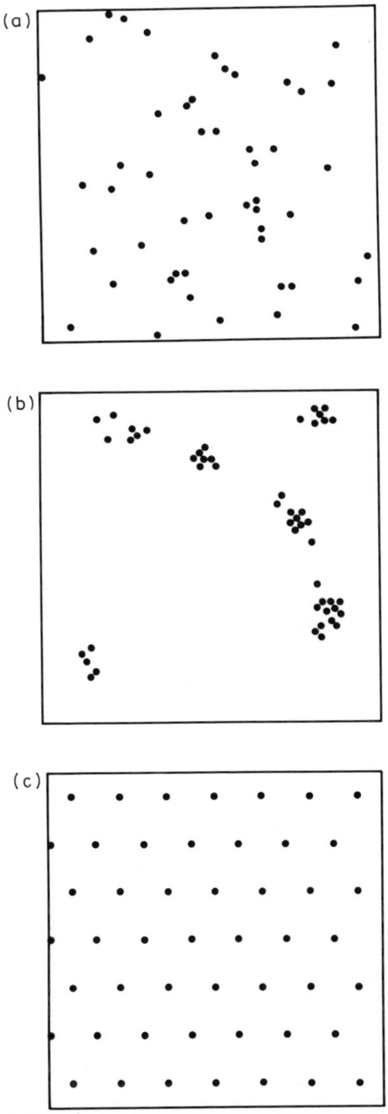

Figure 8.0.1 Classes of point patterns: (a) complete spatial randomness; (b) clustered; (c) regular.

patterns, are not unusual in point pattern analysis. When they occur, if the only evidence we have is the pattern itself, further analysis is necessary to determine the conditions responsible for the observed clustering. Instead of points attracting each other, in some instances, such as diffusion or competition, points may repel each other. Repulsion will likely produce regular patterns. The types of models that produce non-CSR outcomes are reviewed by Haggett *et al.* (1977, Chapter 13), Getis and Boots (1978), Cliff and Ord (1981), Ripley (1981), Diggle (1983) and Upton and Fingleton (1985).

In general, the hypothesis of CSR (or any other hypothesis concerning the spatial nature of the empirical point pattern) is tested by comparing measures of selected characteristics of the empirical point pattern with those of the hypothesized pattern. Two types of measure are possible. *Arrangement* measures characterize the pattern by describing the locations of points relative to other points, while *dispersion* measures take into account the locations in relation to the area included within B. More precisely, arrangement measures are concerned with those characteristics of the pattern which are invariant under the three rigid motions of Euclidean geometry (translation, rotation and reflection) and under changes of scale, while dispersion measures relate to pattern characteristics which may change under such conditions. Figure 8.0.2 shows four patterns with identical arrangement characteristics but differing dispersion characteristics.

The most frequently used approaches involving dispersion measures are quadrat analysis, nearest neighbour analysis and second-order analysis (see Boots and Getis, 1988, for details). *Quadrat analysis* involves defining a set of spatial sampling units (quadrats) of consistent size and shape which, depending on the specific approach adopted, may or may not overlap and which, collectively, may or may not completely cover B. A frequency array $f(x)$ is generated by counting the number of quadrats containing x = 0, 1,

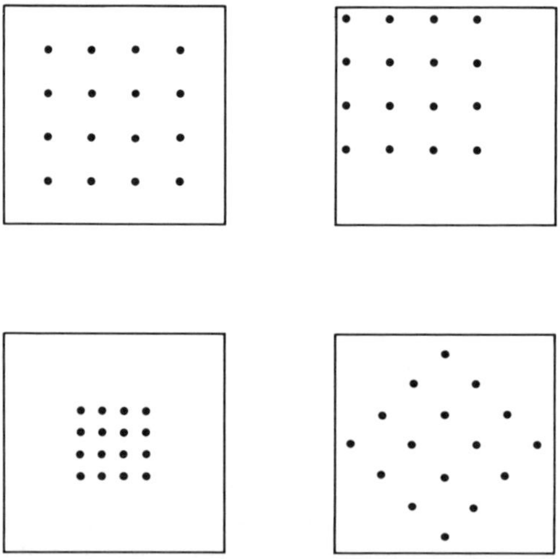

Figure 8.0.2 Four patterns with identical arrangement characteristics but differing dispersion characteristics.

2, ... points. $f(x)$ can then be compared with the corresponding distribution $p(x)$ expected under CSR with the same density as the empirical pattern, using a chi-squared goodness-of-fit test.

Nearest neighbour analysis involves measuring the distance between each point in S and the nearest other point. The nearest neighbour distances so obtained are compared with those expected under CSR with the same intensity as the empirical pattern using a variety of tests. *Second-order analysis* extends this approach by examining the distances between all pairs of points in S. These interpoint distances are compared with the CSR expectations, again taking into consideration the intensity of the points in the empirical pattern.

Some (e.g. Upton and Fingleton, 1985, p. 86) argue that measures such as these three fail to capture the true spatial nature of the pattern. Note also that they, like all measures of dispersion, require that the parameters of the theoretical model be estimated from the empirical pattern. In the case of CSR it is λ, the intensity of Θ_P, which is required. Such estimates can be complicated by the bounded nature of the empirical point pattern which conflicts with the unbounded nature of Θ_P. Such complications are compounded further if B is not regular in form.

In contrast, measures of arrangement such as reflexive nearest neighbour analysis (Boots and Getis, 1988, pp. 69–71) and some of the polygon-based techniques described below, are unaffected by these problems, although as we shall see they are prone to 'edge effects' of a different nature. Also the statistical theory underlying these measures is less well developed than that for measures of dispersion, so that a greater degree of subjectivity and conjecture enters into the interpretation of analyses involving them.

Typically, measures of arrangement emphasize the relationships between 'nearby' points of S. In the discussion of spatial interpolation procedures in Chapter 6 we saw how concepts of both Voronoi diagrams and Delaunay tessellations were used in the definition of 'nearby'. We exploit similar concepts here. Indeed, it is in the area of arrangement measures that these concepts have made their greatest contribution to point pattern analysis. In Section 8.1 we consider approaches involving Voronoi diagrams, while Section 8.2 considers approaches using Delaunay tessellations. In both sections our treatment illustrates the procedures for S in $\mathbb{R}^2$ although the techniques extend naturally to $\mathbb{R}^3$.

8.1 POLYGON-BASED METHODS

8.1.1 Direct approach

If we represent the empirical point pattern as a set $S = \{s_1, s_2, ..., s_n\}$ of n distinct points in a bounded region B of $\mathbb{R}^m$, we can define the bounded Voronoi diagram $\mathcal{V}(S)$ of S (see Section 2.1). For each of the polygons of

$\mathcal{V}(S)$ we can measure the values of a number of characteristics. For $\mathcal{V}(S)$ in $\mathbb{R}^2$ perhaps the most obvious characteristics in terms of dimensionality from zero to two are N, the number of sides (vertices) of a polygon; P, the perimeter; and A, the area. We can compare the values of these characteristics with those of the polygons of the Voronoi diagram defined for a hypothesized point process. Since edge effects influence the values associated with the polygons around the boundary of B, some allowance must be made for this. One approach (already considered in Section 5.3) is to exclude values for any polygon for which a circle, centred at any vertex v of the polygon and passing through the three points of S which are equidistant from v, intersects the boundary of B.

The polygonal approach was originally proposed by Evans (1967) but despite being advocated by a number of researchers since then (Haggett *et al.*, 1977, pp. 436–439; Matern, 1979; Cormack, 1979, pp. 171–175; Cliff and Ord, 1981, p. 110; Ripley, 1981, p. 149; Thomas, 1981), it has not been widely developed. One reason is that the properties of N, by far the most studied and most easily measured characteristic, do not always vary strongly amongst Voronoi diagrams defined for patterns generated by different processes. Recall that in Section 5.4 (equations (5.4.6)) we noted that the average value of N, $E(N)$, is always 6 for any normal, planar tessellation. Furthermore, Boots (1977) demonstrated that the frequency distribution of N for at least one set of clustered patterns is not significantly different from that of CSR. Vincent and Howarth (1982) reached a similar conclusion for a set of regular patterns.

When, as is usually the case initially, the theoretical pattern we are comparing with the empirical one is CSR, the values for the polygons are those reported for the Poisson Voronoi cells (PVCs) in Section 5.4. Recall that these polygons are the members of the tessellation created by defining the Voronoi diagram of a set of points located according to a homogeneous Poisson point process Θ_P. For PVC the distribution of P, the perimeter of a polygon, was estimated by Hinde and Miles (1980) to be approximately normal with mean $4\lambda^{-\frac{1}{2}}$ and variance $0.9472\lambda^{-1}$, where λ is the intensity of Θ_P (see Table 5.4.1). This led Upton and Fingleton (1985, p. 99) to propose the statistic U_k as a test of CSR where

$$U_k = \hat{\lambda} \sum_{i=1}^{k} (P_i - 4\hat{\lambda}^{-\frac{1}{2}})^2/0.9472 \qquad (8.1.1)$$

and P_1, P_2, ..., P_k, are the perimeters of a sample of k of the n polygons of $\mathcal{V}(S)$ and $\hat{\lambda} = n/|B|$ (where $|B|$ is the area of B) is the sample estimate of λ. Clearly, U_k is a measure of dispersion. When the k polygons are selected following certain conditions, the distribution of U_k for CSR will be approximately chi-squared with k degrees of freedom. These conditions involve avoiding both edge effects as discussed above and dependence between the k polygons. As noted by Boots and Murdoch (1983) (see equation 5.4.24) the lengths of the perimeters of neighbouring RVPs are not independent and so the k polygons must be selected so that no two are adjacent.

To illustrate this approach consider Figure 8.1.1 which shows the Voronoi diagram for a set of 65 Japanese black pine saplings (Upton and Fingleton, 1985, Figures 1.39a and 1.42a). In this example k=6 (the shaded polygons of Figure 8.1.1) and U_k is calculated to be 4.17. Since $\chi^2_{0.05,6} = 12.59 > 4.17$, the pattern of pines is not considered significantly different from CSR. No similar tests have been developed using the values of A. However, such values have been used in an alternative approach which borrows from information theory and which is discussed in the next section.

Because the dependence between the lengths of the perimeters of neighbouring PVCs also extends to other polygon characteristics such as the area and the number of sides (see equations (5.4.23) and (5.4.25), respectively), Hutchings and Discombe (1986) proposed an analytical approach using Monte Carlo simulation tests. This involves generating a number of patterns, typically 100, using a known spatial point process (e.g. the homogeneous planar Poisson point process which produces CSR) containing the same number of points as the empirical pattern being investigated. The Voronoi diagram is generated for each simulated pattern and the value of the desired characteristic of all polygons not subject to edge effects (as defined above) is measured. These values are then formed

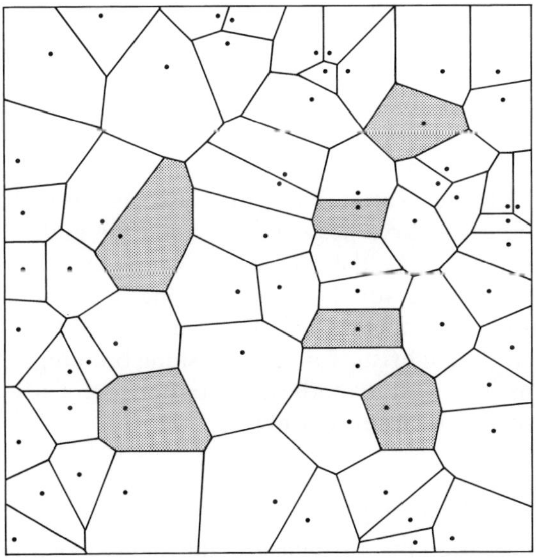

Figure 8.1.1 Voronoi diagram of Japanese black pine saplings (shaded polygons are those used in calculating U_k in equation (8.1.1)). (Redrawn from data in Upton and Fingleton, 1985, Figures 1.39a and 1.42a.)

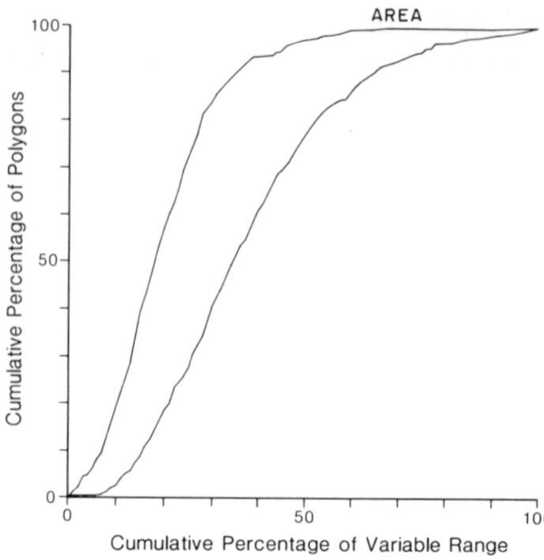

Figure 8.1.2 Standardized cumulative frequency histogram for the areas of Voronoi polygons of points in a CSR pattern. The curves are the limits to an envelope describing the standardized cumulative frequency histograms for a family of simulations of 500 points. (*Source:* Hutchings and Discombe (1986, Figure 2).)

into a cumulative percentage frequency histogram in which the class intervals are set equal to 1% of the range between the smallest and largest values in the set of measurements. This use of percentages avoids the problem of different numbers of polygons in each of the simulations. A curve is produced by plotting the histogram class interval against the cumulative percentage frequency. If the curves for all simulations are superimposed, an envelope of values is created which contains all the values from the simulated patterns (see Figure 8.1.2). A similar frequency curve can be generated for the values of the polygons of the Voronoi diagram associated with the empirical pattern. If this empirical curve lies entirely within the envelope of simulated values, the hypothesis that the process which generated the simulated patterns is also responsible for the empirical one is not rejected.

Using this approach Hutchings and Discombe (1986) suggested that the best single polygon characteristic for distinguishing both regular and clustered empirical patterns from an hypothesized pattern of CSR is area, while perimeter is sensitive to clustering only.

8.1.2 Information theory approaches

Information theory attempts to measure the degree of organization in a given system. It was Shannon (1948) who applied this idea to a set of probabilities by measuring organization using the entropy statistic H given by

$$H = \sum_{i=1}^{n} p_i \ln(1/p_i) \,, \tag{8.1.2}$$

where p_i is the probability of a random variable x taking the value x_i (i=1, ..., n) and

$$\sum_{i=1}^{n} p_i = 1 \quad (p_i \geq 0) \,. \tag{8.1.3}$$

Originally, in communications theory, logarithms to the base 2 were used, but in applications elsewhere these have been replaced by natural logarithms to the base e, as in equation (8.1.2).

If the system being investigated is an empirical point pattern consisting of a set $S = \{s_1, s_2, ..., s_n\}$ of n distinct points in a bounded region B of $\mathbb{R}^m$, one way the level of organization will be reflected is by the variation in the frequency with which points occur in subregions of B. To operationalize this concept in $\mathbb{R}^2$, Chapman (1970) defined the Voronoi diagram of S, $\mathcal{V}(S)$, and set $p_i = A_i/\Sigma_i A_i$ (i=1, ..., n), where A_i is the area of the polygon associated with s_i. Edge effects are avoided by disregarding any member of S which does not meet the criterion described at the beginning of Section 8.1.1.

In fact, H is more properly considered as a measure of disorder since its maximum value, $H_{\max}$, occurs when there is no other information other than n and the area of B, $|B|$. In this case the best estimate of A_i for any member of s_i of S would be $A_i = |B|/n$. This would mean that A_i would equal $1/n$ for each of the n points so that

$$H_{\max} = n\{(1/n)\ln[1/(1/n)]\} = \ln n \,. \tag{8.1.4}$$

However, there are problems in using H directly to evaluate empirical point patterns (Chapman, 1970, pp. 320–321; Lenz, 1979, pp. 376–377). In particular, although H decreases as the extent of clustering in the pattern increases, maximum regularity (members of S located on a triangular grid, see Figure 8.2.2 below) also yields $H_{\max}$. Also, as n gets very large the value of H for CSR also approaches $H_{\max}$, thus making the interpretation of H for an empirical point pattern almost impossible. In addition, since ln n increases as n increases, patterns generated by the same process but with different intensities will not produce the same value of H.

These problems with H can be overcome by using Thiel's redundancy measure R^* given by

$$R^* = \sum_{i=1}^{n} p_i \ln[p_i/(1/n)] = H_{\max} - H \,. \tag{8.1.5}$$

For a regular pattern, $R^* = 0$, while R^* increases with increasing clustering of points. In addition, R^* is less sensitive to variation in n.

If we wish to test an empirical point pattern against a hypothesis of CSR we need both the expected value, $E(R^*)$, and the variance, $Var(R^*)$, of R^* for CSR. Using the conjecture that the area A of a Poisson Voronoi cell (PVC) has a gamma distribution,

$$f(A,\alpha) = [\alpha(\alpha A)^{\alpha-1} e^{-\alpha A}]/\Gamma(\alpha) \tag{8.1.6}$$

for which $\alpha = 3.57$ if the variance of $A = 0.280\lambda^{-2}$, where λ is the intensity of Θ_P (Gilbert, 1962) (see Section 5.4 for a discussion of both the conjecture and the expected value), Lenz (1979) derived estimates of both $E(R^*)$ and $Var(R^*)$ for CSR. These values are reported in Table 8.1.1. Note that for $n \geq 15$ there is little change in $E(R^*)$ but that $Var(R^*)$ is an inverse function of n. The best fit to the data in Table 8.1.1 yields

$$Var(R^*) = 0.2157/^{1.9}\sqrt{n}\,, \tag{8.1.7}$$

which can be used to estimate $Var(R^*)$ for values of n not given in Table 8.1.1.

Lenz (1979) showed that when $n \geq 15$, R^* is approximately normally distributed for CSR so that we may use the variate of the standard normal curve,

$$z = [R^*_{obs} - E(R^*)]/Var(R^*)^{\frac{1}{2}}\,, \tag{8.1.8}$$

where R^*_{obs} is the value of R^* for the empirical point pattern, as a test statistic.

Since R^* is independent of n we can also use it to compare two empirical point patterns, for example of the same phenomenon at different times or different phenomena in the same region. An example is provided by Lenz who compared the pattern of personal robbery locations in the city of Milwaukee in 1975 with the pattern of victims' addresses (see Figure 8.1.3). Values of R^*_{obs} indicate that the former pattern shows more clustering than the latter, which led Lenz to conclude that victims contribute to the occurrence of personal robberies by exposing themselves to high crime risk districts of the city.

Table 8.1.1 Estimates of the expected value and variance of Thiel's redundancy measure R^* for CSR

n	$E(R^*)$	$Var(R^*)$
15	0.124	0.001875
30	0.129	0.001225
60	0.131	0.000660
90	0.132	0.000412
180	0.132	0.000204
360	0.132	0.000090
∞	0.134	0

Source: Lenz (1979).

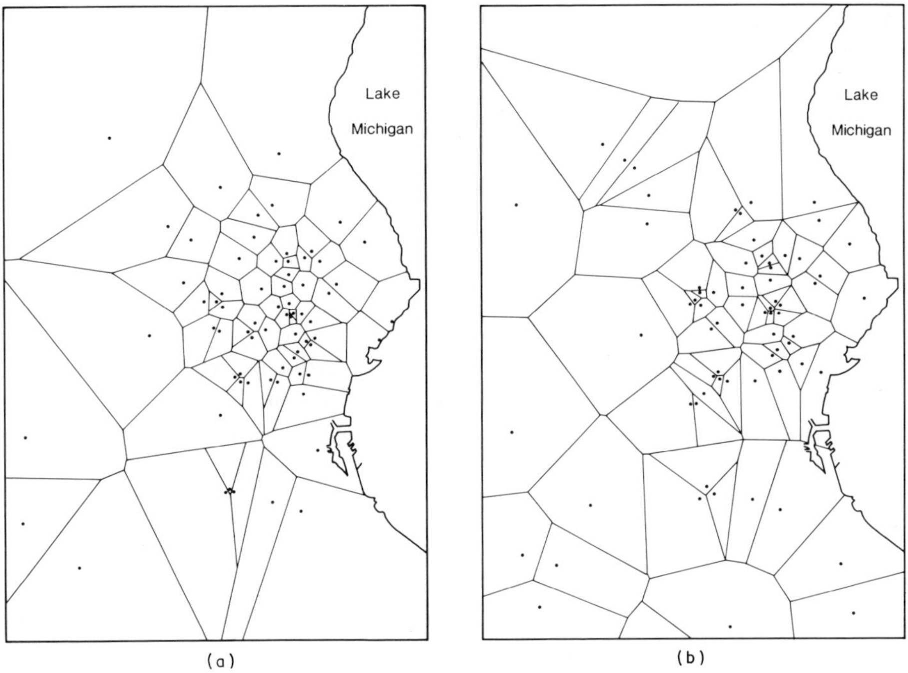

Figure 8.1.3 City of Milwaukee, 1975: (a) Voronoi diagram of locations of personal robberies; (b) Voronoi diagram of locations of victims' addresses. (*Source:* Lenz (1979, Figures 3 and 4).)

8.1.3 Estimating spatial intensity

In the introduction to this chapter a distinction was made between measures of dispersion and measures of arrangement for analysing empirical point patterns. Unlike the latter, the former measures take into account the location of the members of the point set $S = \{s_1, s_2, \ldots, s_n\}$ relative to a bounded study region B, and thus require estimates of the spatial intensity (or density) ρ of the hypothesized point process. For example, in the case of CSR we need to estimate λ, the intensity of the homogeneous planar Poisson Θ_P. However, there are other situations in empirical point pattern analysis where the focus of the investigation is on estimating intensity. Such instances include estimates of the intensity of a tree species in forestry where such estimates may be used in deriving expected timber yields or in metallurgy where intensities of impurities in a given material above a specified level may invalidate its use for a particular purpose.

The obvious way of obtaining an estimate $\hat{\rho}$ of ρ is to use

$$\hat{\rho} = n/|B| , \quad (8.1.9)$$

where $|B|$ is the area of B, or alternatively, to count the number n_b of the members of S in some bounded subregion b of B, of area $|b|$ and use

$$\hat{\rho} = n_b/|b| \,. \tag{8.1.10}$$

But, as pointed out above, intensity estimates are also required for dispersion measures and so the information collected in any of these approaches may also be used to provide estimates of ρ. Thus, ρ may be estimated using any of the techniques mentioned above in the introduction to this chapter such as quadrat, nearest neighbour and second-order methods (see Upton and Fingleton, 1985, Chapter 2, for a review of the use of such methods in estimating intensity). Voronoi polygons may also be used in this role.

If ρ is the intensity or number of points per unit of area, then the area per point is ρ^{-1}. Recognition of this provides an alternative way of estimating ρ. If we generate the Voronoi diagram $\mathcal{V}(S)$ of S we can measure the area A_i of the Voronoi polygon of s_i. If we select a sample of m polygons from $\mathcal{V}(S)$ which satisfy the avoidance of edge effects and independence conditions specified in Section 8.1.1, the mean area $\bar{A}$ of the sample polygons is given by

$$\bar{A} = \sum_{i=1}^{m} A_i/m \,. \tag{8.1.11}$$

Brown (1965) suggests that the intensity may be estimated from

$$\hat{\rho} = 1/\bar{A} \,. \tag{8.1.12}$$

However, this estimate is only asymptotically unbiased. To provide an unbiased estimate of ρ, Ord (1978) used the reciprocal of A_i so that the intensity estimate is given by

$$\hat{\rho} = \sum_{i=1}^{m} (A_i^{-1})/m \,. \tag{8.1.13}$$

If we wish to provide a range for the above estimates, for equation (8.1.12) we need the variance $\mathrm{Var}(A_i)$ of the sample areas given by

$$\mathrm{Var}(A_i) = \left[\sum_{i=1}^{m} (A_i^2) - \left(\sum_{i=1}^{m} A_i\right)^2 \Big/ m\right] \Big/ (m-1) \,, \tag{8.1.14}$$

and for equation (8.1.13) the variance $\mathrm{Var}(A_i^{-1})$ of the reciprocals given by

$$\mathrm{Var}(A_i^{-1}) = \left[\sum_{i=1}^{m} (A_i^{-2}) - \left(\sum_{i=1}^{m} A_i^{-1}\right)^2 \Big/ m\right] \Big/ (m-1) \,. \tag{8.1.15}$$

Thus, a two standard deviation range for $\bar{A}$ in equation (8.1.12) is given by $\{\bar{A} \pm [\mathrm{Var}(A_i)/m^{\frac{1}{2}}\}$, while that for the mean of the reciprocals is given by $\{(\Sigma_{i=1}^{m} A_i^{-1})/m \pm [\mathrm{Var}(A_i^{-1})/m]^{\frac{1}{2}}\}$. In general, the size of the range reflects the nature of the point pattern. If it is regular there is little variation in the areas of the Voronoi polygons and so the variances in equations (8.1.14) and (8.1.15) are small leading to a small range. Conversely, clustered patterns produce Voronoi polygons of greatly varying sizes, thus producing large variances in equations (8.1.14) and (8.1.15) and large ranges.

8.2 TRIANGLE-BASED METHODS

If an empirical point pattern consists of a set $S = \{s_1, s_2, \ldots, s_n\}$ of n distinct points in a bounded region B of $\mathbb{R}^m$, this approach involves defining the Delaunay tessellation of S, $\mathcal{D}(S)$, measuring properties of the polytopes of $\mathcal{D}(S)$ and comparing these with the properties of the polytopes of a Delaunay tessellation defined for a hypothesized point pattern. As with all areas of point pattern analysis the hypothesized pattern which has been emphasized is CSR. Such patterns correspond to those produced when S is generated according to a homogeneous Poisson point process (see Section 1.3.8). Known results relating to this model are reported in Section 5.9 and several of them are used here.

In $\mathbb{R}^2$ the probability density function (pdf) $f(\alpha)$ of α, a random angle of a randomly selected triangle of $\mathcal{D}(S)$ when S is CSR, was derived by Miles (1970a) (see equation (5.9.8)). Boots (1974, 1975b) and later Vincent *et al.* (1976, 1977, 1983) argued that a test for CSR can be obtained by integrating equation (5.9.8) for chosen class intervals to obtain a set of expected frequencies which can be compared with the empirical frequencies using a Pearson goodness-of-fit test. The probability $F(\alpha \leq x)$ that $\alpha \leq x$, derived by integrating equation (5.9.8) is

$$F(\alpha \leq x) = \int_0^x f(\alpha)\,\mathrm{d}\alpha \tag{8.2.1}$$

$$= (1/3)[2\sin^2\alpha + \{\alpha \cos 2\alpha - (3\sin 2\alpha)/2 + 2\alpha\}\,\pi^{-1}]_0^x\,.$$

Both Boots (1975b) and Vincent *et al.* (1977) used this procedure to analyse patterns of urban settlements in various parts of the United States and demonstrated that, in this context, the procedure is superior to the nearest neighbour approach (see Section 8.1) in correctly rejecting a hypothesis of CSR.

However, since this approach requires that we take a random sample of triangles from $\mathcal{D}(S)$, Mardia *et al.* (1977) have argued that it is wasteful of the available information. They proposed instead using the marginal density of the minimum angle, $\alpha_{\min}$, of all triangles in $\mathcal{D}(s)$. The pdf of $\alpha_{\min}$ for CSR is given in equation (5.9.9). Mardia *et al.* assume that the values of $\alpha_{\min}$ are independent for the triangles of $\mathcal{D}(S)$. Integrating equation (5.9.9) we get

$$F(\alpha_{\min} \leq x)$$

$$= [1 + (\tfrac{1}{2}\pi)\{(6\alpha_{\min} - 2\pi)\cos 2\alpha_{\min} - \sin 2\alpha_{\min} - \sin 4\alpha_{\min}\}]_0^x \; (0 < x \leq \pi/3) \tag{8.2.2}$$

as the probability that $\alpha_{min} \leq x$.

Boots (1986b) has proposed that a similar argument may also be made in favour of the use of the maximum angle, α_{max}, of the triangles of $\mathcal{D}(S)$. When S is CSR the marginal density of α_{max} is given by equation (5.9.10) which yields

$$F(\alpha_{max} \leq x) = (\tfrac{1}{2}\pi)\,[\sin 4x + \sin 2x + 2\pi\cos 2x - 6x\cos 2x]$$
$$(\pi/3 \leq x \leq \pi/2)$$

and

$$F(\alpha_{max} \leq x) = (\tfrac{1}{2}\pi)\,[2x\cos 2x - 2\pi\cos 2x - 3\sin 2x + 4x - 2\pi] \quad (\pi/2 \leq x \leq \pi) \qquad (8.2.3)$$

as the probability that $\alpha_{max} \leq x$.

Preliminary investigations by Boots (1986b) using both empirical and simulated patterns containing different degrees of regularity or clustering suggested that α_{min} has the best overall ability to reject a null hypothesis of CSR against both regular and clustered alternatives, although there is some indication that α_{max} is able to detect instances of clustering not discernible by α_{min}. In view of these findings we illustrate this procedure using α_{min} for a pattern of settlements in an area of southeastern Montana, USA, in 1973 (see Figure 8.2.1).

For each triangle in $\mathcal{D}(S)$ in Figure 8.2.1 we identify and measure α_{min}. These values are used to determine the number of values $G(\alpha_{min})$ of α_{min} in a specified interval. In calculating $G(\alpha_{min})$ we must account for edge effects produced by the bounded nature of $\mathcal{D}(S)$. Mardia (1989) (see also the discussion in Kendall, 1989, p. 110) proposed doing this by discounting any triangles for which the circumcentre does not lie in B (recall from Section 2.1 that the circumcentres are vertices of the Voronoi diagram of S). This procedure was followed in this example and such discounted triangles are shaded in Figure 8.2.1. The values of $G(\alpha_{min})$ so obtained are compared with the corresponding values $F(\alpha_{min})$ for CSR obtained from equation (8.2.2) using a chi-squared goodness-of-fit test (see Table 8.2.1). Following usual statistical convention, categories for which the expected values are less than five are combined with adjacent categories. Since the value of $\chi^2 = 11.198$, which results has a probability of less than 0.05 of occurring for a χ^2_4 distribution, we reject the hypothesis of CSR for this settlement pattern.

However, can we proceed further beyond this result? Currently, the answer is a tentative yes. If the points of S are arranged in a perfectly regular way on a triangular grid such as that shown in Figure 8.2.2, all the Voronoi polygons (indicated by dashed lines) will be regular hexagons and all the Delaunay triangles in $\mathcal{D}(S)$ (indcated by solid lines) will be equilateral so that all angles would be equal to $\pi/3$. Thus, if for an empirical point pattern there is an excess of values (relative to the number expected for

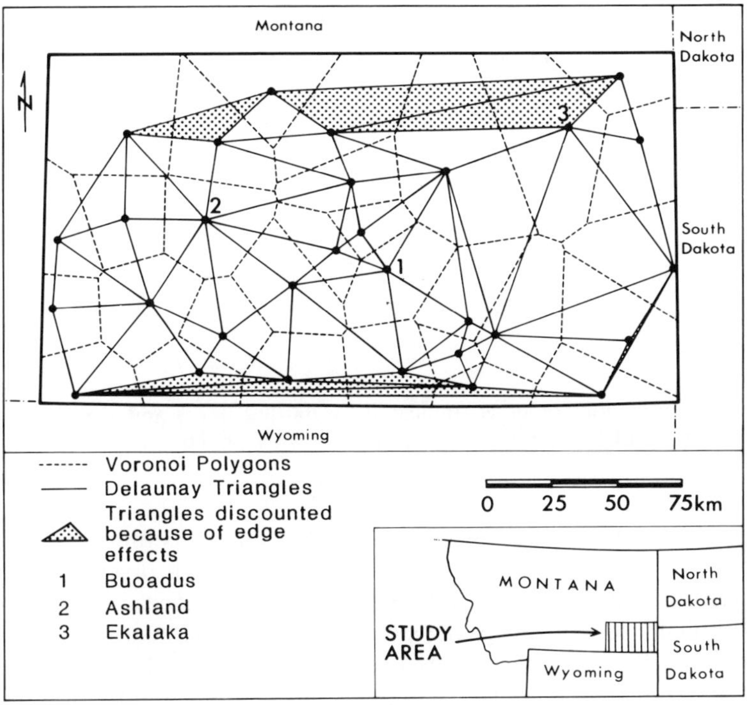

Figure 8.2.1 Voronoi diagram and Delaunay triangulation for pattern of settlements in an area of Montana, USA, in 1973. Re-drawn from data in Boots and Getis (1988, Figure 4.3).

Table 8.2.1 Analysis of locations of settlements in southeastern Montana using the minimum angle technique

Minimum angle, $\alpha_{\min}$ (degrees)	Observed frequency, $G(\alpha_{\min})$		Expected frequency, $F(\alpha_{\min})$	
0–10	1	2	2.468	9.441
11–20	1		6.973	
21–30	15		10.007	
31–40	8		10.521	
41–50	15	16	8.023	11.031
51–60	1		3.008	

$X_4^2 = 11.198$.

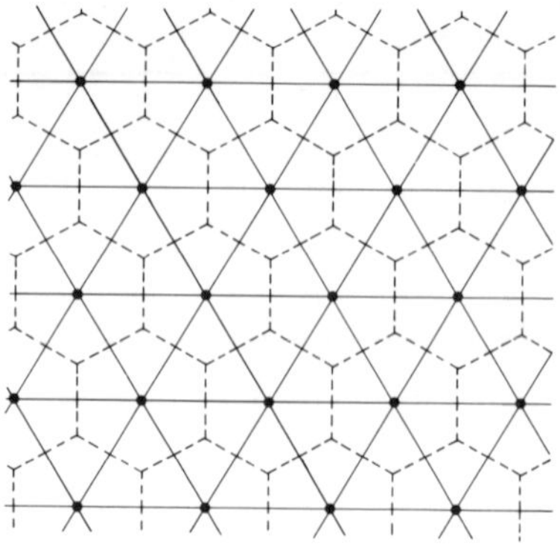

Figure 8.2.2 Voronoi polygons (indicated by dashed lines) and Delaunay triangles (indicated by solid lines) for points located on a triangular grid.

CSR) at the upper tail of $G(\alpha_{\min})$, a regular pattern is indicated. However, in Section 5.9 we noted that there will also be a high proportion of nearly equilateral triangles in $\mathcal{D}(S)$ under CSR so that the excess referred to might prove difficult to detect. If the points are located on a square grid (see Figure 8.2.3) the resulting Voronoi polygons (indicated by dashed lines) will be squares and the Delaunay triangles of $\mathcal{D}(S)$ (indicated by the solid lines) will be right-angled so that $\alpha_{\min} = \pi/4$. A significant excess of such angles would indicate a point pattern with affinities to a square grid. In contrast, a pattern in which points are clustered in distinct groups will have some edges of $\mathcal{D}(S)$ which correspond to links between points on the peripheries of different clusters (see Figure 8.2.4). Many of the triangles formed by such edges will be elongated so that their minimum angles will be small. Thus, a distribution of empirical values of $\alpha_{\min}$ with a significant excess of small angles is suggestive of a pattern of clustered points.

On the basis of these speculations and noting that the greatest differences between the frequencies for the observed angles and the CSR expectations occur in the ranges of 0–20° (where there is a deficit of minimum angles relative to CSR) and 41–50° (where there is an excess) (see Table 8.2), we may conclude that the pattern of settlements in Figure 8.2.1 shows a significant degree of regularity. Furthermore, it appears that this regularity is more consistent with the expectations of a regular square grid than with a triangular one.

Recently, rather than considering individual angles of triangles in $\mathcal{D}(S)$, Mardia (1989) has proposed a test which involves comparing the positions of the triangles of $\mathcal{D}(S)$ in Kendall's shape space (see Section 5.9 and Figure 5.9.2) with the positions of those expected under CSR.

In addition to analysing the angles of the triangles of $\mathcal{D}(S)$, Vincent *et al.* (1976, 1977, 1983) also proposed using two other characteristics of $\mathcal{D}(S)$.

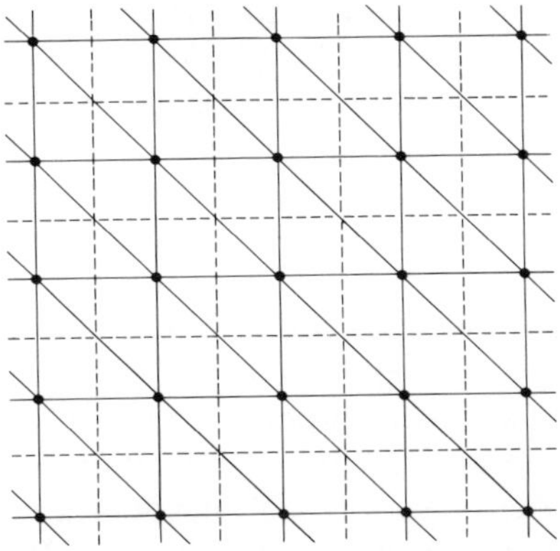

Figure 8.2.3 Voronoi polygons (indicated by dashed lines) and Delaunay triangles (indicated by solid lines) for points located on a square grid.

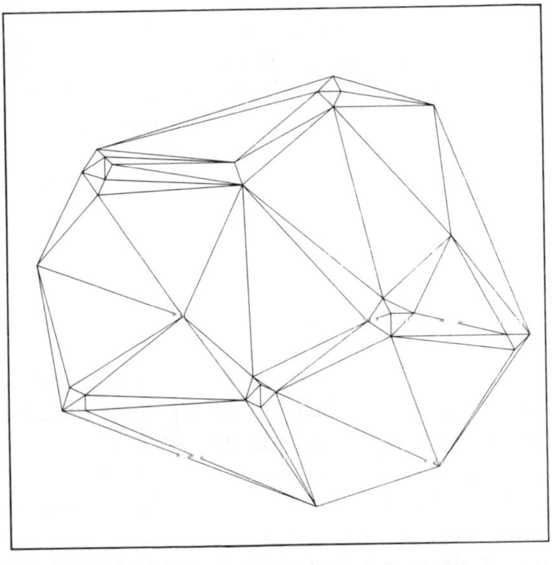

Figure 8.2.4 Delaunay triangulation for a clustered pattern of points.

These are N, the number of edges incident at a vertex in $\mathscr{D}(S)$, and L, the length of an edge of $\mathscr{D}(S)$. Because of the duality between $\mathscr{D}(S)$ and the ordinary Voronoi diagram of S, the former characteristic is equal to the number of edges (vertices) of the polygon associated with the vertex in $\mathscr{D}(S)$. As indicated in Section 8.1.1, there are problems involved in using this characteristic and it is not discussed further here. For CSR the pdf of L, $f(L)$, is given by equation (5.9.11). Unlike tests using α, α_{min}, α_{max} and N, any test using L is a measure of dispersion. The usefulness of this characteristic is essentially untested at present since there is little information on the behaviour of L for non-CSR situations. Using the Monte Carlo simulation test procedure described in Section 8.1.1, Hutchings and Discombe (1986) suggested that L is effective in distinguishing both regular and clustered patterns from those of CSR. Vincent and Howarth (1982), also using a Monte Carlo simulation approach, estimated $f(L)$ for $\mathcal{D}(S)$ associated with S resulting from a hard core process (see Section 5.10). In metallurgy there has also been some preliminary investigation of the behaviour of $\bar{L}_i$, the average length of the edges of $\mathcal{D}(S)$ incident at s_i (referred to in this context as near neighbour distances) for both CSR and some empirical situations (Wray *et al.*, 1983; Spitzig *et al.*, 1985; Burger *et al.*, 1989).

Rather than considering individual characteristics of the triangles of $\mathcal{D}(S)$ independently, both Mardia *et al.* (1977) and Kendall (1989) proposed pursuing them jointly. For CSR a lemma due to Miles (1970a) (see equation (5.9.2)) demonstrates that the size (measured by the circumradius, r) and the shape (measured by the joint orientation of the edges) of an individual triangle are independent with the distribution of both characteristics being known. Using this information Kendall (1989) suggested the following test. First, sort the triangles of $\mathcal{D}(S)$ in terms of their shapes into three groups: those that are 'nearly equilateral', those that are highly 'splinter shaped', and the remainder. Then for the first and last of these groups, examine the departures from independence and from the known distribution of r. By discounting the second group the procedure accounts for edge effects.

8.3 NEAREST NEIGHBOUR DISTANCE METHODS

In this section we show the nearest neighbour distance method, abbreviated to the NND method, as an application of Voronoi diagrams. The NND method is a statistical method which examines the effect of physical objects located in a region upon the distribution of point-like elements over the region. The effect of nuclear power stations on the distribution of abnormal plants, the locational dependency of fast food stores on arterial streets, the spatial association of firms producing integrated circuits around airports, etc. can be investigated with the NND method. We should note that the NND method to be discussed here is different from the ordinary nearest neighbour distance method referred to in biology, geography and ecology (Bartlett, 1975; Pielou, 1977; Getis and Boots, 1978). The ordinary nearest neighbour method also deals with the distribution of point-like elements but it does not directly deal with physical objects affecting the distribution of the point-like elements. In this section we focus on the NND method because this method is an application of the Voronoi diagram which is the major concern of this text. The reader who wishes to study the ordinary nearest neighbour distance method in depth should consult Diggle (1983) and Upton and Fingleton (1985).

To fix a general setting for the analysis, let us consider a region in which m physical objects (such as power plants, arterial streets and airports referred to in the above example) are located in a region, and n point-like elements (such as plants, stores and firms) are distributed over the region excluding the area occupied by the objects. A set of the m objects in the region is represented by a set of points, lines or areas in a bounded region S in $\mathbb{R}^2$, which is denoted by $O = \{O_1, \ldots, O_m\}$ $(1 \leq m < \infty)$, where $O_i \cap O_j = \emptyset$, $i \neq j$, $i, j \in I_n$. A set of the n point-like elements is represented by a set of points distributed over $S_o = S \backslash \cup_{i=1}^{n} O_i$, which is denoted by $P = \{p_1, \ldots, p_n\}$ $(1 \leq n < \infty)$ with the set of their location vectors $\{\mathbf{x}_1, \ldots, \mathbf{x}_n\}$.

Given O and P, we define the distance from a point p_i in $\mathbb{R}^2$ to an object O_j by

$$d(p_i, O_j) = \min_{u_i}\{\|\mathbf{x}_i - \mathbf{u}_j\| \,|\mathbf{x}_j \in O_j\} , \tag{8.3.1}$$

where $\|\mathbf{x}_i - \mathbf{u}_j\|$ is the Euclidean distance between $\mathbf{x}_i \in S_o$ and $\mathbf{u}_j \in O_j$. With $d(p_i, O_j)$, we next define the distance from a point p_i to the nearest object in O by

$$d(p_i, O) = \min_{j}\{d(p_i, O_j), j \in I_n\} . \tag{8.3.2}$$

We call this distance the *nearest neighbour distance*, or briefly the *NN-distance*, from p_i to O. Third, we define the NN-distance averaged across n points, i.e.

$$\bar{t} - \frac{1}{n}\sum_{i=1}^{n} d(p_i, O) . \tag{8.3.3}$$

We call $\bar{t}$ the *average nearest neighbour distance*, or the *average NN-distance* for short.

In the NND method, several indices are proposed to measure or to test the effect of objects O on the distribution of points P. To formulate one of the simplest indices, let us consider the hypothesis that points P are independently and randomly distributed according to the uniform distribution over S_o. This hypothesis implies that objects O do not affect the distribution of points P. Under this hypothesis, the NN-distance $t_i = d(p_i, O)$ is a random variable; consequently $\bar{t}$ of equation (8.3.3) is also a random variable. Let $\mathrm{E}(\bar{t})$ be its expected value and $\bar{\mathrm{t}}$ be an observed value of $\bar{t}$. If the hypothesis is true, we may expect that the observed value $\bar{\mathrm{t}}$ is close to $\mathrm{E}(\bar{t})$. Using this property, we define an index by

$$R = \frac{\bar{t}}{\mathrm{E}(\bar{t})} \tag{8.3.4}$$

(cf. Pielou's, 1977, p. 155, index). If the hypothesis is true, we may expect that R is close to unity. If we observe $R < 1$, we may infer that points P are more closely distributed around objects O than they would be in the random distribution. If we observe $R > 1$, we may infer that points P are more sparsely distributed around objects O than they would be in the random distribution.

The index R can be also used for a statistical test. Since $\bar{t}$ is the average of independent random variables having the same distribution with a finite mean, the central limit theorem guarantees that the distribution of $\bar{t}$ asymptotically approaches the normal distribution with mean $\mathrm{E}(\bar{t}) = \mathrm{E}(t)$ and variance $\mathrm{Var}(\bar{t}) = \mathrm{Var}(t)/n$, where $t = d(p, O)$ is the NN-distance from a random point p to O. It follows from this property that the random variable $R = \bar{t}/\mathrm{E}(\bar{t})$ asymptotically follows the normal distribution with $\mathrm{E}(R) = 1$ and $\mathrm{Var}(R) = \mathrm{Var}(t)/(n\ \mathrm{E}(t)^2)$ as n increases. With this test statistic, we can test the null hypothesis that points P are independently and randomly distributed according to the uniform distribution over S_o, that is, the distribution of points P is not affected by the location of objects O (regarding the method of this statistical test, see Section 1.3.7).

We should make one remark on the boundary effect often discussed in the related literature. The nature of the boundary effect of the NND method is different from that of the ordinary nearest neighbour method. Since the NND method is based upon a uniform distribution over a bounded region and a random NN-distance t_i is the distance from a random point to a fixed object (not a random object), in theory the NND method is not bothered by the boundary effect. In practice, the boundary effect may appear if we apply the NND method to a subregion of a region in which points P are distributed, because we usually neglect objects outside the subregion that affect the distribution of P. In this case we can overcome the boundary effect by considering the objects outside the subregion whose Voronoi regions (generated by O) overlap the subregion.

When we use the index R in practice, we have to compute $d(p_i, O)$, $\mathrm{E}(\bar{t}) = \mathrm{E}(t)$ and $\mathrm{Var}(\bar{t}) = \mathrm{Var}(t)/n$. The latter two values are obtained from the probability distribution function, $F(t)$, of t. Since a point is placed according to the two-dimensional uniform distribution over S_o, the probability of the point being placed in a subregion in S_o is given by the ratio of the area of the subregion to that of the whole region S_o (recall Section 1.3.8). As a subregion, we consider the region $S(t)$ in which the NN-distance from any point p in $S(t)$ to O is less than or equal to t, i.e.

$$S(t) = \{p |\, d(p, O) \leq t, p \in S_o\}\,. \tag{8.3.5}$$

We call this region the *buffer zone* of objects O with distance t. Since the value of $F(t)$ is the probability that the NN-distance from a random point to O is less than or equal to t, $F(t)$ is given by

$$F(t) = \frac{|S(t)|}{|S_o|}, \tag{8.3.6}$$

where $|S(t)|$ indicates the area of $S(t)$. At first glance, the exact form of the function $|S(t)|$ or $F(t)$ is difficult to obtain. Fortunately, however, if we use a Voronoi diagram, we can exactly obtain the function $|S(t)|$. Since the computational method differs according to the shape of objects O, we deal with the NND methods for point-like objects, line-like objects, and area-like objects, separately.

8.3.1 Nearest neighbour distance method for point-like objects

When the area occupied by an object O_i is very small relative to the whole region S_o, we may regard O_i as a point, and O as a set of points. We call the NND method for this set the *NND method for point-like objects*. Using this method, we may examine the effect of point-like objects O on the distribution of point-like elements P. For example, we may examine the effect of firms (O) polluting the air with smoke on the distribution of patients (P) suffering from asthma, and the association of news stands (P) around stations (O). An actual application was carried out by Okabe and Miki (1984) who examined the effect of railway stations (O) on the distribution of several kinds of retail stores (P) in Toshima, Tokyo. Figure 8.3.1 shows the distribution of book stores (filled circles) with respect to railway stations (empty circles).

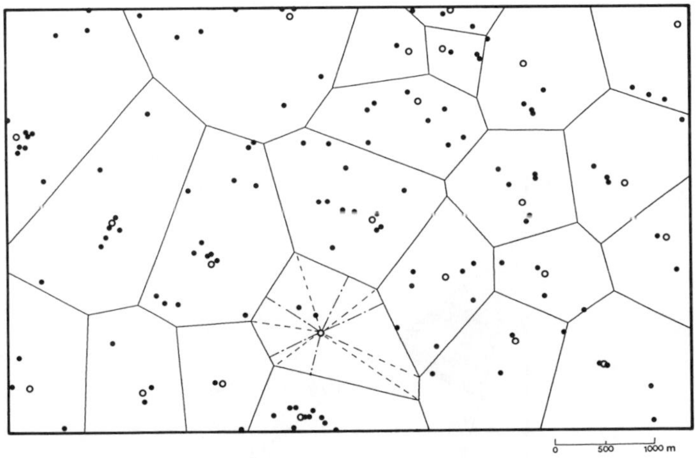

Figure 8.3.1 The location of the railway stations (the unfilled circles), the distribution of book stores (the filled circles) in Toshima and the Voronoi diagram generated by the railway stations. (*Source*: Miki (1983, Figure 3.7).)

As we mentioned above, when we use the index R in practice, we have to calculate the value of $d(p_i, O)$ and the moments of $t = d(p, O)$, i.e. $\mathrm{E}(t)$ and $\mathrm{Var}(t) = \mathrm{E}(t^2) - \mathrm{E}(t)^2$. Recalling Problem P3 in Section 2.3, we notice that $d(p_i, O)$ can be efficiently computed using the ordinary Voronoi diagram. First, we generate the Voronoi diagram $\mathcal{V} = \{V(O_1), \ldots, V(O_n)\}$ with point-like objects O (indicated by the continuous lines in Figure 8.3.1). Second, we find in which Voronoi polygon point p_i is located. This search problem is known as the *point-location problem*, and several efficient computational methods are proposed (see Chapter 4). Once $p_i \in V(O_j)$ is known, $d(p_i, O)$ is readily given by $d(p_i, O_j)$. Substituting this value into equation (8.3.3), we obtain the value of $\bar{t}$ in equation (8.3.4).

The moments of the NN-distance to point-like objects (the moments of the distance from a random point to the nearest generator point) We next show the procedure of calculating the moments of t. First, we generate the Voronoi diagram $\mathcal{V}$ with point-like objects O. We next triangulate each Voronoi polygon by the line segments connecting generator O_i and Voronoi vertices (indicated by the broken lines in Figure 8.3.1). We further partition the resulting triangle by the line passing through O_i perpendicular to the facing edge if the line crosses the edge (indicated by the dash-dotted lines in Figure 8.3.1). As a result, the whole region S is partitioned into a set of obtuse or right angle triangles $S = \{S_1, \ldots, S_k\}$ where $S = S_1 \cup \ldots \cup S_k$.

Figure 8.3.2 depicts a triangle S_i in S, where the vertex O_i is placed on the v-axis and the other two vertices are placed on the u-axis. Let θ_{i1}, θ_{i2}, u_{i1}, u_{i2}, v_{i0}, t_{i1}, t_{i2} be those indicated in Figure 8.3.2, respectively, and $S_i(t)$

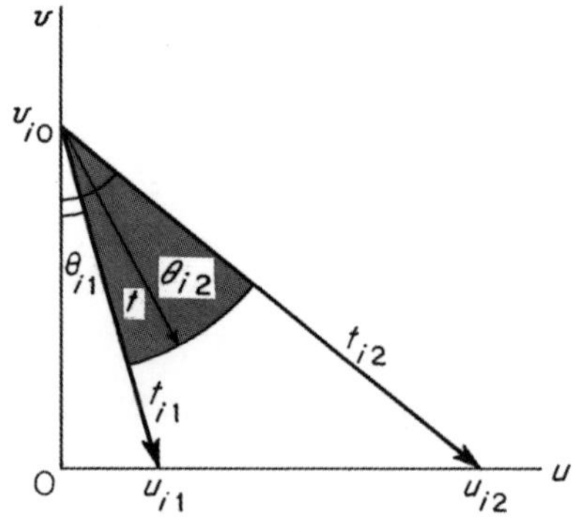

Figure 8.3.2 A triangle in the triangulated Voronoi diagram shown in Figure 8.3.1.

$= S(t) \cap S_i$ (the shaded area in Figure 8.3.2). Then the probability distribution function of equation (8.3.6) is written as

$$F(t) = \frac{1}{|S_0|} \sum_{i=1}^{k} |S_i(t)| . \qquad (8.3.7)$$

With the help of Figure 8.3.2, we obtain $|S_i(t)|$ as

$$|S_i(t)| = \begin{cases} \frac{1}{2} t^2(\theta_{i2} - \theta_{i1}) & \text{for } 0 \le t \le t_{i1} , \\ \frac{1}{2} v_{io}(\sqrt{t^2 - v_{io}^2} - u_{i1}) + \frac{1}{2} t^2 (\theta_{i2} - \arccos \frac{v_{io}}{t}) & \text{for } t_{i1} \le t \le t_{i2} , \\ \frac{1}{2} v_{io} (u_{i2} - u_{i1}) & \text{for } t_{i2} \le t . \end{cases} \qquad (8.3.8)$$

Thus we can explicitly obtain the probability distribution function $F(t)$ by substituting equation (8.3.8) into equation (8.3.7).

The probability density function $f(t)$ of $F(t)$ is obtained from

$$f(t) = \frac{\mathrm{d}F(t)}{\mathrm{d}t} = \frac{1}{|S_o|} \frac{\mathrm{d}|S(t)|}{\mathrm{d}t} = \frac{1}{|S_o|} \sum_{i=1}^{k} \frac{\mathrm{d}|S_i(t)|}{\mathrm{d}t} = \frac{1}{|S_o|} \sum_{i=1}^{k} s_i(t) . \qquad (8.3.9)$$

The rth moment of t is given by

$$\mathrm{E}(t^r) = \int_0^\infty t^r f(t)\mathrm{d}t = \frac{1}{|S_o|} \sum_{i=1}^{k} \int_0^\infty t^r s_i(t) \, \mathrm{d}t . \qquad (8.3.10)$$

After several steps of calculation, we obtain the second integral in equation (8.3.10) as:

if r is even ($r = 2q - 2$, $q \ge 1$),

$$\int_0^\infty t^r s_i(t) \, \mathrm{d}t$$

$$= \frac{(2q-2)!!}{2q(2q-1)!!} \sum_{j=1}^{q-1} \frac{(2j-1)!!}{(2j)!!} [u_{i2} (u_{i2}^2 + v_{io}^2)^j - u_{i1} (u_{i2}^2 + v_{io}^2)^j] \, v_{io}^{2q-2j-1} , \qquad (8.3.11)$$

if r is odd ($r = 2q - 1$, $q \ge 1$),

$$\int_0^\infty t^r s_i(t) \, \mathrm{d}t$$

$$= \frac{(2q-2)!!}{2q(2q-1)!!} \left\{ \sum_{j=1}^{q-1} \frac{(2j-1)!!}{(2j)!!} [u_{i2} (u_{i2}^2 + v_{io}^2)^j - u_{i1} (u_{i2}^2 + v_{io}^2)^j] \, v_{io}^{2q-2j-1} \right.$$

$$\left. + \, v_{io}^{2q+1} \ln \left(\frac{\sqrt{v_{io}^2 + u_{i2}^2} + u_{i2}}{\sqrt{v_{io}^2 + u_{i1}^2} + u_{i1}} \right) \right\} \qquad (8.3.12)$$

(Okabe, 1987; Yoshikawa, 1989). Substituting this equation into equation (8.3.10), we can explicitly obtain the expected value E(t) and the variance Var(t) = E(t^2) − E(t)^2. Okabe and Miki (1984) calculated the value of R for book stores in Figure 8.3.1 with these equations, and obtained R = 0.727. Since this value was smaller than unity with significance level 0.05, they rejected the null hypothesis that book stores were independently and uniformly distributed over Toshima. The book stores tended to gather around railway stations.

Besides the NND method, the moments of the NN-distance is also used in the quantizing problem encountered in applications to analog-to-digital conversion (Conway and Sloane, 1982, 1984, 1988, Chapter 22). This problem asks how to place points in space so that the average second moment of their Voronoi polygons is as small as possible. This problem is discussed with the second moments of the NN-distance of m-dimensional Voronoi diagrams generated by several types of lattice points. These moments are shown in Conway and Sloane (1982, 1984, 1988, Chapter 22) (see also the related discussion in Chapter 7).

8.3.2 Nearest neighbour distance method for line-like objects

When the area occupied by an object O_i is very narrow and long relative to the whole region S_o, we may regard the object as a line segment, and O as a set of line segments. We call the NND method for this set the *NND method for line-like objects*.

Applications of the NND method for line-like objects are shown by Okabe and Fujii (1984) and Okabe and Yoshikawa (1989). Figure 8.3.3 illustrates the distribution of high-class apartment buildings (the dots) and arterial streets (the continuous lines) in Kohto-Sumida, Tokyo. Okabe and Yoshikawa (1989) asked whether or not the distribution of high-class apartment buildings was affected by arterial streets. This question can be examined by the NND method for line-like objects.

To carry out this statistical examination, we have to calculate $d(p_i, O)$ and the moments of $t = d(p, O)$. First, to gain computational tractability, we assume that a set of line-like objects O is represented by a set of chains of straight line segments. A curved line is approximated by the chain of connected small straight line segments. If this modification is made, the procedure is essentially the same as that of the NND method for point-like objects. We construct the Voronoi diagram $\mathcal{V} = \{V(O_1), \ldots, V(O_n)\}$ generated by the decomposed set of the line generator set O (see Section 3.5). We next find in which Voronoi region a point p_i is placed. Once $p_i \in V(O_j)$ is known, the NN-distance $d(p_i, O)$ is given by $d(p_i, O_j)$.

The moments of the NN-distance to line-like objects (the moments of the distance from a random point to the nearest generator line) We next show

Figure 8.3.3 The distribution of high-class apartment buildings (dots) and the arterial streets (continuous lines) in Kohto-Sumida, Tokyo. (*Source*: Okabe and Yoshikawa (1989, Figure 9).)

the procedure of calculating the moments of t. This procedure is similar to that in Section 8.3.1. We first generate the Voronoi diagram $\mathcal{V}$. Second, if a generator is a point O_i, we join the vertices of $V(O_i)$ and O_i by line segments (the broken lines in Figure 8.3.4). Third, if a generator is a line segment O_i, we partition $V(O_i)$ by line passing through a vertex of $V(O_i)$ perpendicular to the generator line O_i if the line crosses the generator line O_i other than its end points (the dash-dotted lines). As a result, S is partitioned into subregions and these subregions are classified into four types, three of which are shown in Figure 8.3.5 and the other one is shown

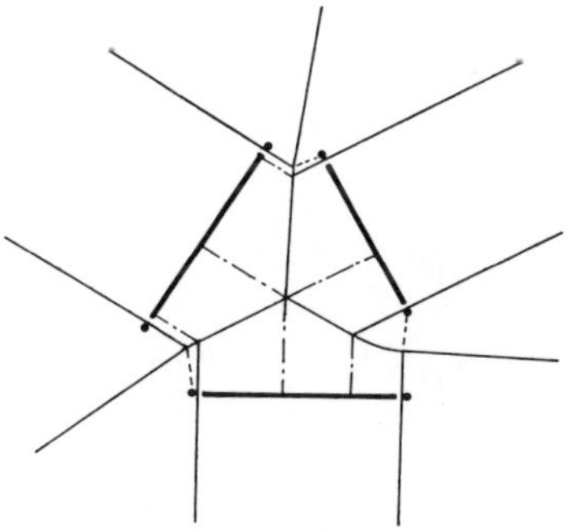

Figure 8.3.4 The partition of a Voronoi diagram generated by a set of lines for calculating the moments of t.

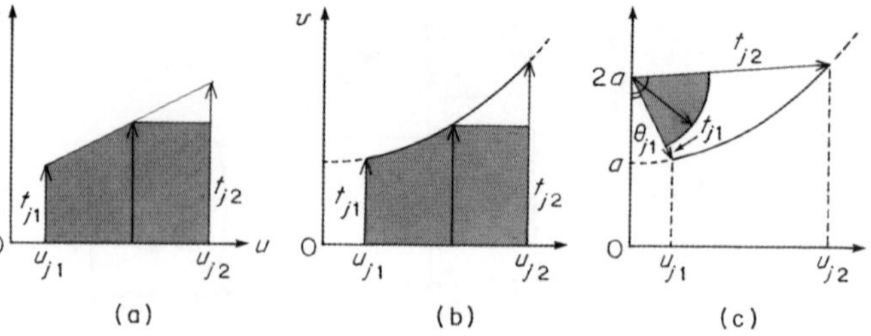

Figure 8.3.5 Four types of subregion constituting a Voronoi diagram generated by a set of line segments (three types are shown here; the fourth is shown in Figure 8.3.2).

in Figure 8.3.2. Let S_{ij} be the jth subregion of type i, $j = 1, \ldots, k_i$, $i = 1, \ldots, 4$, and $S_{ij}(t) = S_{ij} \cap S(t)$, i.e. the buffer zone in S_{ij}. Then the probability distribution function $F(t)$ of equation (8.3.6) is written as

$$F(t) = \frac{|S(t)|}{|S_o|} = \frac{1}{|S_o|} \sum_{i=1}^{4} \sum_{j=1}^{k_i} |S_{ij}(t)| . \qquad (8.3.13)$$

To give the explicit form of $S_{ij}(t)$, let u_{j1}, u_{j2}, t_{j1}, t_{j2}, a, θ_{j1} and θ_{j2} be those indicated in Figure 8.3.5, respectively. Then, $S_{ij}(t)$ is given by

$$|S_{1j}(t)| = \begin{cases} (u_{j2} - u_{j1})t & \text{for } 0 \le t \le t_{j1}, \\ -\dfrac{(u_{j2} - u_{j1})(t - t_{j1})^2}{2(t_{j2} - t_{j1})} + (u_{j2} - u_{j1})t & \text{for } t_{j1} \le t \le t_{j2}, \\ \tfrac{1}{2}(u_{j2} + u_{j1})(t_{j2} - t_{j1}) & \text{for } t_{j2} \le t; \end{cases}$$

$$(8.3.14)$$

$$|S_{2j}(t)| = \begin{cases} (u_{j2} - u_{j1})t & \text{for } 0 \le t \le t_{j1}, \\ -\tfrac{4}{3}\sqrt{a}(t-a)^{3/2} - \tfrac{1}{12}u_{j1}^3 - au_{j1} + u_{j2}t & \text{for } t_{j1} \le t \le t_{j2}, \\ a(u_{j2} - u_{j1}) + \dfrac{u_{j2}^3 - u_{j1}^3}{12a} & \text{for } t_{j2} \le t; \end{cases}$$

$$(8.3.15)$$

$$|S_{3j}(t)| = \begin{cases} \tfrac{1}{2}(\theta_{j2} - \theta_{j1})t^2 & \text{for } 0 \le t \le t_{j1}, \\ \tfrac{1}{3}\sqrt{a}(t + 2a)\sqrt{t-a} - \dfrac{au_{j1}}{2} - \dfrac{u_{j1}^3}{24a} & \\ \quad + \tfrac{1}{2}t^2\left[\theta_{i2} - \arccos\left(\dfrac{2a}{t} - 1\right)\right] & \text{for } t_{j1} \le t \le t_{j2}, \\ \dfrac{u_{j2}^3 - u_{j1}^3}{24a} - \tfrac{1}{2}a(u_{j2} - u_{j1}) & \text{for } t_{j2} \le t. \end{cases} \qquad (8.3.16)$$

Notice that $S_{4j}(t)$ is given by $S_j(t)$ of equation (8.3.8).

The probability density function $f(t)$ of $F(t)$ is obtained from

$$f(t) = \frac{\mathrm{d}F(t)}{\mathrm{d}t} = \frac{1}{|S_o|}\frac{\mathrm{d}|S(t)|}{\mathrm{d}t} = \frac{1}{|S_o|}\sum_{i=1}^{4}\sum_{j=1}^{k_i}\frac{\mathrm{d}|S_{ij}(t)|}{\mathrm{d}t} = \frac{1}{|S_o|}\sum_{i=1}^{4}\sum_{j=1}^{k_i} s_{ij}(t)\,. \tag{8.3.17}$$

The rth moment of t is given by

$$\mathrm{E}(t^r) = \int_0^\infty t^r f(t)\mathrm{d}t = \frac{1}{|S_o|}\sum_{i=1}^{4}\sum_{j=1}^{k_i}\int_0^\infty t^r s_{ij}(t)\mathrm{d}t\,. \tag{8.3.18}$$

Differentiating equations (8.3.8) and (8.3.13)–(8.3.16) with respect to t and evaluating the integral in equation (8.3.18), we obtain

$$\int_0^\infty t^r s_{1j}(t)\mathrm{d}t = \frac{(u_{j2}-u_{j1})(t_{j2}^{r+2}-t_{j1}^{r+2})}{(t_{j2}-t_{j1})(r+1)(r+2)}, \tag{8.3.19}$$

$$\int_0^\infty t^r s_{2j}(t)\mathrm{d}t = \frac{1}{r+1}\sum_{i=1}^{r+1}\binom{r+1}{i}\frac{a^{r+1-2i}(u_{j2}^{2i+1}-u_{j1}^{2i+1})}{2^{2i}(2i+1)}, \tag{8.3.20}$$

$$\int_0^\infty t^r s_{3j}(t)\mathrm{d}t = \frac{1}{r+2}\sum_{i=0}^{r+1}\binom{r+1}{i}\frac{a^{r+1-2i}(u_{j2}^{2i+1}-u_{j1}^{2i+1})}{2^{i}(2i+1)} \tag{8.3.21}$$

(Yoshikawa, 1989). Notice that $\int_0^\infty t^r s_{4j}(t)\mathrm{d}t$ is given by equation (8.3.12). Substituting equations (8.3.12) and (8.3.19)–(8.3.21) into equation (8.3.18), we can explicitly obtain the values of $\mathrm{E}(t)$ and $\mathrm{Var}(t)$.

Through the above procedure, Okabe and Yoshikawa (1989) calculated the statistic R and showed that high-class apartment buildings tended to locate along arterial streets in Kohto-Sumida, Tokyo (Figure 8.3.3).

8.3.3 Nearest neighbour distance method for area-like objects

When we cannot neglect the area of object O_j, we should apply the NND method for area-like objects. Figure 8.3.6 shows its typical application provided by Okabe *et al.* (1988). In the figure, the area-like objects O indicated by the shaded polygons are big parks in Setagaya, Tokyo, and the point-like elements P indicated by the unfilled circles are high-class apartment buildings. They examined whether or not the distribution of high-class apartment buildings were affected by big parks.

The procedure of the NND method for area-like objects is just the same as that for line-like objects, and so we do not repeat it here. Through that procedure, Okabe *et al.* (1988) calculated the statistic R and came to the conclusion that the distribution of high-class apartment buildings being independent of the location of big parks could not be rejected at the 0.05 significance level in Setagaya, Tokyo.

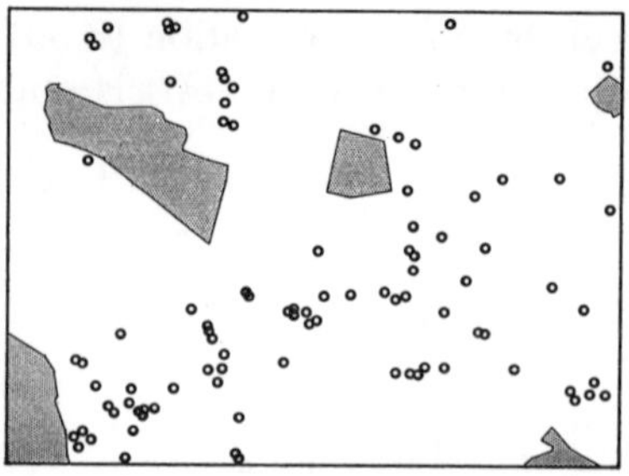

Figure 8.3.6 The distribution of high-class apartment buildings (the unfilled circles) and big parks (the shaded polygons) in Setagaya, Tokyo. (*Source*: Okabe *et al.* (1988, Figure 1).)

8.3.4 Multi nearest neighbour distance method

In the above three subsections we showed methods that examined the effect of only one type of objects O on the distribution of point-like elements P. In the real world, the effect is likely to be compounded; more than one type of objects $O_{(j)}$, $j \in I_{m_o}$, compoundly affect the distribution of point-like objects P. For instance, as is shown in Figure 8.3.3, it is likely that the distribution of high-class apartment buildings is compoundly affected by subway stations (empty circles) and arterial streets (the solid lines).

To examine this compound effect, Okabe and Yoshikawa (1989) generalized the above univariate index R into a multivariate index,

$$R = \left(\frac{\bar{t}_{(1)}}{\mathrm{E}(\bar{t}_{(1)})}, \ldots, \frac{\bar{t}_{(m)}}{\mathrm{E}(\bar{t}_{(m)})} \right), \tag{8.3.22}$$

where $\bar{t}_{(i)}$ be the average NN-distance defined by equation (8.3.2) for objects $O_{(i)}$, $i \in I_m$. With this multivariate index, we can examine the compound effect of several kinds of objects, $O_{(1)}$, ..., $O_{(m)}$, on the distribution of point-like elements P. Using this index, Okabe and Yoshikawa (1989) found that there was a compound effect of arterial streets and subway stations on the distribution of high-class apartment buildings in Kohto-Sumida, Tokyo (see Okabe and Yoshikawa, 1989, for a detailed discussion).

CHAPTER 9

Locational Optimization through Voronoi Diagrams

Voronoi diagrams are useful not only for the spatial analyses shown in the preceding chapters, but also for spatial optimization. In this chapter we discuss four classes of locational optimization problems which can be solved with the aid of Voronoi diagrams. The chapter consists of five sections. Section 9.1 is a preliminary section in which we briefly show a few non-linear, non-convex optimization methods to be used in this chapter. The reader who is familiar with these methods can skip this preliminary section.

In Section 9.2 we deal with the locational optimization of points. In general, the problem is to determine the location of points so that the average distance (more generally, the average cost) to the nearest points is minimized. The problem varies slightly according to the nature of the points. We consider six kinds of points: point-like facilities used by independent individuals (say, public mail boxes); point-like facilities used by groups (say, tennis courts); a hierarchical facility (say, the main library and branch libraries); observation points for estimating some spatial quantities (say, pluviometers); service points of a mobile facility (say, a bookmobile); and nodes through which users go to a central point (say, bus stops for commuters). We show that these locational optimization problems can be solved with the ordinary and weighted Voronoi diagrams.

In Section 9.3 we discuss the locational optimization of lines. We deal with three types of lines. The first type of lines is a service route of a mobile facility (say, an ice-cream vendor). The optimization problem is to determine a service route to minimize the average distance (or cost) to the nearest point on a service route provided that the total length of the service route is given. The second type of lines is a network (say, a railway network). The optimization problem is to determine the location of nodes (stations) and links (railways) to minimize the total flow cost (or transportation cost) between any two points on a plane provided that the total length of a network and the flow density between any two points in a region are given.

The third type of lines is the Steiner tree. The problem is to approximately obtain a big Steiner tree which cannot be obtained from combinatorial methods. We show that the locational optimization problems of these types of lines can be solved with the help of the ordinary Voronoi diagram, the line Voronoi diagram and the Delaunary triangulation.

In Section 9.4 we consider two types of space–time optimization problems. The first type of problem is to optimize the location of point-like facilities that are constructed over multiple time-stages. The optimization is to minimize the distance (or cost) to the nearest facility averaged over users in a region as well as over time. The second type of problem is to determine the location of facilities which open periodically (say, market places open during a certain time interval) so that the average distance (or cost) to the nearest facilities and the waiting time is minimized. We show that these two types of problems can be solved with the aid of the space–time Voronoi diagram and the Voronoi diagram on a cylinder.

In Section 9.5 we show a method of fitting a Voronoi diagram to a given polygonal tessellation. With this method we can see to what extent a Voronoi diagram fits a given tessellation. Moreover, we can apply this method to the locational optimization problem of point-like facilities whose use is spatially restricted.

9.1 PRELIMINARIES

In solving the locational optimization problems mentioned above, we meet a certain type of mathematical optimization problem, called the non-linear, non-convex programming problem. In this section we briefly present a few computationally feasible methods to solve this problem. The reader who wants to know the full details of non-linear programming methods should consult, for example, Fiacco and McCormick (1968), Konno and Yamashita (1978) and Gill *et al.* (1981).

9.1.1 The non-linear non-convex programming problem

The *programming problem* referred to in the literature of mathematical optimization is the problem in which we seek to determine values for n variables $x_1, \ldots, x_n$ which satisfy m_1 inequalities and/or m_2 equations and, in addition, minimize a function of the n variables. Mathematically, the programming problem is written as

Problem 9.1.1 (the general programming problem)

$$\min_{x_1, \ldots, x_n} F(x_1, \ldots, x_n), \qquad (9.1.1)$$

subject to

$$g_i(x_1, \ldots, x_n) \leq 0, \quad i = 1, \ldots, m_1, \tag{9.1.2}$$
$$g_i(x_1, \ldots, x_n) = 0, \quad i = m_1+1, \ldots, m_1+m_2.$$

We call the function in expression (9.1.1) an *objective function*. In this chapter we assume that the objective function has at least the first derivative and it is bounded below, i.e. $F(\mathbf{x}) > K > -\infty$. We call relations (9.1.2) *constraints* and the functions in the constraints *constraint functions*. The programming problems with constraints ($m_1+m_2 \geq 1$) and without constraints ($m_1+m_2 = 0$) are called the *constrained programming problem* and the *unconstrained programming problem*, respectively. For simplicity, $F(x_1, \ldots, x_n)$ and $g_i(x_1, \ldots, x_n)$ are written as $F(\mathbf{x})$ and $g_i(\mathbf{x})$, where $\mathbf{x}^T = (x_1, \ldots, x_n)$ is a vector in the n-dimensional Euclidean space $\mathbb{R}^n$. Note that since the maximization of $F(\mathbf{x})$ is equivalent to the minimization of $-F(\mathbf{x})$, we consider only the minimization problem; since $g_i(\mathbf{x}) \geq 0$ is equivalent to $-g_i(\mathbf{x}) \leq 0$, we consider only the two relations $\leq$ and $=$ in the constraints of relation (9.1.2). In Figure 9.1.1 a simple example is depicted, where the objective function is a univariate function $F(x) = x(x-1)(x-2)(x-4)(x-8)(x-10) + 3000$, and the constraints are $g_1(x) = -x \leq 0$ and $g_2(x) = x \leq 10$ ($m_1 = 2, m_2 = 0$).

The constraints in relation (9.1.2) determine a region in $\mathbb{R}^n$, which is given by

$$S = \{\mathbf{x} | g_i(\mathbf{x}) \leq 0, i = 1, \ldots, m_1; g_i(\mathbf{x}) = 0, i = m_1+1, \ldots, m_1+m_2\}. \tag{9.1.3}$$

We call the set S a *feasible region*, and $\mathbf{x} \in S$ a *feasible solution*. In the above example the feasible region is given by $S = \{x | 0 \leq x \leq 10\}$, which is

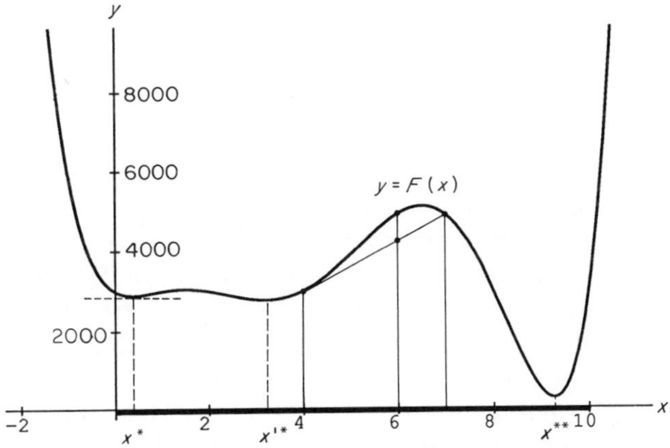

Figure 9.1.1 A feasible region, local minima, and the global minimum $F(x) = x(x-1)(x-2)(x-4)(x-8)(x-10) + 3000$, $g_1(x) = -x \leq 0$, $g_2(x) = x \leq 10$.

indicated by the heavy solid line in Figure 9.1.1. In this feasible region we notice three basins around the points indicated by x^*, x'^* and x^{**}. We call the bottom points in these basins local minima. To be precise, let

$$N_\epsilon(\mathbf{x}^*) = \{\mathbf{x} | \; \|\mathbf{x}-\mathbf{x}^*\| < \epsilon\} \tag{9.1.4}$$

to be an open ball centred at $\mathbf{x}^*$ with radius ϵ. If there exist a positive ϵ and $\mathbf{x}^* \in S$ for which the relation

$$F(\mathbf{x}^*) \leq F(\mathbf{x}) \quad \text{for } \mathbf{x} \in S \cap N_\epsilon(\mathbf{x}^*) \tag{9.1.5}$$

holds, we call $\mathbf{x}^*$ a *local minimum* (*optimum*). In the example of Figure 9.1.1, local minima are x^*, x'^* and x^{**}. Among them, the bottom at x^{**} is the lowest. We call such a bottom point the global minimum. Mathematically, if there exists $\mathbf{x}^{**} \in S$ that satisfies

$$F(\mathbf{x}^{**}) \leq F(\mathbf{x}) \quad \text{for } \mathbf{x} \in S, \tag{9.1.6}$$

we call $\mathbf{x}^{**}$ the *global minimum* (*optimum*).

Since the functions $F(\mathbf{x})$ and $g_i(\mathbf{x})$ are general functions, programming problems subsumed under Problem 9.1.1 are various, but we can classify them into two classes: the linear programming problem and the non-linear programming problem. The *linear programming problem* is the programming problem in which each of the functions $F(\mathbf{x})$ and $g_i(\mathbf{x})$, $i \in I_{m_1+m_2}$, is a linear function, i.e. $F(x_1, \ldots, x_n) = c_1x_1+\ldots+c_nx_n$, and $g_i(x_1, \ldots, x_n) = a_{i1}x_1+\ldots+a_{in}x_n+b_i$. The *non-linear programming problem* is the programming problem in which either $F(\mathbf{x})$, or at least one of the functions $g_i(\mathbf{x})$, $i \in I_{m_1+m_2}$, is a non-linear function, such as $F(x_1, \ldots, x_n) = c_1x_1^2+\ldots+c_nx_n^2$.

We can further classify the non-linear programming problem into two subclasses: the non-linear convex programming problem, and the non-linear non-convex programming problem. If for any two points $\mathbf{x}_1$ and $\mathbf{x}_2$ in $\mathbb{R}^n$ and for all $0 \leq \lambda \leq 1$ the following relation holds

$$F[\lambda\mathbf{x}_1+(1-\lambda)\mathbf{x}_2] \leq \lambda F(\mathbf{x}_1)+(1-\lambda)F(\mathbf{x}_2), \tag{9.1.7}$$

we call function $F(\mathbf{x})$ a *convex function*; otherwise, we call it a *non-convex function* (an alternative definition is given in Section 1.3.3). The function shown in Figure 9.1.1 is a non-convex function, because the relation (9.1.7) does not hold, for example $F(\frac{1}{3}\times4+\frac{2}{3}\times7) > \frac{1}{3}F(4)+\frac{2}{3}F(7)$ (observe the two filled circles in Figure 9.1.1). If every non-linear function in the non-linear programming problem is convex, we call such a programming problem the *non-linear convex programming problem*; if at least one non-linear function in the non-linear programming problem is non-convex, we call it the *non-linear non-convex programming problem*. As will be seen below, the programming problems encountered in the following sections are non-linear non-convex programming problems.

9.1.2 The descent method

Let us first consider the unconstrained case of Problem 9.1.1, i.e.

Problem 9.1.2 (the unconstrainted programming problem)

$$\min_{x_1, \ldots, x_n} F(x_1, \ldots, x_n). \tag{9.1.1}$$

As is indicated by the horizontal broken line in Figure 9.1.1, the necessary condition for a local minimum of a univariate function is that the line tangential to the curve given by $y = F(x)$ at x^* is horizontal. Algebraically, this necessary condition is written as $\mathrm{d}F(x)/\mathrm{d}x = 0$. Similarly, the necessary condition of a multivariate function $F(x_1, \ldots, x_n)$ for a local minimum is that the hyperplane tangential to the surface given by $y = F(\boldsymbol{x})$ at $\boldsymbol{x}^*$ is horizontal. Algebraically, this condition is written as

$$\begin{gathered} \frac{\partial F(\boldsymbol{x}^*)}{\partial x_1} = 0, \\ \vdots \\ \frac{\partial F(\boldsymbol{x}^*)}{\partial x_n} = 0. \end{gathered} \tag{9.1.8}$$

When the function $F(\boldsymbol{x})$ is a non-linear non-convex function, we usually encounter analytical difficulty in solving the simultaneous equations of equation (9.1.8). It is almost impossible to obtain not only the global minimum but also a local minimum with the analytical method, or the so-called classical method. To overcome this difficulty, we may alternatively use a numerical method which gives a solution close to a local minimum with a certain precision level.

The difficulty in solving the non-linear programming problem may be intuitively understood by imagining the difficult situation in which we are lost in a rolling mountainous region (Figure 9.1.2(a)) covered with mist, and we are looking for the bottom of a basin in the region without a contour map (Figure 9.1.2(b)). Since we are surrounded by mist, we can see only the small area around us ($\boldsymbol{x}^{(k)}$ in Figure 9.1.2(b)). What we can do is therefore to walk down observing the slope in this small area. First, we decide in which direction ($\boldsymbol{d}^{(k)}$ in Figure 9.1.2(b)) we should walk down (Step 1). Once the direction is determined (the arrow in Figure 9.1.2(b)), we traverse straight in this direction (the straight line in Figure 9.1.2(b)). Second, we decide at which point we should stop on this line (Step 2); third, when we reach that point ($\boldsymbol{x}^{(k+1)}$ in Figure 9.1.2(b)), we reconsider the direction and repeat the same procedure from that point (Step 3); last, if we reach a flat place, we stop at that point (Step 4). We consider that the point we have reached is the bottom (a local minimum). Since we do not know how many bottoms exist in the region, we only know that the

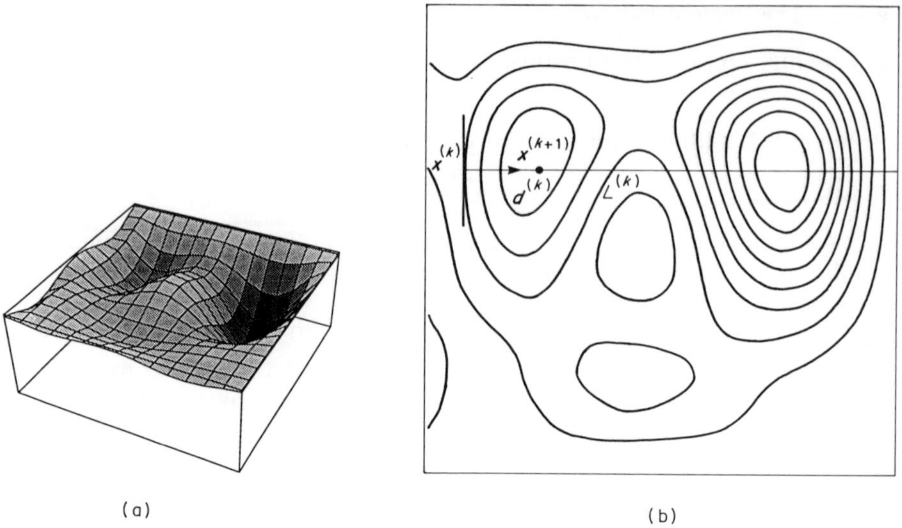

Figure 9.1.2 The descent method: (a) the surface of the function $F(x_1,x_2)$ and (b) its contour lines.

bottom we have reached is one of many bottoms (local minima). If we are lucky enough, the bottom may be the lowest bottom (the global minimum) in the region.

The above search method is called the *descent method*, and its computational procedure is written algorithmically as follows.

Algorithm 9.1.1 (a prototype of the descent method)

Step 0. Choose an arbitrary $\boldsymbol{x}^{(0)} \in \mathbb{R}^n$, and substitute 0 for k (denoted by $k \leftarrow 0$).
Step 1. Determine a direction vector $\boldsymbol{d}^{(k)}$ at the point $\boldsymbol{x}^{(k)}$.
Step 2. Determine a step size $\alpha^{(k)}$ at the point $\boldsymbol{x}^{(k)}$.
Step 3. $\boldsymbol{x}^{(k+1)} \leftarrow \boldsymbol{x}^{(k)}+\alpha^{(k)}\boldsymbol{d}^{(k)}$.
Step 4. If $\|\boldsymbol{x}^{(k+1)}-\boldsymbol{x}^{(k)}\| < \epsilon \approx 0$, report $\boldsymbol{x}^{(k+1)}$; otherwise, $k \leftarrow k+1$ and go to Step 1.

The descent method has several variations in the choice of the direction vector $\boldsymbol{d}^{(k)}$ (Step 1). When the direction vector is given by the steepest direction, i.e.

$$\boldsymbol{d}^{(k)^{\mathrm{T}}} = -\nabla F(\boldsymbol{x}^{(k)}) = -\left(\frac{\partial F(\boldsymbol{x}^{(k)})}{\partial x_1{}^{(k)}}, \ldots, \frac{\partial F(\boldsymbol{x}^{(k)})}{\partial x_n{}^{(k)}}\right), \tag{9.1.9}$$

the descent method is called the *steepest descent method*. In the example of Figure 9.1.3, the steepest direction is shown by the solid arrow (note that

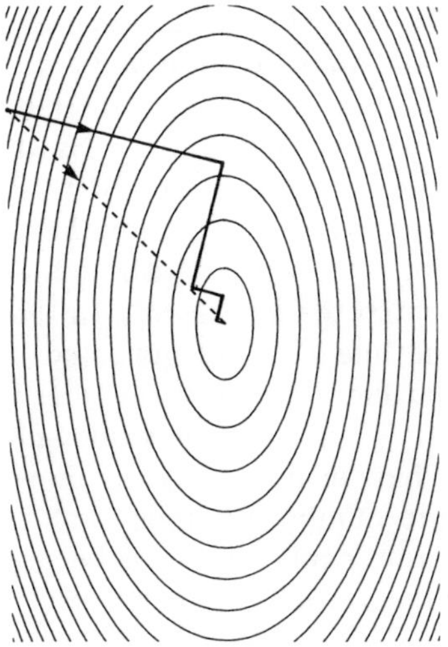

Figure 9.1.3 Direction vectors: the steepest direction (the solid arrow) and the direction used in the Newton method (the broken line).

the steepest direction is orthogonal to the contour line of $F(\boldsymbol{x})$). The steepest descent method has an advantage in that it requires only the first derivatives. On the other hand, it has a disadvantage in that the convergence to a local optimum solution is not always fast (compare the zig-zag solid line with the broken line in Figure 9.1.3). To overcome this disadvantage, the *Newton method* is proposed, in which the direction vector is given by

$$\boldsymbol{d}^{(k)^{\mathrm{T}}} = -[\nabla^2 F(\boldsymbol{x}^{(k)})]^{-1} \nabla F(\boldsymbol{x}^{(k)}), \tag{9.1.10}$$

where $\nabla^2 F(\boldsymbol{x}^{(k)})$ is the Hesse matrix of $F(\boldsymbol{x}^{(k)})$, i.e.

$$\nabla^2 F(\boldsymbol{x}) = \begin{bmatrix} \dfrac{\partial^2 F}{\partial x_1^2} & \cdots & \dfrac{\partial^2 F}{\partial x_1 \partial x_n} \\ \vdots & \ddots & \vdots \\ \dfrac{\partial^2 F}{\partial x_n \partial x_1} & \cdots & \dfrac{\partial^2 F}{\partial x_n^2} \end{bmatrix}. \tag{9.1.11}$$

In the example of Figure 9.1.3, this direction is indicated by the broken arrow. The calculation of the Hesse matrix is, however, sometimes difficult. To diminish this difficulty, we may use a direction vector

$$\boldsymbol{d}^{(k)^{\mathrm{T}}} = -[H^{(k)}]^{-1} \nabla F(\boldsymbol{x}^{(k)}), \tag{9.1.12}$$

where $H^{(k)}$ is a 'simple' approximation of the Hesse matrix in the sense that the matrix operation with $H^{(k)}$ is simpler than that with $\nabla^2 F(\boldsymbol{x})$. For example, $H^{(k)}$ is given by the matrix in which the diagonal elements are given by those of $\nabla^2 F(\boldsymbol{x})$ and the off-diagonal elements are zeros. We call this alternative method the *quasi-Newton method*. Formally, if $H^{(k)} = I$, the quasi-Newton method becomes the steepest descent method; if $H^{(k)} = \nabla^2 F(\boldsymbol{x}^{(k)})$, the method becomes the Newton method.

Having determined the direction vector $\boldsymbol{d}^{(k)}$ at the point $\boldsymbol{x}^{(k)}$ (Step 1), we next consider the straight line $L^{(k)}$ radiating from $\boldsymbol{x}^{(k)}$ in the direction $\boldsymbol{d}^{(k)}$, and search the point $\boldsymbol{x}^{(k+1)}$ on $L^{(k)}$ at which we reconsider the direction (Step 2). We call this search the *line search*. In the example of Figure 9.1.2, direction vector $\boldsymbol{d}^{(k)}$ and line $L^{(k)}$ are indicated by the arrow and the hair line, respectively. In the literature, several rules are proposed for the line search, for example Curry's rule, Altman's rule and Goldstein's rule. Among them, Goldstein's rule has an advantage in that we always can terminate the line search in a finite number of steps.

To show Goldstein's rule explicitly, let α be the distance from $\boldsymbol{x}^{(k)}$ along line $L^{(k)}$ toward the direction $\boldsymbol{d}^{(k)}$, and

$$h^{(k)}(\alpha) = F(\boldsymbol{x}^{(k)} + \alpha \boldsymbol{d}^{(k)}). \tag{9.1.13}$$

The function $h^{(k)}(\alpha)$ shows the value of $F(\boldsymbol{x})$ along the line $L^{(k)}$ (see $\beta = h^{(k)}(\alpha)$ on the α–β plane in Figure 9.1.4 obtained from $L^{(k)}$ in Figure 9.1.2(b)). The line tangential to $\beta = h^{(k)}(\alpha)$ at $\alpha = 0$ is given by

$$\tau^{(k)}(\alpha) = h^{(k)}(0) + \frac{\partial h^{(k)}(0)}{\partial \alpha}\alpha \tag{9.1.14}$$

(see $\beta = \tau^{(k)}(\alpha)$ in Figure 9.1.4). We now consider two lines passing through the point $(0, h^{(k)}(0))$ which are slightly flatter than the line of equation (9.1.14) (see Figure 9.1.4). Mathematically, these lines are written as

$$\eta_i^{(k)}(\alpha) = h^{(k)}(0) + \mu_i \frac{\partial h^{(k)}(0)}{\partial \alpha}\alpha, \quad i = 1,2, \tag{9.1.15}$$

where $0 < \mu_1 \leq \mu_2 < 1$ (in practice, $\mu_1 = 0.4$ and $\mu_2 = 0.6$ are often used). With this function, we define the set

$$M^{(k)} = \{\alpha \mid \eta_2^{(k)}(\alpha) \leq h^{(k)}(\alpha) \leq \eta_1^{(k)}(\alpha)\} \tag{9.1.16}$$

(see the heavy solid line segment on the α-axis in Figure 9.1.4). The set $M^{(k)}$ is always finite because $F(\boldsymbol{x})$ is assumed to be bounded below. *Goldstein's rule* is the rule in which we choose $\alpha^{(k)}$ in $M^{(k)}$. We can find such an $\alpha^{(k)}$ in a finite number of steps using the following algorithm.

Algorithm 8.1.2 (Goldstein's rule)

Step 0. Choose an arbitrary initial value $\alpha_{(0)} > 0$, and $i \leftarrow 0$.
Step 1. $\alpha_{(i)} \leftarrow (i+1)\alpha_{(0)}$.
Step 2. If $\eta_2^{(k)}(\alpha_{(i)}) \leq h^{(k)}(\alpha_{(i)})$, then go to Step 3; otherwise, $i \leftarrow i+1$ and go to Step 1.

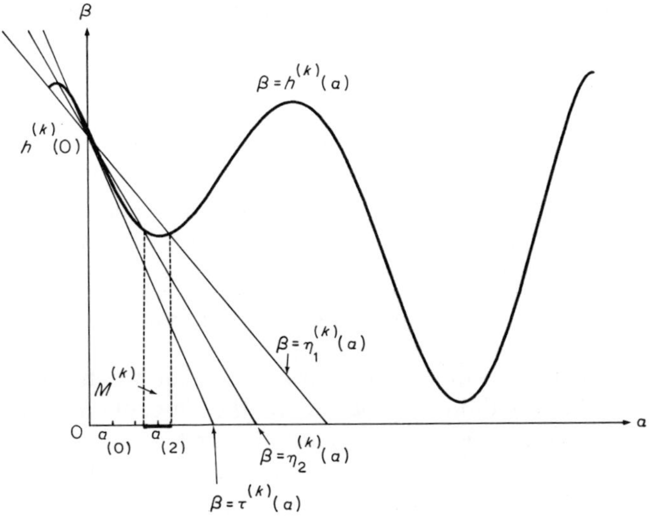

Figure 9.1.4 Goldstein's rule.

Step 3. If $h^{(k)}(\alpha_{(i)} \leq \eta_1^{(k)}(\alpha_{(i)})$, then report $\alpha_{(i)}$; otherwise, $\alpha_{(i)} \leftarrow \{\alpha_{(i)} + \alpha_{(i-1)}\}/2$, and go to Step 2 (where $\alpha_{(-1)} = 0$).

9.1.3 The penalty function method

The descent method is developed for the unconstrained non-linear programming problem (Problem 9.1.2); it cannot be directly applied to the constrained non-linear programming problem (Problem 9.1.1, $m_1 + m_2 \geq 1$). To solve the latter problem, we can use the transformation method. The essential idea is that we first transform a constrained non-linear programming problem into an unconstrained non-linear programming problem (Problem 9.1.2); next we solve this unconstrained programming problem with a method that solves the unconstrained programming problem, say the descent method. The transformation method has several variations, such as the penalty function method and the multiplier method. In this subsection we show the penalty function method, because this method is one of the most basic and many other methods are developed from this method.

In the constrained programming problem of Problem 9.1.1, we have to find a local minimum in the feasible region S determined by the relation (9.1.2). In the process of searching the local minimum, however, we may choose an infeasible solution. In such a case we are supposed to pay a penalty. If this penalty is extremely high, we give up that solution and try to find another solution. In the end, we may reach the local minimum in the feasible region S.

To state this method a little more precisely, suppose that a penalty $R_1(\boldsymbol{x})$ for an infeasible solution is given by

$$R_1(x) = \begin{cases} +\infty & \text{if } \boldsymbol{x} \notin S, \\ 0 & \text{if } \boldsymbol{x} \in S, \end{cases} \tag{9.1.17}$$

and let $L(\boldsymbol{x}) = F(\boldsymbol{x})+R_1(\boldsymbol{x})$. Noticing that $L(\boldsymbol{x})$ coincides with $F(\boldsymbol{x})$ for any feasible solution and that it is $+\infty$ for any infeasible solution, we can restate the constrained programming problem (Problem 9.1.1) as the unconstrained programming problem:

$$\min_{\boldsymbol{x}} L(\boldsymbol{x}) = \min_{\boldsymbol{x}}[F(\boldsymbol{x})+R_1(\boldsymbol{x})]. \tag{9.1.18}$$

In principle, this idea appears to work well, but we have a difficulty in implementing this method in practice. In the descent method, we assume that the objective function has at least the first derivative. The objective function $L(\boldsymbol{x})$ is, however, discontinuous at the boundary of the feasible region S, even if $+\infty$ is replaced by a very large number.

To overcome this difficulty, we approximate the penalty function $R_1(\boldsymbol{x})$ by a smooth function whose value (i.e. the penalty) smoothly increases as $\boldsymbol{x}$ moves farther from the feasible region S. For instance,

$$R_2(\boldsymbol{x}) = \sum_{i=1}^{m_1} [\max\{g_i(\boldsymbol{x}), 0\}]^2 + \sum_{i=m_1+1}^{m_1+m_2} g_i(\boldsymbol{x})^2. \tag{9.1.19}$$

To show that this penalty function works well, let us consider an illustrative example: $F(x) = (x+1)^2$ and $g_1(x) = -x$, $g_2(x) = x-1$, which are depicted in Figure 9.1.5. In this case the penalty function of equation (9.1.19) is written as

$$R_2(x) = [\max\{-x, 0\}]^2 + [\max\{x-1, 0\}]^2, \tag{9.1.20}$$

which is shown in Figure 9.1.5(b). Then the transformed function is given by $L(x) = F(x)+R_2(x)$, which is depicted in Figure 9.1.5(c).

In this example we can analytically obtain the global minimum of $L(x)$, and the solution is $x^* = -0.5$. Obviously, this solution violates the constraint $g_1(x) = -x \leq 0$. This violation suggests that the penalty should be higher. We do not know, however, in advance how high the penalty should be. We thus increase the penalty from low to high according to $2^k R_2(x)$, $k = 0,1 \ldots$. As a result, we obtain a series of solutions, -0.50, -0.33, -0.20, -0.11, -0.05, -0.03, -0.02, -0.01 for $k = 0, \ldots, 8$ (Figure 9.1.5(d)). As is easily noticed in Figure 9.1.5(a), the minimum is achieved at $x = 0$. The above series appear to approach this minimum.

In general, with the penalty function of equation (9.1.19), we can transform the constrained programming problem of Problem 9.1.1 into the unconstrained programming problem

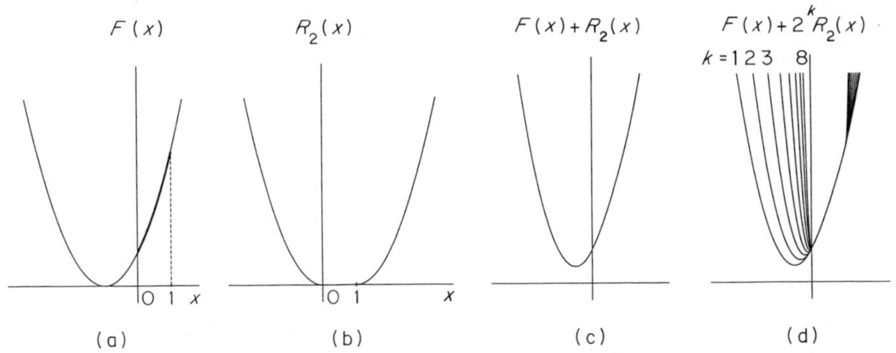

Figure 9.1.5 The penalty function method: (a) $F(x) = (x+1)^2$, $g_1(x) = -x \leq 0$, $g_2(x) = x-1 \leq 0$; (b) $R_2(x) = [\max\{-x,0\}]^2 + [\max\{x-1,0\}]^2$; (c) $L(x) = F(x)+R_2(x)$; (d) $L(x)^{(k)} = F(x)+2^k R_2(x)$.

$$\min_{\mathbf{x}} L^{(k)}(\mathbf{x}) = \min_{\mathbf{x}}[F(\mathbf{x}) + r_o^k R_2(x)], \tag{9.1.21}$$

where $r_o > 1$. It is shown that if there exists a local optimum $\mathbf{x}^{*(k)}$ for this unconstrained programming problem, then there exists a series $\mathbf{x}^{*(0)}$, $\mathbf{x}^{*(1)}$, $\mathbf{x}^{*(2)}$, ..., $\mathbf{x}^{*(k)}$, ..., that converge to a local minimum of the constrained programming problem, Problem 9.1.1. Hence, as we expected, the series in the above example actually converges to 0. This technique is called the *sequential unrestrained minimization technique* (Fiacco and McCormick, 1968).

We call the above method the *penalty function method* or, more specifically, the *exterior penalty function method*. With this method we can solve the constrained programming problems in the following sections. Of course we can use alternative methods which are modified from the exterior penalty function method, such as the *interior penalty function method* and the *mixed penalty function method*. Furthermore, to gain fast convergence, we may use the *multiplier method*. The reader who wishes to use these alternative methods should see, for example, Gill *et al.* (1981, Chapters 4–6).

9.2 LOCATIONAL OPTIMIZATION OF POINTS

Using the computational methods in Section 9.1, we now wish to solve the locational optimization problems mentioned in the introduction. We first discuss the locational optimization of points in a plane or in a space, where the points represent point-like facilities, service points, observation points, nests of birds, crystallites, points from which some influential power is generated, and so forth.

The study of locational optimization of point-like facilities may date back to Launhardt (1882) or Weber (1909) (the so-called the Weberian problem), but the study has been rapidly progressed since Hakimi (1964) in Operations Research. The recent progress can be seen, for example, in Handler and Mirchandani (1979), the *European Journal of Operations Research* (1985, Vol. 20, No. 3), Love *et al.* (1988) and Mirchandani and Francis (1990). It should be noted that the problem dealt with there is locational optimization on a network where the demand for service supplied by facilities is supposed to arise only on nodes of the network.

Compared with this rapid progress, progress was slow in the study of locational optimization in a continuous plane where demand arises at any point in a plane and feasible locations of facilities are at any point in the plane. This slow progress was due to the lack of efficient computational methods for managing complicated geometrical calculation on a plane. In fact, computational methods in the 1960s could deal with less than twenty facilities, for instance Leamer (1968). In the 1980s however, computational geometry (Preparata and Shamos, 1985), in particular efficient computational methods for constructing a Voronoi diagram (Chapter 4) have been rapidly developed and have overcome this difficulty (Iri *et al.*, 1984). We can now in practice solve locational optimizational problems of a large number of points in a continuous plane or space with Voronoi diagrams.

9.2.1 Locational optimization of point-like facilities used by independent users

We first consider the locational optimization of point-like facilities which are used by independent users. A good example of such a facility is a public mail box. For illustrative purposes, we show an actual example in Figure 9.2.1 where the empty circles and the filled circle indicate the locations of public mail boxes in Koganei, Tokyo. Our concern is with whether or not the configuration of those mail boxes is 'optimal'; if not, to what extent we can improve the configuration by relocating the mail boxes?

To answer these questions, we should explicitly define what is 'optimal'. To define it, let us discuss two factors: the cost borne by the post office and the cost borne by users. Regarding the former cost, the major cost is the cost of collecting the mail. In Koganei, the mail boxes are managed by the main post office indicated by the filled circle in Figure 9.2.1. A collector wagon starts from this office, visits all mail boxes successively, and returns to the office. We have to take this cost into account. However, since a public organization is supposed to minimize the user's cost or maximize the user's convenience instead of maximizing profit, the primary factor is the user's convenience rather than the cost borne by the post office. For the moment, let us consider the primary factor. The secondary factor will be taken into account later (see the locational optimization of service points of a mobile facility).

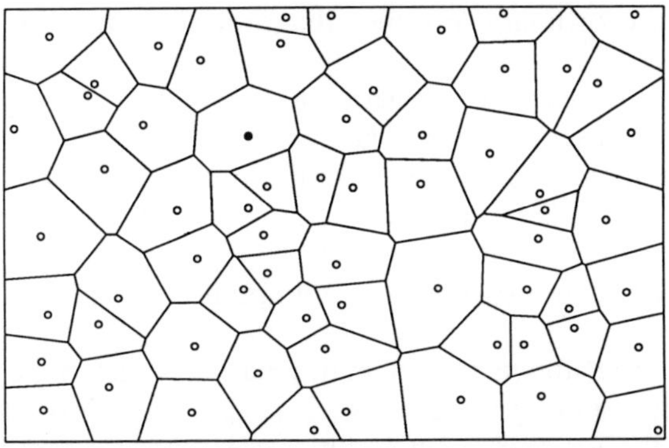
(a)

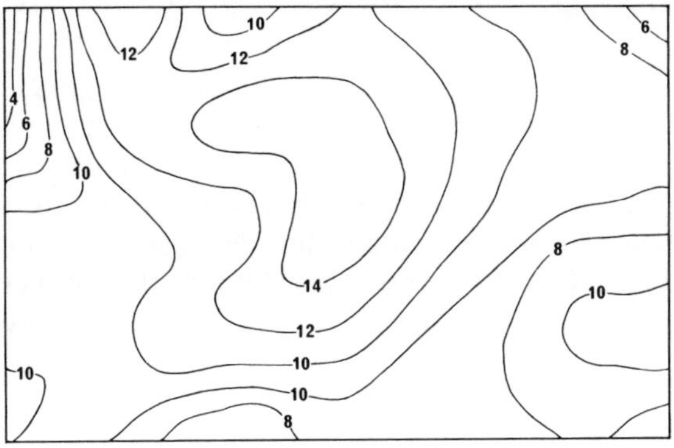

(b)

Figure 9.2.1 (a) Public mail boxes in Koganei district in Tokyo and (b) the distribution of population (× 10/ha).

The user's cost or the user's convenience may be measured in terms of the distance from a user to a mail box. Since mail boxes are indifferent, most users go to their nearest mail box. As a result, every mail box, say mail box $i \in I_n$, has its own catchment area or the area in which the nearest mail box from any point in this area is mail box i. Recalling Property V6 in Chapter 2, we notice that this region is indeed a Voronoi polygon (strictly speaking, since users use streets, this region is given as a Voronoi region of the network area-Voronoi diagram defined in Chapter 3).

Obviously, a street distance is different from a crow-flight distance (the Euclidean distance). This difference is empirically examined by Koshizuka and Kobayashi (1983) and Love *et al.* (1988, Chapter 10). At first glance this difference appears very big, but according to Koshizuka and Kobayashi (1983) this difference is not so large as we might expect. If this difference is allowable in practice, the catchment areas of mail boxes can be approximated by the ordinary Voronoi diagram. The solid lines in Figure

9.2.1 indicate the ordinary Voronoi diagram generated by the mail boxes in Koganei.

As mentioned above, we measure the cost of a user by the Euclidean distance from the user to his/her nearest mail box. The cost of users as a whole is hence measured by the average nearest neighbour distance, i.e. the distance to the nearest mail box averaged over all users in a region. In these terms we can say that the optimal configuration of mail boxes implies the configuration that provides the minimum average nearest neighbour distance.

To formalize the above model mathematically, let S be a finite closed subset of $\mathbb{R}^2$ representing a region; $\mathbf{x}_1, \ldots, \mathbf{x}_n$ be n points (location vectors) in S indicating the locations of n facilities; $\phi(\mathbf{x})$ be the density of users over region S, where the integral of $\phi(\mathbf{x})$ over S is assumed to be unity without loss of generality; and $\mathcal{V}\cap S = \{V_1\cap S, \ldots, V_n\cap S\}$ be the Voronoi diagram generated by $P = (\mathbf{x}_1, \ldots, \mathbf{x}_n\}$ bounded by S (Section 2.1). For simplicity, we write $\mathcal{V} = \{V_1, \ldots, V_n\}$ for $\mathcal{V}\cap S = \{V_1\cap S, \ldots, V_n\cap S\}$.

The Euclidean distance from a user at $\mathbf{x}$ to a mail box at $\mathbf{x}_i$ is written as $\|\mathbf{x}-\mathbf{x}_i\|$. More generally, we can consider the travel cost from a user at $\mathbf{x}$ to a facility at $\mathbf{x}_i$ as a function of the Euclidean distance, or the squared Euclidean distance $f(\|\mathbf{x}-\mathbf{x}_i\|^2)$ (note that we use the squared Euclidean distance only for analytical convenience). One of the simplest cost functions is $f(\|\mathbf{x}-\mathbf{x}_i\|^2) = c\sqrt{(\|\mathbf{x}-\mathbf{x}_i\|^2)} = c\|\mathbf{x}-\mathbf{x}_i\|$. With this cost function the objective function discussed above is written as

$$F(\mathbf{x}_1, \ldots, \mathbf{x}_n) = \sum_{i=1}^{n} \int_{V_i} f(\|\mathbf{x}-\mathbf{x}_i\|^2)\phi(\mathbf{x})\mathrm{d}\mathbf{x}. \tag{9.2.1}$$

Note that since the total number of users is assumed to be unity, the total travel cost is the same as the average cost.

When $f(\|\mathbf{x}-\mathbf{x}_i\|^2) = \|\mathbf{x}-\mathbf{x}_i\|$, and $\phi(\mathbf{x}) = 1/S$ (the density of users is uniform over S), we can calculate the integral in equation (9.2.1). The explicit form is shown in Section 8.3.

In terms of the objective function given by equation (9.2.1), we can formally state the locational optimization problem of public mail boxes as the following programming problem.

Problem 9.2.1 (minimization of the average travel cost to the nearest points without constraints)

$$\min_{x_1, \ldots, x_n} \sum_{i=1}^{n} \int_{V_i} \mathrm{f}(\|\mathbf{x}-\mathbf{x}_i\|^2)\phi(\mathbf{x})\mathrm{d}\mathbf{x}. \tag{9.2.2}$$

The network problem corresponding to Problem 9.2.1 is called the multi-median problem. In this problem permissible locations are only on nodes and a distance is given by the shortest path distance on a network. When a network density is dense, we may expect that the solution of Problem 9.2.1 is close to the solution of the multi-median problem. This approximation is quite useful when a network is huge, because the multi-median problem on a huge network is hard to solve.

The function F of equation (9.2.1) is apparently non-linear, and moreover is non-convex. We can see the latter property from the following fact. If $(\mathbf{x}_1^*, \ldots, \mathbf{x}_i^*, \ldots, \mathbf{x}_j^*, \ldots, \mathbf{x}_n^*)$ is a local minimum of F, then $(\mathbf{x}_1^*, \ldots, \mathbf{x}_j^*, \ldots, \mathbf{x}_i^*, \ldots, \mathbf{x}_n^*)$ is also a local minimum of F, because the equation $F(\mathbf{x}_1^*, \ldots, \mathbf{x}_i^*, \ldots, \mathbf{x}_j^*, \ldots, \mathbf{x}_n^*) = F(\mathbf{x}_1^*, \ldots, \mathbf{x}_j^*, \ldots, \mathbf{x}_i^*, \ldots, \mathbf{x}_n^*)$ holds. This implies that we have multiple local minima unless $\mathbf{x}_1^* = \ldots = \mathbf{x}_n^*$. The solution, $\mathbf{x}_1^* = \ldots = \mathbf{x}_n^*$, means that all facilities are located at the same place. Obviously, this solution does not give a minimum. Therefore the function F has multiple local minima, implying that the function F is non-convex.

Because of this non-convex property, as we mentioned in Section 9.1, we have to use a computational method. If we employ the steepest descent method, we need the first derivatives of F. To derive them, suppose that the ith generator slightly changes its location from $\mathbf{x}_i$ to $\mathbf{x}_i + \delta\mathbf{x}_i$ ($\|\delta\mathbf{x}_i\| \approx 0$) provided that all other generators remain at the same location. Let $\mathcal{V}$ be the Voronoi diagram generated by $\{\mathbf{x}_1, \ldots, \mathbf{x}_i, \ldots, \mathbf{x}_n\}$ and $\mathcal{V}'$ be that generated by $\{\mathbf{x}_1, \ldots, \mathbf{x}_i + \delta\mathbf{x}_i, \ldots, \mathbf{x}_n\}$; V_i be the Voronoi polygon of the ith generator in $\mathcal{V}$ and V_i' be that in $\mathcal{V}'$; and J_i be the set of indices of the Voronoi polygons adjacent to V_i. Owing to the slight move, $\delta\mathbf{x}_i$, the Voronoi diagram slightly changes, but, as is indicated by the broken lines in Figure 9.2.2, this change occurs only in V_i and its adjacent Voronoi polygons, i.e. V_j, $j \in J_i$. Moreover, the Voronoi edges between adjacent Voronoi polygons

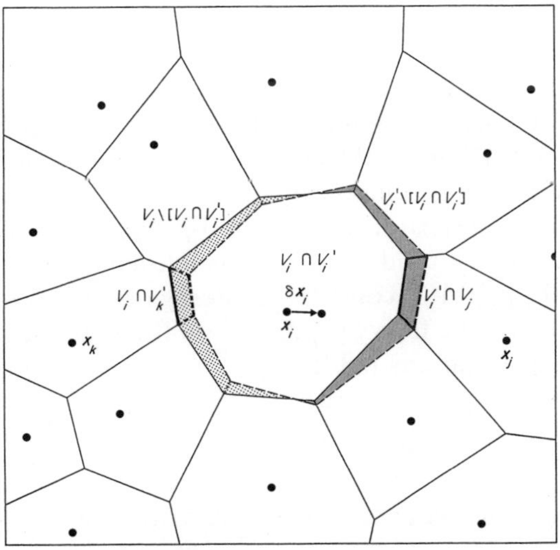

Figure 9.2.2 Variation in a Voronoi polygon.

V_j and V_k, j, $k \in J_i$, and the Voronoi edges between adjacent Voronoi polygons V'_j and V'_k, $j,k \in J_i$, are on the same line. From this property and equation (9.2.1), we obtain

$$\begin{aligned}\triangle F &= F(\mathbf{x}_1, \ldots, \mathbf{x}_i+\delta\mathbf{x}_i, \ldots, \mathbf{x}_n)-F(\mathbf{x}_1, \ldots, \mathbf{x}_i, \ldots, x_n) \\ &= \int_{V'_i} f(\|\mathbf{x}_i+\delta\mathbf{x}_i-\mathbf{x}\|^2)\phi(\mathbf{x})\mathrm{d}\mathbf{x} - \int_{V_i} f(\|\mathbf{x}_i-\mathbf{x}\|^2)\phi(\mathbf{x})\mathrm{d}\mathbf{x} \qquad (9.2.3) \\ &+ \sum_{j\in J_i}\left[\int_{V'_j} f(\|\mathbf{x}_j-\mathbf{x}\|^2)\phi(\mathbf{x})\mathrm{d}\mathbf{x} - \int_{V_j} f(\|\mathbf{x}_j-\mathbf{x}\|^2)\phi(\mathbf{x})\mathrm{d}\mathbf{x}\right].\end{aligned}$$

Since

$$V_i = (V_i\cap V'_i)\cup(V_i\backslash V_i\cap V'_i), \quad V'_i = (V_i\cap V'_i)\cup(V'_i\backslash V_i\cap V'_i), \qquad (9.2.4)$$

$$V_j = (V_j\cap V'_j)\cup(V'_i\cap V_j), \qquad V'_j = (V_j\cap V'_j)\cup(V_i\cap V'_j)$$

(see Figure 9.2.2), equation (9.2.3) is written as

$$\begin{aligned}\triangle F &= \int_{V'_i\cap V_i} [f(\|\mathbf{x}_i+\delta\mathbf{x}_i-\mathbf{x}\|^2) - f(\|\mathbf{x}_i-\mathbf{x}\|^2)]\phi(\mathbf{x})\mathrm{d}\mathbf{x} \\ &+ \int_{V'_i\backslash V_i\cap V'_i} f(\|\mathbf{x}_i+\delta\mathbf{x}_i-\mathbf{x}\|^2)\phi(\mathbf{x})\mathrm{d}\mathbf{x} - \int_{V_i\backslash V_i\cap V pm} f(\|\mathbf{x}_i-\mathbf{x}\|^2)\phi(\mathbf{x})\mathrm{d}\mathbf{x} \qquad (9.2.5) \\ &+ \sum_{j\in J_i}\left[\int_{V_i\cap V'_j} f(\|\mathbf{x}_j-\mathbf{x}\|^2)\phi(\mathbf{x})\mathrm{d}\mathbf{x} - \int_{V'_i\cap V_j} f(\|\mathbf{x}_j-\mathbf{x}\|^2)\phi(\mathbf{x})\mathrm{d}\mathbf{x}\right].\end{aligned}$$

In Figure 9.2.2 the integral domains $V'_i\backslash(V_i\cap V'_i)$ and $V_i\backslash(V_i\cap V'_i)$ are indicated by sparsely and densely shaded regions, respectively; the integral domains $V'_i\cap V_j$ and $V_i\cap V'_k$ are indicated by the regions surrounded by the heavy lines (note that $V_i\cap V'_j$ or $V'_i\cap V_j$ may be empty or may not, but at least one of the two is non-empty). Observing in Figure 9.2.2 that the lightly shaded region $V_i\backslash(V_i\cap V'_i)$ consists of subregions $V_i\cap V'_k$, $(k \in J_i)$ (one of which is indicated by the heavy solid lines on the left-hand side), and the densely shaded region $V'_i\backslash(V_i\cap V'_i)$ consists of subregions $V'_i\cap V_j$, $(j \in J_i)$ (one of which is indicated by the heavy solid lines on the right-hand side), we notice that the following equations hold:

$$V'_i\backslash(V_i\cap V'_i) = \sum_{j\in J_i} V'_i\cap V_j, \qquad V_i\backslash(V_i\cap V'_i) = \sum_{j\in J_i} V_i\cap V'_j. \qquad (9.2.6)$$

Using these equations, equation (9.2.5) is written as

$$\triangle F = \int_{V'_i\cap V_i} \{f(\|\mathbf{x}_i+\delta\mathbf{x}_i-\mathbf{x}\|^2)-f(\|\mathbf{x}_i-\mathbf{x}\|^2)\}\phi(\mathbf{x})\mathrm{d}\mathbf{x}$$

$$+\sum_{j\in J_i}\int_{V_i'\cap V_j}\{f(\|\mathbf{x}_i+\delta\mathbf{x}_i-\mathbf{x}\|^2)-f(\|\mathbf{x}_j-\mathbf{x}\|^2)\}\phi(\mathbf{x})\mathrm{d}\mathbf{x} \tag{9.2.7}$$

$$+\sum_{j\in J_i}\int_{V_i\cap V_j'}\{f(\|\mathbf{x}_j-\mathbf{x}\|^2)-f(\|\mathbf{x}_i-\mathbf{x}\|^2)\}\phi(\mathbf{x})\mathrm{d}\mathbf{x}.$$

As $\delta\mathbf{x}_i$ approaches zero, the integral domains $V_i\cap V_j'$ and $V_i'\cap V_j$ in equation (9.2.7) reduce to the Voronoi edge shared by V_i and V_j, and

$$\lim_{\delta\mathbf{x}_i\to 0}|V_i'\cap V_j|=0, \qquad \lim_{\delta\mathbf{x}_i\to 0}|V_i\cap V_j'|=0. \tag{9.2.8}$$

From this property and the property that equation $\|\mathbf{x}_i-\mathbf{x}\|=\|\mathbf{x}_j-\mathbf{x}\|$ (or $\|\mathbf{x}_i+\delta\mathbf{x}_i-\mathbf{x}\|=\|\mathbf{x}_j-\mathbf{x}\|$) holds if point $\mathbf{x}$ is on the Voronoi edge shared by V_i and V_j (or V_i' and V_j'), we have the relations

$$\|\mathbf{x}_i-\mathbf{x}\|\approx\|\mathbf{x}_j-\mathbf{x}\|, \quad \mathbf{x}\in V_i\cap V_j', j\in J_i, \tag{9.2.9}$$

$$\|\mathbf{x}_i+\delta\mathbf{x}_i-\mathbf{x}\|\approx\|\mathbf{x}_j-\mathbf{x}\|, \quad \mathbf{x}\in V_i'\cap V_j, j\in J_i,$$

for sufficiently small $\delta\mathbf{x}_i$. From this equation and the assumption that the functions $f(\|\mathbf{x}_i-\mathbf{x}_j\|)$ and $\phi(\mathbf{x})$ are continuous, the value of $|\{f(\|\mathbf{x}_i+\delta\mathbf{x}_i-\mathbf{x}\|^2)-f(\|\mathbf{x}_j-\mathbf{x}\|^2)\}\phi(\mathbf{x})|$ is of order $\|\delta\mathbf{x}_i\|$, denoted by $O(\|\delta\mathbf{x}_i\|)$, for $\mathbf{x}\in V_i'\cap V_j$. From equation (9.2.8), the area of the integral domain $V_i'\cap V_j$ is also $O(\|\delta\mathbf{x}_i\|)$ for sufficiently small $\delta\mathbf{x}_i$. Hence, $|V_i'\cap V_j|\{f(\|\mathbf{x}_i+\delta\mathbf{x}_i-\mathbf{x}\|^2) - f(\|\mathbf{x}_j-\mathbf{x}\|^2)\}\phi(\mathbf{x})$ is $O(\|\delta\mathbf{x}_i\|^2)$. Similarly, $|V_i\cap V_j'|\{f(\|\mathbf{x}_i-\mathbf{x}\|^2) - f(\|\mathbf{x}_j-\mathbf{x}\|^2)\}\phi(\mathbf{x})$ is $O(\|\delta\mathbf{x}_i\|^2)$. Therefore equation (9.2.7) is written as

$$\triangle F=\int_{V_i'\cap V_i}\{f(\|\mathbf{x}_i+\delta\mathbf{x}_i-\mathbf{x}\|^2)-f(\|\mathbf{x}_i-\mathbf{x}\|^2)\}\phi(\mathbf{x})\mathrm{d}\mathbf{x}+O(\|\delta\mathbf{x}_i\|^2), \tag{9.2.10}$$

from which we obtain

$$\frac{\partial F}{\partial x_{i\kappa}}=\lim_{\delta x_{i\kappa}\to 0}\frac{\triangle F}{\delta x_{i\kappa}}$$

$$=\lim_{\delta x_{i\kappa}\to 0}\left[\int_{V_i'\cap V_i}\frac{f(\|\mathbf{x}_i+\delta\mathbf{x}_i-\mathbf{x}\|^2)-f(\|\mathbf{x}_i-\mathbf{x}\|^2)}{\delta x_{i\kappa}}\phi(\mathbf{x})\mathrm{d}\mathbf{x}+\frac{O(\|\delta\mathbf{x}_i\|^2)}{\delta x_{i\kappa}}\right], \tag{9.2.11}$$

$\kappa=1,2.$

Therefore we obtain the first derivatives as

$$\frac{\partial F}{\partial x_{i\kappa}}=\int_{V_i}2(x_{i\kappa}-x_\kappa)f'(\|\mathbf{x}_i-\mathbf{x}\|^2)\phi(\mathbf{x})\mathrm{d}\mathbf{x}, \quad \kappa=1,2. \tag{9.2.12}$$

Note that this result holds not only for $\mathbf{x}_i^{\mathrm{T}}=(x_{i1}, x_{i2})$ and $\mathbf{x}^{\mathrm{T}}=(x_1, x_2)\in\mathbb{R}^2$, but also for $\mathbf{x}_i^{\mathrm{T}}=(x_{i1}, \ldots, x_{i\kappa}, \ldots, x_{im})$ and $\mathbf{x}^{\mathrm{T}}=(x_1, \ldots, x_k, \ldots, x_m)\in\mathbb{R}^m$.

In the latter case $\kappa = 1, 2$ is just replaced by $\kappa = 1, \ldots, m$.

Using the steepest descent method with this derivative, we can obtain a local optimal solution. Alternatively, as mentioned in Section 9.1, we may use the Newton method to gain faster convergence. For this use we need the second partial derivatives of F. The derivation of the second partial derivatives is much more lengthy than the above derivation. Since space is limited, we show only the result here. The detailed derivation is given by Suzuki (1988) and Asami (1991). Using the notation $\mathbf{x}_i^{\mathrm{T}} = (x_{i1}, \ldots, x_{i\kappa}, \ldots, x_{i\lambda}, \ldots, x_{im})$ and $\mathbf{x}^{\mathrm{T}} = (x_1, \ldots, x_\kappa, \ldots, x_\lambda, \ldots, x_m) \in \mathbb{R}^m$, the second partial derivatives are written as

$$\frac{\partial^2 F}{\partial x_{i\kappa}\partial x_{i\lambda}} = \int_{V_i} [2\delta(\kappa,\lambda)f'(\|\mathbf{x}_i-\mathbf{x}\|^2)+4(x_{i\kappa}-x_\kappa)(x_{i\lambda}-x_\lambda)f''(\|\mathbf{x}_i-\mathbf{x}\|^2)\phi(\mathbf{x})]\mathrm{d}\mathbf{x}$$
$$-\sum_{j\in J_i}\int_{V_i\cap V_j}\frac{2}{\|\mathbf{x}_j-\mathbf{x}_i\|^2}(x_{i\kappa}-x_\kappa)(x_{i\lambda}-x_\lambda)f'(\|\mathbf{x}_i-\mathbf{x}\|^2)\phi(\mathbf{x})\mathrm{d}\mathbf{x}, \quad (9.2.13)$$

where

$$\delta(\kappa,\lambda) = \begin{cases} 1 & \text{if } \kappa = \lambda, \\ 0 & \text{if } \kappa \neq \lambda,\ \kappa,\ \lambda \in I_m, \end{cases}$$

$$\frac{\partial^2 F}{\partial x_{i\kappa}\partial x_{j\lambda}} = \int_{V_i\cap V_j}\frac{2}{\|\mathbf{x}_j-\mathbf{x}_i\|}(x_{i\kappa}-x_\kappa)(x_{j\lambda}-x_\lambda)f'(\|\mathbf{x}_i-\mathbf{x}\|^2)\phi(\mathbf{x})\mathrm{d}\mathbf{x}, \quad (9.2.14)$$

$$j \neq i, j \in J_i,\ \kappa,\ \beta \in I_m,$$

$$\frac{\partial^2 F}{\partial x_{i\kappa}\partial x_{j\lambda}} = 0, \quad j \neq i, j \notin J_i \quad (9.2.15)$$

(note that $V_i \cap V_j$ is the Voronoi edge shared by V_i and V_j). Although we may use the Newton method with these second derivatives, the calculation of the Hessian becomes time consuming for a large number of facilities when the density $\phi(\mathbf{x})$ is not uniform. To shorten computing time, Iri *et al.* (1984) recommend, on the basis of their numerical experiment, to use the quasi-Newton method with the diagonal matrix H, whose diagonal elements are given by the first term in equation (9.2.13). Using this quasi-Newton method, we can solve Problem 9.2.1 in practice for a large number of facilities. (Note that when $\phi(\mathbf{x})$ is not uniform, Iri *et al.*, 1984, carry out the integration in equations (9.2.12)–(9.2.15) using the numerical integration formula shown in Abramowitz and Stegun, 1964.)

Figure 9.2.3 shows the optimal configuration of $n = 128$ points in a square S for three different density functions $\phi(\mathbf{x})$ (Iri *et al.*, 1984).

Using the same method, we can obtain the local optimal configuration of public mail boxes in Koganei. The result is shown in Figure 9.2.4. It is of interest to compare this local optimal configuration with the present configuration shown in Figure 9.2.1.

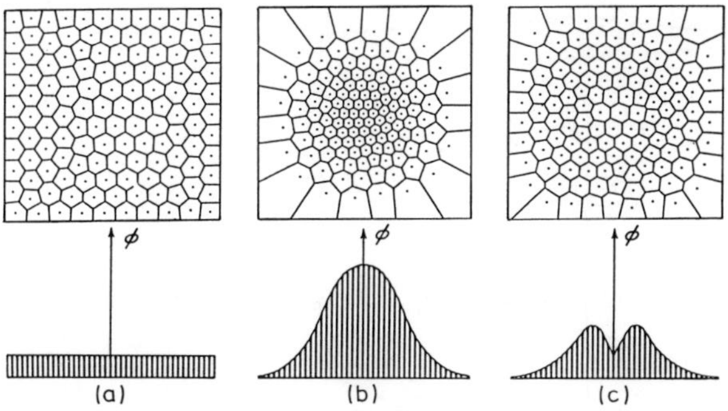

Figure 9.2.3 Optimal location of 128 points in a square for: (a) density function $\phi(\boldsymbol{x})$ given by $\phi(\boldsymbol{x}) = c$, $\boldsymbol{x} \in S$; (b) density function $\phi(\boldsymbol{x})$ given by $\phi(\boldsymbol{x}) = c \exp(-25\|\boldsymbol{x}\|)$, $\boldsymbol{x} \in S$; (c) density function $\phi(\boldsymbol{x})$ given by $\phi(\boldsymbol{x}) = c \exp(-\|\boldsymbol{x}\|(25\|\boldsymbol{x}-10\|))$, $\boldsymbol{x} \in S$. (*Source*: Iri *et al.* (1984, Figures 1 and 2).)

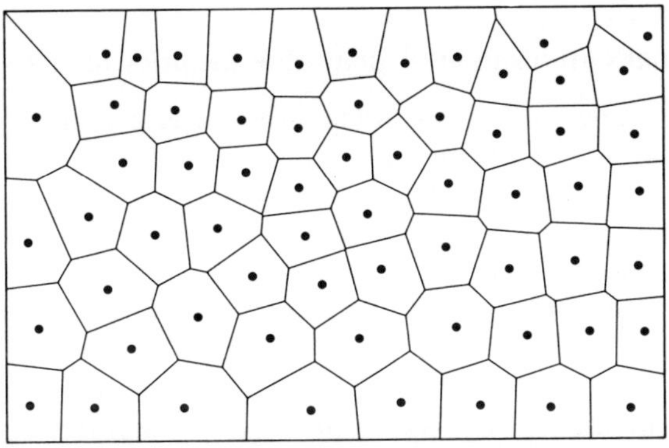

Figure 9.2.4 Locational optimization of public mail boxes in Koganei, Tokyo.

9.2.2 Locational optimization of points in a three-dimensional space

In the above example we formulated Problem 9.2.1 in a plane $\mathbb{R}^2$. We can also formulate the same problem in a space $\mathbb{R}^3$ and can solve the problem almost in the same manner using the three-dimensional Voronoi diagram. Figure 9.2.5 shows the optimal configuration of 16 points in a cube where the density function $\phi(\boldsymbol{x})$ is given by a uniform distribution (Takeda, 1985). This result is quite suggestive for crystalology. It is shown in Rogers (1964) that Voronoi polyhedra become truncated octahedra when generators are given by cubic lattice points in an infinite space (see also Chapter 7).

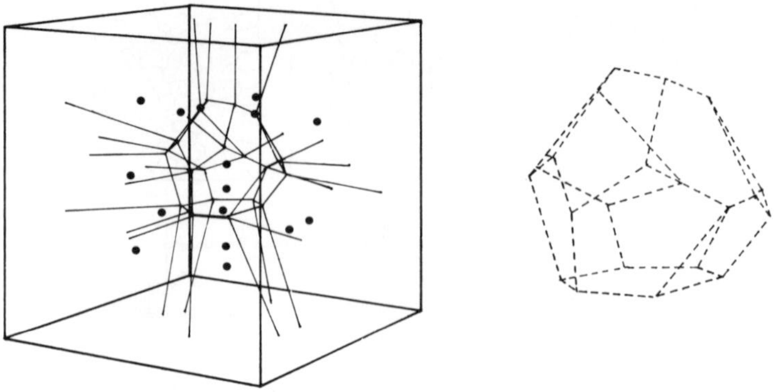

Figure 9.2.5 Optimal location of 16 points in a cube: (a) the Voronoi diagram of the optimized points; (b) a Voronoi polyhedron in (a). (*Source*: Takeda (1985, Figure 3.7).)

9.2.3 Locational optimization of point-like facilities used by groups

Facilities are not always used by independent users; some facilities are used by groups of users. For example, a tennis court is used by at least a pair of players; a convention hall is used by groups of certain members. This type of locational optimization problem is examined by Ohsawa and Suzuki (1987), upon which this subsection is based.

Suppose that there are n facilities located at $\mathbf{x}_1, \ldots, \mathbf{x}_n$ in region S, and that every group consists of m members. Groups are distributed over S according to a density function $\phi(\mathbf{y}_1, \ldots, \mathbf{y}_m)$, where $\mathbf{y}_1, \ldots, \mathbf{y}_m$ indicate the locations of m members. The value of $\phi(\mathbf{y}_1, \ldots, \mathbf{y}_m)\mathrm{d}\mathbf{y}_1, \ldots, \mathrm{d}\mathbf{y}_m$ indicates the number of groups whose members $1, \ldots, m$ are in a small region $(\mathrm{d}\mathbf{y}_1)$ around $\mathbf{y}_1, \ldots,$ and in a small region $(\mathrm{d}\mathbf{y}_m)$ around $\mathbf{y}_m$, respectively. Since m members in a group use the same facility, say facility i, the total travel cost of the group is given by the sum of the travel costs of the m members from their locations to the facility i, i.e. $\Sigma_{j=1}^{m} f(\|\mathbf{y}_j - \mathbf{x}_i\|^2)$.

We make two assumptions. First, every group chooses the facility that gives the minimum total travel cost, i.e. $\min_i \Sigma_{j=1}^{m} f(\|\mathbf{y}_j - \mathbf{x}_i\|^2)$. Second, every group uses the facility α times per unit time regardless of the travel cost to their nearest facility. For convenience, $\alpha = 1$ without loss of generality. Under these assumptions, the total travel cost of groups over region S is given by

$$F(\mathbf{x}_1, \ldots, \mathbf{x}_n) = \int_S \min_i \sum_{j=1}^{m} f(\|\mathbf{y}_j - \mathbf{x}_i\|^2)\phi(\mathbf{y}_1, \ldots, \mathbf{y}_m)\mathrm{d}\mathbf{y}_1 \ldots \mathrm{d}\mathbf{y}_m. \quad (9.2.16)$$

The locational optimization problem is hence to minimize this objective function with respect to $\mathbf{x}_1, \ldots, \mathbf{x}_n$. This problem is difficult to solve for a general travel cost function. The problem, however, becomes tractable in a specific case where the travel cost function is given by $f(\|\mathbf{y}_j-\mathbf{x}_i\|^2) = \|\mathbf{y}_j-\mathbf{x}_i\|^2$. For this travel cost function, the following equation holds:

$$\sum_{j=1}^{m} \|\mathbf{y}_j-\mathbf{x}_i\|^2 = m\|\bar{\mathbf{y}}-\mathbf{x}_i\|^2 + \sum_{j=1}^{m} \|\mathbf{y}_j-\bar{\mathbf{y}}\|^2, \tag{9.2.17}$$

where $\bar{\mathbf{y}}$ is the centroid of m members, i.e. $\bar{\mathbf{y}} = \Sigma_{i=1}^{m} \mathbf{y}_i/m$. Substituting equation (9.2.17) into equation (9.2.16) we obtain

$$\begin{aligned} F(\mathbf{x}_1, \ldots, \mathbf{x}_n) &= \int_S \min_i \sum_{j=1}^{m} \|\mathbf{y}_j-\mathbf{x}_i\|^2 \phi(\mathbf{y}_1, \ldots, \mathbf{y}_m)\mathrm{d}\mathbf{y}_1 \ldots \mathrm{d}\mathbf{y}_m \qquad (9.2.18) \\ &= \int_S \min_i \left[\{m\|\bar{\mathbf{y}}-\mathbf{x}_i\|^2 + \sum_{j=1}^{m} \|\mathbf{y}_j-\bar{\mathbf{y}}\|^2\} \right] \phi(\mathbf{y}_1, \ldots, \mathbf{y}_m)\mathrm{d}\mathbf{y}_1 \ldots \mathrm{d}\mathbf{y}_m. \end{aligned}$$

The first term, $\min_i\|\bar{\mathbf{y}}-\mathbf{x}_i\|^2$, implies the distance to the nearest facility from the centroid of m members. Obviously, if centroid $\bar{\mathbf{y}}$ is in the Voronoi polygon V_i, the nearest facility is the facility at $\mathbf{x}_i$. The second term, $\Sigma_{j=1}^{m} \|\mathbf{y}_j-\bar{\mathbf{y}}\|^2$, is the total cost from m members to their centroid, which is independent of i, and so $\min_i \Sigma_{j=1}^{m} \|\mathbf{y}_j-\bar{\mathbf{y}}\|^2 = \Sigma_{j=1}^{m} \|\mathbf{y}_j-\bar{\mathbf{y}}\|^2$. Thus, equation (9.2.18) is written as

$$F(\mathbf{x}_1, \ldots, \mathbf{x}_n) = m \sum_{i=1}^{n} \int_{V_i} \|\bar{\mathbf{y}}-\mathbf{x}_i\|^2 \phi_g(\bar{\mathbf{y}})\mathrm{d}\bar{\mathbf{y}}+c, \tag{9.2.19}$$

where $\phi_g(\bar{\mathbf{y}})$ is the density function of $\bar{\mathbf{y}}$ obtained from $\phi(\mathbf{y}_1, \ldots, \mathbf{y}_\mathrm{m})$, and c is a constant. Note that when $\phi(\mathbf{y}_1, \ldots, \mathbf{y}_m)$ is uniform over a unit square region, $\phi_g(\bar{\mathbf{y}}) = \phi_g(\bar{y}_1,\bar{y}_2)$ is written as

$$\phi_g(\bar{y}_1,\bar{y}_2) = \frac{m^2}{\{(m-1)!\}^2} \left\{ \sum_{i=1}^{[m\bar{y}_1]} \frac{(-1)^i i!(m\bar{y}_1-i)^{m-1}}{m!(m-i)!} \right\} \left\{ \sum_{j=1}^{[m\bar{y}_2]} \frac{(-1)^j j!(m\bar{y}_2-j)^{m-1}}{m!(m-j)!} \right\}, \tag{9.2.20}$$

where $[m\bar{y}_i]$ is the maximum integer less than or equal to $m\bar{y}_i$ (Mood *et al.*, 1974).

Comparing equation (9.2.19) with equation (9.2.2), we notice that the objective function of equation (9.2.19) is the same as that of equation (9.2.2) except that $f(\|\mathbf{x}-\mathbf{x}_i\|^2)$ is replaced by $\|\mathbf{x}-\mathbf{x}_i\|^2$. Therefore we can apply the same computational method used for Problem 9.2.1 to this locational optimization problem. Actually, using that method Ohsawa and Suzuki (1987) solved the problem for $n = 100$ facilities used by groups of $m = 2$, 4,8,16 members. The results are shown in Figure 9.2.6. We notice from Figure 9.2.6 that facilities tend to gather near the centre as the number of members constituting a group increases.

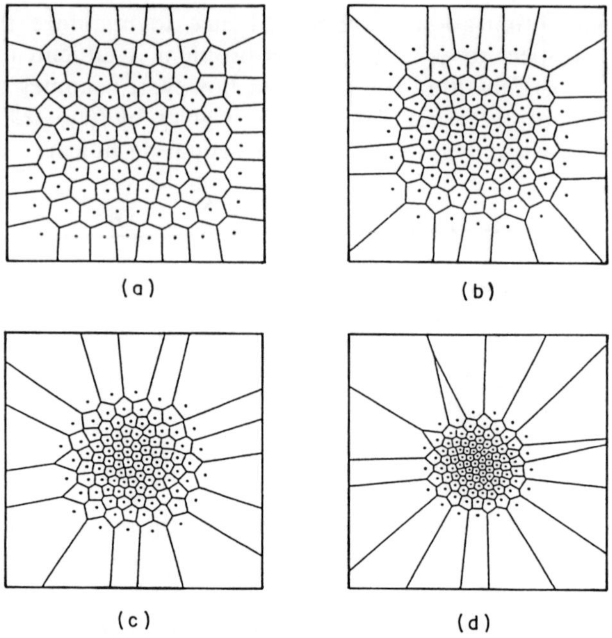

Figure 9.2.6 Optimal locations of 100 facilities used by a group of: (a) $m = 2$ members; (b) $m = 4$ members; (c) $m = 8$ members; (d) $m = 16$ members. (*Source*: Ohsawa and Suzuki (1987, Figure 3).)

9.2.4 Locational optimization of a hierarchical facility

We implicitly assume in the above that all facilities supply the same quality service, or facilities are indifferent, like public mail boxes. This assumption is not always appropriate for some kinds of facilities. For example, in Japan, post offices are ranked from the first to the third according to the level of postal service; a university library system usually consists of the main library and a few branch libraries. We call such a set of facilities a *hierarchical facility*.

Besides a hierarchical facility, we also observe hierarchy in a set of cities in a country. The study of this hierarchy, which has been investigated in spatial economics since Christaller (1933) and Lösch (1940), is worth noting in conjunction with Voronoi diagrams. Christaller (1933) formulates hierarchical systems of cities with Voronoi diagrams generated by triangular lattice points. A few examples are shown in Figures 9.2.7 and 9.2.10 (which will be discussed with the optimal configurations obtained in Figure 9.2.9).

The locational optimization of a hierarchical facility has been studied mainly in Operations Research. Dokmeci (1973) solved this problem for discrete demand with the integer programming. Moore and ReVelle (1982)

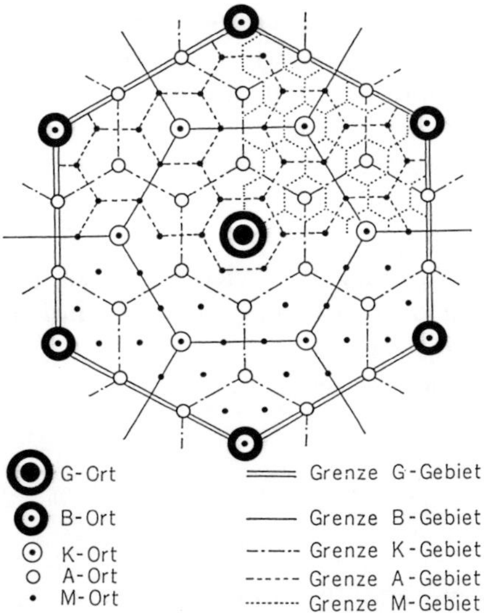

Figure 9.2.7 Hierarchical system of cities. (*Source*: Christaller (1933, Figure 2).)

viewed a hierarchical facility from a covering problem. Tien *et al.* (1983) dealt with the problem of a hierarchical health facility for discrete demand. O'Kelly and Storbeck (1984) took probabilistic allocation into account. Narula (1984) examined the characteristics of several hierarchical facilities and classified the types of hierarchy. Hodgson (1984) solved a multi-level locational problem with the gravity model. Mirchandani (1987) generalized Tien *et al.*'s (1983) model.

As is noticed in those articles, most of the models are formulated on a network; few models are formulated on a continuous plane. Recalling the computational method used in Problem 9.2.1, one may notice that the same method can be applied to the above problem. In fact, T. Suzuki (1989) has applied that method to the locational optimization problem of a hierarchical facility on a continuous plane. In this subsection, we follow T. Suzuki's (1989) model.

Suppose that there are n facilities in region S, and those facilities are ranked from 1 to m. Let n_i be the number of facilities of rank i, ($\Sigma_{i=1}^{m} n_i = n$), and $\mathbf{x}_1, \ldots, \mathbf{x}_n$ be the locations of n facilities. For convenience, the first n_1 locations of $\mathbf{x}_1, \ldots, \mathbf{x}_n$ are those of rank 1 facilities; the next n_2 locations are those of rank 2 facilities, and so forth.

Suppose next that n facilities supply k kinds of services, $s_1, \ldots, s_k$. Service s_j may be supplied by all facilities or may be supplied by facilities of certain ranks. First, we consider the case in which facilities of rank i supply services $s_i, \ldots, s_m$, implying that facilities of rank i supply their proper service as well as the services supplied by facilities of lower ranks $i+1, \ldots, m$. This type of hierarchy is called a *successively inclusive hierarchy* (Tien *et al.*, 1983) or a *nested hierarchy* (Banerji and Fisher, 1974).

We make two assumptions. First, a user who needs service s_j goes to the nearest facility among the facilities supplying service s_j (the facilities of rank 1, ..., j). As a result, service areas of the facilities supplying service s_j are given by the Voronoi diagram $\mathcal{V}^{(i)} = \{V_1^{(j)}, ..., V_{n_1+...+n_j}^{(j)}\}$ generated by $\{\mathbf{x}_1, ..., \mathbf{x}_{n_1+...+n_j}\}$. An example of $m = 3$ is depicted in Figure 9.2.8. Second, every user consumes α_j units of service s_j, $j \in I_k$, per unit time regardless of the Euclidean distance to the facilities. For convenience, α_i's are standardized as $\Sigma_{j=1}^{k} \alpha_j = 1$.

Under these assumptions the average travel cost of users over region S is written as

$$F(\mathbf{x}_1, ..., x_n) = \sum_{i=1}^{k} \alpha_i \sum_{j=n_1+...+n_{i-1}+1}^{n} \int_{V_j^{(i)}} \|\mathbf{x}_j - \mathbf{x}\| \phi(\mathbf{x}) \mathrm{d}\mathbf{x}. \qquad (9.2.21)$$

The computational method for minimizing this objective function is the same as that of Problem 9.2. Using that method, T. Suzuki (1989) obtained the optimal configurations for $n = 37$, $m = 2$ and $\alpha_1 = \alpha_2 = 0.5$. The results are shown in Figure 9.2.9 where $n_1 = 6,7,10$.

As is noticed from Figure 9.2.9, the optimal configuration varies according to the ratio of n_2 to n_1. This ratio K is an important parameter in Christaller's (1933) model. In Figure 9.2.9, the K values are 5.2 in (a), 4.3 in (b) and 2.7 in (c). Christaller (1933) refers to $K = 3,4,7$ (Figure 9.2.10) as the hierarchical systems resulting from the market principle, the transportation principle, and the administrative principle, respectively. Since Christaller (1933) formulated the model on an infinite plane, whereas T. Suzuki (1989) formulated it on a bounded plane, we cannot directly compare their results.

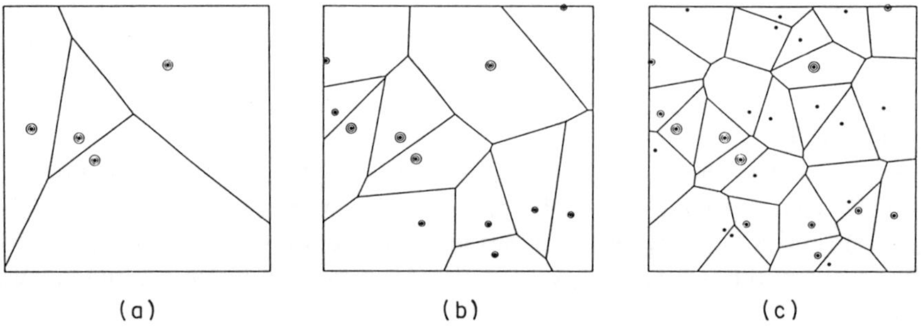

Figure 9.2.8 A nested hierarchical facility: (a) the Voronoi diagram $\mathcal{V}^{(1)}$ generated by the locations of facilities supplying service s_1 (facilities of rank 1); (b) the Voronoi diagram $\mathcal{V}^{(2)}$ generated by the locations of facilities supplying service s_2 (facilities of rank 1 and rank 2); (c) the Voronoi diagram $\mathcal{V}^{(3)}$ generated by the locations of facilities supplying service s_3 (facilities of ranks 1, 2 and 3).

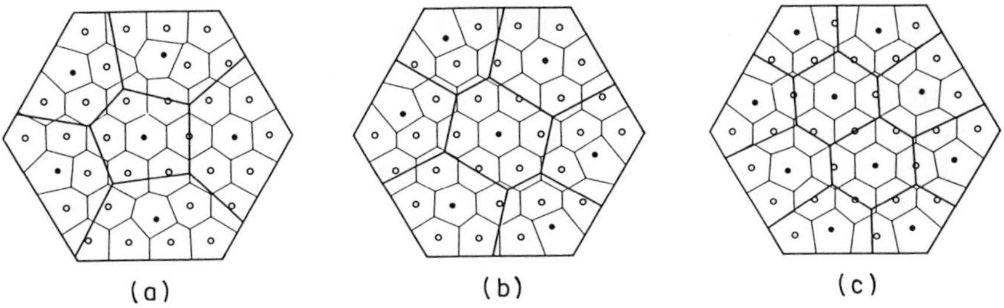

Figure 9.2.9 Optimal configurations of a hierarchical facility, $n = 37$, $m = 2$, $\alpha_1 = \alpha_2 = 0.5$: (a) $n_1 = 6$; (b) $n_1 = 7$; (c) $n_1 = 10$. (*Source*: T. Suzuki (1989, Figure 4.10).)

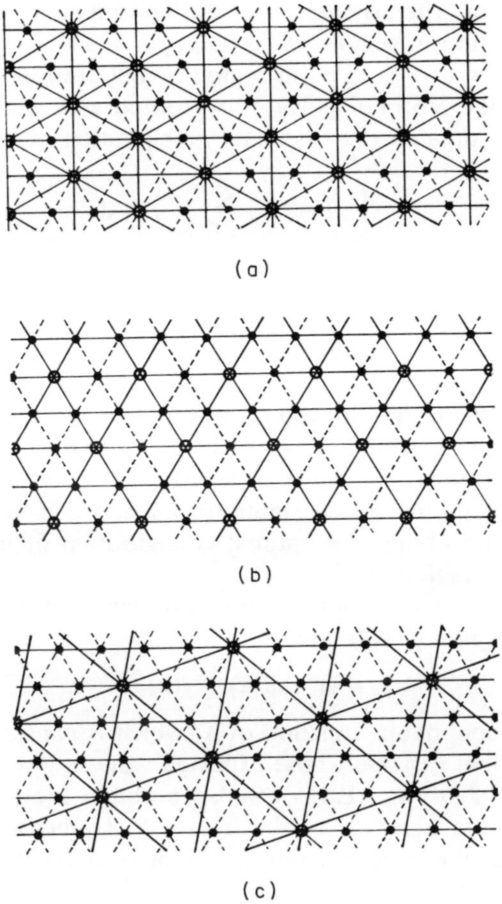

Figure 9.2.10 Christaller's (1933) hierarchical systems of cities: (a) $K = 3$; (b) $K = 4$; (c) $K = 7$.

Figure 9.2.9, however, provides a new look at hierarchical systems of cities from the optimization view point.

Hierarchy is not always nested. Hierarchy may be partially nested or successively exclusive. We can formulate the optimization problems of such a hierarchical facility almost in the same manner as in the above, and can solve them with the descent method. The reader who is interested in this problem should see T. Suzuki (1989).

9.2.5 Locational optimization of observation points for estimating the total quantity of a spatial variable continuously distributed over a plane

As we mentioned in Chapter 1, the Voronoi diagram is alternatively called the Thissen diagram. The original use of the Thissen diagram was to estimate the total precipitation in region S. To be explicit, suppose that there are n pluviometers placed at $\mathbf{x}_1, \ldots, \mathbf{x}_n$ in region S. Let $z(\mathbf{x}_i)$ be the precipitation at $\mathbf{x}_i$, and V_i be the Voronoi polygon associated with a pluviometer at $\mathbf{x}_i$. Thiessen (1911) estimated the true total precipitation in region S, i.e.

$$Z = \int_S z(\mathbf{x})\mathrm{d}\mathbf{x} = \sum_{i=1}^{n} \int_{V_i} z(\mathbf{x})\mathrm{d}\mathbf{x}, \tag{9.2.22}$$

by

$$\hat{Z} = \sum_{i=1}^{n} V_i z(\mathbf{x}_i). \tag{9.2.23}$$

In Thiessen (1911), the locations of pluviometers are fixed, and no locational optimization problem is discussed. However, if we can freely choose those locations, we have a locational optimization problem to minimize the estimation error. A similar problem is considered by Hori and Nagata (1985), who obtain the optimal configuration of monitor points to estimate NO_x in a region.

To formulate the above optimization problem, let $z(\mathbf{x})$ be a random variable at $\mathbf{x}$ in region S as follows:

$$z(\mathbf{x}) = m(\mathbf{x}) + \epsilon(\mathbf{x}), \tag{9.2.24}$$

where $m(\mathbf{x})$ is a continuous function of $\mathbf{x}$, which is constant for a given $\mathbf{x}$, and $\epsilon(\mathbf{x})$ is a random variable whose expected value is given by $\mathrm{E}[\epsilon(\mathbf{x})] = 0$. From equation (9.2.24) and $\mathrm{E}[\epsilon(\mathbf{x})] = 0$, we obtain $\mathrm{E}[z(\mathbf{x})] = m(\mathbf{x})$, which indicates the average trend of variable z over S. To measure the estimation error, we consider the expected squared error over region S, i.e.

$$F(\mathbf{x}_1, \ldots, \mathbf{x}_n) = \mathrm{E}\left[\sum_{i=1}^{n} \int_{V_i} \{z(\mathbf{x}) - z(\mathbf{x}_i)\}^2 \mathrm{d}\mathbf{x}\right]. \tag{9.2.25}$$

Upon substituting equation (9.2.24) into equation (9.2.25), we obtain

$$F(\mathbf{x}_1, \dots, \mathbf{x}_n) = \sum_{i=1}^{n} \int_{V_i} \mathrm{E}[\{m(\mathbf{x})-m(\mathbf{x}_i)\}^2] + \mathrm{E}[\epsilon(\mathbf{x})^2] + \mathrm{E}[\epsilon(\mathbf{x}_i)^2] - 2\mathrm{E}[\epsilon(\mathbf{x})\epsilon(\mathbf{x}_i)] + 2\mathrm{E}[\epsilon(\mathbf{x})\{m(\mathbf{x})-m(\mathbf{x}_i)\}] + 2\mathrm{E}[\epsilon(\mathbf{x}_i)\{m(\mathbf{x})-m(\mathbf{x}_i)\}]\mathrm{d}\mathbf{x}. \quad (9.2.26)$$

The first term, $\mathrm{E}[\{m(\mathbf{x})-m(\mathbf{x}_i)\}^2]$, is written as $\{m(\mathbf{x})-m(\mathbf{x}_i)\}^2$ because $m(\mathbf{x})$ is constant for a given $\mathbf{x}$. The second and third terms are variances, which are denoted by $\sigma(\mathbf{x})$ and $\sigma(\mathbf{x}_i)$. The fourth term is written as $\mathrm{E}[\epsilon(\mathbf{x})\epsilon(\mathbf{x}_i)] = \sigma(\mathbf{x})\,\sigma(\mathbf{x}_i)\,\mathrm{Cov}(\mathbf{x},\mathbf{x}_i)$, where $\mathrm{Cov}(\mathbf{x},\mathbf{x}_i)$ is the covariance. The fifth and sixth terms are zero, because $\mathrm{E}[\epsilon(\mathbf{x})\ \{m(\mathbf{x})-m(\mathbf{x}_i)\}] = \mathrm{E}[\epsilon(\mathbf{x})]\ \{m(\mathbf{x})-m(\mathbf{x}_i)\} = 0$. Thus equation (9.2.26) is written as

$$F(\mathbf{x}_1, \dots, \mathbf{x}_n) = \sum_{i=1}^{n} \int_{V_i} \{m(\mathbf{x})-m(\mathbf{x}_i)\}^2 + \{\sigma(\mathbf{x})-\sigma(\mathbf{x}_i)\}^2 + 2\sigma(\mathbf{x})\sigma(\mathbf{x}_i)\ \{1-\mathrm{Cov}(\mathbf{x},\mathbf{x}_i)\}\mathrm{d}\mathbf{x}. \quad (9.2.27)$$

The locational optimization problem is hence to minimize F of equation (9.2.27) with respect to $\mathbf{x}_i, \dots, \mathbf{x}_n$.

First, let us consider the case in which the variance $\sigma(\mathbf{x})$ is independent of $\mathbf{x}$, and constant for all points in region S, i.e. $\sigma(\mathbf{x}) = \sigma$; the change in the average trend $m(\mathbf{x})$ is small compared with the variance σ. In this case, equation (9.2.27) is approximated by

$$F(\mathbf{x}_1, \dots, \mathbf{x}_n) = \sum_{i=1}^{n} \int_{V_i} 2\sigma^2\{1-\mathrm{Cov}(\mathbf{x},\mathbf{x}_i)\}\mathrm{d}\mathbf{x}. \quad (9.2.28)$$

Empirically, the covariance $\mathrm{Cov}(\mathbf{x},\mathbf{x}_i)$ tends to monotonically decrease as $\|\mathbf{x}-\mathbf{x}_i\|$ increases, and so $1-\mathrm{Cov}(\mathbf{x},\mathbf{x}_i)$ monotonically increases as $\|\mathbf{x}-\mathbf{x}_i\|$ increases. Let $f(\|\mathbf{x}-\mathbf{x}_i\|)$ be this increasing function. Then, the optimization problem reduces to

$$\min_{\mathbf{x}_1, \dots, \mathbf{x}_n} \sum_{i=1}^{n} \int_{V_i} 2\sigma^2\ \{1-\mathrm{Cov}(\mathbf{x},\mathbf{x}_i)\}\mathrm{d}\mathbf{x} = \min_{\mathbf{x}_1, \dots, \mathbf{x}_n} \sum_{i=1}^{n} \int_{V_i} f(\|\mathbf{x}-\mathbf{x}_i\|)\mathrm{d}\mathbf{x}. \quad (9.2.29)$$

We readily notice from equation (9.2.2) that the problem given by expression (9.2.29) is a special case of Problem 9.2.1.

Let us next consider the case in which the variance $\sigma(\mathbf{x})$ is small compared with the change in the average trend $m(\mathbf{x})$. In this case, equation (9.2.27) is approximated by

$$F(\mathbf{x}_1, \dots, \mathbf{x}_n) = \sum_{i=1}^{n} \int_{\mathbf{x}\in V_i} \{m(\mathbf{x})-m(\mathbf{x}_i)\}^2\mathrm{d}\mathbf{x}. \quad (9.2.30)$$

The locational optimization problem is to minimize this function with respect to $\mathbf{x}_1, \ldots, \mathbf{x}_n$. The exact solution of this minimization problem is difficult to obtain for a general function $m(\mathbf{x})$. We may, however, obtain a solution which is possibly close to the optimal solution from the following relation. For a given V_i, the following relation holds:

$$\int_{V_i} \{m(\mathbf{x})-m(\mathbf{x}_i)\}^2 \mathrm{d}\mathbf{x} \geq \int_{V_i} \{m(\mathbf{x})-m_i)\}^2 \mathrm{d}\mathbf{x} \qquad (9.2.31)$$

where

$$m_i = \frac{1}{V_i}\int_{V_i} m(\mathbf{x})\mathrm{d}\mathbf{x}. \qquad (9.2.32)$$

We notice from this relation that if $m(\mathbf{x}_i)$ happens to be equal to m_i, that is, if the value of $m(\mathbf{x}_i)$ at the generator $\mathbf{x}_i$ of the Voronoi polygon V_i happens to be equal to the average value of $m(\mathbf{x})$ over V_i, the value of the left-hand side of equation (9.2.31) achieves the minimum value. This property suggests that the solution may be given by a set $\mathbf{x}_1, \ldots, \mathbf{x}_n$ that satisfies $m(\mathbf{x}_i) = m_i$ (or $m(\mathbf{x}_i) \approx m_i$ in practice). To find such a set, Hori and Nagata (1985) proposed the following heuristic or crude method.

First, we construct the Voronoi diagram $\{V_1^{(0)}, \ldots, V_n^{(0)}\}$ (the heavy solid lines in Figure 9.2.11) generated by an initial generator set $\{\mathbf{x}_1^{(0)}, \ldots, \mathbf{x}_n^{(0)}\}$ (the filled circles in Figure 9.2.11). Hori and Nagata (1985) recommend using the initial generator set given by the optimal solution of Problem 9.2, where $f(\|\mathbf{x}-\mathbf{x}_i\|^2) = \|\mathbf{x}-\mathbf{x}_i\|^2$, $\phi(\mathbf{x}) = m(\mathbf{x})$. Second, we construct the Delaunay triangulation for this Voronoi diagram (the light solid lines in Figure 9.2.11). Let $\mathbf{u}_{ij}^{(0)}, j \in I_{n_i}$ be the Voronoi points of the Voronoi polygon $V_i^{(0)}$, where n_i is the number of Voronoi vertices of $V_i^{(0)}$ (the empty circles in Figure 9.2.11); $T_{ij}^{(0)}$ be a Delaunay triangle which corresponds to vertex $\mathbf{u}_{ij}^{(0)}$; and $\mathbf{x}_{ij1}^{(0)}$ and $\mathbf{x}_{ij2}^{(0)}$ be two vertices of $T_{ij}^{(0)}$ other than $\mathbf{x}_i^{(0)}$. Third, we find

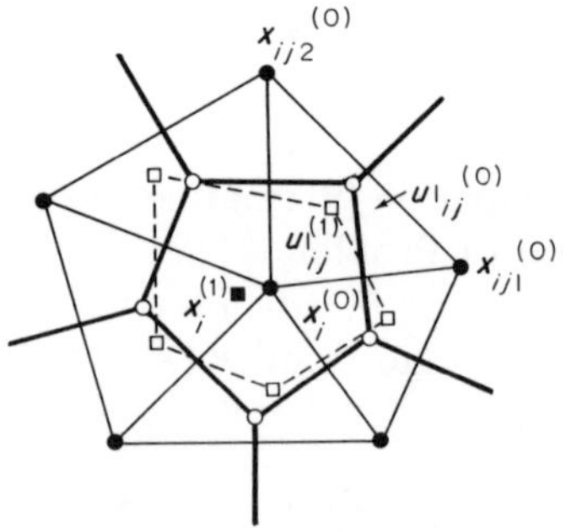

Figure 9.2.11 A heuristic method for searching for $m(\mathbf{x}_i) = m_i$.

a point $u_{ij}^{(1)}$ in $T_{ij}^{(0)}$ or in the circumcircle of $T_{ij}^{(0)}$ that satisfies $m(u_{ij}^{(1)}) = \{m(x_i^{(0)}) + m(x_{ij1}^{(0)}) + m(x_{ij2}^{(0)})\}/3$ (if such a point does not exist, we find a point $u_{ij}^{(0)}$ in such a way that $m(u_{ij}^{(1)})$ is close to $\{m(x_i^{(0)}) + m(x_{ij1}^{(0)}) + m(x_{ij2}^{(0)})\}/3$) (the empty squares in Figure 9.2.11). Fourth, we construct the polygon V'_i whose vertices are given by $u_{ij}^{(1)}, j \in I_{n_i}$ (the broken lines in Figure 9.2.11). Fifth, in this polygon, we find a point $x_i^{(1)}$ that satisfies equation (9.2.32), where $m_i = m(x_i^{(1)})$ and $V_i = V'_i$ (if we cannot find such a point, we find a point $x_i^{(1)}$ in such a way that $m(x_i^{(1)})$ is close to the left-hand side of equation (9.2.32)) (the filled squares in Figure 9.2.11). Last, we generate a Voronoi diagram with $\{x_1^{(1)}, \ldots, x_n^{(1)}\}$. We continue this procedure until $x_i^{(k)}, i \in I_n$, converges. It should be noted, however, that this procedure does not guarantee convergence.

According to Hori and Nagata (1985), the variance in the density of NO_x is fairly small. Hori and Nagata (1985) hence optimized the objective function given by equation (9.2.30) with the above heuristic method. Figure 9.2.12(b) shows the near-optimal configuration of 16 monitor points placed in Kyoto provided that $m(x)$ is given by Figure 9.2.12(a).

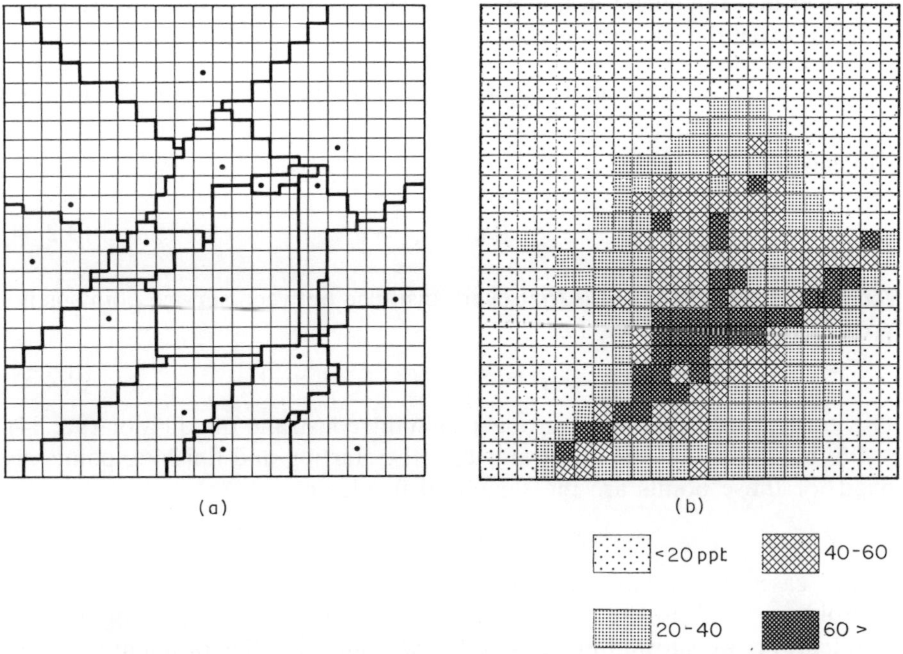

Figure 9.2.12 (a) Density of NO_x in Kyoto, and (b) the optimal configuration of 16 observation points. Redrawn from Hori and Nagata (1985, Figures 6 and 7).

9.2.6 Locational optimization of service points of a mobile facility

Facilities are not always fixed at the same locations. Some facilities are mobile and stop at several points in a region to provide service. An example is a bookmobile (Takeda, 1985). A bookmobile offers a library service at some fixed points in a region for a certain length of time. Library users go to their nearest service points when a bookmobile stops at those points. If a bookmobile could stop at every house, the user's convenience would be maximum. Because of a time constraint, however, the sum of the library service time and travel time should be not greater than a certain limit. The locational optimization problem is hence to minimize the average distance (or cost) of users to their nearest service points provided that the number of service points and the total travel distance of the bookmobile are given.

The mathematical formulation of this locational optimization problem is almost the same as that of Problem 9.2 except for constraints. The objective function is given by equation (9.2.1). If we assume that the total travel distance of the bookmobile is given by the sum of the Euclidean distance between the successive service points, the bookmobile problem is formulated as follows.

Problem 9.2.2 (minimization of the average travel cost to the nearest points provided that the length of a line successively connecting those service points is not greater than a certain limit)

$$\min_{\mathbf{x}_1, \ldots, \mathbf{x}_n} \sum_{i=1}^{n} \int_{V_i} f(\|\mathbf{x}-\mathbf{x}_i\|^2)\phi(\mathbf{x})d\mathbf{x}, \tag{9.2.33}$$

subject to

$$\sum_{i=1}^{n-1} d(\mathbf{x}_i, \mathbf{x}_{i+1}) \le d^*. \tag{9.2.34}$$

An alternative constraint is that every distance between service points is not greater than $d^{**} = d^*/n$, i.e.

$$d(\mathbf{x}_i, \mathbf{x}_{i+1}) \le d^{**}, \quad i = 1, \ldots, n-1. \tag{9.2.35}$$

Obviously, this constraint is stronger than the constraint of equation (9.2.34).

We may also add the constraint that the starting and finishing points are fixed, or those points are the same and fixed, i.e.

$$\mathbf{x}_1 = \mathbf{c}, \quad \mathbf{x}_n = \mathbf{c}' \quad \text{or} \quad \mathbf{x}_1 = \mathbf{x}_n = \mathbf{c}. \tag{9.2.36}$$

In this case, the variables in expression (9.2.33) are $\mathbf{x}_2, \ldots, \mathbf{x}_{n-1}$.

Problem 9.2.2 is the constrained non-linear non-convex programming problem. As we showed in Section 9.1, we can solve this problem using the penalty function method. Note that when n is large, the penalty function method is not efficient. Alternatively, we may use the multiplier method.

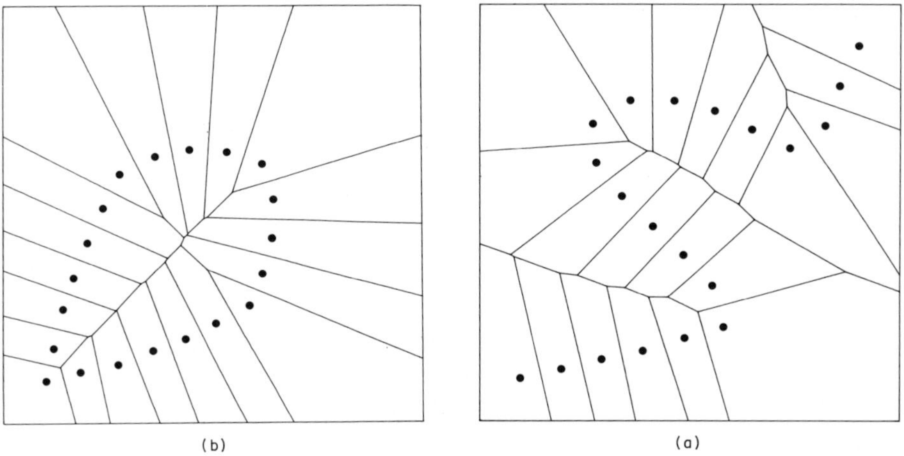

Figure 9.2.13 Locational optimization of $n = 20$ service points: (a) $d^{**} = 0.1$, $\mathbf{x}_1^{\mathrm{T}} = (0.1, 0.1)$ $\mathbf{x}_{20}^{\mathrm{T}} = (0.9, 0.9)$; (b) $d^{**} = 0.1$, $\mathbf{x}_1^{\mathrm{T}} = \mathbf{x}_{20}^{\mathrm{T}} = (0.15, 0.15)$. (*Source*: Takeda (1985, Figure 5.3).)

Figure 9.2.13 shows the optimal configuration of $n = 20$ service points obtained by Takeda (1985).

9.2.7 Locational optimization of terminal points through which users go to a central point

On a network we have terminal facilities, such as stations on a railway network, bus stops on a bus route, access points in a communication network, and so forth. Such networks sometimes have central points. For example, consider a bus route in a residential area. Commuters walk from their homes to bus stops, and take a bus to a railway station, from which they go to their places of work. In this case the central point is the railway station at which a bus terminates. Obviously, commuters want to reach the bus station as fast as possible. The problem is thus to seek a set of locations of bus stops that minimizes the average travel time of commuters. This type of optimization problem is examined by Lesley (1976), Vaughan and Cousins (1977), Wirasinghe and Ghoneim (1981), and T. Suzuki (1987), among others.

To formulate the above problem mathematically, let us consider a bus route L in region S (Figure 9.2.14(a)) on which n bus stops are placed. Those bus stops are indexed from 1 to n from the station. Let l_i be the route distance from bus stop i to bus stop $i+1$, $i = 0, 1, \ldots, n$. Note that bus stop 0 is fixed at the station, and l_{n+1} is the distance from bus stop n to the end of the bus route L from which a bus departs (a bus garage). The location of a point on route L can be represented by a route distance s

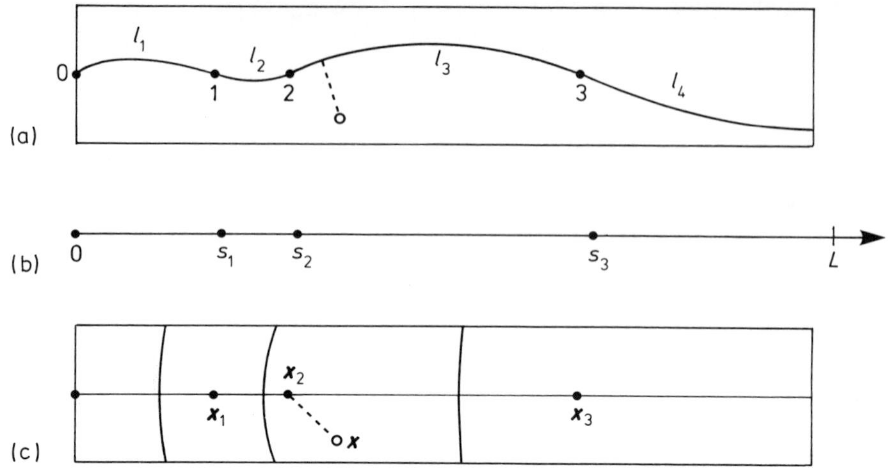

Figure 9.2.14 Bus stops on a bus route.

from the station. The location of bus stop i is hence indicated by $s = s_i = \Sigma_{j=1}^{i} l_j$. Commuters are distributed over S according to the density function $\phi(\mathbf{x})$.

First we assume that every commuter walks from his/her home (the unfilled circle in Figure 9.2.14(a)) to the nearest point (called an access point) on bus route L, and then walks from this access point to a bus stop along bus route L. Under this assumption, the choice of a bus stop depends upon only the distances from the access point to alternative bus stops; the choice does not depend upon the distance from his/her home to the access point. Hence, the two-dimensional problem reduces to a one-dimensional problem (Figure 9.2.14(b)). To be explicit, let $\phi_L(s)$ be the density of commuters whose nearest point on bus route L from their homes is point s. Once $\phi_L(s)$ is obtained from $\phi(\mathbf{x})$ (the density of commuters on a plane), the problem becomes a problem of optimizing n bus stops on a line segment L over which commuters are distributed with density function $\phi_L(s)$.

Let v_b be the speed of a bus, v_w be walking speed and b be the time stopping at every bus stop. We assume that every commuter has a time schedule and arrives at a bus stop when a bus just reaches it. Then, the total travel time from point s to bus stop i on foot and from this bus stop to the station by bus is given by $|s-s_i|/v_w+s_i/v_b+ib$. Obviously, the relation

$$b + \frac{l_i}{v_b} < \frac{l_i}{v_w} \tag{9.2.37}$$

should be satisfied; otherwise, walking is faster than taking a bus. Since every commuter minimizes the travel time, commuters at point s choose the bus stop that gives the minimum travel time, i.e. $\min_i\{|s-s_i|/v_w+s_i/v_b+ib\}$. As a result, every bus stop has its catchment area, which is given by

$$V_i = \left\{ s \,\middle|\, \frac{|s-s_i|}{v_w} + \frac{s_i}{v_b} + ib \le \frac{|s-s_j|}{v_w} + \frac{s_j}{v_b} + jb, \text{ for } j \ne i, j \in I_n \right\}. \quad (9.2.38)$$

As we defined in Chapter 3, this region is the weighted Voronoi region generated on the line L.

From equation (9.2.38), the total travel time of all commuters in region L (or S) is given by

$$F(l_1, \ldots, l_n) = \sum_{i=0}^{n} \int_{V_i} \left\{ \frac{|s-s_i|}{v_w} + \frac{s_i}{v_b} + ib \right\} \phi_L(s) \mathrm{d}s. \quad (9.2.39)$$

The optimization problem is hence written as

$$\min_{l_1, \ldots, l_n} \sum_{i=0}^{n} \int_{V_i} \left\{ \frac{|s-s_i|}{v_w} + \frac{s_i}{v_b} + ib \right\} \phi_L(s) \mathrm{d}s, \quad (9.2.40)$$

subject to

$$b + \frac{l_i}{v_b} < \frac{l_i}{v_w} \quad \text{for } i \in I_n, \quad (9.2.41)$$

$$\sum_{i=1}^{n} l_i \le L, \quad (9.2.42)$$

$$l_i > 0 \quad \text{for } i \in I_{n+1}. \quad (9.2.43)$$

This problem is a constrained non-linear programming problem, which can be solved, as shown in Section 9.1, by use of the penalty function method. When the density function $\phi_L(s)$ is a uniform distribution, the objective function is explicitly written as

$$\begin{aligned} F(l_1, \ldots, l_n) = & \frac{1}{4v_b^2 v_w} (v_b^2+2v_b v_w - v_w^2) \sum_{i=1}^{n} l_i + \frac{1}{v_b} \sum\sum_{i \ne j} l_i l_j \\ & + b \sum_{i=1}^{n} (i-1) l_i + \frac{v_b - v_w}{2v_b} b \sum_{i=1}^{n} l_i - \frac{v_w b^2}{4} n \\ & + \frac{l_{n+1}}{v_b} \sum_{i=1}^{n} l_i + nbl_{n+1} + \frac{l_{n+1}^2}{2v_w}. \end{aligned} \quad (9.2.44)$$

In this case the problem becomes a constrained quadratic programming problem (a special case of the constrained non-linear programming problem).

In the above we assume that commuters first go to the nearest point on route L. Instead, if we assume that commuters take the Euclidean path to a bus stop (the broken line in Figure 9.2.14(c)), the problem cannot be reduced to a one-dimensional problem; the problem should be formulated on a plane. Essentially, the formulation is the same as in the one-dimensional

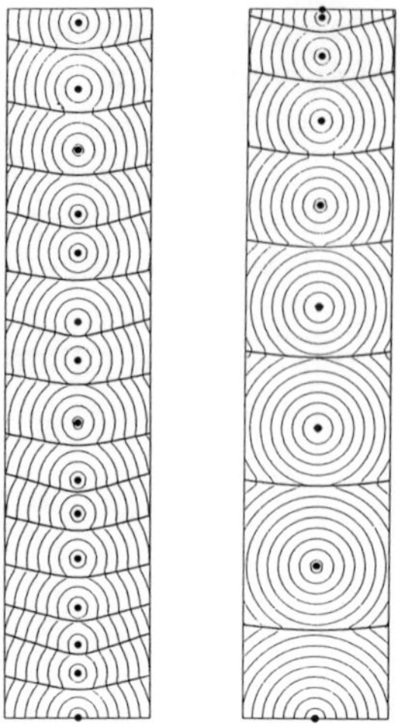

Figure 9.2.15 Locational optimization of bus stops on the Motoyahata–Takatsuka line: (a) the optimal location of bus stops; (b) the present location of bus stops. (*Source*: T. Suzuki (1987, Figure 8).)

case. In place of equation (9.2.38), the catchment area of the bus stop at s_i is given by

$$V_i = \left\{ \mathbf{x} \left| \frac{\|\mathbf{x}-\mathbf{x}_i\|}{v_\mathrm{w}} + \frac{s_i}{v_\mathrm{b}} + ib \leq \frac{\|\mathbf{x}-\mathbf{x}_j\|}{v_\mathrm{w}} + \frac{s_j}{v_\mathrm{b}} + jb, j \neq i, j \in I_n \right. \right\}. \quad (9.2.45)$$

An example is shown in Figure 9.2.14(c). The objective function is thus given by

$$F(l_1, \ldots, l_n) = \sum_{i=0}^{n} \int_{V_i} \left\{ \frac{\|\mathbf{x}-\mathbf{x}_i\|}{v_\mathrm{w}} + \frac{s_i}{v_\mathrm{b}} + ib \right\} \phi(\mathbf{x}) \mathrm{d}s. \quad (9.2.46)$$

The constraints are the same as those given by relations (9.2.37), (9.2.41), (9.2.42) and (9.2.43).

The calculation of the integral in equation (9.2.46) is fairly complex, but we can carry it out in practice. In particular, when the density $\phi(\mathbf{x})$ is uniform, this integral is obtained explicitly. The result is, however, extremely lengthy, and so it is omitted here (see T. Suzuki, 1987, who obtained the results with the help of REDUCE-III).

T. Suzuki (1987) applied the above optimization problem to the Motoyahata–Takatsuka bus line in Ichikawa. This line was chosen because the line is mainly used for commuting to Motoyahata station; the bus route is almost straight; and the population density $\phi(\mathbf{x})$ is fairly uniform. From the empirical survey, the parameter values are observed as $v_\mathrm{b} = 8.00$ m/s,

$v_w = 1.20$ m/s, $b = 30$ s and $L = 4865.7$ m. T. Suzuki (1987) obtained optimal solutions for $n = 2, \ldots, 34$. These solutions indicate that the total travel is minimized when $n = 8$. The optimal locations of eight bus stops are depicted in Figure 9.2.15(a). It is of interest to compare this figure with the actual locations of bus stops shown in Figure 9.2.15(b). The travel time in the optimal solution is shorter than the actual travel time by 2 min on average. Obviously, the optimal solution depends upon an objective function. Instead of minimizing the total travel time, we may maximize the profit of a bus company. This alternative optimization (including another criteria) is discussed in the literature cited above.

9.3 LOCATIONAL OPTIMIZATION OF LINES

In the previous section we discussed the locational optimization of points. In this section, extending the method developed there, we consider the locational optimization of lines. We deal with three types of lines: a service route (Section 9.3.1), a network (Section 9.3.2) and the Steiner tree (Section 9.3.3).

9.3.1 Locational optimization of a service route

We first consider the locational optimization of a service route. Generally, the problem is to seek an optimal service route of a mobile facility that provides some service at any point on the service route. Such mobile services are numerous in Japan. The well-known example is a wagon collecting old newspapers. The wagon travels a route calling with a loudspeaker, 'we are exchanging old newspapers for tissue paper'. Hearing that call, a resident who wants to sell old newspapers goes to the nearest service route to catch the wagon. The wagon then comes to the resident's home and the resident gets tissue paper in exchange for old newspapers. More traditional examples in Japan are an ice-cream vendor riding on a bicycle with a bell, a bean card (tofu) vendor pulling a cart blowing a bugle, a fermented soybeans (natto) vendor calling 'natto' with a resonant voice, and so forth. The problem for these vendors is to find the optimal service route in a region that yields the maximum profit.

To discuss the above problem more precisely, let us consider, for example, an optimal service route of an ice-cream vendor. We suppose, for simplicity, that only one ice-cream vendor exclusively sells ice-cream in a region, and that a service route can be freely placed in a region (a road network is so dense that the route can be almost freely placed in a region). The starting point and the finishing point of the service route may be the same or may not. We assume that every customer buys ice-cream at the nearest point on the service route; the demand for ice-cream is inversely proportional to the average travel cost from all customers to their nearest points on the service route; the ice-cream vendor can satisfy all demand; and the marginal cost of a unit amount of ice-cream is constant. Under these assumptions, the profit-maximization problem reduces to the problem of minimizing the average travel cost. Obviously, as the length of the service route increases, the average travel cost becomes shorter, and the profit increases. Because of a time constraint, however, the vendor cannot walk infinitely; the travel distance in a day is limited. The problem is hence to determine the service route to minimize the average travel cost provided that the length of the service route is not greater than a certain limit.

To formulate the above problem mathematically, we represent a service route in region S by the chain of straight line segments (a curved line is approximated by the chain of small straight line segments). Let $\mathbf{x}_1, \ldots, \mathbf{x}_n$ be the end points of the straight line segments; $\overline{\mathbf{x}_i\mathbf{x}_{i+1}}$ be the open straight line segment connecting points $\mathbf{x}_i$ and $\mathbf{x}_{i+1}$; and $\{V(\mathbf{x}_1), \ldots, V(\mathbf{x}_n), V(\overline{\mathbf{x}_1\mathbf{x}_2}), \ldots, V(\overline{x_{n-1}x_n})\}$ be the line Voronoi diagram generated by $\mathbf{x}_1, \ldots, \mathbf{x}_n, \overline{\mathbf{x}_1\mathbf{x}_2}, \ldots, \overline{\mathbf{x}_{n-1}\mathbf{x}_n}\}$ (Section 3.5). From the definition of the line Voronoi diagram, if point $\mathbf{x}$ is in the Voronoi region $V(\mathbf{x}_i)$, the Euclidean distance from point $\mathbf{x}$ to the nearest point on the service route is given by $\|\mathbf{x}-\mathbf{x}_i\|$; if point $\mathbf{x}$ is in the Voronoi region $V(\overline{\mathbf{x}_i\mathbf{x}_{i+1}})$, the Euclidean distance $d_l(\mathbf{x}, \overline{\mathbf{x}_i\mathbf{x}_{i+1}})$ from point $\mathbf{x}$ to the nearest point on the service route is given by the Euclidean distance from point $\mathbf{x}$ to the line segment $\overline{\mathbf{x}_i\mathbf{x}_{i+1}}$, which is explicitly given by

$$d_l(\mathbf{x}, \overline{\mathbf{x}_i\mathbf{x}_{i+1}}) = \left[\|\mathbf{x}-\mathbf{x}_i\|^2 - \frac{\{(\mathbf{x}-\mathbf{x}_i)^{\mathrm{T}}(\mathbf{x}_{i+1}-\mathbf{x}_i)\}^2}{\|\mathbf{x}_{i+1}-\mathbf{x}_i\|^2}\right]^{1/2}. \tag{9.3.1}$$

Let $\phi(\mathbf{x})$ be the demand density at $\mathbf{x}$ and $f(\|\mathbf{x}-\mathbf{x}_i\|^2)$ be a strictly increasing travel cost function of the squared Euclidean distance $\|\mathbf{x}-\mathbf{x}_i\|^2$. Note that the total demand is assumed to be unity without loss of generality.

In these terms, the average travel cost to the service route is written as

$$\begin{aligned} F(\mathbf{x}_1, \ldots, \mathbf{x}_n) = & \sum_{i=1}^{n} \int_{V(\mathbf{x}_i)} f(\|\mathbf{x}-\mathbf{x}_i\|^2)\phi(\mathbf{x})\mathrm{d}\mathbf{x} \\ & + \sum_{i=1}^{n-1} \int_{V(\overline{\mathbf{x}_i\mathbf{x}_{i+1}})} f(d_l(\mathbf{x}, \overline{\mathbf{x}_i\mathbf{x}_{i+1}})^2)\phi(\mathbf{x})\mathrm{d}\mathbf{x}. \end{aligned} \tag{9.3.2}$$

When $f(\|\mathbf{x}-\mathbf{x}_i\|^2) = \|\mathbf{x}-\mathbf{x}_i\|$, and the demand density is uniform, i.e. $\phi(\mathbf{x}) = 1/S$ over region S, we can explicitly obtain the value of equation (9.3.2) for given $\mathbf{x}_1, \ldots, \mathbf{x}_n$ (see Section 8.3).

With the above objective function, we can formulate the optimal service route problem as follows.

Problem 9.3.1 (minimization of the average travel cost to the nearest point on a service line provided that the length of the service line is not greater than a certain limit)

$$\min_{\mathbf{x}_1,\ldots,\mathbf{x}_n}\left[\sum_{i=1}^{n}\int_{V(\mathbf{x}_i)} f(\|\mathbf{x}-\mathbf{x}_i\|^2)\phi(\mathbf{x})\mathrm{d}\mathbf{x} + \sum_{i=1}^{n-1}\int_{V(\overline{\mathbf{x}_i\mathbf{x}_{i+1}})} f(d_l(\mathbf{x},\overline{\mathbf{x}_i\mathbf{x}_{i+1}})^2)\phi(\mathbf{x})\mathrm{d}\mathbf{x}\right], \tag{9.3.3}$$

subject to

$$\sum_{i=1}^{n-1}\|\mathbf{x}_i-\mathbf{x}_{i+1}\| \le d^*. \tag{9.3.4}$$

We may add the constraints $\mathbf{x}_1 = \mathbf{c}$, $\mathbf{x} = \mathbf{c}'$, or $\mathbf{x}_1 = \mathbf{x}_n = \mathbf{c}$, where $\mathbf{c}$ and $\mathbf{c}'$ are constants. In this case the variables in expression (9.3.3) are $\mathbf{x}_2, \ldots, \mathbf{x}_{n-1}$.

Like Problem 9.2.1, Problem 9.3.1 is a non-linear non-convex programming problem. To solve this problem, as mentioned in Section 9.1, we should use a computational method, such as the penalty function method. If we adopt the steepest descent method in the unconstrained programming problem transformed from the constrained programming problem of Problem 9.3.1 (recall Section 9.1.3), we need the first derivative. The derivation of the first derivative is almost the same as was shown in Problem 9.2.1 (equation (9.2.2)). A slight move of $\mathbf{x}_i$ produces changes in the Voronoi regions $V(\mathbf{x}_i)$, $V(\overline{\mathbf{x}_{i-1}\mathbf{x}_i})$ and $V(\overline{\mathbf{x}_i\mathbf{x}_{i+1}})$ and their adjacent Voronoi regions. Since the changes in the integral domains are cancelled out (recall the derivation from equation (9.2.7) to equation (9.2.12)), the first derivative is obtained from the first derivative of the integrand,

$$\begin{aligned}\frac{\partial F}{\partial x_{i\kappa}} &= \int_{V(\mathbf{x}_i)}\frac{\partial}{\partial x_{i\kappa}} f(\|\mathbf{x}-\mathbf{x}_i\|^2)\phi(\mathbf{x})\mathrm{d}\mathbf{x} \\ &+ \int_{V(\overline{\mathbf{x}_{i-1}\mathbf{x}_i})}\frac{\partial}{\partial x_{i\kappa}} f(d_l(\mathbf{x},\overline{\mathbf{x}_{i-1}\mathbf{x}_i})^2)\phi(\mathbf{x})\mathrm{d}\mathbf{x} \\ &+ \int_{V(\overline{\mathbf{x}_i\mathbf{x}_{i+1}})}\frac{\partial}{\partial x_{i\kappa}} f(d_l(\mathbf{x},\overline{\mathbf{x}_i\mathbf{x}_{i+1}})^2)\phi(\mathbf{x})\mathrm{d}\mathbf{x}, \quad \kappa = 1, 2.\end{aligned} \tag{9.3.5}$$

Since $d_l(\mathbf{x}, \overline{\mathbf{x}_i\mathbf{x}_{i+1}})$ is explicitly given by equation (9.3.1), and the function $f(\mathbf{x})$ is assumed to be differentiable, we can calculate the integrals in equation (9.3.5) once the function $f(\mathbf{x})$ is explicitly given (the first integral is the same as equation (9.2.12)).

Using the penalty function method and the steepest descent method with this derivative, we can solve Problem 9.3.1. Actually, Takeda (1985) solved

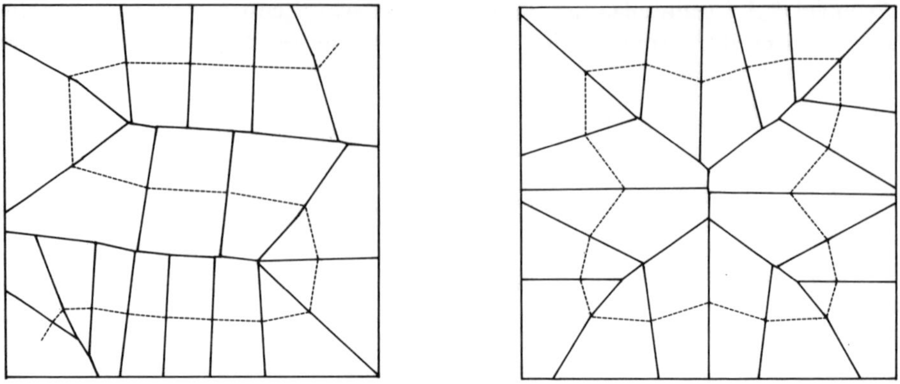

(a) (b)

Figure 9.3.1 Locally optimal service routes in a unit square. Parameter values are: $d^* = \sqrt{2}/10$, $\phi(\mathbf{x}) = 1$, $f(\|\mathbf{x}-\mathbf{x}_i\|^2) = \|\mathbf{x}-\mathbf{x}_i\|^2$; $n = 20$ in (a), $n = 21$ in (b). (*Source*: Takeda (1985, Figure 5.10).)

this problem with a slight modification (the quasi-Newton method) for $f(\|\mathbf{x}-\mathbf{x}_i\|^2) = \|\mathbf{x}-\mathbf{x}\|^2$, $n = 20$, $\phi(\mathbf{x})$ being uniform over a unit square, $d^* = \sqrt{2}/10$ and $\mathbf{x}_1^{\mathrm{T}} = (0.1,0.1)$, $\mathbf{x}_{20}^{\mathrm{T}} = (0.9,0.9)$; $n = 21$, $\mathbf{x}_1^{\mathrm{T}} = \mathbf{x}_{21}^{\mathrm{T}} = (0.15,0.15)$. The results are depicted in Figure 9.3.1.

9.3.2 Locational optimization of a network

In most network optimization problems the location of nodes and links, or at least the location of links, is fixed (for example, recall the bus stop problem in Section 9.2.1). The locational optimization of both nodes and links was rarely studied because of complicated geometrical computation. The recent progress in Voronoi diagrams, however, provides a clue to this difficult problem. In fact, using the Voronoi diagram, Suzuki and Iri (1986c) attempted to solve the optimization problem in which the locations of both nodes and links were to be optimized so that the total travel time was minimized, provided that the connection of links of the network and the traffic volumes between origins and destinations were given.

Let us consider a railway network in region S in which nodes are stations and links are railways connecting the stations. We assume that the speed of a train on this railway network is controlled as follows: first, the speed is accelerated from a station with α until it reaches $v_{\max}$; next, the speed is kept constant at $v_{\max}$; last, the speed is decelerated with $-\alpha$ until it becomes zero at the next station. To avoid unnecessary complicated equations, we assume that the distance between two adjacent stations is greater than $v_{\max}^2/\alpha$, implying that a train achieves the maximum speed between stations.

When a traveller makes a trip from origin $\mathbf{x}_o$ to destination $\mathbf{x}_d$, we assume that the traveller first walks to the nearest station $\mathbf{x}^*(\mathbf{x}_o)$, from origin $\mathbf{x}_o$ at walking speed v_w (since the traveller knows the time schedule of trains, he/she reaches the station just when a train arrives there); next, the traveller takes a train from station $\mathbf{x}^*(\mathbf{x}_o)$ to the station $\mathbf{x}^*(\mathbf{x}_d)$, which is the nearest station from destination $\mathbf{x}_d$; last, the traveller walks from $\mathbf{x}^*(\mathbf{x}_d)$ to destination $\mathbf{x}_d$ at walking speed v_w. To obtain the total travel time of this trip, let $N[\mathbf{x}^*(\mathbf{x}_o), \mathbf{x}^*(\mathbf{x}_d)]$ be the number of stations between station $\mathbf{x}^*(\mathbf{x}_o)$ and station $\mathbf{x}^*(\mathbf{x}_d)$; $d_{\text{net}}(\mathbf{x}^*(\mathbf{x}_o), \mathbf{x}^*(\mathbf{x}_d))$ be the railway distance between station $\mathbf{x}^*(\mathbf{x}_o)$ and station $\mathbf{x}^*(\mathbf{x}_d)$; and b be the time spent at each station. A train runs a distance $v_{\max}^2/\alpha$ while it is accelerating and decelerating, and the train runs a distance $d_{\text{net}}(\mathbf{x}^*(\mathbf{x}_o), \mathbf{x}^*(\mathbf{x}_d)) - N[\mathbf{x}^*(\mathbf{x}_o), \mathbf{x}^*(\mathbf{x}_d)]\, v_{\max}^2/\alpha$ while it keeps a constant speed $v_{\max}$. Since it takes $2v_{\max}/\alpha$ (minutes) for acceleration and deceleration, the total travel time from origin $\mathbf{x}_o$ to destination $\mathbf{x}_d$ is given by

$$
\begin{aligned}
T(\mathbf{x}_o, \mathbf{x}_d) = {} & \frac{1}{v_w}\{\|\mathbf{x}_o - \mathbf{x}^*(\mathbf{x}_o)\| + \|\mathbf{x}_d - \mathbf{x}^*(\mathbf{x}_d)\|\} \\
& + \left(\frac{v_{\max}}{\alpha} + b\right)\{N[\mathbf{x}^*(\mathbf{x}_o), \mathbf{x}^*(\mathbf{x}_d)] + 1\} \\
& + \frac{1}{v_{\max}} d_{\text{net}}(\mathbf{x}^*(\mathbf{x}_o), \mathbf{x}^*(\mathbf{x}_d)).
\end{aligned}
\tag{9.3.6}
$$

To consider the traffic volume between two points in S, suppose that a traveller in a small region around $\mathbf{x}_o$ makes trips to see friends who are distributed over the region S. The number of friends in a small region $(\mathrm{d}\mathbf{x})$ around $\mathbf{x}_d$ is proportional to the number of inhabitants there. In terms of the density function of inhabitants, $\phi(\mathbf{x}_d)$, this number is given by $c_1\phi(\mathbf{x}_d)\mathrm{d}\mathbf{x}$, where c_1 is a constant. We assume that the traveller makes $c_1\phi(\mathbf{x}_d)\mathrm{d}\mathbf{x}$ trips to a small region around $\mathbf{x}_d$ during a unit period of time. We also assume that the number of trip-makers in a small region around $\mathbf{x}_o$ is proportional to the number of inhabitants there, i.e. $c_2\phi(\mathbf{x}_o)\mathrm{d}\mathbf{x}$, where c_2 is a constant. Thus the number of trips from a small region around $\mathbf{x}_o$ to a small region around $\mathbf{x}_d$ is given by $c_1c_2\phi(\mathbf{x}_o)\phi(\mathbf{x}_d)\,\mathrm{d}\mathbf{x}_o\mathrm{d}\mathbf{x}_d = \phi(\mathbf{x}_o)\phi(\mathbf{x}_d)\,\mathrm{d}\mathbf{x}_o\mathrm{d}\mathbf{x}_d$, where we fix $c_1c_2 = 1$ without loss of generality. In region S, the total round travel time of all travellers in a unit period of time is given by

$$
F(\mathbf{x}_1, \ldots, \mathbf{x}_n) = 2\int_S T(\mathbf{x}_o, \mathbf{x}_d)\phi(\mathbf{x}_o)\phi(\mathbf{x}_d)\mathrm{d}\mathbf{x}_o\mathrm{d}\mathbf{x}_d. \tag{9.3.7}
$$

The nearest stations, $\mathbf{x}^*(\mathbf{x}_o)$ and $\mathbf{x}^*(\mathbf{x}_d)$, are readily obtained from the Voronoi diagram $\{V_1, \ldots, V_n\}$ generated by a set of n stations at $\mathbf{x}_1, \ldots, \mathbf{x}_n$. Obviously, if $\mathbf{x}_o \in V_i$, then $\mathbf{x}^*(\mathbf{x}_o) = \mathbf{x}_i$; if $\mathbf{x}_d \in V_j$, then $\mathbf{x}^*(\mathbf{x}_d) = \mathbf{x}_j$. We thus write $N[\mathbf{x}^*(\mathbf{x}_o), \mathbf{x}^*(\mathbf{x}_d)]$ as $N[\mathbf{x}_i, \mathbf{x}_j]$ for $\mathbf{x}_o \in V_i$, $\mathbf{x}_d \in V_j$, and equation (9.3.7) is written as

$$F(\mathbf{x}_1, \ldots, \mathbf{x}_n) = 2\sum_{i=1}^{n}\int_{V_i}\frac{1}{v_\mathrm{w}}\|\mathbf{x}_\mathrm{o}-\mathbf{x}_i\|\phi(\mathbf{x}_\mathrm{o})\mathrm{d}\mathbf{x}_\mathrm{o} + 2\sum_{j=1}^{n}\int_{V_j}\frac{1}{v_\mathrm{w}}\|\mathbf{x}_\mathrm{d}-\mathbf{x}_j\|\phi(\mathbf{x}_\mathrm{d})\mathrm{d}\mathbf{x}_\mathrm{d}$$

$$+ 2\sum_{i=1}^{n}\sum_{\substack{j=1\\ j\neq i}}^{n}\left\{\left(\frac{v_{\max}}{\alpha}+b\right)(N[\mathbf{x}_i, \mathbf{x}_j]+1) + \frac{1}{v_{\max}}d_\mathrm{net}(\mathbf{x}_i, \mathbf{x}_j)\right\} \qquad (9.3.8)$$

$$\times \int_{V_i}\phi(\mathbf{x}_\mathrm{o})\mathrm{d}\mathbf{x}_o \int_{V_j}\phi(\mathbf{x}_\mathrm{d})\mathrm{d}\mathbf{x}_\mathrm{d}.$$

The first term is the round travel time from (to) origins in region V_i to (from) station $\mathbf{x}_i$. The second term is the round travel time from (to) station $\mathbf{x}_j$ to (from) destinations in region V_j. The last term is the round travel time from (to) station $\mathbf{x}_i$ to (from) station $\mathbf{x}_j$. Notice that the first and second terms are the same.

To sum up, the network locational optimization problem is formulated as follows.

Problem 9.3.2 (locational optimization of a network by minimizing the total travel time)

$$\min_{\mathbf{x}_1, \ldots, \mathbf{x}_n}\left[\sum_{i=1}^{n}\int_{V_i}\frac{1}{v_\mathrm{w}}\|\mathbf{x}-\mathbf{x}_i\|\phi(\mathbf{x})\mathrm{d}\mathbf{x}\right.$$

$$+\sum_{i=1}^{n}\sum_{j\geq i+1}^{n}\left\{\left(\frac{v_{\max}}{\alpha}+b\right)(N[\mathbf{x}_i, \mathbf{x}_j]+1) + \frac{1}{v_{\max}}d_\mathrm{net}(\mathbf{x}_i, \mathbf{x}_j)\right\} \qquad (9.3.9)$$

$$\left.\times\int_{V_i}\phi(\mathbf{x})\mathrm{d}\mathbf{x}\int_{V_i}\phi(\mathbf{x})\mathrm{d}\mathbf{x}\right]$$

(note that constant 4 is omitted in expression (9.3.9)). We may add the constraints that $\mathbf{x}_1 = \mathbf{c}$ and $\mathbf{x}_n = \mathbf{c}'$, or $\mathbf{x}_1 = \mathbf{x}_n = \mathbf{c}$.

Again, Problem 9.3.2 is a non-linear non-convex programming problem. We may solve it with the steepest descent method. The first derivative is obtained from

$$\frac{\partial}{\partial x_{i\kappa}}F(\mathbf{x}_1, \ldots, \mathbf{x}_n) = \int_{V_i}\frac{1}{v_\mathrm{w}}\frac{\partial}{\partial x_{i\kappa}}\|\mathbf{x}-\mathbf{x}_i\|\phi(\mathbf{x})\mathrm{d}\mathbf{x}$$

$$+\sum_{i=1}^{n}\sum_{j\geq i+1}^{n}\frac{1}{v_{\max}}\left\{\frac{\partial}{\partial x_{i\kappa}}d_\mathrm{net}(\mathbf{x}_i, \mathbf{x}_j)\right\}\int_{V_i}\phi(\mathbf{x})\mathrm{d}\mathbf{x}\int_{V_j}\phi(\mathbf{x})\mathrm{d}\mathbf{x}$$

$$+\sum_{i=1}^{n}\sum_{j=i+1}^{n}\left\{\left(\frac{v_{\max}}{\alpha}+b\right)(N[\mathbf{x}_i, \mathbf{x}_j]+1) + \frac{1}{v_{\max}}d_\mathrm{net}(\mathbf{x}_i, \mathbf{x}_j)\right\} \qquad (9.3.10)$$

$$\times\left\{\int_{V_j}\phi(\mathbf{x})\mathrm{d}\mathbf{x}\frac{\partial}{\partial x_{i\kappa}}\int_{V_i}\phi(\mathbf{x})\mathrm{d}\mathbf{x} + \int_{V_i}\phi(\mathbf{x})\mathrm{d}\mathbf{x}\frac{\partial}{\partial x_{i\kappa}}\int_{V_j}\phi(\mathbf{x})\mathrm{d}\mathbf{x}\right\},$$

$$\kappa = 1,2.$$

The first term is the same as equation (9.2.12). The second term indicates the change in the railway distance due to the slight move of station $\mathbf{x}_i$. If the railway is assumed to be a chain of lines connecting stations with straight line segments, the value of this term is obtained from

$$\frac{\partial}{\partial x_{i\kappa}} d_{\mathrm{net}}(\mathbf{x}_i, \mathbf{x}_j) = \frac{\partial}{\partial x_{i\kappa}} \{\|\mathbf{x}_{i-1} - \mathbf{x}_i\| + \|\mathbf{x}_i - \mathbf{x}_{i+1}\|\}, \quad \kappa = 1,2. \tag{9.3.11}$$

The last term in equation (9.3.10) treats the change in the area of Voronoi polygons. The derivation is almost the same as that in Section 7.4 (we will also refer to the derivation in Section 9.5).

Using the steepest descent method with this derivative, we can solve Problem 9.3.2. An actual computation is carried out by Suzuki and Iri (1986c) with a modified method (the quasi-Newton method). Figure 9.3.2 shows one of their results, where $n = 16$, $S = \{(x_1,x_2)|-0.5 \leq x_1,x_2 \leq 0.5\}$, $\mathbf{x}_1^{\mathrm{T}} = (-0.9, -0.9)$, $\mathbf{x}_{16}^{\mathrm{T}} = (0.9, 0.9)$, $v_{\max} = 0.01$, $\alpha = 0.002$, $v_{\mathrm{w}} = 0.001$, and

$$\phi(\mathbf{x}) = \begin{cases} \exp(-25\|\mathbf{x}\|) & \text{if } \mathbf{x} \in S, \\ 0 & \text{if } \mathbf{x} \notin S. \end{cases} \tag{9.3.12}$$

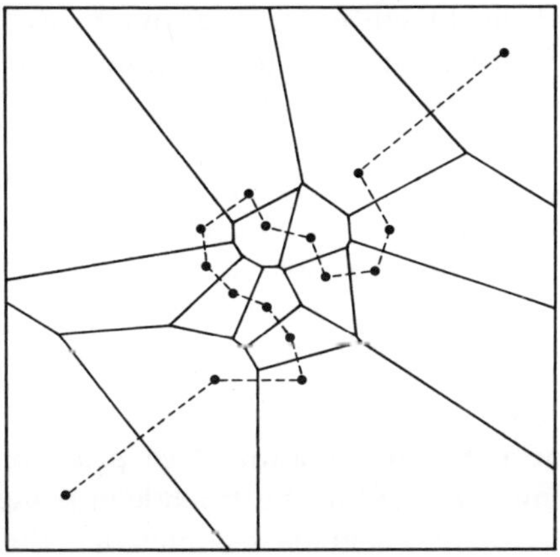

Figure 9.3.2 A locally optimal railway and stations: $n = 16$, $S = \{(x_1, x_2)|-0.5 \leq x_1, x_2 \leq 0.5\}$, $\mathbf{x}_1^{\mathrm{T}} = (-0.9, -0.9)$, $\mathbf{x}_{16}^{\mathrm{T}} = (0.9, 0.9)$, $v_{\max} = 0.01$, $\alpha = 0.002$, $v_{\mathrm{w}} = 0.001$, and $\phi(\mathbf{x})$ is given by equation (9.3.12). (*Source*: Suzuki and Iri (1986c, Figure 1).)

9.3.3 Euclidean Steiner tree

In Section 2.5 we referred to the Euclidean minimum spanning tree (EMST). As is noticed from the term 'minimum', the EMST problem may be regarded as a locational optimization problem of lines, and this problem can be solved with the Delaunay triangulation (Property D21). In this subsection we show another locational optimization problem of lines, called the Steiner problem, which also can be solved with the Delaunay triangulation.

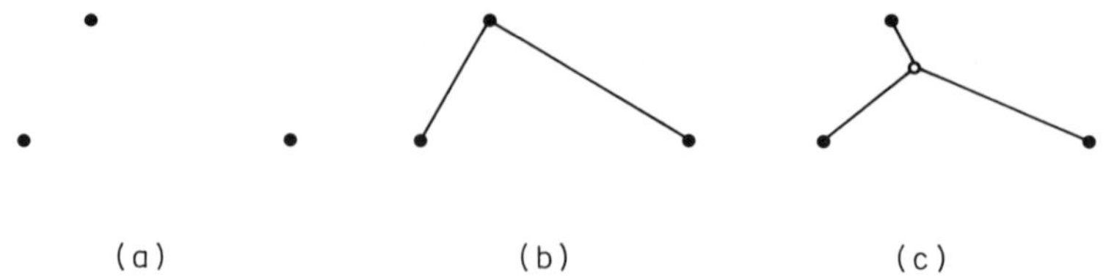

Figure 9.3.3 An example of the Euclidean minimum spanning tree (b) and the Euclidean Steiner minimal tree (c).

To give an example of the Steiner problem, suppose that there are three factories on a plane where we can freely lay out roads (Figure 9.3.3(a)). The problem is to construct a network of roads connecting these factories with the minimum length of roads, provided that the roads are directly connected between these factories. Recalling the definition of the Euclidean minimum spanning tree in Section 2.5, we realize that the solution is given by the Euclidean minimum spanning tree for the three points at which the factories are located (Figure 9.3.3(b)). This solution, however, is not optimal if we are allowed to add a new point (the empty circle in Figure 9.3.3(c)). The problem is to find the fourth point that minimizes the length of roads connecting the three factories and the fourth point. The solution is shown in Figure 9.3.3(c). The length of roads in Figure 9.3.3(c) is shorter than in Figure 9.3.3(b).

According to Kuhn (1974), the above problem dates back to Fermat in the seventeenth century, and was introduced by Courant and Robbins (1941, VII, §6) as the Steiner (Steiner's) problem. Since then, the Steiner problem has been generalized in various directions. In this subsection we consider the following generalized problem.

Problem 9.3.3 (Euclidean Steiner minimal tree problem) Suppose that a certain number of points are fixed in the Euclidean plane (space). We wish to minimize the length of line segments connecting the given points. We can add new points (through which the given points are connected) if the additional points reduce the length. The problem is to find the number of additional points and their locations that minimize the length of the line segments connecting the given points and the additional points.

We call the collection of line segments of this solution the *Euclidean Steiner minimal tree*, and the additional points the *Steiner points*.

To discuss the above problem mathematically, let P be a set of m distinct points $\{\mathbf{x}_1, \ldots, \mathbf{x}_m\}$ which are fixed in $\mathbb{R}^2$; Q_n be a set of n additional points where n and their locations $\{\mathbf{x}_{m+1}, \ldots, \mathbf{x}_{m+n}\}$ are variables; $C_{m+1} = [c_{ij}]$ be the $(m+n)\times(m+n)$ adjacency matrix of points in $P \cup Q_n$, where $c_{ij} = 1$ if nodes $\mathbf{x}_i$ and $\mathbf{x}_j$ are connected by a line segment and $c_{ij} = 0$ if they are not connected; $c_i = \sum_{j=1, j\neq i}^{m+n} c_{ij}$ be the degree of node i (the number of links connected to node i); l_{ij} be $\overline{\mathbf{x}_i\mathbf{x}_j}$ (the line segment joining nodes $\mathbf{x}_i$ and $\mathbf{x}_j$) if $c_{ij} = 1$ and $l_{ij} = \varnothing$ if $c_{ij} = 0$; $L_{m+n} = \{l_{ij}, i\neq j, i,j \in I_{m+n}\}$; and $\mathcal{G}$ be all possible connected geometric graphs of $G(P\cup Q_n, L_{m+n})$ for $n = 0,1, \ldots$ and $\mathbf{x}_{m+1}, \mathbf{x}_{m+2}, \ldots \in \mathbb{R}^2$. In these terms, Problem 9.3.3 is written mathematically as

$$\min_{n;\mathbf{x}_{m+1},\ldots,\mathbf{x}_{m+n}} \sum_{i=1}^{m+n} \sum_{j=i+1}^{m+n} c_{ij}\|\mathbf{x}_i - \mathbf{x}_j\|, \text{ subject to } G(P\cup Q_n, L_{m+n}) \in \mathcal{G}. \tag{9.3.13}$$

To solve this problem, we first solve its subproblem. We fix n and the adjacency matrix C_{m+n}, and consider a set, $\mathcal{G}(C_{m+n})$, of all possible geometric graphs whose adjacency matrix is given by C_{m+n}. The subproblem is written as

$$\min_{\mathbf{x}_{m+1},\ldots,\mathbf{x}_{m+n}} \sum_{i=1}^{m+n} \sum_{j=i+1}^{m+n} c_{ij}\|\mathbf{x}_i - \mathbf{x}_j\|, \text{ subject to } G(P\cup Q_n, L_{m+n}) \in \mathcal{G}(C_{m+n}). \tag{9.3.14}$$

We call the tree of this solution the *relatively minimal tree*. Gilbert and Pollack (1968) showed that the relatively minimal tree is uniquely determined for a given C_{m+n}. Figure 9.3.4 depicts relatively minimal trees for different adjacency matrices. Obviously, the Euclidean Steiner minimal tree is one of the relatively minimal trees including points P.

The Euclidean Steiner minimal tree has many interesting properties, which are discussed by Courant and Robbins (1941), Melzak (1961), Gilbert and Pollack (1968), Chang (1972), Megiddo (1978), Smith *et al.* (1981), Winter (1985), among others. We refer to only three of them which are to be used here.

Property ST1 Exactly three links meet at every Steiner point, and the three angles (α_{i1}, α_{i2}, α_{i3}) made by those links at any Steiner point are all 120° (i.e. $c_i = 3$ and $\alpha_{i1} = \alpha_{i2} = \alpha_{i3} = 120°$).

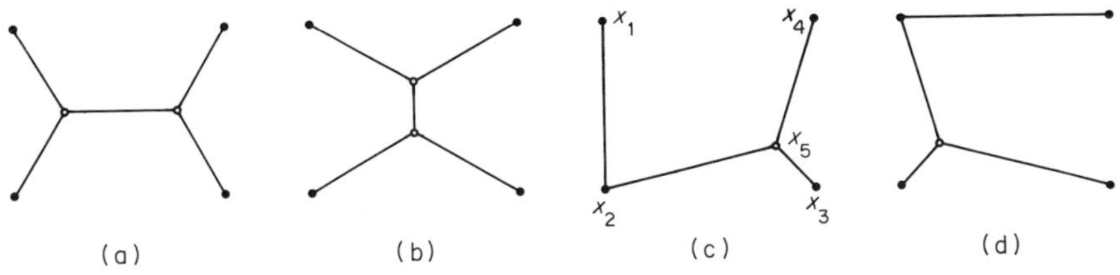

Figure 9.3.4 (a) Euclidean Steiner minimal tree; (a)(b) Euclidean Steiner trees; (a)–(d) relatively minimal trees.

Property ST2 At most three links meet at every point of P ($c_i \leq 3$). If exactly three links meet at a point in P ($c_i = 3$), then $\alpha_{i1} = \alpha_{i2} = \alpha_{i3} = 120°$. If exactly two links meet at a point in P ($c_i = 2$), then $\alpha_{i1} \geq 120°$, where α_{i1} and α_{i2} are angles at a point of degree 2, and $\alpha_{i1} \leq \alpha_{i2}$.

Property ST3 The number n^* of Steiner points is less than or equal to the number of the given points minus 2, i.e. $n^* \leq m-2$.

We call the relatively minimal tree satisfying these three properties the *Euclidean Steiner tree*. Figures 9.3.4(a), (b) are the Euclidean Steiner trees. The Euclidean Steiner minimal tree is the Euclidean Steiner tree whose length is minimum among all the Euclidean Steiner trees (Figures 9.3.4(a)).

To solve the Euclidean Steiner minimal tree, several combinatorial methods are proposed in the literature, for example, Gilbert and Pollack (1968), Change (1972), Meggido (1978) and Winter (1985). In practice, however, their methods can hardly solve the problem with more than twenty points ($m > 20$). In fact, Garey *et al.* (1977) showed that the Steiner problem is NP-hard (the difficulty of the Steiner problem is harder than or equal to the difficulty of problems which can be algorithmically solved with polynomial time). For a large number of points, it is almost impossible to obtain the Euclidean Steiner minimal tree. We may, however, obtain an approximate Euclidean minimal Steiner tree with a computational method. A few heuristic methods are proposed in the literature, for example Smith *et al.* (1981) and Suzuki and Iri (1986b). We show the latter method here, because it is more efficient than the former.

A clue to the problem is shown in Figure 9.3.5. Figure 9.3.4(c) is a relatively minimal tree, but not the Euclidean Steiner tree, because the links at point $\mathbf{x}_2$ do not satisfy Property ST2. To avoid such a point, we *split* (Gilbert and Pollack, 1968) the point as follows. First, we add a new point $\mathbf{x}_6$ in the close neighbourhood of point $\mathbf{x}_2$ (see Figure 9.3.5(a)); second, we connect point $\mathbf{x}_2$ with point $\mathbf{x}_6$, point $\mathbf{x}_6$ with point $\mathbf{x}_1$, and point $\mathbf{x}_6$ with point $\mathbf{x}_5$. As a result (Figure 9.3.5(a)), the adjacency matrix of the resulting network becomes the same as that of the Euclidean Steiner tree (Figure 9.3.4(a)). Solving the optimization problem given by expression (9.3.14), we obtain Figure 9.3.4(a), which is one of the Euclidean Steiner trees.

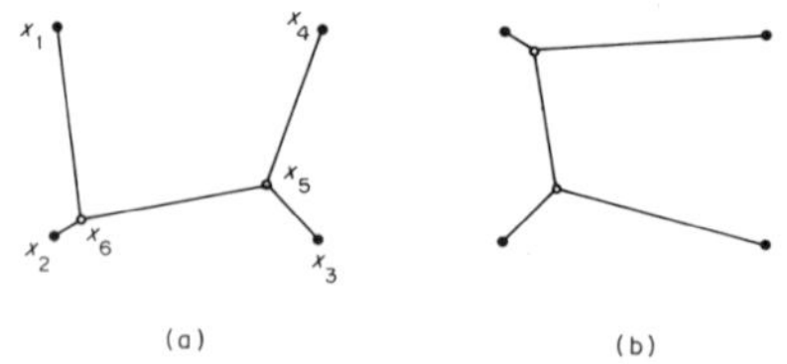

Figure 9.3.5 Splitting a point in relatively minimal trees.

Similarly, we obtain another Euclidean Steiner tree (Figure 9.3.4(b)) from the relatively minimal tree in Figure 9.3.4(d) through splitting shown in Figure 9.3.5(b).

We now formulate the above procedure as the following algorithm.

Algorithm 9.3.1 (Euclidean Steiner tree algorithm)

Step 1. Choose an arbitrary number n larger than $m-2$, and place the n points Q_n randomly in the convex hull of the given points P.

Step 2. Construct the Delaunay triangulation for points $P \cup Q_n$.

Step 3. Obtain the minimum spanning tree from the Delaunary triangulation using, for example, Cheriton and Tarjan's (1976) method.

Step 4. Given the adjacency matrix of this minimum spanning tree, solve the problem of expression (9.3.14) with the descent method.

Step 5. Delete the points of degree $c_i \leq 2$ in Q_n of the resulting relatively minimal tree, and place a new point in the close neighbourhood of each point of degree $c_i \geq 4$ in Q_n (as a result, n changes).

Step 6. For each point of degree $c_i = 2$ in P (denoted by $\mathbf{x}_i^*$), place a new point in the close neighbourhood of each $\mathbf{x}_i^*$ inside the angle $\alpha_{i1} < 120°$ in the direction of the bisector of the angle α_{i1}.

For each point of degree $c_i > 3$ in P and each point of degree $c_i = 3$ in P that does not satisfy $\alpha_{i1} = \alpha_{i1} = \alpha_{i1} = 120°$ (denoted by $\mathbf{x}_i^{**}$), place a new point in the close neighbourhood of each $\mathbf{x}_i^{**}$ inside the smallest angle $\alpha_i^* = \min_j\{\alpha_{ij}, j \in I_{c_i}\}$ in the direction of the bisector of angle α_i^*.

Step 7. If there is no point that can be deleted or added in Steps 5 and 6, report n and $\mathbf{x}_{m+1}, \ldots, \mathbf{x}_{m+n}$. Else, go to Step 2.

Step 5 is the procedure to satisfy Property ST1, and Step 6 is the procedure to satisfy Property ST2. Step 4 is to solve the non-linear non-convex optimization problem of expression (9.3.14). The objective function is given by $F(\mathbf{x}_{m+1}, \ldots, \mathbf{x}_{m+n}) = \sum_{i=1}^{m+n} \sum_{j=i+1}^{m+n} c_{ij} \|\mathbf{x}_i - \mathbf{x}_i\|$. The first-order derivative is given by

$$\frac{\partial F}{\partial x_{i\kappa}} = \sum_{j \in \{j | c_{ij} \neq 0\}} \frac{\partial}{\partial x_{i\kappa}} \|\mathbf{x}_i - \mathbf{x}_j\|, \quad \kappa = 1, 2. \tag{9.3.15}$$

The second derivatives are also obtained from this equation. Using the steepest descent method or the (quasi-)Newtonian method, we can solve the problem in Step 4.

With Algorithm 9.3.1, we can obtain the Euclidean Steiner tree. In the case of Figure 9.3.4, we have only two Euclidean Steiner trees. Hence, comparing those trees, we can obtain the Euclidean Steiner minimal tree (Figure 9.3.4(a)). In the case of a large number of given points, we can

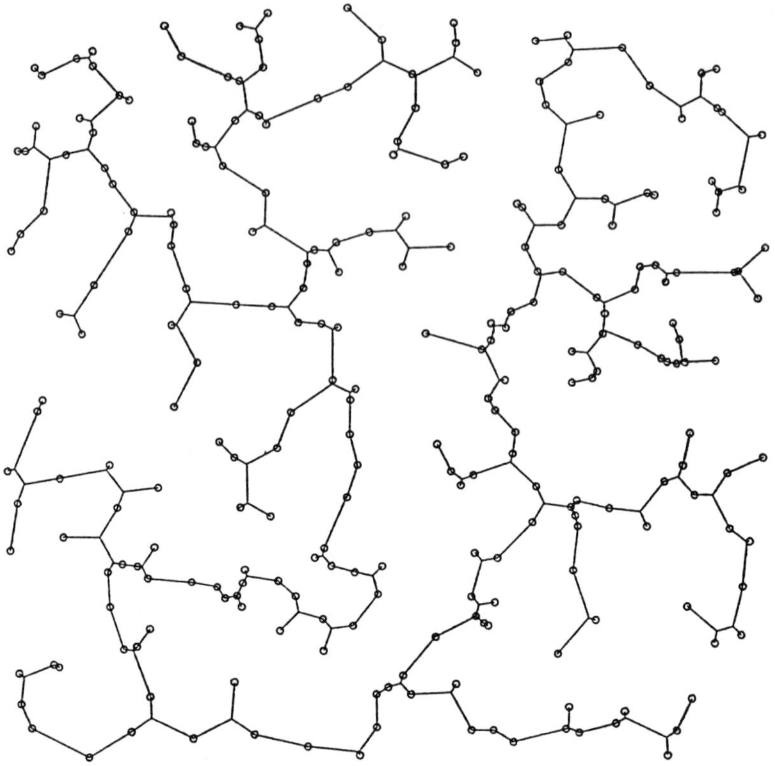

Figure 9.3.6 An approximate Euclidean Steiner tree. (*Source*: Suzuki and Iri (1986b, Figure 9).)

hardly find all the Euclidean Steiner trees. We hence obtain a computationally feasible number of Euclidean Steiner trees and find the minimum tree among them. Obviously, this tree is not guaranteed as the Euclidean Steiner minimal tree, but it may be close to it.

Figure 9.3.6 shows an approximate Euclidean Steiner tree obtained by Suzuki and Iri (1986) for $m = 2^8$ points.

9.4 LOCATIONAL OPTIMIZATION OVER TIME

In the preceding sections we formulated the locational optimization problems without considering the time dimension. In this section we take this dimension into account.

9.4.1 Multi-stage locational optimization

In Section 9.2 we implicitly assumed that all facilities can be constructed at the same time. This assumption, however, is not always acceptable in reality.

One reason is the budget constraint. The number of facilities that can be constructed in a year is sometimes limited because of a budget constraint; we have to construct a required number of facilities over several years. Another reason is inefficiency. When the density of users is expected to increase from low density to high density, it is inefficient to construct all facilities in the early stage; the facilities should be constructed in accordance with the increase in users. Because of these reasons, we sometimes meet the problem in which the location of facilities is to be optimized not only over a region but also over time. The problems related to this locational optimization have been studied by several researchers in Operations Research, for example, Scott (1975), Love (1976), Erlenkotter (1981), Gunawardane (1982) and Van Roy and Erlenkotter (1982). Among them, the problems considered by Scott (1972) and T. Suzuki *et al.* (1991) are the closest to the above problem. The former model is formulated on a network, whereas the latter model is formulated on a continuous plane. We show the latter model here because it is formulated with the Voronoi diagram.

Suppose that we construct n facilities in a region S, and that we can construct only one facility in one stage. Consequently it takes n stages to construct all the facilities. The length of the ith stage (referred to as stage i) is τ_i, which may be the same constant like a year ($\tau_i = 1$) or may not. The n facilities are indexed according to the order of construction; facility i is constructed at the beginning of stage i. Thus, facilities $j = 1, \ldots, i$ are usable in stage i. We assume that once a facility is located, the facility remains there; it cannot be replaced (Figure 9.4.1).

We further assume, as in Section 9.2, that every user uses the nearest facility with respect to the Euclidean distance. Let $\mathcal{V}^{(i)} = \{V_1^{(i)}, \ldots, V_i^{(i)}\}$ be the Voronoi diagram generated by a set of points $\{\mathbf{x}_1, \ldots, \mathbf{x}_i\}$. Obviously $\mathcal{V}^{(1)} = S$. In stage i, users in $V_j^{(i)}$ use facility j; in stage i+1, users in $V_j^{(i+1)}$ use facility j (Figure 9.4.1). Let $f(\|\mathbf{x}-\mathbf{x}_j\|^2)$ be the travel cost function of the squared Euclidean distance, which is assumed to be strictly increasing and differentiable, and $\phi_i(\mathbf{x})$ be the density of users at $\mathbf{x}$ in stage i. We

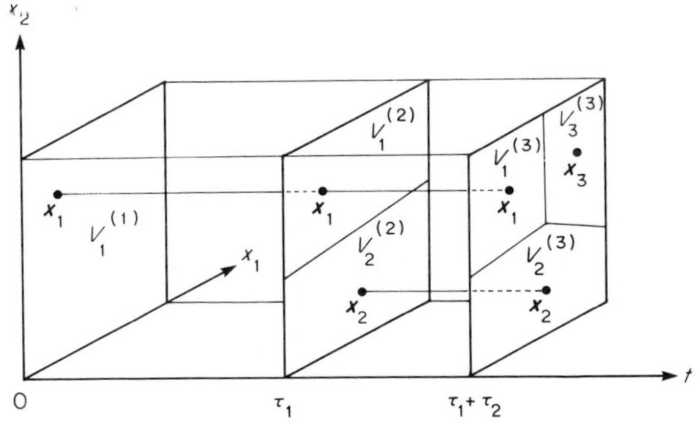

Figure 9.4.1 Locating facilities over multistages.

assume that $\phi(\mathbf{x})$ is constant in stage i (note that $\phi_i(\mathbf{x}) \neq \phi_{i+1}(\mathbf{x})$ in general).

In the above terms, the travel cost averaged over region S and over n stages is written as

$$F(\mathbf{x}_1, \ldots, \mathbf{x}_n) = \sum_{i=1}^{n} \tau_i \sum_{j=1}^{i} \int_{V_j^{(i)}} f(\|\mathbf{x}-\mathbf{x}_j\|^2)\phi_i(\mathbf{x})\mathrm{d}\mathbf{x}. \tag{9.4.1}$$

With this objective function we formulate the above problem as the following mathematical programming problem.

Problem 9.4.1 (minimization of the average travel cost to the nearest points over multistages)

$$\min_{\mathbf{x}_1, \ldots, \mathbf{x}_n} \left[\sum_{i=1}^{n} \tau_i \sum_{j=1}^{i} \int_{V_j^{(i)}} f(\|\mathbf{x}-\mathbf{x}_j\|^2)\phi_i(\mathbf{x})\mathrm{d}\mathbf{x} \right]. \tag{9.4.2}$$

In conjunction with this far-sighted locational optimization problem, it may be of interest to compare a near-sighted locational optimization problem. In the latter problem, we optimize the location of facility 1 in stage 1; in stage 2, provided that the location of facility 1 is fixed at $\mathbf{x}_1$ obtained in stage 1, we optimize the location of facility 2; in stage 3, provided that the locations of facilities 1 and 2 are fixed at $\mathbf{x}_1$ and $\mathbf{x}_2$ obtained in stages 1 and 2, we optimize the location of facility 3; in stage i, provided that the locations of the facilities $j = 1, \ldots, i-1$ are fixed at $\mathbf{x}_1, \ldots, \mathbf{x}_{i-1}$ obtained in stages $1, \ldots, i-1$, we optimize the location of facility i without considering the facilities $j = i+1, \ldots, n$ to be located in the future. Mathematically, this optimization problem is written as

Problem 9.4.2 (myopic locational optimization problem)

$$\min_{\mathbf{x}_i} \left[\sum_{j=1}^{i} \int_{V_j^{(i)}} f(\|\mathbf{x}-\mathbf{x}_j\|^2)\phi_i(\mathbf{x})\mathrm{d}\mathbf{x} \right], \tag{9.4.3}$$

where $\mathbf{x}_1, \ldots, \mathbf{x}_{i-1}$ are fixed and obtained from

$$\min_{\mathbf{x}_{k-1}} \left[\sum_{j=1}^{k-1} \int_{V_j^{(k-1)}} f(\|\mathbf{x}-\mathbf{x}_j\|^2)\phi_{k-1}(\mathbf{x})\mathrm{d}\mathbf{x} \right], \tag{9.4.4}$$

for $k = 2, 3, \ldots, i$, where $V_j^{(0)} = S$.

The computational methods for Problems 9.4.1 and 9.4.2 are almost the same as that for Problem 9.2.1. Regarding Problem 9.4.2, the first derivative of the objective function in expression (9.4.3) is just the same as that of equation (9.2.12). Regarding Problem 9.4.1, the first derivative of equation (9.4.1) is given by

$$\frac{\partial F}{\partial x_{j\kappa}} = \sum_{i=j}^{n} \tau_i \int_{V_j^{(i)}} \frac{\partial}{\partial x_{j\kappa}} f(\|\mathbf{x}-\mathbf{x}_j\|^2)\phi_i(\mathbf{x})\mathrm{d}\mathbf{x}, \quad \kappa = 1,2. \tag{9.4.5}$$

Each term in the summation is the same as equation (9.2.12). We can thus solve Problem 9.4.1 using the descent method with the derivatives obtained from equation (9.4.5). (Note that the Voronoi diagrams $V^{(i)}$, $i = 1, \ldots, n$, are efficiently obtained from the incremental method shown in Chapter 4.) Figure 9.4.2 shows the locally optimal locations of $n = 3, \ldots, 8$ facilities obtained by T. Suzuki *et al.* (1991) for the uniform distribution over a unit square and $f(\|\mathbf{x}-\mathbf{x}_i\|^2) = \|\mathbf{x}-\mathbf{x}_i\|$. The filled circles in panel (a) show the locally optimal locations of Problem 9.4.1 for equal time intervals ($\tau_i = 1$, $i = 1, \ldots, n$), and the unfilled circles in the same panel show those for linearly increasing time intervals ($\tau_i = i$, $i = 1, \ldots, n$). The numbers associated with the filled and unfilled circles indicate the time of construction. As is noticed from these circles, the difference appears quite small. Panel (b) shows the locally optimal locations of facilities under the myopic location policy. Compared with panel (a), the configurational difference is large for a small number of facilities, but becomes small as the number of facilities increases. In fact, T. Suzuki *et al.* (1991) show that the difference in the average nearest neighbour distance over time is within only 3% for $n = 8$, implying that the myopic location policy is not as bad as we expected for this case.

9.4.2 Periodic locational optimization

A few hundred years ago, market places were opened periodically. In Japan, we can find such traces in the names of places, for instance 3rd-Day-Market, 7th-Day-Market, etc. In 3rd-Day-Market, a market opened on the 3rd, 13th, and 23rd of every month. Nowadays, there are few such an old style markets, but we still have such periodic services, for example a monthly mobile clinic

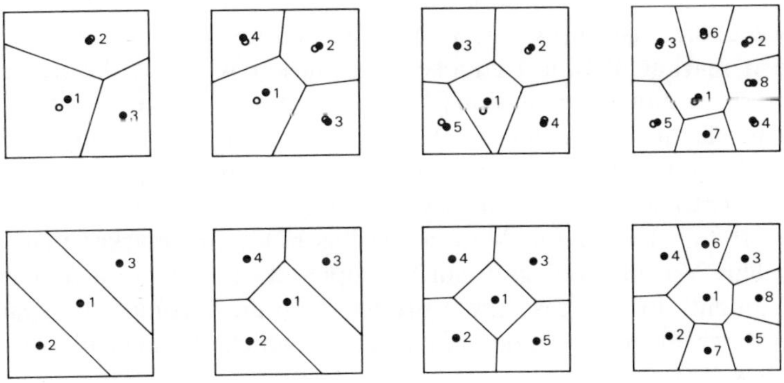

Figure 9.4.2 Locally optimal locations of $n = 1, \ldots, 8$ firms in a unit square region with a uniform distribution and $f(\|\mathbf{x}-\mathbf{x}_i\|^2) = \|\mathbf{x}-\mathbf{x}_i\|$: (a) locally optimal locations for Problem 9.4.1 (optimized over time) $\tau_i = 1$ for filled circles and $\tau_i = i$ for unfilled circles; (b) locally optimal locations for Problem 9.4.2 (optimized at each time stage). (*Source*: T. Suzuki *et al.* (1991, Figure 4).)

visiting several places in a region. In this subsection we consider the locational optimization of facilities open periodically.

Suppose that there are n market places, $i = 1,\ldots,n$, in a region $S \subset \mathbb{R}^2$ and these market places are located at $\mathbf{x}_1, \ldots, \mathbf{x}_n \in S$, respectively. Market place i is open at t_i+jT, $j = 0,1,\ldots,$ where $t_1 < t_2 < \ldots < t_n < T$ is assumed without loss of generality. A market opens for a certain time length, but we assume that this length of time is very short compared with $t_i - t_{i-1}$. Under this assumption, market i located at $\mathbf{x}_i$ and open at t_i is represented by the point $(\mathbf{x}_i, t_i)$ in $\mathbb{R}^3$. Let $t^*_i(t)$ be the length of time from time t to the nearest time when market i opens, i.e.

$$t^*_i(t) = \min_j \{t_i+jT | t_i+jT \geq t\}. \tag{9.4.6}$$

Suppose that at time t a consumer at location $\mathbf{x}$ plans to go to a market. If the consumer chooses market place i, he/she has to wait for $t^*_i(t)$ days and walk $\|\mathbf{x}-\mathbf{x}_i\|$ km; if the consumer chooses market place k, he/she has to wait for $t^*_k(t)$ days and walk $\|\mathbf{x}-\mathbf{x}_k\|$ km. If the waiting time for market i is shorter than that for market k and the walking distance to market i is shorter than market k, the consumer's decision is simple: he/she decides to go to market i. If the former inequality holds, but the latter inequality does not hold (or vice versa), the consumer's decision becomes slightly complex; he/she makes a decision by trading off between waiting time and walking distance. To be explicit, let α be this trade-off rate, i.e. the patience of waiting for one day is equivalent to that of walking α km, and

$$d((\mathbf{x}, t), (\mathbf{x}_i, t^*_i(t))) = \sqrt{(\|\mathbf{x}-\mathbf{x}_i\|^2+\alpha^2(t-t^*_i(t))^2)} \tag{9.4.7}$$

(recall that this distance is equivalent to the space–time distance defined in Chapter 3). If a consumer does not mind waiting at all, then $\alpha = 0$. If a consumer is impatient of waiting, α is very large. In terms of this distance, the above choice assumption is written as: a consumer at $\mathbf{x}$ who considers going to a market at time t chooses the market which is the nearest with respect to the distance of equation (9.4.7), i.e. $\min_i[\sqrt{(\|\mathbf{x}-\mathbf{x}_i\|^2+\alpha^2(t-t^*_i(t))^2)}]$.

With the distance of equation (9.4.7), we can define a Voronoi diagram $\{V_1, \ldots, V_n\}$ generated by points $(\mathbf{x}_i, t_i)$, $i \in I_n$, in the space given by $\{(\mathbf{x}, t) | \mathbf{x} \in S, 0 \leq t \leq T\}$. We can see consumers' choice from this Voronoi diagram. If $(\mathbf{x}, t)$ is in the Voronoi polygon V_i, the market place at i is chosen. This Voronoi polygon can be depicted on a plane if S is given by a line segment. In this case the Voronoi diagram becomes the space–time Voronoi diagram on a cylinder (Chapter 3), which is extensible on a plane (Figure 9.4.3(a)).

The optimization problem is to place n market places so that the average space–time distance to the nearest market places is minimized. Mathematically, this problem is written as follows.

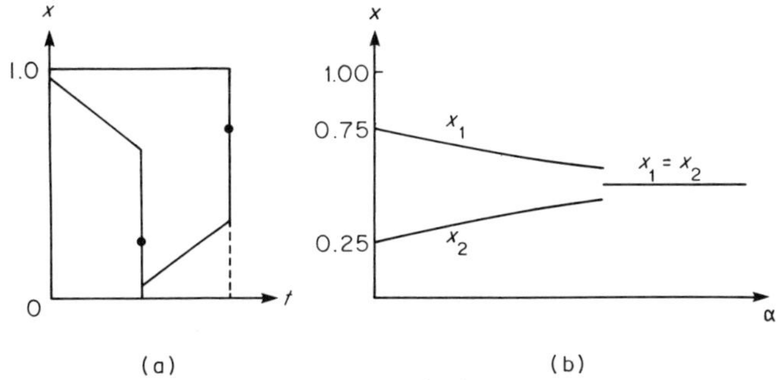

Figure 9.4.3 Local optimal locations of two market places ($n = 2$) for $\phi(\mathbf{x}, t) = 1$, $t_1 = 0.5$, $t_2 = 1.0$: (a) $\alpha = 1$; (b) local optimal locations with respect to α.

Problem 9.4.2 (minimization of the average space–time distance to the nearest points)

$$\min_{\mathbf{x}_1, \ldots, \mathbf{x}_n} \sum_{i=1}^{n} \int_{V_i} \sqrt{(\|\mathbf{x}-\mathbf{x}_i\|^2 + \alpha^2(t-t_i^*(t))^2)}\,\phi(\mathbf{x}, t)\mathrm{d}\mathbf{x}\mathrm{d}t. \tag{9.4.8}$$

The computational method for this problem is almost the same as that for Problem 9.2.1. Figure 9.4.3 shows a locally optimal solution in the simplest case in which two market places are placed in a unit line segment with $\phi(\mathbf{x},t) = 1$, $t_1 = 0.4$, $t_2 = 0.8$ and $\alpha = 1$ (Seoung, 1990). Obviously, the solution changes according to the value of α. When $\alpha = 0$ (implying that consumers do not mind waiting at all), the optimization problem in $\{(\mathbf{x},t)|\mathbf{x} \in S, 0 \le t \le T\}$ reduces to the optimization problem in S (Problem 9.2.1). The solution is given by $\mathbf{x}_1 = 0.25$, $\mathbf{x}_2 = 0.75$. When α is very large (implying that customers do not mind walking to a market place at all), the solution is given by $\mathbf{x}_1 = \mathbf{x}_2 = 0.5$. As is shown in Figure 9.4.3(b), the locations of market places approach the centre as the value of α increases. Note that this change is not continuous; after α exceeds a certain value, both market places locate at the centre.

The case of a two-dimensional region S is studied by Takeda and Iri (1984) with a slightly different distance, i.e. equation (9.4.7) is replaced by

$$t_i^*(t) = \min_j \{|t_i + jT - t|, j = 0,1, \ldots\}. \tag{9.4.9}$$

An example of the Voronoi diagram with this distance is depicted by the solid lines in Figure 9.4.4. In the same figure, the space–time Voronoi diagram is shown by the broken lines. The difference is, as is seen in Figure 9.4.4, distinct, and so their implications of time should be different. To consider a possible implication, suppose that a consumer wants to eat fresh fish every Friday. If a fish market opens on every Friday, the consumer is

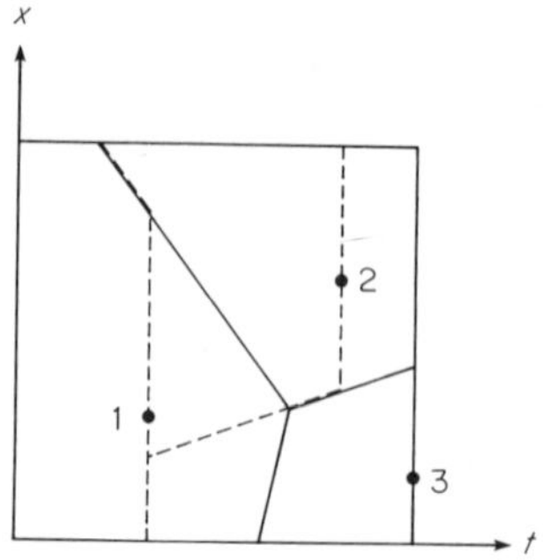

Figure 9.4.4 The difference between the space–time Voronoi diagram (dashed lines) and the ordinary Voronoi diagram (solid lines).

happy. If a fish market opens on every Sunday, the consumer eats fish on Sunday, because he or she does not want to eat refrigerated fish. In this case, the consumer has to wait two days, and hence the degree of dissatisfaction may be measured as two days. If the market opens on every Thursday, the degree of dissatisfaction may be measured as one day. Under this understanding, the distance equation (9.4.9) may be regarded as the measure of dissatisfaction with respect to time.

Figure 9.4.5 shows the local optimal locations of six market places in a unit square obtained by Takeda (1985). The numbers associated with the filled circles indicate t_i. For $\alpha \leq 1$, market places open at six different places, but for $\alpha = 2.0$, 3.0 and 4.0, they open at three, two and one place, respectively.

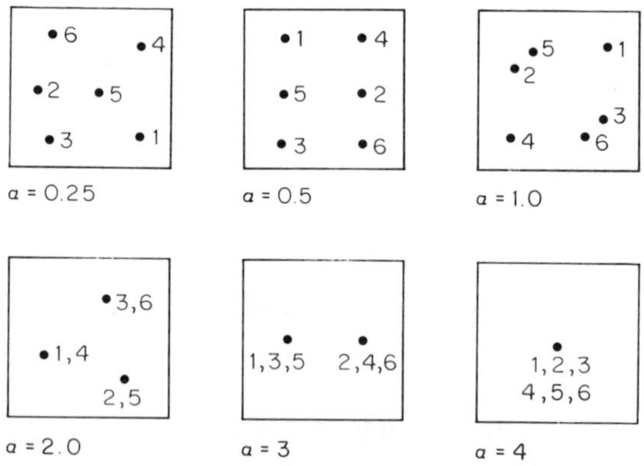

Figure 9.4.5 Locally optimal locations of six market places in a unit square with a uniform distribution. (*Source*: Takeda (1985, Figure 4.5).)

9.5 VORONOI FITTING AND ITS APPLICATION TO LOCATIONAL OPTIMIZATION PROBLEMS

In Chapter 7 we observed that the territories of mouthbreeder fish looked like a Voronoi diagram, and discussed how to measure the fitness of a polygonal tessellation to a Voronoi diagram. In this section we show another method in which we seek to find a Voronoi diagram that gives the 'best' fit to a given polygonal tessellation. We also show that this method is applicable to some locational optimization problems.

9.5.1 Method of fitting a Voronoi diagram to a polygonal tessellation

Let S_1, ..., S_n be a finite number of closed polygons in $S \subset \mathbb{R}^2$ satisfying $[S_i \backslash \partial S_i] \cap [S_j \backslash \partial S_j] \neq \emptyset$ for $i \neq j$, $i, j \in I_n$ and $\cup_{i=1}^{n} S_i = S$. Then $\mathcal{S} = \{S_1, ..., S_n\}$ forms a tessellation of S (the dash-dot lines in Figure 9.5.1). Our objective is to fit a Voronoi diagram to the tessellation $\mathcal{S}$. To this end, we first choose an arbitrary point $\mathbf{x}_i$ in S_i and generate the Voronoi diagram $\mathcal{V} = \{V_1, ..., V_n\}$ by the set of points $\{\mathbf{x}_1, ..., \mathbf{x}_n\}$ (the solid lines in Figure 9.5.1). Since we are concerned with the Voronoi diagram in S, the Voronoi diagram should be bounded by S, i.e. $\{V_1 \cap S, ..., V_n \cap S\}$, which is written as, for notational simplicity, $\{V_1, ..., V_n\}$. In general, we can treat non-convex S using the shortest-path Voronoi diagram (Section 3.4), but we assume, for simplicity, that S is convex.

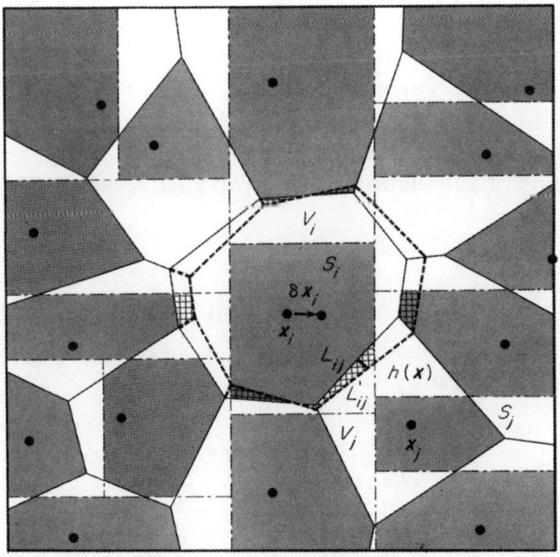

Figure 9.5.1 Change in $V_i \cap S_j$ produced by a slight move of $\mathbf{x}_i$.

The similarity between the Voronoi diagram $\mathcal{V}$ and the tessellation $\mathcal{S}$ may be measured in terms of the intersection area $\Sigma_{i=1}^{n} |S_i \cap V_i|$ (the shaded region in Figure 9.5.1). Obviously, if this value is large (i.e. close to $|S|$) we can say that the tessellation $\mathcal{S}$ is close to the Voronoi diagram $\mathcal{V}$. Mathematically, this intersection area is written as

$$F(\mathbf{x}_1, \ldots, \mathbf{x}_n) = \sum_{i=1}^{n} \int_{V_i \cap S_i} \mathrm{d}\mathbf{x}. \tag{9.5.1}$$

This function may be slightly generalized as

$$F(\mathbf{x}_1, \ldots, \mathbf{x}_n) = \sum_{i=1}^{n} \int_{V_i \cap S_i} \phi(\mathbf{x})\mathrm{d}\mathbf{x}, \tag{9.5.2}$$

where $\phi(\mathbf{x})$ indicates a weight (an explicit meaning will be given in Section 9.5.2). With this generalized objective function, the Voronoi fitting problem is formulated as follows.

Problem 9.5.1 (maximization of the common area between a given tessellation and a Voronoi diagram)

$$\max_{{}^*\mathbf{x}_1, \ldots, \mathbf{x}_n} \sum_{i=1}^{n} \int_{V_i \cap S_i} \phi(\mathbf{x})\mathrm{d}\mathbf{x}. \tag{9.5.3}$$

The computational method for this maximization problem is almost the same as that in expression (7.3.1) in Section 7.3. As is seen in Figure 9.5.1, a slight move, $\delta\mathbf{x}_i$, of $\mathbf{x}_i$ produces the changes in $V_i \cap S_j$, $j \in J_i = \{j | V_i \cap S_j \neq 0, j \in I_n\}$ (the shaded area in Figure 9.5.1). To measure these changes, let V'_i be the Voronoi polygon of the point $\mathbf{x}_i + \delta\mathbf{x}_i$; $L_{ij} = (V_i \cap V_j) \cap S_i$ (the Voronoi edge shared with the Voronoi polygons V_i and V_j included in S_i); and $L'_{ij} = (V'_i \cap V_j) \cap S_i$; $h(\mathbf{x})$ be the length of the perpendicular line segment from a point on L'_{ij} to the foot point, $\mathbf{x}^{\mathrm{T}} = (x_1, x_2)$, on L_{ij} (see Figure 9.5.1). After a few steps of calculation (Suzuki and Iri, 1986a), $h(\mathbf{x})$ is obtained as

$$h(\mathbf{x}) = \frac{(x_{i1} - x_1)\delta x_{i1}}{\|\mathbf{x}_i - \mathbf{x}_j\|} + \frac{(x_{i2} - x_2)\delta x_{i2}}{\|\mathbf{x}_i - \mathbf{x}_j\|}. \tag{9.5.4}$$

The change in the area of $V_i \cap S_j$ (the grid area in Figure 9.5.1) is obtained from the integral of $h(\mathbf{x})$ along the line L_{ij}. Thus, the first derivative of equation (9.5.2) is given by

$$\frac{\partial F}{\partial x_{i\kappa}} = \sum_{j \in J_i} \left\{ \int L_{ij}\, h(\mathbf{x})\phi(\mathbf{x})\mathrm{d}\mathbf{x} - \int_{L_{ji}} h(\mathbf{x})\phi(\mathbf{x})\mathrm{d}\mathbf{x} \right\} \tag{9.5.5}$$

$$= \sum_{j \in J_i} \left\{ \int_{L_{ij}} \frac{x_{i\kappa} - x_\kappa}{\|\mathbf{x}_i - \mathbf{x}_j\|} \phi(\mathbf{x})\mathrm{d}\mathbf{x} - \int_{L_{ji}} \frac{x_{i\kappa} - x_\kappa}{\|\mathbf{x}_i - \mathbf{x}_j\|} \phi(\mathbf{x})\mathrm{d}\mathbf{x} \right\}, \quad \kappa = 1,2.$$

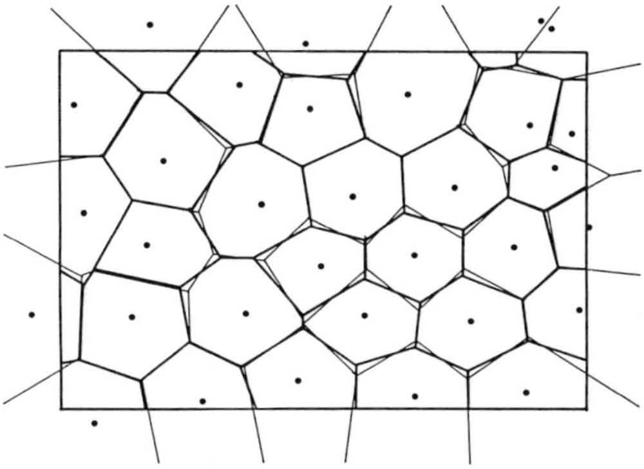

Figure 9.5.2 Fitting a Voronoi diagram to the territories of mouthbreeder fish. (*Source*: Suzuki and Iri (1986, Figure 13).)

Using the steepest descent method with this derivative, we can solve Problem 9.5.1. Figure 9.5.2 illustrates an actual application carried by Suzuki and Iri (1986a) who fitted a Voronoi diagram (the heavy solid lines) to the territories of mouthbreeder fish studied by Barlow (1974) (followed by Hasegawa and Tanemura, 1976, and Honda, 1978). Suzuki and Iri (1986a) also fitted a Voronoi diagram to basaltic columnar jointing studied by Koch (1974),

Figure 9.5.3 School districts in Tsukuba: (a) school districts and the present sites of schools (filled circles), the students in the shaded region cannot go to their nearest schools; (b) school sites (filled squares) that minimize the area in which students cannot go to their nearest schools. (*Source*: Suzuki and Iri (1986a, Figure 14).)

Stoyan and Stoyan (1980), Stoyan and Hermann (1986). Judging by their examination, the Voronoi diagrams appear to give a good fit to these phenomena.

9.5.2 Locational optimization for minimizing restricted areas

The method of fitting a Voronoi diagram to a given tessellation is also useful to solve the locational optimization of facilities whose use is spatially restricted. An example is the locational optimization of schools constrained by school districts. Figure 9.5.3(a) shows the districts of junior high schools in Tsukuba (the filled circles indicate the locations of the schools). In Japan, students in a school district are supposed to go to the school assigned to that district. As a result, students in some areas have to go to the school which is farther than the nearest school. If we assume that students can take the Euclidean path in Figure 9.5.3, the area in which students cannot go to their nearest schools is given by the hatched region in Figure 9.5.3(a), where the polygons are the Voronoi diagram generated by the school sites. The locational optimization problem is to relocate the schools so that the number of students who cannot go to their nearest schools is minimized. This minimization problem is equivalent to the maximization problem of expression (9.5.2), where $\phi(\boldsymbol{x})$ is given by the density of students. Figure 9.5.3 shows the locally optimal locations obtained by Suzuki and Iri (1986a). The area in which students cannot go to their nearest schools reduces from 20% to 10% by this relocation.

References

Aboav, D.A. (1970) The arrangement of grains in a polycrystal *Metallography*, **3**, 383–390. [5.4]

Aboav, D.A. (1980) The arrangement of cells in a net *Metallography*, **13**, 43–58. [5.4]

Aboav, D.A. (1983) The arrangement of cells in a net II *Metallography*, **16**, 265–273. [5.4]

Aboav, D.A. (1984) The arrangement of cells in a net III *Metallography*, **17**, 383–396. [5.4]

Aboav, D.A. (1985) The arrangement of cells in a net IV *Metallography*, **18**, 129–147. [5.4]

Aboav, D.A. (1987) Contiguous polygons in a random Voronoi tesselation—an approximate topological formula *Acta Sterologica*, **6**(3), 371–373. [5.4]

Abramowitz, M. and I.A. Stegun (eds) (1964) *Handbook of Mathematical Functions with Formulas, Graphs, and Mathematical Tables* New York: Dover. [9.2.1]

Aggarwal, A., L.J. Guibas, J. Saxe and P.W. Shor (1989) A linear-time algorithm for computing the Voronoi diagram of a convex polygon *Discrete and Computational Geometry*, **4**, 591–604. [2.4, 4.1, 4.7]

Aho, A.V., M.R. Garey and F.K. Hwang (1976) Rectilinear Steiner trees: efficient special case algorithms *Networks*, **7**, 37–58. [3.9]

Aho, A.V., J.E. Hopcroft and J.D. Ullman (1974) *The Design and Analysis of Computer Algorithms* Reading: Addison-Wesley. [4.1, 4.2, 4.4, 4.5]

Ahuja, N., B. An and B. Schachter (1985) Image representation using Voronoi tessellation *Computer Vision, Graphics and Image Processing* **29**(3), 286–295. [5.2]

Ahuja, N. and B.J. Schachter (1983) *Pattern Models* New York: John Wiley. [1.0, 5.2, 6.1.1]

Akima, H. (1978) A method of bivariate interpolation and smooth surface fitting for irregularly distributed data points *ACM Transactions on Mathematical Software* **4**(2), 148–159. [6.2, 6.3]

Alonso, W. (1964) Location theory. In: J. Friedmann and A.W. Alonso (eds) *Regional Development and Planning: A Reader* Cambridge: Cambridge University Press. [7.3.1]

An, B.H., N. Ahuja and B. Schachter (1983) Representation of images using Voronoi tessellation *Proceedings of the IEEE Computer Society Conference on Computer Vision and Pattern Recognition*, **2**, 188–189. [5.2]

Anjoumani, A. (1981) Market-area analysis with a rectilinear network *Environment and Planning A*, **13**, 943–954. [3.9]

Aoyagi, M. and A. Okabe (1992) Spatial competition of firms in a bounded two-dimensional market *Regional Science and Urban Economics* (forthcoming) [7.3.1]

Armstrong, R.A. (1974) Dynamics of expanding inhibitory fields *Science*, **183**, 444–445. [5.6]

Arnold, D.B. and W.J. Milne (1984) The use of Voronoi tesellations in processing soil survey results *IEEE Computer Graphics and Applications*, **4**, 22–28. [6.1.1]

Arnov, B. (1987) On the geodesic Voronoi diagram of point sites in a simple polygon *Proceedings of the ACM Symposium on Computational Geometry, Waterloo, Ontario* 39–49. [3.4.2]

Asami, Y. (1991) A note on derivatives of objective functions in geographical optimization problems *Journal of the Faculty of Engineering, University of Tokyo*, **41**(1), 1–13. [9.2.1]

Asano, T. and T. Asano (1987) Voronoi diagram for points in a simple polygon In: D.S. Johnson, T. Nishizeki, A. Nozaki and H.S. Wilf (eds) *Discrete Algorithms and Complexity, Perspectives in Computing*, Volume 15 New York: Academic Press. [3.4.1, 3.4.2, 4.8]

Asano, T., T. Asano, L. Guibas, J. Herschberger and H. Imai (1986) Visibility of disjoint polygons *Algorithmica*, **1**, 49–63. [3.9]

Ash, P.F. and E.D. Bolker (1985) Recognizing Dirichlet tessellations *Geometriae Dedicata*, **19**, 175–206. [2.3, 3.7.4, 7.5.1]

Ash, P.F. and E.D. Bolker (1986) Generalized Dirichlet tessellations *Geometriae Dedicata*, **20**, 209–243. [3.1.1, 3.1.4, 3.1.5]

Ashcroft, N.W. and N.D. Mermin (1976) *Solid State Physics* New York: Holt, Reinhart and Winston. [7.1]

Aurenhammer, F. (1987) Power diagrams: properties, algorithms and applications *SIAM Journal of Computing*, **16**, 78–96. [3.2.1, 3.9]

Aurenhammer, F. (1988) Voronoi diagrams—a survey *Institute for Information Processing, Technical University of Graz, Report 263.* [1.0, 3.1.1, 3.1.4, 3.2.1, 3.4.3, 3.7.3]

Aurenhammer, F. (1990) A new duality result concerning Voronoi diagrams *Discrete and Computational Geometry*, **5**, 243–254. [4.8]

Aurenhammer, F. and H. Edelsbrunner (1984) An optimal algorithm for constructing the weighted Voronoi diagram in the plane *Pattern Recognition*, **17**, 251–257. [3.1.1, 4.1, 4.7, 4.8]

Avis, D. and B.K. Bhattacharya (1983) Algorithms for computing *d*-dimensional Voronoi diagrams and their duals *Advances in Computing Research*, **1**, 159–180. [4.7]

Avrami, M. (1939) Kinetics of phase change: general theory *Journal of Chemical Physics*, **7**, 1103–1112. [7.2]

Babuska, I. and A.K. Aziz (1976) On the angle condition in the finite element method *SIAM Journal on Numerical Analysis*, **13**, 214–226. [6.2]

Baker, T.J. (1989) Automatic mesh generation for complex three-dimensional regions using a constrained Delaunay triangulation *Engineering with Computers*, **5**, 161–175. [3.9]

Bajaj, C.L. and W.J. Bouma (1990) Dynamic Voronoi diagrams and Delaunay triangulations *Abstracts of the 2nd Canadian Conference on Computational Geometry, Ottawa, August 1990*, 273–277. [4.1]

Ballard, D.H. and C.M. Brown (1982) *Computer Vision* Englewood Cliffs, N.J.: Prentice-Hall. [4.2]

Banerji, A. and H.B. Fisher (1974) Hierarchical location analysis for integrated area planning in rural India *Papers of the Regional Science Association*, **33**, 177–194. [9.2.4]

Barlow, G.W. (1974) Hexagonal territories *Animal Behaviour*, **22**, 876–878. [7.2, 7.3, 9.5.1]

Barnhill, R.E. (1977) Representation and approximation of surfaces. In: J.R. Rice (ed.) *Mathematical Software III* New York: Academic Press, pp. 69–120. [6.0, 6.2]

Bartlett, M.S. (1975) *The Statistical Analysis of Spatial Pattern* London: Chapman and Hall. [1.3.7, 8.3]

Bartlett, V. (1976) The ordering of multivariate data (with discussion) *Journal of the Royal Statistical Society, Series A*, **139**, 318–354. [6.5]

Baumgart, B. (1975) A polyhedron representation for computer vision *IFIPS Conference Proceedings*, **44**, 589–596. [4.2]

Beckman, M. (1971) Market share, distances, and potential *Regional Science and Urban Economics*, **1**, 1–18. [3.9]

Bellagamba, V., R. Ercoli, A. Gamba and M. Simonett (1985) An introduction to the use of Voronoi polyhedra in the study of organic reactions in solution *Theochem (Netherlands)*, **24**(3–4), 383–389. [7.1]

Bengtsson, B-E. and S. Nordbeck (1964) Construction of isorithms and isorithmic maps by computers *BIT*, **4**, 87–105. [6.2]

Bentley, J.L. and H.A. Maurer (1979) A note on Euclidean near neighbor searching in the plane *Information Processing Letters*, **8**, 133–136. [3.2]

Bentley, J.L., B.W. Weide and A.C. Yao (1980) Optimal expected-time algorithms for closest point problems *ACM Transactions on Mathematical Software*, **6**, 563–580. [2.3, 4.1]

Bernal, J.D. (1937) An attempt at a molecular theory of liquid structure *Transactions of the Faraday Society*, **33**, 27–45. [7.1]

Bernal, J.D. (1959) A geometrical approach to the structure of liquids *Nature*, **183** (4655), 141–147. [7.1]

Besag, J. (1974) Spatial interaction and the statistical analysis of lattice systems (with discussion) *Journal of the Royal Statistical Society, Series B*, **36**, 192–236. [6.1.2]

Besag, J. (1975) Statistical analysis of non-lattice data *The Statistician*, **24**, 179–196. [6.1.2]

Blanc, M. and A. Mocellin (1979) Grain coordination in plane sections of polycrystal *Acta Metallurgica*, **27**, 1231–1237. [5.4]

Bland, R.G. (1977) New finite pivoting rules for the simplex method *Mathematics of Operations Research*, **2**, 103–107. [4.6.2]

Blum, H. (1967) A transformation for extracting new descriptors of shape. In: W. Wathen-Dunn (ed.) *Symposium on Models for the Perception of Speech and Visual Form* Cambridge, Mass.: M.I.T. Press, pp. 362–380. [3.9]

Blum, H. (1973) Biological shape and visual science (Part I) *Journal of Theoretical Biology*, **38**, 205–287. [3.9]

Boggs, S.W. (1951) Delimitation of seaward areas under national jurisdiction *American Journal of International Law*, **45**, 240–266. [3.9]

Bogue, D.J. (1949) *The Structure of the Metropolitan Community: A Study of Dominance and Subdominance* Ann Arbor: Horace M. Rackham School of Graduate Studies, University of Michigan. [1.2, 7.1]

Boots, B.N. (1973) Some models of the random subdivision of space *Geografiska Annaler*, **55B**, 34–48. [5.2, 5.6, 7.2]

Boots, B.N. (1974) Delaunay triangles: an alternative approach to point pattern analysis *Proceedings, Association of American Geographers*, **6**, 26–29. [1.2, 8.2]

Boots, B.N. (1975a) Some observations on the structure of socioeconomic cellular networks *The Canadian Geographer*, **19**, 107–120. [5.2, 5.6, 7.2]

Boots, B.N. (1975b) Patterns of urban settlements revisited *The Professional Geographer*, **27**, 426–431. [8.2]

Boots, B.N. (1977) Contact number properties in the study of cellular networks *Geographical Analysis*, **9**(4), 379–387. [5.10, 8.1.1]

Boots, B.N. (1980) Weighting Thiessen polygons *Economic Geography*, **56**, 248–259. [3.9]

Boots, B.N. (1982) The arrangement of cells in 'random' networks *Metallography*, **15**, 53–62. [5.10]

Boots, B.N. (1984) Comments on 'Aboav's Rule' for the arrangement of cells in a network *Metallography*, **17**, 411–418. [5.10]

Boots, B.N. (1986) Using angular properties of Delaunay triangles to evaluate point patterns *Geographical Analysis*, **18**(3), 250–260. [5.9, 8.2]

Boots, B.N. (1987) Edge length properties of random Voronoi polygons *Metallography*, **20**, 231–236. [5.3, 5.4]

Boots, B.N. and A. Getis (1988) *Point Pattern Analysis*. Sage Scientific Geography Series, Volume 8 Newbury Park, Calif.: Sage Publications. [8.0]

Boots, B.N. and D.J. Murdoch (1983) The spatial arrangement of random Voronoi polygons *Computers and Geosciences*, **9**, 351–365. [4.1, 5.3, 5.4, 8.1.1]

Bowyer, A. (1981) Computing Dirichlet tessellations *The Computer Journal*, **24**, 162–166. [2.1, 4.7]

Brakke, K.A. (1986) Plane Voronoi tessellation second order statistics *Abstracts of the American Mathematical Society*, **7**(2), 235–236. [5.4]

Bramble, J.H. and M. Zlamal (1970) Triangular elements in the finite element method *Mathematics of Computation*, **24**(112), 809–820. [6.2]

Brassel, K.E. and D. Reif (1979) A procedure to generate Thiessen polygons *Geographical Analysis*, **11**, 289–303. [4.1]

Brillouin, L. (1953) *Wave Propagation in Periodic Structures: Electric Filters and Crystal Lattices* (Second edition) New York: Dover Publications. [3.9]

Brostow, W., J-P. Dussault and B.L. Fox (1978) Construction of Voronoi polyhedra *Journal of Computational Physics*, **29**, 81–92. [4.7]

Brostow, W. and Y. Sicotte (1973). Some properties of the informational model of the liquid state *Journal of Statistical Physics*, **9**(4), 339–349. [7.1]

Brostow, W. and Y. Sicotte (1975) Coordination number in liquid argon *Physica*, **80A**, 513–522. [7.1]

Brown, G.S. (1965) Point density in stems per acre *New Zealand Forestry Service Research Notes*, **38**, 1–11. [1.2, 7.1, 8.1.3]

Brown, K.Q. (1979) Voronoi diagrams from convex hulls *Information Processing Letters*, **9**(5), 223–228. [2.4, 4.1, 4.7, 4.8]

Brown, K.Q. (1980) Geometric transforms for fast geometric algorithms, Ph.D. Thesis, Report CMU-CS-80-101, Carnegie-Mellon University, Pittsburgh, Pennsylvania. [2.4, 3.7.4, 4.1, 4.7]

Brumberger, H. and J. Goodisman (1983) Voronoi cells: an interesting and potentially useful cell model for interpreting the small-angle scattering of catalysts *Journal of Applied Crystallography (Denmark)*, **16**, 83–88. [5.2]

Buckley, C.E. (1988) A divide-and-conquer algorithm for computing 4-dimensional convex hulls *Lecture Notes in Computer Science*, **333** (International Workshop on Computational Geometry, Wurzburg, March 1988). Berlin: Springer-Verlag, pp. 113–135. [4.1, 4.7, 4.8]

Buckley, P.A. and F.G. Buckley (1977) Hexagonal packing of Royal Tern nests *The Auk*, **94**, 36–43. [5.2]

Bulow-Olsen, A., N.R. Sackville Hamilton and M.J. Hutchings (1984) A study of growth form in genets of *Trifolium repens L.* as affected by intra- and interplant contacts *Oecologia (Berlin)*, **71**, 383–387. [7.1]

Burger, G., E. Koken, D.S. Wilkinson and J.D. Embury (1988) The influence of spatial distribution on metallurgical processes. In: J.D. Embury and G.R. Purdy (eds) *Advances in Phase Transition* Oxford: Pergamon Press, pp. 247–262. [5.10, 7.1, 7.2]

Burger, G., J.D. Embury and D.S. Wilkinson (1989) The characterization of microstructures using tessellations and their application to deformation processes. Unpublished paper. [5.10, 8.2]

Busacker, R.G. and T.L. Saaty (1965) *Finite Graphs and Networks: An Introduction with Applications* New York: McGraw-Hill. [1.3]

Calabi, L. and W.E. Hartnett (1968) Shape recognition, prairie fires, convex deficiencies and skeletons *American Mathematical Monthly*, **75**, 335–342. [3.9]

Callen, J.S. and G.E. Stephan (1975) Siwai line-villages: Thiessen polygons and the size-density hypothesis *Field Museum of Natural History, Solomon Island Studies in Human Biogeography*, **4**, [7.1]

Cannings, C. and L.M. Cruz Orive (1975) On the adjustment of the sex ratio and the gregarious behaviour of animal populations *Journal of Theoretical Biology*, **55**, 115–136. [7.4]

Canny, J. and B. Donald (1988) Simplified Voronoi diagram *Discrete and Computational Geometry*, **3**, 219–236 [3.5.4]

Cape, J.N., J.L. Finney and L.V. Woodcock (1981) An analysis of crystallization by homogeneous nucleation in a 4000-atom soft-sphere model *Journal of Chemical Physics*, **75**(5), 2366–2373. [7.1]

Carter, G.M., J.M. Chaiken and E. Ignall (1972) Response areas for two emergency units *Operations Research*, **20**, 571–594. [3.7.1]

Cavendish, J.C. (1974) Automatic triangulation of arbitrary planar domains for the finite element method *International Journal for Numerical Methods in Engineering*, **8**, 679–696. [6.2]

Chand, D.R. and S.S. Kapur (1970) An algorithm for convex polytopes *Journal of the ACM*, **17**, 78–86. [4.7]

Chang, R.C. and R.C.T. Lee (1984) On the average length of Delaunay triangulations *BIT*, **24**, 269–273. [6.2]

Chang, S-K. (1972) The generation of minimal trees with a Steiner topology *Journal of the Association for Computing Machinery*, **19**, 699–711. [9.3.3]

Chapman, G.P. (1970) The application of information theory to the analysis of population distributions in space *Economic Geography*, **46**, 317–331. [8.1.2]

Chazelle, B. (1985) How to search in history *Information and Control*, **64**, 77–99. [3.9]

Chazelle, B., R. Cole, F.P. Preparata and C.K. Yap (1986) New bounds for neighbor searching *Information and Control*, **68**, 105–124. [3.2]

Chazelle, B. and L.J. Guibas (1985) Visibility and intersection problems in plane geometry *Proceedings of the ACM Symposium on Computational Geometry, Baltimore, Maryland*, 135–146. [3.4.1]

Chen, Z-T. and J.A. Guevara (1987) Systematic selection of very important points

(VIP) from digital terrain model for constructing triangular irregular networks *Proceedings of Auto-Carto*, **8**, 50–56. [6.4]
Cheriton, D. and R.E. Tarjan (1976) Finding minimum spanning trees *SIAM Journal on Computing*, **5**, 724–742. [2.5, 9.3.3]
Chew, L.P. (1989) Guaranteed-quality triangular meshes *Technical Report, TR-89-983*, Cornell University, Ithaca. [3.9]
Chew, L. and R. Drysdale, III (1985) Voronoi diagrams based on convex distance functions *Proceedings of the ACM Symposium on Computational Geometry, Baltimore, June 1985*, 235–244. [4.8]
Christ, N.H., R. Friedberg and T.D. Lee (1982) Random lattice field theory: general formulation *Nuclear Physics*, **B202**, 89–125. [1.2, 5.2]
Christaller, W. (1933) *Die Zentralen Orte in Süddeutschland* Jena: Fischer. (Translated by C. Baskin, *The Central Places of Southern Germany* Englewood Cliffs: Prentice-Hall, 1966.) [7.1, 9.2.4]
Christensen, A.H.J. (1987) Fitting a triangulation to contour lines *Proceedings of Auto-Carto*, **8**, 57–67. [6.3]
Cliff, A.D. and J.K. Ord (1981) *Spatial Processes: Models and Applications* London: Pion. [1.3.8, 8.0, 8.1.1]
Collins, R. (1967) Statistical geometry and thermodynamics of liquids. In: P.S. Rudman, J. Stringer and R.J. Jaffee (eds.) *Phase Stability in Metals and Alloys* New York: McGraw-Hill, pp. 499–522. [7.1]
Collins, R. (1968) A geometrical sum rule for two-dimensional fluid correlation functions *Journal of Physics C: Solid State Physics (London), Series 2*, **1**, 1461–1471. [5.9, 7.1]
Collins, R. (1972) Melting and statistical geometry of simple liquids. In: C. Domb and M.S. Green (eds). *Phase Transitions and Critical Phenomena* New York: Academic Press, pp. 271–303. [7.1]
Conway, J.H. and N.J.A. Sloane (1982) Voronoi regions of lattices, second moments of polytopes, and quantization *IEEE Transactions on Information Theory*, **28**(2), 211–226. [8.3.1]
Conway, J.H. and N.J.A. Sloane (1984) On the Voronoi regions of certain lattices *SIAM Journal on Algorithms and Discrete Mathematics*, **5**(3), 294–305. [8.3.1]
Conway, J.H. and N.J.A. Sloane (1988) *Sphere Packings, Lattices and Groups* New York: Springer-Verlag. [6.1.1, 7.1, 8.3.1]
Cormack, R.M. (1979) Spatial aspects of competition between individuals. In: R.M. Cormack and J.K. Ord (eds.) *Spatial and Temporal Analysis in Ecology* Fairland, Maryland: International Co-operative Publishing House, pp. 151–212. [8.1.1]
Correc, Y. and E. Chapuis (1987) Fast computation of Delaunay triangulations *Advances in Engineering Software*, **9**(2), 77–83. [4.1]
Courant, R. and Robins, H. (1941) *What is Mathematics?* New York: Oxford University Press. [9.3.3]
Cover, T.M. and P.E. Hart (1967) Nearest neighbor pattern classification *IEEE Transactions, Information Theory*, **IT-13**, 21–27. [3.9]
Cowan, R. (1978) The use of ergodic theorems in random geometry *Advances in Applied Probability*, **10**, Supplement, 47–57. [5.1]
Cowan, R. (1980) Properties of ergodic random mosaic processes *Mathematische Nachrichten*, **97**, 89–102. [5.1]
Cox, K.R. and J.A. Agnew (1974) Optimal and non-optimal territorial partitions: a possible approach toward conflict *Papers of the Peace Science Society (International)*, **23**, 123–138. [3.9, 7.5.2]

Cox, K.R. and J.A. Agnew (1976) The spatial correspondence of theoretical partitions *Syracuse University, Department of Geography, Discussion Paper Series*, **24**. [7.1, 7.5.2]

Crain, I.K. (1972) Monte-Carlo simulation of the random Voronoi polygons: preliminary results *Search*, **3**(6), 220–221. [5.3, 5.4]

Crain, I.K. (1976) Statistical analysis of geotechtonics. In: D.F. Merriam (ed.) *Random Processes in Geology* New York: Springer-Verlag, pp. 3–15. [5.2, 5.3]

Crain, I.K. (1978) The Monte-Carlo generation of random polygons *Computers and Geosciences*, **4**, 131–141. [4.1, 5.3, 5.4]

Crain, I.K. and R.E. Miles (1976) Monte Carlo estimates of the distribution of the random polygons determined by random lines in a plane *Journal of Statistical Computation and Simulation*, **4**, 293–325. [5.7]

Cromley, R.G. and D. Grogan (1985) A procedure for idenifying and storing a Thiessen diagram within a convex boundary *Geographical Analysis*, **17**(2), 167–175. [4.1]

Cruz Orive, L-M. (1979) Distortion of certain Voronoi tessellations when one particle moves *Journal of Applied Probability*, **16**, 95–103. [7.4]

Cunliffe, B. (1971) Some aspects of hillforts and their cultural environments. In: D. Hill and M. Jesson (eds.) *The Iron Age and its Hillforts* Southampton: Millbrook Press, pp. 53–69. [3.9, 7.1]

Dacey, M.F. (1965) The geometry of central place theory *Geografiska Annaler*, **47B**, 111–124. [1.2, 7.1]

Dahlberg, R.E. (1967) Towards the improvement of the dot map *Internationales Jarbuch für Kartographie*, **7**, 157–166. [6.1.1]

David, C.W. (1984) Voronoi polyhedra and cleft recognition in aquated macromolecules *Computers and Chemistry*, **8**(3), 225–226. [7.1]

David, E.E. and C.W. David (1982a) Voronoi polyhedra and cluster recognition *Journal of Chemical Physics*, **77**(6), 3288–3289. [7.1]

David, E.E. and C.W. David (1982b) Voronoi polyhedra as a tool for studying solution structures *Journal of Chemical Physics*, **76**(9), 4611–4614. [7.1]

David, E.E. and C.W. David (1982c) Voronoi polyhedra and solvent structure for aqueous solutions (III) *Journal of Chemical Physics*, **77**(12), 6251–6254. [7.1]

David, E.E. and C.W. David (1983) Voronoi polyhedra for studying solvation structure (IV) *Journal of Chemical Physics*, **78**(3), 1459–1464. [7.1]

De Floriani, L. (1989) Two algorithms for radial ordering of Delaunay triangles *Abstracts of the 1st Canadian Conference on Computational Geometry, Montreal, Quebec*, 40. [2.4]

De Floriani, L., B. Falcidieno and C. Pienovi (1982) Triangulated irregulator networks in geographical data processing. In: S. Rinaldi (ed.) *Environmental Systems Analysis and Management* Amsterdam: North-Holland, pp. 801–811. [6.2, 6.3]

De Floriani, L., B. Falcidieno and C. Pienovi (1983) A Delaunay-based method for surface approximation. In: P.J.W. ten Hagen (ed.), *Eurographics '83* Amsterdam: Elsevier Science Publishers B.V., pp. 333–350. [6.2, 6.4]

De Floriani, L., B. Falcidieno and C. Pienovi (1985) Delaunay-based representation of surfaces defined over arbitrarily shaped domains *Computer Vision, Graphics, and Image Processing*, **32**, 127–140. [6.2, 6.3, 6.4]

Dehne, F. and R. Klein (1987) An optimal algorithm for computing the Voronoi diagram on a cone *Report SCS-TR-122*, School of Computer Science, Carleton University, Ottawa, Canada. [3.7.6, 4.5, 4.8]

Delaunay, B. (1934) Sur la sphère vide *Bulletin of the Academy of Sciences of the U.S.S.R., Classe des Sciences Mathématiques et Naturelles*, Series **7**(6), 793–800. [1.2, 2.2]

Delone, B.N. (1961) Proof of the fundamental theorem in the theory of stereohedra *Soviet Mathematics*, **2**, 812–815. [7.1]

Delone, B.N., N.P. Dolbilin, and M.I. Stogrin (1980) Combinatorial and metric theory of planigons *Proceedings of the Steklov Institute of Mathematics*, Issue 4, 111–141. [7.1]

Devijver, P.A. and M. Dekesel (1982) Insert and delete algorithms for maintaining dynamic Delaunay triangulations *Pattern Recognition Letters*, **1**, 73–77. [4.1]

Devijver, P.A. and M. Dekesel (1983) Computing multidimensional Delaunay tessellations *Pattern Recognition Letters*, **1**, 311–316. [4.7]

Dey, T., C. Bajaj and K. Sugihara (1991) On good triangulations in three dimensions *Proceedings of the ACM Symposium on Solid Modeling Foundations and CAD/CAM Applications, Austin, Texas*, 431–441. [3.9]

DiCenzo, S.B. and G.K. Wertheim (1989) Monte Carlo calculation of the size distribution of supported clusters *Physical Review B*, **39**(10), 6792–6796. [5.2, 5.3, 5.4]

Diggle, P.J. (1983) *Statistical Analysis of Spatial Point patterns* New York: Academic Press. [1.3.8, 8.0, 8.3]

Dijkstra, E. (1959) A note on two problems in connection with graphs *Numerische Mathematik*, **1**, 269–271. [3.4.1]

Dillencourt, M.B. (1990a) Toughness and Delaunay triangulations *Discrete and Computational Geometry*, **5**, 573–601. [2.4]

Dillencourt, M.B. (1990b) Realizability of Delaunay triangulations *Information Processing Letters*, **33**, 283–287. [2.4]

Dirichlet, G.L. (1850) Über die reduction der positeven quadratischen formen mit drei unbestimmten ganzen zahlen *Journal für die Reine und Angewandte Mathematik*, **40**, 209–227. [1.2]

Dobkin, D.P., S.J. Friedman and K.J. Supowit (1987) Delaunay graphs are almost as good as complete graphs *Proceedings of the 28th IEEE Symposium on Foundations of Computer Science, Los Angeles, California*, 20–26. [2.4]

Dobkin, D.P., S.J. Friedman and K.J. Supowit (1990) Delaunay graphs are almost as good as complete graphs *Discrete and Computational Geometry*, **5**, 399–407. [2.4]

Dodge, C.W. (1972) *Euclidean Geometry and Transformations* Reading: Addison-Wesley. [2.4]

Dokmeci, V.F. (1973) An optimization model for a hierarchical spatial system *Journal of Regional Science*, **13**(3), 439–451. [9.2.3]

Drouffe, J.M. and C. Itzykson (1984) Random geometry and the statistics of two-dimensional cells *Nuclear Physics*, **B235**, 45–53. [5.2, 5.3, 5.4, 5.9]

Drysdale, R.L. (1979) Generalized Voronoi diagrams and geometric searching Ph.D Thesis, Department of Computer Science, Stanford University. [3.5]

Drysdale, R.L. and D-T. Lee (1978) Generalized Voronoi diagram in the plane *Proceedings of the 16th Annual Allerton Conference on Communications, Control and Computing*, 833–842. [3, 3.5, 4.1]

Düppe, R.D. and H.J. Gottschalk (1970) Automatische interpolation von isolinien bei willkuerlich verteilten stuetzpunkten *Allgemeine Vermessungs-Nachrichten*, **129**(10), 423–426. [6.2]

Dwyer, R.A. (1987) A faster divide-and-conquer algorithm for constructing Delaunay

triangulations *Algorithmica*, **2**, 137–151. [4.1, 4.4]
Eaton, B.C. and R.G. Lipsey (1975a) The principle of minimum differentiation reconsidered: some new developments in the theory of spatial competition *Review of Economic Studies*, **42**, 27–49. [7.3.1]
Eaton, B.C. and R.G. Lipsey (1975b) The non-uniqueness of equilibrium in the Löschan location model *The American Economic Review*, **66**, 77–93. [7.3.1]
Eaton, B.C. and R.G. Lipsey (1980) The block metric and the law of markets *Journal of Urban Economics*, **7**, 337–347. [3.9]
Edelsbrunner, H. (1987) *Algorithms in Combinatorial Geometry* Berlin: Springer-Verlag. [2.4, 4.7]
Edelsbrunner, H. (1988) The computational geometry column *EATCS (European Association for Theoretical Computer Science) Bulletin*, **37**, 109–116. [2.4, 4.1]
Edelsbrunner, H. and E.P. Mücke (1987) Simulation of simplicity—a technique to cope with degenerate cases in geometric algorithms *Proceedings of the 4th Annual ACM Symposium on Computational Geometry*, Urbana-Champaign, IL, pp. 118–133. [4.6.2]
Edelsbrunner, H. and R. Seidel (1986) Voronoi diagrams and arrangements *Discrete Computational Geometry*, **1**, 25–44. [3, 4.1, 4.7, 4.8]
Edelsbrunner, H., T.S. Tan and R. Waupotitsch (1990) An $O(n^2 \log n)$ time algorithm for the minmax angle triangulation *Proceedings of the 6th Annual Symposium on Computational Geometry, Berkeley, California*, 44–52. [6.2]
Egginton, S. (1990) Morphometric analysis of tissue capillary supply. In: R.G. Boutilier (ed.) *Advances in Comparative and Environmental Physiology*, Volume 6 Berlin: Springer-Verlag. pp. 73–141. [7.1]
Egginton, S. and H.F. Ross (1989a) Quantifying capillary distribution in four dimensions *Advances in Experimental Medicine and Biology*, **248**, 271–280. [7.1]
Egginton, S. and H.F. Ross (1989b) Influence of muscle phenotype on local capillary supply *Advances in Experimental Medicine and Biology*, **248**, 281–291. [7.1]
Egginton, S., Z. Turek and L. Hoofd (1987) Morphometric analysis of sparse capillary networks. In: I.A. Silver and A. Silver (eds.) *Oxygen Transport to Tissue IX* New York: Plenum Publishing Corp., pp. 1–12. [7.1]
Egginton, S., Z. Turek and L.J.C. Hoofd (1988) Differing patterns of capillary distribution in fish and mammalian skeletal muscle *Respiration Physiology*, **74**, 383–396. [7.1]
Ehrlich, P.E. and H.C. Im Hof (1979) Dirichlet regions in manifolds without conjugate points *Commentarii Mathematici Helvetici*, **54**, 642–658. [3.7.8]
El Gindy, H.A. and D. Avis (1981) A linear algorithm for computing the visibility polygon from a point *Journal of Algorithms*, **2**, 186–197. [3.4.1]
Elliott, H.M. (1981) Macrocalifornia and the urban gradient *The California Geographer*, **21**, 1–17. [6.1.2]
Elliott, H.M. (1982) The Pacific nearest larger neighbor gradient *Yearbook, Association of Pacific Coast Geographers*, **44**, 29–45. [6.1.2]
Elliott, H.M. (1983) Surrounding larger neighbors and the Atlantic Coast cardinal neighbor gradient *Economic Geography*, **59**(4), 426–444. [6.1.2]
Elliott, H. (1985) Cardinal place geometry *Geographical Analysis*, **17**(1), 16–35. [6.1.2]
Engel, P. (1981) Über wirkungsbereichsteilungen von kubischer symmetrie *Zeitschrift für Kristallographie*, **154**, 199–215. [7.1]
Erlenkotter, D. (1981) A comparative study of approaches to dynamic location problems *European Journal of Operations Research*, **6**, 133–143. [9.4.1]

Evans, D.G. and S.M. Jones (1987) Detecting Voronoi (area-of-influence) polygons *Mathematical Geology*, **19**(6), 523–537. [7.5.1, 7.5.2]

Evans, I.S. (1967) The properties of patterns of points, measured by space filling and angular relationships *Geographical Articles (Cambridge)*, **8**, 63–77. [1.2, 8.1.1]

Evans, U.R. (1945) The laws of expanding circles and spheres in relation to the lateral growth of surface films and the grain-size of metals *Transactions of the Faraday Society*, **41**, 365–374. [5.6, 7.2]

Feller, W. (1957) *An Introduction to Probability Theory and its Applications* (Second edition) New York: John Wiley. [1.3]

Fetter, F.A. (1924) The economic law of market areas *Quarterly Journal of Economics*, **38**, 520–529. [3.9]

Fiacco, A.V. and G.P. McCormick (1968) *Nonlinear Programming: Sequential Unconstrained Minimization Technique* New York and Toronto: John Wiley. [9.1, 9.1.3]

Finney, J.L. (1970a) Random packings and the structure of simple liquids. I. The geometry of random close packing *Proceedings of the Royal Society of London A*, **319**, 479–493. [5.2, 5.10, 7.1]

Finney, J.L. (1970b) Random packings and the structure of simple liquids. II. The molecular geometry of simple liquids *Proceedings of the Royal Society of London A*, **319**, 495–507. [5.10]

Finney, J.L. (1975a) The structures of laboratory and computer built random packing *Journal de Physique, Colloque C2*, **36**(4 Supplement), C2-1–C2-11. [5.10, 7.1]

Finney, J.L. (1975b) Volume occupation, environment and accessibility in proteins *Journal of Molecular Biology*, **96**, 721–732. [7.1]

Finney, J.L. (1979) A procedure for the construction of Voronoi polyhedra *Journal of Computational Physics*, **32**, 137–143. [4.7]

Firbank, L.G. and A.R. Watkinson (1987) On the analysis of competition at the level of the individual plant *Oecologia (Berlin)*, **71**, 308–317. [7.1]

Fischer, R.A. and R.E. Miles (1973) The role of spatial pattern in the competition between crop plants and weeds. A theoretical analysis *Mathematical Biosciences*, **18**, 335–350. [7.4]

Fischer, W. and E. Koch (1979) Geometrical packing analysis of molecular compounds *Zeitschrift für Kristallographie*, **150**, 245–260. [3.9, 7.1]

Fortune, S. (1986) A sweepline algorithm for Voronoi diagrams *Proceedings of the 2nd Annual ACM Symposium on Computational Geometry, Yorktown Heights, June 1986*, 313–322. [3.5, 4.1, 4.5, 4.8]

Fortune, S. (1987) A sweepline algorithm for Voronoi diagrams *Algorithmica*, *2*, 153–174. [4.1, 4.5, 4.8]

Fowler, R.S. and J.J. Little (1979) Automatic extraction of irregular network digital terrain models *Computer Graphics*, **13**, 199–207. [6.2, 6.3, 6.4]

Frank, F.C. and J.S. Kasper (1958) Complex alloy structures regarded as sphere packings. I. Definitions and basic principles *Acta Crystallographica*, **11**, 184–190. [1.2, 7.1]

Franke, R. (1982) Scattered data interpolation: tests of some methods *Mathematics of Computation*, **38**(157), 181–200. [6.0, 6.3]

Fraser, A.R. (1977) Triangle based probability polygons for forest sampling *Forest Science*, **23**, 111–121. [6.2]

Fraser, A.R. and P. Van den Driessche (1972) Triangles, density and pattern in point populations *Proceedings of the Third Conference of the Advisory Group of Forest Statisticians, Jouy en Josas, France*, 277–286. [6.2]

Fraser, D.P. (1991) Voronoi statistics of hard-core systems. *Materials Characterization*, **26**, 73–83. [5.10]

Frost, H.J. and C.V. Thompson (1987a) The effect of nucleation conditions on the topology and geometry of two-dimensional grain structures *Acta Metallurgica*, **35**(2), 529–540. [5.6, 5.10, 7.2]

Frost, H.J. and C.V. Thompson (1987b) Development of microstructure in thin films. SPIE's 31st Annual International Technical Symposium on Optical and Optoelectronic Applied Science and Engineering, San Diego, Calif. [5.6, 5.10]

Fujii, A. (1976) Activity contour lines, Ph.D. thesis, Department of Architecture, University of Tokyo [in Japanese]. [3.9]

Fujii, A. (1983) Geometrical models for regional analysis *Review of City Planning*, **126**, 93–98 [in Japanese]. [3.9]

Fujii, A. and K. Oikawa (1986) Simulation models of domain division based on the allocation of urban facilities *Monthly Journal of Institute of Industrial Science, University of Tokyo*, **38**, 418–421 [in Japanese]. [3.9]

Fukunaga, K. and P.M. Narendra (1975) A branch and bound algorithm for computing *k*-nearest neighbor *IEEE Transactions on Computers*, **C-24**, 750–753. [3.9]

Gabriel, K.R. and R.R. Sokal (1969) A new statistical approach to geographic variation analysis *Systematic Zoology*, **18**, 259–278. [2.5]

Gabszewicz, J.J., J-F. Thisse, M. Fujita and U. Schweizer (1986) *Location Theory* Switzerland: Harwood Academic Publishers. [7.3.1]

Gambini, R., D.L. Huff and G.F. Jenks (1967) Geometric properties of market areas *Papers and Proceedings of the Regional Science Association*, **20**, 85–92. [3.9]

Garey, M.R. and D.S. Johnson (1977) The rectilinear Steiner tree problem is NP-complete *SIAM Journal of Applied Mathematics*, **32**, 826–834. [3.9]

Garey, M.R., R.L. Graham and D.S. Johnson (1977) The complexity of computing Steiner minimal trees *SIAM Journal on Applied Mathematics*, **32**, 835–859. [9.3.3]

Gellatly, B.J. and J.L. Finney (1982a) Characterisation of models of multicomponent amorphous metals: the radical alternative to the Voronoi polyhedron *Journal of Non-Crystalline Solids*, **50**, 313–329. [3.9]

Gellatly, B.J. and J.L. Finney (1982b) Calculation of protein volumes: an alternative to the Voronoi procedure *Journal of Molecular Biology*, **161**(2), 305–322. [3.9]

Gen, E. (1983) Algorithms for high dimensional Voronoi diagrams, Graduation thesis, Department of Mathematical Engineering and Information Physics, University of Tokyo [in Japanese]. [2.1, 2.2]

Getis, A. and B. Boots (1978) *Models of Spatial Processes: An Approach to the Study of Point, Line and Area Patterns* Cambridge: Cambridge University Press. [1.3.8, 8.0, 8.3]

Gilbert, E.N. (1962) Random subdivisions of space into crystals *Annals of Mathematical Statistics*, **33**, 958–972. [5.4, 5.5, 8.1.2]

Gilbert, E.N. and H.O. Pollack (1968) Steiner minimum trees *SIAM Journal of Applied Mathematics*, **16**(1), 1–29. [9.3.3]

Gill, P.E., W. Murry and M.H. Wright (1981) *Practical Optimization* London: Academic Press. [9.1, 9.1.3]

Glass, L. (1973) Stochastic generation of regular distributions *Science*, **180**, 1061–1063. [5.6]

Glass, L. (1974) Dynamics of expanding inhibitory fields *Science*, **183**, 446. [5.6]

Gnedenko, B.V. and A.N. Kolmogorov (1954) *Limit Distributions for Sum of Independent Random Variables* (translated from the 1949 Russian edition by K.L. Chung; with an appendex by J.L. Doob) Reading, Mass.: Addison-Wesley. [1.3.7]

Gold, C.M. (1980) Triangulation-based terrain modelling—where are we now? *Proceedings of Auto-Carto IV*, Volume II, pp. 104–111. [6.2, 6.5]

Gold, C.M. (1989) Surface interpolation, spatial adjacency and GIS. In: J. Raper (ed.) *Three Dimensional Applications in Geographical Information Systems* London: Taylor and Francis, pp. 21–35. [6.0]

Gold, C.M., T.D. Charters and J. Ramsden (1977) Automated computer mapping using triangular element data structures and an interpolant over each irregular triangular domain *Computer Graphics*, **11**, 170–175. [6.2]

Gold, C. and S. Cormack (1987) Spatially ordered networks and topographic reconstructions *International Journal of Geographical Information Systems*, **1**(2), 137–148. [6.5]

Goodchild, M.F. (1986) *Spatial Autocorrelation. Concepts and Techniques in Modern Geography* **47**, Norwich, U.K.: Geo Books. [6.1.2]

Goodchild, M.F. and B.H. Massam (1969) Some least-cost models of spatial administrative systems in southern Ontario *Geografiska Annaler*, **52B**, 86–94. [7.1]

Gowda, I.G., D.G. Kirkpatrick, D.T. Lee and A. Naamad (1983) Dynamic Voronoi diagrams *IEEE Transactions on Information Theory*, **IT-29**(5), 724–731. [4.1]

Gower, J.C., and G.J.S. Ross (1969) Minimum spanning trees and single linkage cluster analysis *Applied Statistics*, **18**, 54–64. [2.5]

Green, F.H.W. (1955) *Local Accessibility* Chessington, Surrey: Director General of the Ordnance Survey. [7.2]

Green, P.J. (1981) Peeling bivariate data. In: V. Barnett (ed.) *Interpreting Multivariate Data*. New York: John Wiley, pp. 3–19. [6.5]

Green, P.J. and R. Sibson (1978) Computing Dirichlet tessellations in the plane *The Computer Journal*, **21**, 168–173. [4.1]

Greub, W. (1975) *Linear Algebra* (Fourth edition) New York: Springer-Verlag. [4.6.2]

Griffith, D.A. (1982) Dynamic characteristics of spatial economic systems *Economic Geography*, **58**, 177–196. [6.1.2]

Griffith, D.A. (1987) *Spatial Autocorrelation: A Primer* Washington, D.C.: Association of American Geographers. [6.1.2]

Grigorovici, R. (1969) Short-range order in amorphous semiconductors *Journal of Non-Crystalline Solids*, **1**, 303–325. [7.1]

Grigorovici, R. and R. Manaila (1967/8) Structural model for amorphous germanium layers *Thin Solid Films* **1**, 343–352. [7.1]

Grigorovici, R. and R. Manaila (1969) Short-range order in amorphous germanium *Journal of Non-Crystalline Solids*, **1**, 371–387. [7.1]

Grünbaum, B. and G.C. Shephard (1980) Tilings with congruent tiles *Bulletin (New Series) of the American Mathematical Society*, **3**(3), 951–973. [1.2, 7.1]

Grünbaum, B. and G.C. Shephard (1987) *Tilings and Patterns* New York: W.H. Freeman. [7.1]

Guibas, L. and J. Stolfi (1985) Primitives for the manipulation of general subdivisions and the computation of Voronoi diagrams *ACM Transactions on Graphics*, **4**(2), 74–123. [4.1, 4.4]

Gunawardane, G. (1982) Dynamic versions of set covering type public facility location problems *European Journal of Operations Research*, **10**, 190–195. [9.4.1]

Haggett, P., A.D. Cliff and A. Frey (1977). *Locational Methods* London: Edward Arnold. [1.3.8, 8.0, 8.1.1]

Haining, R., D.A. Griffith and R. Bennett (1984) A statistical approach to the problem of missing spatial data using a first-order Markov model *The Professional Geographer*, **36**, 338–345. [6.1.2]

Hakimi, S.L. (1964) Optimal locations of switching centers and the absolute centers and medians of a graph *Operations Research*, **12**, 450–475. [3.9, 9.2]

Hamilton, W.D. (1971) Geometry for the selfish herd *Journal of Theoretical Biology*, **31**, 295–311. [5.2, 5.3, 7.4]

Hammond, N.D.C. (1972) Locational models and the site of Lubaantun: a classic Maya centre. In: D.L. Clarke (ed.) *Models in Archaeology* London: Methuen, pp. 757–799. [7.1]

Hammond, N. (1974) The distribution of Late Classic Maya major ceremonial centres in the Central Area. In: N. Hammond (ed.) *Mesoamerican Archaeology: New Approaches* Austin, Texas: University of Texas Press, pp. 313–334. [7.1]

Handler, G.Y. and P.B. Mirchandani (1979) *Location on Networks, Theory and Algorithms* Cambridge, Mass.: The MIT Press. [3.9, 9.2]

Hanjoul, P., H. Beguin and J-C. Thill (1989) Advances in the theory of market areas *Geographical Analysis*, **21**(3), 185–196. [3.1.2, 3.9]

Hanjoul, P. and J-C. Thisse (1987) Elements of planar analysis in spatial competition *Regional Science and Urban Economics*, **17**, 423–439. [3.9]

Hanson, H.G. (1983) Voronoi cell properties from simulated and real random spheres and points *Journal of Statistical Physics (USA)*, **30**(3), 591–605. [5.3, 5.4, 5.10]

Harding, J.E. (1920–21) Calculation of ore tonnage and grade from drill-hole samples *Transactions of the American Institute of Mining Engineers*, **66**, 117–126. [1.2, 6.2]

Harding, J.E. (1923) How to calculate tonnage and grade of an ore body *Engineering and Mining Journal Press*, **116**(11), 445–448. [1.2, 5.2]

Hasegawa, M. and M. Tanemura (1976) On the pattern of space division by territories *Annals of the Institute of Statistical Mathematics*, **28**, Part B, 509–519. [5.2, 5.3, 7.2, 7.3.2, 9.5.1]

Hasegawa, M. and M. Tanemura (1980) Spatial patterns of territories. In: K. Matusita (ed.) *Recent Developments in Statistical Inference* Amsterdam: North-Holland, pp. 73–78. [5.2, 5.10, 7.3.2]

Hasegawa, M. and M. Tanemura (1986) *Ecology of Territories: Statistics of Ecological Models and Spatial Patterns* Tokyo: Tokai University Press [in Japanese]. [7.3.2]

Hasegawa, M., M. Tanemura and S.Takiguchi (1981) Spatial patterns in ecology *International Roundtable Congress—The 50th Anniversary of the Japan Statistical Society*, 146–161. [5.10, 7.1]

Hayes, W.B. and G.S. Koch (1984) Constructing and analyzing area-of-influence polygons by computer *Computers and Geosciences*, **10**(4), 411–430. [1.2, 6.1.1]

Haynes, K.E. and A.S. Fotheringham (1984) *Gravity and Spatial Interaction Models*, Sage Scientific Geography Series, Volume 2 Beverly Hills, Calif.: Sage Publications. [6.1.2]

Henry, N.F.M. and K. Lonsdale (eds) (1965) *International Tables for X-ray Crystallography. Volume 1: Symmetry groups* Birmingham: Kynoch Press. [7.1]

Hermann, H., H. Wendrock and D. Stoyan (1989) Cell-area distributions of planar

Voronoi mosaics *Metallography*, **23**, 189–200. [5.3, 5.4, 5.10]
Hinde, A.L. and R.E. Miles (1980) Monte Carlo estimates of the distributions of the random polygons of the Voronoi tessellation with respect to a Poisson process *Journal of Statistical Computation and Simulation*, **10**, 205–223. [5.3, 5.4, 8.1.1]
Hodder, I.R. (1972) Locational models and the study of Romano-British settlement. In: D.L. Clarke (ed.) *Models in Archaeology* London: Methuen. [7.1]
Hodder, I. and C. Orton (1976) *Spatial Analysis in Archaeology* Cambridge: Cambridge University Press. [3.9]
Hodgson, M. (1984) Alternative approaches to hierarchical location—allocation systems *Geographical Analysis*, **16**, 275–281. [9.2.4]
Hoffmann, C.M. (1989) *Geometric and Solid Modeling—An Introduction* San Mateo, Calif.: Morgan Kaufmann Publishers. [4.2]
Hogg, A.H.A. (1971) Some applications of surface fieldwork. In: D. Hill and M. Jesson (eds), *The Iron Age and Its Hillforts* Southampton: Millbrook Press, pp. 105–125. [3.9]
Honda, H. (1978) Description of cellular patterns by Dirichlet domains: the two-dimensional case *Journal of Theoretical Biology*, **72**, 523–543. [7.2, 7.5.2, 9.5.1]
Honda, H. (1983) Geometrical models for cells in tissues *International Review of Cytology*, **81**, 191–248. [7.5.2]
Honda, H., T. Morita and A. Tanabe (1979) Establishment of epidermal cell columns in mammalian skin: computer simulation *Journal of Theoretical Biology*, **81**, 745–759. [7.1]
Hoofd, L., Z. Turek, K. Kubat, B.E.M. Ringnalda and S. Kazda (1985) Variability of intercapillary distance estimated on histological sections of rat heart *Advances in Experimental Medicine and Biology*, **191**, 239–247. [1.2, 7.1]
Horalek, V. (1988) A note on the time-non-homogeneous Johnson–Mehl tessellation *Advances in Applied Probability*, **20**, 684–685. [5.6]
Horalek, V. (1989) The Johnson–Mehl tessellation with time dependent nucleation intensity in view of basic 3-d tessellations *Mathematical Research*, **51**, 111–116. [5.6]
Horalek, V. (1990) On ASTM grain-size model and related random tessellation models *Metallography*, **25**, 263–284. [5.6]
Hori, H. and M. Nagata (1985) Examples of optimization methods for environment monitoring systems *Report B-266-R-53-2*, Environmental Sciences, Ministry of Education, Japan, 18–29 [in Japanese]. [9.2.5]
Horspool, R.N. (1979) Constructing the Voronoi diagram in the plane. *Technical Report SOCS 79.12*, School of Computer Science, McGill University. [4.1]
Horton, R.E. (1917) Rational study of rainfall data makes possible better estimates of water yield *Engineering News-Record*, **79**, 211–213. [1.2, 6.1.1]
Hotelling, H. (1929) Stability in competition *Economic Journal*, **39**, 41–57. [7.3.1]
Hubbard, R. (1970) The spatial pattern and market areas of urban settlement in Jamaica *University of the West Indies, Department of Geography, Research Notes*, **1**. [3.9]
Huff, D.L. (1973) The delineation of a national system of planning regions on the basis of urban spheres of influence *Regional Studies* **7**, 323–329. [3.9]
Huff, D.L. and C.F. Jenks (1968) A graphic interpretation of the friction of distance in gravity models *Annals, Association of American Geographers*, **58**, 814–824. [3.9]
Huff, D.L. and J.M. Lutz (1979) Ireland's urban hierarchy *Economic Geography*,

55, 196–212. [3.9]

Hutchings, M.J. and R.J. Discombe (1986) The detection of spatial pattern in plant populations *Journal of Biogeography*, **13**, 225–236. [5.3, 5.10, 7.1, 8.1.1, 8.2]

Hutchings, M.J. and S. Waite (1985) Cohort behaviour and environmental determination of life histories within a natural population of *Plantago coronopus L.* In: J. Haeck and J.W. Woldendorp (eds) *Structure and Functioning of Plant Populations 2.* Amsterdam: North-Holland. [7.1]

Hwang, F.K. (1976) On Steiner minimal trees with rectilinear distance *SIAM Journal of Applied Mathematics*, **30**, 104–114. [3.9]

Hwang, F.K. (1979) An O(n log n) algorithm for rectilinear minimal spanning trees *Journal of the Association of Computing Machinery*, **26**, 177–182. [3.7.1]

Hyson, C.D. and W.P. Hyson (1950) The economic law of market areas *Quarterly Journal of Economics*, **64**, 319–327. [3.9]

Icke, V. and R. Van de Weygaert (1987) Fragmenting the universe. I Statistics of two-dimensional Voronoi foams *Astronomy and Astrophysics*, **184**, 16–32. [1.2, 5.2, 5.4, 5.10, 7.1]

Illeris, S. (1967) Funktionelle regioner i Danmark ommkring 1960 *Geografisk Tidsskrift*, **66**, 225–251. (English translation reprinted in B.J.L. Berry and F.E. Horton (eds) *Geographic Perspectives on Urban Systems with Integrated Readings* Englewood Cliffs, N.J.: Prentice-Hall, pp. 200–207.) [3.9]

Imai, T. (1989) Numerically robust algorithm for constructing Voronoi diagrams for line segments, Master's Thesis, Department of Mathematical Engineering and Information Physics, Faculty of Engineering, University of Tokyo [in Japanese]. [4.8]

Imai, H., T. Asano and T. Asano (1985) Visibility of disjoint polygons Research Memorandum RMI 85-02, Department of Mathematical Engineering and Instrumentation Physics, University of Tokyo. [3.4.1]

Imai, H., M. Iri and K. Murota (1985) Voronoi diagram in the Laguerre geometry and its applications *SIAM Journal of Computing*, **14**(1), 93–105. [3.1.4, 3.5, 3.5.3, 3.7.8, 4.8]

Imai, T. and K. Sugihara (1989) A combinatorial-structure oriented algorithm for Voronoi diagrams of line segments *Notes of the Special Interest Group on Algorithms of Information Processing Society of Japan*, SIGAL 11-2, October 1989 [in Japanese]. [4.8]

Iri, M. and T. Koshizuka (eds) (1986) *Computational Geometry and Geographical Information Processing* Tokyo: Kyoritsu [in Japanese]. [1.0]

Iri, M., K. Murota and T. Ohya (1984) A fast Voronoi-diagram algorithm with applications to geographical optimization problems. In: P. Throft-Christensen (ed.) *Lecture Notes in Control and Information Science. Volume 59: System Modelling and Optimization*, Proceedings of the 11th IFIP Conference Copenhagen Springer-Verlag, Berlin, pp. 273–288. [4.1, 9.2, 9.2.1]

Jack, W.H. (1967) Single tree sampling in even-aged plantations for survey and experimentation *International Union of Forestry Research Organisations, XIV. IUFRO-Kongress, Munchen, Papers* VI, Section 25, 379–403. [1.2, 7.1]

Jackson, J.L. (1974) Stochastic generation of regular distributions: some analytical results *Science*, **183**, 445–446. [5.6]

Jerauld, G.R., J.C. Hatfield and H.T. Davis (1984a) Percolation and conduction on Voronoi and triangular networks: a case study in topological disorder *Journal of Physics C: Solid State Physics (London)*, **17**(9), 1519–1529. [5.2, 7.1]

Jerauld, G.R., L.E. Scriven and H.T. Davis (1984b) Percolation and conduction on the 3D Voronoi and regular networks: a second case study in topological disorder *Journal of Physics C: Solid State Physics (London)*, **17**(19), 3429–3439. [5.2, 5.3]

Joe, B. (1986) Delaunay triangular meshes in convex polygons *SIAM Journal on Scientific and Statistical Computing*, **7**(2), 514–539. [6.2]

Johnson, N.L. and S. Kotz (1969) *Distributions in Statistics: Discrete Distributions* Boston: Houghton Mifflin. [1.3.7]

Johnson, W.A. and R.F. Mehl (1939) Reaction kinetics in processes of nucleation and growth *Transactions of the American Institute of Mining, Metallurgical and Petroleum Engineers*, **135**, 410–458. [3.9, 5.6]

Jones, K.G. and D.R. Mock (1984) Evaluating retail trading performance. In: R.L. Davies and D.S. Rogers (eds) *Store Location and Store Assessment Research* New York: John Wiley, pp. 333–360. [7.1]

Jones, K. and J. Simmons (1987) *Location, Location, Location. Analyzing the Retail Environment* Toronto: Methuen. [7.1]

Jones, T.A. (1989) The three faces of geological computer contouring *Mathematical Geology*, **21**(3), 271–283. [6.2]

Jones, T.A., D.E. Hamilton and C.R. Johnson (1986) *Contouring Geologic Surfaces with the Computer* New York: Van Nostrand Reinhold. [6.2]

Kaler, E.W. and S. Prager (1982) A model of dynamic scattering by microemulsions *Journal of Colloid and Interface Science*, **86**(2), 359–369. [5.2]

Karlin, S. (1969) *First Course in Stochastic Processes* New York: Academic Press. [1.3.8]

Katajainen, J. and M. Koppinen (1988) Constructing Delaunay triangulations by merging buckets in quadtree order *Fundamenta Informaticae*, **11**, 275–288. [4.1, 4.4]

Keeney, R.L. (1972) A method for districting among facilities *Operations Research*, **20**, 613–618. [7.1]

Keil, J.M. and C.A. Gutwin (1989) The Delaunay triangulation closely approximates the complete Euclidean graph *Proceedings of the First Workshop on Algorithms and Data Structures*, 47–56. [2.4]

Kendall, D.G. (1981) The statistics of shape. In: V. Barnett (ed.) *Interpreting Multivariate Data* Chichester: John Wiley. [5.9]

Kendall, D.G. (1983) The shape of Poisson–Delaunay triangles. In: M.C. Demetrescu and M. Iosifescu (eds) *Studies in Probability and Related Topics* Montreal: Nagard, pp. 321–330. [5.3, 5.9]

Kendall, D.G. (1987) Random Delaunay simplices in R^m *Proceedings of the Blacksburg Conference to Honor I.J. Good.* [5.3]

Kendall, D.G. (1988) Delone simplexes in many dimensions, Paper presented at 18th EMS, Berlin, DDR, 24 August 1988. [5.3]

Kendall, D.G. (1989) A survey of the statistical theory of shape *Statistical Science*, **4**(2), 87–120. [5.3, 5.9, 8.2]

Kenkel, N.C., J.A. Hoskins and W.D. Hoskins (1989) Edge effects in the use of area polygons to study competition *Ecology*, **70**(1), 272–274. [5.3]

Khachian, L.G. (1980) Polynomial algorithms in linear programming *USSR Computational Mathematics and Mathematical Physics*, **20**, 53–72. [4.6.2]

Kiang, T. (1966) Random fragmentation in two and three dimensions *Zeitschrift für Astrophysik*, **64**, 433–439. [5.3, 5.4]

King, L.J. (1984) *Central Place Theory* Scientific Geography Series, Volume 1 Beverly Hills, Calif.: Sage Publications. [7.1]

Kirkpatrick, D.G. (1979) Efficient computation of continuous skeletons *Proceedings of the 20th Annual IEEE Symposium on Foundations of Computer Science*, 18–27. [3.5]

Kirkpatrick, D.G. (1980) A note on Delaunay and optimal triangulations *Information Processing Letters*, **10**(3), 127–128. [6.2]

Kishimoto, K. (1978) An algorithm for the optimal triangulation *Information Processing*, **19**, 211–218 [in Japanese]. [2.4]

Klee, V. (1980) On the complexity of *d*-dimensional Voronoi diagrams *Archiv de Mathematik*, **34**, 75–80. [2.3, 4.7]

Klein, R. (1988) Abstract Voronoi diagrams and their applications *Lecture Notes in Computer Science*, **333** (International Workshop on Computational Geometry, Wurzburg, March 1988) Springer-Verlag, Berlin, pp. 148–157. [3.7.2, 3.7.6, 4.8]

Klein, R. (1989) Concrete and abstract Voronoi diagrams *Springer Lecture Notes in Computer Sciences*, **400**. [4.8]

Klein, R., K. Melhorn and S. Meiser (1990) On the construction of abstract Voronoi diagrams, II. Algorithms. International Symposium SIGAL'90, Tokyo, Japan, August 1990. Proceedings, *Lecture Notes in Computer Science*, **450** Berlin: Springer-Verlag, pp. 138–154. [4.8]

Klein, R. and D. Wood (1988) Voronoi diagrams based on general metrics in the plane, *Lecture Notes in Computer Science*, **294**. (Proceedings of the 5th Annual Symposium on Theoretical Aspects of Computer Science, Bordeaux, February 1988) Berlin: Springer-Verlag, pp. 281–291. [3.7.6, 4.8]

Kleinstreuer, C. and J.T. Holdeman (1980) A triangular finite element mesh generation for fluid dynamic systems of arbitrary geometry *International Journal for Numerical Methods in Engineering*, **15**, 1325–1334. [6.2, 6.3]

Knuth, D.E. (1968) *The Art of Computer Programming. Volume I: Fundamental Algorithms* Reading, Mass.: Addison-Wesley. [4.1]

Knuth, D.E. (1973) *The Art of Computer Programming. Volume III: Sorting and Searching* Reading, Mass.: Addison-Wesley. [2.3, 4.4]

Koch, R.A. (1974) Zur Frage der quantitaven Klufterfassung von saulig absonderndem Eruptivgestein (Nephelinbasalt von stolpen) *Abhandlungen des Staatlichen Museums fuer Minerallogie und Geologie zu Dresden*, **21**, 1–37. [9.5.1]

Koken, E., N. Chandrasekaran, J.D. Embury and G. Berger (1988) The role of particle distribution in recrystallization in two-phase alloys *Materials Science and Engineering*, **A104**, 163–168. [7.1]

Kokubo, I. (1985) Computational methods of generalized Voronoi diagrams, Master's thesis, Department of Mathematical Engineering and Information Physics, University of Tokyo [in Japanese]. [3.5, 4.8]

Kolars, J.F. and J.D. Nystuen (1974) *Human Geography* New York: McGraw-Hill. [2.5]

Konno, H. and H. Yamashita (1978) *Nonlinear Programming Methods* Tokyo: Nikka Giken [in Japanese]. [9.1]

Kopec, R.J. (1963) An alternative method for the construction of Thiessen polygons *The Professional Geographer*, **15**, 24–26. [1.2]

Koshizuka, T. and J. Kobayashi (1983) Road distances and straight line distances *Papers of the Annual Conference of the City Planning Institute of Japan*, **18**, 43–48 [in Japanese]. [9.2.1]

Koshizuka, T. and O. Kurita (1986) A theoretical study on the role of a circular part of radial-circular networks *Papers of the Annual Conference of the City Planning Institute of Japan*, **21**, 217–222 [in Japanese]. [3.7.2]

Kreps, D.M. and R. Wilson (1982) Reputation and imperfect information *Journal of Economic Theory*, **27**, 253–279. [7.4]

Kuhn, H.W. (1974) Steiner's problem revisited. In: G.B. Dantzig and B.C. Evans (eds) *Studies in Optimization* **10**, pp. 52–70. [9.3.3]

Lachenbruch, A.H. (1962) Mechanics of thermal contraction cracks and ice-wedge polygons in permafrost *Geological Society of America, Special Paper*, **70**. [7.5.2]

Lam, N.S-N. (1983) Spatial interpolation methods: a review *The American Cartographer*, **10**(2), 129–149. [6.0]

Lambert, C.J. and D. Weaire (1983) Order and disorder in two-dimensional random networks *Philosophical Magazine B*, **47**(4), 445–450. [5.4]

Lantuejoul, C. (1980) Skeletonization in quantitative metallography. In: R.M. Haralick and J.C. Simon (eds) *Issues in Digital Image Processing* Alphen aan der Rijn, The Netherlands: Sijthoff & Noordhoff, pp. 107–135. [7.1]

Lantuejoul, C. and F. Maisonneuve (1984) Geodesic methods in quantitative image analysis *Pattern Recognition*, **17**, 177–187. [3.9]

Launhardt, W. (1882) Die Bestimmung des zweckmässigsten Standortes einer gewerblichen Anlage *Zeitschrift des Deutscher Ingenieure*, **26**, 768. [9.2]

Laves, F. (1930) Ebeneneinteilung in wirkungsbereich *Zeitschrift für Kristallographie*, **76**, 277–284. [7.1]

Lawson, C.L. (1977) Software for C^1 surface interpolation. In: J. Rice (ed.), *Mathematical Software III* New York: Academic Press, pp. 161–194. [2.4, 4.1, 6.2]

Leamer, E.E. (1968) Locational equilibria *Journal of Regional Science*, **8**(2), 229–242. [9.2]

Le Caer, G. (1990) Do Swedish pines diagonalize complex random matrices? Unpublished paper. [5.10]

Le Caer, G. and J.S. Ho (1990) The Voronoi tessellation generated from eigenvalues of complex random matrices *Journal of Physics A: Mathematical and General*, **23**, 3279–3295. [5.3, 5.4, 5.10]

Lee, D-T. (1978) Proximity and Reachability in the Plane, Ph.D. dissertation, University of Illinois at Champaign-Urbana. [3.9]

Lee, D-T. (1980) Two-dimensional Voronoi diagrams in the L_p-metric *Journal of the Association for Computing Machinery*, **27**, 604–618. [3.7.1]

Lee, D-T. (1982a) Medial axis transformation of a planar shape *IEEE Transactions on Pattern Analysis and Machine Intelligence*, **4**, 363–369. [3.5.4, 3.9]

Lee, D-T. (1982b) On *k*-nearest neighbor Voronoi diagrams in the plane *IEEE Transactions on Computers*, **C31**, 478–487. [3.2, 3.2.1, 3.5.4, 4.8]

Lee, D-T. (1983) Visibility of a simple polygon *Computer Vision, Graphics, and Image Processing*, **22**, 207–211. [3.4.1]

Lee, D-T. and I.M. Chen (1985) Display of visible edges of a set of convex polygons. In: G.T. Toussaint (ed.) *Computational Geometry* Amsterdam: North-Holland, pp. 249–266. [3.4.1]

Lee, D-T. and R.L. Drysdale, III (1981) Generalization of Voronoi diagrams in the plane *SIAM Journal of Computing*, **10**, 73–87. [3.5, 3.5.3, 4.8]

Lee, D.T. and A.K. Lin (1986) Generalized Delaunay triangulations for planar graphs *Discrete and Computational Geometry*, **1**, 201–217. [4.8, 6.3]

Lee, D-T. and B.J. Schachter (1980) Two algorithms for constructing the Delaunay triangulation *International Journal of Computer and Information Sciences*, **9**(3), 219–242. [2.4, 4.1, 4.4]

Lee, D-T. and C.K. Wong (1980) Voronoi diagrams in L_1 (L_∞) metrics with 2-

dimensional storage applications *SIAM Journal of Computing*, **9**, 200–211. [3.7.1, 3.9]

Lenstra, A.K. (1984) Polynomial factorization by root approximation. *Lecture Notes in Computer Science*, **174** Berlin: Springer-Verlag, pp. 272–276. [4.6.2]

Lenz, R. (1979) Redundancy as an index of change in point pattern analysis *Geographical Analysis*, **11**, 374–388. [8.1.2]

Leonyuk, N.I. and L.I. Mal'tseva (1980) Use of the Wulff–Delaunay method to assess the habitats of crystals *Soviet Physics—Crystallography*, **25**(5), 524–527. [7.2]

Lesley, L.J.S. (1976) Optimum bus-stop spacing: part 1 *Traffic Engineering Control*, **17**, 399–401. [3.9, 9.2.7]

Leven, D. and M. Sharir (1987) Planning a purely translational motion for a convex object in two-dimensional space using generalized Voronoi diagrams *Discrete and Computational Geometry*, **2**, 9–31. [3.5.4, 4.8]

Lewis, B.A. and J.S. Robinson (1977) Triangulation of planar regions with applications *The Computer Journal*, **21**(4), 324–332. [6.2, 6.3]

Lewis, F.T. (1928) The correlation between cell division and the shapes and sizes of prismatic cells in the epidermis of *Cucumis, Anatomical Record*, **38**, 341–376. [5.4]

Lewis, F.T. (1930) A volumetric study of growth and cell division in two types of epithelium—the longitudinally prismatic cells of *Tradescantia* and the radially prismatic epidermal cells of *Cucumis, Anatomical Record*, **47**, 59–99. [5.4]

Lewis, F.T. (1931) A comparison between the mosaic of polygons in a film of artificial emulsion and in cucumber epidermis and human amnion *Anatomical Record*, **50**, 235–265. [5.4]

Lewis, F.T. (1943) The geometry of growth and cell division in epithelial mosaics *American Journal of Botany*, **30**, 766–776. [5.4]

Lewis, F.T. (1944) The geometry of growth and cell division in columnar parenchyma *American Journal of Botany*, **31**, 619–629. [5.4]

Liddle, M.J., C.S.J. Budd and M.J. Hutchings (1982) Population dynamics and neighbourhood effects in establishing swards of *Festuca rubra, Oikos*, **38**, 52–59. [7.1]

Lingas, A. (1986a) An approximation behaviour and implementation of the greedy triangulation for convex planar point sets *Proceedings of the Second Symposium of Computational Geometry, Yorktown Heights, New York*. [6.2]

Lingas, A. (1986b) The greedy and the Delauney [sic] triangulations are not bad in the average case *Information Processing Letters*, **22**, 25–31. [6.2]

Linhart, J. (1981) Die Beleuchtung von Kugeln *Geometriae Dedicata*, **10**, 145–154 [3.9]

Lloyd, E.L. (1977) On triangulations of a set of points in the plane *Proceedings of the 18th Annual IEEE Conference on the Foundations of Computer Science, Providence, R.I.*, 228–240. [6.2]

Loeb, A.L. (1976) *Space Structures—Their Harmony and Counterpoint* Reading, Mass.: Addison-Wesley. [7.1]

Loftsgaarden, D.O. and D.P. Quesenberry (1965) A nonparametric density function *Annals of Mathematical Statistics*, **36**, 1049–1051. [3.9]

Lorz, U. (1990a) Cell-area distributions of planar sections of spatial Voronoi mosaics *Materials Characterization*, **25**(3), 297–309. [5.5]

Lorz, U. (1990b) Personal Communication. [5.4, 5.5]

Lorz, U. (1991) Distributions of cell characteristics of the spatial POISSON–

VORONOI tessellation and plane sections. In: U. Eckhardt, A. Hubler, W. Nagel and G. Werner (eds) *Geometrical Problems of Image Processing. Research in Information Series*, Volume 4 Berlin: Akademie-Verlag, pp. 171–178. [5.3, 5.4, 5.5]

Lösch, A. (1940) Die raumliche Ordnung der Wirtschaft. Jena: Fischer. (Translated by W.H. Woglom and W.F. Stolper, *The Economics of Location* New Haven: Yale University Press, 1954). [7.1, 9.2.4]

Lotwick, H.W. (1982) Simulation of some spatial hard core models, and the complete packing problem *Journal of Statistical Computation and Simulation*, **15**, 295–314. [5.10]

Love, R.F. (1976) One-dimensional facility location-allocation using dynamic programming *Management Science*, **22**, 614–617. [9.4.1]

Love, R.F., J.Q. Morris and G.O. Wesolowsky (1988) *Facilities Location* New York: North-Holland. [9.2, 9.2.1]

Lozano-Pérez, T. and M.A. Wesley (1979) An algorithm for planning collision free paths among polyhedral obstacles *Communications of the ACM*, **22**(10), 560–570. [3.9]

Mackay, A.L. (1972) Stereological characteristics of atomic arrangements in crystals *Journal of Microscopy*, **95**(2), 217–227. [7.1]

Mahin, K.W., K. Hanson and J.W. Morris, Jr (1976) The computer simulation of homogeneous nucleation and growth processes *Computer Simulation for Materials Applications, Nuclear Metallurgy*, **20**, 39–50. [5.5, 5.6]

Mahin, K.W., K. Hanson and J.W. Morris, Jr (1980) Comparative analysis of the cellular and Johnson–Mehl microstructures through computer simulation *Acta Metallurgica*, **28**, 443–453. [5.5, 5.6, 7.1]

Mahoney, M.S. (1979) *Rene Descartes. Le Monde, ou Traité de la lumière* (Translation and introduction) New York: Abaris Books. [1.1]

Manacher, G.K. and A.L. Zobrist (1979) Neither the Greedy triangulation nor the Delaunay triangulation of a planar point set approximates the optimal triangulation *Information Processing Letters*, **9**(1), 31–34. [6.2]

Mardia, K.V. (1989) Shape analysis of triangles through directional techniques *Journal of the Royal Statistical Society, Series B*, **51**(3), 449–458. [8.2]

Mardia, K.V., R. Edwards and M.L. Puri (1977) Analysis of central place theory *Bulletin of the International Statistical Institute*, **47**, 93–110. [5.9, 8.2]

Mark, D.M. (1975) Computer analysis of topography: a comparison of terrain storage methods *Geografiska Annaler*, **57A** 179–188. [6.2]

Mark, D.M. (1987) Recursive algorithm for determination of proximal (Thiessen) polygons in any metric space *Geographical Analysis*, **19**(3), 264–272. [4.1, 4.9]

Martin, J.L. (1965) Solution to problem 62-8 *SIAM Review*, **7**, 132–134. [5.9]

Massam, B.H. (1972) The spatial structure of administrative systems *Association of American Geographers, Commission on College Geography, Resource Paper*, **12**. [7.1]

Massam, B.H. (1975) *Location and Space in Social Administration* London: Edward Arnold. [7.1]

Matérn, B. (1979) The analysis of ecological maps as mosaics. In: R.M. Cormack and J.K. Ord (eds) *Spatial and Temporal Analysis in Ecology* Fairland, Maryland: International Co-operative Publishing House, pp. 271–278. [8.1.1]

Matheron, G. (1975) *Random Sets and Integral Geometry* New York: John Wiley. [5.3]

Matlack, G.R. and J.L. Harper (1986) Spatial distribution and the performance of individual plants in a natural population of *Silene dioica, Oecologia (Berlin)*,

70, 121–127. [7.1]

Matschinski, M. (1954) Considération statistique sur les polygones et les polyèdres *Université de Paris, Publications de l'Institut Statistique*, **3**, 179–201. [5.4]

Matsuda, T. and E. Shima (1984) Topology of supercluster–void structure *Progress in Theoretical Physics*, **71**(4), 855–857. [7.2]

Matula, D.W. and R.R. Sokal (1980) Properties of Gabriel graphs relevant to geographic variation research and the clustering of points in the plane *Geographical Analysis*, **12**, 205–222. [2.5]

Maus, A. (1984) Delaunay triangulation and the convex hull of *n* points in expected linear time *BIT*, **24**, 151–163. [4.1]

Maxwell, J.C. (1864) On reciprocal diagrams and diagrams of forces *Philosophical Magazine*, **4**(27), 250–261. [2.4]

McCullagh, M.J. (1981) Creation of smooth contours over irregularly distributed data using local surface patches *Geographical Analysis*, **13**(1), 51–63. [6.2, 6.3]

McCullagh, M.J. and C.G. Ross (1980) Delaunay triangulation of a random data set for isarithmic mapping *The Cartographic Journal*, **17**(2), 93–99. [6.2, 6.3]

McLain, D.H. (1976) Two dimensional interpolation from random data *The Computer Journal*, **19**(2), 178–181. [2.4, 6.2]

Mead, R. (1966) A relationship between individual plant-spacing and yield *Annals of Botany, N.S.*, **30**, 301–309. [1.2]

Mead, R. (1967) A mathematical model for the estimation of inter-plant competition *Biometrics*, **23**, 189–205. [7.1]

Mead, R. (1971) Models for interplant competition in irregularly distributed populations. In: G.P. Patil, E.C. Pielou and W.E. Waters (eds) *Statistical Ecology. Volume 2: Sampling and Modeling Biological Populations and Population Dynamics* University Park, Penn.: The Pennsylvania State University Press, pp. 15–30. [7.1]

Mecke, J. (1980) Palm methods for stationary random mosaics. In: R.V. Ambartzumian (ed.) *Combinatorial Principles in Stochasic Geometry* Erevan: Armenian Academy of Sciences, pp. 124–132. [5.1]

Mecke, J. (1984) Parametric representation of mean values for stationary random mosaics *Mathematische Operationsforschung und Statistik. Series Statistik*, **15**, 437–442. [5.1]

Megiddo, N. (1978) Cost allocation for Steiner trees *Networks*, **8**, 1–6. [9.3.3]

Meijering, J.L. (1953) Interface area, edge length, and number of vertices in crystal aggregates with random nucleation *Philips Research Reports*, **8**, 270–290. [5.4, 7.2]

Melzak, Z.A. (1961) On the problem of Steiner *Canadian Mathematical Bulletin*, **4**, 143–148. [9.3.3]

Meshkat, S.N. and C.M. Sakkas (1987) Voronoi diagram for multiply-connected polygonal domains, II—Implementation and application *IBM Journal of Research and Development*, **31**, 373–381. [4.8]

Miki, F. (1983) A study of the distribution of convenience goods stores, Unpublished Ph.D. Thesis, Department of Urban Engineering, University of Tokyo [in Japanese]. [8.3.1]

Miles, R.E. (1964a) Random polygons determined by random lines in a plane: I *Proceedings of the National Academy of Sciences (USA)*, **52**, 901–907. [5.7]

Miles, R.E. (1964b) Random polygons determined by random lines in a plane: II *Proceedings of the National Academy of Sciences (USA)*, **52**, 1151–1160. [5.7]

Miles, R.E. (1970a) On the homogeneous planar Poisson point process *Mathematical Biosciences*, **6**, 85–127. [3.3, 5.3, 5.4, 5.7, 5.9, 8.2]

Miles, R.E. (1970b) A synopsis of 'Poisson flats in Euclidean spaces' *Izvestilia Akademii Nauk Armlianskoi SSR Matematika*, **5**(3), 263–285. [5.4, 5.9]

Miles, R.E. (1971) Random points, sets and tessellations on the surface of a sphere *Sankhya: The Indian Journal of Statistics, Series A*, **33**(2), 145–174. [3.7.4, 5.8, 5.9]

Miles, R.E. (1972a) The random division of space *Advances in Applied Probability*, **4**(Supplement), 243–266. [5.1, 5.5, 5.6]

Miles, R.E. (1972b) Multi-dimensional perspectives on stereology *Journal of Microscopy*, **95**, 181–195. [5.1]

Miles, R.E. (1973) The various aggregates of random polygons determined by random lines in the plane *Advances in Mathematics*, **10**, 256–290. [5.7]

Miles, R.E. (1974) A synopsis of 'Poisson flats in Euclidean spaces'. In: E.F. Harding and D.G. Kendall (eds) *Stochastic Geometry*. New York: John Wiley, pp. 202–227. [5.4, 5.9]

Miles, R.E. (1984) Sectional Voronoi tessellations *Revista de la Union Mathematica Argentina*, **29**, 310–327. [5.5, 5.7]

Miles, R.E. (1986a) Spatial tessellations and their stereology. In: Y. Kato, R. Takaki and J. Toriwaki (eds) *Science on Form: Proceedings of the First International Symposium for Science on Form* Tokyo: KTK Scientific Publishers, pp. 147–155. [5.5, 5.7, 5.9]

Miles, R.E. (1986b) Random tessellations. In: S. Kotz and N.L. Johnson (eds), *Encyclopedia of Statistical Sciences*, Volume 7. New York: John Wiley, pp. 567–572. [5.7]

Miles, R.E. and R.J. Maillardet (1982) The basic structures of Voronoi and generalized Voronoi polygons *Journal of Applied Probability*, **19A**, 97–111. [3.2, 3.2.1, 5.4, 5.7]

Mirante, A. and N. Weingarten (1982) The radial sweep algorithm for constructing triangulated irregular networks *IEEE Computer Graphics and Applications*, **2**(3), 11–21. [6.2, 6.3]

Mirchandani, P.B. (1987) Generalized hierarchical facility locations *Transportation Science*, **21**(2), 123–125. [9.2.4]

Mirchandani, P.B. and R.L. Francis (1990) *Discrete Location Theory* New York: John Wiley. [9.2]

Mithen, R., J.L. Harper and J. Weiner (1984) Growth and mortality of individual plants as a function of 'available area' *Oecologia, (Berlin)*, **62**, 57–60. [7.1]

Møller, J. (1989) Random tessellations in R^d *Advances in Applied Probability*, **21**, 37–73. [5.1, 5.4, 5.5, 5.9]

Møller, J., E.B. Jensen, J.S. Petersen and H.J.G. Gundersen (1989) Modelling an aggregate of space-filling cells from sectional data *University of Aarhus, Department of Theoretical Statistics, Research Report*, 182 [5.0]

Monmonier, M.S. (1982) *Computer-Assisted Cartography: Principles and Prospects* Englewood Cliffs, N.J.: Prentice-Hall. [6.2]

Montarani, U. (1968) A method of obtaining skeletons using a quasi-Euclidean distance *Journal of the ACM*, **15**, 600–624. [3.9]

Moore, G.C. and C. ReVelle (1982) The hierarchical service location problem *Management Science*, **28**(7), 775–780. [9.2.4]

Mood, A.M., F.A. Grayhill and D.C. Boes (1974) *Introduction to the Theory of Statistics* New York: McGraw-Hill [9.2.3]

Mount, D.M. (1985) Voronoi diagrams on the surface of a polyhedron *Report 1496*, University of Maryland. [3.7.7, 4.8]

Muche, L. and D. Stoyan (1990) Contact and chord length distributions of the

Poisson Voronoi tessellation, Paper presented at the Uppsala Conference. [5.5]
Narula, S.C. (1984) Hierarchical location–allocation problems: a classification scheme *European Journal of Operational Research*, **15**, 93–99. [9.2.4]
Niggli, R. (1927) Die topologische strukturanalyse *Zeitschrift für Kristallographie*, **65**, 391–415. [1.2, 7.1]
Nowacki, V.W. (1976) Über allgemeine eigenschaften von wirkungsbereichen *Zeitscrhift für Kristallographie*, **143**, 360–385. [1.2, 3.9]
Odland, J. (1988) *Spatial Autocorrelation*. Scientific Geography Series, **9**, Newbury Park, Calif.: Sage Publications. [6.1.2]
O'Dunlaing, C., M. Sharir and C.K. Yap (1986) Generalized Voronoi diagrams for moving a ladder—topological analysis *Communications on Pure and Applied Mathematics*, **39**(4), 423–483. [3, 3.5.4]
O'Dunlaing, C. and C.K. Yap (1985) A 'reaction' method for planning the motion of a disk *Journal of Algorithms*, **6**, 104–111. [3.9]
Ogawa, T. and M. Tanemura (1974) Geometrical considerations on hard core problems *Progress of Theoretical Physics*, **51**(2), 399–417. [7.1]
Ohji, T. (1986) Normative approach to the study of periodic markets in South India *Human Geography*, **38**(4), 289–315 [in Japanese]. [7.1]
Ohsawa, T. and Y. Kijima (1991) On the triangulations for constructing domes *Journal of Architecture, Planning and Environmental Engineering*, No. 425, 87–89 [in Japanese]. [3.9]
Ohsawa, Y. and A. Suzuki (1987) Location–allocation problem of multi-person facility *Journal of the Operations Research Society of Japan*, **30**, 368–395 [in Japanese]. [9.2.3]
Ohya, T. (1983) On efficient algorithms for the Voronoi diagram, Master's thesis, Department of Mathematical Engineering and Information Physics, University of Tokyo [in Japanese]. [4.1, 4.3]
Ohya, T., M. Iri and K. Murota (1984a) Improvements of the incremental method for the Voronoi diagram with computational comparisons of various algorithms *Journal of the Operations Research Society of Japan*, **27**, 306–336. [4.1, 4.3, 4.4]
Ohya, T., M. Iri and K. Murota (1984b) A fast Voronoi-diagram algorithm with quaternary tree bucketing *Information Processing Letters*, **18**(4), 227–231. [4.3]
Oikawa, K. (1986) A study on the optimal partitioning model with distributed facilities *Papers of the Annual Conference of the City Planning Institute of Japan*, **21**, 235–240 [in Japanese]. [3.9]
Okabe, A. (1987) Exact higher moments of the nearest neighbor distance *Journal of the Faculty of Engineering, University of Tokyo*, **39**, 1–6. [8.3.1]
Okabe, A. and M. Aoyagi (1991) Existence of equilibrium configurations of competitive firms on an infinite two-dimensional space *Journal of Urban Economics*, **29**, 349–370. [7.3.1]
Okabe, A. and A. Fujii (1984) The statistical analysis through a computational method of a distribution of points in relation to its surrounding network *Environment and Planning A*, **16**, 107–114. [3.9, 8.3.2]
Okabe, A. and F. Miki (1984) A conditional nearest-neighbor spatial association measure for the analysis of conditional locational interdependence *Environment and Planning A*, **16**, 163–171. [8.3.1]
Okabe, A. and A. Suzuki (1987) Stability of spatial competition for a large number of firms on a bounded two-dimensional space *Environment and Planning A*, **19**, 1067–1082. [7.3.1]
Okabe, A. and H. Yomono (1988) Statistical methods for evaluating the geometrical hierarchy of a network *Geographical Analysis*, **20**(2), 122–139. [3.8.3]

Okabe, A. and T. Yoshikawa (1989) The multi nearest neighbor distance method for analyzing the compound effect of infrastructural elements on the distribution of activity points *Geographical Analysis*, **21**(3), 216–235. [8.3.2, 8.3.4]
Okabe, A., T. Yoshikawa, A. Fujii and K. Oikawa (1988) The statistical analysis of a distribution of activity points in relation to surface-like elements *Environment and Planning A*, **20**, 609–620. [3.9, 8.3.3]
O'Kelly, M.E. and H.J. Miller (1989) A synthesis of some market area delimitation models *Growth and Change*, **20**, 14–33. [3.9]
O'Kelly, M.E. and J.E. Storbeck (1984) Hierarchical location models with probabilistic allocation *Regional Studies*, **18**(2), 121–129. [9.2.4]
Ooishi, Y. (1990) Numerically robust implementation of the divide-and-conquer algorithm for constructing Voronoi diagrams, Batchelor's Thesis, Department of Mathematical Engineering and Information Physics, Faculty of Engineering, University of Tokyo [in Japanese]. [4.1]
Ooishi, Y. and K. Sugihara (1991) Numerically robust divide-and-conquer algorithm for constructing Voronoi diagrams *Transactions of Information Processing Society of Japan,* **32**(6), 709–720 [in Japanese]. [4.1]
Ord, J.K. (1975) Estimation methods for models of spatial interaction *Journal of the American Statistical Society*, **70**, 120–126. [6.1.2]
Ord, J.K. (1978) How many trees in a forest? *Mathematical Scientist*, **3**, 23–33. [8.1.3]
O'Rourke, J., S.R. Kosaraju and N. Megiddo (1986) Computing circular separability *Discrete and Computational Geometry*, **1**, 105–113. [2.4, 4.1, 4.7]
Osteen, R.E. and P.P. Lin (1974) Picture skeletons based on eccentricities of points of minimum spanning trees *SIAM Journal on Computing*, **3**, 23–40. [2.5]
Overmars, M.H. and J. van Leeuwen (1981) Maintenance of point configurations in the plane *Journal of Computers and System Sciences,* **23**, 116–204.
Paschinger, I. (1982) Konvexe Polytope und Dirichletsche Zellenkomplexe, Dissertation, Institut für Mathematik, University of Salzburg, Austria. [3.7.4]
Pedoe, D. (1970) *Geometry* New York: Dover Publications. [4.2]
Peucker, T.K. and N. Chrisman (1975) Cartographic data structures *The American Cartographer*, **2**(1), 55–69. [6.1.1]
Peucker, T.K. and D.H. Douglas (1975) Detection of surface-specific points by local parallel processing of discrete terrain elevation data *Computer Graphics and Image Processing*, **4**, 375–387. [6.3]
Peucker, T.K., R.J. Fowler, J.J. Little and D.M. Mark (1976) Digital representation of three-dimensional surfaces by triangulated irregular networks (TIN) *Technical Report 10*, Department of Geography, Simon Fraser University. [6.2]
Peucker, T.K., R.J. Fowler, J.J. Little and D.M. Mark (1978) The triangulated irregular network *Proceedings of the Digital Terrain Models Symposium, American Society of Photogrammetry, St. Louis*, 516–540. [6.2]
Philbrick, O. (1968) Shape recognition with the medial axis transform. In: G.C. Cheng, R.S. Ledley, D.K. Pollock and A. Rosenfeld (eds) *Pictorial Pattern Recognition* Washington, D.C.: Thompson Book Company. [3.9]
Philip, G.M. and D.F. Watson (1982) A precise method for determining contoured surfaces *Journal of the Australian Petroleum and Exploration Association*, **22**, 205–212. [6.2]
Pielou, E. (1977) *Mathematical Ecology* New York: Wiley–Interscience. [1.3.8, 2.5, 5.2, 8.3]
Popoff, C.C. (1966) Computing reserves of mineral deposits: principles and conventional methods *U.S. Department of the Interior, Bureau of Mines,*

Information Circular, 8283. [1.2, 6.1.1, 6.2]

Preparata, F.P. (1977) The medial axis of a simple polygon. In: *Proceedings of the 6th Symposium, Mathematical Foundation of Computer Science*, **53**. New York: Springer-Verlag, pp. 443–450. [3.5.4]

Preparata, F.P. and M.I. Shamos (1985) *Computational Geometry—An Introduction* New York: Springer-Verlag. [1.0, 2.2, 3.4.3, 4.1, 4.5, 4.7, 8.2, 9.2]

Prewitt, J.M.S. (1980) Contemporary medical microscopy: the advent of intelligent microscopes *IEEE Transactions on Nuclear Science*, **NS-27**(3), 1207–1217. [7.5.2]

Prim, R.C. (1957) Shortest connection networks and some generalizations *The Bell System Technical Journal*, **36**, 1389–1401. [2.5]

Quine, M.P. and D.F. Watson (1984) Radial generation of *n*-dimensional Poisson processes *Journal of Applied Probability*, **21**, 548–557. [4.1, 5.3, 5.4]

Radecke, W. (1980) Some mean value relations on stationary random mosaics in the space *Mathematische Nachrichten*, **97**, 203–210. [5.1]

Rahman, A. (1966) Liquid structure and self diffusion *Journal of Chemical Physics*, **45**(7), 2585–2592. [5.2, 5.3, 7.1]

Rapaport, D.C. (1983) Density fluctuations and hydrogen bonding in supercooled water *Molecular Physics*, **48**(1), 23–31. [7.1]

Rau, K. (1841) Report to the Bulletin of the Belgian Royal Society. In: W. Baumol and S. Goldfield (eds) *Precursors in Mathematical Economics: An Anthology* London School of Economics and Political Science, pp. 181–182. [3.9]

Reedman, J.H. (1979) *Techniques in Mineral Exploration* London: Applied Science Publishers Ltd. [6.1.1, 6.2]

Renfrew, C. (1973a) Monuments, mobilization and social organization in neolithic Wessex. In: C. Renfrew (ed.) *The Explanation of Culture Change: Models in Prehistory* Pittsburgh: University of Pittsburgh Press, pp. 539–558. [7.1]

Renfrew, C. (1973b) *Before Civilization. The Radiocarbon Revolution and Prehistoric Europe* London: Jonathan Cape. [7.1]

Rhind, D. (1975) A skeletal overview of spatial interpolation techniques *Computer Applications*, **2**, 293–309. [6.0]

Rhynsburger, D. (1973) Analytic delineation of Thiessen polygons *Geographical Analysis*, **5**, 133–144. [4.1]

Ricketts, P. (1986) Geography and international law: the case of the 1984 Gulf of Maine boundary dispute *The Canadian Geographer*, **30**(3), 194–205. [3.9]

Ripley, B.D. (1981) *Spatial Statistics* New York: John Wiley. [3.9, 6.0, 6.2, 6.3, 8.0, 8.1.1]

Ripley, B.D. (1987) *Stochastic Simulation* New York: John Wiley. [5.3]

Ripley, B.D. (1988) *Statistical Inference for Spatial Processes* Cambridge: Cambridge University Press. [1.3.8]

Rivier, N. (1982) Recent results on the ideal structures of glasses *Journal de Physique (Colloque C9)*, **43**(12, Supplement), C9-91–C9-95. [5.4]

Rivier, N. (1983a) Topological structure of glasses. In: V. Vitek (ed.) *Amorphous Materials: Modeling of Structure and Properties* New York: American Institute of Mining, Metallurgical and Petroleum Engineers, pp. 81–98. [5.4]

Rivier, N. (1983b) On the structure of random tissues or froths, and their evolution *Philosophical Magazine B*, **47**(5) L45–L49. [5.4]

Rivier, N. (1985a) Statistical crystallography: structure of random cellular networks *Philosophical Magazine B*, **52**(3), 795–819. [5.4]

Rivier, N. (1985b) Theory of glass *Revisita Brasiliera de Fisca*, **15**. [7.1]

Rivier, N. (1986a) Structure of random cellular networks and their evolution *Physica*, **23D**, 129–137. [5.4]

Rivier, N. (1986b) Distributions of shapes and sizes by maximum entropy methods *Annals, Israel Physical Society*, **8**, 560. [5.4]
Rivier, N. (1986c) A botanical quasicrystal *Journal de Physique, Colloque C3*, **47**(7 Supplement), C3-299–C3-308. [7.1]
Rivier, N. (1988) Crystallography of spiral lattices *Modern Physics Letters B*, **2**(8), 953–960. [7.1]
Rivier, N. and D.M. Duffy (1982) On the topological entropy of atomic liquids and the latent heat of fusion *Journal of Physics C: Solid State Physics*, **15**, 2867–2874. [7.1]
Rivier, N., E. Guyon and E. Charlaix (1985) A geometrical approach to percolation through random fractured rocks *Geological Magazine*, **122**(2), 157–162. [5.2]
Rivier, N. and A. Lissowski (1982) On the correlation between sizes and shapes of cells in epithelial mosaics *Journal of Physics A: Mathematical and General*, **15**, L143–L148. [5.4]
Rivier, N., R. Occelli, J. Pantaloni and A. Lissowksi (1984) Structure of Benard convection cells, phyllotaxis and crystallography in cylindrical symmetry *Journal de Physique*, **45**, 49–63. [7.1]
Rogers, C.A. (1964) *Packing and Covering* Cambridge Tracts in Mathematics and Mathematical Physics No.54. Cambridge: Cambridge University Press. [1.2, 9.2.2]
Rohlb, F.J. (1973) Hierarchical clustering using the minimum spanning tree *The Computer Journal*, **16**, 93–95. [2.5]
Rosenkrantz, D.J., R.E. Stearns and P.M. Lewis II (1977) An analysis of several heuristics for the traveling salesman problem *SIAM Journal on Computing*, **6**, 563–581. [2.5]
Royden, H.L. (1963) *Real Analysis* New York: Macmillan. [1.3]
Saaty, T.L. (1970) *Optimization in Integers and Related Extremal Problems* New York: McGraw-Hill. [2.5]
Sabin, M.A. (1976) Errata *The Computer Journal*, **19**, 384. [6.2]
Sabin, M.A. (1982) Contouring—a review of methods for scattered data. In: K.W. Brodie (ed.) *Mathematical Methods in Computer Graphics and Design* London. Academic Press, pp. 63–86. [6.0]
Sadeck, E.A. (1980) A scheme for automatic generation of triangular finite elements *International Journal for Numerical Methods in Engineering*, **15**, 1813–1822. [6.3]
Saetre, T.O., O. Hunderi and E. Nes (1986) Computer simulation of primary recrystallization microstructures. The effects of nucleation and growth kinetics *Acta Metallurgica*, **34**(6), 981–987. [5.5, 5.6, 7.2]
Saito, N. (1982) Asymptotic regular pattern of epidermal cells in mammalian skin *Journal of Theoretical Biology*, **95**, 591–599. [7.1, 7.3.2]
Sakamoto, M. and M. Takagi (1987) Weighted Voronoi tessellations Denshi Joho Tsushin Gakkai, *PRU87-49*, pp. 21–30 [in Japanese]. [3.1]
Santalo, L.A. (1976) *Integral Geometry and Geometric Probability* Reading, Mass.: Addison-Wesley. [5.7]
Saxena, S., P.C.P. Bhatt and V.C. Prasad (1990) Efficient VLSI parallel algorithm for Delaunay triangulation on orthogonal tree network in two and three dimensions *IEEE Transactions on Computers*, **C-39**, 400–404. [4.1]
Schachter, B. and N. Ahuja (1979) Random pattern generation processes *Computer Graphics and Image Processing*, **10**, 95–114. [5.2]
Schumaker, L.L. (1976) Fitting surfaces to scattered data. In: G.G. Lorentz, C.K. Chui and L.L. Schumaker (eds) *Approximation Theory II* New York: Academic

Press, pp. 203–268. [6.0]

Schut, G.H. (1976) Review of interpolation methods for digital terrain models *The Canadian Surveyor*, **30**(5), 389–412. [6.0]

Schwartz, J. and C.K. Yap (1986) *Advances in Robotics* Hillside, N.J.: Lawrence Erlbaum Associates. [3.9]

Schwarz, H.R. (1988) *Finite Element Methods* London: Academic Press. [6.2]

Scott, A.J. (1971) *Combinatorial Programming, Spatial Analysis and Planning* London: Methuen. [7.1]

Scott, A.J. (1975) Discrete dynamic location systems: notes on a structural framework. In: A. Karlqvist, L. Lundqvist, F. Snickars (eds), *Dynamic Allocation of Urban Space* Westmead, Farnborough, Hants.: Saxon House, pp. 121–160. [9.4.1]

Seidel, R. (1982) On the size of closest point Voronoi diagrams *Report, F94, Institute for Information Processing, Technical University of Graz*. [3.3.1]

Seidel, R. (1986) Constructing higher-dimensional convex hulls at logarithmic cost per face *Proceedings of the 18th Annual ACM Symposium on Theory of Computing*, 404–413. [4.7]

Seidel, R. (1987) Output-size sensitive algorithms for constructional problems in computational geometry, Ph.D. Thesis, Cornell University. [4.7]

Seoung, Y.M. (1990) A study on optimal location problems of facilities using the Voronoi diagram, Unpublished Master's Thesis, Department of Urban Engineering, University of Tokyo, Tokyo [in Japanese]. [3.9, 9.4.2]

Seoung, Y.M. and T. Asano (1987) Facility location problems with obstacles *Proceedings of the 1987 Joint Technical Conference on Circuits and Systems* (JTC-CAS'87), 237–242. [3.4.1, 3.9]

Shaked, A. (1975) The non-existence of equilibrium for the two-dimensional three-firms problem *Review of Economic Studies*, **42**, 51–55. [7.3.1]

Shamos, M.I. (1978) Computational geometry, Ph.D. Thesis, Yale University. [3.2]

Shamos, M.I. and D. Hoey (1975) Closest-point problems *Proceedings of the 16th Annual IEEE Symposium on Foundations of Computer Science*, 151–162. [1.2, 2.3, 3.2, 4.1, 4.4, 6.5]

Shannon, C.E. (1948) A mathematical theory of communication *Bell System Technical Journal*, **27**, 379–423 and 623–656. [8.1.2]

Shapiro, M. (1981) A note on Lee and Schachter's algorithm for Delaunay triangulation *International Journal of Computer and Information Sciences*, **10**, 413–418. [4.1]

Sharir, M. (1985) Intersection and closest-pair problems for a set of planar discs *SIAM Journal on Computing*, **14**, 448–468. [3.5, 3.5.3]

Sharir, M. and A. Baltsan (1986) On shortest paths amidst convex polyhedra *Proceedings of the 2nd Annual ACM Symposium on Computational Geometry, Yorktown Heights, New York*, 193–206. [3.7.7]

Sharir, M. and A. Schorr (1984) On shortest paths in polyhedral spaces *The 16th ACM Symposium on Theory of Computing*, 144–153. [3.7.7]

Shieh, Y-N. (1985) K.H. Rau and the economic law of market areas *Journal of Regional Science*, **25**(2), 191–199. [1.2, 3.9]

Sibson, R. (1978) Locally equiangular triangulations *The Computer Journal*, **21**, 243–245. [2.4, 4.1, 6.2]

Sibson, R. (1980a) A vector identity for the Dirichlet tessellation *Mathematical Proceedings of the Cambridge Philosophical Society*, **87**, 151–155. [3.2, 3.2.2, 4.1, 5.1.2]

Sibson, R. (1980b) The Dirichlet tessellation as an aid in data analysis *Scandinavian Journal of Statistics*, **7**, 14–20. [3.2.2, 4.1]

Sibson, R. (1981) A brief description of natural neighbour interpolation. In: V. Barnett (ed.) *Interpreting Multivariate Data* New York: John Wiley, pp. 21–36. [6.0, 6.1.2]

Singh, R.L. and R.P.B. Singh (1975) Shape analysis in rural settlement geography: qualitative vs. quantitative approach. In: R.L. Singh (ed.) *Readings in Rural Settlement Geography* Varanasi: National Geographical Society of India, pp. 444–452. [7.1]

Singh, R.L. and R.P.B. Singh (1977) Spatial and structural analysis of the rural market network and its relation to society in an Indian environment *National Geographical Journal of India*, **23**, 151–159. [7.1]

Singh, R.P.B. (1976) A note on the transformation of village shapes into Thiessen polygons and hexagons *Geographical Review of India*, **38**, 50–53. [7.1]

Singh, R.P.B. (1979) Poisson process and cell model properties of village shape in Middle Ganga valley, India: a comparative study of locational pattern *National Geographical Journal of India*, **25**, 154–166. [5.2, 7.1]

Singh, R.P.B. and U.P. Singh (1978) Structural characteristics and transformation of village shape in spatial matrix (case of Gorakpur district) *National Geographer*, **13**, 143–154. [5.2, 7.1]

Small, C.G. (1988) Techniques of shape analysis on sets of points *International Statistical Review*, **56**(3), 243–257. [5.9]

Smalley, I.J. (1966) Contraction crack networks in basalt flows *Geological Magazine*, **103**, 110–114. [5.10, 7.2]

Smith, F.W. (1964) The structure of aggregates and the molecular kinematics of the viscosity of a Bernal liquid *Canadian Journal of Physics*, **42**, 304–320. [1.2, 7.1]

Smith, F.W. (1965) The structure of aggregates—a class of 20-sided space-filling polyhedra *Canadian Journal of Physics*, **43**, 2052–2055. [1.2, 7.1]

Smith, J.M., D.T. Lee and J.S. Liebman (1981) An O(*n* log *n*) heuristic for Steiner minimal tree problems on the Euclidean metric *Networks*, **11**, 23–39. [9.3.3]

Snyder, D.E. (1962) Urban places in Uruguay and the concept of a hierarchy *Festschrift: C.F. Jones, Northwestern University, Studies in Geography*, **6**, 29–46. [1.2, 7.1]

Spitzig, W.A., J.F. Kelly and O. Richmond (1985) Quantitative characterization of second-phase populations *Metallography*, **18**, 235–261. [7.1, 8.2]

Srinivasan, V. and L.R. Nackman (1987) Voronoi diagram for multiply-connected polygonal domains, I—Algorithm *IBM Journal of Research and Development*, **31**, 361–372. [4.8]

Steele, M. (1982) Optimal triangulation of random samples in the plane *The Annals of Probability*, **10**(3), 548–553. [6.2]

Stiny, J. (1929) *Technische Gesteinkunde für Bauingenieure, Kulturtechniker Land- und Forstwirte, Sowie für Steinbruchtechniker* Vienna: Julius Springer. [7.2]

Stogrin, M.I. (1968) Regular Dirichlet–Voronoi partitions for the second triclinic group *Soviet Mathematics Doklady*, **9**, 1479–1482. [7.1]

Stouffer, S.A. (1940) Intervening opportunities: a theory relating mobility and distance *American Sociological Review*, **5**, 845–867. [6.1.2]

Stoyan, D. (1986) On generalized planar random tessellations *Mathematische Nachrichten*, **128**, 215–219. [5.3]

Stoyan, D. and H. Hermann (1986) Some methods for statistical analysis of planar random tessellations *Statistics*, **17**(3), 407–420. [5.3, 5.4, 9.5.1]

Stoyan, D., W.S. Kendall and J. Mecke (1987) *Stochastic Geometry and its Applications* New York: John Wiley. [1.0, 1.3.8, 5.1, 5.4, 5.5, 5.9]

Stoyan, D. and H. Stoyan (1980) Gedanken zur Entstehung der Säulenformen bei

Basalten *Zeitschrift für Geologischen Wissenschaften*, **8**, 1529–1537. [7.2, 7.5.1, 9.5.1]

Stoyan, D. and H. Stoyan (1990) Exploratory data analysis for planar tessellations: structural analysis and point process models *Applied Stochastic Models and Data Analysis*, **6**, 13–25. [5.3, 5.4, 7.5.2]

Sugihara, K. (1986) *Machine Interpretation of Line Drawings* Cambridge, Mass.: MIT Press. [2.4]

Sugihara, K. (1992a) Voronoi diagrams in a river *International Journal of Computational Geometry and Applications,* **2**(1), 29–48. [4.5]

Sugihara, K. (1992b) A simple method for avoiding numerical errors and degeneracy in Voronoi diagram construction *IEICE (Institute of Electronics, Information and Communication Engineers of Japan) Transactions on Fundamentals of Electronics, Communications and Computer Sciences,* **E75-A**(4), 468–477.

Sugihara, K. and M. Iri (1988) Geometric algorithms in finite-precision arithmetic *Abstracts of the 13th International Symposium on Mathematical Programming, Tokyo, August 1988*, 196. [4.1, 4.6.3]

Sugihara, K. and M. Iri (1989a) A solid modelling system free from topological inconsistency *Journal of Information Processing*, **12**, 380–393. [4.6.2]

Sugihara, K. and M. Iri (1989b) Two design principles of geometric algorithms in finite-precision arithmetic *Applied Mathematics Letters*, **2**, 203–206. [4.1, 4.6.3]

Sugihara, K. and M. Iri (1989c) VORONOI2 reference manual—Topology-oriented version for the incremental method for constructing Voronoi diagrams Research Memorandum, RMI 89-04, Department of Mathematical Engineering and Information Physics, Faculty of Engineering, University of Tokyo, Tokyo. [4.6.3]

Sugihara, K. and M. Iri (1989d) Construction of the Voronoi diagram for one million generators in single precision arithmetic, Paper presented at the First Canadian Conference on Computational Geometry, 21–25 August 1989, Montreal, Canada. [4.1]

Sugihara, K., Y. Ooishi and T. Imai (1990) Topology-oriented approach to robustness and its applications to several Voronoi-diagram algorithms *Abstracts of the 2nd Canadian Conference on Computational Geometry, Ottawa, August 1990*, 36–39. [4.1]

Supowit, K.J. (1983) The relative neighborhood graph with an application to minimum spanning trees *Journal of ACM*, **30**, 428–447. [2.5]

Suzuki, A. (1989) On the average complexity of the incremental algorithm for the Voronoi diagrams, Preprint, Department of Electrical Engineering and Computer Science, Northwestern University, Evanston. [4.8]

Suzuki, A. and M. Iri (1986a) Approximation of a tessellation of the plane by a Voronoi diagram *Journal of the Operations Research Society of Japan*, **29**, 69–96. [7.1, 7.5.2, 9.5.1, 9.5.2]

Suzuki, A. and M. Iri (1986b) A heuristic method for the Euclidean Steiner problems as a geometrical optimization problem *Asia-Pacific Journal of Operational Research*, **3**(2), 109–122. [9.3.2, 9.3.3]

Suzuki, A. and M. Iri (1986c) Location problems for railroad—minimizing total travel times *Abstract of Operations Research Society of Japan*, Spring, 2-C-9, 210–211. [9.3.2]

Suzuki, T. (1987) Optimum locational patterns of bus-stops for many-to-one travel demand *Papers of the Annual Conference of the City Planning Institute of Japan*, **22**, 247–252 [in Japanese]. [9.2.7]

Suzuki, T. (1989) Study on locational optimization of facilities on a plane with socially optimal criteria, Unpublished Master's Thesis, Department of Urban Engineering, University of Tokyo [in Japanese]. [9.2.4]

Suzuki, T., Y. Asami and A. Okabe (1991) Sequential allocation of public facilities in one- and two-dimensional space: comparison of several policies *Journal of Mathematical Programming Series B*, **52**, 125–146. [9.4.1]

Swart, G. (1985) Finding the convex hull facet by facet *Journal of Algorithms*, **6**, 17–48. [4.7]

Takeda, S. (1985) On geographical optimization and dynamic facility location problem, Unpublished Master's Thesis, Department of Mathematical Engineering and Information Physics, University of Tokyo [in Japanese]. [3.9, 9.2.2, 9.2.6, 9.3.1, 9.4.2]

Talmon, Y. and S. Prager (1977) Statistical mechanics of microemulsions *Nature*, **267**, 333–335. [5.2]

Talmon, Y. and S. Prager (1978) Statistical thermodynamics of phase equilibria in microemulsions *Journal of Chemical Physics*, **69**(7), 2984–2991. [5.2]

Tanemura, M. (1988) Random packing and random tessellation in relation to the dimension of space *Journal of Microscopy*, **151**(3), 247–255. [5.3, 5.4]

Tanemura, M. and M. Hasegawa (1980) Geometrical models of territory, I. Models for synchronous and asynchronous settlement of territories *Journal of Theoretical Biology*, **82**, 477–496. [5.3, 5.10, 7.1]

Tanemura, M., Y. Hiwatari, H. Matsuda, T. Ogawa, N. Ogita and A. Ueda (1977) Geometrical analysis of crystallization of the soft core model *Progress of Theoretical Physics*, **58**(4), 1079–1095. [7.1]

Tanemura, M., T. Ogawa and N. Ogita (1983) A new algorithm for three-dimensional Voronoi tessellation *Journal of Computational Physics*, **51**, 191–207. [4.7]

Thiel, H. (1967) *Economics and Information Theory* Amsterdam: North-Holland. [7.5.2]

Thiessen, A.H. (1911) Precipitation averages for large areas *Monthly Weather Review*, **39**, 1082–1084. [1.2, 6.1.1, 9.2.5]

Thomas, R.W. (1981) Point pattern analysis. In: N. Wrigley and R.J. Bennett (eds), *Quantitative Geography* Henley-on-Thames, Oxon: Routledge and Kegan Paul, pp. 166–176. [8.1.1]

Tien, J.M., K. El-Tell and G.R. Simons (1983) Improved formulations to the hierarchical health facility location-allocation problem *IEEE Transactions on Systems, Man, and Cybernetics*, **13**, 6, 1128–1132. [9.2.4]

Tobler, W.R. (1975) Linear operators applied to real data. In: J.C. Davis and M.J. McCullagh (eds) *Display and Analysis of Spatial Data* London: John Wiley, pp. 14–37. [6.1]

Tokuyama, T. (1988) Deformation of merged Voronoi diagrams with translation *TRL Research Report TR-88-0049*, IBM Tokyo Research Laboratory. [4.1]

Toriwaki, J., K. Mase, Y. Yashima and T. Fukumura (1982) Modified Voronoi diagram and relative neighbors on a digitized picture and their applications to tissue image analysis *Proceedings of the IEEE 1st International Symposium on Medical Imaging and Image Interpretation*, 362–367. [4.1, 4.9]

Tóth, L.F. (1972) *Lagerungen in der Ebene auf der Krugel und im Raum* Heidelberg: Springer-Verlag. [3.9]

Toussaint, G.T. (1980a) The relative neighbourhood graph of a finite planar set *Pattern Recognition*, **12**, 261–268. [2.5]

Toussaint, G.T. (1980b) Pattern recognition and geometrical problems *Proceedings of the Fifth International Conference on Pattern Recognition, Miami Beach*, Volume 2, 1324–1347. [2.5]

Tuominen, O. (1949) Das Einflussgebeit der Stadt Turku im System der Einflussgeibiete SW-Finnlands *Finnia*, **71**, 1–38. [3.9]

Turek, Z., L. Hoofd and K. Rakusan (1986) Myocardial capillaries and tissue oxygenation *Canadian Journal of Cardiology*, **2**, 98–103. [7.1]

Turnball, D. and D.E. Polk (1972) Structure of amorphous semiconductors *Journal of Non-Crystalline Solids* **8–10**, 19–35. [7.1]

Upton, G. and B. Fingleton (1985) *Spatial Data Analysis by Example. Volume 1: Point Pattern and Quantitative Data* Chichester: John Wiley. [1.3.8, 3.7.8, 8.0, 8.1.1, 8.1.3, 8.3]

Urquhart, R. (1982) Graph theoretical clustering based on limited neighbourhood sets *Pattern Recognition*, **15**, 173–187. [2.5]

Van de Weygaert, R. and V. Icke (1989) Fragmenting the universe II. Voronoi vertices as Abell clusters *Astronomy and Astrophysics*, **213**, 1–9. [7.2]

Van Roy, T.J. and Erlenkotter, D. (1982) A dual-based procedure for dynamic facility location *Management Science*, **28**, 1091–1105. [9.4.1]

Vaughan, R.J. and E.A. Cousins (1977) Optimum location of stops on a bus route *Proceedings of the Seventh International Symposium on Transportation and Traffic Theory*, 697–716. [3.9, 9.2.7]

Vincent, P., R. Collins, J. Griffiths and J. Howarth (1983) Statistical geometry of geographical point patterns *Geographia Polonica*, **45**, 109–125. [5.3, 8.2]

Vincent, P.J. and J. Howarth (1982) A theoretical note on spatial pattern of discs and properties of simplicial graph *Forest Science*, **28**(1), 181–186. [5.10, 8.1.1, 8.2]

Vincent, P.J., J.M. Howarth, J.C. Griffiths and R. Collins (1976) The detection of randomness in plant patterns *Journal of Biogeography*, **3**, 373–380. [5.3, 8.2]

Vincent, P.J., J. Howarth, J. Griffiths and B. Collins (1977) Urban settlement patterns and the properties of the simplicial graph *The Professional Geographer*, **29**, 21–25. [8.2]

Von Hohenbalken, B. and D.S. West (1984a) Predation among supermarkets: an algorithmic locational analysis *Journal of Urban Economics*, **15**, 244–257. [7.1, 7.4]

Von Hohenbalken, B. and D.S. West (1984b) Manhattan versus Euclid: market areas computed and compared *Regional Science and Urban Economics*, **14**(1), 19–35. [3.9]

Voronoi, G. (1908) Nouvelles applications des paramètres continus à la théorie des formes quadratiques, deuxième memoire, recherches sur les parallelloèdres primitifs *Journal für die Reine und Angewandte Mathematik*, **134**, 198–287. [1.2]

Walters, P. (1982) *An Introduction to Ergodic Theory* New York: Springer-Verlag. [1.3.8]

Watkinson, A.R., W.M. Lonsdale and L.G. Firbank (1983) A neighbourhood approach to self-thinning *Oecologica (Berlin)*, **56**, 381–384. [7.1]

Watson, D.F. (1981) Computing the *n*-dimensional Delaunay tessellation with application to Voronoi polytopes *The Computer Journal*, **24**, 167–172. [4.1, 4.7]

Watson, D.F. and G.M. Philip (1984a) Triangle based interpolation *Journal of the International Association for Mathematical Geology*, **16**(8), 779–795. [6.2]

Watson, D.F. and G.M. Philip (1984b) Systematic triangulations *Computer Vision, Graphics and Image Processing*, **26**, 217–233. [6.2, 6.5]

Watson, D.F. and G.M. Philip (1985) A refinement of inverse distance weighted interpolation *Geo-Processing*, **2**, 315–327. [6.2]

Watson, D.F. and G.M. Philip (1987) Neighbourhood-based interpolation *Geobyte*, **2**(2), 12–16. [6.1.2]

Watson, D.F. and F.G. Smith (1975) A computer simulation and study of grain shape *Computers and Geosciences*, **1**, 109–111. [7.2]

Weaire, D. (1974) Some remarks on the arrangement of grains in a polycrystal *Metallography*, **7**, 157–160. [5.4]

Weaire, D., J.P. Kermode and J. Wejchert (1986) On the distribution of cell areas in a Voronoi network *Philosophical Magazine B*, **53**(5), L101–L105. [5.3, 5.4]

Weaire, D. and N. Rivier (1984) Soap, cells and statistics—random patterns in two dimensions *Contemporary Physics*, **25**, 59–99. [3.9]

Weber, A. (1909) *Über den Standort der Industrien* Tubingen: Erster Teil, Reine Theorie des Standorts. (English translation by C.J. Friedrich (1957) *Theory of the Location of Industries* Chicago: Chicago University Press.) [9.2]

Wells, A.F. (1984) *Structural Inorganic Chemistry* (fifth edition) Oxford: Clarendon Press. [7.1]

Welzl, E. (1985) Constructing the visibility graph for n line segments in $O(n^2)$ time *Information Processing Letters*, **20**, 167–171. [3.9]

West, D.S. (1981) Testing for market preemption using sequential location data *Bell Journal of Economics*, **12**, 129–143. [7.1]

West, D.S. and B. Von Hohenbalken (1984) Spatial predation in Canadian retail oligopoly *Journal of Regional Science*, **24**(3), 415–427. [7.1, 7.4]

Whitney, E.N. (1929) Areal rainfall estimates *Monthly Weather Review*, **57**, 462–463. [1.2, 6.2]

Wigner, E. and F. Seitz (1933) On the constitution of metallic sodium *Physical Review*, **43**, 804–810. [1.2, 7.1]

Wilkinson, D.S. (1988a) The effect of a non uniform void distribution on grain boundary void growth during creep *Acta Metallurgica*, **36**(8), 2055–2063. [5.3, 5.10, 7.1]

Wilkinson, D.S. (1988b) Modelling the effect of material inhomogeneity on grain boundary creep fracture *Proceedings of Ashby Symposium, World Materials Congress*, TMS-AIME. [7.1]

Wilks, S.S. (1962) *Mathematical Statistics* New York: John Wiley. [1.3.7]

Williams, P.W. (1972) The analysis of spatial characteristics of karst terrain. In: R.J. Chorley (ed.), *Spatial Analysis in Geomorphology* New York: Harper and Row, pp. 135–163. [7.2]

Winter, P. (1985) An algorithm for the Steiner problem in the Euclidian plane *Networks*, **15**, 323–345. [9.3.3]

Winterfield, P.H., L.E. Scriven and M.T. Davis (1981) Percolation and connectivity of random two-dimensional composites *Journal of Physics C: Solid State Physics*, **14**, 2361–2376. [5.2, 5.3]

Wirasinghe, S.C. and N.S. Ghoneim (1981) Spacing of bus-stops for many travel demand *Transportation Science*, **15**(3), 210–221. [3.9, 9.2.7]

Wood, L.J. (1974) Spatial interaction and partitions of rural market space *Tijdschrift voor Economische en Sociale Geografie*, **65**(1), 23–34. [3.9]

Wray, P.J., O. Richmond and H.L. Morrison (1983) Use of the Dirichlet tessellation for characterizing and modeling nonregular dispersions of second-phase particles *Metallography*, **16**, 39–58. [7.1, 8.2]

Yap, C.K. (1987) An $O(n \log n)$ algorithm for the Voronoi diagram of a set of simple curve segments *Discrete and Computational Geometry*, **2**, 365–393. [3.5, 4.8]

Yap, C.-K. (1988) A geometric consistency theorem for a symbolic perturbation scheme *Proceedings of the 4th Annual ACM Symposium on Computational Geometry, Urbana-Champaign, June 1988*, 134–142. [4.6.2]

Yoeli, P. (1977) Computer executed interpolation of contours into arrays of randomly distributed height-points *The Cartographic Journal*, **14**(2), 103–108. [6.3]

Yomono, H. (1991) The Voronoi diagram on a network *Technical Report, 91-09-02*, Nippon System Co., Tokyo [in Japanese]. [3.9]

Yonezawa, F. and T. Ninomiya (1983). Introduction. In: F. Yonezawa and T. Ninomiya (eds) *Topological Disorder in Condensed Matter* Berlin: Springer-Verlag. [7.1]

Yoshikawa, T. (1989) Moments of the nearest neighbor distance from a random point to line segments or polygons *Comprehensive Urban Studies, Special Edition, Tokyo Metropolitan University*, 25–33 [in Japanese]. [8.3.1, 8.3.2]

Yoshikawa, T. and A. Okabe (1991) A statistical analysis of the compound effect of urban infra-structures on the spatial distribution of apartment buildings with the multi-nearest neighbor distance method *Papers of the Annual Conference of the City Planning Institute of Japan*, **27**. (in press) [in Japanese]. [3.9]

Yoshikawa, T., A. Okabe, A. Fujii and K. Oikawa (1987) A statistical method of analyzing the distribution of urban activities in relation to the locations of infrastructural elements *Papers of the Annual Conference of the City Planning Institute of Japan*, **22**, 241–246. [in Japanese]. [8.3]

Yoshioka, S. and S. Ikeuchi (1989) The large scale structure of the universe and the division of space *The Astrophysical Journal*, **341**, 16–25. [5.2, 5.3, 7.1]

Yoshitake, T., K. Sugihara and M. Sugie (1987) Recognition of three-dimensional objects by means of regular pattern projection *Proceedings of the IEEE International Workshop on Industrial Applications of Machine Vision and Machine Intelligence, Tokyo*, 229–234. [4.1, 4.9]

Zallen, R. (1979) Stochastic geometry: aspects of amorphous solids. In: E.W. Montroll and J.L. Lebowitz (eds) *Studies in Statistical Mechanics. Volume VII: Fluctuation Phenomena* Amsterdam: North-Holland. [7.1]

Zaninetti, L. (1989) Dynamical Voronoi tessellation. I The two-dimensional case *Astronomy and Astrophysics*, **224**, 345–350. [5.6]

Zaninetti, L. (1990) Dynamical Voronoi tessellation. II The three-dimensional case *Astronomy and Astrophysics*, **233**, 293–300. [5.2, 5.6]

Zienkiewicz, O.C. and D.V. Phillips (1971) An automatic mesh generation scheme for plane and curved surfaces by isoparametric co-ordinates *International Journal for Numerical Methods in Engineering*, **3**, 519–528. [6.2]

Zinman, J.M. (1979) *Models of Disorder* Cambridge: Cambridge University Press. [7.1]

Index

2–3 tree 245
Aboav's law 305
abstract graph 36
access distance 204
access point 204
additional generator 224
additively weighted (AW) distance 133
additively weighted (Poisson (Voronoi diagram)) 315
additively weighted (power diagram) 138
additively weighted (power distance) 138
additively weighted (Voronoi diagram) 133, 134, 315
additively weighted (Voronoi region) 134
adjacency matrix 40
adjacent 68
adjacent link 39
adjacent node 39
adjustment model 392, 394
adjustment process 394
algorithm 209, 246
alloys 8
amorphous structures 367
amorphous substrates 279
ancestor 328
angle vector 112
APA (area potentially available) 8, 377
Apollonius circle 130
approximation algorithm 267
archaeology 380
area
  market 9, 205, 208, 379
  restricted 488
  service 281, 318, 379, 385, 403
  trade 378
  Voronoi diagram for 183
area-like object 431
area of a (Voronoi polygon) 388
area of influence polygons 8, 343
area potentially available (APA) 8, 377
area-preserving transformation 60
area (Voronoi diagram) 183, 207
area (Voronoi region) 183
arrangement measure 408
assignment model 366
  Voronoi 366
assignment rule 123
  V- 125
  Voronoi generation 125
astronomy 382
astrophysics 318
atomic volume 368
associative law 11
augmented geometric graph 217
AVL tree 245
AW (additively weighted) 133, 206

basalt 383, 384, 398, 404
binomial distribution 46
binomial (point process) 56, 57
bisector 69, 125, 263
bisector (node set) 200
bound
  lower 25
  upper 25
boundary 19
  lower 99
  village 380
boundary growing procedure 224, 253
boundary point 19
boundary (Voronoi polygon) 69
boundary (Voronoi region) 69
bounded above 25
bounded below 25
bounded distance 170
bounded (Voronoi diagram) 69
Brillouin zone 207
bucket 228
buffer zone 207, 424

capillary domain 8, 376
Cartesian coordinates 12
Cartesian product 12
  m-dimensional 12
Cartesian space 12
ccw (counterclockwise) 218
CDT (constrained Delaunay triangulation) 359
cell 252
  Poisson Delaunay 273
  Poisson Voronoi 273
cell model 382
central limit theorem 53
central place theory 378
central point 463
chiefdom territory 380
chain 40
children 228
chi-square distribution 50
choropleth map 336
circle
  Apollonius 130
  empty 82
  largest empty 82
  smallest enclosing 161
circle criterion 354
circular (Dirichlet tessellation) 131
circular shear 373
circuit 40

city-block distance 185
classification 207
closed 19
closed ball 20
closed disk 20
closed half plane 23
closed half space 23
closed (Voronoi polygon) 67
closest pair problem 80
cluster amplification process 333
column vector 13
common support 234
commutative law 11
complement 11
complete random packing 330
complete spatial random pattern 61
complete spatial randomness (CSR) 61, 406
component 12, 41
compound event 43
compound (point process) 332
compound (Poisson (point process)) 63
compoundly weighted (CW) distance 136
compoundly weighted (Voronoi diagram) 136, 137
compoundly weighted (Voronoi region) 137
computational geometry 9
commutative law 11
conduction 278, 279
connected 20, 40
conic (Voronoi diagram) 197
constrained (Delaunay triangulation) 207, 359
constrained programming problem 435
constraint 435
constraint function 435
continuous 28, 29
continuous random variable 47
contour map 336
convex 23, 201, 204
  strictly 23
convex function 25, 436
convex hull 24, 115, 221, 260, 357, 363
convex spherical polygon 193
coordinate 12
coordination number 368
coordination polyhedron 368
coordination shell 368
correlation function 305, 306
cosmic fragmentation 6
countable set 10
crack pattern 383
criterion 354
  circle 354
  empty circumcircle 94
  equiangular 354
  local max-min angle 107
  Pitteway 354, 356
crystal aggregates 315
crystal growth 381
crystalline structure 206
crystalline 366
crystallization 373
crystallography 6, 7, 369, 372
CSR (complete spatial randomness) 61
cumulative distribution function 45
CW (compoundly weighted) 218
cw (clockwise)
cycle 40
cylindrical (Voronoi diagram) 195, 208

data structure 215
decomposed generator set 174
degeneracy 215, 246, 249
degenerate 68
degenerate (Delaunay triangulation) 74, 75
degree 39
  negative 39
  positive 39
Delaunay edge 75
  external 75
  internal 75
Delaunay graph 101
Delaunay pretessellation
  m-dimensional 76
Delaunay pretriangulation 74
Delaunay simplex 10, 285
  m-dimensional 76
Delaunay tessellation 1, 2, 3, 10, 72
  m-dimensional 76
  Poisson 273
Delaunay tetrahedrization 76
Delaunay triangle 75
Delaunay triangulation 72, 74, 75, 115
  constrained 207, 359
  degenerate 74
  modifying 357
  properties of 89
Delaunay vertex 76
derivative 28, 29, 30
descendant 228
descent method 437, 438
diagonal matrix 14
diameter 161
dictionary 245
differentiable 31
digital image 268
digital terrain model (DTM) 351, 358, 361
direct approach 409
directed graph 38
Dirichlet tessellation 6
  circular 131
  hyperbolic 134
disconnected 20, 40
discontinuous 29
discrete distribution 56
discrete random variable 45
discrete sample space 44
disjoint 11
disordered 367
dispersion measures 408
displaced lattice process 332
distance
  access 204
  additively weighted 133
  additively weighted power 138
  bounded 170
  city-block 185
  compoundly weighted 136
  Euclidean 19
  exponential 128
  great circle 193
  Hausdorff 190
  logarithmic 128
  Laguerre 267
  multiplicatively weighted 129
  near neighbour 422
  nearest neighbour 423
  network 200
  power 128, 138
  quadratic form 199
  shortest-path 165

distance (*cont.*)
space-time 199, 483
taxi-cab 185
V- 127, 184
visibility-shortest-path 170
Voronoi generation 127
weighted 129
distance metric 127
distribution
binomial 46
chi-square 50
discrete 56
multinomial 46
normal 49
Palm 277, 286, 308
Poisson 47
standard normal 49
uniform 48
distribution function 45
distributive law 11
divide-and-conquer method 214, 232, 261, 266
domain 13
dominance metric 185
dominance (node set) 200
dominance region 69, 125, 263
domain of an atom 8, 369
doubly stochastic (Poisson (point process)) 63
DTM (digital terrain model) 351, 358, 361
dual graph 43

ecology 8
economic predation theory 396
edge 26
Delaunay 75
external Delaunay 75
internal Delaunay 75
Voronoi 67, 71
edge effect 281, 409, 410, 411, 418
efficiency 210
distance (*cont.*)
empirical point pattern 405
eight-neighbour set 268
empty circle 82
empty circumcircle criterion 94
empty circumcircle theorem 94
empty set 10
EMST (Euclidean minimum spanning tree) 118, 120
end point 21, 39
entropy 289, 400, 412
epigraph 25
equiangular criterion 354
equilibrium state 277
ergodic 61, 277, 281, 286, 307, 319
ergodic theorem 61
error-free world 247, 251
estimating spatial intensity 415
Euclidean distance 19
Euclidean minimum spanning tree 118
Euclidean space 19
Euclidean Steiner tree algorithm 477
Euclidean Steiner minimal tree 474
Euclidean Steiner tree 208, 473
Euler's formula 42, 217, 290
Euler–Schaefli formula 42
event 43
compound 43
simple 43
expectation 50
expected value 50

experiment 43
exponential distance 128
exterior (penalty function method) 443
external Delaunay edge 75

face 26, 276, 286, 287
facet 26
lower 99
Voronoi 71
facility
hierarchical 454
mobile 462
point-like 452
far-side triangle 96, 162
farthest pair 161
farthest-point (Voronoi diagram) 158, 159, 160, 206
farthest-point (Voronoi polygon) 159
feasible region 435
feasible solution 435
finite element method 207, 351, 356, 358
finite graph 36
finitely countable set 10
finite region 25
forestry 415
four-neighbour set 268
FP (farthest point) 160
full neighbour 289
function 12
inverse 13
multivariate 28
univariate 28
function inverse to 13
fundamental theorem of integral calculus 34

Gabriel graph (GG) 116
Gabriel neighbour 116, 287
galaxies 280, 383
generalized additively weighted distance 134
generalized gamma function 296, 323
generalized (Voronoi diagram) 124, 262
generalized (Voronoi region) 124
general Poisson (point process) 62
general programming problem 434
general rule of addition 44
generator 67
generator point 67
generator recognition problem 85
generator set 67, 123, 273, 276, 322, 335
geology 280, 383
geometric graph 37
geometric realization 38
geometry 16
germanium 370
GG (Gabriel graph) 116, 120
global equilibrium 387
global minimum 436
global optimum 436
globular protein 206
golden mean 376
Goldstein's rule 440
goodness-of-fit test 54
granular composites 278
graph 35, 36, 359
abstract 36
augmented geometric 217
Delaunay 101
directed 38
dual 43
finite 36

graph (*cont.*)
  Gabriel 116
  geometric 37, 216, 217
  nearest neighbour 119
  non-planar 38
  planar 38
  primal 43
  relative neighbourhood 117
  simplicial 10
  undirected 38
  visibility 166, 359
  Voronoi 83
great circle distance 193
greedy triangulation (GT) 354
growth model 381
  Voronoi 381
GT (greedy triangulation) 354

Hadamard inequality 248
half hyperbola 242
half line 21
half plane 23
half space 23
hard core process 330, 383, 422
Hausdorff distance 190
Hausdorff line (Voronoi diagram) 192
heap 245
hierarchical facility 454
higher-dimensional Voronoi diagram 257
higher order (Poisson (Voronoi diagram)) 318
higher-order (Voronoi diagram) 142
hill shading 337, 352
homogeneous 59
homogeneous coordinate 220
homogeneous (normal tessellation) 310
homogeneous planar Poisson (point process) 354, 355, 411, 415, 417
homogeneous Poisson (point process) 63, 273, 276, 368, 406, 410
homogeneous tessellation 276
Hotelling model 385, 386
Hotelling process 387
hull
  convex 24
hyperbolic (Dirichlet tessellation) 134
hyperplane 22
hypersphere 258

ice wedge polygon 401
ideal gas 368
identity matrix 14
illumination of ball 206
image 12, 280
image analysis 341
image processing 207
imaginary data site 358
inconsistency 246
incremental method 213, 223, 261, 264
incidence mapping 39
incident 39
  negatively 39
  positively 39
independent 44
independent scattering 58
independent user 444
independently identically distributed (i.i.d.) 285
infinitely countable set 10
infinite region 25
infimum 25
information retrieval system 207
information theory 399, 412
information theory approach 412
inhomogeneous (Poisson (point process)) 309
initial node 39
inner product 15
inorganic chemistry 367
integrable 33
integral 33
integration 32
intensity 58
  estimating spatial 415
  spatial 415
interior point 20
intermediate node 228
internal (Delaunay edge) 75
interpolant 335
  approximate 335
  exact 335
  global 335
  local 335
interpolation
  natural neighbour 343
  nearest neighbour 340
  proximal polygon 343
  spatial 335
  triangle based 351
intersection 11
intervening opportunities model 346
invariant set 61
inverse function 13
inverse matrix 15
inversion 261
inversion transformation 95
invertible 60
Ireland 380, 399
isarithmic map 336
isolated node 39
isomorphic 38
isotropic 60
isotropic Poisson line process 319
isotropic tessellation 276, 307

Johnson–Mehl (J–M) model 206, 315, 385
  one- and two-dimensional sections of 318
  time inhomogeneous 317
J–M (Johnson–Mehl) 206, 315, 385

Karlsruhe metric 188
Karlsruhe-metric (Voronoi diagram) 190
Karlsruhe-metric (Voronoi region) 190
Karlsruhe (Voronoi diagram) 190
karst 384
kernel 26
KNP (*k*th nearest point) 163, 206
*k*th nearest neighbour 164
*k*th nearest-point (Voronoi diagram) 158, 162, 163, 206, 207
*k*th nearest-point (Voronoi region) 163

L_1-metric 271
L-metric 271
Laguerre distance 267
Laguerre (Voronoi diagram) 181
lattice 8, 278, 332, 366, 368, 372
  body centred cubic 370
  diamond 371
  primitive cell of a 369
  random 278
  square 420
  triangular 378, 383, 400, 413, 418

lattice complex 369
law
  associative 11
  commutative 11
  distributive 11
leaf 228
left generator 233
left (Voronoi diagram) 233
Lewis' law 289, 291
lexicographically greater 112
lexicographically maximum triangulation theorem 112
lift-up 87, 112, 261
lift-up method 214, 257
lift-up transformation 87, 111, 258
line 21
  Voronoi 71
linearly dependent 17, 18
linearly independent 18
linear programming problem 436
linear time 210
linear time algorithm 210
line-like object 428
line search 440
line segment 21
  straight 172
line (Voronoi diagram) 171, 207
line (Voronoi region) 171
link 39
  adjacent 39
  parallel 39
liquid 280, 373
  monatomic 368, 372
local coordinate property 158
local coordinates 349
local equilibrium 390
locally equiangular triangulation 111
locally finite 57
local(ly) optimal procedure 110
local max-min angle criterion 107, 353
local max-min angle theorem 109
local minimum 436
local optimum 436
locational optimization of a network 472
locational optimization of lines 467
locational optimization over time 478
location vector 16
logarithmic distance 128
loop 39
lower bound 25
lower bound of the time complexity 222
lower boundary 99
lower common support 234
lower facet 99
lower one-sided (statistical test) 54
lower m-face 260
L_p-metric 267

$(m-1)$-dimensional (Voronoi) face 71
$(m-2)$-dimensional (Voronoi) face 71
Manhattan metric 185
Manhattan-metric (Voronoi diagram) 186
Manhattan-metric (Voronoi polygon) 186
Manhattan (Voronoi diagram) 186, 208
map
  choropleth 336
  contour 336
  isarithmic 336
  soil 341
  slope 337
mapping 13
marked (point process) 306
market area 9, 205, 208, 379
  economic law of 9
matrix 13
  adjacency 40
  diagonal 14
  identity 14
  inverse 15
  random 332
  square 13
  symmetric 14
matrix addition 14
matrix multiplication 15
matrix subtraction 15
maximum angle (min-max) condition 356
m-dimensional (Cartesian product) 12
m-dimensional (Delaunay pretessellation) 76
m-dimensional (Delaunay simplex) 76
m-dimensional (Delaunay tessellation) 76
m-dimensional ordinary (Voronoi diagram) 71
m-dimensional ordinary (Voronoi polyhedron) 71
mean 50
measure 60
measure space 60
measurable 60
measurable space 60
measure-preserving transformation 60
medial axis 182, 207
merger 236
metal 127
  oxide supported 280
  second order particles in 375
  surface film on 315, 381
metal vapour 279
metallic glass 206, 368, 373
metallurgy 375, 381, 415, 422
meteorology 7
method A 281
method R 284
method T 284
metric 127
  distance 127
  dominance 185
  Karlsruhe 188
  L_1- 271
  L- 271
  L_p- 267
  Manhattan 185
  Minkowski power 185
  Moscow 188
  supremum 185
microemulsions 279
minimum length triangulation 354
Minkowski metric L_p
Minkowski power metric 185
missing value 345
mixing 62
  strongly 62
mixing condition 277
MLT (minimum length triangulation) 354
mobile facility 462
modification method 213
modifying (Delaunay triangulation) 357
moment about the mean 51
moment about the origin 51
moments of the nearest neighbour distance to line-like objects 428

moments of the nearest neighbour distance to point-like objects 426
monatomic liquid 368, 372
monotone with respect to ray shooting 112
mosaic 279
  k-phase 336
  Poisson Voronoi random 279
  random 279
  S- 279
Moscow metric 188
motion invariant 60, 276, 278, 305
motion invariant tessellation 276
mouthbreeder fish 384, 403
moving generator (particle) problem 388
multi-median problem 446
multi nearest neighbour distance method 432
multinomial distribution 46
multiplicatively weighted distance 129
multiplicatively weighted (Voronoi diagram) 129, 130
multiplicatively weighted (Voronoi region) 131
multiplier method 443
multi-stage 480
multi-stage locational optimization 478
multivariate function 28
multivariate density estimation 206
muscle tissue 375
mutual information 400
mutual visibility 359
MW (multiplicatively weighted) 129, 206
myopic locational optimization 480

naive method 210, 213, 261
natural neighbour interpolation 343
nearest neighbour (NN) 423
nearest neighbour analysis 409
nearest neighbour distance (NND) 423
nearest neighbour distance method 422
nearest neighbour distance method for area-like objects 431
nearest neighbour distance method for line-like objects 428
nearest neighbour distance method for point-like objects 425
nearest neighbour graph (NNG) 119
nearest neighbour interpolation 340
nearest neighbour search 225
nearest-point (Voronoi diagram) 159
near neighbour distance 422
near-side triangle 96
negative degree 39
negatively incident 39
neighbourhood 20
neighbour set 268
nested hierarchy 456
network 200, 470
  topologically random 278
  triangulated irregular (TIN) 351
network distance 200
network Voronoi area-diagram 204, 205
network Voronoi diagram 200
network Voronoi link-diagram 202, 208
network Voronoi node-diagram 200, 201
Newton method 439
NN (nearest neighbour) 423
NND (nearest neighbour distance) 422
NNG (nearest neighbour graph) 119, 120
node 39, 226
  adjacent 39
  initial 39
  intermediate 228
  isolated 39
  terminal 39
node set
  bisector 200
  dominance 200
  Voronoi 201
non-cocircularity 82
non-collinearity 72
non-convex 23
non-convex function 25, 436
non-cosphericity 82
non-crystalline 367
non-degenerate 68, 176, 277
non-degenerate tessellation 277
non-empty set 10
non-linear convex programming problem 436
non-linear non-convex programming problem 434, 436
non-linear programming problem 436
non-periodic 367
non-planar graph 38
norm 16
normal 277
normal distribution 49
normal tessellation 277, 310, 319
  homogeneous 310
nucleation site 279, 381, 382
null hypothesis 54

object
  area-like 431
  line-like 428
  point-like 425
objective function 435
observation point 458
one-sided (statistical test) 54
one-to-one 13
open 19
open ball 19, 436
open disk 19
open (Voronoi polygon) 67
ordered minimum angle vector 112
ordered order-2 (Voronoi diagram) 152
ordered order-*k* (Voronoi diagram) 151, 154, 207, 347
ordered order-2 (Voronoi polygon) 152, 153
ordered order-*k* (Voronoi polygon) 154
ordering multivariate data 362, 363
order-*k* (power diagram) 147
order-2 (Voronoi diagram) 143
order-*k* (Voronoi diagram) 142, 144
order-2 (Voronoi polygon) 143
order-*k* (Voronoi polygon) 144
ordinary (Voronoi diagram) 65, 70, 71
ordinary (Voronoi polygon) 66, 67, 70
ore reserves
  estimation of 8, 351
orthogonal projection 18
orthographic projection 19

packing
  complete random 330
  random 330
  Voronoi polygonal areal random 331
packing density 331
packing fraction 368
Palm distribution 277, 286, 308
parallel link 39
parent 228

partial derivative 31
partition 32
path 40
pattern recognition 207
PDC (Poisson Delaunay cell) 273
pdf (probability density function) 48
peeling data 363
penalty function method 441, 443
percolation 278, 279
periodic boundary condition 281
periodic locational optimization 481
perpendicular projection 18
perspective view 337
phase transition 381
phyllotaxis 376
Pielou's index 424
Pitteway criterion 354, 356
Pitteway triangulation 90
Pitteway triangulation theorem 91
pixel 268
  positive 268
planar graph 38
planar stochastic (point process) 57
planar tessellation 27
planar ordinary (Voronoi diagram) 66, 67, 70
plane 22
plane sweep method 214, 239
plant polygon 8, 377
plastic flow 375
plesiohedron 7, 369
point 16
point-like facility 452
point-like object 425
point-location problem 81, 426
point pattern 281, 405
point process
  binomial 56, 57
  compound 332
  compound Poisson 63
  doubly stochastic Poisson 63
  general Poisson 62
  homogeneous Poisson 63, 273, 354, 355, 368, 406, 410, 411, 415, 417
  inhomogeneous Poisson 309
  marked 306
  planar stochastic 57
  Poisson 58
  spatial stochastic 56
  spatial-temporal stochastic 56
  stochastic 57
point vector 16
Poisson cluster process 332
Poisson Delaunay cell (PDC) 273, 323
  properties of 323
  shape of 285
  simulating 281
Poisson (Delaunay tessellation) 273
  s-dimensional sections of 329
Poisson (Delaunay tessellation) on the surface of a sphere 329
Poisson distribution 47
Poisson (point process) 58
  compound 63
  doubly stochastic 63
  general 62
  homogeneous 63, 273, 354, 355, 368, 406, 410, 411, 415, 417
  inhomogeneous 309
Poisson Voronoi cell (PVC) 273
  properties of 286
  simulating 281
Poisson (Voronoi diagram) 273, 276
  additively weighted 315
  higher order 318
  properties of 276
  s-dimensional section of higher order 322
  sectional 307
Poisson Voronoi diagram on the surface of a sphere 322
Poisson Voronoi random mosaic (PVRM) 279
polygon 25
  convex spherical 193
  simple 26
  spherical 193
polygon based method 409
polygonal method 337
polygon-based method 409
polyhedral surface 197
polyhedral (Voronoi diagram) 198
polyhedron 26
  coordination 368
  m-dimensional ordinary Voronoi 71
  Voronoi 71
polytope 26, 276, 286
porous oxides 279
positive degree 39
positively incident 39
positive pixel 268
power diagram 138, 139, 206
  additively weighted 138
  order-*k* 147
power distance 128, 138
  additively weighted 138
power (Voronoi polygon) 138
predation process 396
predecessor 218
prey–predator relations 395, 396
primal graph 43
priority queue 245
probability 44
probability density function (pdf) 48
probability function 45
probability measure 44
probability space 60
product
  Cartesian 12
  inner 15
  scalar 15
programming problem 434
projection
  orthogonal 18
  orthographic 19
  perpendicular 18
properties of the Delaunay triangulation 89
properties of the Voronoi diagram 76
proximal polygon 341
PVC (Poisson Voronoi cell) 273
PVRM (Poisson Voronoi random mosaic) 279

quadrat analysis 408
quadratic form distance 199
quantization 343
quantizing problem 428
quantum field theory 278
quasi-Newton method 440
quaternary incremental method 231
quaternary initial guessing 230
quaternary reordering 228
quaternary tree 228

random matrix
random matrix (Voronoi diagram) 332
random variable 45
  continuous 47
  discrete 45
range 13
ray-shoot monotone 112
real Cartesian space 12
real line 1
real plane 1
re-arrangement process 330
reciprocal farthest-pair 161
reciprocal figure 101
reciprocal pair 80
redundancy measure 413
refinement 33, 35
regular 126, 276, 330
regular pattern 406
regular tessellation 276
relative neighbour graph (RNG) 117
relatively minimal tree 475
relocation decision 387
relocation timing 386
restricted area 488
retail stores 9, 379
Riemann integrable 33, 35
Riemann integral 33, 35
Riemann sum 33, 35
Riemannian manifold 199
right generator 233
right (Voronoi diagram) 117, 120, 233
RNG (relative neighbourhood graph) 117
robustness 211
root 226
rooted tree 226
rotation 59, 276
row vector 13

saddle point problem 352
scalar multiplication 15
scalar product 15
second order analysis 409
second order product density 305
second partial derivative 31
sectional (Poisson (Voronoi diagram)) 307
sectional (Voronoi diagram) 140, 206
sectional (Voronoi polygon) 140
selfish herd model 395
separate 149
sequential clustering process 333
sequential unconstrained minimization
    technique 443
service area 281, 318, 379, 385, 403
service point 462
service route 467
set 10
  countable 10
  decomposed generator 174
  eight-neighbour 268
  empty 10
  finitely countable 10
  four-neighbour 268
  generator 67, 273, 276, 322, 335, 397
  infinitely countable 10
  invariant 61
  non-empty 10
  starter 265
  uncountable 11
settlement pattern 418
shape
  theory of 327
shape space 327, 420
shelling method 261
shortest-path (SP) distance 165
shortest-path (Voronoi diagram) 164
shortest-path (Voronoi region) 166
significance level 54
simple 57
simple circuit 40
simple cycle 40
simple chain 40
simple event 43
simple path 40
simple polygon 26
simple sequential inhibition (SSI) process 330
simplicial graph 10
simultaneous linear equations 15
sinkholes 384
site saturation model 382
simplex 24
  m-dimensional Delaunay 76
simultaneous linear equation 15
slope map 337
smallest enclosing circle 161
S-mosaic 279
solid state physics 369
solution structures 373
solvent sheath 373
SP (shortest-path) 166
space–time distance 199, 483
space–time Voronoi diagram 208
spatial autocorrelation 345
spatial competition model 385, 386
spatial intensity 415
spatial interaction 346
spatial interpolation 7, 207, 335
spatial–temporal process 385
spatial–temporal (stochastic (point process))
    56
special rule of addition 44
spherical polygon 193
spherical (Voronoi diagram) 193, 208
spherical (Voronoi polygon) 193
split 9
square matrix 13
SSI (simple sequential inhibition) 330
stable 394
standard deviation 51
standard normal distribution 49
star-shaped 26, 201, 204
starter set 265
stationary 59, 276, 277, 352
stationary tessellation 276, 277
statistical test 54
  lower one-sided 54
  two-sided 54
  upper one-sided 54
steepest descent method 438
Steiner point 474
stereohedron 7, 369
stochastic (point process) 57
straight (line segment) 21
strictly convex 23
strongly connected 41
strongly mixing 62
subgraph 36
subset 11
subtree 208
successively inclusive hierarchy 456
successor 218

supremum 25, 208
supremum metric 185
supremum-metric (Voronoi diagram) 187
surrounding larger neighbour 347
swapping procedure 110
sweepline 239
symbolic perturbation 251
symmetric matrix 14

taxi-cab distance 185
terminal node 39
terminal point 463
territory 9, 206, 280
  animal 280, 378, 384
  chiefdom 380
  village 281
tessellation 27
  homogeneous normal 310
  m-dimensional Delaunay 76
  motion invariant 276
  normal 227, 319
  planar 27
  regular 276
  stationary 276, 277
Thiessen polygon 8, 343
thinning process 330
TIJ–M (time inhomogeneous J–M) 317, 318
time complexity 210
TIN (triangulated irregular network) 351
tissue 8, 375, 403
topology-oriented approach 251, 255
torus 199, 281
trade area 378
transformable 128
transformation 4, 7
  area-preserving 60
  inversion 95
  lift-up 2, 259
  measure-preserving 60
  volume-preserving 60
translation 59, 276
transpose 14
transportation problem 379
  unconstrained 379
tree 41, 245, 266
  2–3 245
  AVL 245
  crown projection diagram of a 378
  Euclidean minimum spanning 118
  Euclidean Steiner 208, 473, 476
  Euclidean Steiner minimal 474
  quaternary 228
  relatively minimal 475
  rooted 226
tree expansion 254
triangle
  far-side 96, 162
  Delaunay 75
  near-side 96
triangle based interpolant 351
triangle-based method 417
triangular method 351
triangulated irregular network (TIN) 351
triangulation 28
  constrained Delaunay (CDT) 359
  degenerate Delaunay 74, 75
  Delaunay 74, 75, 115
  greedy 354
  lexicographically maximum 112
  locally equiangular 111
  minimum length 354
  minimum weight 354
  modifying Delaunay 357
  optimal 354
  Pitteway 90
trivial generalization 128
two-dimensional (Voronoi face) 71
two-sided statistical test 54
two-species model 394

unbounded 25
unconstrained programming problem 435, 437
unconstrained (transportation problem) 379
uncountable set 11
undirected graph 38
uniform distribution 48
union 11
univariate function 28
unordered product 36
upper bound 25
upper common support 234
upper *m*-face 260
upper one-sided (statistical test) 54

value 12
variance 51
vector
  angle 112
  column 13
  location 16
  ordered minimum angle 112
  point 16
  row 13
vertex 26
  Delaunay 76
  Voronoi 67, 71
V-assignment rule 125
V-distance 127, 184
village boundary 380
virtual generator 217
visibility graph 166, 359
visibility polygon 165, 207
visibility-shortest-path (VSP) distance 170
visibility-shortest-path (Voronoi diagram) 168, 170
visibility-shortest-path (Voronoi polygon) 170
volume-preserving transformation 60
Voronoi assignment model 366
Voronoi diagram 1, 6, 66, 71, 273
  additively weighted Poisson 315
  additively weighted 133, 134, 246, 263, 315
  area 183, 207
  bounded 69
  compoundly weighted 136, 137
  conic 197
  cylindrical 195, 208
  farthest-point 158, 159, 160, 206, 261, 264
  fitting a 485
  generalized 124, 262
  Hausdorff line 192
  higher-dimensional 257
  higher-order 142
  higher order Poisson 318
  Karlsruhe (-metric) 190
  *k*th nearest-point 158, 162, 163, 206, 207
  Laguerre 181
  left 233
  line 171, 207, 264
  line and area 208
  Manhattan (-metric) 186

m-dimensional 260
m-dimensional ordinary 71
multiplicatively weighted 129, 130, 263
nearest-point 159
network 200
network node- 208
order-2 143
ordered order-2 152
ordered order-k 151, 154, 207, 347
order-k 142, 144
ordinary 65, 71
planar ordinary 66, 67, 70
Poisson 273, 276
polyhedral 198
properties of 76, 276
properties of Poisson 276
random 4
random matrix 332
right 233
sectional Poisson 307
sectional 140, 206
shortest-path 164, 166
space-time 280
spherical 193, 208
supremum 208
supremum-metric 187
visibility-shortest-path (VSP) 168, 170
weighted 128, 129, 205
weighted area 207
Voronoi diagram for area 183
Voronoi diagram for line 170
Voronoi diagram for Γ with respect to d 262
Voronoi diagram in a river 240
Voronoi diagram in a simple polygon 168
Voronoi diagram on a cone 196
Voronoi diagram on a cylinder 194
Voronoi diagram on a polyhedral surface 197
Voronoi diagram on a quasi-torus 199
Voronoi diagram on a sphere 192
Voronoi diagram with obstacles 164, 207
Voronoi edge 67, 71
Voronoi face 71
($m-1$)-dimensional 71
($m-2$)-dimensional 71
two-dimensional 71
zero-dimensional 71
Voronoi facet 71
Voronoi fitting 485
Voronoi generation (assignment rule) 125
Voronoi generation distance 127
Voronoi graph 85
Voronoi growth model 381
Voronoi line 71
Voronoi link set 202
Voronoi (node set) 201
Voronoi polygon 66
area of a 388
boundary 69
closed 67
farthest-point 159
Manhattan-metric 186
open 67
ordered order-2 152, 153
ordered order-k 154
order-2 143
order-k 144
ordinary 70
power 139
sectional 140
spherical 193
visibility-shortest-path 170
Voronoi polygonal random packing 331
Voronoi polyhedron 71
Voronoi polytope 76
Voronoi region
additively weighted 134
area 183
boundary 69
compoundly weighted 137
generalized 124
Karlsruhe (-metric) 190
kth nearest-point 163
line 171
multiplicatively weighted 131
shortest-path 166
weighted 129
Voronoi vertex 67, 71
VSP (visibility shortest path) 170

walking method 213, 261
Weberian problem 444
weighted distance 129
weighted (Voronoi diagram) 128, 129, 205
weighted area (Voronoi diagram) 207
weighted (Voronoi region) 129
well-behaving bisector 126
well-behaving dominance region 126
Wigner–Seitz region 8, 369
winged-edge data structure 216, 220
Wirkungsbereich 6, 369

x-ray scattering 279

zero conjectural variation 386
zero-dimensional (Voronoi face) 71

Applied Probability and Statistics (Continued)

JOHNSON and KOTZ • Distributions in Statistics
Discrete Distributions
Continuous Univariate Distributions—1
Continuous Univariate Distributions—2
Continuous Multivariate Distributions

JUDGE, GRIFFITHS, HILL, LÜTKEPOHL, and LEE • The Theory and Practice of Econometrics, *Second Edition*

JUDGE, HILL, GRIFFITHS, LÜTKEPOHL, and LEE • Introduction to the Theory and Practice of Econometrics, *Second Edition*

KALBFLEISCH and PRENTICE • The Statistical Analysis of Failure Time Data

KASPRZYK, DUNCAN, KALTON, and SINGH • Panel Surveys

KISH • Statistical Design for Research

KISH • Survey Sampling

LAWLESS • Statistical Models and Methods for Lifetime Data

LEBART, MORINEAU, and WARWICK • Multivariate Descriptive Statistical Analysis: Correspondence Analysis and Related Techniques for Large Matrices

LEVY and LEMESHOW • Sampling of Populations: Methods and Applications

LINHART and ZUCCHINI • Model Selection

LITTLE and RUBIN • Statistical Analysis with Missing Data

MAGNUS and NEUDECKER • Matrix Differential Calculus with Applications in Statistics and Econometrics

MAINDONALD • Statistical Computation

MALLOWS • Design, Data, and Analysis by Some Friends of Cuthbert Daniel

MANN, SCHAFER, and SINGPURWALLA • Methods for Statistical Analysis of Reliability and Life Data

MASON, GUNST, and HESS • Statistical Design and Analysis of Experiments with Applications to Engineering and Science

MILLER • Survival Analysis

MONTGOMERY and PECK • Introduction to Linear Regression Analysis, *Second Edition*

NELSON • Accelerated Testing, Statistical Models, Test Plans, and Data Analyses

NELSON • Applied Life Data Analysis

OCHI • Applied Probability and Stochastic Processes in Engineering and Physical Sciences

OKABE, BOOTS and SUGIHARA • Spatial Tessellations: Concepts and Applications of Voronoi Diagrams

OSBORNE • Finite Algorithms in Optimization and Data Analysis

PANKRATZ • A User's Guide to Principal Components

PANKRATZ • Forecasting with Dynamic Regression Models

PANKRATZ • Forecasting with Univariate Box-Jenkins Models: Concepts and Cases

RACHEV • Probability Metrics and the Stability of Stochastic Models

RENYI • A Diary on Information Theory

RIPLEY • Spatial Statistics

RIPLEY • Stochastic Simulation

ROSS • Introduction to Probability and Statistics for Engineers and Scientists

ROUSSEEUW and LEROY • Robust Regression and Outlier Detection

RUBIN • Multiple Imputation for Nonresponse in Surveys

RUBINSTEIN • Monte Carlo Optimization, Simulation, and Sensitivity of Queueing Networks

RYAN • Statistical Methods for Quality Improvement

SCHUSS • Theory and Applications of Stochastic Differential Equations

SEARLE • Linear Models

SEARLE • Linear Models for Unbalanced Data

SEARLE • Matrix Algebra Useful for Statistics